Advances in Intelligent and Soft Computing 123

Editor-in-Chief: J. Kacprzyk

Advances in Intelligent and Soft Computing

Editor-in-Chief

Prof. Janusz Kacprzyk
Systems Research Institute
Polish Academy of Sciences
ul. Newelska 6
01-447 Warsaw
Poland
E-mail: kacprzyk@ibspan.waw.pl

Further volumes of this series can be found on our homepage: springer.com

Vol. 108. Y. Wang (Ed.)
Education and Educational Technology, 2011
ISBN 978-3-642-24774-3

Vol. 109. Y. Wang (Ed.)
Education Management, Education Theory and Education Application, 2011
ISBN 978-3-642-24771-2

Vol. 110. L. Jiang (Ed.)
Proceedings of the 2011 International Conference on Informatics, Cybernetics, and Computer Engineering (ICCE 2011) November 19-20, 2011, Melbourne, Australia, 2011
ISBN 978-3-642-25184-9

Vol. 111. L. Jiang (Ed.)
Proceedings of the 2011 International Conference on Informatics, Cybernetics, and Computer Engineering (ICCE 2011) November 19-20, 2011, Melbourne, Australia, 2011
ISBN 978-3-642-25187-0

Vol. 112. L. Jiang (Ed.)
Proceedings of the 2011 International Conference on Informatics, Cybernetics, and Computer Engineering (ICCE 2011) November 19-20, 2011, Melbourne, Australia, 2011
ISBN 978-3-642-25193-1

Vol. 113. J. Altmann, U. Baumöl, and B.J. Krämer (Eds.)
Advances in Collective Intelligence 2011, 2011
ISBN 978-3-642-25320-1

Vol. 114. Y. Wu (Ed.)
Software Engineering and Knowledge Engineering: Theory and Practice, 2011
ISBN 978-3-642-03717-7

Vol. 115. Y. Wu (Ed.)
Software Engineering and Knowledge Engineering: Theory and Practice, 2011
ISBN 978-3-642-03717-7

Vol. 116. Yanwen Wu (Ed.)
Advanced Technology in Teaching - Proceedings of the 2009 3rd International Conference on Teaching and Computational Science (WTCS 2009), 2012
ISBN 978-3-642-11275-1

Vol. 117. Yanwen Wu (Ed.)
Advanced Technology in Teaching - Proceedings of the 2009 3rd International Conference on Teaching and Computational Science (WTCS 2009), 2012
ISBN 978-3-642-25436-9

Vol. 118. A. Kapczynski, E. Tkacz, and M. Rostanski (Eds.)
Internet - Technical Developments and Applications 2, 2011
ISBN 978-3-642-25354-6

Vol. 119. Tianbiao Zhang (Ed.)
Future Computer, Communication, Control and Automation, 2011
ISBN 978-3-642-25537-3

Vol. 120. Nicolas Loménie, Daniel Racoceanu, and Alexandre Gouaillard (Eds.)
Advances in Bio-Imaging: From Physics to Signal Understanding Issues, 2011
ISBN 978-3-642-25546-5

Vol. 121. Tomasz Traczyk and Mariusz Kaleta (Eds.)
Modeling Multi-commodity Trade: Information Exchange Methods, 2011
ISBN 978-3-642-25648-6

Vol. 122. Yinglin Wang and Tianrui Li (Eds.)
Foundations of Intelligent Systems, 2011
ISBN 978-3-642-25663-9

Vol. 123. Yinglin Wang and Tianrui Li (Eds.)
Knowledge Engineering and Management, 2011
ISBN 978-3-642-25660-8

Yinglin Wang and Tianrui Li (Eds.)

Knowledge Engineering and Management

Proceedings of the Sixth International Conference on Intelligent Systems and Knowledge Engineering, Shanghai, China, Dec 2011 (ISKE2011)

Editors

Prof. Yinglin Wang
Department of Computer Science
and Engineering
Shanghai Jiao Tong University
800 Dongchuan Road, Shanghai 200240,
China
E-mail: ylwang@sjtu.edu.cn

Prof. Tianrui Li
School of Information Science
and Technology
Southwest Jiaotong University
Chengdu, Sichuan Provincc, 610031,
China
E-mail: trli@swjtu.edu.cn

ISBN 978-3-642-25660-8 e-ISBN 978-3-642-25661-5

DOI 10.1007/978-3-642-25661-5

Advances in Intelligent and Soft Computing ISSN 1867-5662

Library of Congress Control Number: 2011942322

Typeset by Scientific Publishing Services Pvt. Ltd., Chennai, India

Printed on acid-free paper

5 4 3 2 1 0

springer.com

Preface

We would like to extend our warmest welcome to each conference attendee. The 2011 International Conference on Intelligent Systems and Knowledge Engineering (ISKE2011) is the sixth in a series of ISKE conferences, which follows the successful ISKE2006 in Shanghai, ISKE2007 in Chengdu, and ISKE2008 in Xiamen, China, ISKE2009 in Hasselt, Belgium, and ISKE2010 in Hangzhou, China. ISKE2011 will be held in Shanghai, China, during December 15–17, 2011. It has been our pleasure as Program Committee Co-Chairs and Conference Co-Chair to organize this impressive scientific and technical program and the technical proceedings. ISKE2011 emphasizes current practices, experiences and promising new ideas in the broad area of intelligent systems and knowledge engineering. It provides a forum for researchers and practitioners around the world to present their latest results in research and applications and exchange new ideas in this field. ISKE 2011 is technically organized by Shanghai Jiao Tong University, and co-sponsored by California State University, Southwest Jiaotong University, Belgian Nuclear Research Centre (SCK•CEN).

We received 605 submissions from 26 countries and regions. We are very pleased with this level of submission and international participation. From these 605 submissions, the program committee selected 262 papers (including 109 full papers and 153 short papers), based on their originality, significance, correctness, relevance, and clarity of presentation, to be included in the proceedings. The acceptance rate of full papers is 18%, which we are proud of. The acceptance rate of short papers is 25%. Besides the papers in the conference proceedings, we also selected 44 papers from the submissions to be published in the Journal of Shanghai Jiao Tong University and the Journal of Donghua University. All the accepted papers will be presented or posted at the conference. Each of them was reviewed by two or more reviewers and the authors were asked to address each comment made by the reviewers for improving the quality of their papers. The acceptance rate of all the papers in the proceedings is 43%.

The accepted papers in the proceedings are contained in three volumes respectively based on the topics of the papers. The proceedings include “Volume I: Foundations of Intelligent Systems”, “Volume II: Knowledge Engineering and Management” and “Volume III: Practical Applications of Intelligent Systems”. Topics covered by the accepted papers in each volume of the proceedings are as follows:

Volume 1: Foundations of Intelligent Systems

Artificial Intelligence	46
Pattern Recognition, Image and Video Processing	40
Cognitive Science and Brain-Computer Interface	1

Volume 2: Knowledge Engineering and Management

Optimization and Biological Inspired Computation	12
Distributed and Parallel Intelligence	6
Robotics	11
Knowledge Engineering and Management	13
Data Mining, NLP and Information Retrieval	23
Data Simulation and Information Integration	11
Formal Engineering & Reliability	10

Volume 3: Practical Applications of Intelligent Systems

Social Computing, Mobile and Service Computing	8
Intelligent Game and Human Computer Interaction	6
Intelligent Engineering Systems	46
Intelligent Control Systems	9
Intelligent GIS, Networks or the Internet of Things	5
Social Issues of Knowledge Engineering	15

Accepted papers come from 23 countries, which shows that ISKE 2011 is a well-represented major international event, and their statistics (only papers of the proceeding, not include 44 papers which will be published in two journals) in terms of country are as follows:

China	212	Iran	2
Spain	8	Belgium	2
Australia	7	France	1
Brazil	5	Serbia	1
UK	6	Japan	1
USA	4	Vietnam	1
Russia	3	Sweden	1
Finland	3	Germany	1
Turkey	3	Saudi Arabia	1
Canada	2	Pakistan	1
Algeria	2	Korea	1
Poland	2		

ISKE 2011 consists of a three-day conference which includes paper and poster tracks, three invited keynote talks and two tutorials. The keynotes, tutorials and technical sessions cover a wide range of topics in intelligent systems and knowledge engineering.

The three invited speakers are Witold Pedrycz, University of Alberta, Canada; Ronald R. Yager, Iona College, New Rochelle, USA; and Zhi-Hua Zhou, Nanjing University, China. Witold Pedrycz will give a talk on granular models of time series and spatiotemporal data under the title of "User-Centric Models of Temporal and

Spatiotemporal Data: A Perspective of Granular Computing." He will discuss a new category of models in which the mechanisms of description, processing, and predicting temporal and spatiotemporal data are expressed in the language of information granules, especially fuzzy sets and intervals. Ronald R. Yager's talk is entitled "Intelligent Social Network Modeling." He will discuss an approach to enrich the social network modeling by introducing ideas from fuzzy sets and related granular computing technologies. Zhi-hua Zhou will discuss the current research results of his group in the machine learning area.

The two invited tutorial speakers are Gio Wiederhold, Stanford University and Jie Lu, University of Technology, Sydney (UTS), Australia. Gio Wiederhold's tutorial entitled "What is Your Software Worth?" will describe how the value of software can be estimated, and emphasize that awareness of the value of the product of one's knowledge and effort can help in making decisions on the design and the degree of effort to be made. Jie Lu's tutorial entitled "Personalized Recommender Systems for e-Government and e-Business Intelligence" will introduce several recommendation approaches, including case-based recommendation, ontology-based recommendation, fuzzy measure based recommendation, trust social networks-based recommendation related approaches and, in particular, present the recent developments made by her group in recommender systems and their applications in e-government and e-business intelligence.

As Program Committee Co-chairs and Conference Co-chair, we are grateful to all the authors who chose to contribute to ISKE2011. We want to express our sincere appreciation to the Program Committee Members listed below and to the additional reviewers for their great and quality work on reviewing and selecting the papers for the conference. We also would like to thank the webmasters, the registration secretary and financial secretary for their hard work. Last but certainly not the least, we would like to thank all the people involved in the organization and session-chairing of this conference. Without their contribution, it would not have been possible to produce this successful and wonderful conference. At this special occasion, we would especially like to acknowledge our respects and heartfelt gratitude to Professor Da Ruan, the Conference Co-chair of ISKE 2011 and the leading initiator of the ISKE conference series, for his hard work to prepare for this year's conference. Professor Da Ruan worked tirelessly for the conference until he suddenly passed away on July 31. Our thoughts and prayers are with his family. Besides of the above, we also thank all the sponsors of the conference, the National Science Foundation of China (No. 60873108, No. 60773088) and the Springer Publishing Company for their support in publishing the proceedings of ISKE 2011.

Finally we hope that you find ISKE2011 programs rewarding and that you enjoy your stay in the beautiful city of Shanghai.

December 15–17, 2011

Tianrui Li
Program Committee Chair

Yinglin Wang
Program Committee Co-chair

Du Zhang
Conference Co-chair

Organizing Committee

Honorary Co-chairs

L.A. Zadeh	University of California, Berkeley, USA
Gio Wiederhold	Stanford University, USA
H.-J. Zimmermann	Aachen Institute of Technology, Germany
Etienne E. Kerre	Ghent University, Belgium

Conference Co-chairs

Da Ruan	Belgian Nuclear Research Centre, Belgium
Du Zhang	California State University, USA
Athman Bouguettaya	RMIT University, Australia
Javier Montero	Complutense University of Madrid, Spain
Fuchun Sun	Tsinghua University, China

Steering Committee Co-chairs

Ronald R. Yager	Iona College, New Rochelle, USA
Jyrki Nummenmaa	University of Tampere, Finland
Wensheng Zhang	Chinese Academy of Sciences, China
Weiming Shen	National Research Council of Canada, Canada
Koen Vanhoof	University of Hasselt, Belgium

Organization Chair

Yinglin Wang	Shanghai Jiao Tong University, China

Local Organization Co-chair

Hongming Cai	Shanghai Jiao Tong University, China

Program Chair

Tianrui Li	Southwest Jiaotong University, China

Program Co-chairs

Yinglin Wang	Shanghai Jiao Tong University, China
Luis Martinez Lopez	University of Jaén, Spain
Hongtao Lu	Shanghai Jiao Tong University, China
Hongming Cai	Shanghai Jiao Tong University, China
Xuelong Li	Chinese Academy of Sciences, China

Publication Chair

Min Liu	Tongji University, China

Special Session Co-chairs

Jie Lu	University of Technology, Sydney, Australia
Cengiz Kahraman	Istanbul Technical University, Turkey
Victoria Lopez	Complutense University of Madrid, Spain
Zheying Zhang	University of Tampere, Finland

Poster Session Co-chairs

Guangquan Zhang	University of Technology, Sydney, Australia
Jun Liu	University of Ulster at Jordanstown, UK

Publicity Co-Chairs

Wujun Li	Shanghai Jiao Tong University, China
Xianyi Zeng	ENSAIT Textile Institute, France
Jiacun Wang	Monmouth University, USA
Michael Sheng	The University of Adelaide, Australia
Dacheng Tao	University of Technology, Sydney, Australia

Program Committee Members

Abdullah Al-Zoubi (Jordan)
Andrzej Skowron (Poland)
Athena Tocatlidou (Greece)
B. Bouchon-Meunier (France)
Benedetto Matarazzo (Italy)
Bo Yuan (USA)
Cengiz Kahraman (Turkey)
Chien-Chung Chan (USA)
Cornelis Chris (Belgium)
Dacheng Tao (Australia)
Davide Ciucci (Italy)
Davide Roverso (Norway)
Du Zhang (USA)
Enrico Zio (Italy)
Enrique Herrera-Viedma (Spain)
Erik Laes (Belgium)
Etienne E. Kerre (Belgium)
Francisco Chiclana (UK)
Francisco Herrera (Spain)
Fuchun Sun (China)
Gabriella Pasi (Italy)
Georg Peters (Germany)
Germano Resconi (Italy)
Guangquan Zhang (Australia)
Guangtao Xue (China)
Gulcin Buyukozkan (Turkey)
Guolong Chen (China)
Guoyin Wang (China)
H.-J. Zimmermann (Germany)
Hongjun Wang (China)
Hongming Cai (China)
Hongtao Lu (China)
I. Burhan Turksen (Canada)
Irina Perfilieva (Czech Republic)
Jan Komorowski (Sweden)
Janusz Kacprzyk (Poland)
Javier Montero (Spain)
Jer-Guang Hsieh (Taiwan, China)
Jesús Vega (Spain)
Jiacun Wang (USA)
Jianbo Yang (UK)
Jie Lu (Australia)
Jingcheng Wang (China)
Jitender S. Deogun (USA)
Jouni Jarvinen (Finland)
Juan-Carlos Cubero (Spain)
Jun Liu (UK)
Jyrki Nummenmaa (Finland)
Koen Vanhoof (Belgium)
Krassimir Markov (Bulgaria)
Liliane Santos Machado (Brasil)
Lisheng Hu (China)
Luis Magdalena (Spain)
Luis Martinez López (Spain)
Lusine Mkrtchyan (Italy)
Madan M. Gupta (Canada)
Martine De Cock (Belgium)
Masoud Nikravesh (USA)
Michael Sheng (Australia)
Mihir K. Chakraborty (India)
Mike Nachtegael (Belgium)
Mikhail Moshkov (Russia)
Min Liu (China)
Peijun Guo (Japan)
Pierre Kunsch (Belgium)
Qi Wang (China)
Qingsheng Ren (China)
Rafael Bello (Cuba)
Richard Jensen (UK)
Ronald R. Yager (USA)
Ronei Marcos de Moraes (Brasil)
Ryszard Janicki (Canada)
S. K. Michael Wong (Canada)
Shaojie Qiao (China)
Shaozi Li (China)
Sheela Ramanna (Canada)
Su-Cheng Haw (Malaysia)
Suman Rao (India)
Sushmita Mitra (India)
Takehisa Onisawa (Japan)
Tetsuya Murai (Japan)
Tianrui Li (China)
Tzung-Pei Hong (Taiwan, China)
Ufuk Cebeci (Turkey)
Victoria Lopez (Spain)
Vilem Novak (Czech Republic)
Weiming Shen (Canada)
Weixing Zhu (China)

Additional Reviewers

Volunteers

Sponsors

Shanghai JiaoTong University, China

The California State University, USA

Southwest Jiaotong University, China

Belgian Nuclear Research Centre, Belgian

Contents

Invited Talks

Part I: Optimization and Biological Inspired Computation

Part III: Robotics

Part IV: Knowledge Engineering and Management

Part V: Data Mining, NLP and Information Retrieval

User-Centric Models of Temporal and Spatiotemporal Data: A Perspective of Granular Computing

Witold Pedrycz

Department of Electrical & Computer Engineering
University of Alberta, Edmonton Canada
and
Systems Research Institute, Polish Academy of Sciences
Warsaw, Poland
pedrycz@ee.ualberta.ca

Abstract

One of the ultimate objectives of intelligent data analysis is to develop models of data that are user-centric. The human centricity of such pursuits means that a process of analysis along with the obtained results are made transparent to the user and come with a significant degree of flexibility, which helps achieve a sound tradeoff between accuracy and interpretability of results. The perception of data, as realized by humans, inherently invokes information granules (realized through numerous formal approaches including fuzzy sets, interval analysis, and rough sets) and their further processing. This helps establish a suitable level of abstraction at which the data are perceived, analyzed and their models are being formed. By casting the problem in the setting of Granular Computing, we develop a new category of models in which the mechanisms of description, processing, and predicting temporal and spatiotemporal data are expressed in the language of information granules, especially fuzzy sets and intervals.

In this talk, we show how a principle of justifiable information granularity leads to the realization of granular models of time series in which a construction of information granules is viewed as a certain optimization problem.

With regard to spatiotemporal data where their temporal facet as well as their spatial characteristics play a pivotal role, it is demonstrated how information granules are formed through an augmented collaborative clustering. The grouping is completed in the temporal and spatial domain in such a way an identity of relationships present in these two domains is retained. An auxiliary mechanism of information granulation is developed through an optimization of relational constraints (granular codebook) realized through a collection of information granules.

"(The full content will be available during the conference)"

Intelligent Social Network Modeling

Ronald R. Yager

Machine Intelligence Institute, Iona College
New Rochelle, NY 10801
yager@panix.com

Abstract

Web 2.0 has provided for a rapid growth of computer mediated social networks. Social relational networks are becoming an important technology in human behavioral modeling. Our goal here is to enrich the domain of social network modeling by introducing ideas from fuzzy sets and related granular computing technologies. We approach this extension in a number of ways. One is with the introduction of fuzzy graphs representing the networks. This allows a generalization of the types of connection between nodes in a network from simply connected or not to weighted or fuzzy connections. A second and perhaps more interesting extension is the use of Zadeh's fuzzy set based paradigm of computing with words to provide a bridge between a human network analyst's linguistic description of social network concepts and the formal model of the network. Another useful extension we discuss is vector-valued nodes. Here we associate with each node a vector whose components are the attribute values of the node. Using the idea of computing with words we are then able to intelligently query the network with questions that involve both attributes and connections. We see this as a kind of social network database theory. We shall look at some dynamic structures of network particularly the small worlds network.

"(The full content will be available during the conference)"

What Is Your Software Worth?

Gio Wiederhold

Professor Emeritus, Stanford University and MITRE Corporation
gio@cs.stanford.edu

Abstract

Much has been written about the cost of producing software, but that literature largely ignores the benefits of using that software. While software creators believe that their products are valuable, they are rarely called upon to quantify its benefits. Evaluation of software and its benefits in commerce is left to lawyers, economists, software vendors, or promoters. The results are often inconsistent.

This tutorial describes how the value of software can be estimated. The problem being addressed is that the value of software is essentially independent of the cost and effort spent to create it. A few brilliant lines of code can have a very high value, whereas a million lines of code that generate a report that nobody uses have little value. Awareness of the value of the product of one's knowledge and effort can help in making decisions on the design and the degree of effort to be made.

The tutorial will survey methods for valuing software based on the income it can generate. A principal approach is based on software growth, caused by needed maintenance. The valuation is with the accepted framework for valuing intellectual property (IP) in general.

My paper on that topic appeared in the Communications of the ACM, September 2006, but could not cover all of the issues. More material is available at http://infolab.stanford.edu/pub/gio/inprogress.html#worth. Software valuation is also covered in a course at Stanford University, CS207, https://cs.stanford.edu/wiki/cs207/ Participants in the tutorial are encouraged to read the available information and engage in discussion of this challenging topic.

"(The full content will be available during the conference)"

Personalized Recommender Systems for e-Government and e-Business Intelligence

Jie Lu

Decision Systems & e-Service Intelligence Lab
Centre for Quantum Computation & Intelligent Systems
School of Software
Faculty of Engineering and Information Technology
University of Technology, Sydney
P.O. Box 123, Broadway, NSW 2007, Australia
jielu@it.uts.edu.au

Abstract

Web personalisation is an interdisciplinary topic that has been discussed in the literature about information systems, web intelligence, customer relationship management and marketing. Web personalisation is defined as any set of actions that tailor the web experience to a specific user or set of users, anticipating user needs to provide them with what they want or require without having to ask for it explicitly. A number of e-business and e-government development stage models have been proposed in the literature that focuses on classifying functions and features offered by current e-business and e-government. Most of these models have a common final stage which concentrates on providing fully integrated and personalised e-services for their constituents. Recommender systems have gained considerable attention in recent years and are the most successful implementation of web personalisation. Recommender systems use justifications to generate recommended products or services to customers and to ensure the customers like these products or services. These justifications can be obtained either from preferences directly expressed by customers, or induced, using data representing the customer experience. Recommender systems are achieving widespread success and have attracted researchers' attention in the field of e-business and e-government applications.

Recommender systems use different types of information filtering techniques to automatically identify and predict a set of interesting items on behalf of the users according to their personal preferences. The most notable classes of recommender system approaches include: (1) Content-based filtering--mainly depends on items' descriptions to generate personalised recommendations; (2) Collaborative Filtering (CF)-- mainly depends on users ratings of items in a given domain, and works by computing the similarities between the profiles of several users on the basis of their provided ratings and generates new recommendations based on comparisons of user ratings; (3) Knowledge-based filtering--suggests items based on logical inferences about a user's needs and preferences; (4) Semantic-based filtering--exploits the semantic information associated with user and item descriptions to generate recommendations; (5) Trust-based filtering--exploits the level of trust between users in a social trust network and uses that knowledge to generate trustworthy recommendations; (6)

Hybrids-based filtering--combines two or more recommendation approaches to exploit their strengths and reduce their weaknesses.

This tutorial will introduce these recommendation approaches and, in particular, present the recent developments made by our Decision Systems and e-Service Intelligence (DeSI) lab in recommender systems and their applications in e-government and e-business intelligence , including case-based recommendation, ontology-based recommendation, fuzzy measure based recommendation, trust social networks-based recommendation related approaches and their applications in telecom companies and government-to-business services.

"(The full content will be available during the conference)"

Part I

Optimization and Biological Inspired Computation

A Novel TinyOS 2.x Routing Protocol with Load Balance Named CTP-TICN

Yang Song[1], Yaqiong Chai[2], Fengrui Ye[3], and Wenqiang Xu[1]

[1] Mobile Communication Planning and Designing Institute, Jiangsu Posts & Telecommunications Planning and Designing Institute, Nanjing, 210006, China
[2] Center for Earth Observation and Digital Earth, Chinese Academy of Sciences, Beijing, 100094, China
[3] College of Engineering Science and Technology, Shanghai Ocean University, Shanghai, 201306, China
{songyang,xuwenqiang}@jsptpd.com, yqchai@ceode.ac.cn, yefengrui@hotmail.com

Abstract. Due to the limitation of power supply in wireless sensor nodes, the paper represented a new routing protocol which could be applied in TinyOS 2.x operating system named CTP-TICN. The new routing protocol realizes the preliminary achievement of load balance in wireless sensor network. CTP-TICN protocol introduces the "intensity of transmission" and "numbers of one hop up node" declaration, which could help nodes choose the suboptimal parent node for data forwarding on the premise of transmission stability. The simulation shows that CTP-TICN is more effective on load-balance than the CTP routing protocol and it helps the wireless sensor network to live for a longer time.

Keywords: CTP-TICN, TinyOS, Load Balance, Simulation in TOSSIM.

1 Introduction

Wireless sensor network is a kind of distributed computing network which consisted from large amount of nodes which integrate sensors, data processing units and short-distance communication models. It is a special Ad Hoc network and the application of WSN is widely spread in recent years.

Right now, the nodes in WSN network is mainly two series, one is MICA series and the other is TELOS series. These nodes usually supported by the battery, however, charging the battery is difficult in most situations. Thus, compared to the traditional wire network, the wireless sensor network routing protocol not only need to ensure the stability of transmission but also need to consider the living time of the whole network, avoiding some nodes deplete their battery too early because of burdened with a heavy transmission task[1].

With the development of new technology, there are many routing protocol presented in TinyOS 1.x and TinyOS 2.x, such as CTP(collection tree protocol), MultiHopLQI, MintRoute and so on[2]. The packets forwarding of these routing protocol are all based

Y. Wang and T. Li (Eds.): Knowledge Engineering and Management, AISC 123, pp. 3–9.
springerlink.com

on the judgment of link quality. Though the link quality mechanism could ensure the link free, it also let some of the parent node undertake more forwarding task. The result is that the battery of some parent node run down quickly and the living time of the entire network become short.

This passage introduces a CTP-TICN routing protocol after analysis the CTP routing protocol of TinyOS 2.x carefully. It can be applied in Mica and Telos series platforms and it has pragmatic value. CTP-TICN routing protocol introduces the "Intensity of Transmission" and "Numbers of One-hop before Child node" declaration which can control the data flow of parent nodes, realize the load balance of wireless sensor network. At the end of this passage, there exists the result of simulation and the vision for general direction of the WSN in the future.

2 CTP Routing Protocol Analysis

CTP is a normal multi-hop routing protocol. It has already been applied in real project. The basic idea of CTP is to construct a collection tree which regards gather node as its root. Nodes in network transmit information to the gather node through the tree. Every node maintains the bidirectional link quality evaluation of itself and its neighbors and sends data to the neighbors which have the best link quality. At this time, the neighbor becomes the node's parent node[3].

CTP has two mechanisms to evaluate the link quality. The first one is based on the LEEP frame. The basic idea of LEEP frame protocol is the node collect information about the percentage of received broadcast package from its neighbor over the forwarding package sent to the neighbor by the node itself. For example, there are nodes α and β, Rα represent the amount of package α received from β. Sβ represent the amount of package β sent to α. In order so, the received link quality of α from β is: InQuality α-β= Rα/Sβ. So, the InQuality β-α= Rβ/Sα. As the result: OutQuality α-β= Rβ/Sα, OutQuality β-α= Ra/Sβ, InQuality α-β= OutQuality β-α, InQuality β-α= OutQuality α-β. The bidirectional link quality from α to β is represented by EETX (Extra Expected Transmission): EETX= InQuality α-β* OutQuality α-β.

The second method for evaluate the link quality in CTP is the evaluation based on the data package. The overall transmission packages between α and β is TotalData and the successful data package is SuccessData. The EETX= (TotalData/ SuccessData-1)*10 at this time. In order to reduce the evaluation flutter, the link estimator combines both methods above with exponential weighting, calculating the final link quality. The original EETXold is percentage 1-A, the EETXnew is percentage A. EETX= EETXold*(1-A) + EETXnew*A[4].

CTP uses ETX to represent the routing gradient. The root's ETX is 0, other node's ETX is its parent node's ETX plus the EETX of the link quality to the parent node. The nodes broadcast information of routing state periodically, update the ETX of the route from its neighbor nodes to gather nodes. While the route table updated, the route engine choose the parent node depend on the minimum ETX value.

From the link quality evaluation mechanism and the route engine of CTP, we can find that CTP uses the link cost function to determine the relaying node and root node. So, it is impossible to avoid communication tasks on some special nodes, as the figure shows below:

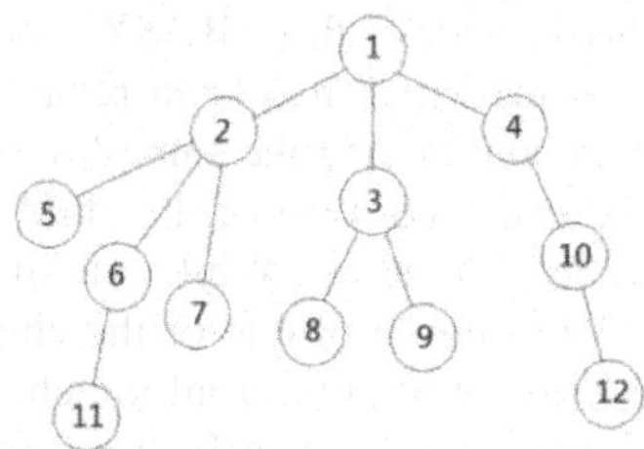

Node 1 is the root node, because of node 2 burden with more reliable data transmission than other same level node such as node 3 and node 4. Node 2 will consume more energy than node 3 and node 4. The data message of node 2 is also easily to be congested. In order to solve such problem, in chapter 3, the paper put forward the CTP-TICN protocol to modulate the data flown dynamically. Thus, other same level nodes will share the data transmission of the "BUSY node" and balance the load of the network.

3 Idea and Realization of CTP-TICN

3.1 Network Model

CTP could be regarded as a collection tree with root node. Define model C (N, L, D), N represent the set of all nodes, L represents the set of all links, D represents the biggest hop between nodes and root. The$(i, j)^n$, $i, j \in C$ represents the node j jump n times reach to node i[5].

1. Define the same level of nodes as: "SameLevelSet".
   $SLSn = \{j \mid \forall (ROOT, i)^n, j \in C, 1 \leqslant n \leqslant D\}$
2. Define the one hop up nodes as: "OneHopUpSet".
   $OHUS = \{ j \mid \forall j \in (i, j)^1, i, j \in C\}$
3. Define "intensity of data transmission" as: U= data flow/ parent node's data flow.
4. Define mean value of the intensity of data transmission among same level nodes: MVI.
5. Define mean value of number of child nodes among the same level nodes: MVN.
6. Define BE (Biggest ETX). BE is the maximum value of ETX value which could ensure the normal working.
7. Define the "BUSY node" state.

3.2 Description of Protocol CTP-TICN

CTP-TICN introduces the declaration of intensity of data transmission and the declaration of one hop up nodes. At first, every same level node "SLSn" broadcast the intensity of its own intensity of data transmission "U" and the number of one hop up nodes "OHUS". While all the intensity of data transmission and the number of one hop up nodes were collected, mean value such as "MVI" and "MVN" should be calculated. If the intensity of data transmission of one of the same level nodes exceeds the average intensity, the node would be remarked as "BUSY node", if the nodes which had been remarked as "BUSY node" has less child nodes than the average amount among the same level nodes, the "BUSY node" avoiding mechanism wouldn't be executed. If the nodes which had been remarked as "BUSY" has more child nodes than the average amount among the same level nodes, the "BUSY node" avoiding mechanism would be executed. One of the child nodes of this parent node would be chosen to avoid the "BUSY node". While one of the parent nodes satisfied the avoiding condition above, the chosen of one of the child node as follows: 1) the value of U of the child node is the minimum of all the child nodes. 2) In the case of condition 1, BE $\geqslant$ ETX. The chosen of suboptimal parent node for the child nodes as follows: 1) BE $\geqslant$ ETX. 2) In the case of condition 1, the original intensity of data transmission of suboptimal parent node is the lowest of all the possible choices.

As the figure 1 shows below, node2, node 3 and node 4 are the same level nodes of level 1. First, in level 1, node 2, 3, 4 broad cast the intensity of their own intensity of data transmission "U" and the number of one hop up nodes "OHUS". As figure 2 shows below, node 2 has a U=50% and OHUS= 3.

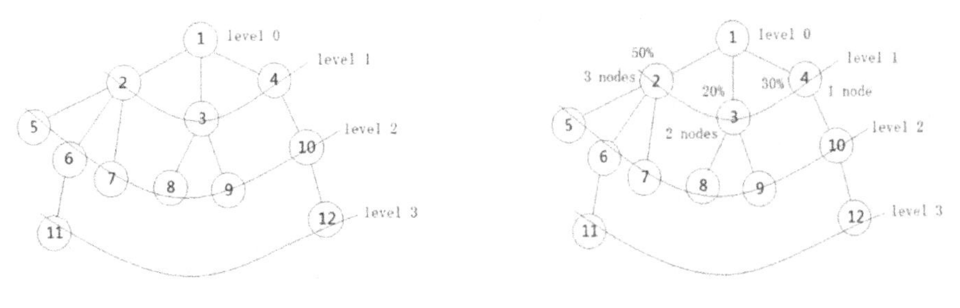

Fig. 1. Fig. 2.

Thus, every level 1 node will know other nodes' condition in level 1. At this time in level 1, mean value should be calculated. MVI= 33.3% and MVN= 2.

While node 2's U > MVI, the node would be remarked as "BUSY node", as figure 3 shows.

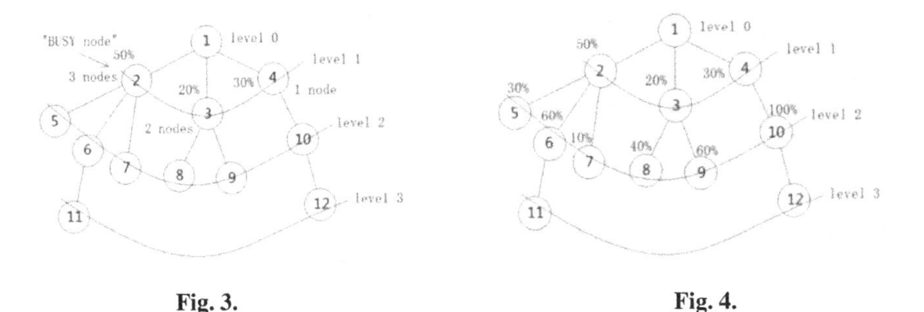

Fig. 3. Fig. 4.

Also, the child node of node 2 > MVN, the "BUSY node" avoiding mechanism would be executed and one of the child nodes of this parent node would be chosen to avoid the "BUSY node". In level 2, the nodes are all the child nodes of the nodes in level 1. The value of U shows in figure 4.

Because of node 7 has the minimum U among node 5, node 6 and node 7, node 7 would be chosen as the alternative node to execute the avoiding mechanism. (Here we SUPPOSE that the link quality between node 3 and node 7, node 4 and node 7 are all satisfied the transmission requirement, which means BE $\leqslant$ ETX.)

Because of node 3 has the lowest transmission intensity in level 1. Thus, node 3 would be chosen as the suboptimal parent node for node 7. Finally, the topology of the net work shows below:

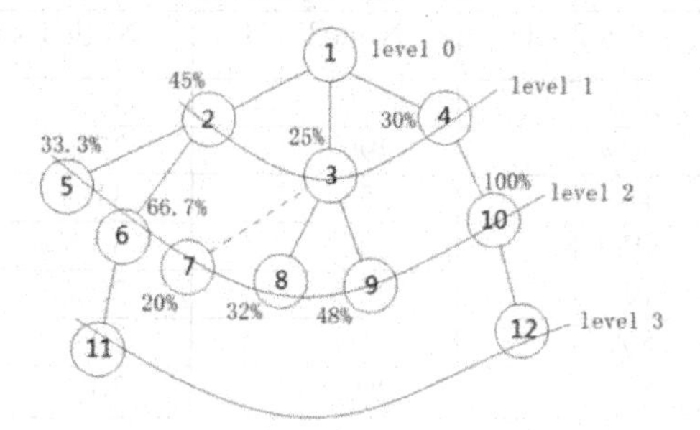

3.3 The Performance Evaluation Index of Load Balance for the Same Level Nodes

In order to judge the performance of the CTP-TICN protocol, we need to have an evaluation index for the load balance of the network. Thus, we put forward a new evaluation index named PEI-LB:

$$\text{PEI-LB}=\sqrt{\frac{\Sigma(\text{U}-\text{MVI})^2}{|\,\text{SLSn}\,|}}$$

From the formula we can find that PEI-LB means the standard deviation of the intensity of the data transmission in the same level. The more the PEI-LB close to 0, the better the network has load balance performance, vice versa.

4 Simulation

In order to test whether the idea of CTP-TICN protocol function well, there exist the need to simulate for this new protocol. Next, we will focus on the simulation of the CTP-TICN protocol. We still use the topology of the network in chapter 3, Figure 2. The SameLevelSet0 = {1}, SameLevelSet1 = {2, 3, 4}, SameLevelSet2 = {5, 6, 7, 8, 9, 10}, SameLevelSet3 = {11, 12}.

The simulation was finished in the TOSSIM[6,7] simulation platform. The type of the simulation nodes were chosen as Mica[9] Z wireless sensor node. Mica Z node uses ATMega128L as its central processing unit. The radio frequency chip of Mica Z was CC2420, it supports 802.15.4/Zigbee technology. Thus, no more telecommunication protocol was needed to develop. In software simulation, the nodes broadcast the routing protocol information to other nodes in the network every 7500ms. From the

TOSSIM flow information, the figure below shows the detailed information of the intensity of data transmission "U" in level 1. Every node has defined initial transmission intensity. At the beginning, Node2, Node3 and Node4 didn't know other nodes' information, it means the topology has not been formed and the three nodes only have their own data flow. So their transmission intensities are nearly the same. After a while, the topology began to form and the transmission intensity differs due to the EETX of the link quality. In this condition, the CTP protocol needs almost 30 second to form the topology in 3 levels. The simulation of CTP protocol shows below:

CTP Routing Protocol

Node ID / Times	Node2's U	Node3's U	Node4's U	SUM of U
7500ms	35%	33%	32%	100%
15000ms	39%	29%	32%	100%
22500ms	44%	25%	31%	100%
30000ms	49.5%	20%	30.5%	100%
37500ms	49.5%	20.5%	30%	100%
Stable	50%	21%	29%	100%
Stable	49.5%	20.5%	30%	100%

The table below shows the information of CTP-TICN protocol in simulation. CTP-TICN routing protocol need more time in topology.

CTP-TICN Routing Protocol

Node ID / Times	Node2's U	Node3's U	Node4's U	SUM of U
7500ms	35%	33.5%	31.5%	100%
15000ms	37%	32%	31%	100%
22500ms	39.5%	30.5%	30%	100%
30000ms	42%	27%	31%	100%
37500ms	45%	25%	30%	100%
45000ms	45%	25.5%	30.5%	100%
Stable	44.5%	25%	30.5%	100%

Table & figure below shows the difference of PEI-LB between CTP and CTP-TICN.

Value of PEI-LB

PEI-LB / Times	PEI-LB for CTP	PEI-LB for CTP-TICN
7500ms	1.53	1.76
15000ms	5.13	3.21
22500ms	9.71	5.35
30000ms	14.95	7.77
37500ms	14.8	10.41
45000ms	14.98	10.13
52500ms	14.78	10.1

Figure of PEI-LB

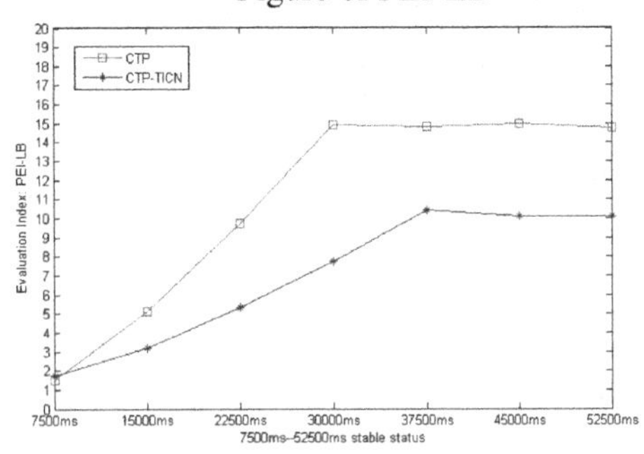

Obviously, the CTP-TICN has less evaluation index for the load balance, which means the CTP-TICN routing protocol has a better ability to balance the load of the network and the energy consumption is more equilibrium than the CTP routing protocol.

5 Ending

In the network, especially the wireless sensor network, huge quantity of aspects in routing protocol needed to be discussed carefully. This passage introduces a new TinyOS 2.x routing protocol named CTP-TICN which had already achieved the preliminary success in load balance for the WSN system. The "intensity of transmission" and "numbers of one hop up node" are put forward after consideration carefully. The simulation shows that these mechanisms solve the load balance in WSN system successfully and the PEI-LB evaluation index could present the degree of load balance in network perfectly. However, there still exist some problems in CTP-TICN. Introducing the load balance mechanism into the network system means the nodes must waste more resources in arithmetic for the routing selection, also, the function of the CTP-TICN is still limited because of its judging condition. Sometimes there exists better topology for the entire network in load balance. These problems are the tendency of further discussion and research.

References

1. Akyildiz, I.F., Su, W., Sankarasubramaniam, Y., et al.: Wireless sensor networks: a survey. Computer Networks, 393–422 (2002)
2. Yi, A., Wang, B.-W., Hu, X.-Y., Yang, W.-J.: Mint-Route-HNLB: A Novel Wireless Sensor Network Routing Protocol with Load Balancing. Computer Science 37(5) (May 2010)
3. Omprakash, G., Rodrigo, F., Kyle, J., et al.: CTP: robust and efficient collection through control and data plane integration. Stanford Information Networks Group, Tech Rep: SING-08-02 (2008)
4. Omprakash, G., Rodrigo, F., Kyle, J., et al.: Collection tree protocol. Stanford Information Networks Group, Tech Rep: SING-09-01 (2009)
5. Chen, J., Yang, Z.-Y., Wang, M.: Research and Simulation of CTP Protocol for wireless Sensor Network. Modern Electronics Technique 34(6) (March 2011)
6. Levis, P., Madder, S., Polastre, J., et al.: TinyOS: An operating system for wireless seneor networks. In: 6th International Symposium Information Processing in Sensor Networks, pp. 254–263 (2007)
7. Ren, F.-Y., Lin, C.: Wireless sensor network. Journal of Software 14(07), 1282–1291 (2003)
8. TinyOS: a component-based OS for the networked sensor regime [EB/OL], http://webs.cs.berkeley.edu/tos
9. Mica motes specifications [EB/OL], http://www.Xbow.com/Products/WirelessSensorNet-work.htm
10. TinyOS Homepage [EB/OL], http://www.tinyos.net
11. David, R., Ignas, G.N.: Ad Hoc networking in future wireless communications. Computer Communications 26(1), 36–40 (2003)

Adaptive Nonlinear Guidance Law Considering Autopilot Dynamics

Dan Yang and Liang Qiu

Department of Automation and Key Laboratory of System Control and Information Processing, Ministry of Education of China
Shanghai Jiao Tong Univercity, Shanghai, China
j.kyu125@163.com

Abstract. A third-order state equation with consideration of autopilot dynamics is formulated. A nonlinear coordinate transformation is used to change the state equation into the normal form. Adaptive sliding mode control theory is depicted and its stability analyses are proved. Applying this theory on the normal form equation, an adaptive nonlinear guidance law is proposed. The presented law adopts the sliding mode control approach can effectively solve the guidance problem against target maneuver and effects caused by autopilot dynamics. Simulation results show that under the circumstance of target escaping with high acceleration and big autopilot dynamics, the proposed guidance law still has high precision.

Keywords: adaptive nonlinear guidance law, autopilot dynamics, sliding mode control.

1 Introduction

In the guidance area, there are, in general, two approaches based on either the classical approach or on modern control theory. The proportional navigation guidance (PNG) [1] used in classical approach need a feedback with constant gain from the angle rate of line of sight. This approach is easy to implement and efficient. Nevertheless, owing to its degradation with target maneuvering and inefficient in some situations, many improved methods have been proposed, such as augmented proportional navigation (APN) [2], generalized true proportional navigation (GTPN) [3], and realistic true proportional navigation (RTPN) [4].All these approaches have been approved to improve control performance. It's evident that they also result in the complexity in designing and analysis of the system simultaneously.

Research on modern methods relative to guidance law is increasingly active. Lots of papers have been proposed in this research area. The optimal control theory has been used to develop the proportional navigation [5]. Sliding mode control method has been proposed applying in guidance of homing missile [6]. Using these methods, the integrated missile guidance and control system can be designed easily. Compared with traditional methods, these approaches obtain excellent robustness, when the system exists disturbance and parameter perturbation, and are adaptive to the target maneuvering and guidance parameter changes.

Y. Wang and T. Li (Eds.): Knowledge Engineering and Management, AISC 123, pp. 11–17.
springerlink.com

In this paper, an adaptive nonlinear approach considering autopilot dynamics is proposed. In the guidance area, missile autopilot dynamics is one of the main factors affecting Precision-guided. In the proposed paper, the target maneuvering is taken into account as bounded disturbances. The angle of LOS acts as zero output state variable. Lyapunov stability theory is used to design an adaptive nonlinear guidance (ANG) law.

The rest of this paper is organized as follows. Section 2 derives the model considering autopilot dynamics. In Section 3, an adaptive guidance law considering target uncertainties and autopilot dynamics is presented. Also the analysis of the stability of the design is derived. In Section 4, the simulation for the presented guidance law is provided. The conclusions are given in Section 5.

2 Model Derivation

The three-dimensional pursuit geometry is depicted in Fig.1.The Line of Sight (LOS) coordinate system ($OX_4Y_4Z_4$) is chosen as reference coordinate system. The original point O is at the missile's mass center, $OXYZ$ is the inertial coordinate system. OX_4 is the line of sight of the initial moment of terminal guidance, the missile to the target direction is positive[7].

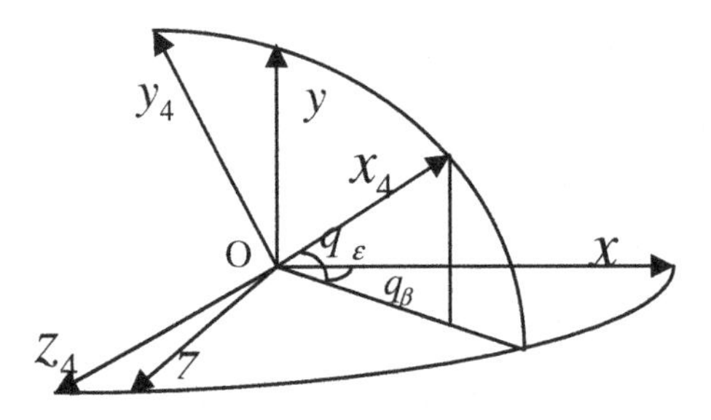

Fig. 1. Three-dimensional pursuit geometry

In the homing process, the acceleration vectors exert on missile and target only change the direction rather than the magnitude of the speed. If missile doesn't rotate, the relative motion between missile and target can be decoupled into two relative independent movements. In this paper, the vertical plane OX_4Y_4 is chosen as an example. Suppose that, during Δt, the incremental of the LOS angle is q. If the time interval is small enough, q is very small. There exists an approximate equation

$$q(t) \approx \sin q(t) = \frac{y_4(t)}{R(t)} \tag{1}$$

In this equation, $R(t)$ represents the relative distance between missile and target.

$y_4(t)$ represents the relative displacement on OY_4 direction during Δt.

The quadratic differential form of (1) can be expressed as

$$\ddot{q} = -a_{g1}q - a_{g2}\dot{q} - b_g a_M + b_g a_T \tag{2}$$

where $a_{g1} = \ddot{R}(t)/R(t), a_{g2} = 2\dot{R}(t)/R(t)$, $b_g = 1/R(t)$, a_M and a_T are the missile and target acceleration.

The first-order autopilot's dynamic characteristic is briefly described as follows

$$\dot{a}_M = -\frac{1}{\tau}a_M + \frac{1}{\tau}a_{MC} \tag{3}$$

where a_{MC} is guidance command provided to autopilot, τ is time-constant of autopilot.

A state space equation is formulated by considering equation (2) and (3) as follows

$$\dot{X}_M = \begin{pmatrix} 0 & 1 & 0 \\ -a_{g1} & -a_{g2} & -b_g \\ 0 & 0 & -\frac{1}{\tau} \end{pmatrix} X_M + \begin{pmatrix} 0 \\ 0 \\ \frac{1}{\tau} \end{pmatrix} u + \begin{pmatrix} 0 \\ f_M \\ 0 \end{pmatrix} \tag{4a}$$

$$Y_M = X_{M2} = \dot{q} \tag{4b}$$

where $X_M = (q \quad \dot{q} \quad a_M)^T$, $u = a_{MC}$, $f_M = b_g a_T$.

In order to apply nonlinear control theory in this problem, the state equation must be transformed into normal form. The output is

$$y = Y_M =: x_1 \tag{5}$$

Differentiating the equation (5) yields

$$\dot{x}_1 = -a_{g1}X_{M1} - a_{g2}x_1 - b_g X_{M3} + b_g a_T \triangleq x_2 + \bar{\Delta}_1 \tag{6}$$

where

$$x_2 = -a_{g1}x_{M1} - a_{g2}x_1 - b_g x_{M3}, \bar{\Delta}_1 = b_g a_T \ . \tag{7}$$

Differentiating x_2 yields

$$\dot{x}_2 = -(a_{g1} + \dot{a}_{g2})x_1 - a_{g2}x_2 - \dot{a}_{g1}x_{M1} - (\dot{b}_g - b_g \frac{1}{\tau})x_{M3} b_g \frac{1}{\tau} u + \bar{\Delta}_2 \tag{8}$$

where $\bar{\Delta}_2 = -a_{g2}\bar{\Delta}_1$.

Assign $x_3 = x_{M1} = q$. From equation (7), x_{M3} can be expressed in terms of x_1, x_2 and x_3 ,therefore x_{M3} can be eliminated from the equation (8)

$$\dot{x}_2 = a_1 x_1 + a_2 x_2 + a_3 x_3 - b_g \frac{1}{\tau} u + \overline{\Delta}_2 \tag{9}$$

where $a_1 = -(a_{g1} + \dot{a}_{g2} - \frac{\dot{b}_g}{b_g} a_{g2} + \frac{1}{\tau} a_{g2})$, $a_2 = -(a_{g2} - \frac{\dot{b}_g}{b_g} + \frac{1}{\tau})$, $a_3 = -(\dot{a}_{g1} - \frac{\dot{b}_g}{b_g} a_{g1} + \frac{1}{\tau} a_{g1})$.

Denoting $\tilde{X} = (x_1 \quad x_2 \quad x_3)^T$, a state space equation is formulated. Then assign $X = (x_1 \quad \dot{x}_1 \quad x_3)^T$, a feedback linearization technique can be applied to have

$$\dot{X} = \begin{pmatrix} 0 & 1 & 0 \\ a_1 & a_2 & a_3 \\ 1 & 0 & 0 \end{pmatrix} X + \begin{pmatrix} 0 \\ -\frac{1}{\tau} b_g \\ 0 \end{pmatrix} u + \begin{pmatrix} 0 \\ f \\ 0 \end{pmatrix} \tag{10a}$$

$$y = x_1 \tag{10b}$$

where $f = b_g \dot{a}_T + \frac{b_g}{\tau} a_T$.

3 Adaptive Nonlinear Guidance Law

In this section, a general approach of adaptive sliding mode control is depicted for multi-variable nonlinear system. Then its application on guidance law is formulated.

3.1 Adaptive Sliding Mode Control

The state space equation of the nonlinear system is

$$\dot{X} = F(X,t) + \Delta F(X,t) + G_1(X,t)U(t) + G_2(X,t)W(t) \tag{11}$$

where $X \in R^n, U \in R^m, W \in R^l$, they are the state variables, control input and disturbance separately. $\Delta F(X,t)$ is the structure perturbation of the system.

Assume the above nonlinear equation satisfies the following conditions

1) $0 < \| W(t) \| < a$, a is a positive constant.

2) $\Delta F(X,t) = E(X,t)\delta(X,t), E(X,t) \in R^{n \times n}$, $\| \delta(X,t) \| \le b, b > 0$, $E(X,t)$ is known, b is a constant.

Define the sliding surface

$$s = CX \tag{12}$$

where $C \in R^{l \times n}$ is a constant matrix.

Then control law is designed based on Lyapunov stability theory. First a Lyapunov function is proposed

$$V = \frac{1}{2}s^T s + \frac{1}{2\gamma_1}\tilde{a}^2 + \frac{1}{2\gamma_2}\tilde{b}^2 \tag{13}$$

where $\tilde{a}$ and $\tilde{b}$ are the error estimates of a and b separately. $\tilde{a} = a - \hat{a}, \tilde{b} = b - \hat{b}$, $\gamma_1, \gamma_2 > 0$ are design parameters.

Take time derivative of (13) and consider equation (11) and (12),

$$\dot{V} = s^T\dot{s} - \frac{1}{\gamma_1}\tilde{a}\dot{\hat{a}} - \frac{1}{\gamma_2}\tilde{b}\dot{\hat{b}} = s^T(CF + C\Delta F + CG_1U + CG_2W) - \frac{1}{\gamma_1}\tilde{a}\dot{\hat{a}} - \frac{1}{\gamma_2}\tilde{b}\dot{\hat{b}} \tag{14}$$

If CG_1 is nonsingular, define U as follows

$$U = (CG_1)^{-1}(-CF - Ks - \varepsilon sign(s)) \tag{15}$$

Substituting (15) to (14) and combining the assumptions yields

$$\dot{V} \le -Ks^T s - \varepsilon \| s \| + \| s \| \, \| CE \| b + \| s \| \, \| CG_2 \| a - \frac{1}{\gamma_1}\tilde{a}\dot{\hat{a}} - \frac{1}{\gamma_2}\tilde{b}\dot{\hat{b}} \tag{16}$$

Denote $\varepsilon = \| CE \| \hat{b} + \| CG_2 \| \hat{a}$, it can be derived that

$$\dot{V} \le -Ks^T s + \| s \| \, \| CG_2 \| \tilde{a} + \| s \| \, \| CE \| \tilde{b} - \frac{1}{\gamma_1}\tilde{a}\dot{\hat{a}} - \frac{1}{\gamma_2}\tilde{b}\dot{\hat{b}} \tag{17}$$

Take the adaptive law as

$$\dot{\hat{a}} = \gamma_1 \| s \| \, \| CG_2 \| \tag{18}$$

$$\dot{\hat{b}} = \gamma_2 \| s \| \, \| CE \| \tag{19}$$

Thus $\dot{V} \le -Ks^T s < 0$, that means $\lim_{t\to\infty} V(t) = 0$, and then $s \to 0$, $\tilde{a}, \tilde{b} \to 0$ the sliding mode is asymptotically reached. The final control law can be expressed as

$$U = -(CG)^{-1}[CF + Ks + (\| CE \| \hat{b} + \| CG_2 \| \hat{a}) sign(s)] \tag{20}$$

3.2 Application on Guidance Law

Select the sliding model s as

$$s = c_1 x_1 + c_2 \dot{x}_1 + c_3 x_3 \tag{21}$$

Assume $|f| \leq M$, apply the above theory on equation (12) yields

$$u = \frac{\tau}{b_g c_2}[(a_1c_2 + c_3)x_1 + (a_2c_2 + c_1)\dot{x}_1 + a_3c_2x_3 + Ks + \varepsilon sign(s)] \tag{22}$$

where $\varepsilon = c_2\widehat{M}$, and the adaptive law is $\dot{\widehat{M}} = c_2\gamma_2 |s|$.

If the guidance law and adaptive law are designed as the above equations, missile can hit the target in a uniform speed. The angle of LOS and acceleration can be stable.

Actually, in the process of terminal guidance, $\dot{R} \approx \dot{R}_0, \ddot{R} \approx 0$, so the final guidance law can be simplified by substituting these two equations.

4 Simulation

In this section, a simulation comparison between the proposed guidance law and PNG in the vertical plane is presented. The target maneuvering is $a_T = 20g\sin(0.4\pi t)$.

During the terminal guidance, the velocity of target and missile are invariable. $V_T = 300m/s$,$V_M = 500m/s$ .The initial distance of missile and target is $6000m$. The angle between the missile velocity and LOS is 60° , the angle between the target velocity and LOS is 120° .Design parameters of guidance are $K = -2\dot{R}_0/R$, $c_1 = 2$, $c_2 = 1$, $c_3 = 0$, $\tau = 0.5s$, $\gamma_2 = 6$ [8]. If the time-constant of autopilot is small enough, the guidance law can be simplified as $a_M = -N\dot{R}_0\dot{q}$. That is PNG.

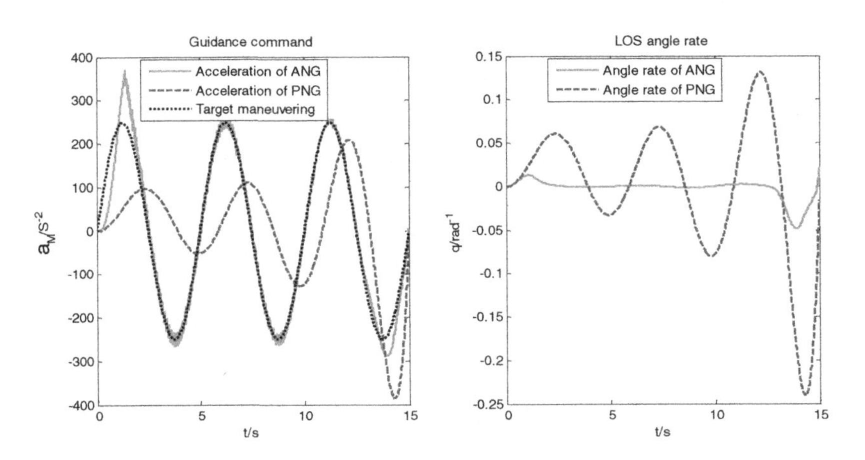

Fig. 2. Acceleration of ANG and PNG), LOS angle rate of ANG and PNG

The first figure of Fig. 2 shows that the normal overload of proportional navigation guidance (PNG) law lags behind target maneuvering significantly. The trend of response property is up. That's because that the PNG reaction to the target maneuvering is slow and indecisive. What's more, the time delay existing in the system also has an effect on the response speed of PNG. However, the normal overload of adaptive nonlinear guidance (ANG) law can keep up with the target maneuvering in real-time after stable. The result shows that ANG has the ability to predict the acceleration of the target. Therefore the proposed guidance law has good adaptability. The results in the second figure of Fig.2 shows that the LOS angle rate become different between ANG and PNG. From the second figure, it is found that the LOS angle rate of ANG stays closely with zero, thus the guidance accuracy is improved. Compared with ANG, the LOS angle rate's amplitude of PNG changes greatly because of the impact of target maneuvering and inertia of autopilot. It can be confirmed that the proposed guidance law can effectively compensate for the target maneuvering and can improve the robustness of the system.

5 Conclusion

In this paper, a guidance law with consideration of autopilot dynamics is proposed. The target acceleration is considered as uncertainties which have bounded disturbances. As the proposed guidance law adopts sliding mode control, the disturbances can be come over. The adaptive law based on Lyapunov stability theory is an estimation of uncertainties. This adaptive law can not only make the system stable but also improve the control performance and the precision of the missile guidance.

References

1. Pastric, H.J., Setlzer, S., Warren, M.E.: Guidance laws for short range homing missile. Journal of Guidance, Control, and Dynamics, 98–108 (1981)
2. Siouris, G.M.: Comparison between proportional and augmented proportional navigation. Nachritentechnische Zeitschrift, 278–228 (1974)
3. Yang, C.D., Yeh, E.B., Chen, J.H.: The closed-form solution of generalized proportional navigation. Journal of Guidance and Control AES-10(2), 216–218 (1987)
4. Yuan, P.J., Chern, J.S.: Solutions of True Proportional Navigation For Maneuvering And Nonmaneuvering Targets. Journal of Guidance, Control, and Dynamics 15(1), 268–271 (1992)
5. Yang, C.D., Yeh, F.B.: Optimal proportional navigation. Journal of Guidance, Control, and Dynamics 11(4), 375–377 (1988)
6. Choi, J.Y., Chwa, D., Kim, M.S.: Adaptive Nonlinear Guidance Law Considering Control Loop Dynamics. IEEE Transactions on Aerospace and Electronic Systems 39(4), 1134–1143 (2003)
7. Zhou, D.: New Guidance Law for Homing Missile. National Defence Industrial Press, Beijing (2002)
8. She, W.X., Zhou, J., Zhou, F.Q.: An Adaptive Variable Structure Guidance Law Considering Missile's Dynamics of Autopilot. Journal of Astronautics 24(3), 245–249 (2003)

An Immune-Enhanced Unfalsified Controller for a High-Speed Spinning Process

Lezhi Wang[1], Yongsheng Ding[1,2,*], Kuangrong Hao[1,2], and Xiao Liang[1]

[1] College of Information Sciences and Technology, Donghua University, 201620 Shanghai, China
[2] Engineering Research Center of Digitized Textile & Fashion Technology, Ministry of Education, Donghua University, 201620 Shanghai, China
ysding@dhu.edu.cn

Abstract. One key process in data-driven methodology is how to use the data easily and efficiently. In this paper, an immune-enhanced unfalsified controller (IEUC) is proposed to act as an efficient process to deal with data. The IEUC consists of two parts, the first part is a unfalsified controller deriving from data-driven methodology; the second part is an immune feedback controller inspired from biologic intelligent methodology. In order to examine control effectiveness of the IEUC, we apply it to a complex plant in high-speed spinning model and compare it with a simple unfalsified control scheme. Simulation results demonstrate that the IEUC can decrease system overshoot as well as reduce rising time successfully and effectively.

Keywords: data-driven, unfalsified control, immune feedback control, high-speed spinning process.

1 Introduction

Data-driven, the beginning and the end of control process are totally based on data, is a more precise illustration of the relationship between data and control system. Data-driven control is a method which entitles data with the power of describing a control system and requires no mathematic model about control process [1].

In the past few years, data-driven methodology has aroused the interests of many researchers. Spall proposed a direct approximation control method using simultaneous perturbation stochastic approximation (SPSA) [2]. Hou and Han built the theory of model free adaptive control (MFAC) [3]. Safonov and Tsao proposed the unfalsified control (UC) [4]. Guadabassi and Savaresi proposed Virtual reference feedback tuning (VRFT) [5]. From the existing studies, we can see that present ameliorative methods could be roughly divided into three parts: the improvement on certain algorithm's performance [6-8]; the mixture of two or more control methodologies [9]; and the combination of intelligent algorithm and data-driven controller [10]. Meanwhile, an interdisciplinary method of model-based control—biologic intelligent control, a

* Corresponding author.

Y. Wang and T. Li (Eds.): Knowledge Engineering and Management, AISC 123, pp. 19–24.
springerlink.com

sophisticated process of simulate human body, is a perfect data processing system—has impressed many researchers [11].

In this paper, we present an immune-enhanced unfalsified controller (IEUC) from biologic principle of immune feedback system and data-driven principle of unfalsified control. In the IEUC, unfalsified control unit is used to select suitable controllers and immune control unit is used to adjust system performance. In order to examine the effectiveness of the IEUC, we apply it to a plant of high-speed spinning model.

This paper is organized as follows. Design of the IEUC is presented in Section 2. Then, the IEUC algorithm is presented in Section 3. We apply the IEUC to a high-speed spinning process in Section 4. Finally, summary and concluding remarks are given in Section 5.

2 Design of the Immune-Enhanced Unfalsified Controller

A novel immune-enhanced unfalsified controller (IEUC) is presented as shown in Fig. 1. The controller uses unfalsified control to evaluate preset candidate controllers and immune control to regulate control performance.

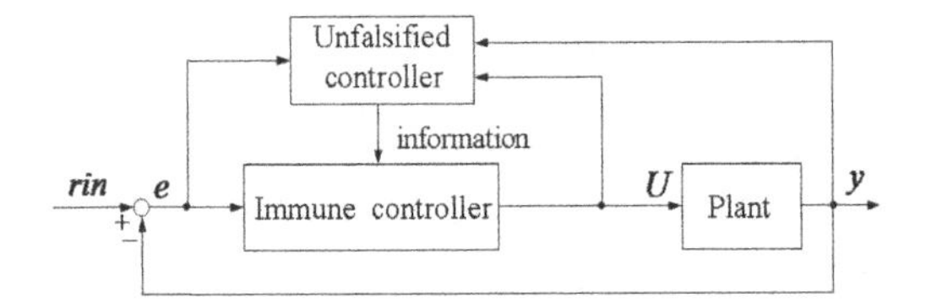

Fig. 1. The structure of the IEUC

The unfalsified control part is based on output ($yout$), control signal (U), and error (e). It transmits selected control's information to immune control part. Based on P-type immune feedback mechanism [11], the immune controller uses this information as well as error signal of system to modulate the control signal of system. In the IEUC, the unfalsified controller and immune controller can control the plant harmoniously, and hold both the advantage of data-driven methodology and biologic intelligent methodology.

For the scope of control system, the unfalsified control scheme (or the unfalsified controller, UC) is a way to build satisfactory controllers with experimental data rather than rely on feigned hypotheses or prejudicial assumptions about the plant [12]. The establishment of a qualified UC is to evaluate each candidate controller from a predefined controller set before put into feedback system which can be visually expressed as "controller sieve". Then, the "controller sieve" makes a strict evaluation of candidate controllers based on input (goal) as well as error (e). After the sieve process, candidate controllers are divided into two parts: reject controllers and accept controllers. Then, the reject controllers turn to falsified controllers as well as the accept controllers turn to unfalsified controllers. And the first unfalsified controller is selected to system loop as a current controller.

This paper builds a model based on immune feedback mechanism to adapt system control performance. The main cells involved in the model are antigen (Ag), antibody

(Ab), B cells (B), help T cells (T_H), suppressor T cells (T_s), and antigen presenting cells (APC) [11]. When Ag invades organisms, it is firstly recognized by APC. Then, the APC sends recognition message to T_H, and T_H secretes the interleukin to activate T_s and B cells. As the T_s receives stimuli, it secretes another kind of interleukin to inhibit the active of T_H. Meanwhile, as the activated B cells begin to divide, their offsprings (plasma cells) secrete Ab to recognize Ag. After a while, when Ag is prevented by Ab, a dynamic relative equilibrium point can be reached, and the immune response is finished.

3 The IEUC Algorithm

As the IEUC cannot be easily implemented without a basic control algorithm, a conventional PID control algorithm mixed with the IEUC is proposed. The basic algorithm flowchart of the IEUC is shown in Fig. 2. The left part of Fig. 2 is unfalsified control algorithm and the right part is immune control algorithm.

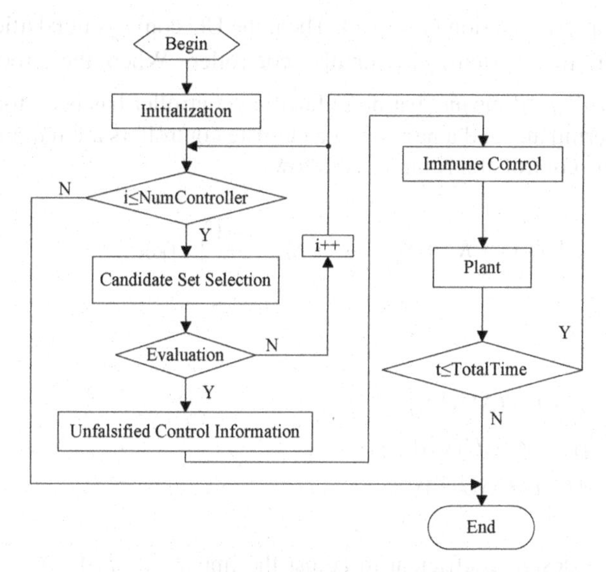

Fig. 2. The basic algorithm of the IEUC

The selection and evaluation process for the candidate set is based on the algorithm of unfalsified control [12]: $r_i(t)$ is the fictitious reference signal, combing with a basic PID-type algorithm:

$$r_i(t) = yout_i(t) + \frac{s}{s \cdot K_p + K_i} \cdot \left(U_i(t) + \frac{s \cdot K_d}{\varepsilon \cdot s + 1} \cdot yout_i(t) \right), \tag{1}$$

where i is current controller number, ε is a small enough value in PID approximate algorithm, K_p, K_i, K_d denotes proportional, integral, and derivative parameter, respectively. Performance specification set T_{spec} is

$$T_{spec}(r_i(t), yout(t), u(t)) = |\omega_1 * (r_i(t) - y(t))|^2 + |\omega_2 * u(t)|^2 - |r_i(t)|^2, \tag{2}$$

where ω_1 and ω_2 are weighting filters depends on user's demand, $*$ denotes the convolution operator.

Moreover, performance evaluation standard E_{value} is

$$\begin{aligned} E_{value}(i, k\cdot t_s) &= E_{value}(i,(k-1)\cdot t_s) + \frac{1}{2}\cdot ts\cdot\{T_{spec}\left(r_i(k\cdot t_s), y(k\cdot t_s), u(k\cdot t_s)\right) \\ &+ T_{spec}(r_i((k-1)\cdot t_s), y((k-1)\cdot t_s), u((k-1)\cdot t_s))\}, \end{aligned} \tag{3}$$

If $E_{value} > 0$, the current controller is falsified, the algorithm will discard the controller and continue the iteration ($i = i + 1$). Then, the UC conveys unfalsified information (K_p, K_i, K_d in this system) to immune controller. When the process comes to $i > NumController$, it means that no unfalsified controller has been found. The algorithm would terminate, and a new set of candidate controllers are required.

The immune control algorithm is as follows:

$$U(k) = \left(K_P + K_I \cdot \frac{1}{z-1} + K_D \cdot \frac{z-1}{z}\right) \cdot e(k), \tag{4}$$

where,

$$\begin{cases} K_P = K_p \cdot \{1 - \eta \cdot f(\Delta U(k))\} \\ K_I = K_i \cdot \{1 - \eta \cdot f(\Delta U(k))\} \\ K_D = K_d \cdot \{1 - \eta \cdot f(\Delta U(k))\} \end{cases}.$$

Where, η is a design coefficient to adjust the impact of $f(\cdot)$, $f(\cdot)$ is a nonlinear function for considering the effect of the reaction of B cells and the antigens. Also, when η=0, the IEUC controller is equal to UC controller. In this paper, the fuzzy controller with two inputs and one output is employed here to approximate $f(\cdot)$. The immune controller signal $U(k)$ and change of the controller signal $\Delta U(k)$ are two inputs variables; the output of system $yout(k)$ is the output variable. The fuzzy control rules and fuzzy functions are shown in [11].

4 Simulation Results

In order to examine the control performance of the IEUC, we consider a three-order plant in the high-speed spinning model,

$$G(s)=\frac{1.059}{0.000942s^3+0.3316s^2+1.1988s} \tag{5}$$

We set the initial parameters as, $NumController = 45$, $t_s = 0.1s$, $TotalTime$=30s, candidate controllers: $K_p = \{5,10,30,110,150\}$, $K_i = \{50,75,100\}$, $K_d = \{20,10,4\}$. Performance evaluation: $W_1(s)=\frac{s+20}{2(s+3)}$, $W_2(s)=\frac{0.01}{1.2(s+1)^3}$. Immune control parameters: $\eta = 0.6$, $l = 1$. Additionally, based on the IEUC algorithm discussed in Section 4, when $\eta = 0$, the IEUC controller is equal to UC controller.

Seen from Fig. 1, in the unfalsified control element, if the mathematical method is available, the unfalsified set of controllers can be easily obtained. However, based on Eqs. (2) and (3), as the progress of sieving, many candidate controllers could be evaluated to be falsified and rejected and it is possible that the finite set Kr of $NumController$ candidates would become empty. When the algorithm is being terminated due to this reason, argument the set Kr with additional candidate controllers or reset of the performance specification need to be done. In this case, a carefully designed set of PID parameters has been employed to act as candidate controllers and a relatively comprehensive performance specification has been adopted. In the immune fuzzy control element, by adjusting η, l in Eq. (4) and fuzzy rules, the fuzzy immune controller can quickly drive the system output to the desired level [11].

Using the IEUC, we obtain desirable control performance for the system as shown in Figs. 3 and 4. The control effectiveness is also illustrated by comparing the performance of the IEUC and the UC. In order to show the contrast effectiveness of the influence on $yout$ with rin changing, the set points of rin at the 10-th , 20-th , 30-th seconds are changed. The simulation results demonstrate that the IEUC is robust and has better control performance than that of the UC.

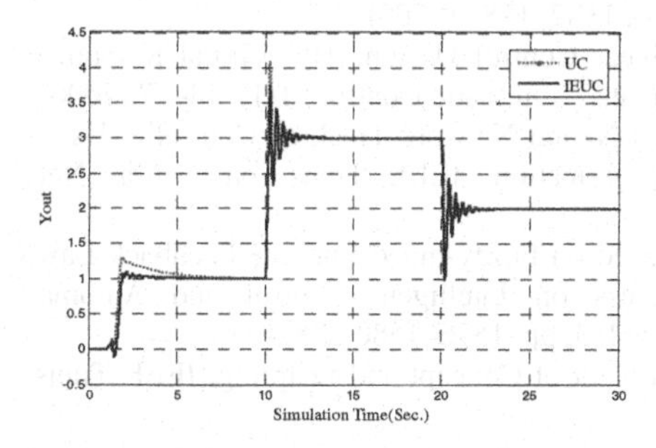

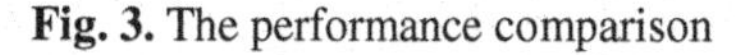
Fig. 3. The performance comparison

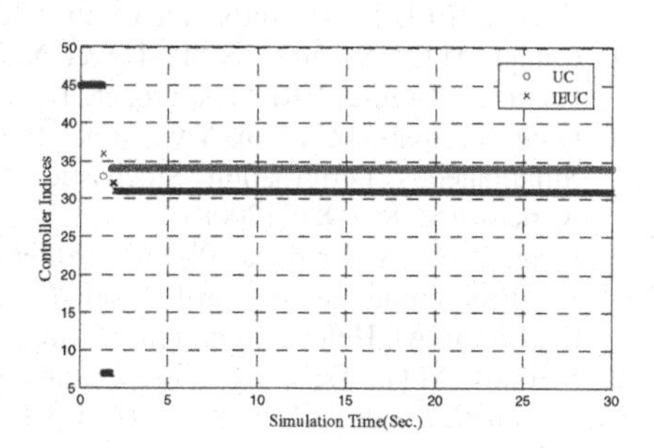

Fig. 4. The change of controller

5 Conclusions

In this paper, an immune-enhanced unfalsified controller and its control scheme are presented and applied to control a complex three-order plant in high-speed spinning model. Simulation results demonstrate that the IEUC can rapidly response to the changing of desired level. Moreover, compared with single unfalsified control algorithm, the IEUC can decrease system overshoot as well as reduce rising time easily and successfully.

Acknowledgments. This work was supported in part by the Key Project of the National Nature Science Foundation of China (No. 61134009), the National Nature Science Foundation of China (No. 60975059), Specialized Research Fund for the Doctoral Program of Higher Education from Ministry of Education of China (No. 20090075110002), and Project of the Shanghai Committee of Science and Technology (Nos. 11XD1400100, 10JC1400200, 10DZ0506500, 09JC1400900).

References

1. Hou, Z.-S., Xin, X.-J.: On Data-driven Control Theory: the Sate of the Art and Perspective. Acta Automatica Sinica 35(16), 350–667 (2009)
2. Spall, J.C.: Multivariate Stochastic Approximation Using A Simultaneous Perturbation Gradient Approximation. IEEE Trans. Automation Control 37(3), 332–341 (1992)
3. Hou, Z.-S., Han, Z.-G.: Parameter Estimation Algorithm of Nonlinear Systems And Its Dual Adaptive Control. Acta Automatica Sinica 21(1), 122–125 (1995)
4. Safonov, M.G., Tsao, T.C.: The Unfalsified Control Concept: A Direct Path From Experiment to Controller. Feedback Control, Nonlinear Systems, and Complexity 202, 196–214 (1995)
5. Guadabassi, G.O., Savaresi, S.M.: Virtual Reference Direct Design Method: An Off-line Approach to Data-based Control System Design. IEEE Trans. Automatic Control 45(5), 954–959 (2000)
6. van Helvoort, J.J.M., de Jager, B., Steinbuch, M.: Direct Data-driven Recursive Controller Unfalsification with Analytic Update. Automatica 43(12), 2034–2046 (2007)
7. Previdi, F., Schauer, T., Savaresi, S.M., et al.: Data-driven Control Design for Neuroprotheses: A Virtual Reference Feedback Tuning (VRFT) Approach. IEEE Trans. Control Systems Technology 12(1), 176–182 (2004)
8. Hildebrand, R., Lecchini, A., Solari, G., et al.: Asymptotic Accuracy of Iterative Feedback Tuning. IEEE Trans. Automatic Control 50(8), 1182–1185 (2005)
9. Campi, M.C., Savaresi, S.M.: Direct Nonlinear Control Design: The Virtual Reference Feedback Tuning (VRFT) Approach. IEEE Trans. Automatic Control 51(1), 14–27 (2006)
10. Qing, S., Spall, J.C., Soh, Y.C., et al.: Robust Neural Network Tracking Controller Using Simultaneous Perturbation Stochastic Approximation. IEEE Trans. on Neural Networks 19(5), 817–835 (2008)
11. Ding, Y.-S.: A Nonlinear PID Controller Based on Fuzzy-tuned Immune Feedback Law. In: Proceeding of the 3rd World Congress on Intelligent Control and Automation(WCICA), Hefei, China, June 28-July 2, vol. 3, pp. 1576–1580 (2000)
12. Safonov, M.G., Tsao, T.C.: The Unfalsified Control Concept and Learning. IEEE Trans. Automatic Control 42(6), 843–847 (1997)

Real-Time Adaptive Task Allocation Algorithm with Parallel Dynamic Coalition in Wireless Sensor Networks

Chengyu Chen, Wenzhong Guo, and Guolong Chen

College of Mathematics and Computer Science, Fuzhou University, Fuzhou 350108, China
chenchengyu53@gmail.com, guowenzhong@fzu.edu.cn, fzucgl@163.com

Abstract. In this paper, we develop a real-time adaptive task allocation algorithm based on parallel dynamic coalition in WSNs. The algorithm gives a priority level to each task according to the idea of EDF. And the task with relatively higher priority will be scheduled firstly. When coalitions are parallel generated through PSO algorithm, the corresponding task of coalition will be allocated according to the current load of sensors and the remaining energy balance degree. The experimental results show that the proposed algorithm has strong capability to meet deadline constraint and it can prolong the lifetime of the whole network significantly.

Keywords: wireless sensor networks, task allocation, dynamic coalition, particle swarm optimization, earliest deadline first.

1 Introduction

Wireless sensor networks (WSNs) are usually made up of a large number of sensor nodes with limited communication and computation abilities. Each sensor is energy-constrained and it always works in a dynamical network environment without human participation. The application of WSNs confronts many challenges.

The issue on task allocation is important for the study of WSNs. In [1], an EBSEL algorithm to extend the network's lifetime through balancing energy consumption among the sensors is developed. But it just considers a single-hop cluster of homogeneous nodes where the communication cost is asymmetric. An energy balance DAG task scheduling algorithm to acquire lowest energy consumption and highest balanced use of energy is presented in [2], however, it mainly focuses on global tasks allocation and will need to consume too much energy to communicate. In [3], a localized cross-layer real-time task mapping and scheduling solutions for DVS-enabled WSNs is proposed, but the scale of nodes is small in simulation experiment.

If task allocation is unreasonable or the network does not make use of these limited resources rationally, the deadline of task may miss; and what's more, it may lead to the imbalance of energy distribution and reduce the lifetime of the whole networks. Therefore, in order to maximize the utilization of resources, prolong the lifetime of the networks under the constraint of task deadline, it is effective that sensor nodes cooperate with one another. Based on these factors mentioned above, an effectively

Y. Wang and T. Li (Eds.): Knowledge Engineering and Management, AISC 123, pp. 25–32.
springerlink.com

real-time adaptive task allocation algorithm based on parallel dynamic coalition (ERATA) is proposed in this paper to solve these issues.

2 Problem Description

There is a WSN, which consists of n heterogeneous sensors in an area of $100*100m^2$, and there are m independent tasks with deadline to be finished. Each task is composed of l attributes and each sensor has different ability for different attributes. In order to meet deadline constraint and prolong the lifetime of WSNs, the target is assigning the m tasks to the n sensor nodes rationally to execute. For simplicity, some assumptions are made as follows:

1) All sensors have different limited remaining energy, which are generated randomly in our simulation.

2) The remaining energy state and ability for different sub-tasks of each sensor in this area can be acquired by the sink node through dynamic topology or route control of WSNs. In this paper, we just focus on task allocation. And nodes can be aware of its residual energy and their location. The sink node is responsible for running the tasks allocation algorithm (proposed in Section 3), and the node can sense tasks and split a task into l different attributes (l sub-tasks) at most.

Some definitions are as follows:

The total energy consumption:

$$E_{sum} = \sum_{i=1}^{m} E_T(i) + \sum_{i=1}^{m} E_C(i) \ . \tag{1}$$

Where, $E_T(i)$ is communication consumption of i-th task; and our communication energy consumption model is based on the first order radio model[4]; $E_c(i)$ is total computation consumption of i-th task. The value of total energy consumption with lower value is better.

Remaining energy balance of networks:

$$Ba = \frac{\sum_{i=1}^{n} | E_i - E_{ave} |}{n} \ . \tag{2}$$

Where, E_i expresses the remaining energy of i-th node. E_{ave} is average remaining energy of the WSNs; Ba denotes the remaining energy balance degree. The smaller value of Ba is, the better. Through adjusting Ba properly, we can prolong the lifetime of WSNs effectively.

Task finish time factor:

$$Gap_i = \begin{cases} F_i - D_i, & if \ F_i \geq D_i \\ 0, & else \end{cases}, \quad (0 \leq i < m) \ . \tag{3}$$

Where, D_i and F_i represent the deadline and finish time of i-th task respectively; if Gap_i is not equal to 0, it shows that i-th task will not be finished before its deadline. The value of Gap_i denotes the precise shortage between deadline and practical finish time of i-th task. The smaller the value of Gap_i, the better. Of course, it would be best that each value of Gap_i is *0*.

3 Algorithm Description

3.1 Dynamic Coalition

As a typical research field of the distributed artificial intelligence, agent theories and technologies play an important role in modern computer science and application. The characteristic of agent is similar to that of sensor. Therefore, some theories of multi-agent would be applied to the algorithm design of WSNs[5], If a sensor node is regarded as an agent, a WSN is a kind of multi-agent system. Based on the characteristic of WSNs, it is not suitable that we adopt the method of dynamic coalition directly, because agents in MAS need to consult with one another repeatedly, which will lead to excessive energy consumption. There are two forms of complicated coalition: multi-tasks coalition and overlapping coalition[6]. In our method, we employ parallel dynamic coalition based on overlapping coalition; and a task is corresponding to a coalition and agents in a coalition cooperate with each other.

3.2 Discrete Particle Swarm Optimization

Particle swarm optimization (PSO) algorithm is a population based on stochastic optimization technique. In this paper, we utilize discrete particle swarm optimization (DPSO) [7] algorithm to generate coalition in parallel.

An $m*n$ matrix X and V represent the position coding and corresponding velocity respectively [8], where:

$$x_{ij} = \begin{cases} 1, & \text{if } j^{th} \text{ node is in } i^{th} \text{ coalition} \\ 0, & \text{otherwise} \end{cases}, \quad (0 \le i < m; 0 \le j < n) . \tag{4}$$

The update equation of the velocity for each particle is described as the follows:

$$V^i_{dj}(t+1) = wV^i_{dj}(t) + c_1 rand_1()(P^i_{dj} - X^i_{dj}(t)) + c_2 rand_2(P_{gd} - X^i_{dj}(t)) . \tag{5}$$

$$X^i_{dj}(t+1) = \begin{cases} 1, & \text{if } r^i_{dj}(t+1) < sigmoid(V^i_{dj}(t+1)) \\ 0, & else \end{cases} . \tag{6}$$

Where, i denotes i-th particle; d represents the d-dimensional vector; j denotes the j-th element of the d-dimensional vector. r_{dj} is a random variable, whose range is [0,1], and sigmoid(V)=1/(1+exp(-V)); in addition, $c_1 = c_2 = 2$. As we known, suitable selection of inertia weight w provides a balance between global and local exploration. The value of w can be set according to the following formula [9]:

$$w = w_{\max} - \frac{w_{\max} - w_{\min}}{t_{\max}} * t \ . \tag{7}$$

Where t_{max} is the maximum iteration number; w_{max} and w_{min} are the maximum and minimum inertia weight respectively. Therefore, w decreases linearly during a run, which is beneficial to the convergence of the algorithm. In our simulation, some parameters of PSO algorithm are set as follows: $w_{max} = 0.9$, $w_{min} = 0.4$.

3.3 Fitness Function

In our model, we make tasks finish before their deadline as far as possible. At the same time, remaining energy balance of networks and less energy consumption are taken into consideration as well. According to the discussion above, we transfer the three factors to a single-objective optimization issue simplistically by adopting three weights. Therefore, fitness function is shown as follows:

$$Fitness = w_1 * \sum_{i=1}^{m} Gap_i + w_2 * E_{sum} / n + w_3 * Ba \ . \tag{8}$$

Where, $w_1 = 0.6$, $w_2 = 0.2$, $w_3 = 0.2$ and $w_1+w_2+w_3 = 1$; during each iteration of DPSO algorithm, we use this function to evaluate a particle. The smaller the value of fitness is, the better the particle is.

3.4 Real-Time Sub-tasks Allocation Algorithm Based on Load and Energy Balance

Step 1: According to formulation (11), select a sensor whose value of *U(i)* is smallest from sensors, which have been in the same task coalition.

Step 2: If the remaining energy of the node is below the average energy of sensors in this coalition, the sensor should be neglected and repeat step 1.

Step 3: For this sensor node, the sub-task corresponding to the strongest ability, which the sensor has, will be picked up. The selected sub-task will be allocated to this sensor node, if the sub-task has not been allocated yet. Otherwise, the sub-task corresponding to the second strongest ability, which the sensor has, will be picked up until the sensor can take over a sub-task of this coalition.

Step 4: The value of *U(i)*, *B(i)* and *EP(i)* corresponding to the sensor should be updated. Then, repeat Step 1, until all the sub-tasks are allocated.

$$B(i) = Bzy_i \Big/ \sum_{j=1}^{nc} Bzy_j \ . \tag{9}$$

$$EP(i) = \frac{1}{e^{-(\frac{ENC_{ave}}{E_i}-1)}+1} \ . \tag{10}$$

$$U(i) = wt_1 * B(i) + wt_2 * EP(i) \ . \tag{11}$$

Where, i denotes i-th sensor; nc and ENC_{ave} represent the sensor number of current coalition and the average remaining energy of sensors in current coalition respectively. Bzy_i ($0 \le i < n$) denotes the busy time of i-th node. In other words, because i-th node is busy handling some task, it is unavailable for i-th sensor to handle other task unless the time Bzy_i goes pass from now on. $EP(i)$ reflects the remaining energy degree of i-th sensor node in current coalition; $B(i)$ signifies the busy degree of i-th node, which shows the load degree of current sensor; wt_1 and wt_2 are weight coefficients set as 0.9 and 0.1 respectively; and $wt_1 + wt_2 = 1$. What's more, Both of $B(i)$ and $EP(i)$ are mapped into the range [0,1].

In this paper, The smaller the value of $U(i)$ is, the better performance of node is. Therefore, we can see that the idle node and the node with more energy are more likely to be picked up.

3.5 Framework of Algorithm

Earliest Deadline First (EDF) is one of the most important dynamic priority algorithms. Besides, it have been proved to be the best dynamic priority algorithm[10]. The priority of a task is inversely proportional to its absolute deadline. In other words, the highest priority task is the one with the earliest deadline.

In this paper, we adopt the idea of EDF to our algorithm. Once all the coalitions are generated by PSO, the coalition with relatively higher priority has higher priority run sub-tasks allocation algorithm. The task allocation algorithm based on parallel dynamic coalition is as follows:

Step 1: Sort m tasks into ascending order according to their deadlines.

Step 2: A swarm of particles are initialized randomly. The position and velocity of each particle are generated randomly. And we get an initial scheme of parallel dynamic coalitions.

Step 3: All coalitions of each particle, runs real-time sub-tasks allocation algorithm respectively according to the result of Step 1.

Step 4: Update previous fitness of each particle and global fitness according to the current value of fitness function; update the position and velocity of every particle. And then, repeat Step 2, if current situation does not meet the ending condition.

4 Simulation Results and Analysis

We will compare the algorithm proposed in this paper with random sub-task allocation algorithm based on parallel dynamic coalition (RSAA)[11] and sub-task allocation algorithm based on parallel dynamic coalition with minimum completion time[12](MCTSAA). Where, the algorithm of MCTSAA is improved by us, load of each sensor has also been taken into consideration.

In this group experiments, both the number of task and sensor node are set as 100. In order to investigate the effect of task deadline, the different deadlines with the same tasks set are given, which are uniformly distributed over [0, 0.5], [0.5, 1], [1, 1.5], [1.5, 2], [2, 2.5], [2.5, 3] respectively.

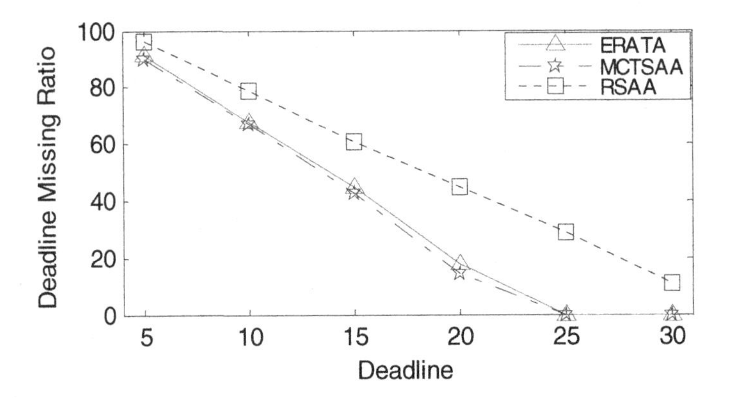

Fig. 1. Deadline missing ratio

As shown in Fig.1, both ERATA and MCTSAA have better capability to meet the deadline constraint compared with RSAA. When deadlines are extremely small, both of algorithms are nearly out of work, because, indeed, most of tasks are impossible at this moment. When deadline increases, the algorithm of ERATA and MCTSAA drops much faster than RSAA, because load of sensors have been taken into consideration in ERATA and MCTSAA. But, ERATA also considers the factor of remaining energy balance. Therefore, ERATA is worse than MCTSAA in the aspect of deadline missing ratio. We can also see that the performance of ERATA is close to that of MCTSAA, which is superior algorithm, because it just considers the factor of deadline. For RSAA, it always randomly chooses nodes to execute tasks; it means that this algorithm hardly thinks of how to improve the performance of parallel coalition to meet the deadline constraint. As a result, the RSAA gets the worst results.

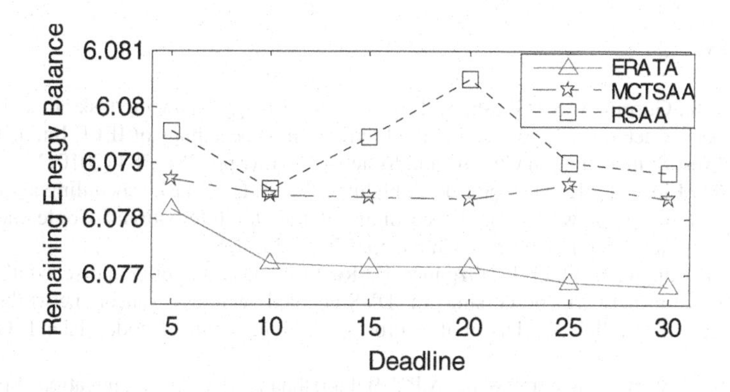

Fig. 2. Remaining energy balance

Regarding remaining energy balance degree of all sensors in Fig.2, ERATA has better performance than MCTSAA and RSAA, because the algorithm of ERATA has not only considered the load of sensors, but also considered the remaining energy balance factor. In ERATA, w_3 and wt_2, which are both corresponding to remaining energy balance weight in equation (8) and equation (11), are just set as 0.2 and 0.1, therefore, when many tasks would not be finished before their deadlines, the algorithm will focus on finding a solution to meet the deadline constraint firstly instead of finding a solution to optimize the energy balance. This can explain that with deadline increasing, the value of remaining energy balance is dropping. As to MCTSAA, it also has better performance than RSAA. MCTSAA is designed to finish tasks as soon as possible and the sensor with small busy time will be considered firstly. As we known, busy time can reflect the current load of sensor. Therefore, the strategy of MCTSAA is beneficial to load balance of sensors; and the load balance is beneficial to the balance of remaining energy of the whole network. So, the strategy of MCTSAA has an indirect effect in keeping remaining energy balance.

5 Conclusion

In this paper, we develop a real-time task allocation based on parallel dynamic coalition in WSNs. In ERATA algorithm, when coalitions are generated through PSO algorithm in parallel, corresponding tasks of coalitions will be allocated according to the current load and remaining energy of sensors. The experimental results show that the proposed algorithm has strong capability to meet deadlines and it can prolong the lifetime of the whole network significantly.

Acknowledgments. This work is supported by the National Natural Science Foundation of China under Grant No. 61103175, the Technology Innovation Platform Project of Fujian Province under Grant No.2009J1007, the Key Project Development Foundation of Education Committee of Fujian province under Grand No.JA11011, the project development foundations of Fuzhou University under Grant No. 2010-XQ-21 and XRC-1037.

References

1. Abdelhak, S., Gurram, C.S., Ghosh, S., Bayoumi, M.: Energy-balancing task allocation on wireless sensor networks for extending the lifetime. In: Proceedings of IEEE International 53rd Midwest Symposium on Circuits and Systems, Seattle, pp. 781–784 (2010)
2. Zeng, Z.W., Liu, A.F., Li, D., Long, J.: A highly efficient DAG task scheduling algorithm for wireless sensor networks. In: Proceedings of the 9th International Conference for Young Computer Scientists, Hunan, China, pp. 570–575 (2008)
3. Tian, Y., Boangoat, J., Ekici, E., Ozguner, F.: Real-time task mapping and scheduling for collaborative in-network processing in DVS-enabled wireless sensor networks. In: Proceedings of Parallel and Distributed Processing Symposium, Rhodes Island, Greece (2006)
4. Heinzelman, W.B., Chandrakasan, A.P., Balakrishnan, H.: An Application Specific Protocol Architecture for Wireless Micro-sensor Networks. IEEE Transactions on Wireless Communications 1, 660–670 (2002)
5. Younis, M., Munshi, P., Al-Shaer, E.S.: Architecture for efficient monitoring and management of sensor networks. In: Marshall, A., Agoulmine, N. (eds.) MMNS 2003. LNCS, vol. 2839, pp. 488–502. Springer, Heidelberg (2003)
6. Zhang, G.F., Jiang, J.G., Xia, N., Su, Z.P.: Solutions of Complicated Coalition Generation Based on Discrete Particle Swarm Optimization. Acta Electronica Sinica 35 (2007)
7. Kennedy, J., Eberhart, R.C.: A discrete binary version of the particle swarm optimization algorithm. In: Proceedings of the IEEE Conference on Systems, Man, and Cybernetics, Orlando, vol. 5, pp. 4104–4109 (1997)
8. Guo, W.Z., Gao, H.L., Chen, G.L., Yu, L.: Particle Swarm Optimization for the Degree-constrained MST problem in WSN Topology Control. In: The International Conference on Machine Learning and Cybernetics, Baoding China, vol. 7, pp. 1793–1798 (2009)
9. Shi, Y., Eberhart, R.C.: A modified particle swarm optimizer. In: Proceedings of the IEEE International Conference on Evolutionary Computation, Piscataway, pp. 69–73 (1998)
10. Chetto, H., Chetto, M.: Some Results of the Earliest Deadline Scheduling Algorithm. IEEE Transaction on Software Engineering 15 (1989)
11. Lesser, V., Ortiz, C.L., Tambe, M.: Distributed sensor networks: a multiagent perspective. Kluwer Academic Publishers (2003)
12. Armstrong, R., Hensgen, D., Kidd, T.: The relative performance of various mapping algorithms is independent of sizable variances in run-time predictions. In: Proceedings of the 7th IEEE Heterogeneous Computing Workshop, pp. 79–87 (1998)

Optimization of Seedlings Transplanting Strategy Based on Greedy Algorithm

Junhua Tong, Huanyu Jiang*, Xiaolong Xu, and Chao Fang

College of Biosystems Engineering and Food Science, Zhejiang University, Hangzhou 310012, China
hyjiang@zju.edu.cn

Abstract. As labor costs rising, development of agriculture robot to alter labor-intensive traditional agriculture is becoming more and more important .Task of transplanting is very heavy and important in bedding plants system in facility agriculture. In this paper, control strategy of transplanting healthy seedlings from high density plug tray to low density growing tray are analyzed. Cause of difficult to get the optimal solution of the whole combination route strategy in the real time, list four typical transplanting scheme, then introduced greedy algorithm to optimize control strategy. Through the simulation, greedy algorithm make the path distance shortened. Controller get approximate optimal solution at the situation of unknown optimal solution, improved the efficiency of transplanting.

Keywords: Greedy algorithm, Seedlings transplanting, Facility agriculture.

1 Introduction

In facility agriculture, seedlings transplanting equipment has many advantages, it can take the place of the traditional manual, reduction of labor intensity, improve the production efficiency, and make seedling transplanting well consistency, solve labor shortage problems, and so on[1]. Healthy seedlings in high density plug tray need to be transplanted into lower density tray for further growth[2]. In this processing, transplanting order of seedlings in the two plug tray need to be plan, combination reasonable transplanting path, all these for the aim of shortest movement distance of mechanical arm with end-effector, and the whole execution cost the least time. This problem is similar knapsack problem and traveling salesman problem, typical and difficult to solve the problem of combination and optimization. The greedy algorithm is a kind of improved grading method, optimization by greedy algorithm, trying to find the solution of the transplanting which approximate to the optimal.

The greedy algorithm is not optimization from the global, its choice is local optimal in a sense, that is the best choice in current situation[5]. Hope that the end result is an optimal solution through do choose by greedy algorithm every time. In some cases, even if we can't get whole optimal solution by greedy algorithm, the result is the approximation solution to optimal.

* Corresponding author.

Y. Wang and T. Li (Eds.): Knowledge Engineering and Management, AISC 123, pp. 33–39.
springerlink.com

2 Materials and Methods

Seedlings transplanting automation equipment in facility agriculture mainly comprise four parts: plug tray conveying system, machine visual identity system, control system and transplanting system[3]. Physical of the transplanter are shown in figure 1.

Fig. 1. Physical map of the robotic transplanter in greenhouses

Identify the healthy state of seedlings by machine vision system, get the position of seedlings in plug tray, then plug tray conveying system transport the plug tray to the transplanting system with three coordinate system. At last, control system drive the mechanism operation according to the transplanting strategy.

2.1 Analysis of Transplanting Control Strategy

The seedling transplanting equipment operate with mechanical arm in rectangular type coordinate, and linear motion as the foundation unit. Movement mechanism act in Cartesian coordinate system with three degrees of freedom in X, Y, Z space.

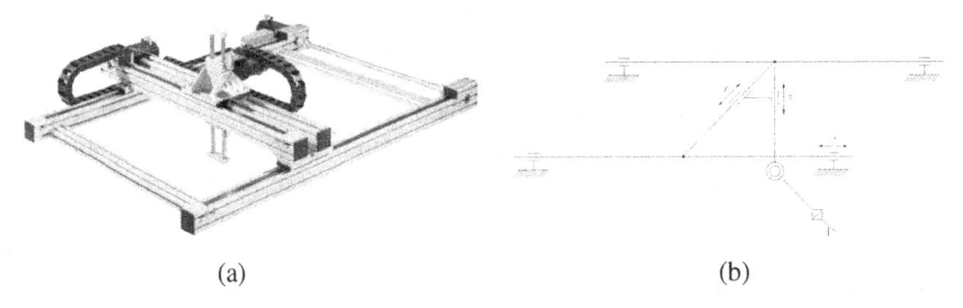

(a) (b)

Fig. 2. Institutions and joint composition of Cartesian type manipulator
(a) Mechanism arm with Rectangular coordinate (b) Schematic diagram of institutions

All these are driving by servo-motor, can realize the movement in all three directions at the same time. The work space is a rectangular space, arm can take the end-effector to any point in this area, the schematic are shown in figure 2.

In the process of transplanting, seedlings transplanted from transplanting to aim tray, it need to make a reasonable control strategy, make sure the path planning. Simplified model are shown in figure 3, (a) aim tray is empty, waiting for transplanting seedlings, (b) small circle in transplanting tray stand for seedlings need to be transplanted, and the vacancy stand for no sprout or no healthy seedling point suit for transplanting. Mechanism arm move from transplanting tray to the aim tray, and then move back to transplanting tray. Such a reciprocating process, path planning is essential, save time and improve efficiency.

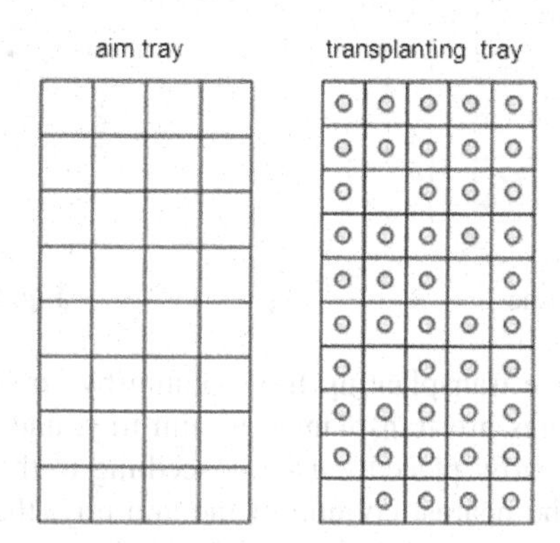

Fig. 3. The position of transplanting seedlings

In the model, size of plug tray is 250 mm x 500 mm, the distance between the two plug tray is 100 mm. the specification of transplanting tray is 5 x 10, and the aim tray is 4 x 8, suppose there are 0 to 9 vacancy plug which is not suitable for transplanting in each simulation.

In rectangular coordinate, each plug seedling has its own coordinate which is unique. (Ax_i, Ay_i) represent point of aim tray, (Bx_i, By_i) represent point of transplanting tray. The distance between two transplanting path can be expressed as:

$$d_i = \sqrt{(Bx_i - Ax_i)^2 + (By_i - By_i)^2}$$

The path of transplanting has many choices, and there are more than $32 \times 40\,!$. Our transplanting strategy's objective is to get the minimum $\sum d_i$, and the constraint conditions is transplanting less than 32 times. If we get the optimal transplanting strategy by hundred million times calculate using computer, obviously, it does not meet the requirement of transplanting in real-time, also, most of these calculation are no using.

2.2 Typical Transplanting Strategy

Compare four typical transplanting strategy which represent the most of reasonable strategy.

Strategy one: scan the transplanting tray row by row, from left to right, from up to down, seedlings are transplanted to aim tray, and also placed in the aim tray row by row, the process are shown in figure 4.

Strategy two: scan the transplanting tray row by row, from left to right, from up to down, seedlings are transplanted to aim tray, and also placed in the aim tray follow the principle of nearby placed and row by row which is different to strategy one, the process are shown in figure 5.

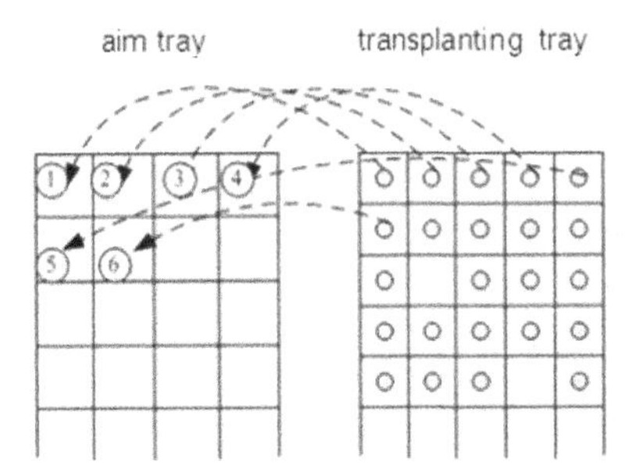

Fig. 4. Strategy one

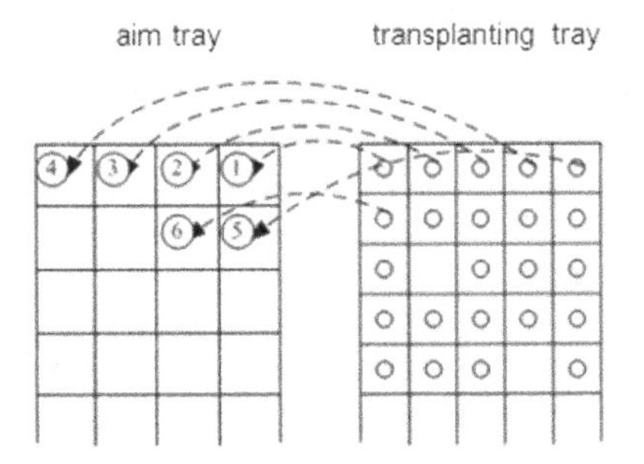

Fig. 5. Strategy two

Strategy three: scan the transplanting tray column by column, from up to down, from left to right, seedlings are transplanted to aim tray, and placed in the aim tray follow the principle of nearby placed, namely seedling in the first column of transplanting tray placed to the nearest column of the aim tray, the process are shown in figure 6.

Strategy four: scan the transplanting tray column by column, from up to down, from left to right, seedlings are transplanted to aim tray, and placed to corresponding position in the aim tray, namely seedling in the first column of transplanting tray placed to the first column of the aim tray, the process are shown in figure 7.

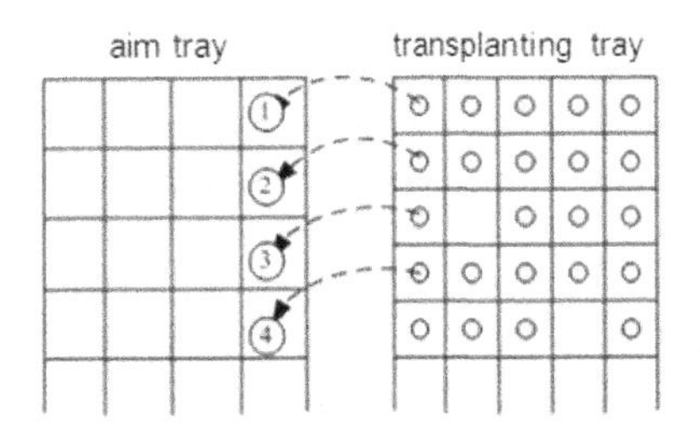

Fig. 6. Strategy three

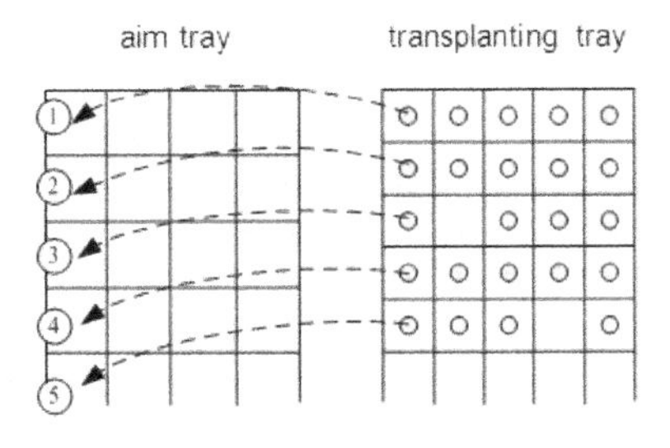

Fig. 7. Strategy four

Generated the transplanting tray with 9 vacancy hole randomly, and the distribution of the vacancy hole are not certain, the number of vacancy hole from 0 to 9. Simulate the trajectory of end-effector by computer, calculated the movement distance of the mechanical arm which seedlings fill all the aim tray, the results are shown in the table 1.

From the table, we found the corresponding distance of strategy one and strategy two is equal, and the corresponding distance of strategy three and strategy four is equal, too. The distance of transplanting is change along with the increasing of vacancy hole in the transplanting tray and the different location. The advantage of transplanting path scanning by row or by column is not obvious. But in the case of litter vacancy hole or no vacancy hole, scanning by row of the transplanting strategy superior to the scanning by column.

Table 1. The distance of transplanting (unit: mm)

Hole number	Strategy 1	Strategy 2	Strategy 3	Strategy 4
0	27600	27600	31412	31412
1	27250	27250	31062	31062
2	26850	26850	30712	30712
3	26387	26387	30362	30362
4	25887	25887	30212	30212
5	25837	25837	27375	27375
6	25887	25887	24937	24937
7	26237	26237	24287	24287
8	26237	26237	25287	25287
9	26237	26237	30550	30550

From the above, we decide to take the strategy of scanning by row, and attempt to improve the effective of the transplanting strategy by greedy algorithm.

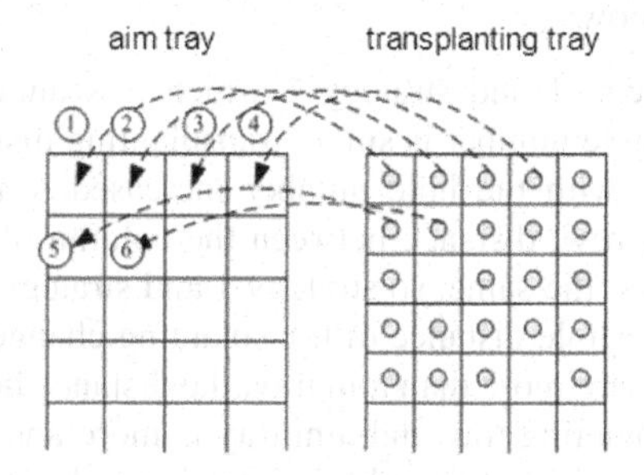

Fig. 8. Improvement Strategy

After optimization, transplanting strategy as follows:

1. Scanning transplanting tray row by row, record the location of hole which have the seedlings;
2. Transplanting the seedling to the aim tray one by one;
3. Judge the corresponding row in the aim tray fill up or not, if the row in the aim tray are filled up nearest the transplanting tray, end-effector take seedling in the next row.
4. Follow the regulation until the aim tray is filled up.

Improvement strategy process are shown in figure 8.

Use the greedy algorithm test the above 10 transplanting tray with 0-9 vacancy hole, distance of transplanting is 23800, 24100, 24300, 24500, 24600, 25000, 25300, 25700, 25700, 25700, the unit is mm.

3 Results and Discussion

Gathering the data of the five transplanting strategy, and drawing the graph, they are shown in figure 9.

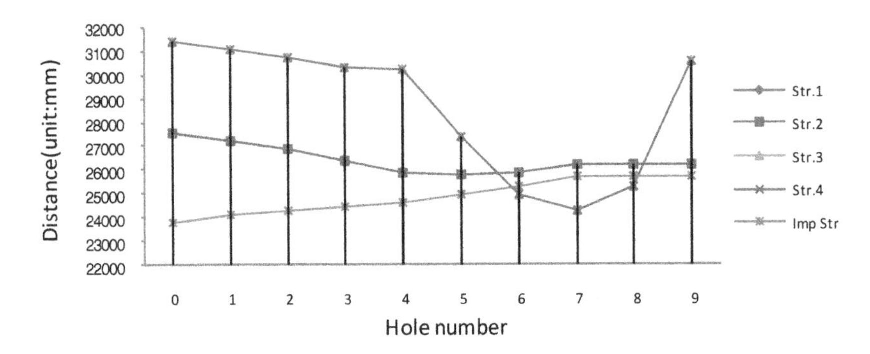

Fig. 9. Graph of the five transplanting strategy test

From the graph, we know:

1. Distance of strategy 1 and strategy 2 which is scanning by row are almost equal; when the hole number is small, transplanting distance change litter, and it decrease along with the hole number increased. Cause the phenomena is that: scanning by row, distance between the column of transplanting tray and aim tray are almost the same, so strategy 1 and strategy 2 is just different with combination, the whole distance of transplanting change litter; and the rows of transplanting tray are more than aim tray, the distance between the corresponding row of transplanting tray and aim tray is more and more large along with the rows increased, the hole number's increasing, decrease the cumulative distance, so the whole distance decrease along with the hole number increase.
2. Distance of strategy 3 and strategy 4 which is scanning by column are almost equal; when the hole number is small, transplanting distance change litter, and it decrease greatly along with the hole number increased. Cause the phenomena is that: the first reason is the same with the strategy 1 and strategy 2; the cumulative distance transplanted by column are bigger than by row, so when the hole number increase, the decreased range of the whole distance is big.
3. Contrast strategy1,2 and strategy3,4, the whole distance of scanning by column is bigger than scanning by row when hole number is litter, and the decreased range is also bigger when hole number is big. Cause the phenomena is that: when every column transplant to the corresponding column in the aim tray, the cumulative distance are bigger than transplanting by row, so when the hole number is bigger, the corresponding range of decrease is bigger.

4. Base on the advantage of the scanning by row, using greedy algorithm optimize strategy 1. The whole distance is shorter than strategy 1, and the advantage is more obvious when the hole number is litter, but the advantage disappear gradually along with the whole number is bigger, Cause the phenomena is that: in this transplanting problem, the advantage of greedy algorithm is weakened along with the hole number increase.

In this thesis, using the greedy algorithm realize the optimization of transplanting strategy, the whole distance of transplanting is shorter than the other four strategy.

4 Conclusions

Using the greedy algorithm to solute the problem is very common, and the greedy strategy is the most closing the strategy of common thinking to the people. The final solution calculated by greedy algorithm can't guarantee the optimal, but it can provide a range feasibility for some problem. For seedling transplanting in facility agriculture, optimizing transplanting strategy with greedy algorithm instead of the method of exhaust algorithm can also get a good control strategy, save the operation time, and further improve efficiency of the equipment. In a word, making full use of the advantage of greedy algorithm, optimal the local case, greedy strategy is easy to constitution, and simple to realize it.

Acknowledgments. The research presented in this paper was partially supported by the New Century Excellent Talent of the Ministry of Education in China (NCET-10-0689), Zhejiang province of China "new century 151 talents project".

References

1. Choi, W.C., Kim, D.C., Ryu, I.H.: Development of a seedling pick-up device for vegetable transplanters. Transactions of the ASAE 45, 13–19 (2002)
2. Ting, K.C., Giacomelli, G.A., Shen, S.J.: Robot work cell for transplanting of seedlings. Part I: Layout and materials flow. Transactions of the ASAE 33, 1005–1010 (1990)
3. Jiang, H., Shi, J., Ren, Y., Ying, Y.: Application of machine vision on automatic seedling transplanting. Transactions of the CSAE 25, 127–131 (2009)
4. Chang, Y., Xiao, G., Zeng, M.: Exploration of greedy algorithm. Journal of Chongqing Electric Power College 13, 40–47 (2008)
5. Temlyakov, V.N.: Weak greedy algorithms. Advances in Computational Mathematics 12, 213–227 (2000)
6. De Vore, R.A., Temlyakov, V.N.: Some remarks on greedy algorithms. Advances in Computational Mathematics 5, 173–187 (1996)
7. Edwin Romeijn, H., Morales, D.R.: A class of greedy algorithms for the generalized assignment problem. Discrete Applied Mathematics 103, 209–235 (2000)

An Investigation of Squeaky Wheel Optimization Approach to Airport Gate Assignment Problem

Xueyan Song and Yanjun Jiang

School of Computer Science & Technology, Tianjin University
Tianjin, China
{songxy,jiangyanjun}@tju.edu.cn

Abstract. This paper investigates a squeaky wheel optimization (SWO) approach to the airport gate assignment problem (AGAP). A graph coloring method is incorporated into the SWO procedure to construct solutions in our approach. Some initial experimental results are presented towards the validation of this approach.

Keywords: Airport Gate Assignment Problem, Squeaky Wheel Optimization, Graph Coloring Heuristic.

1 Introduction

Airport gate assignment problem involve scheduling a number of arriving and departing flights into a set of airport gates with the aim to satisfy certain objectives and constraints. A typical objective is to minimize the number of the delayed gates. A number of constraints need to be considered in the procedure of constructing solutions. Such as: only one flight is allowed at one gate within the same time interval; each flight should not be assigned to more than one gate at the same time; the time interval between the flight arriving and departing should not be longer than the minimum ground time;

This paper addresses the AGAP problem with the dispersion of gate idle time period as the objective. The purpose is to improve the robustness of the schedule, which expects that slight time variation won't cause too much disruption in the original plan.

It has been proved that AGAP is a combinatorial problem. Heuristic method is generally applied to obtain approximate optimal solutions with reasonable cost. In the literature, there are a number of state of art works on AGAP. Ahmet Bolat [1] used the branch and bound method to assign commercial service aircrafts to the available airport gates ; Loo Hay Lee [2] et al. developed a multi-objective genetic algorithm to schedule the flights to gates; Rosenberger [3] et al. constructed a fleet assignment model that can be used to improve robustness based on the structure of a hub-and-spoke flight network to create a partial rotation with many short cycles.

The contribution of this paper is to present a Squeaky Wheel Optimization approach to the AGAP. The Squeaky Wheel Optimization [6] algorithm can be divided into three steps—Construct / Analyze / Prioritize. Firstly, an initial elements ordering can be

Y. Wang and T. Li (Eds.): Knowledge Engineering and Management, AISC 123, pp. 41–45.
springerlink.com

acquired by evaluating the difficulty of these elements. According to this ordering, an initial solution is constructed element by element. Then, this solution is analyzed to find those elements that can't work well and a strategy is used to increase their priorities while others' stay unchanged. Finally, a new sequence of elements ordering is obtained by prioritizing the elements in descending order of their priorities. The SWO algorithm cycles around these three steps until certain stop criteria are met. Some initial experimental results show that the proposed SWO approach to AGAP can be used to improve the schedule quality.

This paper is organized as follows. The proposed SWO algorithm to AGAP is given in Section 2. Section 3 describes the experimental data and discusses the results. A conclusion is provided in Section 4.

2 The Proposed Method

In this section, A Squeaky Wheel Optimization algorithm incorporated with the graph coloring method as the basic heuristic is provided to solve the airport gate assignment problem.

The three steps of SWO are termed constructor / analyzer / prioritizer respectively. On each iteration, the constructor gives a solution (may violate some hard constraints), and the analyzer finds those "difficult points" ("squeaky wheels") and at the same time increases their priorities, the prioritizer reprioritizes these elements in descending order of their priorities (measured by difficulty). Iteration after iteration, these difficult points are given more and more attention and move to the front of the elements sequence gradually ("It is the squeaky wheel that gets the oil."). In the gate assignment problem domain, these elements are the flights to be assigned.

A graph coloring method derived from the literature [7],[8] and [9] is applied as the basic heuristic in the constructor to obtain a solution in each cycle. In this work, two graph coloring heuristics, Largest Degree First (LDF) and Saturation Degree (SD), are used to construct a solution.

The difficulty which measures how difficult a flight can be assigned, and the heuristic modifier which helps to dynamically alter the overall difficulty of a flight, are applied in the analyzer. The difficulty of flight f at iteration i is given in the following equation:

$$difficulty(f,i)=heuristic(f,i)+heurmod(f,i) \quad (1)$$

The $heuristic(f,i)$ is measured by the conflict number of flight f with the other flights at iteration i. The $heurmod(f,i)$ is a variable which is altered dynamically in each iteration. Four types of heuristic modifier proposed in [9] are applied in our work. They are: custom (C), additive (AD), Multiplicative (MP), Exponential (MP).

After the prioritizer has set an order of flights, a shuffle strategy is applied on the sorted flight sequence. The sorted flights are divided into fixed size blocks, and within each block, the flights are shuffled randomly. The idea behind that is that the diversity of the schedule might be improved.

The pseudo code of the proposed SWO algorithm to AGAP is presented in Figure 1.

3 Experiments

The test data used in the experiments is generated by the following method: the integer arriving time of each flight is drawn from a uniform distribution between 0 and 48. Similarly, the ground time for each flight is from a uniform distribution between 6 and 18. Each gate has two attributes: the gate number and gate type value. The gate number is the same as the sequence number generating the gate. While, the gate type value is drawn from a uniform distribution between 1 and 3.

Since a random function is used to shuffle the ordering of flights, the results of each run are different. Thus, the best result of 10 runs is recorded for each method. We have carried out two group experiments. One group use LDF heuristic in SWO, and the other use SD heuristic. For each group experiment, four type of modifier with different shuffle strategies are tested. Table 1 presents the two group experimental results, which are by LDF and SD respectively.

```
Algorithm 1. Pseudo code of the SWO methodology to AGAP
Do
 Applying graph color method to construct a solution;
 calculate the fitness value of the current solution;
 if current solution is better than the best solution ever
found
 then replace best solution with current solution;
 sort all the flights by the value of difficulty in the
descending order;
 partition all the ordered flights into fixed size bolcks;
 shuffle all flights within each block randomly;
 while stop critera aren't met;
 return best solution;
```

Fig. 1. Pseudo-code of the SWO algorithm to AGAP

It can be seen from table 1 that the performance by SD is generally better than by LDF. For each group experiment, there is a possibility to have the improved performance by applying different shuffle strategy.

Table 1. Experimental results

LDF			*SD*		
run number	method	fitness value	run number	method	fitness value
1	C--0	238925	1	C—0	144200
2	C--2	238925	2	C--2	90875
3	C--3	238925	3	C--3	91275
4	C--4	238925	4	C--4	89700
5	C--5	238925	5	C--5	72325
6	C--6	238925	6	C--6	81775
7	AD--0	235875	7	AD--0	144200
8	AD--2	228350	8	AD--2	90875
9	AD--3	235875	9	AD--3	93075
10	AD--4	193350	10	AD--4	85525
11	AD--5	235875	11	AD--5	71700
12	AD--6	235875	12	AD--6	79575
13	MP--0	238925	13	MP--0	144200
14	MP--2	221950	14	MP--2	90875
15	MP--3	194475	15	MP--3	94750
16	MP--4	186150	16	MP--4	89700
17	MP--5	103700	17	MP--5	67050
18	MP--6	186950	18	MP--6	76300
19	EX--0	235875	19	EX--0	144200
20	EX--2	228350	20	EX--2	90875
21	EX--3	194475	21	EX--3	93075
22	EX--4	191925	22	EX--4	87100
23	EX--5	142525	23	EX--5	64275
24	EX--6	186150	24	EX--6	79000

4 Conclusions

In this paper, we initially apply the Squeaky Wheel Optimization algorithm to the airport gate assignment problem. And preliminary experimental results show that our method can be used to improve the robustness of the flight schedule. In the future, we will try to investigate how to combine the SWO with other intelligent algorithm

(Evolutionary algorithm e.g.) so as to solve the gate assignment problem more effectively.

Acknowledgements. This work was supported by Tianjin University Innovation Fund reference 2010XJ-0091 and The National Natural Science Foundation of China reference 61039001.

References

[1] Bolat, A.: Procedures for providing robust gate assignments for arriving aircrafts. European Journal of Operational Research 120, 63–80 (2000)
[2] Lee, L.H., Lee, C.U., Tan, Y.P.: A multi-objective genetic algorithm for robust flight scheduling using simulation. European Journal of Operational Research 177, 1948–1968 (2007)
[3] Rosenberger, J.M., Johnson, E.L., Nemhauser, G.L.: A robust fleet-assignment model with hub isolation and short cycles. Transportation Science, ProQuest Science Journals, 357 (August 2004)
[4] Wei, D., Liu, C.: Optimizing gate assignment at airport based on genetic-tabu algorithm. Automation and Logistics. 2007 IEEE International Conference on Publication Date 18-21 (2007)
[5] Burke, E.K., De Causmaecker, P., De Maere, G., Mulder, J., Paelinck, M., Vanden Berghe, G.: A multi-objective approach for robust airline scheduling. Computers and Operations Research 37, 822–832 (2010)
[6] Joslin, D.E., Clements, D.P.: Squeaky wheel optimization. In: AAAI-1998 Proceedings, vol. 10, pp. 353–373 (1999)
[7] Burke, E.K., Newall, J.P.: Solving examination timetabling problems through adaption of heuristic orderings. Annals of Operations Research 129, 107–134 (2004)
[8] Qu, R., Burke, E.K., McCollum, B.: Adaptive automated construction of hybrid heuristics for exam timetabling and graph coloring problems. European Journal of Operational Research 198, 392–404 (2009)
[9] Rahman, S.A., Bargiela, A., Burke, E.K., Özcan, E., McCollum, B.: Construction of examination timetables based on ordering heuristics. In: 24th International Symposium on ISCIS 2009, September 14-16, pp. 680–685 (2009)

A Suppression Operator Used in TMA

Jungan Chen, Qiaowen Zhang, and Zhaoxi Fang

Electronic Information Department, Zhejiang Wanli University, No.8 South Qian Hu Road
Ningbo, Zhejiang, 315100, China
friendcen21@hotmail.com, cn_hnzqw@yahoo.com.cn,
zhaoxifang@gmail.com

Abstract. In T-detector Maturation Algorithm with Overlap Rate, the parameter Omin is proposed to control the distance among detectors. But Omin is required to be set by experience. To solve the problem, T-detector Maturation Algorithm with NS operator is proposed. The results of experiment show that the proposed algorithm can achieve the same effect with TMA-OR when 2-dimensional synthetic data and iris data are used as the data set.

Keywords: Artificial immune system, suppression operator, TMA.

1 Introduction

Nowadays, Artificial Immune System (AIS) has been applied to many areas such as computer security, classification, learning and optimization [1]. Negative Selection Algorithm, Clonal Selection Algorithm, Immune Network Algorithm and Danger Theory Algorithm are the main algorithms in AIS [2][3].

A real-valued negative selection algorithm with variable-sized detectors (V-detector Algorithm) is proposed to generate detectors with variable r, which are applied in abnormal detection. A statistical method (naïve estimate) is used to estimate detect coverage [4]. But as reported in paper[5], the performance of V-detector algorithm on the KDD Cup(1999) data is unacceptably poor. So a new statistical approach (hypothesis testing) is used to analyze the detector coverage [6]. But hypothesis testing requires np>5, n(1-p)>5 and n>10. When p is set to 90%, n must be set to at least 50. While the number of detectors have an important effect on the detect performance, the hypothesis has poor performance. Actually in naïve estimate method, V-detector algorithm tries to maximize the distance among valid detectors. With the number of valid detectors increasing, it is difficult to find valid detector. To choose the appropriate distance among valid detectors and achieve less number of detectors generated, a parameter overlap rate (Omin) in T-detector Maturation Algorithm (TMA) is proposed to control the distance among detectors [7].But the optimized Omin is required to be set by experience. To solve this problem, a suppression operator called Negative Selection operator (NS operator) is used in TMA. NS operator is first proposed to eliminate those network cells which are recognized by others in optaiNet [8]. So there is no parameter Omin in TMA with NS operator(TMA-NS).

Y. Wang and T. Li (Eds.): Knowledge Engineering and Management, AISC 123, pp. 47–52.
springerlink.com

2 Algorithm

2.1 Match Range Model

$U=\{0,1\}^n$,n is the number of dimensions. The normal set is defined as selves and the abnormal set is defined as nonselves. selves ∪ nonselves=U. selves∩nonselves=Φ.There is two point $x=x_1x_2\ldots x_n$, $y=y_1y_2\ldots y_n$. The Euclidean distance between x and y is:

$$d(x,y)=\sum_{i=1}^{n}(x_i-y_i)^2 \tag{1}$$

The detector is defined as dct = {<center, selfmin, selfmax > | center ∈ U, selfmin, selfmax∈N}. 'center' is one point in U. 'selfmax' is the maximized distance between dct.center and selves. 'selfmin' is the minimized distance. The detector set is definined as DCTS. Selfmax and selfmin is calculated by setMatchRange(dct, selves), dct.center∈U, i∈[1, |selves|], $self_i$∈selfves:

$$setMatchRange=\begin{cases}\text{selfmin}=\min(\{d(\text{self}_i,\text{dct.center})\})\\ \text{selfmax}=\max(\{d(\text{Self}_i,\text{dct.center})\})\end{cases} \tag{2}$$

[selfmin,selfmax] is defined as self area. Others is the nonself area. Suppose there is one point x∈U and one detector dct ∈ DCTS. When d(x,dct) ∉ [dct.selfmin, dct.selfmin], x is detected as abnormal.

2.2 NS Operator

NS operator is first proposed to eliminate those network cells which are recognized by others in optaiNet [8]. In this work, it is defined as following:

$$IsValidAnd=\begin{cases}false, NSMatchAnd(dctx,dct_k)=true, \exists dct_k\in DCTS\\ true, \quad others\end{cases} \tag{3}$$

$$NSMatchAnd=\begin{cases}true, d<dctx.self\min\wedge d<dctk.self\min\\ false, \quad others\end{cases} \tag{4}$$

$$d=d(dctx.center, dctk.center) \tag{5}$$

$$IsValidOR=\begin{cases}false, NSMatchOR(dctx,dct_k)=true, \exists dct_k\in DCTS\\ true, \quad others\end{cases} \tag{6}$$

$$NSMatchOR=\begin{cases}true, d<dctx.self\min\vee d<dctk.self\min\\ false, \quad others\end{cases} \tag{7}$$

As there are two logic operators including AND and OR, two type NS operators in equation 3 and 6 are provided.

2.3 The Model of Algorithm

The algorithm, called TMA-NS (TMA with Negative Select operator), is shown in Fig.1. Step 2~4 is used to generate candidate detector which does not covered by self with rs. Step 10 is used to estimate the detect coverage. Step 5 is used to decide whether candidate detector is a valid detector according equation 5 or equation 8. As AND or OR operator is used in step 5, there are two algorithms called TMA-NS-AND or TMA-NS-OR.

```
1.  Set the desired coverage pc, Self radius rs
2.  Generate one candidate detector dctx randomly
3.  setMatchRange(dctx,selves) // equation 2
4.  if dctx.selfmin< rs² then Go to 2;
5.  if isvalidAnd(dctx,DCTS) then// equation 3
//  isvalidOR(dctx,DCTS)// equation 6
6.         dctx is added to detector set DCTS
7.     covered=0
8.  Else
9.     covered ++
10. If covered <1/(1- pc) then goto 2
```

Fig. 1. TMA-NS algorithm model

3 Experiments

For the purpose of comparison, experiments are carried out using every data set list in table 1.In table 1, 2-dimensional synthetic data is described in Zhou's paper[9]. Over the unit square $[0,1]^2$,various shapes are used as the self region. In every shape, there are training data (self data) of 1000 points and test data of 1000 points including both self points and nonself points. In the famous benchmark Fisher's Iris Data, one of the three types of iris is considered as normal data, while the other two are considered abnormal [4]. As for KDD data, 20 subsets were extracted from the enormous KDD data using a process described in [5]. Self radius from 0.01 up to 0.2 and Omin used in TMA-OR from 0 up to 0.7 is conducted in these experiments. All the results shown in these figures are average of 100 or 20 (see table 1) repeated experiment with coverage rate 99%.

Table 1. Data set and parameters used in experiments

Data set		Parameters		
		r_s	*Omin*	Repeated times
2-dimensional synthetic data	Comb	0.01 ~ 0.2	0 ~ 0.7	100
	Cross			
	Ring			
	Triangle			
	Stripe			
	Intersection			
	Pentagram			
Iris data	Setosa as self data			
	Versicolor as self data			
	Virginica as self data			
KDD data		0.05~0.2		20

3.1 The Optimized rs Value

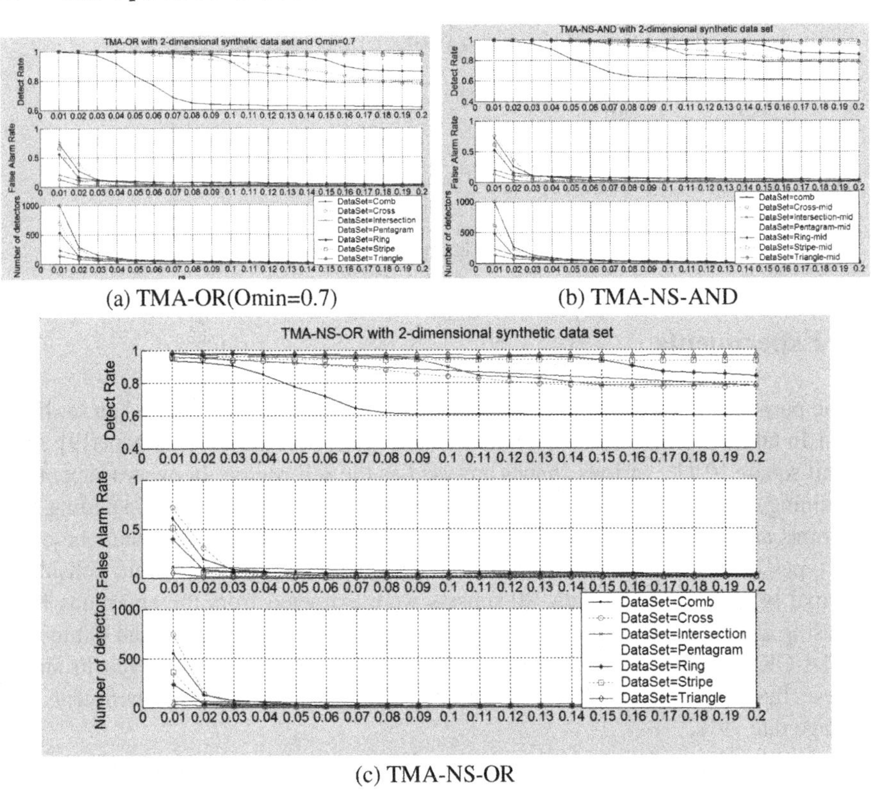

(a) TMA-OR(Omin=0.7)

(b) TMA-NS-AND

(c) TMA-NS-OR

Fig. 2. Results with 2-dimensional synthetic data and different rs

In Fig.2, it shows that these algorithms achieve the optimized value at rs=0.03. So 0.03 is taken the optimized value in following discussion.

3.2 Comparison

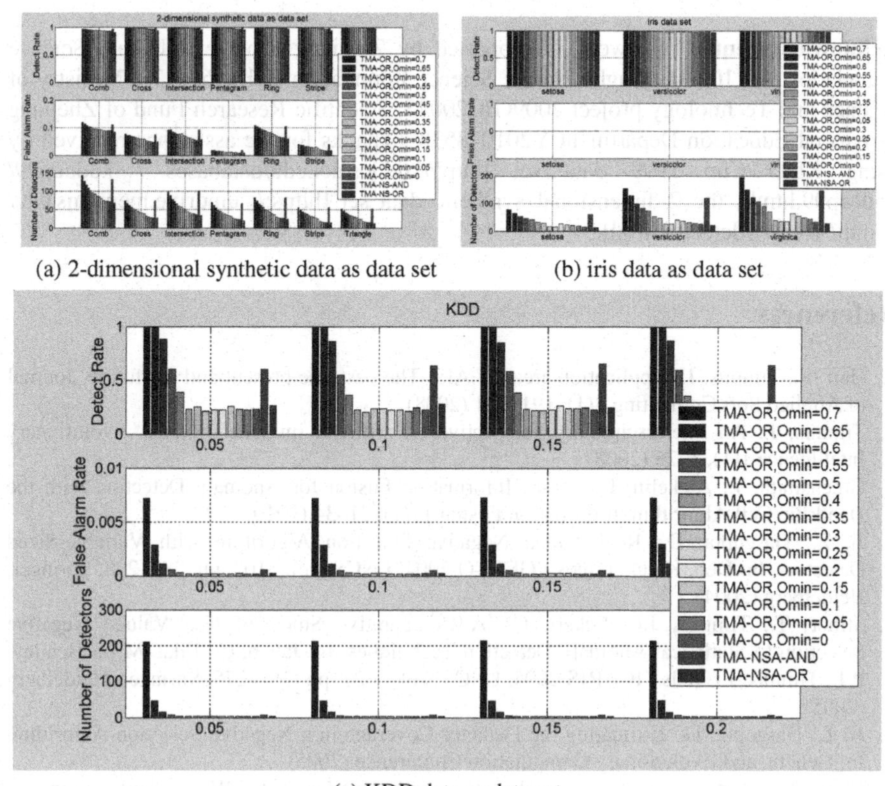

(a) 2-dimensional synthetic data as data set

(b) iris data as data set

(c) KDD data as data set

Fig. 3. Results with rs=0.03

In fig.3(a)(b),the results of TMA-NSA-AND and TMA-OR(Omin=0.7) have almost the same effect So does between TMA-NSA-OR and TMA-OR(Omin=0). According the equation 4 and 7, TMA-NSA-AND requires AND operator and TMA-NSA-OR requires OR operator. So TMA-NSA-OR leads more detectors to be removed because OR operator is easier to become true than AND operator. As a result, TMA-NSA-OR generates less valid detectors in the third figure.

In Fig.3(c), when KDD data set is used, TMA-NSA-AND shows less effective than TMA-OR(Omin>=0.6) because TMA-NSA-AND generates less valid detectors in the third figure.

4 Conclusion

As the parameter Omin in TMA-OR is required to be set by experience. To solve the problem, T-detector Maturation Algorithm with NS operator (TMA-NS) is proposed. TMA-NS, where there is no Omin required to be set , can achieve the same effect in

2-dimensional synthetic data and iris data. But TMA-NS shows less effective than TMA-OR when KDD is as the data set.So further research is required to be done.

Acknowledgments. This work is supported by Zhejiang Provincial Nature Science Foundation Y1110200, Ningbo Nature Science Foundation 2010A610173, Ministry of Science and Technology project 2009GJC20045, Scientific Research Fund of Zhejiang Provincial Education Department Y201018538 . Thanks for the assistance received by using KDD Cup 1999 data set [http://kdd.ics.uci.edu/databases / kddcup99/ kddcup99.html], the 2-dimensional synthetic data set [https:// umdrive.memphis.edu/ zhouji/ www/ vdetector.html].

References

1. Hart, E., Timmis, J.: Application areas of AIS: The past, the present and the future. Journal of Applied Soft Computing 8(1), 191–201 (2008)
2. Timmis, J.: An interdisciplinary perspective on artificial immune systems. Evolutionary Intelligence 1(1), 5–26 (2008)
3. Greensmith, J., Aickelin, U., et al.: Information Fusion for Anomaly Detection with the Dendritic Cell Algorithm. Information Fusion 11(1), 21–34 (2010)
4. Ji, Z., Dasgupta, D.: Real-Valued Negative Selection Algorithm with Variable-Sized Detectors. In: Deb, K., et al. (eds.) GECCO 2004. LNCS, vol. 3102, pp. 287–298. Springer, Heidelberg (2004)
5. Stibor, T., Timmis, J.I., Eckert, C.: A Comparative Study of Real-Valued Negative Selection to Statistical Anomaly Detection Techniques. In: Jacob, C., Pilat, M.L., Bentley, P.J., Timmis, J.I. (eds.) ICARIS 2005. LNCS, vol. 3627, pp. 262–275. Springer, Heidelberg (2005)
6. Ji, Z., Dasgupta, D.: Estimating the Detector Coverage in a Negative Selection Algorithm. In: Genetic and Evolutionary Computation Conference (2005)
7. Chen, J.: T-detector Maturation Algorithm with Overlap Rate. Wseas Transactions on Computers 7(8), 1300–1308 (2008)
8. Chen, J.: A novel suppression operator used in optaiNet. BSBT 57, 17–23 (2009)
9. Ji, Z.: Negative Selection Algorithms: from the Thymus to V-detector. PhD Dissertation. University of Memphis (2006)

Existence and Simulations of an Impulsive Appropriate Pest Management SI Model with Biological and Chemical Control

Yan Yan, Kaihua Wang, and Zhanji Gui*

School of Mathematics and Statistics Hainan Normal University
Haikou, Hainan, 571158
zhanjigui@sohu.com

Abstract. Using the continuation theorem of coincidence degree theory and analysis techniques, we establish criteria for the existence of periodic solutions of predator-prey models and impulsive perturbations. It is more appropriate to add the density-dependent term to these models in this paper. Further, computer simulation shows that our models can occur in many forms of complexities including periodic oscillation and gui chaotic strange attractor.

Keywords: Periodic Solution, SI, Impulses, Coincidence degree theory.

1 Introduction

The predator-prey competitive and cooperative models have been studied by many authors (see[1,2,3]). The permanence and extinction are significant concepts of those models which also show many interesting results. However, the stage structure of species has been considered very little. In the real world, almost all animals have the stage structure. Recently, papers ([3,4,5,6,7,8]) studied the stage structure of species with or without time delays. In this paper, we shall explore the dynamics of impulsive differential equation modeling the process of releasing infective pests and spraying pesticides in a more general form

$$\begin{cases} \left.\begin{array}{l} \frac{ds(t)}{dt} = r(t)s(t)\left(1 - \frac{s(t)+Q(t)I(t)}{K(t)}\right) - \beta(t)s(t)I(t) \\ \frac{dI(t)}{dt} = \beta(t)s(t)I(t) - I(t) \end{array}\right\} t \neq t_k, \\ \left.\begin{array}{l} \Delta s(t_k) = p_k^1 s(t_k^-) \\ \Delta I(t_k) = p_k^2 I(t_k^-) \end{array}\right\} t = t_k. \end{cases} \quad (1)$$

where $s(t), I(t)$ represents the density of species at time t; $r(t)$ denotes the intrinsic growth rate of species; $K(t)$ means the environment carrying capacity of species in the absence of competition; $\beta(t)$ measures the amount of competition between the species $s(t)$ and $I(t)$; p_k^i are constants. $(i = 1, 2)$

* Corresponding author.

Y. Wang and T. Li (Eds.): Knowledge Engineering and Management, AISC 123, pp. 53–60.
springerlink.com

In system (1), we give two hypotheses as follows:

(H_1) $r(t), K(t)$ and $\beta(t)$ are all nonnegative ω-periodic functions defined on R.

(H_2) $1 + p_k^i > 0$ and there exists a positive integer q such that $t_{k+q} = t_k + \omega, p_{k+q}^i = p_k^i, (i = 1, 2)$.

For the sake of convenience, we introduce the following notations: $\bar{g} = \frac{1}{\omega}\int_0^\omega g(t)dt$, here $g(t)$ is a ω-periodic functions.

2 Existence of Positive Periodic Solutions

To prove our results, we need the notion of the Mawhin's continuation theorem formulated in [4]

Lemma 1. *Let X and Y be two Banach spaces. Consider an operator $Lx = \lambda Nx$ where $L : \mathrm{Dom}\, L \cap X \to Y$ is a Fredholm operator of index zero. Let P and Q denote two projectors such that $P : X \to \mathrm{Ker}\, L$ and $Q : Y \to Y/\mathrm{Im}\, L$. Assume that $N : \overline{\Omega} \to Y$ is L-compact on $\overline{\Omega}$, where Ω is open bounded in X. Furthermore, assume that*
(a) for each $\lambda \in (0,1), x \in \partial\Omega \cap \mathrm{Dom}\, L, Lx \neq \lambda Nx$,
(b) for each $x \in \partial\Omega \cap \mathrm{Ker}\, L$, $QNx \neq 0$,
(c) for each $\deg\{JQN, \Omega \cap \mathrm{Ker}\, L, 0\} \neq 0$.
Then equation $Lx = Nx$ has a solution on $\overline{\Omega} \cap \mathrm{Dom}\, L$.

Now we are ready to state and prove the main results of the present paper.

Theorem 1. *Assume that the following conditions*
$(H_3) 1 - \frac{1}{\omega}\sum_{k=1}^{q} \ln(1 + p_k^2) > 0$,
$(H_4) \bar{r} + \frac{1}{\omega}\sum_{k=1}^{q} \ln(1 + p_k^1) > \overline{(\frac{r}{K})} \exp\{D_1\}$,
where $D_1 = \ln\left[\frac{1 - \frac{1}{\omega}\sum_{k=1}^{q}\ln(1+p_k^2)}{\bar{\beta}}\right] - 2\bar{r}\omega$,
then system (1) has at least one ω-periodic solution.

Proof. Make the change of variables $\exp\{s(t)\} \to s(t)$, $\exp\{I(t)\} \to I(t)$, then (1) can be reformulated as

$$\begin{cases} \frac{ds(t)}{dt} = r(t)\left(1 - \frac{\exp\{s(t)\}+Q(t)\exp\{I(t)\}}{K(t)}\right) - \beta(t)\exp\{I(t)\}, \\ \frac{dI(t)}{dt} = \beta(t)\exp\{s(t)\} - 1, \\ \Delta s(t_k) = \ln(1 + p_k^1), \\ \Delta I(t_k) = \ln(1 + p_k^2). \end{cases} \quad (2)$$

Let

$$PC(J, R) = \left\{ \begin{array}{l} x : J \to R |\, x(t) \textit{ is continuous with} \\ \textit{respect to } t \neq t_1, ..., t_q; x(t^+) \textit{ and} \\ x(t^-) \textit{ exsit at } t_1, ..., t_q; \textit{and} \\ x(t_k) = x(t_k^+), k = 1, 2, ..., q \end{array} \right\}.$$

To complete the proof, we only need to search for an appropriate open bounded subset $\Omega \subset X$ verifying all the requirements in Lemma 1.

Note $u = (x, y)^T$, define $X = \{u \in PC(R, R^2) : x(t+\omega) = x(t)\}, Y = X \times R^{2q}$ show that both X and Y are Banach space when they are endowed with the norms $||u||_c = \sup_{t\in[0,\omega]} |u(t)|$ and $||(u, c_1, ..., c_q)|| = (||u||_c^2 + |c_1|^2 + \cdots + |c_q|^2)^{1/2}$. Let L: Dom$L \to Y, Lx = (x', \Delta x(t_1), ..., \Delta x(t_q))$. It is easy to prove that L is a Fredholm mapping of index zero.
Let

$$N : X \to Y$$
$$Nu = N\begin{pmatrix} s(t) \\ I(t) \end{pmatrix} = \left(\begin{pmatrix} f_1(t) \\ f_2(t) \end{pmatrix}, (\Phi_1(u(t_1)), ...\Phi_q(u(t_q))\right).$$

where $f_1(t) = r(t)\left(1 - \frac{\exp\{s(t)\}+Q(t)\exp\{I(t)\}}{K(t)}\right) - \beta(t)\exp\{I(t)\}$,

$f_2(t) = \beta(t)\exp\{s(t)\} - 1$ and $\Phi_k(u(t_k)) = \begin{pmatrix} \ln(1+p_k^1) \\ \ln(1+p_k^2) \end{pmatrix}, k = 1, 2, \cdots, q.$

Consider the operator equation

$$Lx = \lambda Nx, \quad \lambda \in (0, 1). \tag{3}$$

Integrating (3) over the interval $[0, \omega]$, we obtain

$$\begin{aligned} \overline{r}\omega = &-\sum_{k=1}^{q} \ln[1+p_k^1] + \int_0^\omega \beta(t)\exp\{I(t)\}\, dt \\ &+ \int_0^\omega \frac{r(t)[\exp\{s(t)\} + Q(t)\exp\{I(t)\}]}{K(t)}\, dt, \end{aligned} \tag{4}$$

$$\omega = \sum_{k=1}^{q} \ln[1+p_k^2] + \int_0^\omega \beta(t)\exp\{s(t)\}\, dt. \tag{5}$$

We can derive that

$$\int_0^\omega |s'(t)|\, dt \leq 2\overline{r}\omega + \sum_{k=1}^{q} \ln(1+p_k^1),$$

$$\int_0^\omega |I'(t)|\, dt \leq 2\omega.$$

Since $s(t), I(t) \in PC([0, \omega], R)$, there exits $\xi_i, \eta_i \in [0, \omega] \cup [t_1^+, t_2^+, ..., t_q^+], i = 1, 2.$ such that $s(\xi_1) = \inf_{t\in[0,\omega]} s(t)$, $s(\eta_1) = \sup_{t\in[0,\omega]} s(t)$, $I(\xi_2) = \inf_{t\in[0,\omega]} I(t)$, $I(\eta_2) = \sup_{t\in[0,\omega]} I(t)$.

From (4) and (5), we have

$$\int_0^\omega \beta(t)\exp\{s(\eta_1)\} \geq \omega - \sum_{k=1}^{q} \ln(1+p_k^2).$$

It follows that

$$s(\eta_1) \geq \ln\left[\frac{1 - \frac{1}{\omega}\sum_{k=1}^{q}\ln(1+p_k^2)}{\overline{\beta}}\right] = A_1.$$

$$\int_0^\omega \frac{r(t)}{K(t)}\exp\{s(\xi_1)\} \leq \overline{r}\omega + M \leq \overline{r}\omega + \sum_{k=1}^{q}\ln(1+p_k^1).$$

where $M = \sum_{k=1}^{q}\ln(1+p_k^1) - \int_0^\omega [\frac{Q(t)r(t)}{K(t)} + \beta(t)]\exp\{I(t)\}\ dt$.
It follows that

$$s(\xi_1) \leq \ln\left[\frac{\overline{r} + \frac{1}{\omega}\sum_{k=1}^{q}\ln(1+p_k^1)}{\overline{(\frac{r}{K})}}\right] = B_1.$$

Then for $\forall t \in [0, \omega]$, we have

$$s(t) \leq s(\xi_1) + \sum_{k=1}^{q}\ln(1+p_k^1) + \int_0^\omega |s'(t)|\ dt \leq B_1 + 2\overline{r}\omega + 2\sum_{k=1}^{q}\ln(1+p_k^1) = C_1. \quad (6)$$

$$s(t) \geq s(\eta_1) - \sum_{k=1}^{q}\ln(1+p_k^1) - \int_0^\omega |s'(t)|\ dt \geq A_1 - 2\overline{r}\omega = D_1. \quad (7)$$

Then we can derive

$$|s(t)| \leq \max\{\ |C_1|, |D_1|\ \} := N_1.$$

Similarly, for $\forall t \in [0, \omega]$, we have

$$I(\xi_2) \leq \ln\left[\frac{r + \frac{1}{\omega}\sum_{k=1}^{q}\ln(1+p_k^1)}{\overline{(\frac{Qr}{K})}}\right] = B_2,$$

$$I(\eta_2) \geq \ln\left[\frac{r + \frac{1}{\omega}\sum_{k=1}^{q}\ln(1+p_k^1) - \overline{(\frac{r}{K})}\exp\{D_1\}}{\overline{(\frac{Qr}{K})}}\right] = A_2.$$

Then for $\forall t \in [0, \omega]$, we have

$$I(t) \leq I(\xi_2) + \sum_{k=1}^{q}\ln(1+p_k^2) + \int_0^\omega |I'(t)|\ dt \leq B_2 + 2\omega + \sum_{k=2}^{q}\ln(1+p_k^1) = C_2. \quad (8)$$

$$I(t) \geq I(\eta_2) - \sum_{k=1}^{q}\ln(1+p_k^2) - \int_0^\omega |I'(t)|\ dt \geq A_2 - 2\omega - \sum_{k=1}^{q}\ln(1+p_k^2) = D_2. \quad (9)$$

Then we can derive

$$|I(t)| \leq \max\{\,|C_2|, |D_2|\,\} := N_2.$$

Obviously, N_1, N_2 are independent of λ. Choosing $r > N_1 + N_2, \Omega = \{u \in X : ||u||_c < r\}$, then N is L-compact on $\overline{\Omega}$. So, for $\forall u = (\bar{s}, \bar{I})^T \in \partial\Omega \cap \operatorname{Ker} L$, we have $QNx \neq 0$. Let $J : \operatorname{Im}Q \to x, (d, 0, ..., 0) \to d$. Then when $x \in \Omega \cap \operatorname{Ker} L$, in view of the assumptions in Mawhin's continuation theorem [4], one obtains, $\deg\{JQN, \Omega \cap \operatorname{Ker}L, 0\} \neq 0$. By now we have proved that Ω satisfies all the requirements in Mawhin's continuation theorem [4]. Hence, (2) has at least one ω-periodic solution $x^*(t) = (s^*(t), I^*(t))^T$ in $\operatorname{Dom} L \cap \overline{\Omega}$. The proof is completes. □

3 An Illustrative Example

In this section, we shall discuss an example to illustrate main results. In (1), we take H_5 : $t_k = kT$, $r(t) = 3 + \sin t$, $K(t) = 2 + \cos t$, $\beta(t) = 1 + \sin t$, $Q(t) = 2 - \cos t$, $p_k^1 = 0.4$, $p_k^2 = 0.8$. Clearly, all conditions of Theorem 1 are satisfied.

If $T = \pi$, then system (1) under the conditions H_5 has a unique 2π-periodic solution (See Fig.1-Fig.3), where $[s(t),\ I(t)]^T = [0.1,\ 0.1]^T$). Because of the influence of the period pulses, the influence of pulse is obvious.

If $T = 5$, then H_2 is not satisfied. Periodic oscillation of system (1) under the conditions H_5 will be destroyed by impulsive effect. Numeric results (see Fig.4-Fig.6) show that system (1) under the conditions H_5 has Gui chaotic strange attractors [6].

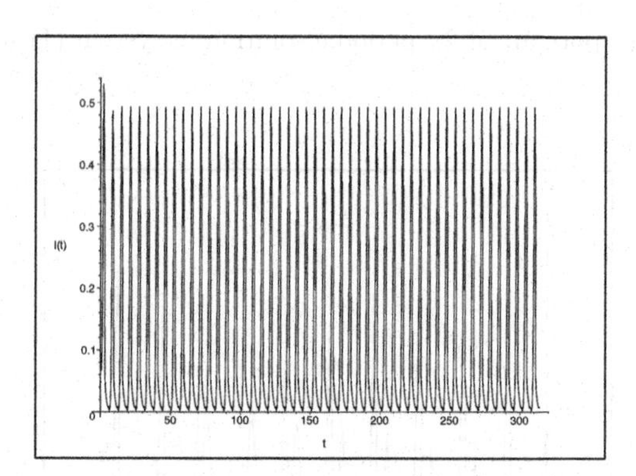

Fig. 1. Time-series of $I(t)$ evolved in system (1) with $T = \pi$

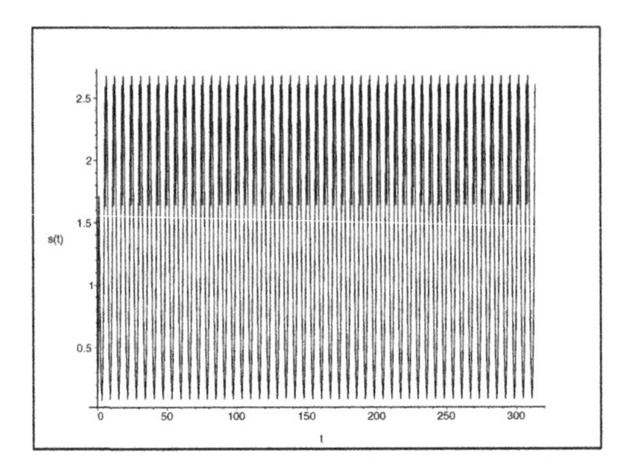

Fig. 2. Time-series of $s(t)$ evolved in system (1) with $T = \pi$

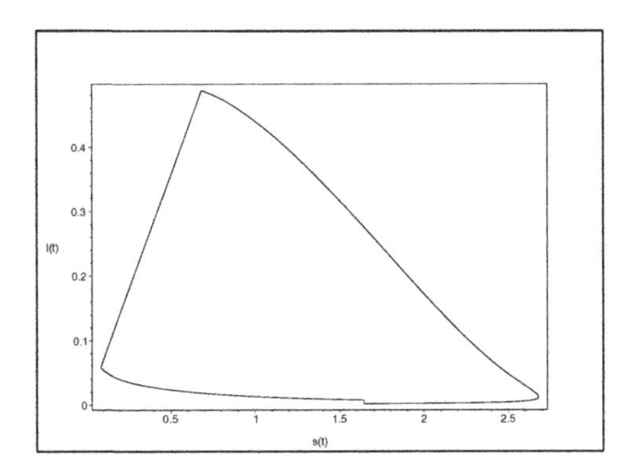

Fig. 3. Phase portrait of 2π-periodic solution of system (1) with $T = \pi$

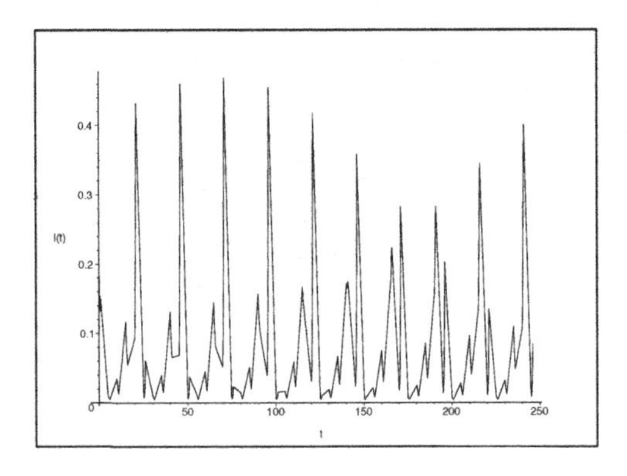

Fig. 4. Time-series of $I(t)$ evolved in system (1) with $T = 5$

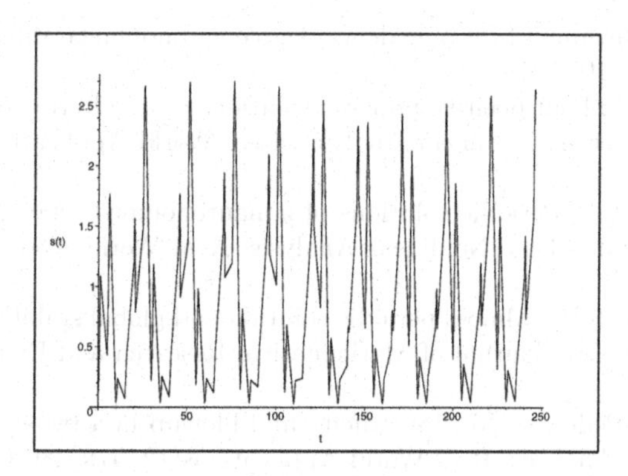

Fig. 5. Time-series of $s(t)$ evolved in system (1) with $T = 5$

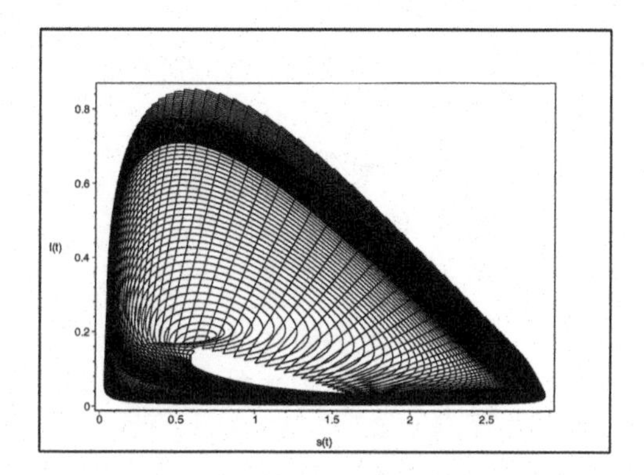

Fig. 6. Phase portrait of 2π-periodic solution of system (1) with $T = 5$

Acknowledgement. This work is supported by the National Natural Science Foundation of China (60963025); Hainan Nature Science Foundation (110007, 609009) in China; the Start-up fund of Hainan Normal University (00203 020201); the Foundation of the Office of Education of Hainan Province(Hj2009-36,Hnjg2011-11) and the Foundation of Hainan College of Software Technology.

References

1. Hofbauer, J., Sigmund, K.: Evolutionary Games and Population Dynamics. Cambridge University, Cambridge (1998)
2. Meng, X.Z., Chen, L.S.: Permance and global stability in an impulsive LotkaC-Volterra N-species competitive system with both discrete delays and continuous delays. Int. J. Biomath. 1, 179–196 (2008)
3. Jiao, J.J., Chen, L.S.: Global attractivity of a stage-structure variable coefficients predator-prey system with time delay and impulsive perturbations on predators. Int. J. Biomath. 1, 187–208 (2008)

4. Gaines, R.E., Mawhin, J.L.: Coincidence degree and nonlinear diferential equations. Springer, Berlin (1977)
5. Hu, D., Zhang, Z.: Four positive periodic solutions to a Lolterra cooperative system with harvesting terms. Nonlinear Analysis: Real World Applications 11, 1115–1121 (2010)
6. Zhang, J., Gui, Z.J.: Periodic solutions of nonautonomous cellular neural networks with impulses and delays. Nonlinear Analysis: Real World Applications 10, 1891–1903 (2009)
7. Wei, F.Y., Wang, S.H.: Almost periodic solution and global stability for cooperative L-V diffusion system. Journal of Mathematical Research and Exposition 30, 1108–1116 (2010)
8. Lin, Z., Liu, J., Pedersen, M.: Periodicity and blowup in a two-species cooperating model. Nonlinar Analysis: Real World Applications 12, 479–486 (2011)

Existence and Simulations of Periodic Solution of a Predator-Prey System with Holling-Type Response and Impulsive Effects

Wenxiang Zhang, Zhanji Gui, and Kaihua Wang*

School of Mathematics and Statistics Hainan Normal University
Haikou, Hainan, 571158
kaihuawang@qq.com

Abstract. The principle aim of this paper is to explore the existence of periodic solution of a predator-prey model with functional response and impulsive perturbations. Sufficient and realistic conditions are obtained by using Mawhin's continuation theorem of the coincidence degree. Further, some numerical simulations show that our model can occur in many forms of complexities including periodic oscillation and chaotic strange attractor.

Keywords: Periodic Solution, Predator-prey system, Impulses, Coincidence degree theory.

1 Introduction

In this paper, we will consider the following T-periodic Holling-type functional response predator-prey system [1,2,3,4] with diffusion and impulsive effects:

$$\begin{cases} \left.\begin{array}{l} \dot{x}_1(t) = x_1(t)(b_1(t) - d_1(t)x_1(t)) - a_1(t)y(t)\frac{\alpha(t)x_1(t)}{N(t)+x_1(t)} \\ \qquad\quad +D_1(t)(x_2(t) - x_1(t)), \\ \dot{x}_2(t) = x_2(t)(b_2(t) - d_2(t)x_2(t)) + D_2(t)(x_1(t) - x_2(t)), \\ \dot{y}(t) \;\;= y(t)(b_3(t) - d_3(t)y(t)) + a_2(t)y(t)\frac{\alpha(t)x_1(t)}{N(t)+x_1(t)}, \end{array}\right\} t \neq t_n, \\ \left.\begin{array}{l} x_1(t_n^+) = (1 + h_{1n})x_1(t_n), \\ x_2(t_n^+) = (1 + h_{2n})x_2(t_n), \\ y(t_n^+) = (1 + g_n)y(t_n), \end{array}\right\} t = t_n, n \in Z^+. \end{cases} \tag{1}$$

where $x_1(t)$ and $y(t)$ are the densities of prey species and predator species in patch I at time t, $x_2(t)$ is the density of prey species in patch II, prey species $x_1(t)$, $x_2(t)$ can diffuse between two patches while the predator species $y(t)$ is confined to patch I. $a_1(t)$ is the maximum of prey that can be eaten by a predator per unit of time, $a_2(t)$ a conversion efficiency, $b_i(t)\,(i = 1, 2, 3)$ intrinsic growth rate, $d_i(t)\,(i = 1, 2, 3)$ the rate of intra-specific competition, $D_i(t)\,(i = 1, 2)$ the dispersal rate of prey species, and h_{in} and g_n represent the annual birth pulse of population $x_i(t)$, $y(t)$ at t_n $(i = 1, 2)$, $n \in Z^+$. In this paper, we will assume that the following conditions are fulfilled:

* Corresponding author.

Y. Wang and T. Li (Eds.): Knowledge Engineering and Management, AISC 123, pp. 61–70.
springerlink.com

(A1) $a_i(t)$, $D_i(t)$, $b_i(t)$, $d_i(t)$ $(i = 1, 2, 3)$ and $\alpha(t)$, $N(t)$ are continuous positive T-periodic functions;

(A2) h_{1n}, h_{2n}, g_n are constants and there exists a positive integer q such that $h_{1(n+q)} = h_{1n}$, $h_{2(n+q)} = h_{2n}$, $g_{n+q} = g_n$, $t_{n+q} = t_n + T$.

With model (1) we can take into account the possible exterior effects under which the population densities change very rapidly. For instance, impulsive reduction of the population density of a given species is possible after its partial destruction by catching, a natural constraint in this case is $1+h_{1n} > 0$, $1+h_{2n} > 0$, $1+g_n > 0$, $n \in Z^+$.

2 Notations and Preliminaries

Let $J \subset R$, denote by $PC(J, R)$ the set of functions $\psi : J \to R$, which are piecewise continuous in $[0, T]$, have points of discontinuity $t_n \in [0, T]$, where they are continuous from the left. Let $PC^1(J, R)$ denote the set of functions ψ with derivative $\dot{\psi}(t) \in PC(J, R)$. Throughout this paper we deal with the Banach space of T-periodic functions

$$PC_T = \{\psi \in PC([0, T], R) | \psi(0) = \psi(T)\}$$

with the supremum norm:

$$\|\psi\|_{PC_T} = \sup\{|\psi(t)| : t \in [0, T]\}$$

and

$$PC_T^1 = \{\psi \in PC^1([0, T], R) | \psi(0) = \psi(T)\}$$

with the supremum norm:

$$\|\psi\|_{PC_T^1} = \max\{\|\psi\|_{PC_T}, \|\dot{\psi}\|_{PC_T^1}\}.$$

we will also consider the product space $PC_T \times PC_T$ which is also a Banach space with the norm

$$\|(\psi_1, \psi_2)\|_{PC} = \|\psi_1\|_{PC} + \|\psi_2\|_{PC}.$$

Moreover, for any $y \in C_T$ or $y \in PC_T$, define average value of y as follows:

$$\overline{y} := \frac{1}{T} \int_0^T y(t)dt$$

and the minimum, maximum of y respectively are:

$$y^L := \min_{t \in [0, T]} y(t), \qquad y^M := \max_{t \in [0, T]} y(t).$$

Give $\alpha, \beta \in PC_T$, $\beta > 0$, we consider the following Logistic equation with impulsive effects.

$$\begin{cases} \dot{\omega}(t) = \alpha(t)\omega(t) - \beta(t)\omega^2(t), & t \neq t_n, n \in Z^+, \\ \omega(t^+) = (1 + c_n)\omega(t_n), & t = t_n, n \in Z^+. \end{cases} \quad (2)$$

where $c_n (n \in Z^+)$ is constant, there exists an integer $q > 0$ such that $c_{n+q} = c_n$, $t_{n+q} = t_n + T$, and assume that $1 + c_n > 0$ $(n \in Z^+)$.

Lemma 1. *System (2) admits a unique positive solution if and only if* $\bar{\alpha} + \frac{1}{T}\sum_{n=1}^{q} \ln(1 + c_n) > 0$.

Let $\theta_{[\alpha,\beta]}$ denote the unique positive periodic solution to (2). Dividing $\dot{\theta}_{[\alpha,\beta]} = \alpha\theta_{[\beta,\beta]} - \beta\theta^2_{[\alpha,\beta]}$ by $\theta_{[\alpha,\beta]}$ and integrating over intervals $(0, T]$, we have

$$\bar{\alpha} + \frac{1}{T}\sum_{n=1}^{q} \ln(1 + c_n) = \frac{1}{T}\int_0^T \beta\theta_{[\alpha,\beta]} dt = \overline{\beta\theta_{[\alpha,\beta]}}.$$

To shorten notation, we rewrite $\theta_\alpha := \theta_{[\alpha,\beta]}$.

We denote by $\Phi_{[a,b]}(t, t_0, \omega_0)$ the unique solution of Cauchy problem

$$\begin{cases} \dot{\omega}(t) = \alpha(t)\omega(t) - \beta(t)\omega^2(t), & t \geq t_0 (t \neq t_n), \\ \omega(t_n^+) = (1 + c_n)\omega(t_n), & t = t_n, \\ \omega(t_0^+) = \omega_0. \end{cases} \quad (3)$$

Lemma 2. *Give* $\alpha, \beta \in PC_T$, *with* $\beta > 0$, *for any* $\omega_0 > 0$ *we have*

$$\lim_{t\to\infty} |\Phi_{[a,b]}(t, t_0, \omega_0) - \theta_\alpha| = 0$$

provided that $\bar{\alpha} + \frac{1}{T}\sum_{n=1}^{q} \ln(1 + c_n) > 0$ *and* $1 + c_n > 0$ *for* $n \in Z^+$.

Lemma 3. *Given a positive* $x_0 \in R$, *consider two functions* $a, b \in PC((t_0, \infty), R)$ *with* $b > 0$, *suppose that* $x(t) \in PC^1_T$ *such that*

$$\begin{cases} \dot{x}(t) \geq ax(t) - bx^2(t), & t \geq t_0 (t \neq t_n), \\ x(t_n^+) \geq (1 + c_n)x(t_n), & t = t_n, \\ x(t_0^+) \geq x_0. \end{cases} \quad (4)$$

Then $x(t) \geq \Phi_{[a,b]}(t, t_0^+, x_0)$ *for all* $t \geq t_0$. *Similarly* $x(t) \leq \Phi_{[a,b]}(t, t_0^+, x_0)$ *for all* $t \geq t_0$ *if all the sign of inequalities in (4) are converse.*

In order to obtain the existence of positive T-periodic solution to system (1), we must use the following lemma, named as the continuation theorem of coincidence degree theory [5].

Let X, Z be normed vector spaces, $L : \text{Dom}L \subseteq X \to Z$ be a linear mapping, $N : X \to Z$ be a continuous mapping. If $\text{dimKer}L = \text{comdimIm}L < +\infty$ and $\text{Im}L$ is closed in Z, then the mapping L will be called a Fredholm mapping of index zero. If L is a Fredholm mapping of index zero, there exist continuous projects $P : X \to X$ and $Q : Z \to Z$ such that $\mathrm{Im}P = \mathrm{Ker}L$, $\mathrm{Im}L = \mathrm{Ker}Q = \mathrm{Im}(I - Q)$. It follows that $L|_{\mathrm{Dom}L \bigcap \mathrm{Ker}P} : (I - P)X \to \mathrm{Im}L$ has an inverse which is denoted by K_P. If Ω is an open bounded subset of X, the mapping N will be called L-compact on $\overline{\Omega}$ provided that $QN(\overline{\Omega})$ is bounded and $K_p(I - Q)N : \overline{\Omega} \to X$ is compact. Since $\mathrm{Im}Q$ is isomorphic to $\mathrm{Ker}L$ there exists an isomorphism $F : \mathrm{Im}Q \to \mathrm{Ker}L$.

Lemma 4. *Let L be a Fredholim mapping of index zero and N be L-compact on $\overline{\Omega}$. Suppose that*

(a) For each $\lambda \in (0, 1)$, every solution x of $Lx = \lambda Nx$ such that $x \notin \partial\Omega$;
(b) $QNx \neq 0$ for each $x \in \mathrm{Ker}L \bigcap \partial\Omega$;
(c) deg $\{FQN, \Omega \bigcap \mathrm{Ker}L, 0\} \neq 0$.

Then the equation $Lx = Nx$ has at least one solution lying in $\mathrm{Dom}L \bigcap \overline{\Omega}$.

3 Existence of Positive Periodic Solution

In this section, we study the existence of positive periodic solution to (1).

Theorem 1. *If system (1) satisfies*

1. $b + \frac{1}{T}\sum_{n=1}^{q} \ln(1 + m_n) > 0$, $\bar{b}_3 + \frac{1}{T}\sum_{n=1}^{q} \ln(1 + g_n) > 0$,
2. $b_i(t) > D_i(t) (i = 1, 2)$, $\bar{b}_2 - \bar{D}_2 + \frac{1}{T}\sum_{n=1}^{q} \ln(1 + h_{2n}) > 0$.

here $b = \max\{b_1^M, b_2^M\}$, $m_n = \max\{1 + h_{1n}, 1 + h_{2n}\}$, $d = \min\{d_1^L, d_2^L\}$. Then system (1) has at least one T-periodic positive solution.

Proof. Let $x_1(t) = e^{u_1(t)}$, $x_2(t) = e^{u_2(t)}$, $y(t) = e^{u_3(t)}$ then system (1) is reformulated as

$$\begin{cases} \left.\begin{array}{l} \dot{u}_1(t) = b_1(t) - D_1(t) - d_1(t)e^{u_1(t)} - \frac{a_1(t)\alpha(t)e^{u_3(t)}}{N(t)+e^{u_1(t)}} + D_1(t)e^{u_2(t)-u_1(t)}, \\ \dot{u}_2(t) = b_2(t) - D_2(t) - d_2(t)e^{u_2(t)} + D_2(t)e^{u_1(t)-u_2(t)}, \\ \dot{u}_3(t) = b_3(t) - d_3(t)e^{u_3(t)} \frac{a_2(t)\alpha(t)e^{u_1(t)}}{N(t)+e^{u_1(t)}}, \end{array}\right\} t \neq t_n, \\ \left.\begin{array}{l} u_1(t_n^+) = u_1(t_n) + \ln(1 + h_{1n}), \\ u_2(t_n^+) = u_2(t_n) + \ln(1 + h_{2n}), \\ u_3(t_n^+) = u_3(t_n) \ln(1 + g_n), \end{array}\right\} t = t_n. \end{cases} \quad (5)$$

If system (5) has a T-periodic solution $(u_1(t), u_2(t), u_3(t))^T$, then

$$(e^{u_1(t)}, e^{u_2(t)}, e^{u_3(t)})^T = (x_1^*(t), x_2^*(t), y^*(t))$$

is a positive T-periodic solution to system (1). So, in the following,we discuss the existence of T-periodic solution to system (5).

In order to use Lemma 2.4, we set $\mathbf{u} = (u_1(t), u_2(t), u_3(t))^T$. Define $X = \{x \in PC(R, R^3) : x(t + T) = x(t)\}$, $Z = X \times R^{3q}$, then it is standard to show both X and Z are Banach space when they are endowed with the

norms $\|x\|_c = \sup_{t\in[0,\omega]} |x(t)|$ and $\|(x, c_1, c_2, c_3)\| = (\|x\|_c^2 + |c_1|^2 + |c_2|^2 + |c_3|^2)^{1/2}$.
Let Dom$L \subset X = \{x \in C^1[0,\omega; t_1, ..., t_m] | x(0) = x(\omega)\}$, L: Dom$L \to Z$, $L\mathbf{u} = (\mathbf{u}', \Delta\mathbf{u}(t_1), ..., \Delta\mathbf{u}(t_q))$; $N : X \to Z$, N: Dom$L \to Z$, $N\mathbf{u} = (\mathbf{u}', \Delta\mathbf{u}(t_1), ..., \Delta\mathbf{u}(t_q))$. It is easy to prove that L is a Fredholm mapping of index zero.

Consider the operator equation

$$L\mathbf{u} = \lambda N\mathbf{u}, \quad \lambda \in (0, 1). \tag{6}$$

Integrating (6) over the interval $[0, T]$, we obtain

$$\begin{cases} B_1 = \int_0^T [d_1(t)e^{u_1(t)} + \frac{a_1(t)\alpha(t)e^{u_3(t)}}{N(t)+e^{u_1(t)}} - D_1(t)e^{u_2(t)-u_1(t)}]dt, \\ B_2 = \int_0^T [d_2(t)e^{u_2(t)} - D_2(t)e^{u_1(t)-u_2(t)}]dt, \\ B_3 = \int_0^T [d_3(t)e^{u_3(t)} - \frac{a_2(t)\alpha(t)e^{u_1(t)}}{N(t)+e^{u_1(t)}}]dt, \end{cases} \tag{7}$$

here $B_i = \bar{b}_i T - \bar{D}_i T + \sum_{n=1}^q \ln(1 + h_{in})(i = 1, 2)$,
$B_3 = \bar{b}_3 + \sum_{n=1}^q \ln(1 + g_n)$.
From (6) and (7), we have

$$\int_0^T |\dot{u}_1(t)|dt \leq 2(\bar{b}_1 - \bar{D}_1)T + \left|\sum_{n=1}^q \ln(1 + h_{1n})\right|, \tag{8}$$

$$\int_0^T |\dot{u}_2(t)|dt \leq 2(\bar{b}_2 - \bar{D}_2)T + \left|\sum_{n=1}^q \ln(1 + h_{2n})\right|, \tag{9}$$

$$\int_0^T |\dot{u}_3(t)|dt \leq 2\bar{b}_3 T + \left|\sum_{n=1}^q \ln(1 + g_n)\right|. \tag{10}$$

Since $u_i(t) \in PC_T$, there exist $\xi_i, \eta_i \in [0, T](i = 1, 2, 3)$ such that

$$u_i(\xi_i) = \min_{t\in[0,T]} u_i(t), \qquad u_i(\eta_i) = \max_{t\in[0,T]} u_i(t). \tag{11}$$

Let $v(t) = \max\{u_1(t), u_2(t)\}$, then $v(t) \in PC_T$, moreover

1. if $u_1(t) \geq u_2(t)$ but $\dot{u}_1(t) \geq \dot{u}_2(t)$, then $v(t) = u_1(t)$ and $\dot{u}_1(t) \leq \lambda(b_1(t) - d_1(t)e^{u_1(t)}) \leq \lambda(b_1^M - d_1^L e^{u_1(t)})$,
2. if $u_2(t) \geq u_1(t)$ but $\dot{u}_2(t) \geq \dot{u}_1(t)$, then $v(t) = u_2(t)$ and $\dot{u}_2(t) \leq \lambda(b_2(t) - d_2(t)e^{u_2(t)}) \leq \lambda(b_2^M - d_2^L e^{u_2(t)})$.

Denote $b = \max\{b_1^M, b_2^M\}$, $d = \min\{d_1^L, d_2^L\}$, $m_n = \max\{h_{1n}, h_{2n}\}$, then

$$\begin{cases} D^+ v(t) \leq \lambda(b - de^{v(t)}), \ t \neq t_n, \\ \Delta v(t_n) \leq \lambda \ln(1 + m_n), \ t = t_n. \end{cases} \tag{12}$$

Integrating (12) over $[0, T]$, we have

$$-\sum_{n=1}^q \ln(1 + m_n) \leq bT - d\int_0^T e^{v(t)}dt,$$

therefore,

$$\int_0^T e^{u_i(\xi_i)}dt \leq \frac{bT + \sum\limits_{n=1}^{q} \ln(1+m_n)}{d} \quad (i=1,2),$$

so

$$u_i(\xi_i) \leq \ln \left[\frac{bT + \sum\limits_{n=1}^{q} \ln(1+m_n)}{dT} \right] (i=1,2),$$

then

$$\begin{aligned} u_i(t) \leq & u_i(\xi_i) + \int_0^T |\dot{u}_i(t)|dt + \left| \sum_{n=1}^{q} \ln(1+h_{in}) \right| \\ \leq & \ln \left[\frac{bT + \sum\limits_{n=1}^{q} \ln(1+m_n)}{dT} \right] + 2(\bar{b}_i - \bar{D}_i)T \\ & +2 \left| \sum_{n=1}^{q} \ln(1+h_{in}) \right| = M_i \ (i=1,2). \end{aligned} \tag{13}$$

From (7) and (11), we have

$$\int_0^T d_2(t)e^{u_2(\eta_2)}dt \geq \int_0^T d_2(t)e^{u_2(t)}dt \geq B_2,$$
$$\int_0^T d_3(t)e^{u_3(\eta_3)}dt \geq \int_0^T d_3(t)e^{u_3(t)}dt \geq B_3.$$

that is

$$u_2(\eta_2) \geq \ln\left(\frac{B_2}{\bar{d}_2T}\right);\ u_3(\eta_3) \geq \ln\left(\frac{B_3}{\bar{d}_3T}\right).$$

Then

$$\begin{aligned} u_2(t) \geq & \ u_2(\eta_2) - \textstyle\int_0^T |\dot{u}_2(t)|dt - \left| \sum\limits_{n=1}^{q} \ln(1+h_{2n}) \right| \\ \geq & \ \ln\left(\tfrac{B_2}{\bar{d}_2T}\right) - 2(\bar{b}_2 - \bar{D}_2)T - 2\left| \sum\limits_{n=1}^{q} \ln(1+h_{2n}) \right| = M_3, \\ u_3(t) \geq & \ u_3(\eta_3) - \textstyle\int_0^T |\dot{u}_3(t)|dt - \left| \sum\limits_{n=1}^{q} \ln(1+h_{3n}) \right| \\ \geq & \ \ln\left[\tfrac{B_3}{\bar{d}_3T}\right] - 2(\bar{b}_3 - \bar{D}_3)T - 2\left| \sum\limits_{n=1}^{q} \ln(1+h_{3n}) \right| = M_4. \end{aligned}$$

So we have

$$B_3 \geq \int_0^T [d_3(t)e^{u_3(\xi_3)} - \frac{a_2(t)\alpha(t)e^{M_1}}{N(t)+e^{M_1}}]dt = \bar{d}_3 T e^{u_3(\xi_3)} - \overline{\left(\frac{a_2\alpha e^{M_1}}{N(t)+e^{M_1}}\right)}T,$$

that is

$$u_3(\xi_3) \leq \ln\left[\frac{B_3 + \overline{\left(\frac{a_2(t)\alpha(t)e^{M_1}}{N(t)+e^{M_1}}\right)}T}{\bar{d}_3 T}\right],$$

Then

$$\begin{aligned} u_3(t) &\leq u_3(\xi_3) + \int_0^T |\dot{u}_3(t)|dt + \left|\sum_{n=1}^{q} \ln(1+g_n)\right| \\ &\leq \ln\left[\frac{B_3 + \overline{\left(\frac{a_2(t)\alpha(t)e^{M_1}}{N(t)+e^{M_1}}\right)}T}{\bar{d}_3 T}\right] + 2\bar{b}_3 T + 2\left|\sum_{n=1}^{q} \ln(1+g_n)\right| \\ &= M_5. \end{aligned}$$

Similarly,

$$B_3 \geq \int_0^T [d_3(t)e^{M_5} - \frac{a_2(t)\alpha(t)e^{u_1(\eta_1)}}{N(t)}]dt,$$

that is

$$u_1(\eta_1) \geq \ln\left[\frac{\bar{d}_3 e^{M_5} T - B_3}{\overline{(\frac{a_2\alpha}{N})}T}\right],$$

then

$$\begin{aligned} u_1(t) &\geq u_1(\eta_1) - \int_0^T |\dot{u}_1(t)|dt - \left|\sum_{n=1}^{q} \ln(1+h_{1n})\right| \\ &\geq \ln\left[\frac{\bar{d}_3 e^{M_5} T - B_3}{\overline{(\frac{a_2\alpha}{N})}T}\right] - 2(\bar{b}_1 - \bar{D}_1)T - 2\left|\sum_{n=1}^{q} \ln(1+h_{1n})\right| = M_6. \end{aligned}$$

Thus, we have

$$\sup_{t\in[0,T]} |u_1(t)| \leq \max\{|M_1|, |M_6|\} = N_1,$$
$$\sup_{t\in[0,T]} |u_2(t)| \leq \max\{|M_2|, |M_3|\} = N_2,$$
$$\sup_{t\in[0,T]} |u_1(t)| \leq \max\{|M_4|, |M_5|\} = N_3.$$

Obviously, there exists a constant $N_4 > 0$ such that $\max\{|u_1|, |u_2|, |u_3|\} < N_4$. Take $r > N_1+N_2+N_3+N_4$, $\Omega = \{x \in X | \|x\|_c < r\}$, then N is L-compact on $\overline{\Omega}$. So, for $\forall\ \mathbf{u} = (u_1, u_2, u_3)^T \in \partial\Omega \bigcap \mathrm{Ker}L$, we have $QN\mathbf{u} \neq 0$. Let $J : \mathrm{Im}Q \to x$, $(d, 0, \ldots, 0) \to d$. Then when $\mathbf{u} \in \Omega \bigcap \mathrm{Ker}L$, in view of the assumptions in Mawhin's continuation theorem, one obtains, $deg\{FQN, \Omega \bigcap \mathrm{Ker}L, 0\} \neq 0$. By now we have proved that Ω satisfies all the requirements in Mawhin's continuation theorem. Hence, (5) has at least one T-periodic solution in $\mathrm{Dom}L \cap \overline{\Omega}$. □

4 Some Simulations

In this section, we shall discuss an example to illustrate main results. For system (1), we take: $t_n = n\omega$, $b_1(t) = 1 + 0.2\sin(t)$, $d_1(t) = 0.8 + 0.2\sin(t)$, $a_1(t) = 1 + 0.8\cos(t)$, $\alpha(t) = 1 + 0.2\sin(t)$, $N(t) = 2 + \sin(t)$, $D_1(t) = 0.2 + 0.1\sin(t)$, $b_2(t) = 1 + 0.1\cos(t)$, $d_2(t) = 0.9 + 0.1\cos(t)$, $D_2(t) = 1 + 0.2\sin(t)$, $b_3(t) = 1+0.2\cos(t)$, $d_3(t) = 0.8+0.2\cos(t)$, $a_2(t) = 1+0.6\sin(t)$, $h_{1n} = 0.2$, $h_{2n} = 0.2$, $g_n = 0.2$. Obviously, all conditions of Theorem 1 are satisfied.

If $\omega = \pi/2$, then system (1) under the above conditions has a unique 2π-periodic solution (In Fig.1-Fig.4, we take $[x_1(0), x_2(0), y(0)]^T = [0.5, 0.6, 1]^T$). We find the occurrence of sudden changes in the figures of the time-series and phase portrait. The influence of pulse is obvious.

If $\omega = 2$, then (A2) is not satisfied. Periodic oscillation of system (1) under the above conditions will be destroyed by impulsive effect. Numeric results (see Fig.5) show that system (1) under the above conditions has Gui chaotic strange attractor [6].

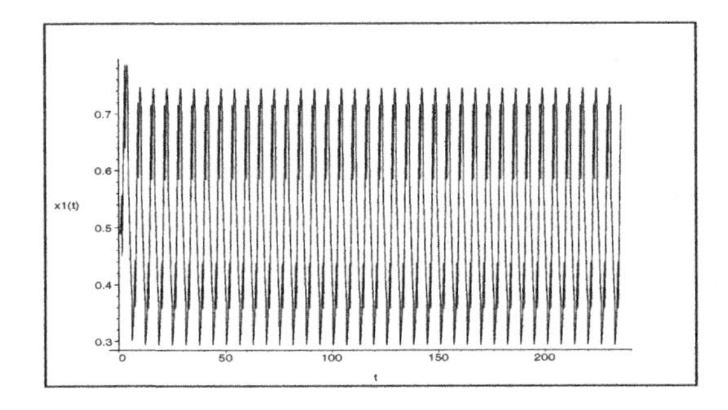

Fig. 1. Time-series of $x_1(t)$ evolved in system (1) with $\omega = \pi/2$

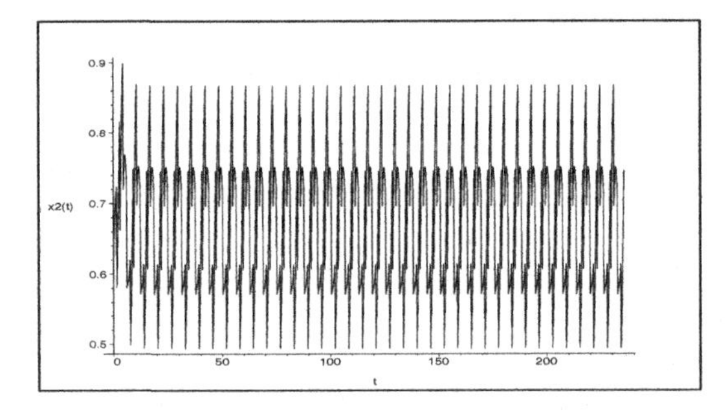

Fig. 2. Time-series of $x_2(t)$ evolved in system (1) with $\omega = \pi/2$

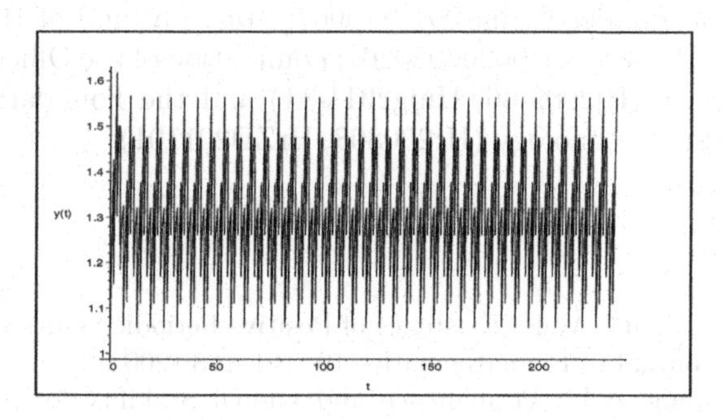

Fig. 3. Time-series of $y(t)$ evolved in system (1) with $\omega = \pi/2$

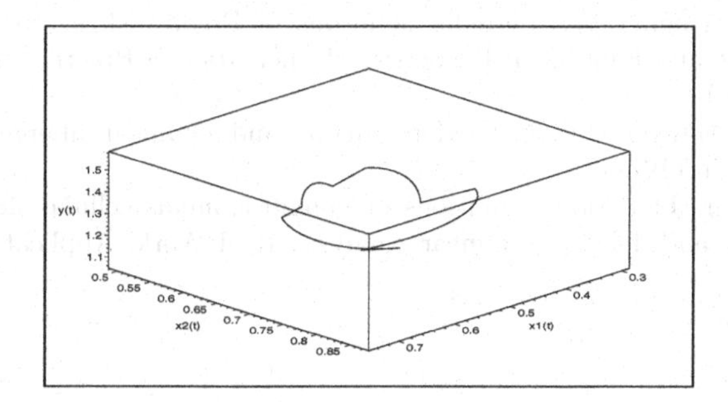

Fig. 4. Phase portrait of 2π-periodic solution of system (1) with $\omega = \pi/2$

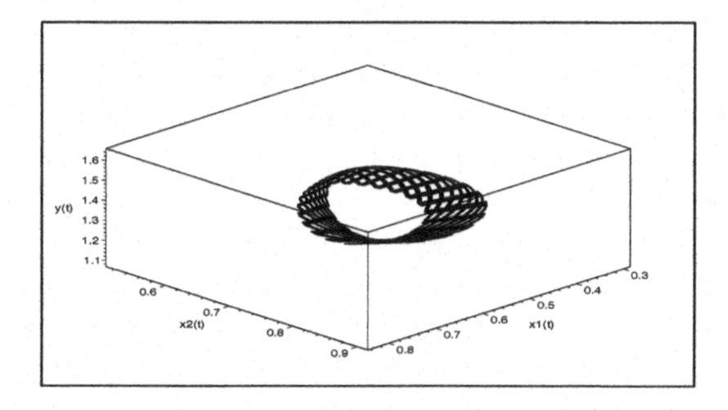

Fig. 5. Phase portrait of chaotic strange attractor of system (1) with $\omega = 2$

Acknowledgements. This work is supported jointly by the Natural Sciences Foundation of China under Grant No. 60963025; Natural Sciences Foundation of Hainan Province under Grant No. 110007; Start-up fund of Hainan Normal University under Project No. 00203020201; Foundation of the Office of Education of Hainan Province (Hj2009-36, Hnjg2011-11) and the Foundation of Hainan College of Software Technology (Hr200808, HrZD201101).

References

1. Ye, D., Fan, M., Zhang, W.P.: Existence of Positive Periodic Solution of a Predator-Prey System. Journal of Biomathematies 19, 161–168 (2004)
2. Meng, X.Z., Wang, X.L.: Permanence and Global Stability for Nonautonomous Predator-Prey Systems with Peridic Coefficients and Type Holling Functional Response. College Mathematics 19, 14–19 (2003)
3. Gao, Y.L., Gui, G.Z.: The Qualitative Analysis of a Model of Predator-Prey System with Holling III Functional Response. Journal of Information Engineering University 6, 19–22 (2005)
4. Zhang, B.L., Yuan, L.W.: Periodic Solutions of Predator-Prey System with Impulsive Effects and Functional Response. Mathematics in Practice and Theory 39, 109–115 (2009)
5. Gaines, R.E., Mawhin, J.L.: Coincidence degree and nonlinear differential equations. Springer, Berlin (1977)
6. Zhang, J., Gui, J.: Periodic solutions of nonautonomous cellular neural networks with impulses and delays. Nonlinear Analysis: Real World Applications 19, 1891–1903 (2009)

A Novel Selection Operator of Cultural Algorithm

Xiaowei Xue, Min Yao, and Ran Cheng

College of Computer Science and Technology, Zhejiang University,
Hangzhou, China
{xwxue,myao,rancheng}@zju.edu.cn

Abstract. Cultural Algorithms (CAs) are a series of new algorithms which depict cultural evolution as a process of dual inheritance. In this paper, cultural algorithm using Genetic Algorithms (GAs) and the knowledge in belief space to guide the evolution of population space is introduced. GAs simply use the fitness to evaluate the quality of solutions, however, it may lose the diversity of population and even lead to premature convergence. To solve this problem, we put forward a novel selection operator. Compared with conventional CA based on GA, CA with our selection operator performs better in the global convergence.

Keywords: Cultural Algorithm, Genetic Algorithm, selection operator, premature convergence.

1 Introduction

In human society, culture can be seen as a set of knowledge that has been involved from individuals' past experience. New individuals can learn knowledge and be guided to solve some problem by it. Inspired by the process of cultural evolution, in 1994, American scholar Reynolds proposed a new evolutionary computation called cultural algorithm (CA)[1].

In this paper, cultural algorithm using the Genetic Algorithms (GAs) and the knowledge in belief space to guide the evolution of population space is introduced. GAs simply use the fitness to evaluate the quality of solutions, however, it may lose the diversity of population and even lead to premature convergence. To solve this problem, we put forward a novel selection operator.

In section 2 the basic components of the CA is introduced. In section 3 we analyze why premature convergence may happen in the multi-peak optimization problem. On the basis of the analysis, we put forward a novel selection operator, this algorithm is abbreviated as (NSCA). By using the knowledge of the belief space and inserting new individuals, more diversity is acquired and the premature convergence is prevented to some extent. In section 4, an experiment with three benchmark test functions is implemented and the result is analyzed.

Y. Wang and T. Li (Eds.): Knowledge Engineering and Management, AISC 123, pp. 71–77.
springerlink.com

2 Cultural Algorithm

The frame of CA can be seen as a process of dual inheritance[1] from both a micro-evolutionary level (population space) and a micro-evolutionary level(belief space).

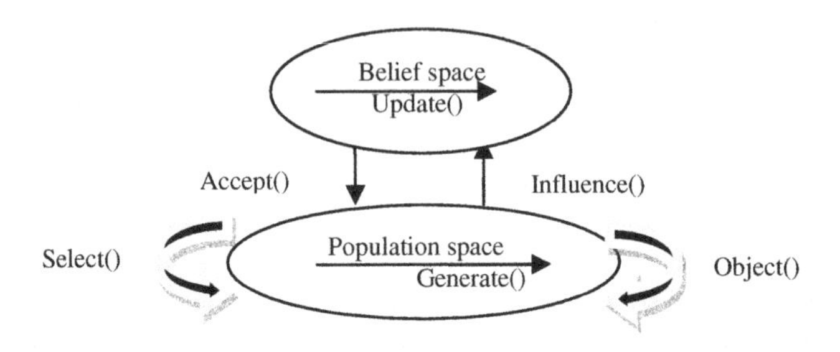

Fig. 1. The framework of Cultural Algorithm

The population space simulates the individual evolution, as a result it can support any population-based computing models, such as evolutionary computation models, including GA, evolutionary programming[3]-[4], evolution strategies. In this paper we select GA as population space.

The belief space extracts various types of implicit information and stores the information in form of knowledge. Knowledge in the belief space is different when computing strategies or applications are different. Generally, knowledge in belief space is divided into five categories: Situational Knowledge[5], Normative Knowledge[5], Topographical Knowledge[6], History Knowledge[7] and Domain Knowledge[7].

Aim at solving multi-peak optimization problems, we select situation knowledge and Normative Knowledge. Situational knowledge provides a set of exemplary cases and leads individuals to "move toward the exemplars". Normative knowledge is a set of promising variable ranges that provides standards for individual behavior and guidelines and eventually leads individuals to "jump into the good range" if they are not already there.

Communications protocol is used to connect population space and belief space. It includes accept function, influence function and update function.

3 A Novel Selection Operator of CA

In the process of evolution, if an individual has a better fitness, its genes may spread rapidly in the population space. As a result, global optimal solution may not be found. When the premature convergence happens, the belief space also can't make guidelines for population space. Aiming at solving these problems, researchers have put forward various methods. For example, [8] uses multi-belief space, [9] changes the form of the knowledge. Their improvement performs well, but they ignore that the root causes of premature convergence is the loss of the population diversity. Research has already shown that only the selection operator reduces the diversity of the population. Our novel selection operator prevents the premature convergence to some extent.

3.1 The Basic Selection Operator

The basic selection operator uses the victory value to select individuals as the parentes for the next iteration. N parents generate N children by crossover, mutation and the influence of the knowledge in belief space. P individuals are selected randomly from the 2N individuals and compared with the 2N individuals. If the fitness of the 2N individuals is better, the individual's victory value pluses one. At last the 2N individuals are sorted based on the victory value and the top N individuals are selected as parents for the next iteration.

3.2 The Influence of Selection Operator to Algorithm Efficiency

Initial population is generated randomly. The probability that there is a global optimal solution in the Initial population is like this:

$$p_0 = 1-(1-r/s)^N \tag{1}$$

for s is the size of the individual space, r is the number of best solutions in the population. N is the size of the population space. Generally $r << s$, N is a finite number, so $p_0 \approx 0$. GA generates offspring mainly by crossover, the average number of new individuals generated by N parents' crossover in t th generation is: $N_t = N p_e$, for p_e is crossover rate.

Considering the new individuals may repeat, so the number of new individuals generated by 2 parents' crossover in t th generation is :

$$N_t = p_e g_1 g_2 N \tag{2}$$

for g_1 is the probability that new individuals are not as same as those in t th generation. g_2 is the probability that new individuals are not as same as those that generated in the past generations. The probability that there is a global optimal solution in t th generation is:

$$p_t = (1-(1-r/s)^{p_e g_1 g_2 N})\, c_t \tag{3}$$

for c_t is the increase of the probability that a iteration after selection operator's influence generates a best individual than the former one.

In order to improve the algorithm efficiency, we need to maximize p_t . From formula(3), it is clear that we can improve p_e , g_1 , g_2 , N and c_t ,where p_e is close to 1. Improving N will increase the computing time and be negative to algorithm efficiency. g_1 , g_2 will decline with the genetic iteration, so for improving p_t we mainly focus on improving c_t and slow down the decline rate of g_1 , g_2 in the iteration.

3.3 A Novel Selection Operator

By using of the knowledge in belief space and inserting new individuals, the novel selection operator improves c_t and slows down the decline rate of g_1 ,g_2.

A Making Use of Knowledge in Belief Space

The belief space uses knowledge to guide the evolution of the population space. One of the knowledge is a series of variable ranges. The variable ranges provide standards for individual behavior. New individuals generated by the variable ranges are more easy to become the best individual.

The improved method is: before selecting the novel selection operation generates M individuals by the variable ranges in the belief space and puts them with 2N individuals together to take victory value judgments. At last the operator sorts the 2N+M individuals by victory value and the top N individuals are selected as parents for the next generation.

In this way, we have to consider two cases.

- **The knowledge in belief space plays a guiding role in evolution:**

  Compared with individuals generated by crossover and mutation, individuals randomized by the variable ranges can more easily become the best individuals. After the victory value selection, there will be more excellent individuals. As a result, we improve c_t .

- **The knowledge in belief space can't play a guiding role in evolution:**

  If the knowledge can't guide the population's evolution, the individuals randomized by the knowledge have no more advantage. But they can increase the diversity of the population space. As a result we improve g_1 , g_2 .

If the individuals' fitness randomized by the knowledge are generally better than parents and offspring, most of the random individuals may be selected as parents for the next iteration and it may lead to a new premature convergence. In order to prevent this phenomenon, the number of random individuals can't be too large. Summarizing the experiments, the value of M sets at 10%N-20%N.

B Inserting New Individuals

The evolution of population space is the basis of the algorithm. If premature convergence of individuals in population space happens, it will lead to the loss of the guiding ability of the knowledge space. At last, premature convergence of algorithm happens.

As is mentioned above, N individuals are selected from 2N+M individuals. In order to improve the diversity of population space, we replace the worst Q individuals of the N individuals. The random range is the range for initializing the individuals of the population space. The reason why we use the initialized range is to ensure the reliability of the information source and to improve the information diversity of the population space. At the same time, inserting random individuals can avoid the situation that the fitness of the individual near the best individual is always worse than the other local optimal solution's fitness. It can also avoid the situation that the global optimal solution is eliminated because of the random of genetic manipulation.

Since the random range of the inserted individuals is the whole solution space, the probability of generating the global optimal solution is very small. So the inserted number should be not too many, at the same time we need to improve the information diversity by them. Summarizing the experiments, the value of Q sets at 10%N or so.

3.4 A Novel Selection Operator

Step 1: Initialize population space and belief space.
Step 2: Evaluate the individuals' fitness values in population space.
Step 3: Breed new individuals through crossover, mutation. The knowledge induce individuals via influence function. We use the variable range of the knowledge to influence the individuals.
Step 4: Use the novel selection operation to select N individuals. First, random M individuals by the knowledge of belief space, use the victory value to select N individuals from 2N+M individuals. Random Q individuals by the range of whole solution space. Remove the last Q individuals of the N individuals selected by the selection operator with the new Q individuals. Now the selection operation select N individuals as parents for the next iteration.
Step 5: Use the accept function to select the top individuals of the parents. Then judge whether the update moment is met or not. If the moment meets, update the knowledge of the belief space.
Step 6: Judge if the termination condition is met or not.If not meets, goes back to step 2, otherwise, stops.

4 Experimental Results

In our experiments, three standard test functions[10] for benchmark test are selected. For each function, the dimension number is 30 and for each dimension i, $x_i \in [-100,100]$ and their global optimal solutions are all 0.The first is a unimodal function, but it is a good function for test.

During the test, in order to ensure the accuracy of the test, we set the same parametersand collect the experimental data for the comparative experiment.

Table 1. Parameters for CA and NSCA

Cross rate	Mutate rate	Pop size	Evolution period	Precision
0.9	0.05	20	1000	10-10

Table 2. Result comparisons of NSCA with CA

		CA	NSCA
Sphere function f1: $f_1(x)=\sum_{i=1}^{n} x_i^2$	Best value	0.000000	0.000000
	Average value	0.000000	0.000000
	Worst value	0.000000	0.000000
Rosenbrock function f2: $f_2(x)=\sum_{i=1}^{n}[100(x_{i+1}-x_i^2)^2+(x_i-1)^2]$	Best value	27.036260	0.000000
	Average value	27.540140	8.5173222
	Worst value	28.015277	28.433610
Rastrigrin function f3 $f_3(x)=\sum_{i=1}^{n}(x_i^2-10\cos(2\pi\cdot x_i)+10)$	Best value	0.000000	0.000000
	Average value	0.000000	0.000000
	Worst value	0.000000	0.000000

In Table 2, both f1 and f3 get the global optimal solution. By comparing the iteration , however, it is clear that the NSCA's convergence is significantly better than the conventional CA and it proves that the novel selection operator has played a positive role in preventing the premature convergence. The test function f3 does not reach the global optimum, but the best solution of NSCA is much better than that of conventional CA and the average of optimal solution obtained in the 20 experiments is also better than that obtained in conventional CA. In Fig 2, we show the comparison chart of the convergence of the NSCA and CA. We can see that in about 150 iterations, premature convergence happens in CA and CA gets the solution of 27 or so. But NSCA keeps on evolution, about 850 iterations, it evolutes at the precision of 10^{-10}. The novel selection operator prevents the premature convergence successfully.

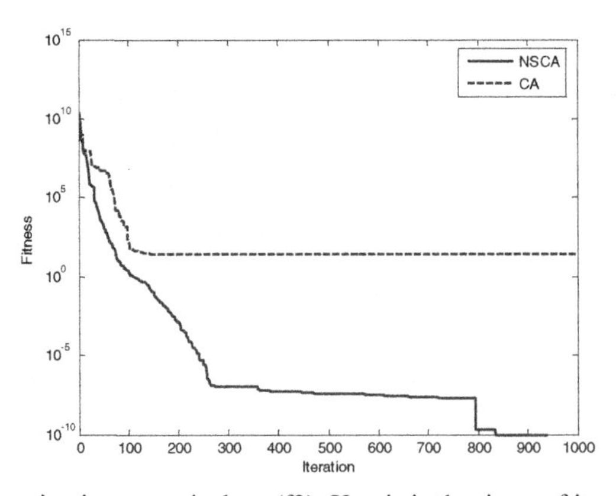

Fig. 2. The 1000 iteration in one typical run(f3). X-axis is the times of iterations , Y –axis is the fitness of individuals and the standard interval of it is 10^5 , so in the late evolution, a small fluctuation may be very obvious.

5 Conclusions

In this paper ,we propose a novel selection operator of the cultural algorithm based on genetic algorithm(NSCA). NSCA takes a good advantage of the knowledge of the belief space to improve algorithm efficiency and inserts random individuals to slow down the spread of losing the information diversity of the population space. Experiments have proved that NSCA performs well in preventing the premature convergence and improves the quality of solutions.

Though the results show that our improvement works, future study for cultural algorithm still requires. In future work we will be interested in proposing a new knowledge of the belief space.

Acknowledgments. This paper is the partial achievement of project Y1110152 supported by Natural Science Fund of Zhejiang Province.

References

1. Reynolds, R.: An Introduction to Cultural Algorithms. In: Proceedings of the 3rd Annual Conference on Evolutionary Programming, pp. 131–139. World Scientific Publishing (1994)
2. Holland, J.H.: Adaptation in natural and artificial systems. In: Ann. Arbor, The University of Michigan Press, MI (1975)
3. Reynolds, R.G., Sverdlik, W.: Problem Solving Using Cultural Algorithms. In: Proceeding of the First IEEE Conference on Evolutionary Computation, vol. 2, pp. 645–650 (1994)
4. Franklin, B., Bergerman, M.: Cultural Algorithms: Concepts and Experiments. In: Proceedings of the 2000 Congress on Evolutionary Computation, USA, vol. 2, pp. 1245–1251 (2000)
5. Chung, C.: Knowledge-based approaches to self-adaptation in cultural algorithms. In Ph.D. thesis, Wayne State University, Detroit, Michigan (1997)
6. Xi, D.J., Reynolds, R.G.: Using knowledge-based evolutionary computation to solve non-linear constraint optimization problems. In: A Cultural Algorithm Approach (1999)
7. Saleem, S.M.: Knowledge-based solution to dynamic optimization problems using cultural algorithms. In: ETD Collection for Wayne State University (2001) AAI3010120
8. Xue, Z., Guo, Y.: Improved Cultural Algorithm based on Genetic Algorithm. In: IEEE International Conference on Integration Technology, ICIT 2007 (2007)
9. Carlos, A., Coello, C., Becerra, R.L.: Adding Knowledge And Efficient Data Structures To Evolutionary Programming: A Cultural Algorithm For Constrained Optimization. In: Proceedings of the Genetic and Evolutionary Computation Conference, pp. 201–209
10. Cheng, R., Yao, M.: A Modified Particle Swarm Optimizer with a Novel Operator. Artificial Intelligence and Computational Intelligence, 293–301 (2010)

An Approach to Analyzing LOH Data of Lung Cancer Based on Biclustering and GA

Jun Wang, Hongbin Yang, Yue Wu, Zongtian Liu, and Zhou Lei

School of Computer Engineering & Science Shanghai University
200072 Shanghai, China
{wangjuntz,hbyoungshu,ywu,ztliu,leiz}@shu.edu.cn

Abstract. There is a close relation between the phenomenon of LOH and malignant tumor. Bicluster algorithms have been applied to the data of loss of heterozygosity analysis and can find the submatrix which is composed by SNPs loci related to cancer. But the conventional Cheng and Church method requires experience values as a threshold, and discovered results must be randomized. In this paper, we use k-means and GA to overcome this shortcoming. The experimental results demonstrate the effectiveness and accuracy of our method in discovering chromosome segments related to suppressor genes of lung cancer.

Keywords: LOH, Tumor, SNPs, Biclustering.

1 Introduction

Lung cancer is a human malignant tumor that uniquely belongs to the human. Its mortality rate is the highest in malignant tumors. Loss of Heterozygosity (LOH) is the loss of one allele at a specific locus, caused by a deletion mutation; or loss of chromosome from a chromosome pair, resulting in abnormal hemizygosity [1]. LOH regions often contain tumor suppressor gene, and is closely related to the incidence of tumor [2]. When one or a pair of alleles change or lose, the incidence of cancer will be large. Therefore, the analysis of LOH in the cancer research has an important significance.

Micro satellites (Short Tandem Repeat, STR) are widely used in the research of gene expressions. But it cannot be expressed in chromosomes. At the same time, there are some shortcomings, such as higher mutation rate and insufficient accuracy [3]. The new genetic sign Single Nucleotide Polymorphisms (SNPs), which is the most abundant genetic variation of human and has high quantity, great density, wide distribution and low mutation, is very suitable for gene analysis. Therefore, in this paper we use SNPs to replace the traditional methods in the research of lung cancer.

Clustering is a very important analytical method, which has been widely used in the analysis of gene expressions. The results of cluster analysis are different from each other. But biological systems' internal are interrelated. In order to overcome this shortcoming, Cheng and Church proposed biclustering algorithm in 2000 [4]. In their approach, rows/columns were deleted from gene expression data matrix to find a bicluster using a greedy row/column removal algorithm. Repeating this approach cannot get different biclusters, unless discovered results are masked. This method can find num sub-matrices by num iterations [5]. In order to discover other biclusters, discovered biclusters must be

Y. Wang and T. Li (Eds.): Knowledge Engineering and Management, AISC 123, pp. 79–84.
springerlink.com

randomized. These random numbers will interfere with the future discovery of biclusters, especially those that overlap with the discovered ones [6].

The biclustering algorithm has been applied to analysis of SNPs data. The purpose is to use the Cheng and Church algorithm to find the relations between the SNPs of LOH and the samples of lung cancer based the supposition that the SNPs of LOH are related to the cancer suppressor genes. But the implementation of the Cheng and Church method requires experience value as a threshold, and the discovered results must be randomized. In this paper, we use k-means and GA to overcome this shortcoming. We use the same mean squared residue score given by Cheng and Church. First we use the k-means algorithm to generate the initial population. We then use GA to search the solution space. Finally the bicluster with high mean squared residue will be selected as our answer.

2 Bicluster Algorithms

In the gene expression data set, let X be a set of genes and Y be a set of conditions. Gene expression data set A= (a_{ij}), where a_{ij} is the element of gene expression matrix, and A represents the logarithm of the relative abundance of the mRNA of the *ith* gene under the *jth* condition. Let I⊆X and J⊆Y, the pair (I, J) specifies the sub matrix.

A bicluster with coherent values identifies coherent values on both columns and rows. If we use additive model to find biclusters, the values of each row or column can be calculated by shifting values of other rows or columns by common offset [4]. In this case, each element a_{ij} can be defined through row mean a_{iJ}, column mean a_{Ij} and the matrix mean a_{IJ}, as follows:

$$a_{ij} = a_{iJ} + a_{Ij} - a_{IJ} \tag{1}$$

Due to the noise of gene expression data, the result of bicluster is not necessarily ideal. Therefore the concept of residues is introduced, the definition is as follows:

$$RS_{IJ}(i, j) = a_{ij} - a_{iJ} - a_{Ij} + a_{IJ} \tag{2}$$

In order to evaluate the quality of bicluster, the pair (I, J) specifies a sub-matrix A_{IJ} with the following mean squared residue score:

$$H(I, J) = \frac{1}{|I||J|} \sum_{i \in I, j \in J} \left(a_{ij} - a_{Ij} - a_{iJ} + a_{IJ}\right)^2 \tag{3}$$

Where $a_{Ij} = \frac{1}{I} \sum_{i \in I} a_{ij}$ $\quad a_{iJ} = \frac{1}{J} \sum_{j \in J} a_{ij}$ $\quad a_{IJ} = \frac{1}{|I||J|} \sum_{i \in I, j \in J} a_{ij}$.

The mean squared residue score is used to measure whether the data is correlated in the matrix. If H has a high value, it means the matrix is uncorrelated. On the contrary, a low H value means the matrix is correlated. A matrix of equally spread random values over the range [a, b] has an expected H score $(b-a)^2/12$. A low H score means that there is a correlation in the matrix.

3 Genetic Algorithm Based on Biclustering

3.1 Generating the Initial Population

In 1967, J.B. MacQueen proposed the k-means algorithm [7]. It is a classical clustering algorithm and is widely used in scientific research and industrial applications. The main process is as follows:

1) Randomly choose k data points from the datasets, and set the k data points as the initial cluster center;
2) Calculate the distance from each data to each cluster center, and assigned the data to the nearest cluster;
3) After all the data be assigned, recalculate each cluster center;
4) Compare with the previous k cluster centers, if any cluster center changes, go to 2); otherwise, end the algorithm;

Firstly we use the k-means algorithm to generate the initial population. The k-means algorithm is used to cluster the LOH data on rows and columns respectively. After clustering, the LOH data is partitioned into n gene clusters and m sample clusters. Thus we have partitioned the initial LOH data matrix into n*m disjoint sub-matrices.

A binary string is used to encode each bicluster. The length of this binary string is x + y, x and y represent the number of genes and samples. The first x bits encode the genes and the next y bits encode the samples. If a gene or sample is in the bicluster, the corresponding bit is set to 1; otherwise it is set to 0.

3.2 Genetic Algorithm

A genetic algorithm [8] is proposed firstly by John Holland, Michigan University, in 1975. It is a global optimization random search algorithm.

When using bicluster to cluster the LOH data of lung cancer, we want to select the best individual into the next generation. The best individual specifies as a matrix with a low H value and contains as much data. So we take the ratio that the mean squared residue divided by the size of a bicluster as our fitness. The smaller H value and the larger bicluster size, the better the clustering result. The fitness function is defined as follows:

$$fitness(b) = \frac{H}{|X||Y|} \tag{4}$$

We set Pc as the select rate, and Pm as the mutation rate. The number of iteration is N. The size of the population is S. The main process of the algorithm is defined as follows:

1) Use k-means algorithm to generate the population, and set this population as initial population B. set counter to 1;
2) Use the fitness function formula (4) to calculate the fitness of each chromosome f(b);
3) If the termination condition is satisfied, return the bicluster with the biggest fitness as the result;
4) Choose S*Pc chromosomes from B into the next generations and get the population B1. The probability of selecting bicluster bi is given by $P(b_i) = \frac{fitness\ (b_i)}{\sum_{j=1}^{s} fitness\ (b_j)}$;

5) Select (1-Pc)*S/2 pairs of bicluster from B. For each pair, use crossover operation and add the new chromosomes to B1;
6) We get the number of variation m by mutation rate Pm. Randomly select m chromosomes from B1. One bit is randomly selected and assigned with 1 or 0. Get population B2;
7) Update population. Use B2 instead of B as new population, t=t+1, go to 2);

4 Result and Analysis

4.1 LOH Data of Lung Cancer

The human lung cancer dataset was downloaded from *http://research2.dfci.harvard.edu/dfci/snp/index.php?dir=Hind%20array*. The samples used in this experiment include 12 normal tissue samples and 101 cancer samples. Lung cancer patients can be classified 4 categories: 51 primary tumor samples of NSCLC, 26 cell line samples of NSCLC, 19 primary tumor samples of SCLC, 5 cell line samples of SCLC[9].

We use the software Dchip to get the probability of LOH. The software uses a Hidden-Markov model to deduce a score between 0 and 1 as the probability of LOH given the genotypes [10]. The processed data formed a large matrix consisting of 62982 loci of SNPs as line and 113 samples as columns. In the matrix, the data are the probabilities of LOH. The threshold value is set to 0.5. A value that is above this threshold is considered to be LOH; a value that is below this threshold is considered to retain the heterozygosity.

4.2 Data Preprocessing

Randomly select part of a lung cancer sample and construct the coordinates, setting the SNP as X-axis and the LOH values of SNP as Y-axis. Shown as Figure 1:

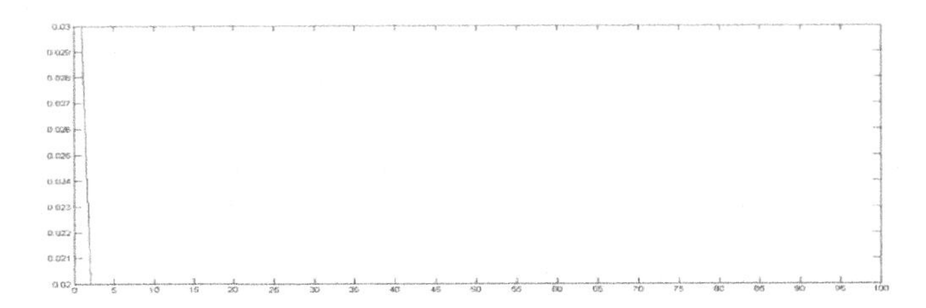

Fig. 1. SNP-LOH Score of Sample AD157T

As shown in Figure 1, the majority of retaining heterozygosity SNP-LOH score is 0. Because the value of retaining heterozygosity is almost the same, the SNPs of retaining heterozygosity will be clustered together in the processing of clusters. Through biology we know that LOH data is more meaningful in the study of tumor. To prevent the SNPs retaining heterozygosity from being clustered together, we replace the data with random data between 0 and 0.5 and retaining the value of SNPs of LOH.

4.3 Result and Analysis

The algorithm proposed in this paper was applied to the preprocessed data. We obtained a bicluster which was composed of 20 samples. The 20 samples consist of 8 normal samples and 12 lung cancer samples and each sample contains 5579 SNPs. Then we establish the relationship between SNP loci and LOH-Score and display the bicluster in the dimensional plane. We construct the coordinates, setting the SNPs as X-axis and the LOH values of SNPs as Y-axis. In the coordinates there are 20 curves each of which represents a sample shown in Figure2.

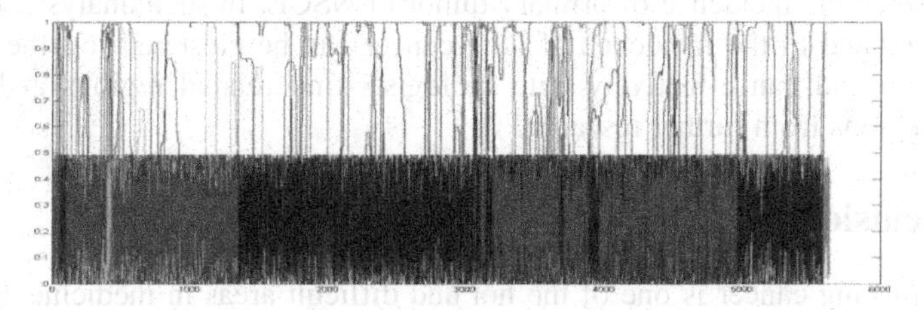

Fig. 2. SNP-LOH Score of 20 Samples

From Figure 2, we see that the bicluster result includes both the LOH SNPs and the retaining heterozygosity SNPs. Due to the excessive amount of data, Figure 2 cannot clearly respond to the relationship between lung cancer and LOH. For further analysis of the bicluster result, 8 normal samples and 12 lung cancer samples were extracted from the bicluster result respectively and shown in Figure 3 and Figure 4.

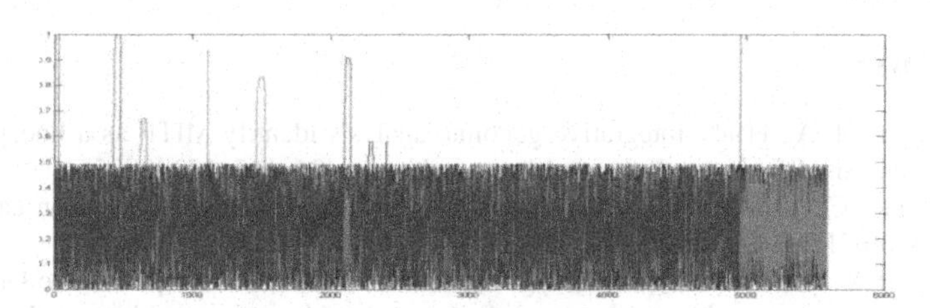

Fig. 3. SNP-LOH Score of 8 Normals

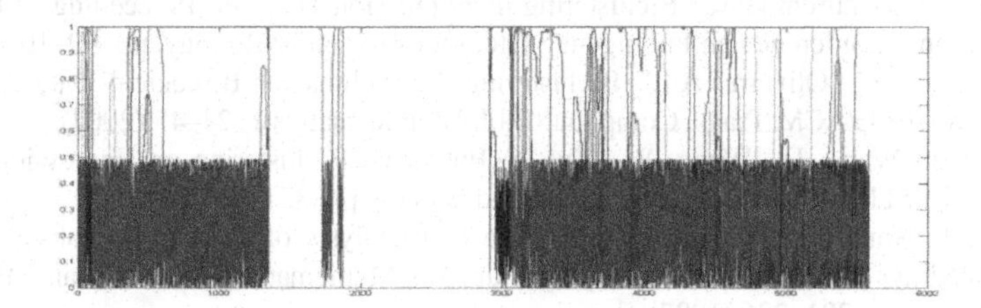

Fig. 4. SNP-LOH Score of 12 lung cancer samples

It can be seen in Figure 3, most LOH values of the 8 normal samples were below the threshold. This indicates that the majority of 8 normal samples do not have LOH in these 5579 SNPs.

As shown in Figure 4, the 12 lung cancer samples all have LOH in these 1400-2800 SNPs except the sample S0453T, which has few retaining heterozygosity SNPs. The 12 lung cancer samples have the same loss of heterozygosity status and the expression is consistent in this region. It is worth noting that the SNPs in this region are related to the incidence of lung cancer.

As can be seen intuitively through Figures 3 and 4, in the bicluster result, the normal samples have retaining heterozygosity in these 1400-2800 SNPs loci, but the lung cancer samples have LOH in this region. According to biological knowledge all the lung cancer samples are from primary tumor samples of NSCL. So these 1400-2800 SNPs influence the incidence of primary tumor of NSCL. In such analysis, the SNPs with LOH related to the incidence of lung cancer can be clustered together through the algorithm and can effectively help biologists find related regions and exclude irrelevant regions from further research.

5 Conclusions

The study of lung cancer is one of the hot and difficult areas in medicine and bioinformatics research. There is a close relation between the phenomenon of LOH and malignant tumor, so the study of LOH data has great significance. The conventional Cheng and Church method requires experience values as a threshold, and discovered results must be randomized. In order to overcome these adverse effects, we combined a bicluster algorithm with a GA algorithm. This algorithm is an intelligent bicluster algorithm, which avoids setting the threshold and randomizing the discovered matrix. The experiments have proven the validity and effectiveness of the method.

References

1. Garraway, L.A., et al.: Integrative genomic analysis identify MITF as a lineage survival oncogene amplified in malignant melanoma. Nature 436, 117–122 (2005)
2. Lengauer, C., Kinzler, K.W., Vogelstein, B.: Genetic instabilities in human cancers. Nature 396(6712), 117–122 (1998)
3. Irving, J.A., et al.: Loss of heterozygosity in childhood acute lymphoblastic leukemia detected by genome-wide microarray single nucleotide polymorphism analysis. Cancer Res. 65, 3053–3058 (2005)
4. Cheng, Y., Church, G.M.: Biclustering of Expression Data. In: Proceeding of the 8th International Conference on Intelligent System for Molecular Biology, pp. 99–103 (2000)
5. Madeira, S.C., Oliveira, A.L.: Biclustering Algorithms for Biological Data Analysis: A Survey. IEEE/ACM Trans. Computat. Biol. Bioinformatics 1, 24–45 (2004)
6. Yang, J., Wang, H., Wang, W., Yu, P.: Enhanced Biclustering on Expression Data. In: Proc. Third IEEE Conf. Bioinformatics and Bioeng, pp. 321–327 (2003)
7. Mac, J.: Some methods for classification and analysis of multivariate observations. In: Proceedings of 15th Berkeley Symposium on Mathematics, Statistics and Probability, Berkeley, pp. 281–296 (1997)
8. Holland, J.H.: Adaption in Natural Artificial System, pp. 1–7. MIT Press (1975)
9. http://research2.dfci.harvard.edu/dfci/snp/index.php?dir=Hind%20array LOH dataset
10. Li, C., Wong, W.H.: DNA-Chip Analyzer (dchip). In: Parmi Giani, G., Garrett, E.S., Irizarry, R.A., Zeger, S.L. (eds.) The Analysis of Gene Expression Data: Methods and Software. Springer, Heidelberg (2003)

Using Simulated Binary Crossover in Particle Swarm Optimization

Xiaoyu Huang[1,*], Enqiang Lin[1], Yujie Ji[1], and Shijun Qiao[2]

[1] School of Software Engineering, Northeastern University, Shenyang, China
xy.huang@yahoo.cn, linenqiang1988@gmail.com, realones@163.com
[2] School of Science, Northeastern University, Shenyang, China
greengreenseye@sina.com

Abstract. Simulated binary crossover (SBX) operator is widely used in real-coded genetic algorithms. Particle swarm optimization (PSO) is a well-studied optimization scheme. In this paper, we combine SBX together with particle swarm optimization (PSO) procedures to prevent possible premature convergence. Benchmark tests are implemented and the result turns out that such modification enhances the exploitation ability of PSO.

1 Introduction

The particle swarm optimization (PSO) is a well studied nature inspired optimization scheme[1]. Soon after PSO was proposed, it was found working well with nonlinear function optimization. Afterwards, however, more and more study has shown that PSO has no ability to perform a fine grain search to improve the quality of solutions with the increase of iterations (generations) [2]. This is considered to be caused by the lack of exploitation[3]. Therefore many researchers have proposed different operators to enhance its exploitation ability[4,5,6].

In genetic algorithms (GAs), crossover performs most exploitation work . Since the exploitation ability directly influences the quality of search results, most work about GAs focus on crossover operators, especially for real-coded genetic algorithms (RCGAs)[7]. Among the most widely used real-coded crossover operators[8,7], the simulated binary crossover (SBX) is a competitive one. It uses a parent-centric probability model to perform recombination between selected parents which allows a large probability of creating a solution near each parent, rather than near the centroid of the parents.

Strictly speaking, both PSO and RCGA belong to the real-coded family. PSO lacks in exploitation ability, and the SBX in RCGA provides exploitation enhancement. So it's a intuitive idea to combine them together to construct a modified PSO, at least, it's worth trying that.

In the remainder of this paper, for convenience, we name such a modification as SBX-PSO. We first give a brief introduce to PSO and SBX. Thereafter, the SBX-PSO modification is presented. At last, to show the performance of SBX-PSO, optimization on four well studied benchmark test functions is implemented with both PSO and SBX-PSO to give a comprehensive comparison.

* Corresponding author.

Y. Wang and T. Li (Eds.): Knowledge Engineering and Management, AISC 123, pp. 85–90.
springerlink.com

2 The Particle Swarm Optimization

In particle swarm optimization (PSO), individuals which record solution information are regarded as particles in space. For each particle, its target is to find the global optimal point in search space. To reach its garget, a particle moves under a velocity and its position presents solution vector. The velocity and position evolve in each iteration. In the standard PSO [9], the information of each particle i in the swarm is recorded by the following variables: (1) the current position X_i, (2) the current velocity V_i, (3)the individual best position $pbest_i$, and (4) the swarm best position $gbest$. In each iteration, the positions and velocities are adjusted by the following equations:

$$v_{ij}(t+1) = \omega v_{ij}(t) + c_1 rand()[p_{ij}(t) - x_{ij}(t)] + c_2 Rand()[p_{gj}(t) - x_{ij}(t)] \tag{1}$$

$$x_{ij}(t+1) = x_{ij}(t) + v_{ij}(t+1) \tag{2}$$

for $j \in 1..d$ where d is the dimension number of the search space, for $i \in 1..n$ where n is the number of particles, t is the iteration number, ω is the inertia weight, $rand()$ and $Rand()$ are random numbers uniformly distributed in the range $[0,1]$, c_1 and c_2 are accelerating factors.To control the flying step size of the particles, v_{ij} is constrained in the range $[-v_{max}, v_{max}]$ where v_{max} is commonly set as $10\% - 20\%$ of each search dimension size [10].

3 The SBX-PSO

Deb proposed the simulated binary crossover (SBX) in 1995[7]. Like most crossover operators, it works with two parent individuals and produce two children individuals. The name indicates that this operator simulates the binary operator in conventional genetic algorithms. Studies show that this operator respects common interval schemata between parents. The SBX is based on a spread factor:

$$\beta_i = \frac{x_i^{2,t+1} - x_i^{1,t+1}}{x_i^{2,t} - x_i^{1,t}} \tag{3}$$

where $x_i^{1,t+1}$ and $x_i^{2,t+1}$ are two children of parents $x_i^{1,t}$ and $x_i^{2,t}$. With this spread factor, a probability distribution density function is defined as follows:

$$P(\beta_i) = \begin{cases} 0.5(\eta+1)\beta_i^{\eta}, & \text{if} \beta \le 1 \\ 0.5(\eta+1)\frac{1}{\beta_i^{\eta+2}}, & \text{otherwise} \end{cases} \tag{4}$$

As is shown by Fig. 1, η determines the crowding degree of the above probability distribution. A large η indicates higher probability to create near parent solutions, and a small η allows children solutions distant from their parents. To make use of such a distribution, a random number u_i within 0 in 1is created. Thereafter, a corresponding β_{ki} is calculated so that the area under the probability curve from 0 to β_{ki} is equal to u_{ki}:

$$\beta_{ki} = \begin{cases} (2u_i)^{\frac{1}{\eta+1}}, & \text{if} u_i \leq 0.5 \\ (\frac{1}{2(1-u_i)})^{\frac{1}{\eta+1}}, & \text{otherwise} \end{cases} \tag{5}$$

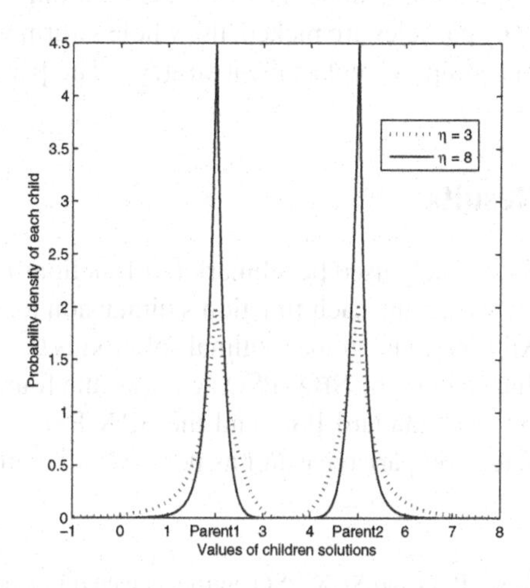

Fig. 1. The children solution value probability density distribution with parents values of 2 and 5 respectively

After getting β_{ki}, the children solutions are calculated as follows:

$$x_i^{(1,t+1)} = 0.5[(1+\beta_{ki})x_i^{(1,t)}) + (1-\beta_{ki})x_i^{(2,t)})], \tag{6}$$

$$x_i^{(2,t+1)} = 0.5[(1-\beta_{ki})x_i^{(1,t)}) + (1+\beta_{ki})x_i^{(2,t)})]. \tag{7}$$

The SBX operator has bias to create children near their parents. This is a good property since near parent solutions have more chance to inherit the valuable information from their parents so that the whole population would evolve in a relatively stable evolutionary process. For PSO, if adding too much variation in the evolutionary process, the algorithm might degenerate to a random search process. In this sense, SBX is so suitable a variation operator for PSO. All we need to do is just insert the SBX operator into the PSO procudure, thereafter, we get our modification:

The SBX-PSO Procedure

STEP 1: Initialize a swarm, including random positions and velocities.
STEP 2: Evaluate the fitness of each particle, $t \leftarrow t+1$.
STEP 3: For each particle, compare its fitness with the individual best position *pbest*. If better, update *pbest* with current position.

STEP 4: For each particle, compare its fitness with the swarm best position *gbest*. If better, replace *gbest* with current position.
STEP 5: Update the velocities and positions with (1) and (2).
STEP 6: Randomly pick up two particles and do SBX, replace the picked parents with their children.
STEP 7: Repeat steps 2–6 until an expected number of iterations is completed or a stop criterion is satisfied.

In STEP 6, the SBX operation is done. Here we just randomly pick 2 particles for crossover, since if too many particles are picked, the whole swarm would be messed up. Besides, we have used the simplest replacement strategy. The parents are replaced by their children.

4 Experimental Results

Here we have selected four widely used benchmark test functions to do our experiment [5,6,8,10]. For simplicity, we set the each function's dimension to 30 ($x_i \in [-100, 100]$ for each dimension i). All functions' global optimal solution is 0.

To see how much enhancement the SBX-PSO acquires, the four functions above are respectively optimized by the standard PSO and the SBX-PSO so that a comparison could be made. To make the comparison as fair as possible, all common parameters are

Table 1. Comparisons between PSO and SBX-PSO on four selected functions. It is the statistical result of 25 runs.

Function name	Presentation		PSO	SBX-PSO
Sphere	$f_1(x) = \Sigma_{i=1}^{n} x_i^2$	Best	3.3278E-81	**3.6342E-82**
		Worst	3.3278E-76	5.0794E-76
		Mean	1.2711E-77	1.6583E-77
		Std	4.9074E-77	7.3302E-77
Rosenbrock	$f_2(x) = \sum_{i=1}^{n} [100(x_{i+1} - x_i^2)^2 + (x_i - 1)^2]$	Best	3.0360E-1	**2.0271E-2**
		Worst	9.5598E+1	9.3768E+1
		Mean	3.3572E+1	**2.6955E+1**
		Std	2.8819E+1	2.4668E+1
Rastrigin	$f_3(x) = \sum_{i=1}^{n} (x_i^2 - 10\cos(2\pi x_i) + 10)$	Best	1.5919E+1	**9.9500E-1**
		Worst	5.1738E+1	1.4917E+1
		Mean	3.2793E+1	**3.2481E+0**
		Std	8.2707E+0	2.0135E+0
Griewank	$f_4(x) = \frac{1}{4000} \sum_{i=1}^{n} x_i^2 - \prod_{i=1}^{n} \cos(\frac{x_i}{\sqrt{i}}) + 1$	Best	0.0000E+0	**0.0000E+0**
		Worst	2.2700E-2	3.9512E-2
		Mean	8.6000E-3	1.9912E-2
		Std	6.8000E-3	1.1829E-2

Table 2. The parameter settings of PSO and SBX-PSO in our experiment

	c_1	c_2	ω	v_{max}	p_c	η
PSO	2	2	0.4	4	-	-
SBX-PSO	2	2	0.4	4	0.05	2

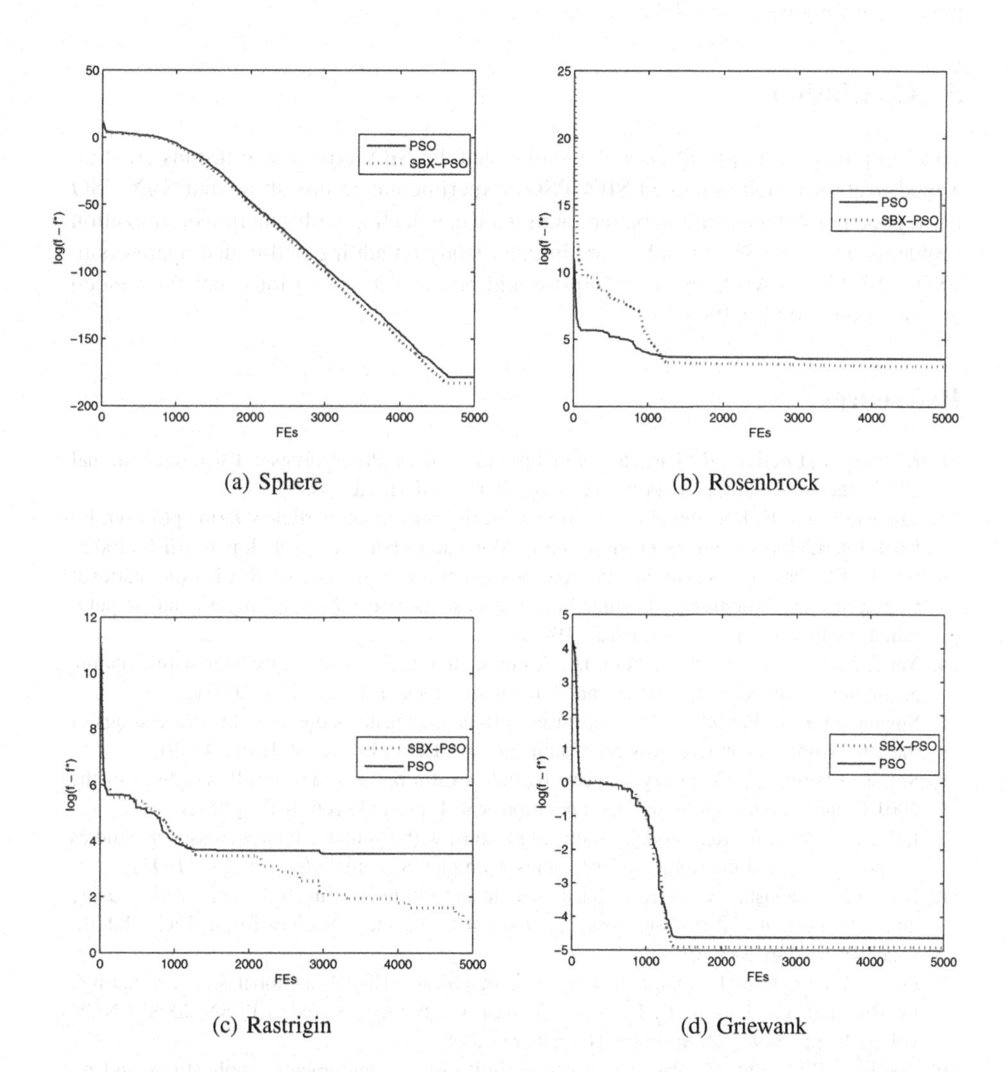

(a) Sphere (b) Rosenbrock

(c) Rastrigin (d) Griewank

Fig. 2. The convergence curves of optimization procedures. To clearly see the convergence tendency, the y axis is $log(f - f^*)$ rather than $f - f^*$, where f is the fitness value and f^* is the real global optimum.

set with the same values. The η in SBX and the mutation probability p_m are set with suggested values[11]. (See Table 2)

The population size for both PSO and SBX-PSO is set to 30. All functions are optimized for 25 runs, and 5000 FEs(Fitness Evaluations) for each run. The result is listed in Table 1. As we can see, SBX-PSO performs much better when optimizing function Rastrigin. For the other 3 functions, it also has potential in finding better solutions. Furthermore, the convergence curves are shown in Fig. 2. Generally, SBX-PSO has more mature convergence than PSO.

5 Conclusion

To add more variation in PSO, we've embedded the SBX operator in it, thus creating a modification which is named SBX-PSO. Experimental results show that SBX-PSO has a good potential in finding better solutions when dealing with function optimization problems. However, this is only a preliminary study on adding real-coded operators in PSO, so in future work, more attempts could be done to try various such real-coded crossover operators in PSO.

References

1. Kennedy, J., Eberhart, R.: Particle swarm optimization. In: Proceedings of IEEE International Conference on Neural Networks, vol. 4, pp. 1942–1948. IEEE (1995)
2. van den Bergh, F., Engelbrecht, A.P.: A new locally convergent particle swarm optimiser. In: IEEE International Conference on Systems, Man and Cybernetics, vol. 3, p. 6. IEEE (2002)
3. Shi, Y., Eberhart, R.: A modified particle swarm optimizer. In: The 1998 IEEE International Conference on Evolutionary Computation Proceedings, IEEE World Congress on Computational Intelligence, pp. 69–73. IEEE (1998)
4. Yang, X., Yuan, J., Yuan, J., Mao, H.: A modified particle swarm optimizer with dynamic adaptation. Applied Mathematics and Computation 189(2), 1205–1213 (2007)
5. Suganthan, P.N.: Particle swarm optimiser with neighbourhood operator. In: Proceedings of the 1999 Congress on Evolutionary Computation, CEC 1999., vol. 3, IEEE (1999)
6. Shi, Y., Eberhart, R.C.: Fuzzy adaptive particle swarm optimization. In: Proceedings of the 2001 Congress on Evolutionary Computation, vol. 1, pp. 101–106. IEEE (2001)
7. Deb, K., Kumar, A.: Real-coded genetic algorithms with simulated binary crossover: Studies on multimodel and multiobjective problems. Complex Systems 9(6), 431–454 (1995)
8. Ono, I., Kobayashi, S.: A real-coded genetic algorithm for function optimization using unimodal normal distribution crossover. Journal of Japanese Society for Artificial Intelligence 14(6), 246–253 (1997)
9. Pošík, P.: Preventing Premature Convergence in a Simple EDA Via Global Step Size Setting. In: Rudolph, G., Jansen, T., Lucas, S., Poloni, C., Beume, N. (eds.) PPSN 2008. LNCS, vol. 5199, pp. 549–558. Springer, Heidelberg (2008)
10. Eberhart, R.C., Shi, Y.: Particle swarm optimization: developments, applications and resources. In: Proceedings of the 2001 Congress on Evolutionary Computation, vol. 1, pp. 81–86. IEEE, Piscataway (2001)
11. Deb, K., Beyer, H.: Self-adaptive genetic algorithms with simulated binary crossover. Evolutionary Computation 9(2), 197–221 (2001)

Part II
Distributed and Parallel Intelligence

A Parallel Cop-Kmeans Clustering Algorithm Based on MapReduce Framework

Chao Lin, Yan Yang, and Tonny Rutayisire

School of Information Science & Technology, Southwest Jiaotong University, Chengdu, 610031, P.R China

Abstract. Clustering with background information is highly desirable in many business applications recently due to its potential to capture important semantics of the business/dataset. Must-Link and Cannot-Link constraints between a given pair of instances in the dataset are common prior knowledge incorporated in many clustering algorithms today. Cop-Kmeans incorporates these constraints in its clustering mechanism. However, due to rapidly increasing scale of data today, it is becoming overwhelmingly difficult for it to handle massive dataset. In this paper, we propose a parallel Cop-Kmeans algorithm based on MapReduce- a technique which basically distributes the clustering load over a given number of processors. Experimental results show that this approach can scale well to massive dataset while maintaining all crucial characteristics of the serial Cop-Kmeans algorithm.

Keywords: Parallel Clustering, Cop-Kmeans Algorithm, MapReduce.

1 Introduction

Clustering algorithms are often useful in applications in various fields such as data mining, machine learning and pattern recognition. They conduct a search through a space of a dataset, grouping together similar objects while keeping dissimilar objects apart, as much as possible. Normally this search proceeds in an entirely unsupervised manner. For some domains, however, constraints on which instances must (ML) or cannot (CL) reside together in the same cluster either are known or are computable automatically from background knowledge [1]. Cop-Kmeans is one popular clustering algorithm which has incorporated these instance based constraints in its clustering mechanism [2]. The approach in this algorithm has been shown to be successful in guiding the clustering process toward more accurate results. However this algorithm only works satisfactorily with relatively small dataset, when the size of the dataset is very large it becomes terribly slow.

With the rapid development of information technology, data volumes processed by typical business applications are very high today which in turn pushes for high computational requirements. To process such massive data, a highly efficient, parallel approach to clustering needs to be adopted. Recently, several attempts have been made to improve the applicability of K-Means algorithm for massive applications

Y. Wang and T. Li (Eds.): Knowledge Engineering and Management, AISC 123, pp. 93–102.
springerlink.com

through parallelization [3-5]. To our knowledge though, widely used semi-supervised clustering algorithms are serial and can only run on a single computer. This greatly hinders their capability to handle very large dataset.

Cop-Kmeans is probably the most popular semi-supervised clustering algorithm that has been used in a variety of applications [2]. Due to its popularity, we believe that proposing a parallel version based on MapReduce can be of significant use to the clustering community. MapReduce is a parallel programming model for processing huge dataset using large numbers of distributed computers (nodes), collectively known as a cluster [6]. The major contributions of this work are two folds. First is to address the issue of constraint-violation in Cop-Kmeans by emphasizing a sequenced assignment of cannot-link instances after conducting a Depth-First Search of the cannot-link set. Second is to reduce the computational complexities of Cop-Kmeans by adopting a MapReduce Framework.

The rest of the paper is organized as follows; Section 2 introduces the underlying mechanism of Depth-First Search and further illustrates its role in solving constraint-violation in Cop-Kmeans. Section 3 presents our approach and the proposed parallel Cop-Kmeans algorithm based on MapReduce framework. Section 4 presents our experimental results as well as evaluation for the proposed algorithm. Finally we draw our conclusions and future work in Section 5.

2 Parallel Cop-Kmeans Algorithm Based on MapReduce

2.1 Depth-First Search

Depth-First Search (DFS) is a general technique for traversing a graph whose principal is "going forward (in depth) while there is such possibility, otherwise backtrack". The mechanism is about choosing a starting vertex and then explore as far as possible along each branch before backtracking. Fig. 1 illustrates the process of Depth-First Search.

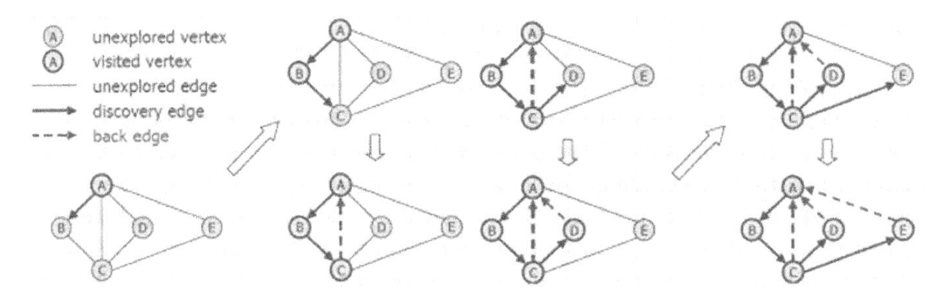

Fig. 1. Depth-First Search

The overall depth first search algorithm will simply initializes a set of markers so we can tell which vertices are visited, chooses a starting vertex A, initializes a tree T to A, and calls DFS(A). Then, we traverse the graph by considering an arbitrary edge (A, B) from the current vertex A. If the edge (A, B) takes us to a visited vertex B, then we back down to the vertex A. On the other hand, if edge (A, B) takes us to an unvisited vertex B, then we paint the vertex B and make it our current vertex, and repeat the above computation. When we get to a point where all the edges from our current vertex take us to visited vertices, we then backtrack along the edge that brought us to that point. We take the immediate visited vertex that we find on the way back, make it our current vertex and start the computations for any edge that we missed earlier. When the depth-first search has backtracked all the way back to the original source vertex, A, it has built a DFS tree of all vertices reachable from that source.

2.2 Solving Constraint-Violation in Cop-Kmeans Algorithm

Whereas clustering with pairwise constraints is generally proven to enhance the conventional K-Means, it is often associated with a problem of constraint-violation; this brings about failure in hard-constrained clustering algorithms like Cop-Kmeans [2] when an instance has got at least a single cannot-link in every cluster. For this to occur, it can be caused by two situations: either there is no feasible solution for that particular clustering or a wrong decision about the assignment order of instances to clusters was made.

Definition 1. Feasibility Problem [7]**:** Given a dataset X, a collection of constraints C, a lower bound $K\ell$ and an upper bound Ku on the number of clusters, does there exists a partition of X into k groups such that $K\ell \leq k \leq Ku$ and all the constraints in C are satisfied?

Definition 2 [7]**:** A feasibility problem instance is β-easy if a feasible solution can be found by an algorithm β given the ordering of instances in the training set which determines the order they will be assigned to clusters.

From definitions 1 and 2, we can see that much as there could be a feasible solution for a particular clustering, it may not be necessarily easy to find one. As it is interpreted in Brook's theorem [7], that it is only when the number of CL constraints involving on instance is less than K (number of clusters), that one is assured of a feasible solution regardless of the order in which instances are assigned to clusters. Since this condition is not always the case, previous work [8, 9] have studied the problem of constraint-violation and used different approaches to solve it in Cop-Kmeans Algorithm.

The solution in this work capitalizes on re-arranging the CL-set (before assignment) in a sequence such that any CL instance will have at least one cluster for assignment. This kind of sequence is produced by the Depth-First Search function in Algorithm 1 below. In the Depth-Search mechanism, clustering under CL constraints is looked at as traversing a graph. The vertices represent the CL-Instances, edges representing the CL between any two instances while the path of traversing it represents the order of assigning the CL-Instances to clusters.

For simplicity let's illustrate this process using Fig. 1 above: where {C ≠ (A, B), C ≠ (A, C), C ≠ (A, D), C ≠ (A, E), C ≠ (B, C), C ≠ (C, D), C ≠ (C, E)}. From this CL-Set we can observe that the maximum number of CL involving one instance, Δ is 4. Assuming K is 4 and taking Brook's theorem into consideration, a conventional Cop-Kmeans could easily fail due to constraint-violation. Since it has no clear mechanism for sequencing CL instances before assignment, it is possible to assign {A, D, B, E, C} in that order, supposing that A, D, B and E are each assigned to a different cluster, then C will have no feasible cluster for assignment. On the other hand, using the depth-first search to traverse this CL-Set (graph), the CL instances will be sequenced in a stack produced by a specific path depending on the first CL instance encountered. For example if we consider the path in Fig. 1 the following stack will be produced {A, B, C, A, D, A, E, A} in which case all red-colored instances represent back edged vertices. Note that since the principal of a stack is "Last-in-First-out", these CL-instances will be assigned in the order {A, E, D, C, B} in which case even if (worst scenarios) A, E, D and C are each assigned to a different cluster, B could still be assigned in either D or E and there could not be any constraint-violation and hence no failure of the algorithm. Although back edged vertices (instances) may appear more than once in the stack, they can only be assigned once in the cluster as shown in the order above. Note that the DFS mechanism ensures that while the first CL instance encountered may be randomly selected, instances with the highest degree (highest number of CL involving a single instance) are always among the first to be assigned. This means that most of those instances involved in CL with these higher degree instances could have a high chance of being assigned together in the same cluster since most of them may not necessarily have CL ties among themselves.

3 Parallel Cop-Kmeans Algorithm Based on MapReduce

In this section, we present the proposed parallel Cop-Kmeans algorithm, but prior to that, we introduce the underlying mechanism of the core approach used in this algorithm namely: MapReduce.

3.1 The MapReduce Framework

MapReduce is a programming model introduced by Google to support distributed computing of large dataset on clusters of computers. The name "MapReduce" was inspired by the "map" and "reduce" functions in the functional programming. Users specify the computations in terms of "map" and "reduce" functions and the underlying runtime system automatically parallelizes the computation across a large cluster of computers, handles machine failures and schedules inter-computer communication to make efficient use of the network and the disks. This enables programmers with no experience with distributed systems to easily utilize the resources of a large distributed system.

In MapReduce [3], the Map function processes the input in the form of key/value pairs to generate intermediate key/value pairs, and the Reduce function processes all intermediate values associated with the same intermediate key generated by the Map function. Fig. 2 below illustrates the different phases of MapReduce model.

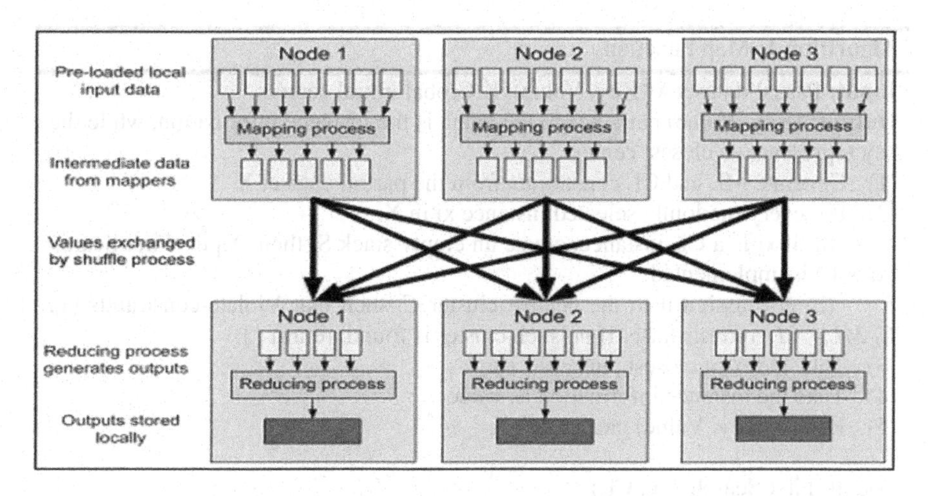

Fig. 2. The MapReduce model

3.2 The Proposed Cop-Kmeans Algorithm Based on MapReduce

The underlying idea behind the original Cop-Kmeans algorithm is to assign each instance in the dataset to the nearest feasible cluster center. We can see that the algorithm involves typically three steps: calculating the shortest distance between each instance and the cluster centers, assigning instance while avoiding constraint-violation and updating cluster centers. From the serial point of view, all these steps are handled by a uniprocessor and all the concerned data is kept in its local memory. However, in order to incorporate these tasks in MapReduce model, some crucial modifications must be made.

Map-Function. Ideally, since the task of distance computations is so bulky and would be independent of each other, it is reasonable to execute it in parallel by the Map function. The challenge about this is that the next task of assigning the instance to the cluster center with which it has the shortest distance takes into consideration constraint-violation. This would mean dependence on assignment of instances in other mappers which could have a ML or CL with the instance in question. To overcome this issue and simplify, we generate constraints from the partial dataset allocated to each mapper. This would ensure that distance computations and considerations of ML and CL are purely independent of other mappers. This map function outputs intermediate data to be used in the Reduce function.

Algorithm 1 shows the map function of our proposed parallel Cop-Kmeans.

Algorithm 1: Map Function

Input: Partial dataset X (Key, Value), K Global initial centers
Output: (Key, Value) pairs where the value is the instance information, while the key represents its closest center
(1) Generate ML and CL constraints from the partial dataset X
(2) For every randomly selected instance xi in X;
(a) if xi is a CL-instance, create an empty stack S, then Depth- First Search (xi, CL) is implemented.
(b) else assign it to the nearest cluster C_j such that Violate-constraints (x_i, C_j, ML, CL) returns false. If no such cluster is found, return { }.
(3) Take index of closest cluster as Key
(4) Take the instance information as value
(5) Return (Key, Value) pair

Depth-First Search (x_i, CL)
(1) Visit x_i;
(2) Insert x_i into S.
(3) For each child w_i of x_i
if w_i is unvisited (not in S)
{
Depth-First Search (w_i, CL);
Add edge (x_i,w_i) to tree T;
}
(4) For every instance S_i in S (Last-in First-out), assign it to the nearest cluster C_j such that Violate-constraints (L_i, C_j,ML, CL) returns false. If no such cluster is found, return { }.

Violate-constraints (object x_i, cluster C, must-link constraints ML, cannot-link constraints CL)
(1) If x_c is already in cluster C and $(x_i, x_c) \in CL$, return true.
(2) If x_m is already assigned to another cluster other than C and $(x, x_m) \in ML$, return true.

Reduce Function. The Reduce Function gets its input from the Map Function of each mapper (host). As shown in algorithm 1 above, in every mapper, the map function outputs a list of instances each labeled with its closest center. Therefore it follows that, in each mapper, all instances labeled with same current cluster center are sent to a single reducer. In the reduce function, the new cluster center can be easily computed by averaging all the instances in the reducer. Algorithm 2 details the Reduce Function of the parallel Cop-Kmeans.

Algorithm 2: Reduce Function (Key, L)

Input: Key is the index of a particular cluster, L is the list of instances assigned to the cluster from different mappers
Output: (Key, Value) where, key is the index of the cluster, Value is the information representing the new centre

(1) Compute the sum of values for each dimension of the instances assigned to the particular cluster.
(2) Initialize the counter to record the number of instances assigned to the cluster
(3) Compute the average of values of the instances assigned to the cluster to get coordinates of the new centre.
(4) Take coordinates of new cluster centre as value.
(5) Return (Key, Value) pair

As shown in the algorithm 2, the output of each reducer is consequently the cluster index (key) and its new coordinates (value). The new centers are then fed back to the mappers for the next iteration and this goes on until convergence.

4 Experiments

4.1 Experimental Methodology

To evaluate the performance of the improved Cop-Kmeans, we have compared it with original Cop-Kmeans with respect to their proportions of failure (constraint-violation) and F-measures. For this purpose, we used four UCI numerical dataset namely: Iris (150, 4, 3), Wine (178, 13, 3), Zoo (101, 16, 7) and Sonar (208, 60, 2).The figures in the brackets indicate the number of instances, number of attributes and number of classes for each dataset respectively. In this set of experiments, for a given number of constraints, each of the two algorithms was run 100 times on a dataset. Note that in both algorithms, initial cluster centers and all pairwise constraints are randomly generated from the dataset. Also note that both algorithms are designed to return an empty partition (fail) whenever constrain-violation arises. With increasing number of constraints as inputs, we generated corresponding average proportions of failure and F-measure for each algorithm running on a given dataset (see Fig. 3 and Fig. 4).

In the second set of experiments, we evaluate the efficiency of the proposed parallel Cop-Kmeans with respect to Speedup and Sizeup characteristics. Note that the idea behind this algorithm is to intelligently distribute the computational workload across a cluster of computer nodes. In this set of experiment therefore we don't evaluate accuracy but rather efficiency of the parallel Cop-Kmeans in processing massive dataset. The experiments were run on Hadoop MapReduce platform of four nodes; each having 2.13 GHz of processing power and 2GB of memory. And in this part, we use USCensus1990 dataset (823MB) which owns 68 categorical attributes. Many of the less useful attributes in the original dataset have been dropped, the few continuous variables have been discretized and the few discrete variables that have a large number of possible values have been collapsed to have fewer possible values. For comparison purposes, we divided the dataset into 4 groups of files, approximately having 200MB, 400MB, 600MB and 800MB in that order.

Speedup of a parallel system is defined by the following formula:

$$\text{Speedup}(p) = T_1/T_P \quad (1)$$

Where p is the number of nodes, T1 is the execution time on one node and TP is the execution time on p nodes. To measure the Speedup of a parallel system, we keep the dataset constant while increasing the number of nodes in the system. In the experiment, we compared the Speedup performances produced when dataset of varying sizes are given as inputs.

On the other hand, the Sizeup metric is defined by formula (2):

$$\text{Sizeup}(D, p) = T_{SP}/T_{S1} \quad (2)$$

where D is the size of the dataset, p is the multiplying factor of the dataset, T_{SP} is the execution time for p * D, T_{S1} is the execution time for D. To measure the Sizeup of a system, we keep the number of nodes in the system constant while growing the size of the dataset by p. For our experiment, we compared the Sizeup performances produced by fixing the number of nodes in the system to 1, 2, 3, and 4.

4.2 Experimental Results

Fig. 3 below depicts the average proportion of failures out of the 100 runs for both the original Cop-Kmeans and the improved Cop-Kmeans at a given number of constraints as input. It can be observed from the Fig. 3 (left) that in Cop-Kmeans, the proportion of failure worsens as more constraints are given. On the other hand, when the Depth-First Search mechanism is used to sequence CL-instances before assignment in the improved Cop-Kmeans, there is no single case of failure as shown in the Fig. 3 (right). The improvement also extends in terms of accuracy where Fig. 4 below shows relatively higher instances of average F-measures in our improved Cop-Kmeans compared to the original Cop-Kmeans on two dataset. The results at each point on the curve are obtained by averaging the F-measures over the 100 runs for a given number of constraints. Note that in both algorithms, initial cluster centers and all pairwise constraints are generated randomly from the dataset.

Fig. 5 reports the Speedup and Sizeup evaluations for our proposed parallel Cop-Kmeans Algorithm. As shown in the Fig. 5 (left), for each dataset, we increase the number of nodes in the system while reporting corresponding Speedup for that dataset. It can be noted that the results show a good Speedup performance for our proposed parallel Cop-Kmeans. Further more, it can be noted that as the size of the dataset increases, also Speedup performance increases; an indication that indeed our parallel algorithm can efficiently handle massive dataset. Sizeup evaluation also shows good performance of the parallel Cop-Kmeans. From the results (see Fig. 5 (right)), we can observe that as we increase the number of nodes in the system, we get better results for Sizeup.

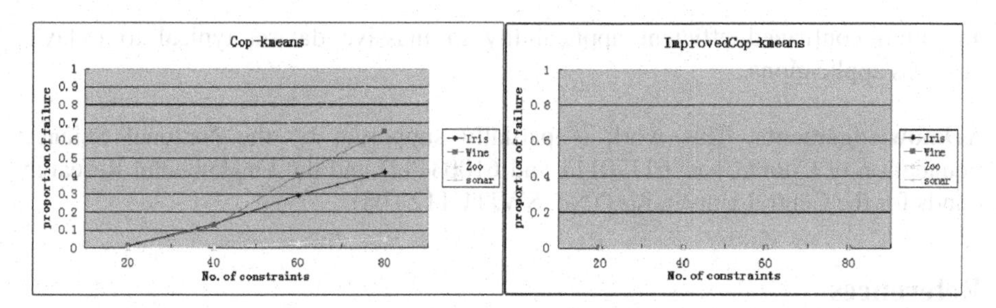

Fig. 3. Proportion of Failure for Cop-Kmeans and Improved Cop-Kmeans

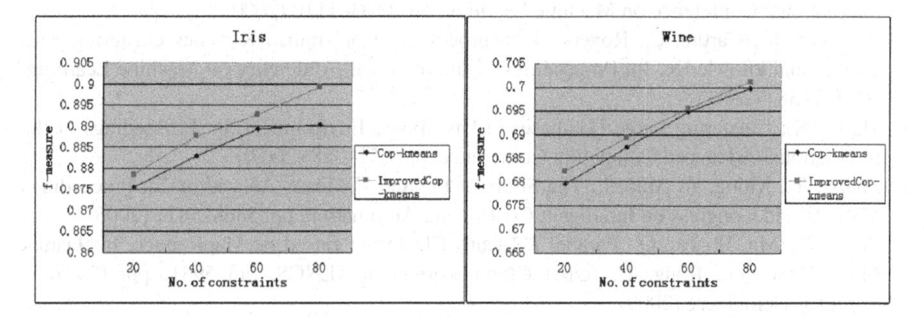

Fig. 4. The Average F-measures for Cop-Kmeans and Improved Cop-Kmeans

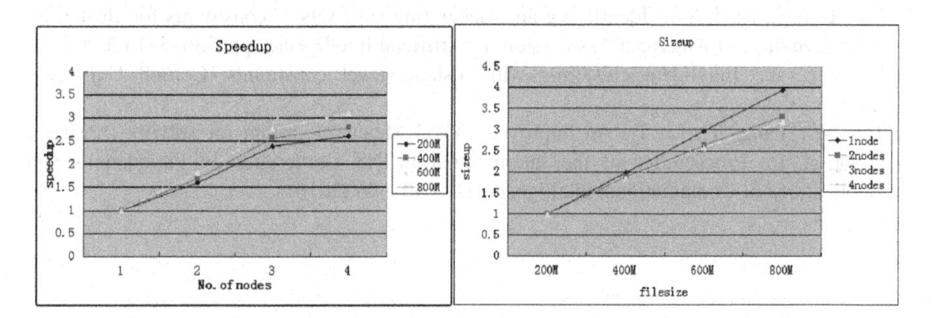

Fig. 5. Speedup and Sizeup evaluations of parallel Cop-Kmeans algorithm

5 Conclusion

In this paper, we proposed a parallel Cop-Kmeans Algorithm based on MapReduce Framework. This Algorithm adopts two crucial mechanisms: Depth-First Search and MapReduce. Inspired by the sensitivity to assignment order of instances, DFS capitalizes on sequencing CL-instances in a way that will not allow constraint-violation to happen. Results from our experiments confirm that this improvement enhances the accuracy of Cop-Kmeans without a single case of failure. Parallelizing the algorithm on a MapReduce Framework also gave positive results in term of speeding up and sizing up the computational processes and hence our proposed algorithm confirmed efficient applicability to massive dataset typical to today's business applications.

Acknowledgements. This work is partially supported by the National Science Foundation of China (Nos. 61170111 and 61003142) and the Fundamental Research Funds for the Central Universities (No. SWJTU11ZT08).

References

1. Wagstaff, K., Cardie, C.: Clustering with instance level constraints. In: Proceedings of the International Conference on Machine Learning, pp. 1103–1110 (2000)
2. Wagstaff, K., Cardie, C., Rogers, S., Schroedl, S.: Constrained K-means clustering with background knowledge. In: Proceedings of International Conference on Machine Learning, pp. 577–584 (2001)
3. Malay, K.: Clustering Large Databases in Distributed Environment. In: Proceedings of the International Advanced Computing Conference, pp. 351–358 (2009)
4. Zhang, Y., Xiong, Z., Mao, J.: The Study of Parallel K-Means Algorithm. In: Proceedings of the World Congress on Intelligent Control and Automation, pp. 5868–5871 (2006)
5. Zhao, W., Ma, H., He, Q.: Parallel *K*-Means Clustering Based on MapReduce. In: Jaatun, M.G., Zhao, G., Rong, C. (eds.) Cloud Computing. LNCS, vol. 5931, pp. 674–679. Springer, Heidelberg (2009)
6. Xuan, W.: Clustering in the Cloud: Clustering Algorithm Adoption to Hadoop Map/Reduce Framework. Technical Reports-Computer Science, paper 19 (2010)
7. Davidson, I., Ravi, S.S.: Identifying and Generating easy sets of constraints for clustering. In: Proceedings of American Association for Artificial Intelligence, pp. 336–341 (2006)
8. Wagstaff, K.: Intelligent clustering with instance-level constraints. Cornell University (2002)
9. Tan, W., Yang, Y., Li, T.: An improved COP-KMeans algorithm for solving constraint violation. In: Proceedings of the International FLINS Conference on Foundations and Applications of Computational intelligence, pp. 690–696 (2010)

Design of Aerospace Data Parallel Pre-processing Ground System Based Grid Workflow

Chaobin Zhu[1,2,3], Baoqin Hei[1], and Jiuxing Zhang[1]

[1] Payload Application Center, Chinese Academy of Sciences, No.9, South Dengzhuang Road, HaiDian District, BeiJing, China
[2] Academy of Opto-Electronics, Chinese Academy of Sciences, No.9, South Dengzhuang Road, HaiDian District, BeiJing, China
[3] Graduate University of Chinese Academy of Sciences, No.80, East ZhongGuanCun Road, HaiDian District, BeiJing, China
{zhuchaobin,hbq,jxzhang}@csu.ac.cn

Abstract. Aerospace data pre-processing ground system play an important role in aerospace exploiting. Today, there're several special aerospace data pre-processing ground system in operation. As experimental activities of space exploration continue to carry out, payload types is increasing, data types and the satellite downlink data are growing on a large scale. For multi-tasks and multi-payloads aerospace massive data processing, this article analyses and designs a aerospace data parallel pre-processing ground system with grid workflow scheduling. It can be used to create appropriate processes flexibility and quickly, adjusting the data pre-processing of multi-missions and multi-payload[2]. Addition, it has perfect scalability in business and computing ability. So, it can meet the future requirements of aerospace data pre-processing.

Keywords: workflow, grid, pre-processing, parallel, scheduling, load balancing.

1 Introduction

Aerospace data processing ground system is an integrated system of aircraft data pre-processing, catalog archiving and management, product processing and product releasing. It is an important support system of the space scientific experiments and the Earth observation satellites. It is the bridge connecting the ground application of space data and aerospace data acquisition. Its main functions are as follows:

(1) Data preprocessing. Data processing as radiometric calibration and geometric correction based on aircraft's track and posture, the sensor's parameters and so on.
(2) Generating and releasing of Standard product of all levels.
(3) Archiving of raw data products of all levels.

Today, there have formed several complete aircraft data pre-processing technologies and systems such as the data processing system of scientific exploration satellite. For example, the Information Power Grid(IPG)[6] of NASA, the space scientific exploration data pre-processing system of Double-Satellite, the Beijing-1 small

Y. Wang and T. Li (Eds.): Knowledge Engineering and Management, AISC 123, pp. 103–111.
springerlink.com

satellite data pre-processing system. They use distributed processing and have high-volume processing capabilities.

With the rapid development of technology, the scope of space technology study is growing day by day. There will be more and more spacecraft be used in scientific experiments and Earth observations. In particular, space station can carry hundreds of different payloads. So, aerospace ground data pre-processing system must meet the requirement of multi-task and multi-preload data parallel processing.

With the rapid development of information technology and sensor technology, the resolution of Earth observation payload instruments is improving. The using of relay satellite makes the space data can be received 24-hour. Further more, X-band and S-band data transmission technology are improving. The ground system will receive more downlink data from space. So, aerospace ground data pre-processing system must meets the requirement of high-speed data processing.

In addition, with the continuous development of remote sensing applications, further processing of remote sensing is gradually within the context of pre-processing system. So, aerospace ground data pre-processing system must meets the new requirement of geometric correction, radiometric correction, Image Fusion and so on.

For the requirement massive data processing, traditional pre-processing systems as minicomputer or workstation platforms are constrained about performance extensions. Then pre- processing system began to use parallel processing based cluster platform.

The parallel processing technology greatly improves the system capacity in some extent to meet the massive data processing performance requirements. But the extensions of the system's performance are still weak. The future space experiments require data pre-processing system not only with high performance, but also have good scalability to adapt to multi-task multi-load handling needs.

Against the problems mentioned above and the current development of aerospace pre-processing systems, this paper designs a workflow-based aerospace ground-based data parallel pre-processing system.

It can redefine[1] the mission processes quickly to adapt different tasks. So it can meet the requirement of multi-task and multi-preload data parallel processing. It has the ability of workflow management and auto scheduling. User can create workflow depend on different task and different data type flexibly. Then it will select the proper computing node in the computing grid to execute though the workflow scheduling engine. Even more, the system can make the load balancing and processing at high speed. It use the flexibly of workflow configuration to achieve the goal of Dynamic Scalability on business and use the Seamless expansion of the computing grid to achieve goal of Dynamic Scalability on computing ability. Thus, it can meet the future requirements of aerospace ground-based data pre-processing.

2 Hardware Platform Design

As is shown in Fig1, the hardware platform consists of the operating management server, the archive management server, the data processing grid, the user service server, the operating management client, the secondary storage system of XIV and the tape library, the network infrastructure and other components.

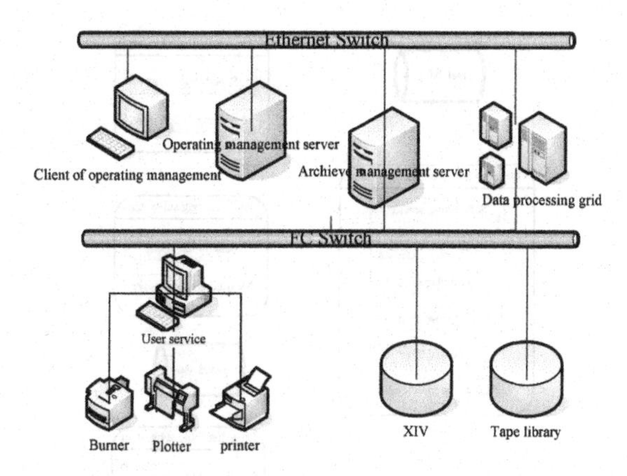

Fig. 1. Hardware platform of aerospace ground-based data pre-processing systems

Operating management server is responsible for the automatic scheduling of the whole system. There is BS structure between it and the operating management server client. Every computer terminal in the task network in the same network segment with the operating management server can use a web browser to login and run as a client. And one can use the web pages to do the Configuration Management of the operating management or monitor the running of the whole system in real-time.

Operating management server schedule the proper computing nodes in the data processing grid parallel in accordance with the pre-configured workflow via Ethernet and monitor the running of the whole system in real-time. When the data processing over, the operating management server will schedule the data archive management server to archive the raw data and data products in different levels. After the archiving, the operating management server will schedule the user service server to distribute the products and release the latest products details through the JSP web pages.

For high-speed mass data processing, storage performance and reliability is essential. The system uses the two tiered storage model base on XIV Storage System and the tape library. The XIV server can meet the fast storing of massive raw data and products. And the data in XIV server can dump to the tape library user the Strategy of incremental backup. The archive management server, the data processing grid and the user service serve connect to the XIV through the optical switch and use the General Parallel File System (GPFS) to Shared the data. Thus, the data transfer speed will reach GB/S and can meet the demand of the high speed reading, writing and storing of mass data.

3 Software Architecture

As is shown in Fig2, the software system consists of the operating management subsystem, the data processing subsystem, data archiving sub-system, data distribution subsystems.

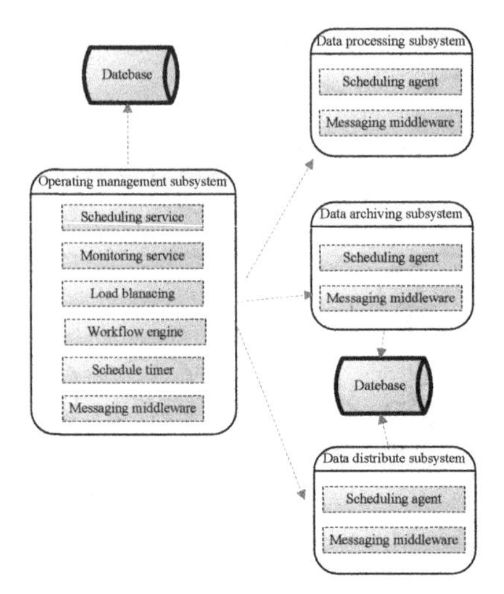

Fig. 2. Software architecture of aerospace ground-based data pre-processing

The operating management subsystem is responsible for scheduling the appropriate process through the scheduling agents that posed on the other subsystems. And other subsystems can feedback the execution information through the scheduling agents. And then the system administrator can monitor the execution information of the whole system through the operating management client. The communication between the operating management server and its client can use local dynamic refreshing base on AJAX[4] to make sure the real-time refreshing of the web client. Further more, the system administrator also can make manual operation on the process. For example, halting, stopping or restarting and so on.

3.1 The Operating Management Subsystem

As is shown in Fig3, the operating management subsystem consists of resource management, loadblanacing, workflow design, job scheduling, Configuration Management, and status monitoring.

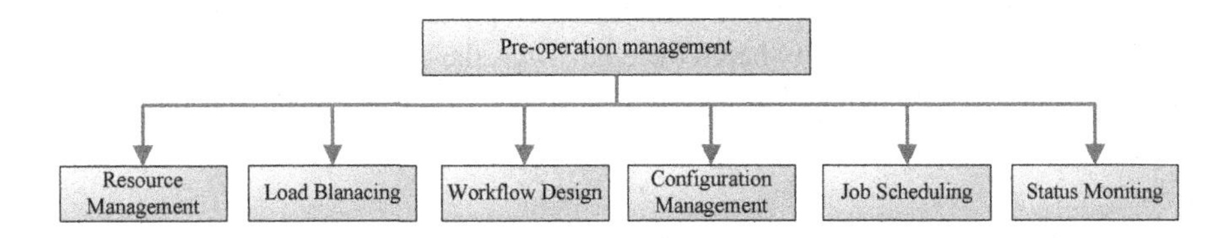

Fig. 3. Pre-operation management software components diagram

The operating management system on charge of the scheduling and configuration of data processing, data archiving, product releasing. And monitor the status of the processes, to achieve centralized controlling and automation running of the whole system. The operating management can be abstracted to a model of three layers, as is shown in Fig4.

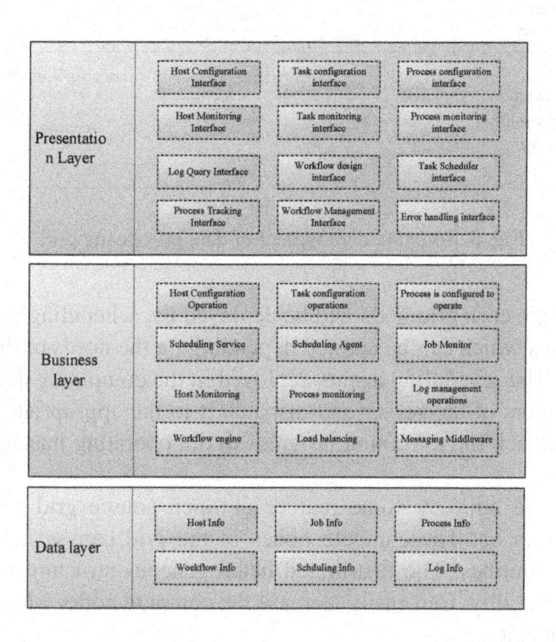

Fig. 4. Three layers model of the operating management system

The presentation layer contains the host configuration interface, tasks configuration interface, the process configuration interface, the host monitoring interface, task monitoring interface, process monitoring interface, the log querying interface, the user management interface, the workflow designing interface, task scheduling interface and so on.

The business layer contains the host configuration operations, the task configuration operations, process configuration operations, scheduling service, scheduling agents, task monitoring, host monitoring, process monitoring, log management operations, user management operations, workflow engine, messaging middleware, scheduling timer, data archiving operations and so on.

The data layer contains the host entity, the task entity, process entity, workflow entity, scheduling entity, log entity, the user entity and so on.

The design of three layers makes the system good versatility and scalability. For different space missions, and for the data processing of different payloads, the changes of the system only related to the presentation layer. Users only need to use the visual workflow design interface to drag the controls and create new workflow process. And dispose the process to the data processing grid next. Then the system will own the function and capability to carry the new task.

3.2 Software Deployment on Data Processing Grid

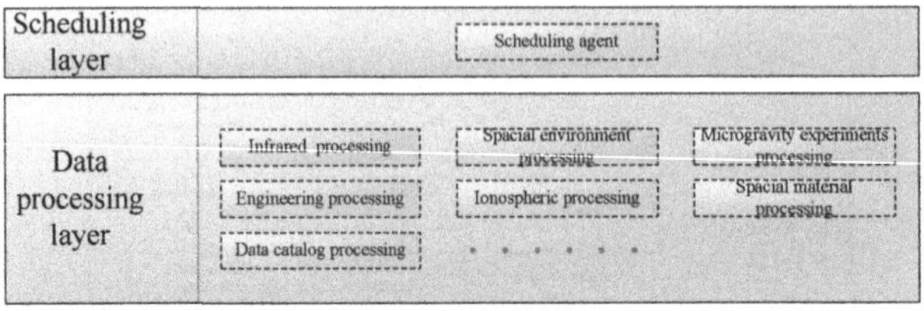

Scheduling layer
Scheduling agent
Data processing layer
Infrared processing
Spacial environment processing
Microgravity experiments processing
Engineering processing
Ionospheric processing
Spacial material processing
Data catalog processing
. . . . . .

Fig. 5. Software deployment of data processing grid

As is shown in Fig5, each grid node will be deployed the scheduling agent and the data processing routines which can be applied depending on the needs of dynamic business logic extensions. The scheduling agents deployed in the computing nodes will respond the scheduling orders of the server and dispatch it to the appropriate data processing routines, and feedback the execution progress to the operating management server in real-time.

The Grid can be built by some mature or open source grid platform, such as GLOBUS[5] and so on. The compute nodes of the grid can be heterogeneous, but deploy the same routines. The distributed heterogeneous structure makes the grid a good dynamic scalability. Can easily increase the compute nodes when need to expand the computing capacity.

3.3 Workflow Designer

The workflow engine can use the mature open source packages, such as osworkflow2.8[7]. The designing view of work-flow scheduling service as follows:

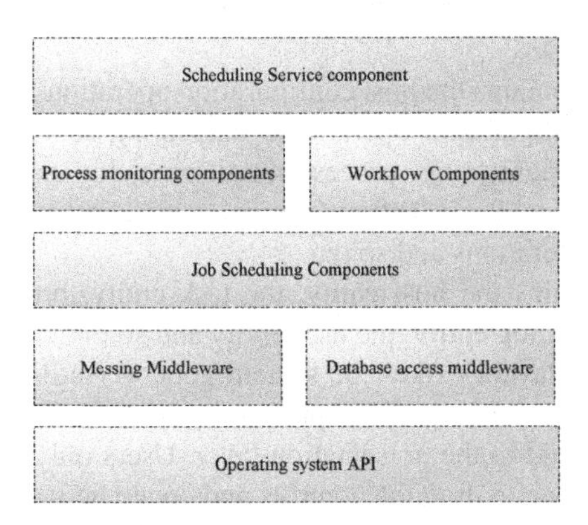

Fig. 6. Designing view of work-flow scheduling service

As is shown in Fig6, the scheduling services are integrated in the dispatch service component and can be developed use EJO[9] framework. It uses scheduling monitoring component which can be carried out by JMX to monitor the status of the scheduling agents. Then schedule the workflow engine by workflow component to achieve the goal of auto scheduling. And the operation related the database can be realized by database component such as Hibernate frame. The same way, the workflow design component can also be developed use EJO framework. The management of workflow templates and the designing of task process can be carried out in workflow component. The presentation layer can be realized by some Visualization technology, for instance, the Flex[9] technology.

The workflow designing component may use visual, interactive technology to make it convenient to create or modify workflows depend on the new task quickly and easily. Such as the use of Flex technology can provide a web-based visual graphic designer. The designer contains steps, branches, joint, movement and other elements which can make the definition of workflow tasks more intuitive and convenient. The system administrator can design the workflow fist, and then configure the task process by certain task, time or workflow templates. At last, the workflow configuration file and the workflow layout file will be generated. They're all stored in the database by the form of XML[9] file. When the task will be scheduled, the workflow scheduling engine will analyze these XML file and make the right scheduling decision as the system administrator set previously.

4 Experimental Verification

Here is an experimental to verify how the visual workflow designer can meet multitasking and multi pre-loads data pre-processing flexibility.

First, drag the control of the visual workflow designer and create a process. Second, select the "Raw data pigeonhole" and the "Product pigeonhole" progresses for the two steps. Third, deploy the two progresses in the data processing grid and add process information (the executable file and configuration file path, process name, process ID and so on) to the operating management database. Then, this process for raw data and products achieving is finished. The layout of workflow designed is shown in Fig7.

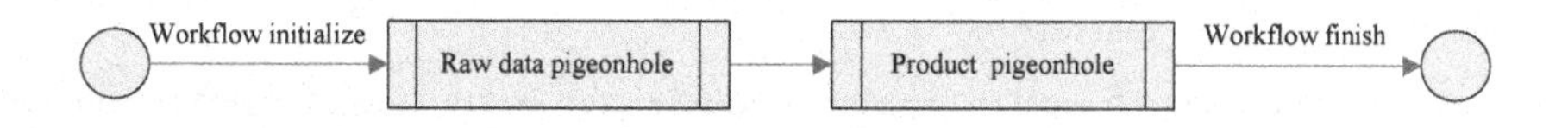

Fig. 7. Workflow layout

When the graphic designing of workflow is completed, the workflow designer will automatically generate the corresponding workflow definition XML file as follows.

```
<?xml version="1.0" encoding="UTF-8"?>
<layout>
  <cell id="0" type="InitialActionCell" height="36" width="36" x="9" y="66" name="begin" />
  <cell id="1" type="EndCell" height="36" width="36" x="501" y="67" name="end" />
  <cell id="2" type="StepCell" height="30" width="12" x="17" y="7" courseId = "2"
name="Raw data pigeonhole"/>
  <cell id="5" type="StepCell" height="30" width="12" x="33" y="7" courseId = "9"
name="Product pigeonhole"/>
 <connector id="8" linewidth="1.0" color="-16777216" labelx="44.5" labely="85"
labeltext="Workflow initialize" from="5" to="4" fromx="44.5" fromy="85"> </connector>
 <connector id="10" linewidth="1.0" color="-16777216" labelx="236.65" labely="86"
labeltext="Workflow finish" from="5" to="4" fromx="236.65" fromy="86"> </connector>
 </layout>
```

When there is a new task to do the products quality control and Product issuance, we can use the workflow process as a template. We only need to modify the progresses selection of the steps of template. And then deploy the new progresses to the grid and add its information to the operating management database. A new process will be designed. The layout of workflow designed is shown in Fig8.

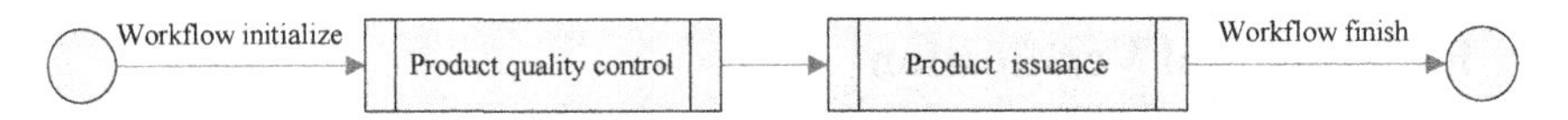

Fig. 8. Workflow layout

And the workflow designer will automatically generate the new workflow definition XML file as follows.

```
<?xml version="1.0" encoding="UTF-8"?>
<layout>
  <cell id="0" type="InitialActionCell" height="36" width="36" x="9" y="66" name="begin" />
  <cell id="1" type="EndCell" height="36" width="36" x="501" y="67" name="end" />
  <cell id="2" type="StepCell" height="30" width="120" x="11.15" y="70" courseId = "29"
name="Product quality control" />
  <cell id="5" type="StepCell" height="30" width="120" x="310.15000003" y="70" courseId
= "910" name="Product issuance" />
 <connector id="8" linewidth="1.0" color="-16777216" labelx="44.5" labely="85"
labeltext="Workflow initialize" from="5" to="4" fromx="44.5" fromy="85"> </connector>
 <connector id="10" linewidth="1.0" color="-16777216" labelx="236.65" labely="86"
labeltext="Workflow finish" from="5" to="4" fromx="236.65" fromy="86"> </connector>
 </layout>
```

5 Summary and Future Work

From the long-term development of the aerospace technology, aerospace ground-based data pre-processing system design and construction should be proactive to make a plan for the next few decades. Data management, the building of hardware and software facilities should have the ability to carry the large-scale space data processing of the past and the future. Meet the requirements of high reliability, high performance, dynamic scalability, security. The workflow-based aerospace ground-based data parallel pre-processing system designed in the paper can accept the future challenge. And the future work of this field will be the high security and the high reliability research.

References

1. Atlasis, A.F., Loukas, N.H., Vasilakos, A.V.: The Use of Learning Algorithms in ATM Networks Call Admission Control Problem: a Methodology. Computers Networks 34, 341–353 (2000)
2. Zhu, X., Xu, J., Liu, H.: On application of workflow technology in satellite ground system. Spacecraft Engineering 20, 1673–8748 (2011) 01—114-06
3. Tham, C.-K., Liu, Y.: Minimizing transmission costs through adaptive marking in differentiated services networks. In: Almeroth, K.C., Hasan, M. (eds.) MMNS 2002. LNCS, vol. 2496, pp. 237–249. Springer, Heidelberg (2002)
4. Galstyan, A., Czajkowski, K., Lerman, K.: Resource Allocation in the Grid Using Reinforcement Learning. In: AAMAS (2004)
5. Vengerov, D.: A Reinforcement Learning Framework for Utility-Based Scheduling in Resource Constrained Systems. Sun Microsystems Inc., Tech. Rep. (2005)
6. http://www.nasa.gov/
7. Vidal, J.M., Buhler, P., Stahl, C.: Multi-agent Systems with Workflows. IEEE Internet Computing 8(1), 76–82 (2004)
8. Cluement, L., Ding, Z., et al.: Providing Dynamic Virtualized Access to Grid Resources via the Web 2.0 Paradigm. In: 3rd Grid Computing Environment Workshop Supercomputing Conference 2007, Reno, Nevada, October 10-16 (2007)
9. Cao, J., et al.: WorkFlow Management for Grid Computing. In: Proc. of the 3rd IEEE/ACM International Symposium on Cluster Computing and the Grid, pp. 198–205 (2003)
10. Amin, K., von Laszewski, G., Hategan, M., Zaluzec, N.J., Hampton, S., Rossi, A.: GridAnt: A Client-Controllable Grid Workflow System. In: Proceedings of the 37th Hawaii International Conference on System Sciences (2004)
11. Deelman, E., et al.: Mapping Abstract Complex Workflows onto Grid Environments. Journal of Grid Computing 1(1), 25–39 (2003)

Development Platform for Heterogeneous Wireless Sensor Networks

Xu Chong

Information Technology Department, Far East Horizon Limited,
36 Floor, Jin Mao Tower, No 88 Century Avenue, Shanghai, China
zephyr.xu@gmail.com

Abstract. A universal platform is developed to provide several interfaces that every sensor can register or communicate with each other and platform (environment). To understand the operation of the platform, sensor nodes simulated by the Blackfin processor and PCs are used. This report presents the performance of the DSN based on an experiment in which different sensor nodes are connected to the platform, send/get data to/from platform and also can communication with each other. The results of this study demonstrate the principles and implementation of distributed sensor networks which can be extended to a complete and operational sensor management environment.

1 Introduction

Smart environments represent the next evolutionary development step in building, utilities, industrial, home, shipboard, and transportation systems automation. Like any sentient organism, the smart environment relies first and foremost on sensory data from the real world. Sensory data comes from multiple sensors of different modalities in distributed locations. The smart environment needs information about its surroundings as well as about its internal workings.

The challenges in the hierarchy of detecting the relevant quantities, monitoring and collecting the data, assessing and evaluating the information, formulating meaningful user displays, and performing decision-making and alarm functions are enormous [1].The information needed by smart environments is provided by Distributed Wireless Sensor Networks, which are responsible for sensing as well as for the first stages of the processing hierarchy. The importance of sensor networks is highlighted by the number of recent funding initiatives, including the DARPA SENSIT program, military programs, and NSF Program Announcements.

Y. Wang and T. Li (Eds.): Knowledge Engineering and Management, AISC 123, pp. 113–119.
springerlink.com

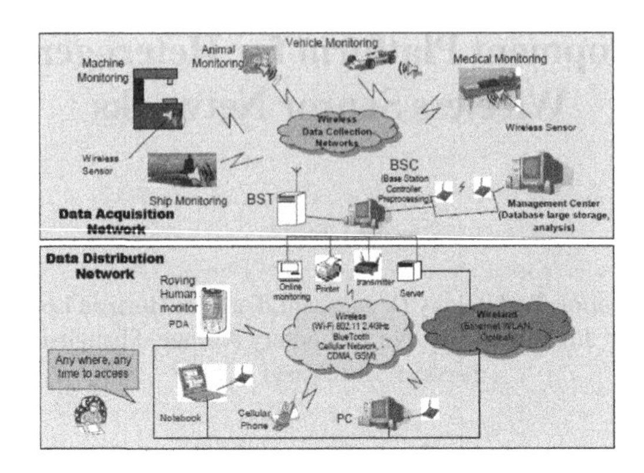

Fig. 1. Distributed sensor network [2]

The complexity of wireless sensor networks, which generally consist of a data acquisition network and a data distribution network, monitored and controlled by a management center.

2 Design and Implement Platform

Sensor networks are the key to gathering the information needed by smart environments, whether in buildings, utilities, industrial, home, shipboard, transportation systems automation, or elsewhere. To cite an example, recent terrorist and guerilla warfare countermeasures require distributed networks of sensors that can be deployed using, e.g. aircraft, and have self-organizing capabilities. In such applications, running wires or cabling is usually impractical. A sensor network is required that is fast and easy to install and maintain.

2.1 Basic Architecture

From a holistic perspective, a service oriented architecture (SOA)-based system is a network of independent services, machines, the people who operate, affect, use, and govern those services as well as the suppliers of equipment and personnel to these people and services. This includes any entity, animate or inanimate, that may affect or be affected by the system. With a system that large, it is clear that nobody is really "in control" or "in charge" of the whole ecosystem; although there are definite stakeholders involved, each of whom has some control and influence over the community [4, 5].

Instead of visualizing a SOA as a single complex machine, it is perhaps more productive to think of it as an ecosystem: a space where people, machines and services inhabit in order to further both their own objectives and the objectives of the larger community. In certain situations this may be a difficult psychological step for owners of so-called enterprise systems to take: after all, such owners may rightly believe that since they own the system they should also have complete control of it.

This view of SOA as ecosystem has been a consistent guide to the development of this architecture.

Taking an ecosystems perspective often means taking a step back: for example, instead of specifying an application hierarchy, we model the system as a network of peer-like entities; instead of specifying a hierarchy of control, we specify rules for the interactions between participants [6]. The three key principles that inform our approach to a SOA ecosystem are:

- An SOA is a medium for exchange of value between independently acting participants;
- Participants (and stakeholders in general) have legitimate claims to ownership of resources that are made available via the SOA;
- The behavior and performance of the participants is subject to rules of engagement which are captured in a series of policies and contracts.

Summing up the above, the distributed sensor platform is implemented in service oriented architecture, the platform is considered as a service, a contractually defined behavior that can be implemented and provided by a component for use by another component.

2.2 Smart Sensors

Wireless sensor networks satisfy these requirements. Desirable functions for sensor nodes include: ease of installation, self-identification, self-diagnosis, reliability, time awareness for coordination with other nodes, some software functions and DSP, and standard control protocols and network interfaces [IEEE 1451 Expo, 2001].

There are many sensor manufacturers and many networks on the market today. It is too costly for manufacturers to make special transducers for every network on the market. Different components made by different manufacturers should be compatible. Therefore, in 1993 the IEEE and the National Institute of Standards and Technology (NIST) began work on a standard for Smart Sensor Networks. IEEE 1451, the Standard for Smart Sensor Networks was the result. The objective of this standard is to make it easier for different manufacturers to develop smart sensors and to interface those devices to networks.

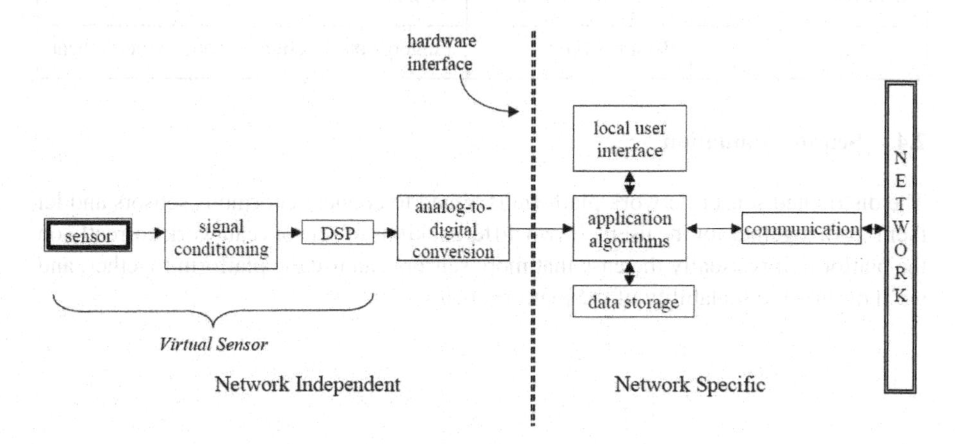

Fig. 2. A general model of smart sensor

2.3 Sensors for Smart Environments

Many vendors now produce commercially available sensors of many types that are suitable for wireless network applications. See for instance the websites of SUNX Sensors, Schaevitz, Keyence, Turck, Pepperl & Fuchs, National Instruments, UE Systems (ultrasonic), Leake (IR), CSI (vibration). The table below shows which physical principles may be used to measure various quantities. MEMS sensors are by now available for most of these measured.

Table 1. Measurements for Wireless Sensor Networks

	Measurand	Transduction Principle
Physical Properties	Pressure	Piezoresistive, capacitive
	Temperature	Thermistor, thermo-mechanical,
Motion Properties	Position	E-mag, GPS, contact sensor
	Velocity	Doppler, Hall effect, optoelectronic
Contact Properties	Strain	Piezoresistive
	Force	Piezoelectric, piezoresistive
	Torque	Piezoresistive, optoelectronic
Presence	Tactile/contact	Contact switch, capacitive
	Proximity	Hall effect, capacitive, magnetic, seismic,
Biochemical	Biochemical agents	Biochemical transduction
Identification	Personal features	Vision
	Personal ID	Fingerprints, retinal scan, voice, heat

2.4 Sensor Simulation

The distributed sensor network platform is aimed to connect enormous sensors and let them work together for the users. If two different kinds of sensors can work correctly on the platform, it is usually the case that more sensors can use the platform together, and we'll discuss the scalability of the platform below.

In this platform, sensor has several attributes:

ID: the id assigned by platform, NAME: the name of sensor,

DESCRIPTION: the description of the sensor,

COORDINATORID: id of the sensor's coordinator,

STATUS: show the status of sensor,

INITIALBATTERYLEVEL: the initial battery level of the sensor,

CURRENTBATTERYLEVEL: the current battery level of the sensor,

REGTIME: the time sensors start up.

For the purpose of simulating sensors, there are two sensors connect to the platform. One is called "Temperature_P" simulated by PC, as a temperature, it sends its data (temperature) to the platform every 20 seconds, at the same time, it want knows the temperature in other places which was collected by one another sensor called "Temperature_B" simulated by Blackfin-537., It get data of "Temperature_B" from platform every 10 seconds. On the contrary, the sensor on Blackfin will send data every 10 seconds, and get data from "Temperatrue_P" every 20 seconds.

"Temperature_P" Sensor:
When the sensor starts up, it will register to the platform and describe itself: "I work on PC, I send my collected data each 20 seconds, and get data each 10 seconds."

"Temperature_B" Sensor:
At the same time, also starts up the sensor on Blackfin, it does the same jobs as the PC sensor does. Its description is: "I work on Blackfin, I collect data each 10 seconds, and get data each 20 seconds."

3 Scalability Measurement

The scalability of the platform was tested by creating multiple threads used to stimulate multi-sensors connected to platform. The performance wasmeasured when the number of sensors is 1, 10, 20, 40…and so on and respectively calculated the time when the sensor does some jobs with the platform.

For instances, one sensor registers to the platform cost about 280ms, and 10 sensors register at same time, the minimum time is 310ms, and maximum time is 368ms, and the average time of ten sensors is 340ms.

This figure blew show the result of measurement that the transformation of time when sensors messaging and register platform with the increase of number. Here, the x axle is the number of sensors; y axle is the average time that the sensor costs when it finish the operation, unit is millisecond (ms). The pink line with circular point is the messaging line, and the purple line with rectangle point is the registration line.

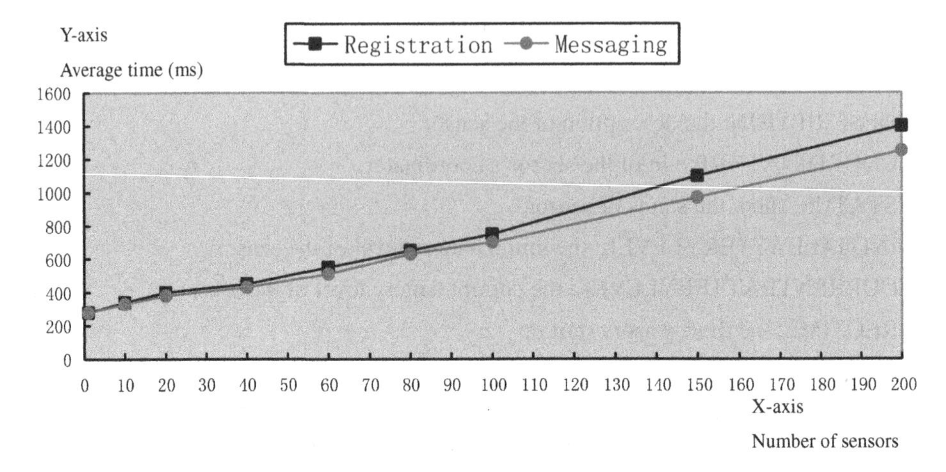

Fig. 3. Scalability Measurement

4 Performances Analysis

From the section 4.3, the performances of the platform can be seen that the time of communication between sensor and platform will increase linearly with the connected sensors enhanced when the number of sensors is not particularly large.

As to the web services, the platform process incoming requests by a process thread pool which within a group of threads. When the request comes, an idle thread in the pool will serve it. The process thread pool will not create an unlimited number of threads to handle a large volume of requests. Therefore, the communication time with the platform of the multi-sensor is larger than single sensors. It depends on the ability of server's CPU that deals with multi-threads jobs.

Is that possible to connect an infinite number of sensors to the platform? The answer is negative. The maximum number of sensors is approximately 500. In the test, when the sensors exceed 500, a few sensors will fail to communicate with platform, when the number is more than 600 or even much more than this, the most of sensors will announce that they can't invoke the service on the platform.

According the response time, from the figure 3, we know that platform will response sensor within 300ms if only a single sensor connected to the platform, but, with the growing number of connected nodes, time has also increased. When the number reaches 1000, the average response time will excess 6500ms, meanwhile, nearly half nodes doesn't get service.

If we used coordinator to manage a series of sensors, in order to send request sequentially, instead of every sensors send request to platform at same time, the system will more stable and efficiently. For example, a Blackfiin connect 1000 sensors as their coordinator, then the Blackfin sends all the registration request of sensors to the platform every 500ms, the platform response each request within the 400ms. That's the advantage of using coordinator to manage sensors, so that we'll discuss coordinator in the next chapter.

Theoretically, every operation of sensor will cost same time because every service in the platform is equal. However, in our implement, it can be seen in the figure 3, the registration takes a little more than the operation of send message. This is due to the database operation, obviously, the register operation in the platform does more work than the sensor send message to the platform.

If server creates threads infinitely, all resources on the server can only be used to manage these threads. By limiting the number of threads can be created, we can make thread management overhead to maintain in a controlled level. If a request arrives for all threads in the thread pool are occupied, then the request will be queued up, only after busy thread completes his task, the idle threads can process the new request. This method is actually more effective than switching to a new thread, because you do not need to be carried over from request to request threads switching. But the problem is, if the thread efficiency is not high (especially in busy server), the waiting request queue will become large.

5 Recommendations for Further Work

Due to the limitation of the time of research, this project has limitations that require some improvement or need to be solved in the future. For instance, the platform should be extended to connect more types of sensors which are implemented by other hardware and other environment. Research on the security aspect of data transmission between the sensors and the platform is also needed.

As discussed in section 3, the number of nodes that connected to the platform has a limitation. Therefore, it is efficient that use coordinator nodes to manage number of sensors, it will improve the performance of platform greatly.

References

1. Altman, E., Basar, T., Jimenez, T., Shimkin, N.: Competitive routing in networks with polynomial costs. IEEE Trans. Automat. Control 47(1), 92–96 (2002)
2. Lewis, F.L.: Smart Environments: Technologies, Protocols, and Applications. In: Cook, D.J., Das, S.K. (eds.). John Wiley, New York (2004)
3. Bronson, R., Naadimuthu, G.: Operations Research, 2nd edn. Schaum's Outlines. McGraw Hill, New York (1997)
4. Bulusu, N., Heidemann, J., Estrin, D., Tran, T.: Self-configuring localization systems: design and experimental evaluation. In: ACM TECS Special Issue on Networked Embedded Computing, pp. 1–31 (August 2002)
5. Cao, J., Zhang, F.: Optimal configuration in hierarchical network routing. In: Proc. Canadian Conf. Elect. and Comp. Eng., Canada, pp. 249–254 (1999)
6. Chen, T.-S., Chang, C.-Y., Sheu, J.-P.: Efficient path-based multicast in wormhole-routed mesh networks. J. Sys. Architecture 46, 919–930 (2000)

Knowledge-Based Resource Management for Distributed Problem Solving

Sergey Kovalchuk, Aleksey Larchenko, and Alexander Boukhanovsky

e-Science Research Institute, National Research University ITMO,
Saint-Petersburg, Russian Federation
sergey.v.kovalchuk@gmail.com, aleksey.larchenko@gmail.com,
avb_main@mail.ru

Abstract. Knowledge-based approach for composite high-performance application building and execution is proposed as a solution to solving complex computational-intensive scientific tasks using set of existing software packages. The approach is based on semantic description of existing software, used within composite application. It allows building applications according to user quality requirements and domain-specific task description. CLAVIRE platform is described as an example of successful implementation of proposed approach's basic principles. Exploration of described software solution performance characteristics is presented.

Keywords: composite application, high-performance computing, e-science, expert knowledge processing, performance estimation.

1 Introduction

Nowadays scientific experiment often requires huge amount of computation during simulation or data processing. Performance of contemporary supercomputers is increasing rapidly. It allows solving computation-intensive scientific problems, processing large arrays of data stored in archives or produced by sensor networks. Today we can speak about a new paradigm for scientific research often called e-Science [1]. This paradigm introduces many issues that have to be solved by collaboration of IT-specialists and domain scientists. These issues become more urgent as appropriate hardware and software turn into large complex systems.

Currently there is a lot of software for solving particular domain problem developed by domain scientists using their favorite programming language and parallel technologies. Thus today we have a great diversity of software for particular problem solving in almost each scientific domain. On the other hand with powerful computational resources we have an ability to solve complex problems that requires the use of different software pieces combined within composite applications. To address this issue problem solving environments (PSE) [2] are introduced as an approach for software composition. But still there are two problems. First, problem of previously developed software integration: with diversity of technologies, data formats, execution platforms it's quite complicated task to join different software even within PSE.

Y. Wang and T. Li (Eds.): Knowledge Engineering and Management, AISC 123, pp. 121–128.
springerlink.com

Second, there is a problem of using third party software with a lack of knowledge about its functionality, and internal features.

Looking at contemporary computational resources we can also see a high level of diversity in architectures, technologies, supported software etc. Moreover today we have an ability to combine different computational resources using metacomputing, Grid, or Cloud approaches. In this case the problem of integration becomes more important as we should care of performance issues in heterogeneous computational environment because of computational intensity of e-Science software.

Today's common approach for solving integration problem is typically based on service-oriented architecture (SOA). This approach allows developing of composite application using sets of services that give access to diverse resources in unified way. But the problem of composite application performance within heterogeneous computational environment still remains. The developer should care of configuring every running service in the way that allows making the whole application faster. In case of a single application implementing simple algorithm there is a lot of approaches for performance estimation and optimization depending on software and hardware features [3, 4]. But in case of composite application using diverse existing software and hardware resources, implementing complex algorithms we are faced with more complex issues of composite application development and execution.

In this paper we present our experience of knowledge-based description of distributed software and using it for performance optimization and support of software composition. Within this approach expert knowledge is used to describe a set of domain-specific services in the way that allows composing application for solving complex simulation problems taking into account technical features of software and hardware available within computational environment.

2 Basic Concepts

2.1 Semantic Description of Resources

As it was described before there is great diversity of software and hardware resources needed to be integrated within composite applications for solving e-Science problem. Description of this software should include the following structure of knowledge typically available to experts:

- ***Software specification.*** Basic statements used for software identification (name, version etc.) are mentioned within this part of description.
- ***Implemented algorithms.*** This part of description allows making composition of services for solving more complex problems within the domain. Also it can be used for searching and comparing alternatives solutions.
- ***Performance model.*** This part of knowledge should allow make estimation of execution time depending on execution parameters: computational environment specification and domain-specific data parameters. These two parts allow estimate execution time of the application with particular hardware and input parameters.
- ***Input and output sets.*** Describing set of incoming and outgoing data this part of knowledge gives information on parameters' structure, data formats and the way of

passing. Using this information it's possible to connect software pieces within composite application automatically, to apply data decomposition and transformation.

- ***Way of running.*** This part describes the procedure of calling the application including low level parameters passing (e.g. files or streams) and environment access procedure (e.g. service invocation, pre- and post-configuration of resources).
- ***Hardware and software dependencies.*** The set of requirements of the application is presented within this part. It is required for appropriate software deployment (in case this procedure is available within the computational environment).

Using this set allows describing a set of existing software available for particular problem domain. But also it is required to describe computational service environment that allows running these software. Service environment description should include the following structure of information:

- ***Hardware characteristics.*** With a set of resources available statically, dynamically or on-demand it is required to have full description of resources.
- ***Available services.*** This part maps the set of services (described as shown before) on the set of resources available within computational environment.

Finally, having the information mentioned above we should describe domain-specific usage process of the software for solving particular tasks. Here and further we use quantum chemistry (QC) as an example of problem domain. The proposed concept and technologies were applied within this domain during the HPC-NASIS project [5] - platform for QC computer simulation using distributed environment with integrated well-known computational software.

- ***Problems.*** This set of knowledge describes known domain problems which can be solved using a set of available software. Typically the problem set is well known for particular field of knowledge (e.g. in the field of quantum chemistry such problems could be a *single point problem* or *geometry optimization*).
- ***Methods.*** This part defines well-known methods of problem domain implemented in software. E.g. concerning QC problem domain there are such method as *Hartree-Fock (HF)* or *Density functional theory (DFT)*.
- ***Domain specific values.*** This part of semantic description contains domain-specific types, their structure and possible values. E.g. for QC domain we should describe concepts like *basis* or *molecular structure*.
- ***Solution quality estimation.*** With expert knowledge it is possible to define quality estimation procedure for known methods and its implementation taking into account domain-specific input values. This procedure is defined for particular quality characteristic space (precision, speed, reliability etc.). For example it is possible to define *precision* of selected method for solving *single point problem* for the defined molecular structure and basis. Using this part of knowledge gives opportunity to estimate quality metrics for each call within composite application and integral quality of whole application.

First of all the description defined above guides composite application building using a) semantic description of particular software pieces and b) available resource definition. In this case set of performance models allows execution time estimation and consequential structural optimization of composite application. But beside of that last part of knowledge extend available facilities with explanation and quality estimation expressed using domain-specific terms. As a result it is possible to build software system that can "speak" with domain specialist using his/her native language.

2.2 Implementation of Knowledge Base

Today semantic knowledge is typically expressed using ontologies. Concerning semantic software description mentioned above we can define ontology structure which integrates all the parts of knowledge. Fig. 1 shows an example of ontology part describing available software using proposed structure. This part of ontology (simplified for illustration) defines five concepts: package (which represents particular software piece), method (implemented within the software), cluster (as a subclass of resource), value (domain-specific) and data format. Important part of the knowledge is mentioned as attributes of individuals and relationships between the individuals. E.g. we can define that *ORCA* package implements *DFT* method with particular quality and performance (defined as constant quality value, function with set of parameters or table with profile values).

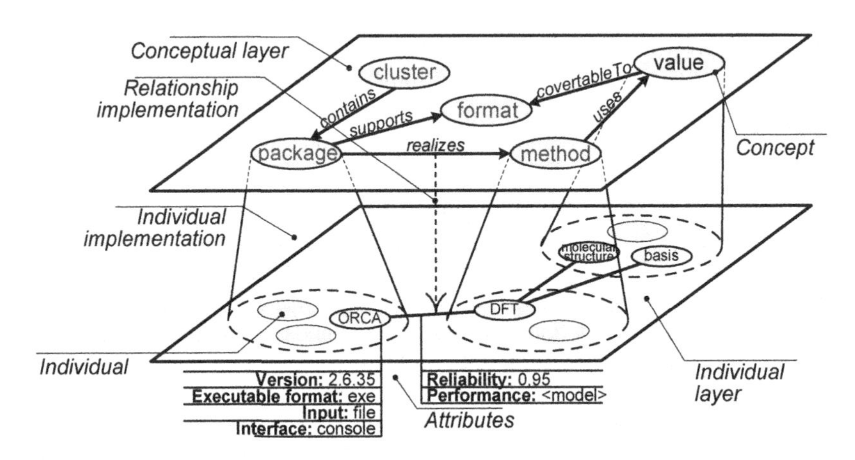

Fig. 1. Basic ontology structure for software semantic description

This approach allows supporting dynamic composition of software and hardware resources within complex application given by set of requirements. For instance it is possible to compose an application that requires shortest time to execute or gives the most precise solution using available parameter dictionary within the desired time.

3 Approach Implementation

3.1 CLAVIRE Platform

The approach described above can be concerned as a part of iPSE concept (Intelligent Problem Solving Environment) [6], which forms a conceptual basis for the set of projects performed by National Research University ITMO within last years. During these projects the infrastructure platform for building and executing of composite application for e-Science was developed. The CLAVIRE (CLoud Applications VIRtual Environment) platform allows building composite applications using domain specific software available within distributed environment. One of important features of the platform is expert knowledge application for solving the following tasks:

- Composite application building using domain-specific knowledge within intelligent subsystem. This process takes into account actual state of computational resources within the environment, available software and data uploaded by the user. Using semantic description of software this system forms an abstract workflow (AWF) definition of composite application. The AWF contains calls of software (using domain-specific values as high-level parameters) without mapping to particular resources.
- Parallel execution of AWF using performance optimization based on models defined as a part of knowledge. During this procedure technical parameters of execution are tuned to reach the best available performance for resources selected for execution parts of AWF. As AWF's elements are mapped to particular resources and low level parameters are defined the workflow turns into a concrete workflow (CWF).
- Data analysis and visualization. These procedures are supported by knowledge about a) data formats using during execution of composite application; b) solving domain-specific problem for automatically selection and presenting the data required by the user.

3.2 Knowledge-Based Solution Composition

The most interesting part of knowledge-based procedures mentioned above is composite application building using used-defined data and requirements (performance, quality etc.). Within the CLAVIRE platform tree-based dialog with the user is presented as a tool for decision support (see Fig. 2).

Passing through the nodes presented by domain-specific concepts (there are four levels of the current tree implementation: problem, method, package (software) and service) the user can define or select given input and required output values for every level of the tree. Passing through the levels 1 (Problem) to 3 (Package) produce AWF, available for automatic execution. But the user can pass further to the level 4 (Service) which allows fine tuning the execution parameters during producing of CWF.

Passing through the tree nodes user can control generation of next-level nodes by defining parameters of auto-generation or by blocking nodes processing for selected nodes. After passing the tree user can compare available solutions which will be estimated by quality values. The performance will be estimated for composite application using performance models of software, used for solving subtasks. It is possible to estimate and optimize the execution time using performance model.

Another performance issue is related to planning process of the whole composite application. The CLAVIRE platform uses well-known heuristics selected depending on performance estimation for current state of computational environment.

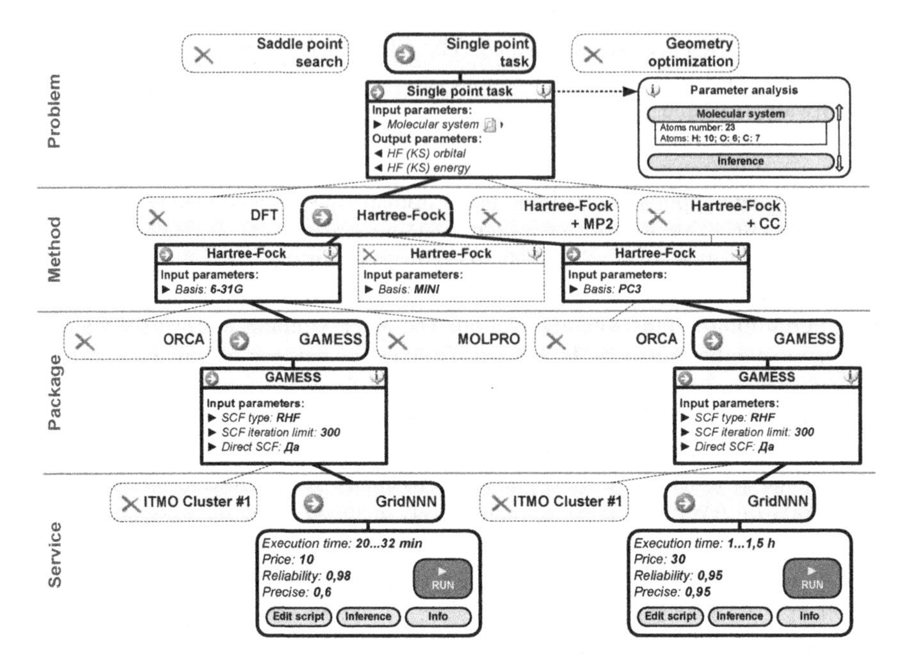

Fig. 2. Tree-based dialog

4 Numerical Results

During exploration the CLAVIRE platform a lot of experiments were performed. Fig. 3 shows the selected results of computational experiments with performance models used by the platform. The experiment was performed using planning simulation system which used performance models, parameterized using experiments with set of packages during test runs with CLAVIRE platform.

A) Comparing alternative implementation of the same computational problem. Three QC packages (GAMESS – 1, ORCA – 2, MOLPRO – 3) were compared during solving single point problem using on the same input data. Performance estimation using parameterized models shows that in case of time optimization the best choice is to use ORCA (I) running on 2 CPU cores (II).

B) Comparing packages with two-dimension models that takes into account shows more complex case of alternative selection. Looking at shown example it can be seen that package 1 (GAMESS) is better in region above dotted line while package 2 (ORCA) is better for parameters in region below the line. This simple case shows that the performance model can combine domain-specific parameters (here is count of basis function) with technical parameters (CPUs count).

C) Comparing CLAVIRE overhead to overhead of underlying computational Grid environment (Russian Grid network for nanotechnologies – GridNNN [7] was used as computational platform) shows that CLAVIRE has quite low overhead level and can be used as distributed computation management system.

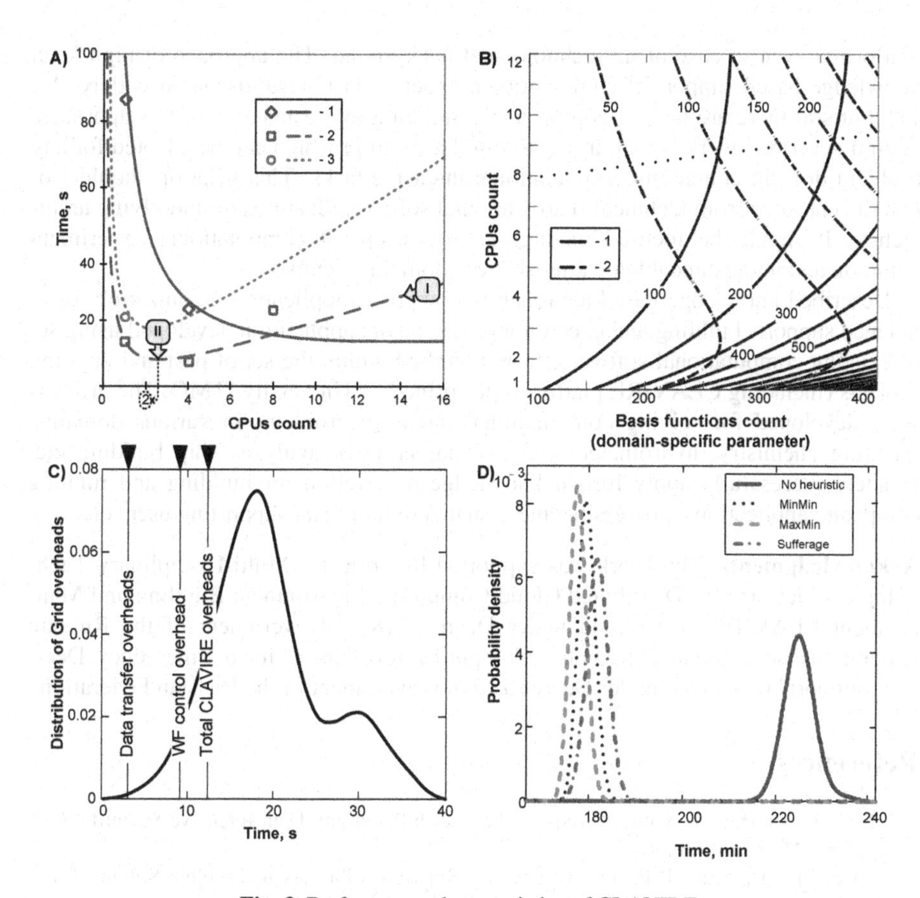

Fig. 3. Performance characteristics of CLAVIRE

D) Experiments with heuristic application for workflow planning show that it can bring us to notable decreasing of execution time for composite application with large amount of subtasks. On the other hand in case of identical (from performance point of view) subtasks using different heuristics gives almost the same results. For instance distributions of estimated execution time for parameter sweep task in case of stochastic behavior of computational environment planned with different heuristics almost overlaps.

5 Discussions and Conclusion

Today there are a lot of solutions trying to build composite applications automatically using knowledge bases (e.g. [8, 9]). Typically they use semantic pattern-based composition of workflows as a composite application description. But the most powerful approach for composite application building should actively involve domain-specific expert knowledge, which describes high-level computational experiment process. This description has to be clear for domain specialists (i.e. end-users of computational platforms) even in case of no technological background. The approach of high-level knowledge-based support of computational experiment is also discussed widely [10, 11], but still there are lack of appropriate common implementations of this approach. Within described work we are trying to build a solution which can be adapted for any problem domain containing computational-intensive tasks. This solution should isolate the end-user from technical (hardware and software) features of underlying architecture. It should be focused on high-level concept of computational experiment common and understandable by almost every domain scientist.

Described knowledge-based approach to composite application organization for e-Science supports building and execution of composite application developed using set of existing computational software. It was applied within the set of past and ongoing projects (including CLAVIRE platform) performed by University ITMO. The projects were developed for solving computational intensive problems in various domains: quantum chemistry, hydrometeorology, social network analysis, ship building etc. Projects successfully apply formal knowledge description for building and running composite applications, processing and visualization of data, supporting users etc.

Acknowledgments. This work was supported by projects "Multi-Disciplinary Technological Platform for Distributed Cloud Computing Environment Building and Management CLAVIRE" performed under Decree 218 of Government of the Russian Federation and "Urgent Distributed Computing for Time-Critical Emergency Decision Support" performed under Decree 220 of Government of the Russian Federation.

References

1. Hey, T., Tansley, S., Tolle, K. (eds.): The Fourth Paradigm. Data-Intensive Scientific Discovery, Microsoft (2009)
2. Rice, J.R., Boisvert, R.F.: From Scientific Software Libraries to Problem-Solving Environments. IEEE Computational Science & Engineering 3(3), 44–53 (1996)
3. Kishimoto, Y., Ichikawa, S.: Optimizing the Configuration of a Heterogeneous Cluster with Multiprocessing and Execution-Time Estimation. Parallel Computing 31(7), 691–710 (2005)
4. Dolan, E.D., Moré, J.J.: Benchmarking Optimization Software with Performance Profiles. Mathematical Programming 91(2), 201–213 (2002)
5. HPC-NASIS, `http://hpc-nasis.ifmo.ru/`
6. Boukhanovsky, A.V., Kovalchuk, S.V., Maryin, S.V.: Intelligent Software Platform for Complex System Computer Simulation: Conception. In: Architecture and Implementation, vol. 10, pp. 5–24. Izvestiya VUZov, Priborostroenie (2009) (in Russian)
7. Start GridNNN, `http://ngrid.ru/ngrid/`
8. Kim, J., et al.: Principles For Interactive Acquisition And Validation Of Workflows. Journal of Experimental & Theoretical Artificial Intelligence 22, 103–134 (2010)
9. Gubała, T., Bubak, M., Malawski, M., Rycerz, K.: Semantic-based grid workflow composition. In: Wyrzykowski, R., Dongarra, J., Meyer, N., Waśniewski, J. (eds.) PPAM 2005. LNCS, vol. 3911, pp. 651–658. Springer, Heidelberg (2006)
10. Gil, Y.: From Data to Knowledge to Discoveries: Scientific Workflows and Artificial Intelligence. Scientific Programming 17(3), 1–25 (2008)
11. Blythe, J., et al.: Transparent Grid Computing: a Knowledge-Based Approach. In: Proceedings of the 15th Annual Conference on Innovative Applications of Artificial Intelligence, pp. 12–14 (2003)

Research on and Realization of an Adaptive Clock Synchronization Algorithm for Distributed Systems

Jia Yang[1] and Piyan He[2]

[1] College of Communication Engineering, Chengdu University of Information Technology, Chengdu, 610225, China
[2] Institute of Aerospace Science and Technology, University of Electronic Science and Technology of China, Chengdu 610054, China
yangjia@cuit.edu.cn, he_py@126.com

Abstract. In distributed systems, the uncertainty of the clock drift and transmission delay is a problem that cannot be ignored, because it influences the precision of the clock synchronization directly. In this paper, considering this problem, an adaptive clock synchronization algorithm which aims the characteristics of enterprise distributed systems is proposed based on the passive algorithm. The adaptive algorithm can automatically determine the time interval of the two adjacent clock synchronization adjustments and choose the optimized measurement times and the clock adjustment value of each clock synchronization adjustment, so that, an optimized plan can be used to achieve the synchronization task. It is proved that the algorithm can inhibit effectively the influence of the uncertainty to meet the accuracy requirements of the clock synchronization in enterprise distributed systems. The algorithm has a good practical value.

Keywords: distributed systems, adaptive algorithm, clock synchronization, transmission delay.

1 Introduction

Clock synchronization is one of core technologies of distributed systems. Its task is to make sure the information, events and each node whose behavior associated with time have a global consistent reference. The key of promoting the precision of the clock synchronization is to offset or inhibit the influence of the uncertainty of the clock drift and transmission delay. So in this paper, according to characteristics of enterprise distributed systems, a network structure for the adaptive algorithm is built. On this basis, the algorithm is discussed how to resolve the influence of the uncertainty. At last, the realization and performance analysis of the algorithm are given to prove its effect.

2 Network Structure for the Adaptive Algorithm

These are three kinds of clock synchronization algorithms: centralized, master-slave, distributed. This paper aims enterprise distributed systems, such as distributed real-time

Y. Wang and T. Li (Eds.): Knowledge Engineering and Management, AISC 123, pp. 129–134.
springerlink.com

data motoring systems. So it pays attention to convenient management, high reliability, low cost, adaptation etc. Firstly considering the convenient management, the designer chooses the passive algorithm of the centralized mode that clients should be managed by at least one server and the server receives clock synchronization message from clients passively. Secondly, considering the high reliability, according to the number of clients, if the number is small, only two servers are needed. One of them manages clients and another as a backup. The two servers synchronize with the external GPS. If the number is large, it is considered to divide the whole area to several sub areas and each sub area can be managed by one server. All of the servers synchronize with external GPS. Once one of servers has fault, there is a server in the near sub area to replace it temporarily.

3 The Analysis of the Adaptive Clock Synchronization Algorithm

According to the passive algorithm, the communication process of the client and server is shown as Fig.1.

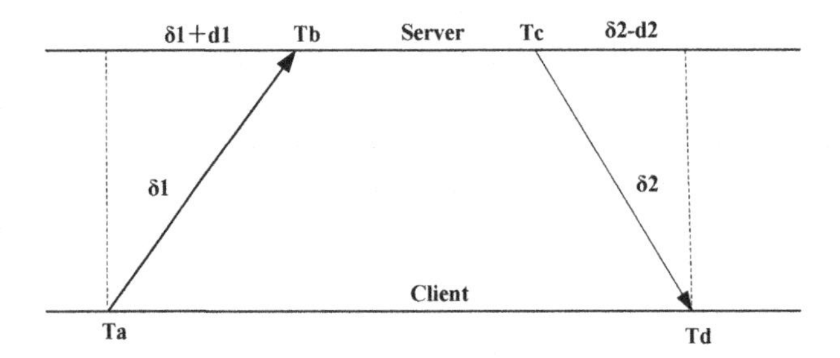

Fig. 1. The clock synchronization based on Client/Server model

Supposed: T_a, T_b, T_c, T_d separately are the time at which the client sends the clock synchronization request message, the time at which the server receives the request message, the time at which the server sends back the response message including the standard time, and the time at which the client receives the response message from the server. δ_1 is the delay from the client sending the request message to the server receiving the request message. δ_2 is the delay from the server sending the response message to the client receiving the response. d_1 is the clock deviation of the client relative to the server at T_a. d_2 is the clock deviation of the Client relative to the server at T_d. The key of the precision of the clock synchronization depends on the communication delay δ_1, δ_2 and the clock deviation d_1, $d2$. In theory, $\delta_1=\delta_2=\delta$, so formula (1) is shown as below:

$$\begin{cases} T_d - T_c = -d_2 + \delta . \\ T_b - T_a = d_1 + \delta . \end{cases} \tag{1}$$

For the computer clock drift is caused by the different frequency of the physical Oscillators in the side of the different nodes. For the specific system, with the advance of the timeline, the clock drift of the client relative to the server approximates a line [2], that $\Delta d = a\Delta t$. a is a constant that can be gotten in the experimental environment. In the actual measurement, because the time that the server processes the message is very shot, so Δt can be approximately considered as 2δ, that $\Delta d = 2a\delta$. So that according to formula (1), the formula (2) can be gotten. It is shown as below:

$$\delta = \frac{1}{2(1-a)}(T_b + T_d - T_a - T_c) . \tag{2}$$

In the actual measurement, because of involving many aspects, so the random error maybe brought to the measurement data. δ should be cumulated n times to calculate the average $\overline{\delta}$ shown as formula (3).

$$\overline{\delta} = \frac{1}{2(1-a)N} \sum_{k=1}^{N} (T_{bk} + T_{dk} - T_{ak} - T_{ck}) . \tag{3}$$

According to formula (3), the formula (4) can be gotten as below:

$$T_{jz} = T_c + \overline{\delta_n} . \tag{4}$$

Notes: Tc is the standard time of the server responding the client at the N time' measurement. T_{jz} is the client time after synchronization adjusting.

According to the analysis above, the synchronization focuses on the selection of the measurement times in each clock synchronization adjustment and the determination of the interval of the adjacent twice clock synchronization adjustment." **The mind of the adaptive algorithm is":** because of the random error, the standard error σ of the average is brought to analyze the jitter of measurement data of N times, shown as formula (5). The value of σ is smaller and the average is closer to the real data value, and it shows that the jitter of N times 'measurement is smaller.

$$\sigma = \sqrt{\frac{\sum_{i=1}^{n} (\delta_i - \overline{\delta})^2}{(n-1)}} . \tag{5}$$

Notes: δ_i is the communication delay of the client and server at the i time.

For the selection of the measurement times in each clock synchronization adjustment, the algorithm will set the upper limit MAX of the measurement times. MAX is the multiple of 10. Each clock synchronization adjustment will do MAX times measurement. When the measurement times N is cumulated to the multiple of 10, the average and average standard error of N times before will be calculated and compared with the standard error of the N-10 times before. The smaller value of the standard error, the measurement times and average corresponding with the standard error will be recorded. So the smallest standard error, the measurement times M and the average $\overline{\delta}$ corresponding with it will be gotten by comparing method. According to the formula (4), the $\overline{\delta}$ and Tc at the MAX time will be used to adjust the client's system time.

For the determination of the time interval of the two adjacent clock synchronization, the algorithm considers the clock drift of the client relative to the server approximates a line, that $\Delta d = a\Delta t$. So according to the actual network, we can set Te as the upper limit of the clock deviation, every one hour calculate the Δd and compared with Te, until $\Delta d = T_e$ .Then the client will send synchronization request message to do another clock synchronization adjustment. Considering the current life and precision of the clock, the time interval T_k when $\Delta d = T_e$ can be recorded. From now on, during a sometime, T_k can be used as the clock synchronization interval. For the higher precision, because of the aging of a clock, the time T_k should be measured periodically.

4 The Realization and Performance Analysis of the Algorithm

The software for the algorithm has two modules. One of them is synchronization module, another is performance analysis module. For the synchronization module, by using Client/Server mode in NTP (Network Time Protocol) clock synchronization measurement is done according to the Fig.2 as below.

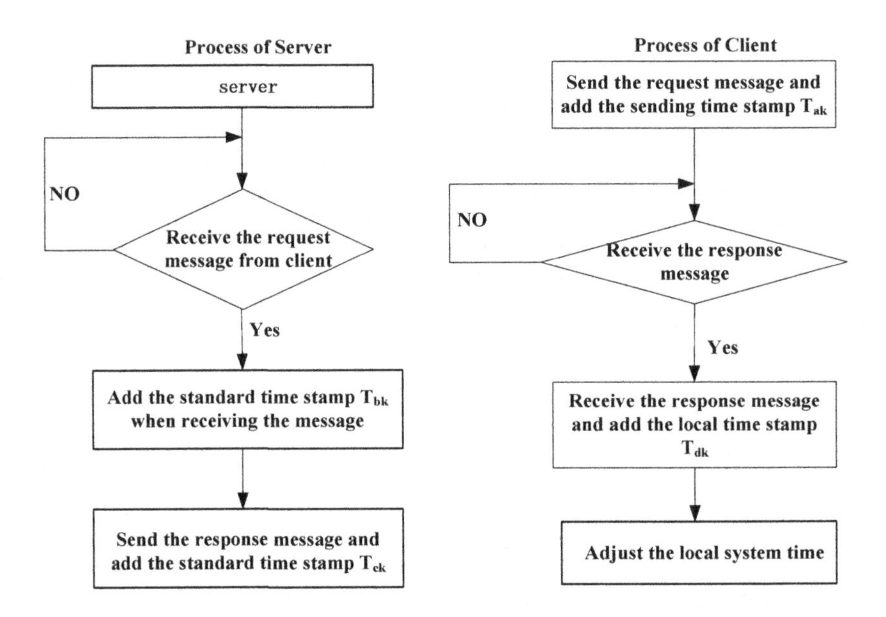

Fig. 2. Clock synchronization process

For getting the system time of the client and server, there are two kinds of plans to be provided. One is 'QueryPerformanceCount' time counter. It can be accurate to microsecond. If it needs a higher precision, and the client use the CPU above the level of Intel Pentium and including Intel Pentium. The timer class of RDTSC instruction: Ktimer class can be read the time stamp of the CPU main frequency. For the CPU whose frequency is above 1GHz, the system time can be accurate to nanosecond by using Ktimer class.

According to the design above, in some enterprise's distributed systems, the adaptive algorithm is tested. 'QueryPerformanceCount' time counter is used to read the system time. The specific performance analysis is shown as Fig.3 and Fig.4.

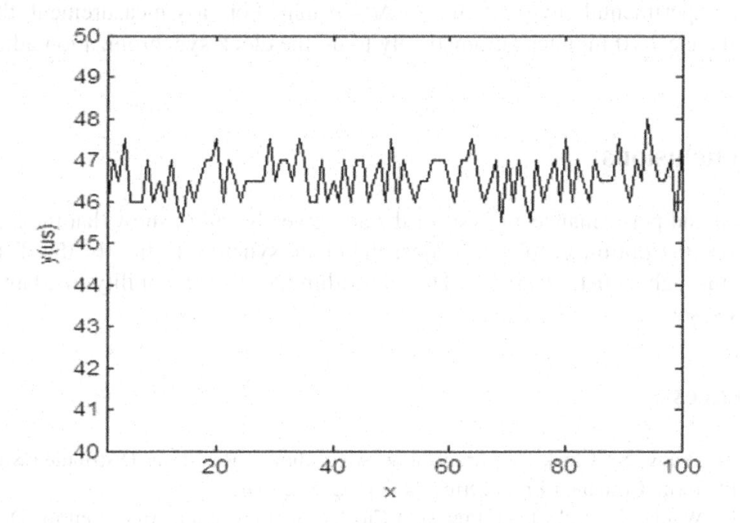

Fig. 3. The data of 100 times measurement in once clock synchronization adjustment

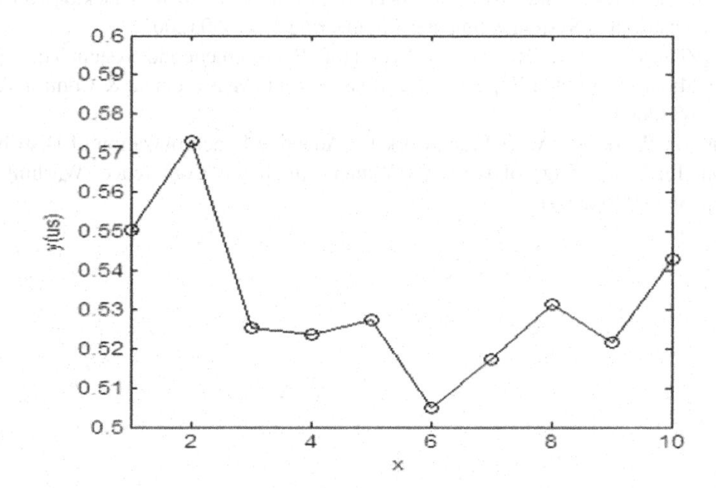

Fig. 4. The average standard error of $x \times 10$ times before

In Fig.3, x coordinate expresses the times of measurement in once clock synchronization adjustment. The times are 100. y coordinate is the communication delay δ between the client and server. Its unit is μs .

In Fig.4, based on the data in Fig.3, for easy watch, the data of 100 times is divided by 10 times as a unit. For example, $x=6$ means 60 measurement times. y coordinate is standard error of average, its unit is μs .

In Fig.4, when $x=60$, the value of y is smallest. $y=0.5050$.So the algorithm will select the average of 60 times before to adjust the client clock. For the determination of the time interval between the two adjacent clock synchronization adjustments, the enterprise stipulates the clock deviation must be smaller than 10 ms, a is 2.1E-6 that is gotten in experimental environment. So, $\Delta t \cong 79\,\text{min}$. For easy measurement, the algorithm will select 70 minutes automatically to do the clock synchronization adjustment again.

5 Conclusions

The result and performance analysis of the adaptive algorithm show that the algorithm can choose the optimized plan to adjust the clock synchronization of the distributed systems through statistic learning. This algorithm has a strong ability to adapt the enterprises system.

References

1. Zhu, B., Yang, S.: Clock Synchronization Algorithm of Real-time Distributed Simulation Test Platform. Computer Engineering 24, 231–235 (2010)
2. Liu, L., Wang, X., et al.: High-Precision Clock Synchronization Algorithm for Distributed System and Its Realization. Journal of Tianjin University 8, 923–927 (2006)
3. Ning, W., Zhang, B., et al.: Statistical Mean Algorithm of Realizing Clock Synchronization Based on Distributed System. Computer Engineering 5, 47–50 (2005)
4. Qin, X., Zhai, Z., Liu, Z.: Research of Real-Time Requirement and System Time Synchronization About Distributed Experiment& Measurement. Measurement & Control Technology 2, 63–65 (2004)
5. Altuntas, B., Wysk, R.: A.: A Framework for Adaptive Synchronization of Distributed Simulation. In: Proceedings of the 2004 Winter Simulation Conference, Washington DC, USA, pp. 363–369 (2004)

Realistic Task Scheduling with Contention Awareness Genetic Algorithm by Fuzzy Routing in Arbitrary Heterogeneous Multiprocessor Systems

Nafiseh Sedaghat[1], Hamid Tabatabaee-Yazdi[2], and Mohammad-R Akbarzadeh-T[3]

[1] Islamic Azad University, Mashhad Branch, Young Researches club, Iran
[2] Islamic Azad University, Quchan Branch, Iran
[3] Ferdowsi University, Mashhad, Iran

nf_sedaghat@yahoo.com, hamid.tabatabaee@gmail.com, akbarzadeh@ieee.org

Abstract. Task scheduling is an essential aspect of parallel processing system. This problem assumes fully connected processors and ignores contention on the communication links. However, as arbitrary processor network (APN), communication contention has a strong influence on the execution time of a parallel application. In this paper, we propose genetic algorithms with fuzzy routing to face with link contention. In fuzzy routing algorithm, we consider speed of links and also busy time of intermediate links. To evaluate our method, we generate random DAGs with different Sparsity value based on Bernoulli distribution and compare our method with genetic algorithm and classic routing algorithm and also with BSA (bubble scheduling and allocation) method that is a well-known algorithm in this field. Experimental results show our method (GA with fuzzy routing) is able to find a scheduling with lower makespan than GA with classic routing and also BSA.

Keywords: Task Scheduling, Fuzzy Routing, Distributed Heterogeneous Systems.

1 Introduction

Distributed heterogeneous systems have become widely used for scientific and commercial applications such as high-definition television, medical imaging, or seismic data process and weather prediction. These systems require a mixture of general-purpose machines, programmable digital machines, and application specific integrated circuits [1]. A distributed heterogeneous system involves multiple heterogeneous modules connected by arbitrary architecture and interacting with one another to solve a problem. The common objective of scheduling is to map tasks onto machines and order their execution so that task precedence requirements are satisfied and there is a minimum schedule length (makespan) [1]. In the earlier researches the processors are assumed to be fully-connected, and no attention is paid to link contention or routing strategies used for communication [2].Actually, most heterogeneous systems cannot meet this condition and their processors are linked by an arbitrary processor network (APN) [1]. In APN algorithms, the mapping of tasks to

Y. Wang and T. Li (Eds.): Knowledge Engineering and Management, AISC 123, pp. 135–144.
springerlink.com

processors is implicit, and messages are also scheduled while considering link contention. Scheduling tasks on is a relatively less explored research topic and very few algorithms for this problem have been designed [2].However, Macey and Zomaya[1] showed that the consideration of link contention is significant for producing accurate and efficient schedules.

In this paper, we propose genetic algorithm for solving the problem of static scheduling and mapping precedence-constrained tasks to anarbitrary network of heterogeneous processors. We consider link contention and just like tasks, messages are also scheduled and mapped to suitable links during the minimization of the finish time of tasks. To reach an optimum makespan, we propose a fuzzy routing algorithm. To evaluate our method, we generate random DAGs with different Sparsity value based P-Method and compare our method (GAFR) with genetic algorithm with classic routing algorithm (GACR) and also with BSA (bubble scheduling and allocation) method that is a well-known algorithm. Experimental results show our method, GAFR (GA with fuzzy routing), is better than GACR and also BSA.

The rest of the paper is organized as follows: Section 2 presents definition of task scheduling problem in classic and contention awareness model. In Section 3, we review some related works and in Section 4 we present our method. In Section 5, the simulation experimental results are presented and analyzed. This paper concludes with Section 6 and in this section, we discuss about future works.

2 Task Scheduling Model

In task scheduling, the program to be scheduled is represented by a directed acyclic graph $G = (\boldsymbol{V}, \boldsymbol{E}, w, c)$representing a program P according to the graph model. The nodes in $\mathbf{V}$ represent the tasks of P and the edges in $\mathbf{E}$ representthe communications between the tasks. An edge$e_{ij} \in \mathbf{E}$ from node $\mathrm{n_i}$to $\mathrm{n_j}$, $\mathrm{n_i}, \mathrm{n_j} \in \mathbf{V}$, represents the communication from node $\mathrm{n_i}$ to node $\mathrm{n_j}$. A task cannot begin execution until all its inputs have arrived and no output is available until the computation has finished and at that time all outputs are available for communication simultaneously [4].

A schedule of a DAG is the association of a start time and a processor with every node of the DAG. To describe a schedule S of a DAG $G = (\boldsymbol{V}, \boldsymbol{E}, w, c)$ on a target system consisting of a set P of dedicated processors, the following terms are defined: $t_s(n, P)$denotes the start time and $w(n, P)$ the execution time of node $n \in \boldsymbol{V}$ on processor $p \in P$. Thus, the node's finish time is given by$t_f(n, P) = t_s(n, P) + w(n, P)$.The processor to which n is allocated is denoted by $proc(n)$.

Most scheduling algorithms employ a strongly idealized model of the target parallel system [5-8]. This model, which shall be referred to as the classic model, is defined in the following, including a generalization toward heterogeneous processors.

Definition 1 (Classic System Model). *A parallel system $M_{classic} = (\mathbf{P}, w)$ consists of a finite set of dedicated processors* $\mathbf{P}$ *connected by a communication network. The* *processor heterogeneity, in terms of processing speed, is described by the execution time function* w*. This dedicated system has the following properties:*

1. *local communication has zero costs,*
2. *communication is performed by a communication subsystem,*
3. *communication can be performed concurrently, and*
4. *the communication network is fully connected.*

Based on this system model, the edge finish time only depends on the finish time of the origin node and the communication time.

The classic scheduling model (Definition 1) does not consider any kind of contention for communication resources. To make task scheduling contention aware, and thereby more realistic, the communication network is modeled by a graph, where processors are represented by vertices and the edges reflect the communication links. The awareness for contention is achieved by edge scheduling, i.e., the scheduling of the edges of the DAG onto the links of the network graph, in a very similar manner to how the nodes are scheduled on the processors.

Thus, communication can overlap with the computation of other nodes, an unlimited number of communications can be performed at the same time, and communication has the same cost $c(e_{ij})$, regardless of the origin and the destination processor, unless the communication is local[4].

The system model is then defined as follows:

Definition 2 (Target Parallel System—Contention Model).

A target parallel system $M_{TG} = (TG, w)$ *consists of a set of possibly heterogeneous processors* $\mathbf{P}$ *connected by the communication network* $TG = (\mathbf{P}, \mathbf{L})$*. This dedicated system has the following properties:*

1. *local communications have zero costs and*
2. *communication is performed by a communication subsystem.*

The notions of concurrent communication and a fully connected network found in the classic model are substituted by the notion of scheduling the edges $\mathbf{E}$ on the communication links$\mathbf{L}$. Corresponding to the scheduling of the nodes, $t_s(e, L)$ and $t_f(e, L)$ denote the start and finish time of edge $e \in \boldsymbol{E}$ on link $L \in \mathbf{L}$, respectively.

When a communication, represented by the edge e, is performed between two distinct processors P_{src} and P_{dst}, the routing algorithm of TG returns a route from P_{src} to P_{dst} : $R =< L_1, L_2, \dots, L_l >$, $L_i \in \mathbf{L}$ for $i = 1, 2, \dots, l$. The edge e is scheduled on each link of the route as below definition:.

Definition 3 (Edge Finish Time—Contention Model).*Let* $G = (\mathbf{V}, \mathbf{E}, w, c)$ *be a DAG and* $M_{TG} = ((\mathbf{P}, \mathbf{L}), w)$ *a parallel system. Let* $R =< L_1, L_2, \dots, L_l >$ *be the route for the communication of* $e_{ij} \in \boldsymbol{E}$, $n_i, n_j \in \boldsymbol{V}$, *if* $proc(n_i) \neq proc(n_j)$*. The finish time of* e_{ij} *is*

$$t_f(e_{ij}) = \begin{cases} t_f(n_i) & if\ proc(n_i) = proc(n_j) \\ t_f(e_{ij}, L_l) & otherwise \end{cases} \tag{1}$$

Thus, the edge finish time$t_f(e_{ij})$ is now the finish time of e_{ij} on the last link of the route, L_l, unless the communication is local.

3 Related Work

There are only a few scheduling algorithms that consider arbitrary topology for processors network and contention on network links. This group of algorithms is called the APN (arbitrary processor network) scheduling algorithms [9]. Two well-known scheduling algorithms for APNs are Dynamic-Level Scheduling (DLS) algorithm and the Bubble Scheduling and Allocation (BSA) algorithm [10].

The DLS Algorithm [10] is a list scheduling heuristic that assigns the node priorities by using anattribute called *dynamic level* (DL). The dynamic level of a task $T_i$on a processor $P_j$is equal to

$$DL(T_i, P_j) = blevel^S(T_i) - EST(T_i, P_j) \tag{2}$$

This reflects how well task $T_i$and processor $P_j$are matched. The *blevel* value of a task T_i, is the length of the longest path from T_i to the exit task including all computation and communication costs on the path. The DLS algorithm uses static *blevel* value, $blevel^S$, which is computed by considering only the computation costs. At each scheduling step, the algorithm selects (ready node, available processor) pair that maximizes the value of the dynamic level. The computation costs of tasks are set with the median values. A new term, $\Delta(T_i, P_j)$, isadded to previous equation for heterogeneous processors, which is equal to the difference between the median execution time of task $T_i$and its executiontime on processor P_j.The DLS algorithm requires a message routing method that is supplied by the user; and no specific routing algorithm is presented in their paper. It should be noted that it does not consider the insertion-based approach for both scheduling tasks onto processors and scheduling messages on the links.

BSA algorithm [2] has two phases. In the first phase, the tasks are all scheduled to a single processor–effectively the parallel program is serialized. Then, each task is considered in turn for possible migration to the neighbor processors. The objective of this process is to improve the finish time of tasks because a task migrates only if it can “bubble up”. If a task is selected for migration, the communication messages from its predecessors are scheduled to the communication link between the new processor and the original processor. After all the tasks in the original processor are considered, the first phase of scheduling is completed. In the second phase, the same process is repeated on one of the neighbor processor. Thus, a task migrated from the original processor to a neighbor processor may have an opportunity to migrate again to a processor one more hop away from the original processor. This incremental scheduling by migration process is repeated for all the processors in a breadth-first fashion.

4 The Proposed Algorithm

In this paper, we consider the problem of scheduling precedence-constrained tasks to an arbitrary network of heterogeneous processors with considering link contentions. Therefore, scheduling algorithms for such systems must schedule the tasks as well as the communication traffic by treating both the processors and communication links as equally important resources. We use genetic algorithm for finding best scheduling in this environment that has minimum finish time of tasks. In the evaluation function, tasks and messages both are scheduled on processors and links, respectively. For mapping messages on links, we use classic and fuzzy routing algorithms to determine the best path between source and destination processors. In the following, we present structure of genetic algorithm and routing algorithms.

4.1 Genetic Algorithm

Encoding is an effective stage in GA. Each chromosome is encoded using decimal numbers which represents a possible schedule [12].

The fitness function of the task-scheduling problem is to determine the assignment of tasks of a given application to processors so that its schedule length is minimized. To compute schedule length, we schedule precedence-constrained tasks (DAG's nodes) and messages (DAG's edges) on processors and links, respectively. While the start time of a node is constrained by the data ready time of its incoming edges, the start time of an edge is restricted by the finish time of its origin node. The scheduling of an edge differs further from that of a node, in that an edge might be scheduled on more than one link. A communication between two nodes, which are scheduled on two different but not adjacent processors, utilizes communication route of intermediate links between the two processors. The edge representing this communication must be scheduled on each of the involved links. For determining involved links we use fuzzy routing algorithm that is explained in the next subsection. We use insertion policy for scheduling nodes and edges.

For selection step in GA, we use tournament selection. After the selection process is completed, we use the cycle crossover method [13] to promote exploration as used in [11]. For mutation, we randomly swap elements of a randomly chosen individual in the population. The initial population is generated randomly. GA will evolve the population until one or more stopping conditions are met. The best individual is selected after each generation and if it doesn't improve for 30 generations, GA stops evolving. The maximum number of generations is set at 200.

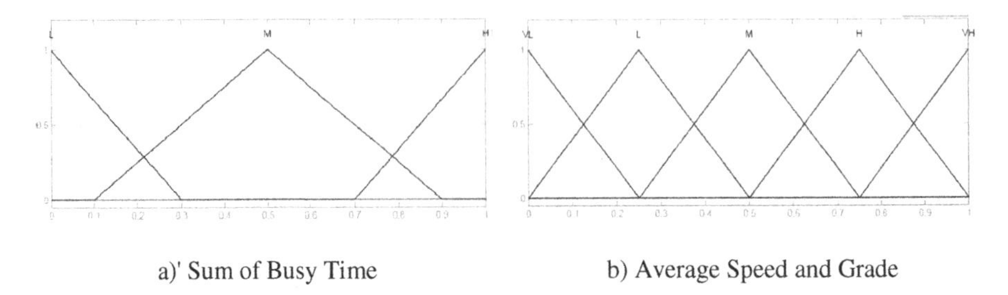

a)' Sum of Busy Time

b) Average Speed and Grade

Fig. 1. Fuzzy variables

4.2 Proposed Routing Algorithm

For transferring a message from a task to another task, scheduled on different processors, we must determine a path between P_{Src} and P_{Dst}, based on network topology graph and busy time of links for preventing contention. We propose all links are full duplex and we use store-and-forward (SAF) switching for message transfer.

In *SAF switching*, when a message traverses on a route with multiple links, each intermediate processor on the route forwards the message to the next processor after it receives and stores the entire message [10]. To find the best route for transferring message from P_{Src} to P_{Dst}, T. Yong et al. [1] examine all of paths between P_{Src} and P_{Dst} and compute edge finish time for each path and then select route that minimized edge finish time. Although, this method selects optimum route, but computing edge finish time for all possible paths between P_{Src} and P_{Dst}has more time complexity. A. Alkaya et al. [10] specify the links to be used for inter-task data communication according to the switching technique (the VCT or the SAF switching) by using the Dijkstra's shortest path algorithm (we called it as classic routing algorithm). The time complexity of this method is low, but it isn't accurate; because they have not considered busy time of links. Sometimes some paths have fastest links, but these links are too busy and transferring messages by these links waste more time on waiting queue. Therefore it is better that use another path with empty waiting queue that results in lower edge finish time. For this reason we propose a fuzzy routing algorithm. In spite of classic routing algorithms that select path only based on speed of links, in fuzzy routing we also consider busy time of intermediate links.

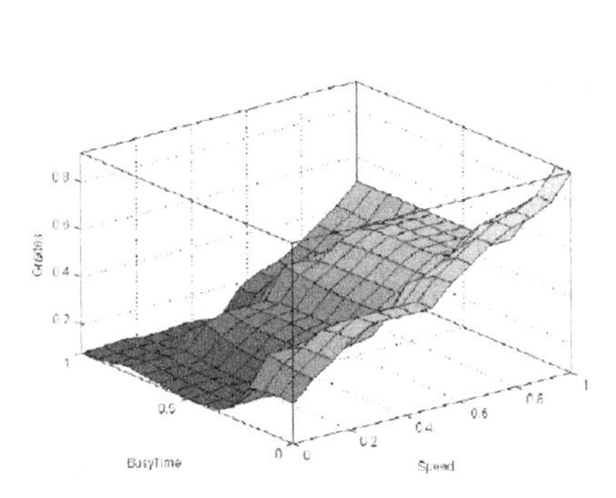

Fig. 2. Control Surface in FIS

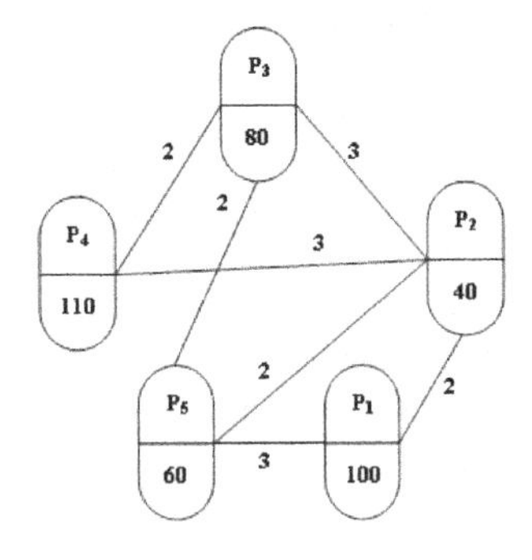

Fig. 3. Unstructured heterogeneous distributed system

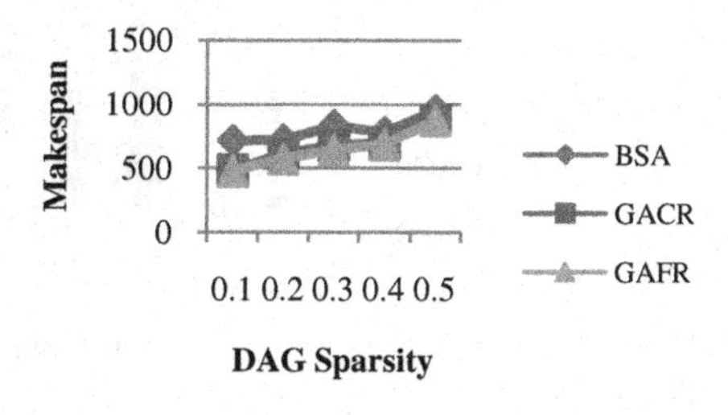

Fig. 4. Makespan

Two important factors effect on selecting a route are: speed of intermediate links and busy time of these links. Speed of each link is known and busy time of each link is equal to the needed time for transferring previous messages through that link. For selecting best path between two processors, we find all possible paths between them and then evaluate each path by a fuzzy inference engine. Fuzzy inference engine has two inputs (average speed of links and sum of busy time of these links) and one output (grade).First input is the average speed of links along path (Figure 1-a).Second input is sum of busy time of links (Figure 1-b).Output is a grade that shows the path goodness. Membership functions of output are the same as those of the second input and they were shown in (Figure 1-b). Figure 2shows control surface of proposed FIS.As shown in Figure 2, in control surface, when speed is close to 1 and busy time is close to 0, goodness is high. Path selection process by fuzzy routing is as followed:

1. Find All Path between $P_{Src}$and P_{Dst}.
2. Calculating Average Speed and Sum of Busy Time for each path.
3. Normalizing each fuzzy inference engine inputs to [0, 1].
4. Assigning grades to paths by fuzzy inference.
5. Selecting path that has most grade.

5 Experimental Results

To evaluate proposed method we generate random DAGs using the P-Method [14]. The parameter p in P-Method can be considered to be the Sparsity of the task graph. With this method, a probability parameter of $p = 1$ creates a totally sequential task graph, and $p = 0$ creates an inherently parallel one. Values of p that lie in between these two extremes generally produce task graphs that possess intermediate structures. The P-method was used to generate 5 DAGs based on parameter p ($p = 0.1, 0.2, 0.3, 0.4, 0.5$). The number of tasks in each task graph was 40. The weight of each node (computation cost of each task) and edge (communication cost between tasks) in task graph was chosen randomly based on normal distribution.

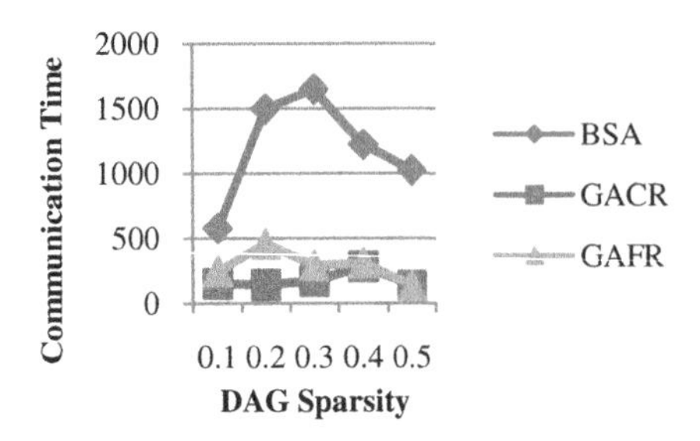

Fig. 5. Total communication time

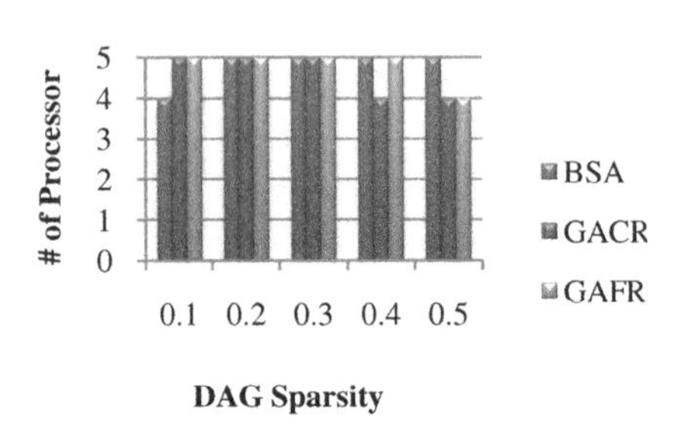

Fig. 6. Number of processor used

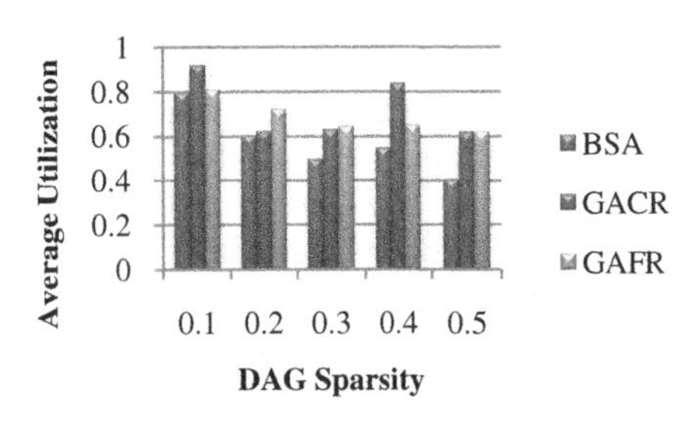

Fig. 7. Average Utilization

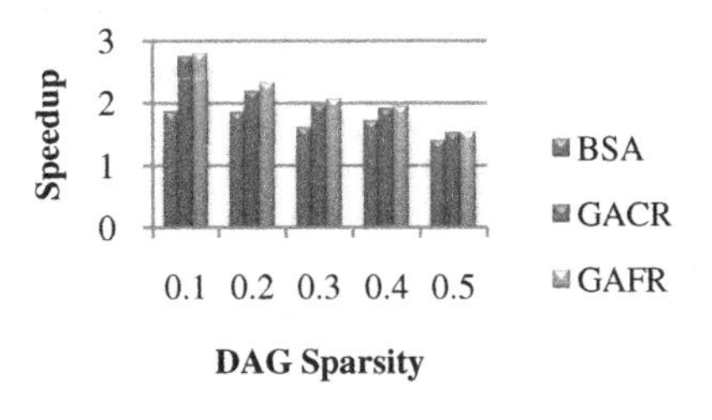

Fig. 8. Speedup

We schedule these task graphs with 5 unstructured heterogeneous distributed processors. Figure 3shows its topology. The weights associated with each processor show processing speed of processor and weights associated with each link show transferring rate of each link. For computing the execution time each task on each processor, it is enough to divide computation cost of task to processing speed of processor. Characteristics of GA are: population size: 50, crossover probability: 0.8, mutation probability: 0.3.Figure 4shows makespan vs. DAG Sparsity. Makespan is an objective that we minimize it with GA. As shown in Figure 4, GAFR and GACR are better than BSA. Also GAFR is better than GACR, however they are close together. As shown in this figure, when the Sparsity is increased, makespan of three methods be closed together. This is because the sequentiality of a task graph is high and it is less difficult to find best found schedules. Figure 5shows the total communication time in scheduling. As shown in this figure, BSA has most total communication time and wastes most time for communicating. GACR has lower communication time than GAFR; this is because in some situation that links are too busy, GAFR prefers transferring messages from another path to waiting for empty time slots in links, to achieve lower edge finish time that results in better makespan. Figure 6shows the number of processors used in scheduling. When the Sparsity is increased, we expect the number of processors used, decrease. This is because the sequentiality of a task graph is high and for reducing communication time, we need to schedule tasks on same processors. As shown in Figure 6 BSA has not good performance; in experiment with Sparsity0.1, it is better to use more processors because the parallelism degree in DAG is high and by using more processors we achieve lower makespan. Also, in experiments with higher Sparsity, BSA uses more processors whereas we need to reduce the number of processors used to decrease communication time. Figure 7 shows average utilization of processors. For dedicated distributed system, we want to increase average utilization.

As shown in Figure 7, average utilization of processors in BSA method in all experiments is lower than GACR and GAFR. The speedup is defined as the ratio of sequential time $seq(G) = \sum_{n \in V} w(n)$ (local communication has zero costs) to the makespan of the produced scheduling [4]; therefore we want to increase this ratio. As shown in Figure 8, BSA has the lowest speedup and GAFR has the highest values.

5 Conclusion

In this paper, we use genetic algorithm for solving the problem of static scheduling precedence-constrained tasks to an arbitrary network of heterogeneous processors for finding the scheduling with minimum finish time. Just like tasks, we schedule messages to suitable links during the minimization of the finish time of tasks. To find a path for transferring a message between processors, we need a routing algorithm. We propose a fuzzy routing algorithm which select path based on speed of links and busy time of intermediate links. We generate random DAGs with different Sparsity to evaluate our method and compare the results of our method with GACR and BSA methods. Experimental results show our method, GAFR, finds scheduling with lower makespan than GACR and BSA. Based on these results, we can say when communication in network is high it is useful to check busy time of links as well as speed of links in paths to find best path that minimize edge finish time. Also we expect in large networks and also with large DAGs the performance of GAFR is very high, while the performance of other method, because of their routing algorithm is low.

References

[1] Tang, X.Y., Li, K.L., Padua, D.: Communication contention in APN list scheduling algorithm. Science in China Series F: Information Sciences 52, 59–69 (2009)

[2] Kwok, Y., Ahmad, I.: Link Contention-Constrained Scheduling and Mapping of Tasks and Messages to a Network of Heterogeneous Processors. In: Cluster Computing, pp. 113–124 (2000)

[3] Sinnen, O.: Task scheduling for parallel systems. JohnWiley & Sons-Interscience (2007)

[4] Sinnen, O., Sousa, L.A., Sandnes, F.E.: Toward a realistic task scheduling model. IEEE Trans. Parallel and Distributed Systems 17, 263–275 (2006)

[5] Cheng, S.-C., Shiau, D.-F., Huang, Y.-M., Lin, Y.-T.: Dynamic hard-real-time scheduling using genetic algorithm for multiprocessor task with resource and timing constraints. Expert Systems with Applications 36, 852–860 (2009)

[6] Yoo, M.: Real-time task scheduling by multiobjective genetic algorithm. Systems & Software 82, 619–628 (2009)

[7] Shin, K., Cha, M., Jang, M., Jung, J., Yoon, W., Choi, S.: Task scheduling algorithm using minimized duplications in homogeneous systems. Parallel and Distributed Computing 68, 1146–1156 (2008)
[8] Yoo, M., Gen, M.: Scheduling algorithm for real-time tasks using multiobjective hybrid genetic algorithm in heterogeneous multiprocessors system. The Journal of Computers & Operations Research 34, 3084–3098 (2007)
[9] Kwok, Y.K., Ahmad, I.: Benchmarking and comparison of the task graph scheduling algorithms. Parallel and Distributed Computing 59, 381–422 (1999)
[10] Alkaya, A.F., Topcuoglu, H.R.: A task scheduling algorithm for arbitrarily-connected processors with awareness of link contention. Cluster Computing 9, 417–431 (2006)
[11] Zomaya, A.Y., Teh, Y.-H.: Observations on Using Genetic Algorithms for Dynamic Load-Balancing. IEEE Trans. Parallel and Distributed Systems 12, 899–911 (2001)
[12] Page, A.J., Naughton, T.J.: Dynamic task scheduling using genetic algorithms for heterogeneous distributed computing. In: Proceedings of the 19th IEEE International Parallel and Distributed Processing Symposium, IPDPS 2005 (2005)
[13] Oliver, I.M., Smith, D.J., Holland, J.: A study of permutation crossover operators on the traveling salesman problem. In: Second International Conference on Genetic Algorithms on Genetic Algorithms and Their Application, pp. 224–230. Lawrence Erlbaum Associates, Inc. (1987)
[14] Al-Sharaeh, S., Wells, B.E.: A Comparison of Heuristics for List Schedules using The Box-method and P-method for Random Digraph Generation. In: Proceedings of the 28th Southeastern Symposium on System Theory, pp. 467–471 (1996)

Part III
Robotics

Tsallis Statistical Distribution in a Completely Open System with OLM Formalism

Heling Li, Bin Yang, and Yan Ma

School of Physics and Electrical Information Science, Ningxia University,
Yinchuan 750021, China
ningxiayclhl@163.com, yang_b@nxu.edu.cn

Abstract. Under the third kind of constraint, using the method of OLM (Optimal Lagrange Multipliers formalism) and maximum entropy principle, we derive the statistical distribution and thermodynamic formulas in completely open system based on Tsallis entropy. Two hardships existed in the method of TMP (Tsallis-Mendes-Plastino formalism) are overcome. We give out the specific expressions of the three Lagrangian multipliers and make their physical significance clear.

Keywords: Tsallis statistics, Statistical distribution, Completely open system.

1 Introduction

Before our latest works[1], both B-G statistical mechanics[2,3,4] and non-extensive statistical mechanics[5-10] take micro canonical, canonical and grand canonical systems as discussion objects, lacking completely open systems because there was no statistical distribution function of a completely open system that can be applied in any macroscopic system. Completely open systems are the most general ones in the natural while others are just their special cases. Though T. L. Hill [11, 12] and others had deduced the statistical distribution of completely open system under the framework of extensive statistical mechanics with the method of ensemble transformation, it was restricted by the "phase rule" [4] in extensive statistical mechanics and was thought to be false. While they proved it can be used in small macroscopic systems (whose N, the particle number of the system, is in the range 10^2—10^3.) [13] and they also successfully discussed the small systems [14] with it. The thermodynamic quantities calculated from this distribution are the same as from canonical and grand canonical distributions, but the relative fluctuations are proportional to 1, an entirely new result, not proportional to $1/\sqrt{N}$ like canonical and grand canonical distributions. This result could account for the large fluctuations observed in critical phenomena, supercooled states and overheated states that can't be explained by the traditional extensive statistical mechanics. A reasonable interpretation of evolution in nature and human society also requires large fluctuations. According to the theory of "dissipative structure", a new order comes into existence through fluctuations that can only be produced in the nonlinear region far from the old order's equilibrium state [15].

In this paper we demonstrate that when a system has long-range interactions, traditional thermodynamic quantities are no longer purely extensive or intensive. The

Y. Wang and T. Li (Eds.): Knowledge Engineering and Management, AISC 123, pp. 147–152.
springerlink.com

"phase rule" of extensive statistical mechanics does not hold. Under the framework of non-extensive statistical mechanics, selecting the third kind of constraint [6] for energy and others, using the method of OLM [16] and basing on the non-extensive Tsallis entropy, we deduce the statistical distribution and thermodynamic formulas in completely open system from the maximum entropy principle, which can be used in any system, not limited to small systems.

2 Statistical Distribution of Completely Open System Based on Tsallis Entropy

2.1 Non-extensive Systems Are Not Restricted by the "Phase Rule" of Extensive Statistical Mechanics

Non-extensive Tsallis entropy can be expressed as [5]:

$$S_q = k\left(1-\sum_{i=1}^{W} p_i^q\right)/(q-1) \tag{1}$$

Where q is non-extensive parameter, k is positive constant (Tsallis usually select k=1 for convenience. Contrasting it with Shannon entropy, if k has no connection with q, Tsallis entropy would tend to Shannon entropy when $q\to 1$, so k should be Boltzmann constant k_B), p_i is the probability of the system at state i, and W is the number of microstates of the system that may be happen. The non-extensive property of Tsallis entropy performs as: when the two subsystems A and B are independent with each other, and the probabilities of A at state i and B at state j are respectively p_i^A and p_j^B, then the probability of the total system is $p_{ij}^{A\cup B} = p_i^A p_j^B$. Inserting it into Eq.(1), we obtain the total entropy of the system:

$$S_q(A\cup B) = S_q(A) + S_q(B) + (1-q)S_q(A)S_q(B)/k. \tag{2}$$

The third term on the right side of Eq.(2) is the non-extensive term. Selecting the second kind of energy constraint [5] $\sum_{i=1}^{W} p_i^q E_i = U_q$ (where E_i is the energy of the system at energy level i, and U_q is the internal energy), we can get the total internal energy from canonical distribution [1] of Tsallis statistics:

$$U_q(A\cup B) = U_q(A)Z^{1-q}(B) + U_q(B)Z^{1-q}(A) + (1-q)U_q(A)U_q(B)/(kT). \tag{3}$$

Where Z is generalized partition function. It's clear that U_q is also non-extensive. When the non-extensive parameter q tends to 1, the non-extensive terms in Eqs. (2) and (3) would disappear automatically and the two equations would restore to the conclusion of extensive statistical mechanics.

Noticing Eq.(3) is the result under the condition $p_{ij}^{A\cup B} = p_i^A p_j^B$. Although the meaningful generalized particle distribution functions were obtained under the second kind of energy constraint, the energy of the total system is not equal to the summation of the energy of the two subsystems that have no interactions, which goes against the law of conservation of energy. There are other deficiencies under the second energy constraint (for example: the statistical distribution changes when the energy level translates). For this, Tsallis and others put forward the third kind of constraint:

$$\sum_{i=1}^{W} p_i^q O_i / \sum_{j}^{W} p_j^q = O_q . \tag{4}$$

Where O_i is the value of a physical quantity O as the system at state i and the corresponding probability is p_i. When $p_{ij}^{A \cup B} = p_i^A p_j^B$, we have:

$$U_q(A \cup B) = U_q(A) + U_q(B). \tag{5}$$

From the well known $\partial S_q / \partial U_q = k\beta = 1/T$ and the obvious discrepancy between Eqs.(2) and (3)[or (5)], we can easily know: the temperature T is not necessarily intensive (this depends on concrete systems). The chemical potential μ and pressure P are also not necessarily intensive. Then the phase rule in extensive statistical mechanics does not hold any more. Any combination of three quantities among entropy S_q, volume V, internal energy U_q, chemical potential μ, temperature T, and pressure P can describe a system completely. That is to say, any three can be the independent variables to describe the system.

Actually, nonextensivity is the natural quality of physical system and only related to the extent and the interactions of the physical system. To select any kind of constraint or entropy function is just to make us study the property of the physical system more tersely and expediently. A system must be non-extensive and has the properties above if only it has long-range interactions.

2.2 Statistical Distribution of Completely Open System Based on Tsallis Entropy

Considering a completely system connected with an enormous source, each microscopic state of the system may have different particle number N_i, energy E_i and volume V_i [This volume V_i is a microscopic quantity, not a pure mechanical quantity like coordinate, momentum, energy and so on, but it has common ground with the energy and particle number. When it is alterable, the system may have more than one corresponding microscopic state for a determined value. So the different microscopic states of a completely open system may have different particle number N_i, energy E_i and volume V_i.]. After the system and the source reach equilibrium, the corresponding average value N_q, E_q=U_q and V_q will be definited. Denote p_i as the probability of the system at state i, then we have the following constraints [5]:

$$\sum_{i}^{W} p_i^q N_i = N_q \sum_{j}^{W} p_j^q , \ \sum_{i=1}^{W} p_i^q E_i = U_q \sum_{j}^{W} p_j^q , \ \sum_{i=1}^{W} p_i^q V_i = V_q \sum_{j}^{W} p_j^q , \ \sum_{i=1}^{W} p_i = 1 \tag{6}$$

Calculating the extreme value of the Tsallis entropy, we can obtain the statistical distribution and generalized partition function [1]:

$$p_i = \exp_q \left\{ -\left[\alpha(N_i - N_q) + \beta(E_i - U_q) + \kappa(V_i - V_q) \right] / \sum_{j}^{W} (p_j)^q \right\} / \overline{Z}_q \tag{7}$$

$$\overline{Z}_q = \sum_{i}^{W} \exp_q \left\{ -\left[\alpha(N_i - N_q) + \beta(E_i - U_q) + \kappa(V_i - V_q) \right] / \sum_{j}^{W} (p_j)^q \right\} \tag{8}$$

And we can prove [1]:

$$S_q = k \ln_q \overline{Z}_q = k\left(\ln_q Z_q + \alpha N_q + \beta U_q + \kappa V_q\right). \tag{9}$$

Where $\ln_q Z_q = \ln_q \overline{Z}_q - \alpha N_q - \beta U_q - \kappa V_q$, then

$$\alpha = \frac{\partial (S_q / k)}{\partial N_q},\ \beta = \frac{\partial (S_q / k)}{\partial U_q},\ \kappa = \frac{\partial (S_q / k)}{\partial V_q}. \tag{10}$$

The functions $\ln_q x$ and e_q^x ($\ln_q x$ and e_q^x are inverse functions of each other) in the above equations can be defined as:

$e_q^x \equiv [1+(1-q)x]^{1/(1-q)}$; $\ln_q x \equiv \left(x^{1-q}-1\right)/(1-q)$.

We have $\ln_q x \to \ln x$, $e_q^x \to e^x$ when $q \to 1$.

Contrasting Eq.(10) with the fundamental thermodynamic equation $TdS = dU + PdV - \mu dN$, we obtain the constants:

$$\alpha = -\mu / kT,\quad \beta = 1/kT,\ \text{and}\ \kappa = P/kT. \tag{11}$$

Where μ is chemical potential, T is absolute temperature and P is pressure. S. Martinez etc. propounded that the process of derivation with the constraints in Eq. (6) exists two hardships [16]:

- The p_i and Z_q expressions are explicitly self-referential.
- The Hessian of L is not diagonal.

The second hardship is more serious, namely, a maximum is not necessarily guaranteed. For this, Martinez etc. advised [16] that the constraints in Eq. (6) should deform to:

$$\sum_i^W p_i^q N_i = N_q \sum_j^W p_j^q,\ \sum_{i=1}^W p_i^q E_i = U_q \sum_j^W p_j^q,\ \sum_{i=1}^W p_i^q V_i = V_q \sum_j^W p_j^q,\ \sum_{i=1}^W p_i = 1. \tag{12}$$

Considering the constraints in Eq. (12) and Tsallis entropy in Eq. (1), we introduce the Lagrangian function L:

$$L = \frac{S_q}{k} - \alpha' \sum_i^W p_i^q (N_i - N_q) - \beta' \sum_i^W p_i^q (E_i - U_q) - \kappa' \sum_i^W p_i^q (V_i - V_q) - \gamma\left(\sum_i^W p_i - 1\right). \tag{13}$$

Where α', β', κ' and γ are all Lagrangian multipliers. We calculate the conditional extreme value and obtain:

$$p_i = \exp_q\left[-\alpha'(N_i - N_q) - \beta'(E_i - U_q) - \kappa'(V_i - V_q)\right] / \overline{Z'}_q. \tag{14}$$

$$\text{Where}\quad \overline{Z'}_q = \sum_i^W \exp_q\left[-\alpha'(N_i - N_q) - \beta'(E_i - U_q) - \kappa'(V_i - V_q)\right]. \tag{15}$$

$$S_q = k \ln_q \overline{Z'}_q. \tag{16}$$

We make Legendre transformation to Eq.(16) as following:

$$\ln_q Z'_q(\alpha, \beta, \kappa) = \ln_q \overline{Z'}_q(N_q, E_q, V_q) - \alpha' N_q - \beta' U_q - \kappa' V_q \tag{17}$$

$$\text{Then}\quad S_q = k\left(\ln_q Z'_q + \alpha' N_q + \beta' U_q + \kappa' V_q\right). \tag{18}$$

$$N_q = -\partial \ln_q Z_q' / \partial \alpha', \quad U_q = -\partial \ln_q Z_q' / \partial \beta', \quad V_q = -\partial \ln_q Z_q' / \partial \kappa' \tag{19}$$

$$\alpha' = \partial (S_q / k) / \partial N_q, \quad \beta' = \partial (S_q / k) / \partial U_q, \quad \kappa' = \partial (S_q / k) / \partial V_q \tag{20}$$

Martinez etc. proved [16]: if the constraints in Eq.(6) are deformed to Eq. (13), the two hardships would be eliminated without losing the beautiful properties of the Tsallis-Mendes-Plastino formalism (TMP), that is :

- Self-reference is avoided.
- The Hessian of L becomes diagonal.

In the derivation above, Martinez etc. used Optimal Lagrange Multipliers (OLM) formalism.

Comparing Eq. (7) with Eq.(13) under considering Eq. (16), we obtain:

$$\alpha' = \alpha / \overline{Z}_q^{1-q}, \quad \beta' = \beta / \overline{Z}_q^{1-q}, \quad \kappa' = \kappa / \overline{Z}_q^{1-q} \quad . \tag{21}$$

While comparing Eq. (10) with Eq. (20), we obtain:

$$\alpha' = \alpha, \quad \beta' = \beta, \quad \kappa' = \kappa \quad . \tag{22}$$

The obvious discrepancy between Eq. (21) and Eq. (22) shows a new hardship, Martinez and others seemed to realize it and selected $k=k(q)$ in the expression of Tsallis entropy in reference [16]. However in the method of OLM they used, the indefinite or incorrect equations like $\partial(S_q/k)/\partial U_q = \overline{Z}_q^{1-q} \beta' = \beta$ [16] still existed. Obviously, Martinez etc. were unaware of the reason and the solution of the problem at that moment. In their later paper [17], they selected $k(q) = k_T = k_B \overline{Z}_q^{q-1}$ to solve the problem. Here k_B is Boltzmann constant.

Our opinion:

- The fundamental thermodynamic equation is an expressive form of energy conservation, which should be held all the time. Namely, both $\partial(S_q/k)/\partial U_q = \beta$ and $\partial(S_q/k)/\partial U_q = \beta'$ should be held.
- The expression of Tsallis entropy should be changed at no time. Namely, the constant k in Tsallis entropy should not be changed for choosing different Lagrangian multipliers. Therefore, it is unreasonable to select the constant of Tsallis entropy $k = k_B \overline{Z}_q^{q-1}$ in the method of TMP and $k = k_B$ in the method of OLM to solve the contradiction between Eq. (21) and Eq. (22). Since the two hardships existed in the method of TMP can be overcome when we use the method of OLM without changing the physical essence of the TMP, so we should select $k = k_B$, which is simple and reasonable and without deviating from the original intention of Tsallis entropy.

It is more reasonable to select the conclusion of OLM [the probability distribution in Eq. (14), the partition function in Eq. (15) and the three constants $\alpha' = -\mu / k_B T$, $\beta' = 1/k_B T$, and $\kappa' = P/k_B T$]. This is also demonstrated in the discussion of classical ideal gas in the reference [17].

3 Conclusions

• We introduce the idea of completely open system into Tsallis statistics based on Tsallis entropy in Eq. (1) and indicate that it is more reasonable to select $k = k_B$ than $k = k_B \overline{Z}_q^{q-1}$ in Tsallis entropy.

• We derive the statistical distribution Eq. (14) and generalized partition function Eqs.(15) and (17) in a completely open system with the method of OLM under the third kind of constraint in Eq. (4), they are not limited to the small systems. The method eliminate the two hardships existed in the method of TMP [1,16].

• We obtain the thermodynamic formulas in Eqs. (18) and (19), and make the physical significance of the three Lagrangian multipliers clear: $\alpha' = -\mu / k_B T$, $\beta' = 1/k_B T$, and $\kappa' = P / k_B T$.

Acknowledgments. The financial support of the natural science foundation of Ningxia (NZ1023) is gratefully acknowledged.

References

1. Li, H., Ying, X., Li, Y.: Statistical distribution in a completely open system. Physica A 390, 2769–277 (2011), doi:10.1016/j.physa.2011.03.023
2. Huang, K.: Statistical Mechanics. John Wiley & Sons, New York (1963)
3. Pathria, R.K.: Statistical Mechanics. Pergamon Press, London (1977)
4. Reichl, L.C.: A Modern Course in Statistical Physics. University of Texas Press, Texas (1980)
5. Tsallis, C.: Possible generalization of Boltzmann-Gibbs statistics. J. Stat. Phys. 52(1-2), 479–487 (1988)
6. Tsallis, C., Mendes, R.S., Plastino, A.R.: The role of constraints within generalized non-extensive statistics. Physica A 261(3-4), 534–554 (1998)
7. Büyükkilic, F., Demirhan, D., Gülec, A.: A statistical mechanical approach to generalized statistics of quantum and classical gases. Phys. Lett. A 197, 209–220 (1995)
8. Hamity, V.H., Barraco, D.E.: Generalized Non-extensive Thermodynamics Applied to the Cosmic Backgroung Radiation in a Robertson-Walker universe. Phys. Rev. Lett. 75, 4664–4666 (1995)
9. Wang, Q.A.: Incomplete statistics: non-extensive generalizations of statistical mechanics. Chaos, Solitons &Fractals 12, 1431–1437 (2001)
10. Lima, J.A.S., Bezerra, J.R., Sliva, R.: On certain incomplete statistics. Chaos, Solitons & Fractals 19, 1095–1097 (2004)
11. Hill, T.L.: Statistical Mechanics. McGraw Hill, New York (1956)
12. Münster, A.: Statistical Thermodynamics. Springer, New York (1969)
13. Hill, T.L.: Thermodynamics of Small Systems. Dover, New York (1994)
14. Hill, T.L., Chamberlin, R.V.: Fluctuations in Energy in Completely Open Small Systems. Nano Lett. 2(6), 609–613 (2002)
15. Nicolis, G., Prigoging, I.: Self-organization in Non-equilibrium Systems. John Wiley & Sons, New York (1977)
16. Martínez, S., Nicolás, F., Pennin, F., Plastino, A.: Tsallis' entropy maximization procedure revisit. Physica A 286, 489–502 (2000)
17. Abe, S., Martínez, S., Pennini, F., Plastino, A.: Ideal gas in non-extensive optimal Lagrange multipliers formalism. Phys. Lett. A 278, 249–254 (2001)

Symbolic Controller with PID Feedback for Locally Linearized System

Bin Xu, Tianyi Wang, and Haibin Duan

Science and Technology on Aircraft Control Laboratory, School of Automation Science and Electrical Engineering, Beihang University, Beijing 100191, P. R. China
helloxubin@gmail.com, skytw.01@163.com, hbduan@buaa.edu.cn

Abstract. In this paper, we proposed a novel symbolic controller with PID (proportional integral derivative) feedback for locally linearized system. The state-space of this system is analyzed in Brunovsky coordinates, and then a practical control structure which combines symbolic controller and PID controller is presented. Series of Experiments on an inverted pendulum are conducted, and the results show the feasibility and efficiency of our proposed structure.

Keywords: symbolic controller, PID controller, inverted pendulum, locally linearized system.

1 Introduction

For a wide class of systems, symbolic control was proved to be an effective method in dealing with linear system with feedback encoding. It can reduce the cost of communication and storage resources [1], and solve the kinematic and dynamic constraints simultaneously [2].

Symbolic control is based on the exact discrete-time linear models of control systems. On the contrary to the linear system, it is not possible to obtain the exact models of nonlinear system. Tabuada proposed methodology worked for nonlinear control systems based on the notion of incremental input-to-state stability [3]. However, as to the locally linearized system, the non-linear effects can be treated as a kind of turbulence and be damped by continuous PID feedback control, as the rest of this paper shows.

2 Symbolic Control

Symbolic control is inherently related to the definition of elementary control events, whose combination allows the specification of complex control actions [1]. Letters from the alphabet $\Sigma = \{\sigma 1, \sigma 2, \ldots\}$ can be utilized to build words of arbitrary length.

A. Bicchi, A. Marigo, and B. Piccoli, introduce a feedback encoding as the Fig.1 shows. In this encoder, an inner continuous (possibly dynamic) feedback loop and an outer discrete-time loop-both embedded on the remote system-are used to achieve richer encoding of transmitted symbols[2].

Y. Wang and T. Li (Eds.): Knowledge Engineering and Management, AISC 123, pp. 153–158.
springerlink.com

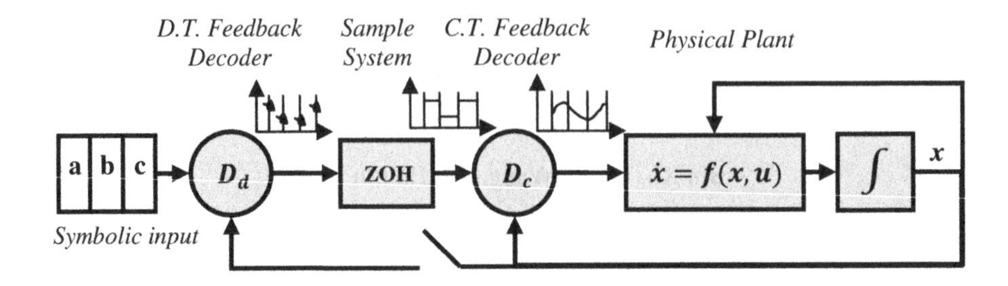

Fig. 1. Nested discrete-time continuous-time feedback encoding

By using nested feedback encoding, all feedback linearized systems are hence additively approachable. The argument can be also directly generalized to multi-input systems by transforming to the Brunovsky form [2].

3 Model of Inverted Pendulum

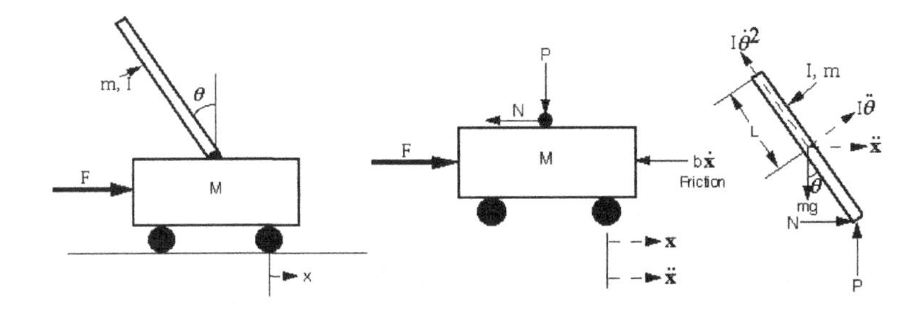

Fig. 2. Inverted pendulum

Consider a nonlinear math model of an inverted pendulum [5]:

$$\frac{d^2x}{dt^2} = \frac{1}{M}\sum F_x = \frac{1}{M}\left(F - N - b\frac{dx}{dt}\right) \quad (1)$$

$$\frac{d^2\theta}{d\theta^2} = \frac{1}{I}\sum \tau = \frac{1}{I}[NLcos(\theta) + PLsin(\theta)] \quad (2)$$

(mass of the cart)	0.5 *kg*
m(mass of the pendulum)	0.2 *kg*
b(friction of the cart)	0.1 *N/m/sec*
l(length to pendulum center of mass)	0.3 *m*
I(inertia of the pendulum)	0.006 *kg*m^2*
F	force applied to the cart
x	cart position coordinate
theta	pendulum vertical angle

By sampling and linearizing the model around its working states, the plant model can be transformed into the state-space form:

$$\begin{bmatrix} x^{(1)} \\ \theta^{(1)} \\ x^{(2)} \\ \theta^{(2)} \end{bmatrix} = \begin{bmatrix} 0 & 0 & 1 & 0 \\ 0 & 0 & 0 & 1 \\ 0 & 2.67 & -0.18 & 0 \\ 0 & 31.18 & -0.45 & 0 \end{bmatrix} \begin{bmatrix} x \\ \theta \\ \dot{x} \\ \dot{\theta} \end{bmatrix} + \begin{bmatrix} 0 \\ 0 \\ 1.82 \\ 4.55 \end{bmatrix} u \tag{3}$$

The x position is the controlled variable and u denotes the F.

4 Symbolic Controller with PID Feedback

4.1 State Space Analyze

If a controller was designed under the control laws mentioned above and applied to the linear model of the cart Eq. (3), the response is identical to the results in Ref. [1] (see Fig. 3(a)).

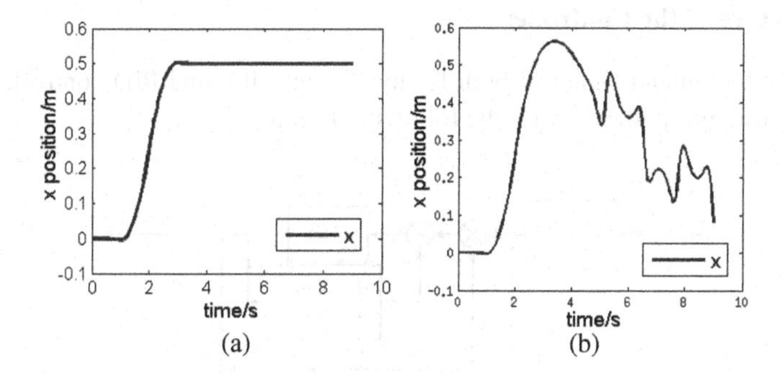

Fig. 3. Responds of the inverted pendulum using symbolic controller on: a)linear model b)nonlinear model

However, if this controller applied to the nonlinear model of the system, the responds of the system is unstable (see Fig. 3(b)), which indicates that the problem should be caused by the nonlinear factor of the plant.

If the states are observed in the continuous time, the result is shown as Fig.4 (a). This shows that in the span of two sampling time, there exists a large range of erratic states in the system states. Since the model is strictly linear, the system arrives at its states according to the sampling time precisely and then updates the new feedback control values with the states according to sampling time. However, the nonlinear factor of the plant caused minor differences against the ideal linear system. Such differences can become larger due to the symbolic feedback controller cannot damp such turbulence, and then the system finally becomes unstable.

A solution to this problem is adding a continuous feedback (PID in this paper) to the system to enhance its robust quality, so the error could be reduced gradually, preventing the feedback of symbolic control being affected.

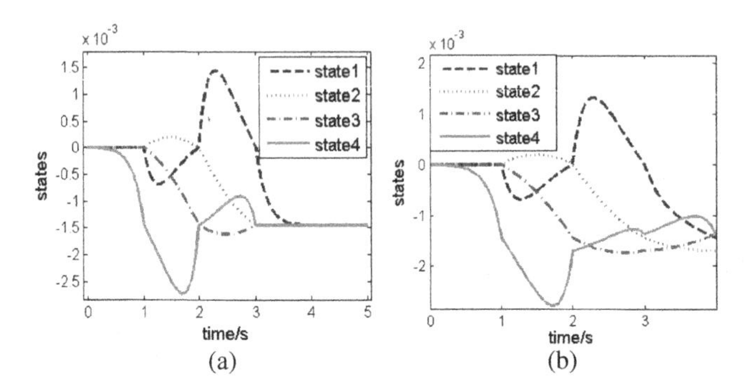

Fig. 4. Brunovskycoordinates states in continuous time of: a) linear model. b)nonlinear model.

4.2 Structure of the Controller

After adding the amend signal of both feedback controller and PID controller, the final system structure is shown with the following Fig. 5.

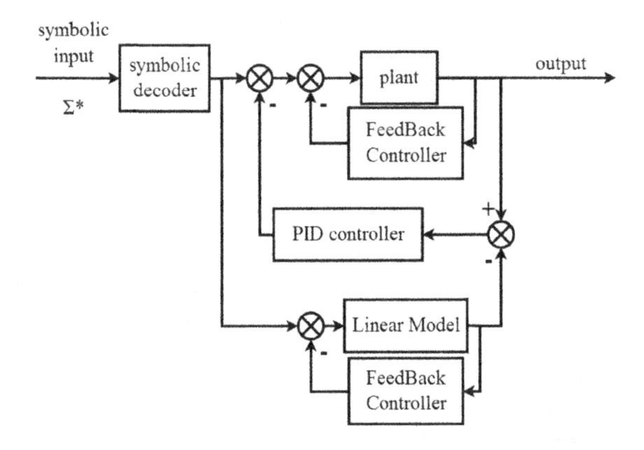

Fig. 5. Structure with symbolic control and PID controller

In this structure, the linear model of the plant is used to generate the output which the symbolic controller is designed for and provide a trace for the output of the plant. Then the difference of the plant and the linear model's outputs is damped by a PID controller to force their behavior to be identical.

5 Experimental Results

In this section, we will illustrate the power of the PID-symbolic hybrid controller by solving the problem of controlling the nonlinear inverted pendulum.

First of all, by observing the linear model of the pendulum (Eq. (3)) and using the methods in section 2, the equilibrium manifold of the system is gained as $\xi = \{x \in \mathbb{R}^4 | x = (\alpha, 0,0,0)\}$ then a symbolic feedback matrix K and the matrices S, V can be calculated according to the linear state-space equation [1]. In here, $K = [-0.000 \;\; 7.0840 \;\; -0.11161 \;\; 1.2846]^T$. The equilibrium manifold can be written into the Brunovsky coordinates form: $\xi = \{\beta \cdot 1_4\}$, and we assume that the desire precision is 0.01 and the corresponding scale factor$\gamma = 0.0029m$.

Then we specify a PID controller for the system:$K_P = 100, T_I = 1, T_D = 20$ and set the input of the PID controller as the difference of *theta*.

Finally, a symbolic control value$x_t$is given to 0.5, then the symbolic control value is $v_i = x_t/\gamma = 0.5/0.0029 = 1724 \approx 172$(*i*=1,2,3,4) and the motion of the cart and the pendulum are shown in Fig. 6 (b).

Compared with the result in Fig. 3(b), the system is stabilized via the PID-symbolic controller (see Fig. 6(a)). Also, under the Brunovsky coordinates (see Fig.6(c)), the figure is similar to the idea linear system states (see Fig. 4(a)) .The static error between two states exits because the input of PID controller is the difference of *theta*, adding an extra PID controller will reduce this error.

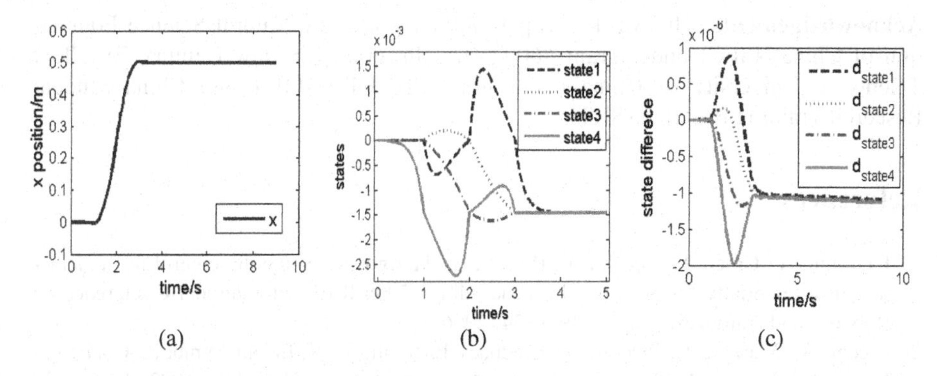

Fig. 6. a) Corresponding Brunovsky coordinates states in continuous time using symbolic-PID controller. b) Corresponding Brunovsky coordinates states in continuous time using symbolic controller. c) Differences between the linear model system states and the nonlinear model states.

Moreover, a larger control value is x_t given to the system to test its ability. A final displacement x_t is set as *15m* and the corresponding symbolic control values are given to the system. The responds of the *x* and *theta* show in the Fig. 7(a) and Fig. 7(b).

According to the figures above, the pendulum arrives at the final position accurately without overflows in a short time. Moreover, the system withstands a 50 degrees oscillation of *theta* which contains a significant part of nonlinear factors and keeps stable. This result shows that the control structure overcomes the nonlinear factors of the system effectively.

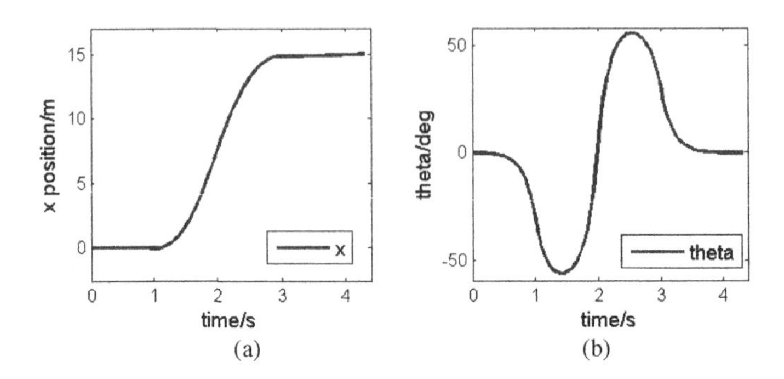

Fig. 7. a) *x* responds for a larger x_t. b): *theta* responds for a larger x_t

6 Conclusion

In this paper, we proposed a structure which combines the continuous PID and symbolic control to control the nonlinear inverted pendulum and obtained stable control results and successfully eliminated the nonlinear factors of the system. However, evaluating the range of applications of this structure on locally linearized systems should be further studied.

Acknowledgements. This work was partially supported by Natural Science Foundation of China(NSFC) under grant #60975072, Program for New Century Excellent Talents in University of China under grant #NCET-10-0021 and China Student Research Training Program (SRTP).

References

1. Fagiolini, A., Greco, L., Bicchi, A., Piccoli, B., Marigo, A.: Symbolic control for underactuated differentially flat systems. In: Proceedings 2006 IEEE International Conference on Robotics and Automation, pp. 1649–1654 (2006)
2. Bicchi, A., Marigo, A., Piccoli, B.: Feedback Encoding for Efficient Symbolic Control of Dynamical Systems. IEEE Transactions on Automatic Control 51(6), 987–1002 (2006)
3. Tabuada, P.: An Approximate Simulation Approach to Symbolic Control. IEEE Transactions on Automatic Control 53(6), 1406–1418 (2008)
4. Tabuada, P.: Symbolic Control of Linear Systems Based on Symbolic Subsystems. IEEE Transactions on Automatic Control 51(6), 1003–1013 (2006)
5. Control Tutorials for MATLAB and Simulink, `http://www.engin.umich.edu/class/ctms/simulink/examples/pend/pendsim.htm`

Multistage Decision Making Based on One-Shot Decision Theory

Peijun Guo and Yonggang Li

Faculty of Business Administration
Yokohama National University, Yokohama, 240-8501 Japan
guo@ynu.ac.jp

Abstract. In this paper, a multistage decision problem for minimizing the total cost with partially known information is considered. In each stage, a decision maker has one and only one chance to make a decision. The optimal decision in each stage is obtained based on the one-shot decision theory. That is, the decision maker chooses one of states of nature (scenario) of each alternative in every stage with considering the satisfaction of the outcome and its possibility. The selected state of nature is called the focus point. Based on the focus points, the decision maker determines the optimal alternative in each stage by dynamic programming problems.

Keywords: One-shot decision theory, focus point, multistage decision problem, dynamic programming problem.

1 Introduction

Decision theories under uncertainty are theories of choice under uncertainty where the objects of choice are probability distributions (for Expected Utility Theory, Subjective Expected Utility and their varieties), or prospects framed in terms of gains and losses (for Prospect Theory [13]), or possibility distributions (regarded as possibilistic lotteries [6]). In fact, for one-shot decision problems, there is one and only one chance for only one state of nature occurring. In the paper [11], the one-shot decision theory is proposed for dealing with such one-shot decision problems. The one-shot decision process is separated into two steps. The first step is to identify which state of nature should be taken into account for each alternative with considering possibility of a state of nature and satisfaction of an outcome. The states of nature focused are called focus points [8]. The second step is to evaluate the alternatives to obtain the optimal alternative. In the paper [9], a duopoly market of a new product with a short life cycle is analyzed within one-shot decision framework. In the paper [10], private real estate investment problem is analyzed within the same framework.

In this research, we apply the one-shot decision theory to multistage decision problems in which for each stage only one chance exists for a decision maker to make a decision. The decision maker chooses one of state of nature (scenario) of each alternative (focus point) in each stage with considering the satisfaction of the outcome and

Y. Wang and T. Li (Eds.): Knowledge Engineering and Management, AISC 123, pp. 159–164.
springerlink.com

its possibility. Based on the focus points, the decision maker determines the optimal alternative in each stage.

It is well known that dynamic programming is a powerful vehicle for deal with multistage decision-making problem [2]. The stochastic dynamic programming problems and fuzzy dynamic programming problems are used for handing the uncertainty existing in the multistage decision problems [4-5, 7, 12, 14-18]. The stochastic dynamic programming problems are formulated as a maximization problem of expected value with the assumption that the system under control is a Markov chain while fuzzy dynamic programming problems introduce a fuzzy constraint and a fuzzy goal in each stage following Bellman and Zadeh's framework [3]. Fuzzy dynamic programming problems are also well used for multiple objective dynamic programming problems [1]. Different from the existing approaches, the proposed approaches consider a possibilistic dynamic system under control and in each stage a decision maker makes a decision according to the chosen scenario (focus point). The payoff in each stage is only associated with the chosen scenarios in the succeeding stages. The proposed method can provide a clear optimal decision path for the multistage decision procedure with possibilistic information.

2 Multistage One-Shot Decision

Let us consider a multistage decision problem. The length of decision period is T. $A_t = \{a_{t,1}, a_{t,2}, \cdots, a_{t,m_t}\}$ is the set of the alternatives available in the stage t, the cardinality of A_t is m_t. $X_{t+1}(x_{t,k}, a_{t,i})$ is the set of a state in the next stage $t+1$ when the state is $x_{t,k}$ and the alternative $a_{t,i}$ is selected in the stage t. $X_{t+1}(x_{t,k}, a_{t,i})$ is as follows:

$$X_{t+1}(x_{t,k}, a_{t,i}) = \{x_{t+1,1}(x_{t,k}, a_{t,i}), x_{t+1,2}(x_{t,k}, a_{t,i}), \cdots, x_{t+1,k_{t+1}(x_{t,k},a_{t,i})}(x_{t,k}, a_{t,i})\}. \tag{1}$$

Each $x_{t+1,j}(x_{t,k}, a_{t,i}), 1 \le j \le k_{t+1}(x_{t,k}, a_{t,i})$ is one of states in the stage $t+1$ resulted by the alternative $a_{t,i}$ when the state is $x_{t,k}$ in the stage t. In other words when we choose $a_{t,i}$ at time t, the next possible state must be an element of $X_{t+1}(x_{t,k}, a_{t,i})$. $X_{t+1}(x_{t,k}, a_{t,i})$ is governed by a possibility distribution defined below.

Definition 1. A possibility distribution is a function $\pi_{t+1} : X_{t+1}(x_{t,k}, a_{t,i}) \to [0,1]$ satisfying $\max_{x \in X_{t+1}(x_{t,k},a_{t,i})} \pi_{t+1}(x) = 1$. $\pi_{t+1}(x)$ is the possibility degree of x in the stage $t+1$. $\pi_{t+1}(x) = 1$ means that it is normal that x occurs. The smaller the possibility degree of x, the more surprising the occurrence of x.

Let $c_{t+1,j}(x_{t,k}, a_{t,i}, x_{t+1,j}) = K(x_{t,k}, a_{t,i}, x_{t+1,j})$ express the transition cost generated from the process $x_{t,k} \to x_{t+1,j}(x_{t,k}, a_{t,i})$ when the state is $x_{t,k}$ and the action we choose is $a_{t,i}$ at time t. The set of the cost corresponding to $X_{t+1}(x_{t,k}, a_{t,i})$ is expressed as

$$C_{t+1}(x_{t,k}, a_{t,i}) = \{c_{t+1,j}(x_{t,k}, a_{t,i}, x_{t+1,j}) \mid 1 \le j \le k_{t+1}(x_{t,k}, a_{t,i})\}. \tag{2}$$

Suppose that in each stage, there is one and only one chance for the decision maker making a decision. Let us consider how to make a decision based on the information A_t, $X_{t+1}(x_{t,k},a_{t,i})$, $\pi_{t+1}(x)$ and $C_{t+1}(x_{t,k},a_{t,i})$. We give the satisfaction function as follows:

Definition 2. The function $u_T : C_T(x_{T-1,k},a_{T-1,i}) \rightarrow [0,1]$ is called a satisfaction function of $a_{T-1,i}$ in the stage T if it satisfies $u_T(w_1) > u_T(w_2)$, for any $w_1 < w_2$, $w_1, w_2 \in C_T(x_{T-1,k},a_{T-1,i})$.

The procedure of one-shot decision in the last stage, $T-1$ is as follows:

Step 1: We seek the focus points for each $a_{T-1,i}$ with the condition that the state in the current stage is $x_{T-1,k}$. The decision maker can choose one of the 12 types of focus points based on his different attitudes about possibility and satisfaction [11]. Now we consider the IX type focus point, which is denoted by

$$\begin{aligned} &x_T^*(x_{T-1,k},a_{T-1,i}) \\ &= \arg \max_{j \in k_T(x_{T-1,k},a_{T-1,i})} \min[\pi_T(x_{T,j}(x_{T-1,k},a_{T-1,i})), u_T(c_{T,j}(x_{T-1,k},a_{T-1,i},x_{T,j}(x_{T-1,k},a_{T-1,i})))] \end{aligned} \quad (3)$$

where $\min[\pi_t, u_t] = [\pi_t \wedge u_t, \pi_t \wedge u_t]$. It can bee seen that the state with higher possibility and higher satisfaction level is chosen for making a one-shot decision in this stage. It means that the decision maker is optimistic.

Step 2: The optimal alternative $a_{T-1}^*(x_{T-1,k})$ is determined as follows:

$$f_1(x_{T-1,k}) = \min_{a \in A_{T-1}} K(x_{T-1,k}, a, x_T^*(x_{T-1,k}, a)) \quad , \quad (4)$$

where $a_{T-1}^*(x_{T-1,k}) = \arg f_1(x_{T-1,k})$.

Next, we consider the decision in the stage $T-2$. Set

$$\begin{aligned} G_{T-1}(x_{T-2,k},a_{T-2,i}) &= \{g_{T-1,j}(x_{T-2,k},a_{T-2,i},x_{T-1,j}) \mid 1 \le j \le k_{T-1}(x_{T-2,k},a_{T-2,i})\} \\ &= \{c_{T-1,j}(x_{T-2,k},a_{T-2,i},x_{T-1,j}) + f_1(x_{T-1,j}) \mid 1 \le j \le k_{T-1}(x_{T-2,k},a_{T-2,i})\} \end{aligned} . \quad (5)$$

We have the satisfaction function $u_{T-1} : G_{T-1}(x_{T-2,k},a_{T-2,i}) \rightarrow [0,1]$ which satisfies $u_{T-1}(w_1) > u_{T-1}(w_2)$, for any $w_1 < w_2$, $w_1, w_2 \in G_{T-1}(x_{T-2,k},a_{T-2,i})$.

The decision in the step $T-2$ is obtained as follows:

Step 1: Finding out the focus points for $a_{T-2,i}$:

$$\begin{aligned} &x_{T-1}^*(x_{T-2,k},a_{T-2,i}) \\ &= \arg \max_{j \in k_{T-1}(x_{T-2,k},a_{T-2,i})} \min[\pi_{T-1}(x_{T-1,j}(x_{T-2,k},a_{T-2,i})), u_{T-1}(g_{T-1,j}(x_{T-2,k},a_{T-2,i},x_{T-1,j}(x_{T-2,k},a_{T-2,i})))] \end{aligned} \quad (6)$$

Step 2: Obtaining the optimal action in the stage $T-2$:

$$f_2(x_{T-2,k}) = \min_{a \in A_{T-2}} [K(x_{T-2,k}, a, x^*_{T-1}(x_{T-2,k}, a)) + f_1(x^*_{T-1}(x_{T-2,k}, a))], \tag{7}$$

where $a^*_{T-2}(x_{T-2,k}) = \arg f_2(x_{T-2,k})$.

For the stage $T-n$, $n \geq 2$, let

$$\begin{aligned} G_{T-n+1}(x_{T-n,k}, a_{T-n,i}) &= \{g_{T-n+1,j}(x_{T-n,k}, a_{T-n,i}, x_{T-n+1,j}) \mid 1 \leq j \leq k_{T-n+1}(x_{T-n,k}, a_{T-n,i})\} \\ &= \{c_{T-n+1,j}(x_{T-n,k}, a_{T-n,i}, x_{T-n+1,j}) + f_{n-1}(x_{T-n+1,j}) \mid 1 \leq j \leq k_{T-n+1}(x_{T-n,k}, a_{T-n,i})\} \end{aligned}. \tag{8}$$

The satisfaction function $u_{T-n+1}: G_{T-n+1} \to [0,1]$ satisfies $u_{T-n+1}(w_1) > u_{T-n+1}(w_2)$, for any $w_1 < w_2$, $w_1, w_2 \in G_{T-n+1}$.

The one-shot decision in the stage $T-n$ is obtained as follows:

Step 1: Finding out the focus points for $a_{T-n,i}$:

$$\begin{aligned} &x^*_{T-n+1}(x_{T-n,k}, a_{T-n,i}) \\ &= \arg \max_{j \in k_{T-n+1}(x_{T-n,k}, a_{T-n,i})} \min[\pi_{T-n+1}(x_{T-n+1,j}(x_{T-n,k}, a_{T-n,i})), u_{T-n+1}(g_{T-n+1,j}(x_{T-n,k}, a_{T-n,i}, x_{T-n+1,j}(x_{T-n,k}, a_{T-n,i})))]. \end{aligned} \tag{9}$$

Step 2: Obtaining the optimal action in the stage $T-2$:

$$f_n(x_{T-n,k}) = \min_{a \in A_{T-n}} [K(x_{T-n,k}, a, x^*_{T-n+1}(x_{T-n,k}, a)) + f_{n-1}(x^*_{T-n+1}(x_{T-n,k}, a))], \tag{10}$$

where $a^*_{T-n}(x_{T-n,k}) = \arg f_n(x_{T-n,k})$.

Repeating (9) and (10), we can obtain the optimal decision sequence $a^*_0, a^*_1, \cdots, a^*_{T-1}$ for the initial state x_0. We can construct the optimal path under the decision sequence $a^*_0, a^*_1, \cdots, a^*_{T-1}$, that is, x_0, $x^*_1(x_0, a^*_0)$, $x^*_2(x^*_1, a^*_1)$, ..., $x^*_T(x^*_{T-1}, a^*_{T-1})$.

3 Numerical Examples

Let us consider a simple numerical example to show how to make multistage one-shot decisions. The set of states is $X = \{0,1,2\}$ and the set of actions is $A_t = \{a_1, a_2\}$ for any *t*. The possibilistic transition matrix associated with a_1 is given by the following matrix

$$\begin{bmatrix} 1 & 0 & 0 \\ 0 & 1 & 0 \\ 0 & 0 & 1 \end{bmatrix}. \tag{11}$$

The possibilistic transition matrix associated with a_2 is

$$\begin{bmatrix} 0 & 1 & 0.4 \\ 0.7 & 0 & 1 \\ 0.8 & 1 & 0 \end{bmatrix}. \tag{12}$$

The elements of (11) and (12) describe the possibility degree of the next state for the corresponding action. The cost generated by a_1 is always 1, that is $c_t = 1$. When taking the action a_2, if $x_{t+1} > x_t$, the cost is $c_t = 0.5$, otherwise the cost is $c_t = 2$.

In each stage, if one decision chance is given, how to make decisions to minimize the total cost?

If $T = 1$, the problem is degenerated as a typical one-shot decision problem. We define the satisfaction function as

$$u_1(x_0, a_i) = \frac{\min_j c_{1,j}(x_0, a_i, x_{1,j}(x_0, a_i))}{c_{1,j}(x_0, a_i, x_{1,j}(x_0, a_i))} \quad . \tag{13}$$

Set the initial state as $x_0 = 0$. Using (3), the focus points of a_1 and a_2 are obtained as 0 an 1, respectively. Using (4), we know the optimal action is a_2. If $x_0 = 1$, the focus points of a_1 and a_2 are 1 and 2, respectively. The optimal action is a_2. If $x_0 = 2$, the focus points of a_1 and a_2 are 2 and 1, respectively. The optimal action is a_1. We can see that $f_1(0) = f_1(1) = 0.5$, $f_1(2) = 1$.

Set $T = 2$ and $x_0 = 0$. We need to know the series of optimal actions a_0^*, a_1^*. Using the above results, we can calculate

$$g_1(0, a_1, 0) = c_1(0, a_1, 0) + f_1(0) = 1.5, \tag{14}$$
$$g_1(0, a_1, 1) = c_1(0, a_1, 1) + f_1(1) = 1.5, \tag{15}$$
$$g_1(0, a_1, 2) = c_1(0, a_1, 2) + f_1(2) = 2, \tag{16}$$
$$g_1(0, a_2, 0) = c_1(0, a_2, 0) + f_1(0) = 2.5, \tag{17}$$
$$g_1(0, a_2, 1) = c_1(0, a_2, 1) + f_1(1) = 1, \tag{18}$$
$$g_1(0, a_2, 2) = c_1(0, a_2, 2) + f_1(2) = 1.5. \tag{19}$$

The satisfaction function in the stage 1 is defined as

$$u_1(x_0, a_i) = \frac{\min_j g_{1,j}(x_0, a_i, x_{1,j}(x_0, a_i))}{g_{1,j}(x_0, a_i, x_{1,j}(x_0, a_i))} \quad . \tag{20}$$

According to (9), we can obtain the focus points of a_1 and a_2 as 0 and 1, respectively. Based on (10), the optimal action is a_2. So that the optimal action sequence is a_2, a_2 and the state will move from 0 to 1 then to 2. In the same way, we can make a decision for any T.

4 Conclusions

In this paper, a new decision approach for multistage one-shot decision problems is proposed. In such decision problems, a decision maker has one and only one chance to make a decision at each stage. The proposed approach lets a decision maker choose one state amongst the all states according to his/her attitude about satisfaction and possibility in each decision-making stage. Based on the selected state (focus point), the optimal alternative is determined. The payoff in each stage is only associated with focus points in the succeeding stages. The proposed method provides an efficient vehicle to deal with multistage one-shot decision problems under uncertainty which are extensively encountered in business, economics, and social systems.

References

1. Abo-Sinna, M.A.: Multiple objective (fuzzy) dynamic programming problems: a survey and some applications. Applied Mathematics and Computation 157(3), 861–888 (2004)
2. Bellman, R.E.: Dynamic Programming. Princeton University Press, Princeton (1957)
3. Bellman, R.E., Zadeh, L.A.: Decision-making in a fuzzy environment. Management Sciences 164, B141–B164 (1970)
4. Cervellera, C., Chen, V.C.P., Wen, A.: Optimization of a large-scale water reservoir network by stochastic dynamic programming with efficient state space discretization. European Journal of Operational Research 171(3), 1139–1151 (2006)
5. Cristobal, M.P., Escudero, L.F., Monge, J.F.: On stochastic dynamic programming for solving large-scale planning problems under uncertainty. Computers & Operations Research 36(8), 2418–2428 (2009)
6. Dubois, D., Prade, H., Sabbadin, R.: Decision-theoretic foundations of possibilty theory. European Journal of Operational Research 128, 459–478 (2001)
7. Esogbue, A.O., Bellman, R.E.: Fuzzy dynamic programming and its extensions. TIMS/Studies in the Management Sciences 20, 147–167 (1984)
8. Guo, P.: Decision analysis based on active focus point and passive focus point. In: Proceedings of the International Workshop of Fuzzy Systems and Innovational Computing, pp. 39–44 (2004)
9. Guo, P.: One-Shot Decision Approach and Its Application to Duopoly Market. International Journal of Information and Decision Sciences 2(3), 213–232 (2010)
10. Guo, P.: Private Real Estate Investment Analysis within One-Shot Decision Framework. International Real Estate Review 13(3), 238–260 (2010)
11. Guo, P.: One-shot decision Theory. IEEE Transactions on SMC: Part A 41(5), 917–926 (2011)
12. Kacprzyk, J., Esogbue, A.O.: Fuzzy dynamic programming: Main developments and applications. Fuzzy Sets and Systems 81, 31–45 (1996)
13. Kahneman, D., Tversky, A.: Prospect Theory: An analysis of decision under risk. Econometrica 47, 263–291 (1979)
14. Kung, J.J.: Multi-period asset allocation by stochastic dynamic programming. Applied Mathematics and Computation 199(1), 341–348 (2008)
15. Li, D., Cheng, C.: Stability on multiobjective dynamic programming problems with fuzzy parameters in the objective functions and in the constraints. European Journal of Operational Research 158(3), 678–696 (2004)
16. Piantadosi, J., Metcalfe, A.V., Howlett, P.G.: Stochastic dynamic programming (SDP) with a conditional value-at-risk (CVaR) criterion for management of storm-water. Journal of Hydrology 348(3-4), 320–329 (2008)
17. Topaloglou, N., Vladimirou, H., Zenios, S.A.: A dynamic stochastic programming model for international portfolio management. European Journal of Operational Research 185(3), 1501–1524 (2008)
18. Zhang, X.B., Fan, Y., Wei, Y.M.: A model based on stochastic dynamic programming for determining China's optimal strategic petroleum reserve policy. Energy Policy 37(11), 4397–4406 (2009)

Speed Control for Under-actuated Planar Biped Robot

Yuanyuan Xiong, Gang Pan, and Ling Yu

Department of Control Science & Control Engineering, Zhejiang University, Hangzhou, 310027, China
20932051@zju.edu.cn, {gpan,lyu}@iipc.zju.edu.cn

Abstract. For the under-actuated planar biped robot, we propose a novel method to change walking speed based on periodic walking. We design reference joint trajectory as Bezier curve. Based on virtual constrain, linear feedback has been used to make joint trajectory asymptotically follow the reference trajectory. To change walking speed, different Bezier parameters are chosen for different walking rate and are fused by fuzzy algorithm. As these fusion outputs are rough, PI controller is designed to regulate some legs' parameters to get a stable walking gait. The whole control system is a two-level control structure. In the first level, a feedforward-feedback controller consisting of fuzzy controller and PI controller is used to generate reference trajectory. While in the second level, linear feedback could be used to achieve trajectory tracking. Simulation confirms the efficiency of this method.

Keywords: virtual constrain, fuzzy algorithm, two-level control structure.

1 Introduction

The control and design of biped robot is always a difficult topic, which has attracted lots of professors to study. In the meantime, the application of biped robot is very broad, which can replace human to complete hazard works. Professors of CNRS designed a bipedal robot RABBIT[1] in the view of low energy consuming, diversity, well anti-interference. From a control design perspective, the challenges in legged robots arise from many degrees of freedom in the mechanisms, intermittent nature of the contact conditions with the environment, and underactuation.

Variety of walking speed is an aspect of robot's walking diversity. D.G.E.Hobbelen, etc [2], have proved that the energy efficiency could become higher as walking speed increases. And they have found that step size, ankle actuate force and trunk angle have main influence on walking speed of periodic walking[3],[4]. E.R.westervelt, etc [5],[6], have designed parameter switch method and PI controller to regulate gait parameter to adjust walking speed. However, parameter switch method can only change walking rate on pre-designed walking speed. C.Sa-bourin, etc [7],[8], have designed Fuzzy-CMAC control strategy based on human walk rules, which could automatically regulate locomotion according to average speed. In addition, it had good anti-interference. However, the CMAC parameters were difficult to confirm and the learning speed was very low.

Y. Wang and T. Li (Eds.): Knowledge Engineering and Management, AISC 123, pp. 165–172.
springerlink.com

The main idea of this paper includes two parts. In the first part, walking gait has been designed for a walking speed. Linear feedback has been used to generate a stable dynamic walking with step length. Because the gait parameters can influence walking gait, it could influence walking speed too. In the second parts, different gait parameters have been designed for different walking speed. Fuzzy algorithm is used to fuse these parameters. The fusion parameters will be output, which allows us to generate new gait and theoretically to carry out an infinity of trajectories only from a limited number of walking references. Simulation confirmed this method is feasible.

2 Biped Walk Dynamic Model

The biped robot model is a planar open kinematic chain consisting of torso, two thighs and two shanks. The end of a shank, whether it has links constituting a foot or not, will be referred to as a foot. As depicted in Fig.1, it has a point foot and there is no actuation between the stance leg tibia and the ground. The biped robot dynamic model includes swing phase model where only one leg is in contact with the ground and impact model where both legs are in contact with the ground.

During the swing phase, assume that the biped has point feet as in Fig.1 and the stance leg end acts as an ideal pivot. So the model of single leg support can be defined by Newton-Lagrange theory. The equations of motion are

$$M(q)\ddot{q}+C(q,\dot{q})\dot{q}+G(q)=Bu \tag{1}$$

Where, $q=[q_1,q_2,q_3,q_4,q_5]^T$ are the joint angles, $\boldsymbol{M}$ is the mass-inertial matrix, $\boldsymbol{C}$ is a matrix relevant to Coriolis and centripetal force, $\boldsymbol{G}$ is the gravity vector, $u=[\tau_1,\tau_2,\tau_3,\tau_4]^T$ are joint torques, $\boldsymbol{B}$ is the linear transformation from current angle space to joint relative angle space. Define $x=(q^T,\dot{q}^T)^T$, the model is written in state-space form as

$$\begin{aligned}\dot{x}&=\begin{bmatrix}\dot{q}\\ M^{-1}(q)(Bu-G(q)-C(q,\dot{q})\dot{q})\end{bmatrix}\\&=f(x)+g(x)u.\end{aligned} \tag{2}$$

Impact occurs when the swing leg touches the walking surface. In the case of a rigid walking surface, the duration of the impact event is very short and it is common to approximate it as being instantaneous. Because the actuator can't produce impulse torque, the influence of it should be ignored in impact process. Under these assumptions and angular momentum conservation, the impact model can be defined as

$$x^+=\Delta(x^-) \tag{3}$$

Where, $(\bullet)^-$ and $(\bullet)^+$ are expressed as instantaneous state before impact and after impact, respectively. The impact switch set $\boldsymbol{S}$ is defined as the horizontal coordinate p_2^h of swing foot greater than zero, while the vertical coordinate p_2^v equal to zero.

$$S = \{(q,\dot{q}) \mid p_2^v(q) = 0, p_2^h(q) > 0\} \tag{4}$$

Walking consists of alternating phases of single and double support. Then the hybrid model includes continuous and discrete state, and it can be expressed as

$$\begin{cases} \dot{x} = f(x) + g(x)u & x^- \notin S \\ x^+ = \Delta(x^-) & x^- \in S \end{cases} \tag{5}$$

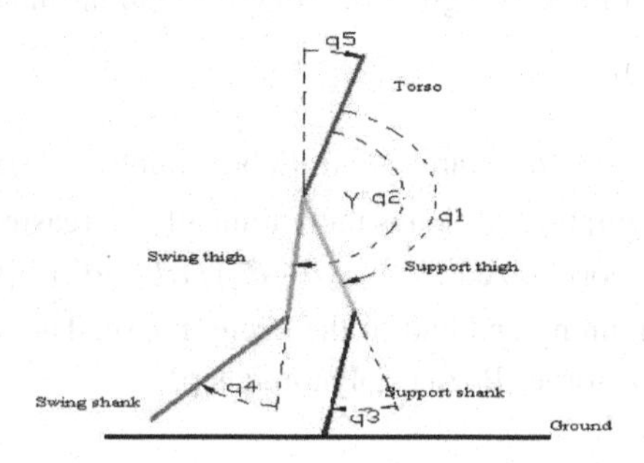

Fig. 1. The N-rigid-link model consists of torso, two thighs and two shanks. During swing phase, only one leg is in contact with ground.

3 Control Strategy for Variable Speed Walking Based on Fuzzy Algorithm

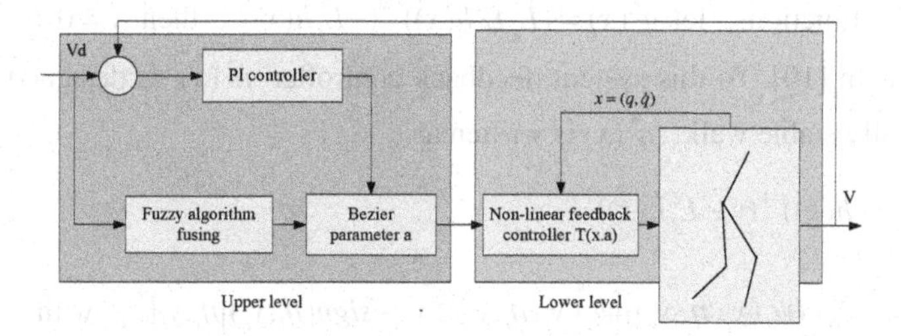

Fig. 2. It is a two-level control structure. In the upper level, PI controller and fuzzy fusion are used to change reference gait parameters. In the lower lever, non-linear feedback controller is used to follow the reference gait.

For speed changing, the control idea includes two parts. Firstly, the nonlinear feedback controller based on virtual constrain[1] is designed to get regular speed walk. The walking speed can be changed when we regulate the parameters of virtual constrain. Secondly, the fuzzy algorithm will be used to fuse these parameters after different parameters have been chosen for different walking rates. The fusion parameters could be obtained according to the given rate. The map of parameters to rate is nonlinear, so the fuzzed results have deviation. Therefore the PI controller is designed to regulate some parameters to achieve stable alterable rate walking. The control strategy is shown as Fig.2.

3.1 Lower Level Non-linear Feedback Controller

Non-linear feedback controller is designed based on output function—virtual constrain, which are only relative to biped configuration. Suppose output function as

$$y = h(q) = h_0(q) - h_d(\theta(q)) \tag{6}$$

Where, $h_0(q) = \begin{bmatrix} q_1 & q_2 & q_3 & q_4 \end{bmatrix}^T$ are controllable joints, $\theta(q) = cq$ is a scalar function relevant to configuration, which is monotonically increasing over the duration of a step. So $\theta(q)$ can be normalized as $s = (\theta - \theta^+)/(\theta^- - \theta^+) \in [0,1]$, θ^+ and θ^- are the values of θ at the beginning and end of the swing phase. The reference trajectory $h_d(\theta(q))$ are designed as M-order Bezier polynomials [9],

$$h_{di}(\theta(q)) = \sum_{k=0}^{M} a_k^i \frac{M!}{k!(M-k)!} s^k (1-s)^{M-k} \qquad i = 1,2,3,4 \tag{7}$$

The Bezier curve parameters can be obtained by manual selection or optimization.

The biped robot dynamic system is a two-order system, so the output function is two-order too. Making second differential for output function, then we can get

$$\ddot{y} = L_f^2 h(q,\dot{q}) + L_g L_f h(q) u \tag{8}$$

From these functions, let $u^*(x) = (L_g L_f h(x))^{-1}(-L_f^2 h(x))$, then $\ddot{y} = 0$ is a double integral system [10]. To this system, feedback controller $u^*(x)$ is designed to realize asymptotically stable walk. $u^*(x)$ is written as

$$u^*(x) = (L_g L_f h(x))^{-1}(v - L_f^2 h(x)) \tag{9}$$

Where $v = -sign(\psi_i(y_i, \eta_t \cdot \dot{y}_i))|\psi_i(y_i, \eta_t \cdot \dot{y}_i)|^{\frac{\alpha}{2-\alpha}} - sign(\eta_t \cdot \dot{y}_i)|\eta_t \cdot \dot{y}_i|^{\alpha}$, with $\psi_i(y_i, \eta_t \cdot \dot{y}_i) = y_i + (\frac{1}{2} - \alpha) sign(\eta_t \cdot \dot{y}_i) . |\eta_t \cdot \dot{y}_i|^{2-\alpha}$, and $0 < \alpha < 1$, η_t is the parameter, which is used to regulate the convergence time of the controller and should be designed to make the output function asymptotically convergence to zero before the swing leg touches the walking surface.

3.2 Fuzzy Algorithm Fusing Parameters

The average speed $\overline{v}$ is defined as the ratio between the step length when the two legs are in contact with the ground and the duration of the step, which can be written as

$$\overline{v} = \frac{p_2^h(q_0^-)}{T(\Delta)} \tag{10}$$

Set several different walking speeds, which are shown in Table.1. Then Bezier curve parameters which satisfied gait suppositions should be obtained by manual selection or optimization. Now fuzzy algorithm is used to fuse parameters of different speeds. Then, when the given speed changes, the fusion parameters could be output as reference parameters. Furthermore, the method designed to change walking speed in literature mostly fix the step length, and just adjust the duration of the step to change speed. But the method designed in this paper can change the step length at the same time. From Table.1, we can see the step length increases with the walking speed .

For output function, Bezier curve is *M*-order curve, so there are *M+1* numbers parameters. In the whole system, there are n ($n=1,2,3,4$) curves, then there are *(M+1)*n* parameters. Parameters a_k^i for different pre-designed speed are fuzzed. The input of fuzzy controller is $\overline{v}$, while the outputs are Bezier parameters which have been fused through fuzzy inference system. The fuzzy principle is shown as follows

- If $\overline{v}$ is small then a_k^i is Y_k^{i1} .
- If $\overline{v}$ is medium then a_k^i is Y_k^{i2} .
- If $\overline{v}$ is big then is a_k^i is Y_k^{i3} .

Where, Y_k^i are Bezier parameters of Gait_i. The average velocity is modeled by fuzzy sets (small, medium, big). μ^l is the weight of Gailt_i. Each a_k^i is computed by using Eq. (14)

$$a_k^i = \frac{\sum_{l=1}^{3} \mu^l Y_k^{il}}{\sum_{l=1}^{3} \mu^l} . \tag{11}$$

3.3 PI Controller Regulates Bezier Parameters

The parameter fusions that have been designed in section 3.2 is nonlinear, so there is error between the desired speed and actual speed. Therefore, PI controller is designed to adjust some parameters of all Bezier curves (or some Bezier curves). PI controller for Bezier parameters $a = \left\{ a_k^i \middle| k = 0,\ldots,M,\quad i=1,2,3,4 \right\}$ is designed as follows:

$$\begin{cases} e(k+1) = e(k) + (v^* - \overline{v}(k)) \\ w(k) = Kp*(v^* - \overline{v}(k)) + Ki*e(k) \\ a(k+1) = a(k) + w(k)\delta a \end{cases} \tag{12}$$

In function (12), $w(k)$ is the deviation of a, which is computed according to the error between the true speed and the desired speed. Here δa can be set to a fixed value [5],[11].

The whole control can be stated as follows, when the given $\bar{v}$ changes, fusion parameters output through fuzzy inference system, then the under level controllers are switched, so the actual speed changes suddenly, however, it can't hop to the given speed but just round it. Then, the fuzzy controller does not work and PI controller is used to adjust some Bezier parameters to follow the given speed gradually.

4 Simulation

Table 1. Gait parameters. θ_{torso} is the angle of torso when collision happened, θ^{-} is the collision angle.

Gait_i	$\bar{v}$ (m/s)	Step L_{step} •m•	θ^{-}	θ_{torso}
Gait_1	0.8	0.6	-7/8*pi	20°
Gait_2	0.7	0.5415	-8/9*pi	20°
Gait_3	0.6	0.4935	-9/10*pi	20°

The parameters of simulation plant come from biped robot RABBIT. The Bezier curves and controller are obtained according to Table.1 and hypothetical conditions. Then the attraction domain of controller can be computed at the same time. According to section 3.2, fuzzy inference system is designed. The output parameters shall be obtained according to the given speed. Then the parameters of the swing leg are regulated by PI controller. Set $\delta a_k^i = 1$ ($i = 2, k = 3$), the rest parameters of a are set to zero, and the parameters of PI controller relevant to w are set to $Kp1 = 0.2$ and $Ki1 = 0.001$.

To verify this walking controller, a 150-step simulation is performed. As show in Fig.3, the given average walking rate starts at 0.6m/s, changes to 0.7m/s at second step, to 0.75m/s at 51st step, to 0.6m/s at 101st step. The actual walking rate follows the given walking rate. Fig.4.shows the phase plane portrait of this walk, which is a limit cycle, implying the stability of the walking gait.

Fig.5.is the simulation result with PI controller. The given walking rate changes to 0.6m/s at 6th step, to 0.8m/s at 36th step. The actual speed follows the given speed after a long time's regulation, but the rise time is long, the oscillation is severe and overshoot is large too. From simulation figures, we can see the walking speed regulate result of this paper is superior to PI controller obviously.

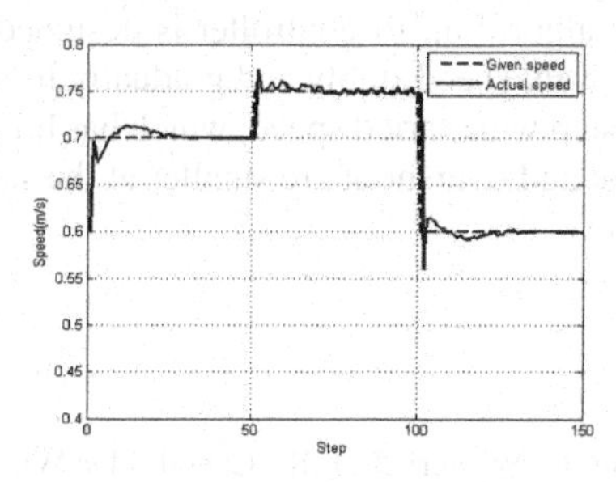

Fig. 3. The desired rate(*dashed line*) and the actual rate(*solid line*)

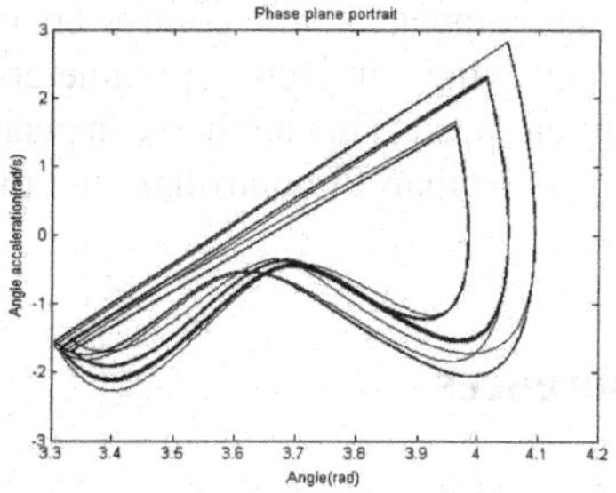

Fig. 4. Phase plane portrait

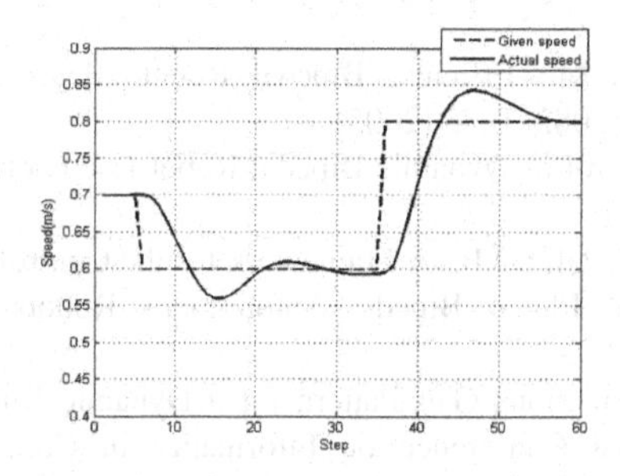

Fig. 5. PI controller to regulate walk speed. The desired rate(*dashed line*) and the actual rate(*solid line*)

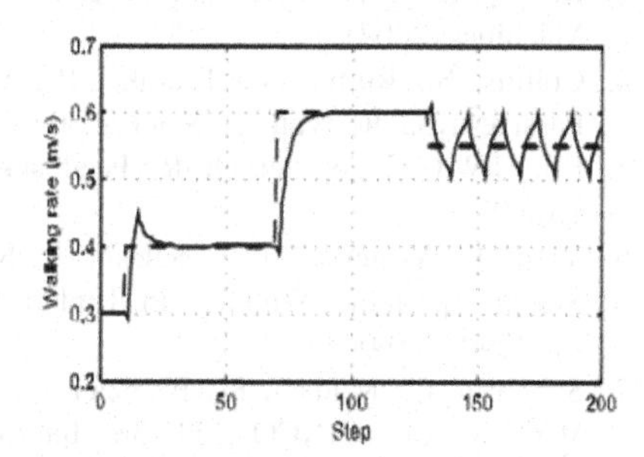

Fig. 6. Step-wise average walking rate(*solid line*) and the desired walking rate(*dashed line*)

The simulation result of literature [7] for direct controller switch is given in Fig.6. In this literature the speed of the reference gait is only pre-designed on 0.5m/s and 0.6m/s without 0.55m/s. So when the desired speed changes to 0.55m/s, the step-wise walking rate oscillates between 0.5m/s and 0.6m/s, while the overall average walking rate is close to the desired walking rate, which indicate this controller can't control the robot to walk at the given speed which is not pre-designed. However, the fuzzy fusion method designed in this paper can let the robot stably walk at the given speed 0.75m/s not pre-designed.

5 Result

In this paper, for biped robot varying speed walking, according to the control experience of stable periodic walk, fuzzy controller is designed to fuse Bezier parameters with pre-designed given speed, and fusion parameters are output according to the given speed. The controller parameters are switched when biped robot leg collide with the ground, which make the actual walking speed follow the given speed more or less. To compensate the control error of fuzzy algorithm, PI controller is designed to regulate swing's trajectory parameters to make walk speed stably and gradually follow the given speed. This method can change walk step with varied speed, which has bigger speed scope than PI controller, and the rise time and overshoot are smaller at the same time.

References

1. Chevallereau, C., Abba, G., Aoustin, Y., Plestan, F., Westervelt, E.R., Canudas De Wit, C., Grizzle, J.W.: RABBIT: A Testbed for Advanced Control Theory. IEEE Control Systems Magazine 23(5), 57–79 (2003)
2. Hobbelen, D.G.E., Wisse, M.: Controlling the Walking Speed in Limit Cycle Walking. International Journal of Robotics Research 27(9), 989–1005 (2008)
3. Daan, M.W., Hobbelen, G.E.: Limit Cycle Walking, Humanoid Robots:Human-like Machines (2007)
4. Collins, S., Ruina, A., Tedrake, R., Wisse, M.: Efficient Bipedal Robots Based on Passive-dynamic Walkers. Science 307(5712), 1082–1085 (2005)
5. Eric, J.W.G.C., Westervelt, R.: Feedback Control of Dynamic Bipedal Robot Locomotion (2007)
6. Yang, T., Westervelt, E.R., Serrani, A., Schmiedeler, J.P.: A Framework for the Control of Stable Aperiodic Walking in Underactuated Planar Bipeds. Autonomous Robots 27, 277–290 (2009)
7. Sabourin, C., Madani, K., Bruneau, O.: Autonomous Gait Pattern for a Dynamic Biped Walking. In: ICINCO 2006-3rd International Conference on Informatics in Control, Automation and Robotics, Proceedings (2006)
8. Sabourin, C., Bruneau, O., Buche, G.: Control Strategy for the Robust Dynamic Walk of a Biped Robot. International Journal of Robotics Research 25(9), 843–860 (2006)
9. Bezier, P.: Numerical Control: Mathematics and Applications. Wiley (1972)
10. Bhat, S.P., Bernstein, D.S.: Continuous Finite-time Stabilization of the Translational and Rotational Double Integrators. IEEE Transactions on Automatic Control 43(5), 678–682 (1998)
11. Westervelt, E.R., Grizzle, J.W., Canudas De Wit, C.: Switching and PI Control of Walking Motions of Planar Biped Walkers. IEEE Transactions on Automatic Control 48(2), 308–312 (2003)

Performance Analytical Model for Interbay Material Handling System with Shortcut and Blocking

Lihui Wu[1], Jianwei Wang[2], and Jie Zhang[3]

[1] School of Mechanical and Electrical Engineering, Henan University of Technology, Zhengzhou 450001, China
wulihui@haut.edu.cn
[2] School of Civil Engineering and Architecture, Henan University of Technology, Zhengzhou 450001, China
lygxywjw@163.com
[3] School of Mechanical Engineering, Shanghai Jiao Tong University, Shanghai 200030, China
zhangjie@sjtu.edu.cn

Abstract. To effectively analyze and evaluate the performances of Interbay material handling system with shortcut and blocking in semiconductor fab, an extended Markov chain model (EMCM) has been proposed, in which the system characteristics such as vehicle blockage and system's shortcut configuration are well considered. With production data from Interbay material handling system of a 300mm semiconductor fab, the proposed EMCM is compared with simulation analytic model. The result demonstrates that the proposed EMCM is an effective modeling approach for analyzing Interlay's performances in the early phase of system design.

Keywords: Performance analysis, Markov model, Interbay material handling system, semiconductor fab.

1 Introduction

The semiconductor wafer fabrication system (SWFS) distinguishes itself from traditional manufacturing systems by many characteristics, such as high reentrant flows, enormous process complexity, shared production tools, long cycle times, and so on [1]. These characteristics lead to high level of wafer lot transportations within the wafer fab [1]. In a typical 300mm semiconductor fabrication line, there can be as many as 5000 wafer lots waiting for about 100 types of tools [2]. Each wafer lot repeatedly goes through deposit film, pattern film, etch film, wafer testing and cleaning process 20-40 times, for a total of about 200 to 600 operations [3]. Each move between operations requires transportation, thus each wafer lot travels approximately 8-10 miles to complete the manufacturing operations. Automated Interbay material handling system (often called Interbay system) becomes critically important to reduce manufacturing cost and maintain high tool utilization [4-5]. During the Interbay system design process, to analyze and evaluate the impact of the critical factors on the Interlay's and SWFS's performance, a practical and feasible performance analytical model is required.

Y. Wang and T. Li (Eds.): Knowledge Engineering and Management, AISC 123, pp. 173–178.
springerlink.com

Recently, there are many researches on performance analytical model for material handling system, including queuing theory model, queuing network model, Markov chain model, and so on. Benjaafar [6], Koo et al. [7], and Smith [8] presented queuing theory models to estimate the performance indexes of automated material handling system (MHS) in production systems. However, the characteristics of material handling system, such as vehicle blockages and empty vehicle's transportation time, are not considered in these queuing theory-based analysis models. Fu [9], Heragu et al. [10], and Jain et al. [11] used queuing network model to analyze the operating performance of MHS in different manufacturing system. In general, these queuing network models can analyze MHS more accurately to compare with queuing theory-based model. However, the time to solve these queuing network models grows exponentially with the system scale and number of vehicles. Nazzal and Mcginnis [12] proposed a modified Markov chain model to estimate the simple closed-loop automated MHS's performance considering the possibility of vehicle blocking. However, this modeling approach is mainly suitable for simple closed-loop material handling system without shortcut rail. This paper proposes an extended Markov chain model (EMCM) for Interbay system with shortcuts and blocking, in which the system characteristics, such as stochastic behavior, vehicle blockage, shortcut rail, and multi-vehicle transportation, are considered. The Interlay's performance is analyzed effectively.

The paper is organized as follows. The next section gives a brief description of the Interbay system. Section 3 discusses the EMCM. Section 4 is dedicated to an experimentation study on Interbay system of a 300mm semiconductor wafer fab. Finally, conclusions are outlined in Section 5.

2 Interbay System Description

The Interbay system typically consists of transportation rails, stockers, automated vehicles, shortcut, and turntable. A detailed description of the Interbay and its' operation can refer to [13]. Let $L(m)$ refer to the Interbay system with m vehicles. Denote S as the set of stocker in the Interbay system, $S=\{s_i, i=1,...,n\}$. s_i is the ith stocker in the Interbay system. Each stocker has two load ports: an input port (called drop-off station) where wafer lots are dropped off by vehicles and an output port (called pick-up station) where wafer lots are picked up by vehicles. Let s_i^p and s_i^d denote the pick-up station and drop-off station, respectively. Let the pick-up and drop-off stations are represented as nodes, and transportation rails and shortcuts are represented as directed arcs, then the layout of Interbay system can be simplified as Fig. 1.

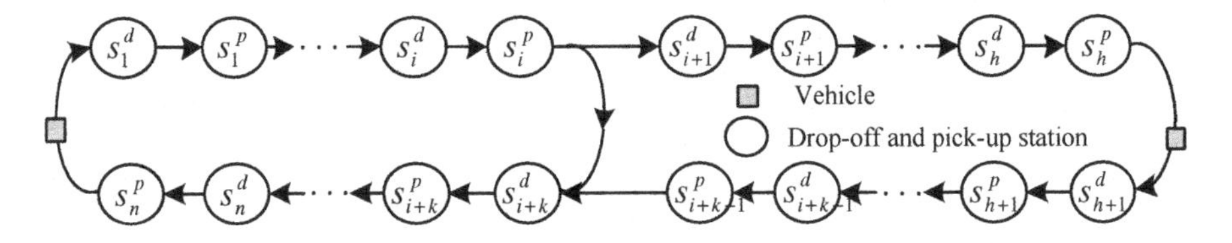

Fig. 1. Layout of simplified Interbay system

3 The Extended Markov Model

During extended Markov model analyzing, there are several system assumptions. (1) Move requests per time period in each stocker is constant. (2) Move request arrived according to Poisson distribution process. (3) Each vehicle transports wafer lot based on FCFS rule. (4) Vehicles travel independently in Interbay system, i.e., the correlation among the vehicles is not considered. (5) The probability that a vehicle is blocked by the downstream stocker is proportional to the number of vehicles in system.

3.1 Notation

r_i : the probability that loaded vehicle drops-off its load at s_i^d .

q_i : probability that an empty vehicle arriving at s_i^p finds a waiting move requests.

p_i^d : the probability that the vehicle at s_{i-1}^p is blocked by another vehicle at s_i^d .

p_i^p : the probability that the vehicle at s_i^d is blocked by another vehicle at s_i^p .

$\overline{\tau}_w$: the probability of loaded vehicle in pick-up station s_w^p gets through the shortcut connected with stocker s_i and enters drop-off station s_{w+k}^d, $\overline{\tau}_w = 1 - \tau_w$. τ_w is the probability of loaded vehicle in pick-up station s_w^p enters drop-off station s_{w+1}^d .

According to assumption 5, the process of m vehicles transporting wafer lots can be equivalent to m independent processes of single-vehicle transporting wafer lots, which can be described as Markov chain. As only those pick-up and drop-off stations are considered, a state of Interbay system with single vehicle is defined by the 3-tuple structure: $\{S, Kp, Ks\} = \{(s_i, i = 1,...,n), (p,d), (e, f, s, b)\}$. Where Kp is the pick-up and drop-off station set where vehicle located. Ks is the state set of vehicle at pick-up and drop-off station. p,d represents vehicle locates at pick-up station and drop-off station, respectively. e, f, s, b means the state of vehicle at pick-up and drop-off station is empty, loaded, receiving service, and blocked, respectively.

3.2 EMCM Analysis

Consider an Interbay system with n stockers, the detail process of state changing can be analyzed and the transition probability of each states of system can be deduced. Based on these probabilities, the states transition matrix $\mathbf{R}$, which specifies the movement of the vehicle between the states of system, can be identified. Denote state visited ratio vector $\mathbf{v} = \{v_w\}$, $w \in W$, where v_w means the steady state probability of visited ratio to state w , W is the state set of EMCM. Without loss of generality, set $v_{(1,p,e)} = 1$. Let $\mathbf{R}$ be the state transition probability matrix of EMCM. For a finite state, positive recurrent Markov chain, the steady-state probabilities can be uniquely obtained by solving the square system of equations [14].

$$\mathbf{R} \cdot \mathbf{v} = \mathbf{v} . \tag{1}$$

$$v_{(1,p,e)} = 1 . \tag{2}$$

In states transition matrix $\mathbf{R}$, some of these probabilities variables are unknown. Specifically, the dropping-off/picking-up probability variables $\mathbf{r} = \{r_i\}$ and $\mathbf{q} = \{q_i\}$, $i = 1,...,n$, the blocking probabilities variables $\mathbf{p^d} = \{p_i^d\}$ and $\mathbf{p^p} = \{p_i^p\}, i = 1,...,n$, and the shortcut-selecting probabilities variables $\boldsymbol{\tau} = \{\tau_i\}$, $i = h_1,...,h_l$.

1) The probability variables $\mathbf{r} = \{r_i\}, i = 1,...,n$. Based on system's stability condition analysis, the vehicle's picking up probability r_i can be described as:

$$\Lambda_i = r_i \cdot a_i^d \Rightarrow r_i = \Lambda_i / a_i^d . \tag{3}$$

Where Λ_i is mean arrival rate of move requests dropped-off to s_i^d, and a_i^d is rate of loaded vehicles arrivals to s_i^d. a_i^d can be get based on material handling flow analyzing of Interbay system.

2) The probability variables $\mathbf{q} = \{q_i\}, i = 1,...,n$. Based on system's stability condition analysis, q_i can be inferred as:

$$q_i = (\lambda_i \times C) / (v_{(i,p,e)} \times m) . \tag{4}$$

Where λ_i is mean arrival rate of move requests picked up from s_i^p, and C is the mean cycle time.

3) The probabilities variables $\mathbf{p^d} = \{p_i^d\}$ and $\mathbf{p^p} = \{p_i^p\}, i = 1,...,n$. Based on system blocking probability analysis, the blocking probability variable p_i^d and p_i^p can be estimated as follows.

$$p_i^d = (m-1) \times v_{(i,d)} / (\sum_{j=1}^{|R|} (v_{(i,d)} + v_{(i,p)}), \forall s_i^d, i = 1,...,n . \tag{5}$$

$$p_i^p = (m-1) \times (v_{(i,p,s)} + v_{(i,p,b)}) / \sum_{j=1}^{|R|} (v_{(i,d)} + v_{(i,p)}), \forall s_i^p, i = 1,...,n . \tag{6}$$

4) The probability variables $\boldsymbol{\tau} = \{\tau_i\}, i = h_1,...,h_l$. Based on system's stability condition analysis, the equation of $\overline{\tau_u}$ can be described as:

$$\overline{\tau_u} = (\sum_{i=1}^{u} (\sum_{j=u+k}^{n} p_{ij} + \sum_{j=1}^{i-1} p_{ij}) + \sum_{i=u+k+1}^{n} \sum_{j=u+k}^{i} p_{ij}) / (\sum_{i=1}^{u} (\sum_{j=u+1}^{n} p_{ij} + \sum_{j=1}^{i-1} p_{ij}) + \sum_{i=u+2}^{n} \sum_{j=u+1}^{i} p_{ij}) . \tag{7}$$

Combing equation sets (1) through (7), the quantity of stability state equations and variables is $|W| + 3n + 1$. We can find the unique solution to the system of equations and calculate the visit ratio to every state, and vehicle-blocking probability.

4 Experimental Study

This section presents a numerical experiment study to demonstrate the effectiveness of the proposed EMCM. The data used for numerical experiment in this study was shared by a semiconductor manufacturer in Shanghai, China. There are 14 stockers in the Interbay system. The distance between adjacent stockers is 20 meters, the distance between pick-up port and drop-off port of a stocker is 5 meters, and the Interbay rail loop is 320-meter long with 2 shortcuts. The system's impact factors considered in the experimental study include vehicle quantity (3-15 vehicles), system's release ratio (9000, 13500, and 15000 wafer /month), and vehicle's speed (1m/s and 1.5 m/s). Totally, 72 scenarios are designed for different combinations of the selected factors.

Based on the experimental data, the proposed EMCM is comprehensively compared with discrete event simulation model in each scenario. A total of three performance indexes - system's throughput capability, vehicle's mean utilization ratio, and mean arrival time interval of empty vehicle are compared. The frequency of relative error percentages of EMCM and simulation analytic model in all scenarios and all performance indexes is analyzed. The Fig. 2 illustrates that 96% of all relative error values belongs to [-8%, 10%]. It's demonstrated that the proposed EMCM performs reasonably well with acceptable error percentages and is an effective modelling approach for performance analyzing of Interbay system with shortcut and blocking.

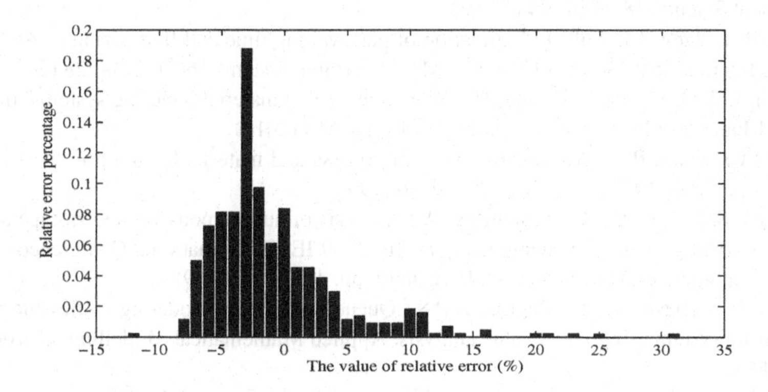

Fig. 2. The frequency of relative error percentages of EMCM and simulation analytic model

5 Conclusion

Interbay system is an indispensable and critical sub-system in 300mm semiconductor wafer fab. In order to analyze and evaluate performances of Interbay system with shortcut and blockage effectively in system's design stage, an EMCM has been proposed. With production data from Interbay of a 300mm semiconductor wafer fabrication line, the proposed EMCM is compared with simulation analytic model. The results demonstrates that the proposed EMCM and simulation analytic model have small relative errors in terms of system's throughput capability, vehicle's mean utilization ratio, and mean arrival time interval of empty vehicle. The 96% of all relative error percentages ranges from -8% to 10%. It means that the proposed EMCM is an effective modelling approach for performances analysing of Interbay system with shortcut and lockage in semiconductor wafer fab.

Acknowledgments. This work was supported by the National Nature Science Foundation of China under Grant No. 60934008 and 50875172.

References

1. Agrawal, G.K., Heragu, S.S.: A survey of automated material handling systems in 300-mm semiconductor Fabs. IEEE Transactions on Semiconductor Manufacturing 19, 112–120 (2006)
2. Chung, S.L., Jeng, M.: An overview of semiconductor fab automation systems. IEEE Robot & Automation Magazine 11, 8–18 (2004)
3. Montoya-Torres, J.R.: A literature survey on the design approaches and operational issues of automated wafer-transport systems for wafer fabs. Production Planning & Control 17, 648–663 (2006)
4. Tompkins, J.A., White, J.A.: Facility Planning. John Wiley, New York (1984)
5. Wang, M.J., Chung, H.C., Wu, H.C.: The evaluation of manual FOUP handling in 300mm wafer fab. IEEE Transactions on Semiconductor Manufacturing 16, 44–55 (2003)
6. Benjaafar, S.: Modeling and analysis of congestion in the design of facility layouts. Management Science 48, 679–704 (2002)
7. Koo, P.H., Jang, J.J., Suh, J.: Estimation of part waiting time and fleet sizing in AGV systems. International Journal of Flexible Manufacturing Systems 16(3), 211–228 (2004)
8. Smith, J.M.: Robustness of state-dependent queues and material handling systems. International Journal of Production Research 48, 4631–4663 (2010)
9. Fu, M.C., Kaku, B.K.: Minimizing work-in-process and material handling in the facilities layout problem. IIE Transactions 29, 29–36 (1997)
10. Heragu, S.S., Cai, X., Krishnamurthy, A.: Analysis of autonomous vehicle storage and retrieval system by open queueing network. In: 2009 IEEE International Conference on Automation Science and Engineering, Bangalore, pp. 455–459 (2009)
11. Jain, M., Maheshwarl, S., Baghel, K.P.S.: Queueing network modeling of flexible manufacturing system using mean value analysis. Applied Mathematical Modelling 32, 700–711 (2008)
12. Nazzal, D., Mcginnis, L.F.: Analytical approach to estimating AMHS performance in 300mm Fabs. International Journal of Production Research 45, 571–590 (2007)
13. Wu, L.H., Mok, P.Y., Zhang, J.: An adaptive multi-parameter based dispatching strategy for single-loop interbay material handling systems. Computer in Industry 62, 175–186 (2011)
14. Ching, W.K., Ng, M.K.: Markov chains, models, algorithms and applications. Springer Science Business Media, New York (2006)

Robotic 3D Reaching through a Development-Driven Double Neural Network Architecture

Fei Chao, Lin Hu, Minghui Shi, and Min Jiang

Cognitive Science Department, Fujian Key Laboratory of the Brain-like Intelligent Systems, Xiamen University, Xiamen, Fujian, P.R. China, 361005
fchao@xmu.edu.cn

Abstract. Reaching ability is a kind of human sensory motor coordination. The objective of this work is to imitate the developmental progress of human infant to create a robotic system which can reach or capture objects. The work proposes to employ a double neural network architecture to implement control a robotic system to learn reaching within 3D experimental environment. A constraint releasing mechanism is applied to implement the development procedure for the robot system. In addition, the experimental results are described and discussed in this paper.

Keywords: Developmental Robotics, Hand-eye Coordination, Double Network Architecture.

1 Introduction

Robotic reaching, which is a sub-item of robotic hand-eye coordination, is one of the most important skills for the intelligent robot to survive and work in the unconstrained environments, and it is also the essential part and significant research topic in the field of autonomous robotic systems. Furthermore, the hand-eye coordination of robot is extensively used in wide range of applications, such as vehicle manufacturing, space exploration, food packaging, etc. [1]. The core technique of the robotic reaching ability is to implement the hand-eye coordination, which is the mapping from robotic visual sensory to robotic actuators. Mostly, the traditional robotic hand-eye coordination systems are calibrated by human engineers. But regretfully, this approach contains a key limitation, which is such systems have to re-calibrate itself when a small change occurs in the system.

To address this problem, this work introduces a new research field of intelligent robotics "Developmental Robotics" to give the autonomous ability to our robotic system. A number of robotic reaching and hand-eye coordination models have been proposed recently. However, Those works have limitations on various facts: most of the robotic hand-eye coordination systems only use one camera as the robotic vision system, merely carry out 2 dimensional experiments which

Y. Wang and T. Li (Eds.): Knowledge Engineering and Management, AISC 123, pp. 179–184.
springerlink.com

might simplify their systems' learning complexity e.g. [2][3][4]; or their robotic learning systems ignore imitating the developmental progress of human infants reaching objects, e.g. [5][6]. The objective of our approach is to extend those works. We apply inspirations from human infant development to implement a developmental robotic learning system, and use a vision system with two cameras to coordinate with a robotic arm to achieve reaching ability in 3-dimensional environment, especially, we emphasize that developmental constraint is the key fact that drives robot to develop reaching behaviors from scratch. This will considerably deepen the research on robotic reaching and hand-eye coordination, and fill up the gap between developmental robotics and developmental psychology.

2 Related Work of Reaching and Development

In order to build a development-driven approach, it is necessary to understand human infant development procedure, and to abstract the significant developmental features from the procedure. Therefore, the following subsections introduce the infant development model and developmental constraint respectively.

2.1 Hand-Eye Coordination Development in Infant

The procedure of infant development demonstrates that the reaching movement starts from coarse reaching movements to precise. After birth, human infants are capable of visually orienting and of making directed hand movements toward visual targets. However, these "pre-reaching" movements are not successful in making contact with targets [7]. Later, towards the end of the first year of life, infants acquire the ability to independently move their fingers, and the vision of the hand gains importance in the configuring and orienting of the hand in anticipation of contact with target objects [8]. Therefore, our robotic system can follow this pattern: a reaching movement consists of the pre-reaching movements and the correction movements. From the aspect of neural processing modules, human infants's reaching is the result of the basal ganglia, and cerebellum working together to generate and control [9]. The "pre-reaching" movements are produced by the basal ganglia-cerebral cortex. As time passes during the first year and the cerebellar network becomes more matures, the cerebellum loops should gradually exert influence on the approximate commands selected by the basal ganglia loops [10].

Inspired by the two types of movements and the two areas of brain cortices, this work proposes to design two neural networks by using the experiences of infant reaching to build the learning system. One neural network is trained to control large amplitude arm movements around the objects, imitating the early infant pre-reaching and learning basal ganglia-cerebral cortex loops taking rough action. The other neural network is designed to behave small amplitude arm movements to make correct reaching movements, imitating the later movement of infant reaching. The design of double neural networks perfectly reflects the two phases of the developmental process of infant reaching.

2.2 Development and Developmental Constraints

Developmental psychology show evidences that lifting constraint can lead infants to develop from one competence to a new, even more complicated competence. Lee et al. [11] introduced the Lift-Constraint, Act, Saturate (LCAS) algorithm. Therefore, we apply the LCAS to simulate a development sequence: when a constraint has been saturated, a new constraint is released into the system. In this paper, we raise two constraints to drive the development of the whole robotic learning system. One parameter is the visual definition, and another is the movement amplitude of robotic arm. The whole procedure can be described as follows: at the beginning, the vision is obscure, the arm can wave in large amplitude only, and the robotic system uses these rough movements to train the network simulating the basal ganglia cortex, during this moment, the system is not saturated until the network has been fully trained. As the change of both constraint parameters, the vision's definition becomes increasing and the robotic starts to train the other network simulating the cerebellum to refine reaching movements until the arm could make correct reaching movements to capture target objects accurately. After that, the whole system becomes stable and mature.

3 An Experimental System for Robotic Hand-Eye Coordination

The experiment aims to achieve the goal that the robotic arm can learn to capture target objects through the robot's random movements. This section describes the double network's architecture the learning algorithms for training the robotic learning system, and the robotic system's configuration.

3.1 Hardware

Fig. 1 illustrates the robotic experimental system: we use an "AS-6 DOF" aluminum alloy robotic arm including 6 Degree-of-Freedoms (DOF), the arm is mounted on a workspace. We lock 2 DOFs of the hand, use the left 4 DOFs to finish reaching movements. This setup can support the robotic arm moves and captures objects in 3D environment. Each rotational joint of the robot arms has a motor drive and also an encoder which senses the joint angle, thus providing a proprioceptive sensor. 2 cameras are applied to build the robotic vision system in this work. One camera is mounted on a frame placed next to the arm; another camera is mounted above and looks down on the work space installed above the workspace.

3.2 Double Neural Network Architecture

The entire computational learning system consists of two neural networks. The first network N_1 is able to behave rough reaching movements, and the second network N_2 can make correction reaching movements after a rough reaching movement has been made. The robotic arm behaves spontaneous movement which

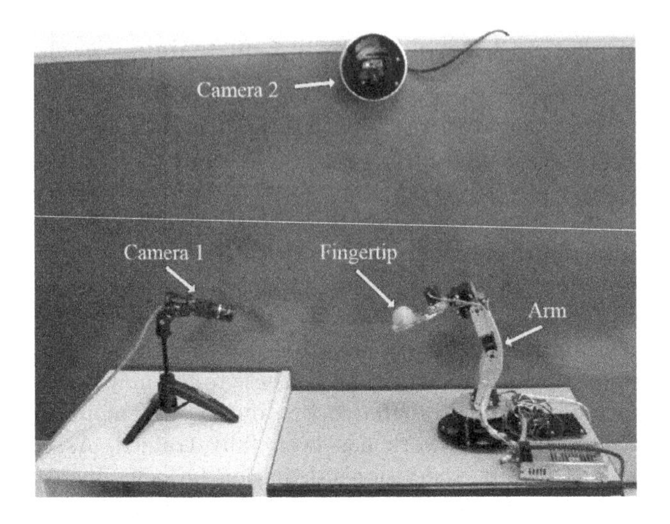

Fig. 1. The experimental system

means the arm moves randomly. Before each movement, the both cameras can calculate the fingertip position, x_1, y_1 indicates the fingertip position in Camera 1, x_2, y_2 gives the position in Camera 2; meanwhile, the joint values $(\theta_1, \theta_2, \theta_3, \theta_4)$ of the robotic motors are acquired from the robotic controller. After one movement, both the fingertip position and the 4 joint values are changed, and we use x'_1, y'_1, x'_2, y'_2 and $\theta'_1, \theta'_2, \theta'_3, \theta'_4$ to identify those new values. Here, the Euclidean distance d is the distance between x_1, y_1, x_2, y_2 and x'_1, y'_1, x'_2, y'_2. $\Delta\theta_{1-4}$ are the different values between θ_{1-4} and θ'_{1-4}. A threshold δ which can be used to determine which network is trained or is used to control the arm. If $d \geqslant \delta$, N_1 network is selected. x'_1, y'_1, x'_2, y'_2 are the input of N_1, and θ'_{1-4} are the network's expected output. The Back-Propagation (BP) algorithm is applied to train N_1 by using x'_1, y'_1, x'_2, y'_2 and θ'_{1-4} as the training pattern. If $d < \delta$, N_2 is trained, its input contains x_1, y_1, x_2, y_2 and $\Delta x_1, \Delta y_1, \Delta x_2, \Delta y_2$, its expected output has $\Delta\theta_{1-4}$.

3.3 Developmental Learning Algorithms

The changing tendencies of the two constraints are: the visual definition slowly becomes better, and the arm amplitude gradually decreases. Thus, at the beginning of the training phase, the captured images are blurry and the arm can merely produce long range spontaneous movements. In this case, only N_1 is trained. While the two constraints gradually change, small range of movements can be made by the arm and can be detected by the vision system. N_2 gains opportunities to have itself trained. When the two constraints achieve their extreme values, only N_2 can be trained.

$$Sat(t) = \begin{cases} true; & \text{for } t = \varphi \cdots n, \text{ if } |e_t - e_{t-\varphi}| < \delta \\ & \text{and } e_t < \psi \\ false; & \text{else} \end{cases} \quad (1)$$

It is very crucial to setup the maturation assessment, which indicates that when a level of constraint is saturated. Equation 1 gives the saturation rule: if the output error of the double network architecture remains low ($e_t < \psi$) for a while $|e_t - e_{t-\varphi}| < \delta$, a new constraint is released into the system, otherwise, the system still acts in the current situation.

4 Experimental Procedure and Results

The experiment follows this procedure: Firstly, no target is put in the workspace, the learning system only generate random movements, according to the changing of two constraints, the robotic arm behaves no small movements at the beginning, but will generate several small movements in the end of the experiment. Fig. 2 illustrates the learning progress: the "N1" line shows how many times N_1 network has been trained after each spontaneous movement. "N2" stands for the training times of N_2 network. "Constraint" means when the constraint is released. We can observe from the figure: The whole procedure behaves 2 developmental stages, the fast increasing stage of N_1 at the beginning and the increasing stage of N_2 in the end; between such two stages, it is the fast constraint releasing phase. All this situations are caused by the changing of the two constraints. Therefore, the entire robotic system's developmental procedure is driven by the two constraints.

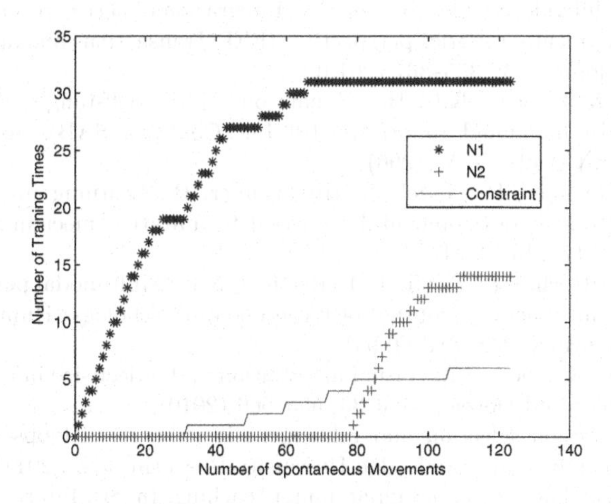

Fig. 2. The learning results

5 Conclusions

This paper has shown how ideas and data from psychology and primate brain science can inspire effective learning algorithms that could have considerable impact on robotics applications. Our experimental system shows two distinct stages of behavior produced from a single method. The constraint changing is able to drive the whole robotic learning system to develop from random movements to make accurate reaching behaviors. There exists a large gap between our psychological understanding of development and our ability to implement working developmental algorithms in autonomous agents. This paper describes one study on the path towards closing that gap and indicates the potential benefits for future robotic systems. In the future work, we propose to use motorized cameras to replace the two static cameras in the system to gain larger view of the vision system.

Acknowledgments. This work is funded by the Natural Science Foundation of China (No. 61003014), and two Natural Science Foundations of Fujian Province of China (No. 2010J01346 and No. 2010J05142).

References

1. Connolly, C.: Artificial intelligence and robotic hand-eye coordination. Industrial Robot: An International Journal 35, 495–503 (2008)
2. Meng, Q., Lee, M.H.: Automated cross-modal mapping in robotic eye/hand systems using plastic radial basis function networks. Connection Science 19, 25–52 (2007)
3. Chao, F., Lee, M.H.: An autonomous developmental learning approach for robotic eye-hand coordination. In: Proceeding of AIA 2009, Innsbruck, Austria (2009)
4. Huelse, M., McBride, S., Law, J., Lee, M.: Integration of active vision and reaching from a developmental robotics perspective. IEEE Transactions on Autonomous and Mental Development 2, 355–367 (2010)
5. Marjanovic, M.J., Scassellati, B., Williamson, M.M.: Self-taught visually-guided pointing for a humanoid robot. In: 4th Int. Conf. on SAB., pp. 35–44. MIT Press/Bradford Books, MA (1996)
6. Andry, P., Gaussier, P., Nadel, J., Hirsbrunner, B.: Learning invariant sensorimotor behaviors: A developmental approach to imitation mechanisms. Adaptive Behavior 12, 117–140 (2004)
7. Clifton, R.K., Rochat, P., Robin, D.J., Berthier, N.E.: Multimodal perception in the control of infant reaching. Journal of Experimental Pschology: Human Perception and Performance 20, 876–886 (1994)
8. Berthier, N.E., Carrico, R.L.: Visual information and object size in infant reaching. Infant Behavior and Development 33, 555–566 (2010)
9. Houk, J.C.: Action selection and refinement in subcortical loops through basal ganglia and cerebellum. Cambridge University Press (in press, 2011)
10. Berthier, N.E.: The syntax of human infant reaching. In: 8th International Conference on Complex Systems, pp. 1477–1487 (2011)
11. Lee, M.H., Meng, Q., Chao, F.: Staged competence learning in developmental robotics. Adaptive Behavior 15, 241–255 (2007)

Automatic C. Elgance Recognition Based on Symmetric Similarity in Bio-Micromanipulation

Rong Zhang[1], Shichuan Tang[2], Jingkun Zhao[1], Shujing Zhang[2], Xu Chen[1], and Bin Zhang[2]

[1] School of Mechanical Engineering and Automation, Beijing University of Aeronautics and Astronautics, 100191 Beijing, China
[2] Beijing Municipal Institute of Labour Protection, 100054 Beijing, China
zhangrong@buaa.edu.cn

Abstract. A new image recognition algorithm for a small creature—Caenorhabditis elgance is developed in this paper. This methods first use edge detection, binarization and other methods to get the body contours of C.elgance. Then using its body symmetry, C.elgance is recognized by comparing similarity of consecutive lines in image. The experimental result shows the high positioning accuracy and rapidity of the proposed algorithm. This method can also be applied to other object recognition from the biological image of creatures with symmetric body.

Keywords: visual, image, symmetry, correlation, recognition.

1 Introduction

Filtering operation is a common bio-engineering research technique. Small creatures' filtering tasks can be divided into two levels -- "soft selection" and "hard selection". “Hard selection” refers to the physical operations of manipulate objects, such as injection, cutting, clamping, adsorption. “Soft selection” refers to the individual recognition, feature extraction (such as length, diameter, life and death, etc.), and characteristic information statistics of manipulate objects. “soft selection” is not only to achieve the position of manipulate objects, but also to understand the characteristics of objects' image. This kind of selection reflects the higher level of cognitive function. Ideal “soft selection” operation should have the capability to identify the identity of small creatures rapidly and accurately in their entire life cycle.

Caenorhabditis elegans, called C. elegans for short, is an important model organism in the biomedical research. Their larvae's length is about 80μm, and their width is about 20μm. Adult C. elegans' length is about 500μm～1200μm, and their width is about 60μm～120μm. At present, the manipulate objects in researches on bio-engineering of micro-operation field are cells or chromosomes [1]-[4], whereas researches on “soft selection” of small creatures in similar size with C. elegans are carried out less.

Y. Wang and T. Li (Eds.): Knowledge Engineering and Management, AISC 123, pp. 185–194.
springerlink.com

Features of C. elegans' image are as follow: they are growing in the medium paste with confusing background, furthermore, the area of bubbles in the medium paste is similar to the area of C. elegans' bodies, which will interfere with determination; there are C. elegans of different ages in a single dish, whose body shapes vary differently and the worms often self-wind, overlap with other individuals, drill into the medium, and other complex situations. So the physical edges are rendered as numbers of non-contiguous segments when handled. As a result, cell recognition methods commonly used, such as area of measurement recognition method, skeleton extraction method, and template matching method and central axis method often used in chromosome identification, are not suitable for C. elegans identification.

In this paper, the feature that C. elegans' body is symmetric in high degree is captured firstly. Secondly, the chain code of C. elegans' body contour on one side is achieved by using image analysis. And then C. elegans can be identified by comparing the similarity coefficient of the chain code value of the adjacent segments. This algorithm can distinguish C. elegans and bubbles in the medium correctly, restore the contours of C. elegans' body, and also can be used for identification of other organisms with symmetrical features.

2 Constitution of C. Elegans' Life Cycle Detection System

The aim of the detection system of C. elegans' life cycle involved in this paper is to assist experimental operators in carrying out the C. elegans RNAi experiment. This system can realize fully automated from the ceiling and injection of Escherichia coli before experiment to C. elegans' life and death judging, corpse sorting and experimental data recording during experiment. By this way, a large number of experimental operators can be rescued from the tedious boring work, and experimental efficiency is improved. Complete flow chart of the system is shown in Figure 1.

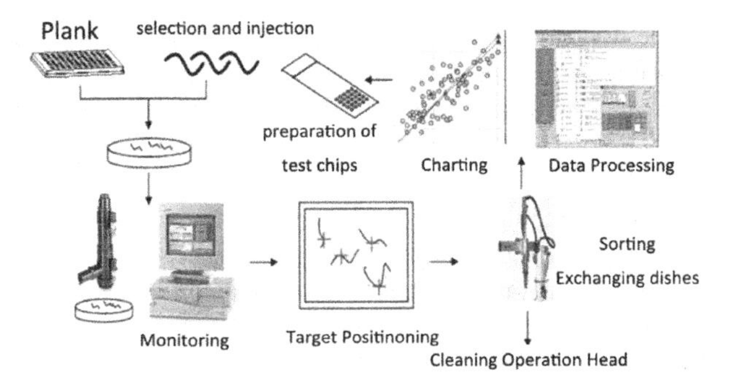

Fig. 1. Flow chart of C.elegans' life cycle detection system

Thereinto, visual system, whose task is to identify the C. elegans (about 100) in the dish, to judge their life condition (dead or alive), and then to get the specific location information of dead C. elegans, is the core of the whole system.

The visual system is made of two main components ---- hardware and software.

A microscope, a CCD video camera, an image acquisition card and an image processing computer are included in the hardware part. The microscope is an inverted biological microscope, OLYMPUS IX70, with 4 times amplification selected; the CCD video camera is JVC TK1270E type, which can output PAL standard video signal (25 frames/ second) with 600×450μm field of view and 0.872μm resolution; the image acquisition card is Matrox-CronosPlus; and the image processing computer (IPC IC-6908P4) is product of Taiwan Advantech Ltd.

Image capture, auto focusing, preprocessing, feature recognition, location marked, information output and some other parts are included in the software part. Introductions to image pre-processing and feature recognition algorithm are described herein.

3 Image Pre-processing

Due to various factors, the C. elegans' image collected was interfered by insufficient ideal contrast or splotchy background. Therefore, image pre-processing was needed in the first place before recognition. Specific steps are as follows.

3.1 Edge Detection and Binarization

By comparison, the best detection results could be achieved by using Gaussian- Laplace (LoG) operator on edge detection. LoG combines the filtering properties of Gaussian operator with the edge extraction characteristics of Laplace operator. Image noise was filtered by Gaussian filter, and then the part which had the maximum gradient value was taken as the edge by the edge detector in order to reduce the impact of edge diffusion. However, the binary image achieved from preprocessing with white background still contained higher interference.

3.2 Detection of Moving Targets and Image Segmentation

Optical flow method and image difference method are the main methods to detect moving targets. Great deal of time would be cost if optical flow method that has poor real-time as well as practicality was used. Whereas the principle of image difference method is relatively simple, which is to do differential computing to the two current continuous images[4]. According to the experiment and measurement, the swimming speed of adult C. elegan was 0.2 mm/s, which means images could be captured by an 1s interval to conduct the differential computing. Since the specific location of the dead C. elegans was the only care of this system, images could be divided into areas, and the areas where only survival C. elegans were located were not our concern (these areas are for target tracking). The determination of existence and location of dead C. elegans in other areas was the focus of this system.

4 Identification of C. Elegans' Features

The biggest factor influence the identification of C.elegans was the bubbles within the culture medium. The length of bubble edges was similar to that of C. elegans' bodies. Moreover, the area of bubbles was also approximately equal to that of C. elegans' bodies. A conclusion that C. elegan' body was completely symmetrical was found by observing. Although the complete envelope curve of C. elegan' edge could not be obtained just by preprocessing, there were still high degree of similarity in the segments on both side of C. elegan' body. However, there were no similar edge curves around bubbles. Through this characteristic, C. elegan' body curve and the edge of bubble were distinguished clearly.

Therefore the recognition algorithm procedures of C. elegans were as follows:

4.1 Image Progressive Scanning

(1) First scanning

Aim of this step was to remove the isolated points and bifurcation points in order to refine edges of image.

There were many bifurcation segments contained in the binary images obtained from preprocessing, so bifurcation points should be removed in the first place to guarantee that only single branch segments were contained in the image in order to facilitate the following operations.

An eight adjacent domain model was defined around each pixel in order to facilitate the description of algorithm. As is shown in Figure 2, there are only three conditions around those 8 points:

(a) Only one black point is around the central one, which means this central point is the beginning one or the ending one of a segment.

(b) Two black points are around the central one, which means this central point is in the middle of a segment.

(c) Three or more than three black points are around the central one, which means this central point is the intersection point or bifurcation point. Thereinto, there are nine kinds of points satisfying the characteristics of bifurcation points, which are shown in Figure 3. So bifurcation points were removed one by one through template matching method in order to make the image edge an entire single pixel wide edge one [2].

P3	P2	P1
P4	P	P0
P5	P6	P7

Fig. 2. An eight adjacent domain model.

Fig. 3. An eEight adjacent domain model

(2) Second scanning

Aim of this step was to detect segments in order to remove the interference short segments and to record the location information and midpoint coordinates of the long segments. In this way, the segment chain code value was determined. Steps were as follows:

(a). Segments were scanned line by line to find out the beginning point of each segment and then a label was given to each point.

(b). As is shown in Figure 4, trend of the segment was determined in turn by the situation of pixel point neighborhood, and the same label was marked along its trend of the segment. When one segment had been marked, the next beginning point would be found out in the same method mentioned in step (a).

1	0	0	2	2	2	2	2	0	0	0	0
0	1	0	0	0	0	0	0	2	0	0	0
0	0	1	1	0	1	0	0	0	2	0	0
3	3	0	1	0	1	0	0	0	0	2	0
0	3	0	1	0	0	1	0	0	0	2	0
0	3	0	0	1	1	0	0	2	2	0	0
0	0	3	0	0	0	0	2	0	0	0	0

Fig. 4. Label for the segment in turn

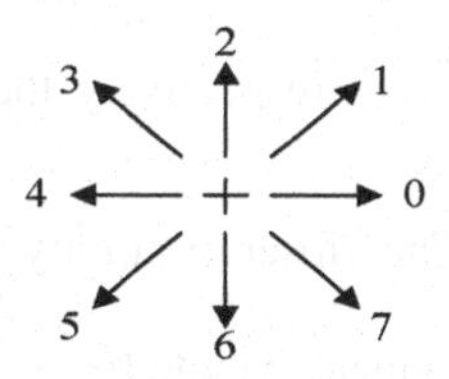

Fig. 5. 8 connected region chain code

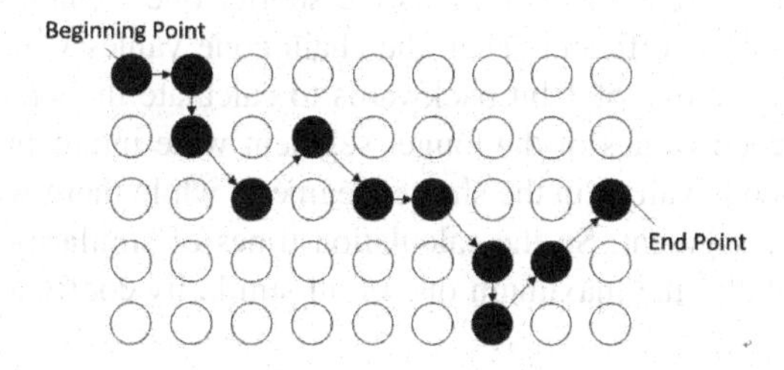

Fig. 6. Edge tracking of image (the chain code is 0671707611)

(c).The aim of this step was to determine the chain code of the segment. The chain code is very crucial to the segment, because not only the length of that segment but also its trend can be expressed by its chain code. Furthermore, some characteristics of the objects should be identified by its chain code. The definition of the 8 connected regional chain code (also called Freeman code) value [2] used in this paper is shown in figure 5. When a beginning point was found and labeled, the rest points along the trend of the segment were found and labeled one by one in turn till the end point, so that the chain code value of this segment was determined. After that, the next beginning point will be found according to the horizontal scanning order to determine the chain code value of next segment. All these steps are shown in Figure 6.

4.2 Similarity Calculating

After calculating the similarity of two closely adjacent segments, information of segments which had the highest similarity was kept for every C. elegan's body checked in order to calculate the midpoint coordinates of C. elegan's body and then mark the midpoint with a cross.

Specific method: Firstly, all the information of points on one segment was extracted; secondly, areas around the beginning point were detected to see if there were other segments, and if there were any then information of points on that segment should also be extracted. The similarity was calculated with the shorter one of the two segments as the benchmark by using their chain code values. Mathematically, the formula to calculate the similarity coefficient is as follow [7]:

$$r = \frac{\sum_{i=1}^{n}\sum_{i=1}^{n}(A_i - \overline{A})(B_i - \overline{B})}{\sqrt{(\sum_{i=1}^{n}\sum_{i=1}^{n}(A_i - \overline{A})^2)(\sum_{i=1}^{n}\sum_{i=1}^{n}(B_i - \overline{B})^2)}} \tag{1}$$

Where $[A_1,\cdots,A_n]$ and $[B_1,\cdots,B_n]$ are respectively the chain code value of the two segments;

$\overline{A} = \sum_{i=1}^{n} A_i/n$, $\overline{B} = \sum_{i=1}^{n} B_i/n$. The closer similarity coefficient was equal to 1, the more similar shapes of the two segments would be.

In general, the length of the two segments in the image studied in this paper was not equal. So in the first place a length of chain code of the longer segment was intercepted from scratch, which was as long as the shorter one's chain code, in order to calculate the similarity coefficient. Then the chain code values with the same length were intercepted by a move of 1 bit backwards to calculate the similarity coefficient until all the chain code values of the longer segment were intercepted. For example, there were n chain code values in the shorter segment, while there were m chain code values in the longer segment. So the calculation times of similarity coefficient were totally (m-n+1). Finally, the maximum one of all similarity coefficients was the final similarity coefficient of the two segments. As the C. elegan's body is symmetric, two contours from one body of a C. elegan were significantly similar, which meant their similarity coefficient was very close to 1. However, similarity coefficient from different C. elegans' body contours or from a C. elegan's body contour and a bubble contour was much smaller than 1, which was why this method could be used to determine if two segments were from one body of a C. elegan. According to the definition of similarity coefficient, when it is greater than 0.7, the degree of similarity is high. So 0.7 was taken as a threshold to determine similarity. Recognition algorithm flow chart is shown in figure 7.

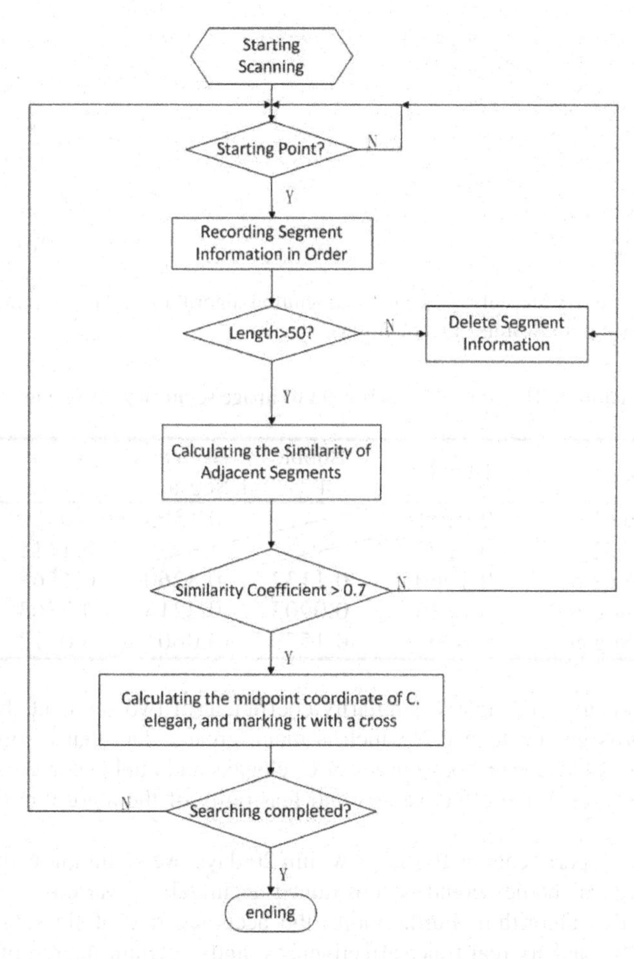

Fig. 7. Flow chart of C. elegan's recognition algorithm

The processing time of algorithm was decided by the complexity of the image. According to estimates, the processing cycle of the single frame image identification and positioning algorithm of C. elegans was less than 0.2 seconds, which was much higher than the speed of manual sorting.

5 The Experiments of C. Elegans' Identification

The image processing results of C. elegans' identification experiments are shown in Figure 8 (6 dead C. elegans are concluded in this figure). The experimental results of similarity calculation of segments are shown in Table 1.

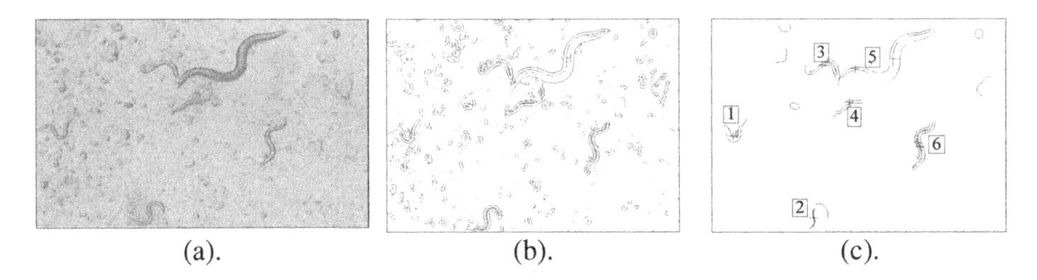

Fig. 8. Result of every step of C. elegans' recognition algorithm: (a) The original image; (b) After preprocessing; (c) Results of recognition

Table 1. The similarity coefficient of image segments comparing

	Bubble 3	Bubble 4	Worm 1 Seg a	Worm 2 Seg a	Worm 3 Seg a
Bubble 1	0.0141		0.1565	0.0180	
Bubble 2	0.0290			0.1111	0.3541
Worm 1 Seg b	0.1960	0.1137	0.8260	0.1565	0.0601
Worm 2 Seg b	0.0336	0.0903	0.2213	0.7363	0.1339
Worm 3 Seg b	0.0903	0.1570	0.0601	0.0212	0.7409

As can be seen from Table 1, similarity coefficient of two segments belonging to one C. elegan is greater than 0.7, which is much greater than that of two segments from different C. elegans or body curves of C. elegans and bubble contours. So results in Table 1 also verify the effectiveness and feasibility of the algorithm described in this paper.

By multiple experiments in 10 dishes within 20 days, we could know that the location of C. elegans' bodies could be positioned accurately in various complex environments by this algorithm. Furthermore, the accuracy rate of this algorithm was more than 98%, and its real-time, effectiveness, and a certain degree of robustness was also good.

Skeleton extraction method and area of measurement recognition method were also used to identify the 7 C. elegans' images for comparison, and the results of these two methods are shown in Table 2. As can be seen in Table 2, there are not good results by these two methods due to the interference of bubbles and influences of other factors, which means this image recognition algorithm based on symmetric similarity proposed in this paper is more accurate.

Table 2. Comparison of the three algorithms

	Exp 1	Exp 2	Exp 3	Exp 4	Exp 5	Exp 6	Exp 7
The Actual Number	3	6	1	2	1	2	4
Method inThis Paper	3	6	1	2	1	2	4
Skeleton Method	5	8	1	3	1	4	4
Area Method	4	9	1	2	1	3	3

Exp - Experient

6 Conclusions

C. elegans' image recognition algorithm based on symmetric similarity method is proposed in this paper. What this algorithm using the symmetry of C. elegans' bodies is needed to do is only to scan the image twice in order to complete several tasks, including labeling all the complex segments, removing the excess segments, and calculating the similarity, etc. Meanwhile, the point information of segments from C. elegans' bodies and the corresponding boundary chain code value is achieved to prepare data for the subsequent processing. Furthermore, the positioning accuracy rate is up to 98%. When time cycle is under 0.2s using this algorithm, the effectiveness is also high.

This algorithm has already been used in the life cycle detection system prototype development of C. elegans, and outline recovery and tracking and positioning algorithm of C. elegans will be added in the future to make this process of system image identification more complete.

Acknowledgments. This research was supported in part by the National Natural Science Foundation of China (Grant No. 61071158), and the Ph.D. Programs Foundation of Ministry of Education of China (Grant No. 20091102120022).

References

1. Zhang, B., Zhao, W., Ma, S.: Banding of Human Chromosome Image Analysis and Recognition System. China Journal of Image and Graphics 1, 148–150 (1996)
2. Sonka, M., Hlavac, V., Boyle, R.: Image Processing, Analysis and Machine Vision, 2nd edn. People Post Press, Beijing (2003)
3. Yang, B., Wang, W., Li, Y.: Image Separation Algorithm of Blood Cells Based On Skeleton. Computer Engineering and Applications 16, 94–97 (2003)

4. Liu, Y., Wang, B., Hu, P.: CD4 Analysis and Recognition Research of Cell Microscopic Image. Chinese Journal of Medical Instrumentation 29, 419–422 (2005)
5. Wang, J.: A Method to Find the Middle Axis of C-band Chromosome. Journal of Sichuan University (Natural Science) 41, 768–773 (2004)
6. Qin, T., Zhou, Z.: A New Way to Detect the Moving Target in the Sequences Image. Computer Application and Software 21, 105–107 (2004)
7. Wang, X., Wu, C.: System Analysis and Design Based on MATLAB-—Image Processing, pp. 57–58. Xidian University Press (2000)

Multi-sensor System of Intellectual Handling Robot Control on the Basis of Collective Learning Paradigm

Nikolay Tushkanov[1], Vladimir Nazarov[2], Alla Kuznetsova[2], and Olga Tushkanova[1]

[1] The North Caucasian State technical university, Umar Aliev St. 91, 357700 Kislovodsk, Russia
{tnb49,olga_tushkanova}@mail.ru

[2] The South Russian State technical university, Prosvesheniya St. 132, 346428 Novocherkassk, Russia
nazarov_v@mail.ru, avk@novoch.ru

Abstract. Approaches based on self-learning and self-organization of collective of interacting intellectual agents (sensor channels, units of neural networks structures and others DAI-forming entities) are recently more often used in development of collective recognition and control systems. Approaches to organization of co-learning and mutual learning on the basis of reliability coefficients of sensor data are described. Approach to development of recognizing multi-sensor system on a basis of two-dimensional two-level neural networks is offered. Scheme of multi-unit manipulators control on a basis of neural approach are described. The results are used for development of robotics systems for operation in extreme (underwater) conditions.

Keywords: control systems, artificial intelligence, multi-sensor systems, collective learning, co-learning, neural networks, multi-unit manipulators.

1 Introduction

Recently approaches based on procedures of self-learning and self-organization of interacting intellectual agents (sensor channels, units of neural networks structures and others DAI-forming entities) collective [3], [10], [11], [12] are used in realization of systems of collective recognition and management even more often.

2 Classification of Collective Learning Methods

The classification of concepts and methods of collective learning in the natural and artificial environments presented in a figure 1, without claiming for completeness, defines a place of the approaches to collective learning on the basis of co-learning and mutual learning paradigms created by author.

Y. Wang and T. Li (Eds.): Knowledge Engineering and Management, AISC 123, pp. 195–200.
springerlink.com

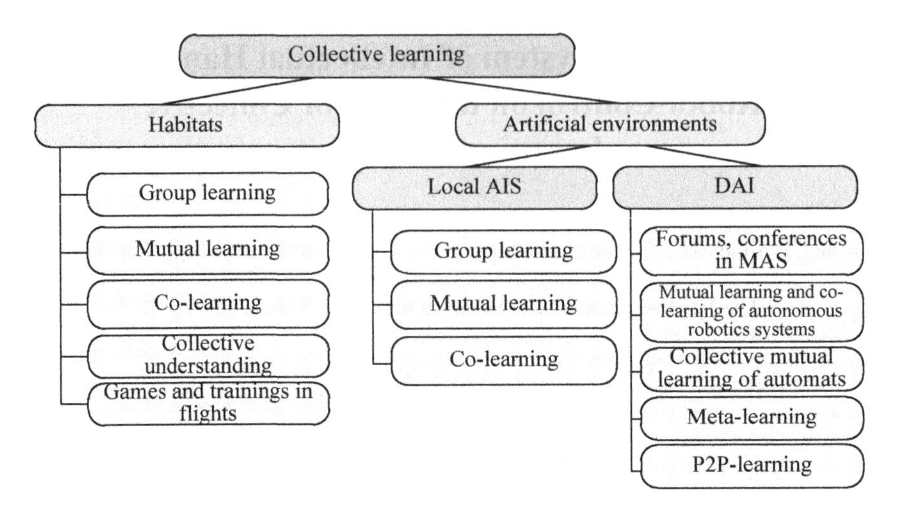

Fig. 1. Classification of collective learning methods: AIS – artificial intelligent systems, DAI – distributed artificial intelligent, MAS – multi-agent systems

3 Methods of Mutual Learning and Co-learning Organization

Let's consider the methods of collective learning laid in basis of created intellectual systems.

3.1 Mutual Learning on the Basis of Reliability Coefficients

Let's understand mutual learning as the process of collective interaction where competent agents learn less competent agents representing itself as the teacher.

Let's define reliability coefficient as the variable K_r accepting values from a range [0..1], and "0" corresponds to absolutely unreliable information arriving from the agent (i.e. video channel operation in utter darkness), "1" corresponds to absolute trustworthy information (i.e. 100 % coincidence of recognized object pattern and standard pattern stored in recognizing system).

Let's consider possible variants of mutual learning organization:

a) agents with value $K_r > 0,5$ are selected as "competent" agents. They form "jury" of agents which classify a situation by voting. The received decision becomes the learning information for all remaining agents;

b) "competent" agent is selected by criterion of a maximum of K_r, i.e.

$$K_{rj} = \max [K_{ri}]. \tag{3.1}$$

This agent in a specific situation becomes, as a matter of fact, the meta-agent.

3.2 Co-learning on the Basis of Reliability Coefficients

Let's define co-learning as one of varieties of the procedures presented below:

1) the procedure corresponding to process of "collective understanding" or procedure of well known Osborn "brainstorming (brainstorming session)" (procedure is implemented when $K_r<1$, i.e. in case of "exact" decision absence). Thus all agents in turn "publish" results of situation recognition (i.e. value of K_r and the recognized object description) in the joint database and analyze similar results of other agents/channels (a phase 1 of "brainstorming" procedure; herewith not only closest "standards" and images are published, but also slightly conceding to it, for example, with $K_r = 0{,}75\ K_{rs}$). Then each agent analyzes a level of coincidence of its results with results of others and declares (for example, on "bulletin board") a level of coincidence (an amount of conterminous descriptions from 50). And the image which has typed the greatest point, is choose as the collective decision;

2) the procedure implemented in a situation when K_{rj} is found out in one of sensor channels (agents):

$$K_{rj} >> K_{ri},\ \ i = [1,2,3,\ldots,j\text{-}1,j+1,\ldots,N]. \tag{3.2}$$

In this situation the channel becomes the teacher and "trains" remaining channels (agents).

3) the procedure organized in the absence of the "strong" superiority of one of agents over another. Herewith agent with K_{rj} is used as the learning agent:

$$K_{rj} = \max[K_{ri}]. \tag{3.3}$$

4 Procedure of Manipulator Control on the Basis of Co-learning Paradigm

Authors offer approach to creation of heterogeneous multi-sensor systems on the basis of a sensor channels co-learning and mutual learning paradigm. One of application variants of co-learning is its usage in creation of multi-sensor control system for intellectual handling robot. For handling robotics complex such channels can be represented by video, tactile (touch imitating), force, ultrasonic, infrared channels etc. Used procedure, as a matter of fact, carries out data fusion according to JDL-model of multi-sensor systems organization [4], [8]. According to the offered approach characteristics and possibilities of sensor subsystems are fused (not data itself) with help of the following procedure:

1) All channels which are not demanding manipulations for their functioning, perceive information about situation (conditions) or about object.

2) The level of received information reliability (using of Kalman filtration is possible here) is defined.

3) Selection of video objects and normalization of selected zones sizes for neural networks structures is produced. Thus, images invariance is provided.

4) Normalized images (visual, ultrasonic, infrared, etc.) submit on inputs of the associative recognizing subsystems (for example, two-dimensional Hopfield network, Hamming network or networks of adaptive resonance theory).

5) The channel with the highest K_r values learns channels with low (but not zero) reliability level.

6) The choice of appropriate (closest) object images from databases is carried out in manipulation zone at the top ontology level (possible situations knowledge bases).

7) Information about location and object type transmits to the gripper control system.

8) Gripper positions in a proper place and does attempt of capture (or palpation) of an object.

9) In a case of successful capture tactile and force sensor information in the form of patterns is processed in specialized recognizing subsystems (they can be neural networks too), where level of proximity to samples stored in data and knowledge bases is determined. In a case of high level of information reliability these channels can represent itself as "teachers" of visual subsystems.

10) In a case of absence of close images in bases escalating (reconfiguration) of neural structures and their learning for new images is produced.

11) Knowledge bases with possible objects transformations (deformation, merging etc.) are entered at ontology level for increasing effectiveness of collective recognition system.

12) Transformed images arrive in recognizing subsystem as secondary templates.

The system applying described approach is implemented in program complex form and currently is being tested on models of situations arising in extreme conditions.

5 Approach to Development of Multi-unit Manipulator Control Systems

One of the most perspective varieties of manipulator for underwater robots working in extreme conditions is so-called flexible or multi-unit manipulator (MUM) [6]. Authors offer the approach to creation of MUM control system on the basis of two-dimensional two-level neural network (figure 2). Here two-dimensional array of forces moments, which are changed in time, created by MUM links drives is submitted on neural network input. $M_{1,H}$ is the moment created by a drive of the first link in initial time moment, $M_{1,K}$ is finish moment value for this drive. Indexes 1...N correspond to drives from the first to the last, where N is an amount of MUM links. So two-dimensional array of moments corresponds to one of MUM relocation variant. Neural networks NN 1...NN N during learning (Learning 1) provide adjustment of local drives by criterion of maximum high-speed performance, and output neural network (NN 2d level) is learned to select "good" motions by criterion of minimum of forces moments with different sign (i.e. opposite directed) created by drives (Learning 2). As a result of learning, the quasi-optimal vector of moments $||M||(t)$ which is stored in control system memory is formed in output of second level neural network. Such approach to some extent implements Bernstein's ideas on synergetic synthesis of biological objects movements in training activity [1]. In classification (figure 1) this approach corresponds to a group paradigm of collective training.

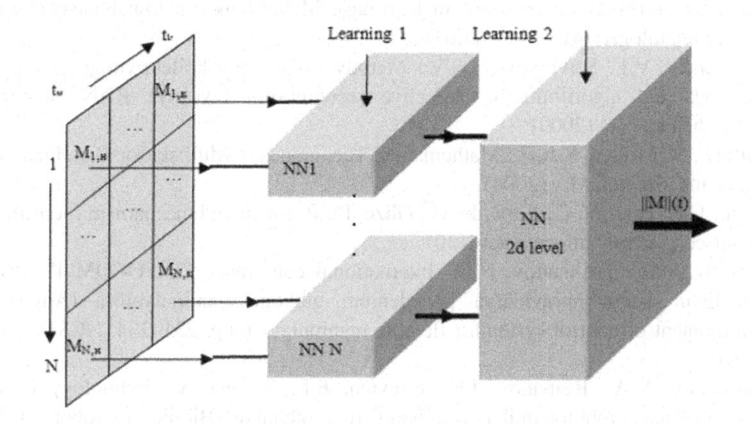

Fig. 2. Control system on the basis of two-level neural network

6 Modeling Results

At the moment the model of MUM dynamics and algorithms of MUM control system learning are developed and implemented. A set of typical movements, one of which is presented in figure 3, is implemented.

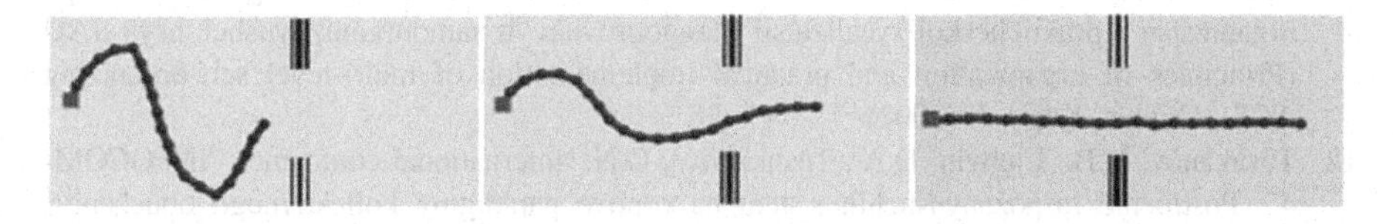

Fig. 3. Driving of type "penetration"

7 Conclusion

The approaches to recognizing MSS organization, technical and algorithmic implementations of manipulator control system offered in this paper are laid in basis of perspective control systems for complex manipulation robots. Work is carried out in collaboration with the Center for Robotics (CRDI RTC), Saint-Petersburg. Results are used for development of robotics systems for operation in extreme (underwater) conditions.

References

1. Bernstein, N.A.: O lovkosti i yee razvitii (About dexterity and its development). Moscow: Fizkul'tura i sport (1991)
2. Burdakov, S.F., Smirnova, N.A.: Metody obucheniya v sistemah upravleniya robotami (Methods of robots control system learning). Mehatronika, avtomatizatsiya, upravlenie (Novye tehnologii) (4), 15–19 (2003)
3. Gorodetsky, V.I., Serebryakov, S.V.: Metody i algoritmy kollektivnogo raspoznavaniya (Methods and algorithms of collective recognition). Izvestiya RAN "Avtomatika i Telemehanika" (1) (2009)
4. Hall, D., McMullen, S.A.H.: Mathematical Techniques in Multisensor Data Fusion. Artech House Inc., Boston,MA (2004)
5. Mano, J.-P., Bourjot, C., Lopardo, G., Glize, P.: Bio-inspared mechanism for artificial self-organised systems. Informatica 1(30)
6. Nazarov, V.A., Tushkanov, N.B.: International conference "INFOCOM-4". Podhody k sozdaniyu sistem upravleniya povedeniem gibkogo manipulyatora (Approaches to development of control systemfor flexible manipulator), pp. 231–234. NCSTU, Stavropol (2010)
7. Smolnikov, V.A., Romanov, I.P., Jurevich, E.I.: Science & Technology Conference: Ekstremal'naya robototehnika. Bionika v robototehnike (Bionics in robotics). SPb, St. Petersburg (2001)
8. Steinberg, Bowman, C.: Revisions to the JDL data fusion process model. In: Proceedings of the National Symposium on Sensor and Data Fusion (NSSDF), Lexington, MA (1998)
9. Terekhov, V.A.: Nyeirosetevye sistemy upravleniya (Neural control). Publishing house of St. Petersburg University, St. Petersburg (1999)
10. Timofeev, A.V: XI Science & Technology Conference: Ekstremal'naya robototehnika. Intellektual'noe i mul'tiagentnoe upravlenie robototehnicheskimi sistemami (Intellectual and multi-agent control of robotics systems). Publishing house of SPbSPU, St. Petersburg (2001)
11. Tushkanov, N.B., Jurevich, E.I.: International conference "Artificial intelligence. Printsipy organizatsii i prakticheskoi ryealizatsii mnogourovnevyh samoorganizuyushchihsya SAU (Principles of organization and practical implementation of multi-level self-organizing ACS). Crimea, Kaciveli (2002)
12. Tushkanov, N.B., Ljubvin, D.A., Tushkanova, O.N.: International conference "INFOCOM-4". Postroenie raspoznayushchih sistem na osnove paradigmy kollektivnogo obucheniya (Development of recognizing systems on the basis of a collective learning paradigm). NCSTU, Stavropol (2010)

A Polynomial-Time Algorithm for the Planar Quantified Integer Programming Problem*

Zhiyao Liang[1] and K. Subramani[2]

[1] Faculty of Information Technology, Macau University of Science and Technology, Macau
zyliang@must.edu.mo

[2] LDCSEE, West Virginia University, Morgantown, WV, U.S.A.
ksmani@csee.wvu.edu

Abstract. This paper is concerned with the design and analysis of a polynomial-time algorithm for an open problem in Planar Quantified Integer Programming (PQIP), whose polynomial-time solution is previously unknown. Since the other three classes of PQIP are known to be in PTIME, we also accomplish the proof that PQIP is in PTIME. Among its practical implications, this problem is a model for an important kind of scheduling. A challenge to solve this problem is that using quantifier elimination is not a valid approach. This algorithm exploits the fact that a PQIP can be horizontally partitioned into slices and that the feasibility of each slice can be checked efficiently. We present two different solutions to implement the subroutine *CheckSlice*: The first one uses IP(3), i.e., Integer Programming with at most three non-zero variables per constraint; the other is based on counting lattice points in a convex polygon. We compare the features of these two solutions.

Keywords: Quantified Integer Programming, Integer Programming, Combinatorial Optimization, Polynomial-time, Algorithm, Scheduling, Counting Lattice Points.

1 Introduction

Let us consider a special kind of scheduling. Suppose we need to schedule the starting o'clock y and the duration hours x of an important flight, for integral y and x, while the following constraints need to be satisfied: There are m crew members; Each crew member i, $1 \leq i \leq m$, has a personal requirement on y and x, described as $a_i y + b_i x \leq c_i$, for some numbers a_i, b_i, and c_i; Especially, the flight can possibly start at any hour between $3am$ and $5pm$. The question is, can we guarantee that such a flight can always be scheduled, no matter what is the starting time? This paper deals with such a problem.

Quantified Integer Programming, which is introduced in [17], is a mathematical programming paradigm which generalizes traditional integer programming (IP). In traditional integer programming, we are given a system of linear inequalities and asked whether the defined polyhedron encloses a lattice point. Integer programming encodes

* This research was supported in part by the Air-Force Office of Scientific Research under contract FA9550-06-1-0050 and in part by the National Science Foundation through Award CCF-0827397.

a number of combinatorial optimization problems and is strongly NP-complete [14]. In the polyhedral system $\mathbf{A} \cdot x \leq b$, the quantifiers are implicit and existential; i.e., we are asked whether there exists at least one x such that $\mathbf{A} \cdot x \leq b$.

In the Quantified Integer Programming problem (QIP), each variable can be quantified either existentially or universally; thus, QIP clearly generalizes traditional integer programming. Indeed QIP is PSPACE-complete; the hardness is established through a trivial reduction from the Quantified Satisfiability problem (SAT) to QIP. The completeness of QIP requires a more involved argument and mirrors the argument used to establish the completeness of IP for the class NP.

QIP can be used to model problems in a number of distinct domains including real-time scheduling [16], constraint databases [11], logical inference [3], computer vision [4], and compiler construction [10]. A number of interesting issues regarding the tractability and complexity of various QIP classes were addressed in [17].

Planar QIPs, or PQIPs, are constituted of precisely two variables. This paper is concerned with establishing the complexity of a specialized and non-trivial class of PQIPs called planar FPQIPs. In [17], a pseudo-polynomial time algorithm was proposed for FPQIP. but its complexity was unresolved. Here we propose an algorithm for FPQIP based on geometric insights, which converges in time polynomial in the size of the input. This paper is the first to present a *strongly polynomial-time* algorithm for the FPQIP problem. Since the other three classes of PQIPs are known to be in PTIME [17], our work finally establishes the PTIME complexity of PQIP.

The Quantified Linear Programming in the plane (PQLP) can be solved in polynomial time, using a quantifier elimination technique such as Fourier-Motzkin (FM) elimination. However, as is well-known, FM is not a valid technique for checking feasibility in integer constraints.

The rest of this paper is organized as follows: Section 2 formally specifies the problem under consideration. Section 3 describes related work in the literature. Our algorithm for the FPQIP problem, with two different solutions to implement the subroutine *CheckSlice*, is presented in Section 4. In Section 5 we discuss the correctness and complexity of the algorithm. We conclude in Section 6 by summarizing our contributions and outlining avenues for future research. More detailed discussion, analysis, and proofs of the algorithm are included in a longer version of the paper, which is available upon request.

2 Statement of Problem

Definition 1. *Let $X = \{x_1, x_2, \cdots x_n\}$ be a set of n variables with integral ranges. A mathematical program of the form*

$$Q_1 x_1 \in \{a_1 - b_1\}\ Q_2 x_2 \in \{a_2 - b_2\} \cdots Q_n x_n \in \{a_n - b_n\}\ \mathbf{A} \cdot \mathbf{x} \leq \mathbf{b},$$

where each Q_i is either $\exists$ or $\forall$, is called a Quantified Integer Program (QIP).

Note that the a_i and b_i are integers; further, we use $\{a_i - b_i\}$ to indicate the set of integers r, such that $a_i \leq r \leq b_i$.

Definition 2. *A planar QIP (PQIP) is the one in which the number of variables is restricted to two.*

Accordingly, a PQIP has the following form:

$$Q_1x_1 \in \{a_1 - b_1\}\ Q_2x_2 \in \{a_2 - b_2\}\ \mathbf{A} \cdot [x_1, x_2]^{\mathsf{T}} \leq b$$

Depending upon the nature of Q_1 and Q_2, each can be either $\exists$ or $\forall$, we have four different classes of PQIPs. Among them, only one class whose PTIME complexity is unknown, which is presented as follows.

When $Q_1 = \forall$, $Q_2 = \exists$, we refer to this class of PQIPs as FPQIPs to denote the fact that the universal quantifer ($\forall$) precedes the existential quantifier ($\exists$) in the quantifier string. Without loss of generality (and for purposes of conformity with the literature in computational geometry), such a PQIP has the following form:

$$\forall y \in \{c-d\}\ \exists x \in \{a-b\}\ \mathbf{A} \cdot [x,\ y]^{\mathsf{T}} \leq \mathbf{b}, \tag{1}$$

where

- $\mathbf{A}$ is a matrix with m rows and two columns, for some positive integer m;
- $\mathbf{b}$ is an integral m-vector,
- a, b, c, and d are integers. We assume $a \leq b$ and $c \leq d$. Note that if $c > d$ then Program (1) has an immediate answer which is **true**. Otherwise, if $c \leq d$ and $a > b$, then the answer is trivially **false**.

For FPQIP, the feasibility query can be described as follows: Is it the case, that for each integer $y \in \{c-d\}$, there exists a corresponding integer $x \in \{a-b\}$ such that $\mathbf{A} \cdot [x,\ y]^{\mathsf{T}} \leq b$. The rest of this paper is devoted to developing a polynomial time algorithm for the class of FPQIPs.

In this paper we refer to FPQIP as Program (1) or Problem (1). We can translate Program (1) to an equivalent form by adding the following four (redundant) constraints to it: $x \leq b$, $-x \leq a$, $y \leq d$, and $-y \leq c$. Such an equivalent program is described as

$$\forall y \in \{c-d\}\ \exists x \in \{a-b\}\ \mathbf{A}' \cdot [x,\ y]^{\mathsf{T}} \leq \mathbf{b}', \tag{2}$$

where $\mathbf{A}'$ is a matrix with two columns and $m+4$ rows, and $\mathbf{b}'$ is a vector with $m+4$ integers. Program (2) can also be described as:

$$\forall y \in \{c-d\}\ \exists x \in \{a-b\}\ \mathbf{g} \cdot x + \mathbf{h} \cdot y \leq \mathbf{b}', \tag{3}$$

where $\mathbf{g}$ is the left column of $\mathbf{A}'$ and $\mathbf{h}$ is the right column of $\mathbf{A}'$.

The ***feasible region*** of $\mathbf{A}' \cdot [x,\ y]^{\mathsf{T}} \leq \mathbf{b}'$ is a two-dimensional area such that $(x,\ y)$ is the Cartesian coordinates of a point in this area, for some real numbers x and y, if and only if $[x,\ y]^{\mathsf{T}}$ satisfies all the constraints of $\mathbf{A}' \cdot [x,\ y]^{\mathsf{T}} \leq \mathbf{b}'$. A point with integral x and y-coordinates is called a ***lattice point***. Program (2) has a graphical interpretation: Deciding true or false that for each integer t, with $c \leq t \leq d$, there is a lattice point whose y-coordinate is t that is covered by the feasible region of the constraints of Program (2).

Each constraint $x \cdot g_i + y \cdot h_i \leq b'_i$ of Program (2), with $1 \leq i \leq m+4$, implies a half-plane bounded by the line $x \cdot g_i + y \cdot h_i = b'_i$. The feasible region of the $m+4$ constraints of Program (2) is the intersection between the $m+4$ implied half-planes, which is a convex polygon.

3 Related Work

Both Quantified Linear Programming and Quantified Integer Programming are mathematical programming paradigms that owe their origin to [15], where they were used to model a number of real-time scheduling specifications. Quantified Integer Programming as an independent abstraction were studied in [17] and a number of special cases were analyzed from the perspective of computational complexity. Besides FPQIP, depending on the the nature of Q_1 and Q_2 described in Definition 1, there are three more classes of PQIP, which are clearly in PTIME, as discussed below:

(i) $Q_1 = \exists$, $Q_2 = \exists$ - In this case, the PQIP is simply a 2-dimensional integer program and can be solved in polynomial time [6].
(ii) $Q_1 = \exists$, $Q_2 = \forall$ - In this case, the PQIP has the form: $\exists x_1 \in \{a_1 - b_1\}\ \forall x_2 \in \{a_2 - b_2\}\ \mathbf{A} \cdot [x_1, x_2]^{\mathsf{T}} \leq b$. As detailed in [17], the discrete interval $\{a_2 - b_2\}$ can be made continuous ($[a_2, b_2]$) without affecting the feasibility of the PQIP. Further, using quantifier elimination, variable x_2 can be removed and the resultant program is an integer program in 1 dimension, which can be easily checked in polynomial time.
(iii) $Q_1 = \forall$, $Q_2 = \forall$ - Such a PQIP is called a Box QIP and can be checked for feasibility in polynomial time using polytope inclusion [17].

[18] shows that the language of reals with universal quantifiers is decidable. It is well-known that a fragment of integer arithmetic called Presburger Arithmetic, in which multiplication and exponentiation are not permitted, is decidable [9].

[8] discusses empirical observations in using the Fourier-Motzkin elimination technique for linear constraints over both rational domains and lattice domains; practical issues in polyhedral projection are further addressed in [7], where certain clausal systems are resolved by constructing the *approximate* convex hull of the feasible space.

[17] discusses the PQIP problem in general and the FPQIP in particular. Observe that the FPQIP problem can be solved in *pseudo-polynomial* time using the following approach: Simply enumerate all integers y in the range $\{c - d\}$. For each value of y, we can check whether the corresponding one dimensional polyhedron (an interval) encloses a lattice point. Note that this algorithm runs in $O(|d - c| \cdot n)$ time and is hence pseudo-polynomial. This paper presents the first *strongly polynomial* time algorithm for the FPQIP problem.

Besides Barvinok's method, there are other algorithms that can count lattices points in a convex polygon in a plane in polynomial time, such as (but not limited to) those presented in [2] and [5]. Different algorithms may have different advantages in terms of easiness of implementation. Barvinok's method [1] has the advantage that it can handle convex polyhedron beyond four dimensions.

4 A Polynomial-Time Algorithm for FPQIP

Our algorithm starts by translating Program (1) to Program (2). Then, the coordinates of the vertices (extreme points) of its feasible region are computed.

For each different y-coordinate, say g, of some vertex of the feasible region, we can draw a horizontal line described by $y = g$. These lines cut the polygon into slices. A

slice is the area in the polygon between two neighboring horizontal lines. The *top slice* is specially defined as the intersection of the line $y = t$ and the polygon, where t is the largest y-coordinate of the vertices of the polygon. So, the top slice is either a single point or a horizontal line segment. The idea of partitioning a polygon into slices is illustrated in Fig. 1.

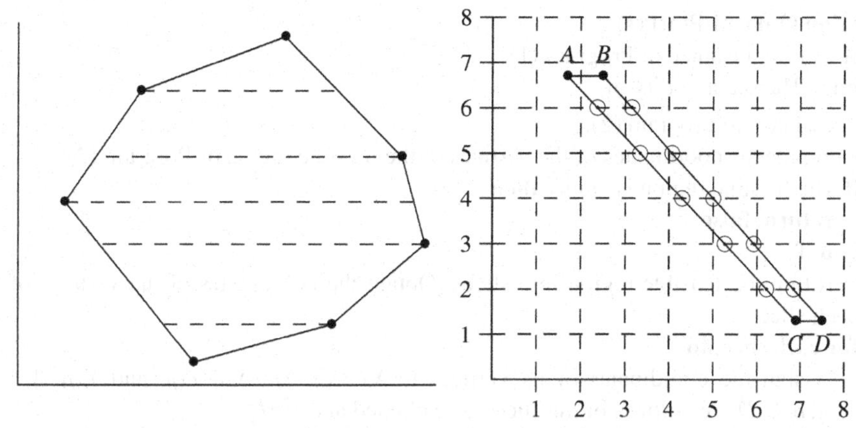

Fig. 1. The feasible region can be horizontally partitioned into slices

Fig. 2. The idea of the function *CheckSlice*

A slice is a trapezoid in general, whose top edge and bottom edge are horizontal. We name the two vertices on the top edge of a slice, from left to right, as A and B, and the two vertices on the bottom edge, from left to right, as C and D. Fig. 2 shows a possible slice with its vertices marked with A B C and D. A slice may look unusual, such as the one in Fig. 2, which is long and narrow and slant. Besides the common case that a slice is a trapezoid with four different vertices there are some special cases:

- A slice can be a triangle, when the top or bottom edge of the trapezoid is a single point. Then we consider that this single point has two names, A and B if it is the top edge of the slice, or C and D if it is the bottom edge of the slice. This case can happen when the slice is the second one (next to the top one) or is the bottom one.
- A slice can be a line segment with two different vertices. When it is non-horizontal, the top vertex is named as A and B, and the bottom vertex is named as C and D. When it is horizontal, the left vertex is named as A and C, and the right vertex is named as B and D. This case can happen when the slice is the top one, or when the whole feasible region is a non-horizontal line segment.
- A slice can be a single point, which has four names, A, B, C, and D. This case can happen when the slice is the top one.

Therefore, a slice can be uniformly described by four vertices named as A, B, C, and D.

Program (2) is satisfiable if and only if each slice of the feasible region is satisfiable, which is defined as follows:

[Slice Satisfiability]: Given a slice, let g' and g be the y-coordinates of its top and bottom borders respectively. The *slice is satisfiable* if and only if for each integer t in the closed interval between g' and g, there is a lattice point covered in this slice with coordinates (x,t), for some integer x.

The function *CheckFPQIP*, which is presented in Algorithm 1 below, is designed to solve FPQIP (Problem 1). For each slice, *CheckFPQIP* checks its satisfiability by calling the subroutine *CheckSlice*, whose implementation is explained later.

Function CHECKFPQIP(s)
Input s is an instance of Program (1).
Output The satisfiability of s.

1: Translate s to Program (2).
2: Compute the coordinates of the vertices of the feasible region of Program (2).
3: **if** The feasible region is empty **then**
4:     **return false**
5: **end if**
6: Partition the feasible region into slices. Obtain the coordinates of the vertices of each slice.
7: **for** each slice **do**
8:     Assign the coordinates of its vertices to X_A, X_B, X_C, X_D, Y_{AB}, and Y_{CD}. {/* A,B,C,D are vertices of the slice, as explained above */}.
9:     **if** $CheckSlice(X_A,X_B,X_C,X_D,Y_{AB},Y_{CD})$ = **false then**
10:         **return false**
11:     **end if**
12: **end for**
13: **return true**

Algorithm 1. *An algorithm that solves FPQIP (Problem 1).*

The simple method of checking the satisfiability of a common slice by vertically partitioning it into a rectangle and one or two right triangles, does work for a slice like the one the one shown in figure 2, whose satisfiability is non-trivial to check.

The subroutine *CheckSlice*, which is called by the *CheckFPQIP* shown above, decides the satisfiability of a slice. We present two different solutions to implement *CheckSlice*; They are the functions *CheckSlice*1 and *CheckSlice*2, which are shown in Algorithm 2 and Algorithm 3, respectively.

Function CHECKSLICE1($X_A,X_B,X_C,X_D,Y_{AB},Y_{CD}$)
Input The parameters are the coordinates of the vertices A B C and D of the slice.
Output The satisfiability of the slice, which is either **true** or **false**.

1: $n := \lfloor Y_{AB} \rfloor - \lceil Y_{CD} \rceil + 1$
    {/* n is the number of integers in the closed interval between Y_{AB} and Y_{CD}. */}
2: **if** n = 0 **then**
3:     **return true** {/* Trivially satisfiable*/}
4: **end if**
5: **if** $Y_{AB} = Y_{CD}$ {/* Y_{AB} and Y_{CD} are the same integer */} **then**
6:     **if** $\lfloor X_B \rfloor - \lceil X_A \rceil + 1 = 0$ {/* No integer is covered by the line segment*/} **then**
7:         **return false**

8: **else**
9: **return true**
10: **end if**
11: **end if**{/* The slice that is a horizontal line segment or a single point is handled.*/}
12: $\alpha := (X_C - X_A)/(Y_{CD} - Y_{AB})$
$\beta := (X_D - X_B)/(Y_{CD} - Y_{AB})$
13: $firstIntY := \lfloor Y_{AB} \rfloor$
14: Compute h_1 and h_2 that are the x-coordinates of the two intersection points of the line $y = firstIntY$ and the two edges AC and BD, respectively. {/* $h_1 \leq h_2$ */}
15: Compute the the following integer program with three variables p, q and r, with the goal to minimize r:
$0 \leq r \leq n-1$
$h_1 + r \cdot \alpha \leq p < h_1 + r \cdot \alpha + 1$ /* $p = \lceil h_1 + r \cdot \alpha \rceil$ */
$h_2 + r \cdot \beta - 1 < q \leq h_2 + r \cdot \beta$ /* $q = \lfloor h_2 + r \cdot \beta \rfloor$ */
$q - p + 1 = 0$ /* No lattice point on the rth cut.
The 0th cut is on the line $y = firstIntY$.*/
16: **if** there is no solution for this integer program **then**
17: **return true**
18: **else**
19: **return false**
20: **end if**

Algorithm 2. *Checking the satisfiability of a slice based on IP(3).*

We use the slice depicted in Fig. 2 as an example to explain the function *CheckSlice*1. We call an intersection of a slice and a line $y = t$, for some integer t, a ***cut***. For the slice (the trapezoid) illustrated in Fig. 2, the vertices of the cuts are marked with circle dots. There are five cuts in the slice. This slice is not satisfiable since on the two cuts along the lines $y = 3$ and $y = 2$ there is no lattice point. If *CheckSlice*1 is called on this slice, then h_1 and h_2 are the x-coordinates of the two vertices of the cut along the line $y = 6$, and $firstIntY = 6$, the minimal r is 3, and the return is **false**.

Note that when a slice is not satisfiable, *CheckSlice*1 finds an integer r such that the cut on the line $y = r$ does not cover any lattice point and this cut is the closest to the top border (or to the bottom border, with a simple adjustment to the algorithm) of this slice. It is a helpful feature for solving other problems related to FPQIP.

We present another method to decide the satisfiability of a slice based on counting lattice points in a convex polygon, which is *CheckSlice*2, presented as Algorithm 3.

The key difference of *CheckSlice*2 is on how to handle a special kind of slice, whose top border and bottom border (parallel to the x-axle) are strictly shorter than 1. Given such a *special slice* $\mathscr{S}$, each cut over it (recall that a cut is the intersection of the slice and a line $y = i$ for some integer i) can cover at most one lattice point. Suppose the range of the y-coordinates of $\mathscr{S}$ covers n integers, then deciding the satisfiability of $\mathscr{S}$ means checking whether the number of lattice points covered in $\mathscr{S}$ is n. There are known algorithms that can count the lattice points in a two-dimensional convex polygon in polynomial time, including those described in [1] [2] [5]. For a slice such that its top border and bottom border are both no shorter than 1, this slice is obviously satisfiable (a *trivial slice*). For a slice such that only one of its top border and bottom border is shorter than 1, we can cut it into one special slice and one trivial slice.

Function CHECKSLICE2($X_A, X_B, X_C, X_D, Y_{AB}, Y_{CD}$)
Input The coordinates of the vertices A B C and D of a slice.
Output The satisfiability of the slice, **true** or **false**.
Require: The algorithm of [1] is implemented as a function *CountLatticePoints*, which computes the lattice points covered in a slice. The 6 parameters of this function are the coordinates of the vertices of a slice, just like the function *CheckSlice*.

1: $n := \lfloor Y_{AB} \rfloor - \lceil Y_{CD} \rceil + 1$
  {/* n is the number of integers in the closed interval between Y_{AB} and Y_{CD}. */}
2: **if** n = 0 **then**
3:   **return true** {/* The slice is trivially satisfied */}
4: **end if**
5: **if** $Y_{AB} = Y_{CD}$ {/* Y_{AB} and Y_{CD} are the same integer */} **then**
6:   **return** $\lfloor X_B \rfloor - \lceil X_A \rceil + 1 \neq 0$ {/* **true** or **false** for this inequality */}
7: **end if**{/* The slice that is a horizontal line segment or a single point is handled. */}

8: **if** $X_B - X_A \geq 1$ and $X_D - X_C \geq 1$ **then**
9:   **return true** {/* A trivial slice */}
10: **end if**
11: **if** $X_B - X_A < 1$ and $X_D - X_C < 1$ **then**
12:   $u :=$ *CountLatticePoints*$(X_A, X_B, X_C, X_D, Y_{AB}, Y_{CD})$.
13:   **return** $u = \lfloor Y_{AB} \rfloor - \lceil Y_{CD} \rceil + 1$ {/* **true** or **false** for this equality */}
14: **end if**{/* A special slice is handled*/}
15: **if** $X_B - X_A \geq 1$ and $X_D - X_C < 1$
  {/* The slice will be cut into two slices—a trivial one beneath a special one. */}
  then
16:   Find the first (smallest) integer t such that $t > Y_{CD}$, and for the length L of the intersection (cut) between the slice and the line $y = t$ it is true that $L \geq 1$. {/* it must be true that $Y_{CD} < t \leq Y_{AB}$ */}
17:   **if** $t - 1 < Y_{CD}$ **then**
18:     **return true**
19:   **else**
20:     Let X_1 and X_2 be the x-coordinates of the left and right intersection points of the line $y = t - 1$ and the slice.
21:     Let $u :=$ *CountLatticePoints*$(X_1, X_2, X_C, X_D, t - 1, Y_{CD})$.
22:     **return** $u = t - 1 - \lceil Y_{CD} \rceil + 1$ {/* **true** or **false** for this equality */}
23:   **end if**
24: **end if**
25: **if** $X_B - X_A < 1$ and $X_D - X_C \geq 1$ **then**
26:   Compute similarly to Lines 16-23. Again the slice is partitioned into two parts; but the special one is on top of the trivial one.
27: **end if**

Algorithm 3. *Checks the satisfiability of a slice based on counting lattice points.*

5 Correctness and Resource Analysis

The correctness of Algorithm 1 clearly follows its design using slices.

In the *arithmetic-complexity model*, each basic arithmetic operation $+$, $-$, $\times$, or $/$ is a unit-cost operation, while in the *bit-complexity model* the complexity of a basic arithmetic operation is counted w.r.t. to the bit length of its parameters. For two integers c and d such that each of them has l bits, $d-c$ or $d+c$ can be computed with a time cost $O(l)$, while $d \times c$ or d/c can be computed with a time cost $O(l^2)$, or even more efficiently $O(l(\log\ l)(\log\ \log\ l))$ using the Schönhage-Strassen algorithm [13]. In order to analyze the time cost of the algorithms, and especially for a clarified comparison between the time cost of computing the feasible region and the time cost of the executions of *CheckSlice*, we use the bit-complexity model in this paper.

Although calculating the precise complexity of the algorithm involves considerable details, doing so is relatively straightforward since the complexity of 3-integer programming and counting lattice points are well-known. A longer version of this paper, which is available upon request, has detailed proofs. We summarize the complexity results of the algorithm as follows.

Given an instance of Program (2) with bit length l, let $M(l)$ be the bit-complexity of l-bit integer multiplication.

- A call of *CheckSlice*1 (Algorithm 2) has a time cost $O(l^2)M(l)$, and an expected time cost $O(\log\ l)M(l)$;
- A call of *CheckSlice*2 (Algorithm 3) has a time cost $O(l \cdot \log\ l)M(l)$.

Therefore, Given an instance of FPQIP, i.e. Program (1), with m constraints and a bit length l, we can show that

- using *CheckSlice*1, i.e, Algorithm 2, the time cost and expected time cost of the computation of *CheckFPQIP* (Algorithm 1) are $O(m \cdot l^2)M(l)$ and $O(m \cdot \log\ l)M(l)$ respectively;
- using *CheckSlice*2, i.e., Algorithm 3, the time cost of *CheckFPQIP* is $O(m \cdot l \cdot \log\ l)M(l)$.

Base on the above discussion, we also have the following complexity results:

- FPQIP is in `PTIME`, since *CheckFPQIP* is a polynomial-time solution.
- Planar Quantified Integer Programming (PQIP) is in `PTIME`, since the other three classes of PQIP beside FPQIP are known to be in PTIME [17].

Asymptotically *CheckSlice*1 is slower than *CheckSlice*2. However, *CheckSlice* has the advantage that when a slice is not satisfiable, it can show an evidence that the slice does not cover any lattice point on a line defined by $y = r$, for a certain integer r, while *CheckSlice*2 cannot. Such a line is also a no-certificate showing that the polygon is not satisfiable, and it can be easily verified. *CheckSlice*1 also has the advantage of being useful for some optimization problems related to FPQIP, since it can compute more than the feasibility of an FPQIP, while *CheckSlice*2 does not.

6 Conclusion

In this paper, we established the polynomial time complexity of FPQIP, which is the only unsolved class of PQIP. The algorithm proposed for FPQIPs is based on our insights into the geometry of the problem and not quantifier elimination. The approach of slicing the feasible region and then check the satisfiability of each slice obviates the need for enumerating all the points of the universally quantified variable.

There are several directions for future research:

(i) Optimization problems of PQIPs - When a given PQIP is deemed infeasible, it is interesting to consider the maximum feasible range of the universally quantified variable, or the maximum subset of feasible constraints.
(ii) An empirical study - Our goal is to integrate a practical implementation of our work into an SMT solver such as Yaices [12]. Several SMT solvers solve integer programs through SAT encoding and it would be interesting to contrast the running times of the SAT approach and the approach proposed in this paper.
(iii) Hardness of PQIPs - We would like to establish the exact complexity of PQIPs. Given that our approach is easily parallelizable, it is likely that this problem belongs to the class NC.
(iv) It is natural to investigate the design and implementation of efficient randomized algorithms for FPQIPs.

Acknowledgments. This research was conducted primarily in the Lane Department of Computer Science and Electrical Engineering in West Virginia University, where Zhiyao Liang worked as a Postdoctoral Research Fellow. Zhiyao Liang wishes to thank Barvinok Alexander for his kind help on the precise complexity of counting lattice points. K. Subramani wishes to thank Natarjan Shankar for friendly discussions when K. Subramani was an Invited Professor in School of Computer Science, Carnegie Mellon University.

References

1. Barvinok, A.: A polynomial time algorithm for counting integral points in polyhedra when the dimension is fixed. In: Symposium on Foundations of Computer Science, pp. 566–572 (1993)
2. Beck, M., Robins, S.: Explicit and efficient formulas for the lattice point count in rational polygons using dedekind - rademacher sums. Discrete & Computational Geometry 27(4), 443–459 (2002), http://front.math.ucdavis.edu/math.CO/0111329
3. Chandru, V., Hooker, J.N.: Optimization Methods for Logical Inference. Series in Discrete Mathematics and Optimization. John Wiley & Sons Inc., New York (1999)
4. Hentenryck, P.V.: Personal Communication
5. Yanagisawa, H.: A Simple Algorithm for Lattice Point Counting in Rational Polygons. Tech. rep., IBM Research Tokyo Research Laboratory, Technical Report RT0622 (2005)
6. Kannan, R.: A polynomial algorithm for the two-variable integer programming problem. JACM 27(1), 118–122 (1980)
7. Lassez, C., Lassez, J.: Quantifier Elimination for Conjunctions of Linear Constraints via a Convex Hull Algorithm. Academic Press (1993)

8. Lassez, J.L., Maher, M.: On fourier's algorithm for linear constraints. Journal of Automated Reasoning 9, 373–379 (1992)
9. Papadimitriou, C.H.: Computational Complexity. Addison-Wesley, New York (1994)
10. Pugh, W.: The omega test: A fast and practical integer programming algorithm for dependence analysis. Comm. of the ACM 35(8), 102–114 (1992)
11. Revesz, P.: Safe query languages for constraint databases. ACM Transactions on Database Systems 23(1), 58–99 (1998)
12. Rushby, J.M.: Tutorial: Automated formal methods with pvs, sal, and yices. In: SEFM, p. 262 (2006)
13. Schönhage, A., Strassen, V.: Schnelle multiplikation großer zahlen. Computing 7(3), 281–292 (1971)
14. Schrijver, A.: Theory of Linear and Integer Programming. John Wiley and Sons, New York (1987)
15. Subramani, K.: Duality in the Parametric Polytope and its Applications to a Scheduling Problem. Ph.D. thesis, University of Maryland, College Park (August 2000)
16. Subramani, K.: A Specification Framework for Real-Time Scheduling. In: Grosky, W.I., Plášil, F. (eds.) SOFSEM 2002. LNCS, vol. 2540, pp. 195–207. Springer, Heidelberg (2002)
17. Subramani, K.: Tractable fragments of presburger arithmetic. Theory of Computing Systems 38(5), 647–668 (2005)
18. Tarski, A.: A Decision Method for Elementary Algebra and Geometry. Univ. of California Press, Berkeley (1951)

Building Computational Model of Emotion Based on Particle System

Yan-jun Yin[1,2] and Wei-qing Li[1]

[1] College of Computer Science and Technology, Nanjing University of Science and Technology, Nanjing 210094

[2] College of Computer and Information Engineering, Inner Mongolia Normal University, Hohhot, 010022

ciecyyj@163.com, li_weiqing@139.com

Abstract. Emotions play a critical role in increasing the believability of virtual agents. The paper develops an emotion model of virtual agent using particle system and OCC model. To better portray emotions of virtual agents, emotional experience is reflected by two aspects, which are the outer emotion defined by the intensity of particles, and the inner feeling by the number of particles. Based on the perspective of psychic energy, the interactional effect of emotions is incarnated through particle motion. Simulation is done by using Matlab software, and the results show that the emotion model can simulate better dynamic process of emotion transferring and change spontaneously.

Keywords: Emotion Model, Particle System, Outer emotion, Inner feeling.

1 Introduction

Emotions play an important role in daily life; they influence how we think and behave, and how we communicate with others. Psychology, neurology and cognitive science have been concerned with modeling the emotions for many years [1]. Inspired by these models, intelligent agents researchers have begun to recognize the utility of computational models of emotions for improving complex, interactive programs [2]. With the growing interest in AI, the research on computational models of emotions is becoming a hot topic in recent years [3]. J.Bates built believable agents for OZ project [4] using OCC model [5].EI-Nasr [6] provided a computational model of emotion based on event appraisal. Wang considered that emotions change in accordance with statistically rules [7]. Hu [8] built an emotion model combined particle system and active field. In the particle system, the particles are eternal and the equilibrium state of emotions was completed by Brownian motion. Based on psychology energy idea, Teng [9] presented not only the changing process of the emotion ,but also the Markov chain model of emotion transferring spontaneously and the Hidden Markov chain model of emotion transferring stimulated based on affective energy. Inspired by psychic energy [10], we present a model for studying the dynamic process of emotion transferring of virtual agent based on his personality and external stimulus.

Y. Wang and T. Li (Eds.): Knowledge Engineering and Management, AISC 123, pp. 213–218.
springerlink.com

2 Basic Concepts of Building Emotion Model

2.1 Emotion Space and Personality Space

Emotion is the complex psychophysiological experience of an individual's state of affect as interacting with biochemical and environmental influences. Its effect on human cognition arouses the attention of researchers. However, Up to now, there is no consensus on the definition about emotion. Researchers understand emotion and define emotion according to their research. zhao[11] defined the basic emotion as four mutual reverse emotions. Ekman provided six basic emotions according to the relationship between emotions and facial expressions. Based on event's consequences and causes, we define four basic emotions: Happy, Anger, Relax and Anxiety.

For emotional intensity, most researchers agreed consistently that human being can produce different intensities of emotions. Ren [12] pointed out that emotions have different intensities, and emotional experiences consume psychic energy. The paper defines emotion intensity is in the range of 0～3 in order to easily reflect the different emotion intensity. An agent's emotion ES is defined as a four-dimensional vector, where each dimension is represented by a emotional factor, $e_i \in [0,3]$.

$$ES=<e_{Happy}, e_{Anger}, e_{Relax}, e_{Anxiety}> \quad (1)$$

Different individuals usually show different emotional responses after they accepted same external stimulations. This difference mainly reflects in their different personalities. For these reasons, personality is defined as one of the main features to make a distinction among individuals in most of emotion models. Currently, most widely accepted models have three-dimensional personality model (PEN) and five-dimension personality model (OCEAN). This paper adopted two-dimensional model provided by Eysenck. An agent's personality (PS) is a two-dimensional vector, where each dimension is represented by a personality factor, $p_i \in [0,1]$ .Two factors of determining personality are extraversion and neuroticism.

$$PS=<P_E, P_N> \quad (2)$$

2.2 Emotion Particle

Emotion particles, as the major component to express the virtual agent's emotion, fill the whole emotion space in discrete way. And they do irregular motion supported by the physical energy which is provided by virtual agent itself. Each emotional particle has a certain lifetime. In this paper, we define the emotional particle as a 3-tuple, Particle= {Px, Py, life}. Px, Py$\in$[0,3] is the position of emotional particle in emotion space, and life$\in$R represents the particles lifetime.

Outer emotion is described as the outer expression degree of virtual agent's emotion and is measured by the emotion intensity of particles. For a certain emotion, the intensity of the outer emotion can be calculated by summing up values gained through projecting particles on the axis of basic emotions.

$$outer\_emotion = \sum_{j=1}^{n} p_j(i) \quad (3)$$

Where i∈ES, $p_i(j)$ is the projection value on the axis of the emotion i, and n is the number of all particles.

Inner feeling is defined as the degree of agent's internal feeling for emotion experience. It measures by the number of particles located in the basic emotional area. By calculating the inner feeling of virtual agent, his emotion experience can be revealed under the outer emotional expression. The inner feeling is defined as follow:

$$Feeling=\sum_{j=1}^{n} N(p_j) \tag{4}$$

Where i∈E , and $N(p_j)$ equals 1 if search_emotion_loc(p_j) is equals to i and 0 otherwise. The primary functions of search_emotion_loc() is to check whether the particles position is in basic emotional area. Basic emotional area is a quarter of a circle, formed by the emotion axis which rotated 45 degree toward its both sides.

3 Emotion Model Based on Particle System

3.1 Emotion Particle Generation

In particle system, generation of emotion particles mainly involves initialization and running stages. In the process of initialization, the particles are randomly generated, and their locations are in the unit circle of the emotion space and are consistent with normal distribution. The lifetime distributions of these particles are in accordance with uniformly probability, and the value range is 1 to L units.

In operating process, emotion particle generation, on the one hand, is an auto-added process in order to recoup the loss particles as the result of death; on the other hand, it is activated due to external stimulation. After the initialization process is completed, the number of dying particles in the ⊿t time is K, where K = N / L *⊿t, N represents the number of particles, and L particles lifetime. As complement of died particles, the way of particles generation, without external stimulation, is the same as that in initialization process. However, under external stimulus, the particle system will activate particles located the unit circle centered at the stimulus point. The corresponding equation of particle generation under external stimulus is

$$p_j(t+1)=sti\_Point(t) + rand() \tag{5}$$

$$p_j(life)= \gamma*(L-1)*rand() + 1 \tag{6}$$

Where j=1,2,…N and the function sti_Point(t) represents the intensity of the stimulation point in the time t. γis an adjustable factor to control emotion experience time. The rand() function generates random numbers between -1.0 and +1.0 in according with normal distribution.

3.2 Emotion Particle Motion

Particle motion represents the state change of human emotions, and relies on the energy from the external stimulation. Particle motion is described below.

$$p_j(t+1)=\varpi * p_j(t)+\lambda * v * P_E \tag{7}$$

Where p_j represents the position of particle j in the emotion space, ω the inertial factor, λ the psychic energy coming from external stimulus, P_E the neuroticism factor of an agent's personality.

In the case of external stimulation, emotional particles move towards the stimulus point, which is affected by distance, between the particle and the stimulus point, and by the angle θ formed by the particle, the stimulus point and coordinate origin in emotion space. The longer the distance and the greater the angle, the less the energy obtained from stimulation point. Based on these analyses, the stimulus energies changed from external stimulus are defined as follow:

$$\lambda = sti\_Point(t) * (\frac{1}{1+r_1^2} + r_2) \tag{8}$$

Where r_1 is the distance between particle and stimulus point, r_2 the value of cosine.

3.3 Emotion Particle Decay

Emotion does not immediately disappear with external stimulus' end, but decays over time. In the particle system, emotion decay is implemented by emotion decreasing in intensity and the lapse of particle life. The degree of lapsing life of particles, located in different positions of emotion space, is the same, but the extent of emotional intensity decay is different. In the first intensity area of emotion space, emotion is in peace state, and particles movement follows Brownian motion without energy consumption. The higher the particles locate in intensity area, the more the energies consume. When the energies of particles are exhausted or their lifetime is end, these particles will die. The lapse of particle life is described below.

$$p_j(life, t+1) = p_j(life, t) - \delta \tag{9}$$

where δ is an adjustment factor in order to control the lapse of particles' life-time. In the paper, we set it to 1.The intensity decay of particle can be described as follow:

$$P_j(t+1) = p_j(t) * (1 - abs(\lfloor p_j o - 1 \rfloor) * P_E * rand()) \tag{10}$$

Where P_jO represents the Euler distance between p_j and coordinate origin in the emotion space. The basic meaning of the formulae of the intensity decay is: first, in the peace state, no energy is consumed. Second, even in the same emotion state, the energy consumptions are different because of the different personality of virtual agent. Third, the rand () function reflects that, in the same emotion state, the emotional energy consumption for specific agent is also not always exactly the same.

4 Stimulation and Analysis

A large number of simulations have been performed using Matlab software, drawing a variety of rational conclusions. In this section some of the simulation results are discussed. Simulations shown here is using a circle of radius 3 represented the emotion space filled with particles. We do the experiments with the particle number from 100 to 10,000.The experiment results have no significant improvement after the particle number was set to 2000. So the number of particles we chose is 2000, and the lifetime of particles in system is 5 unit-long. We describe the emotion reactions of virtual agent under no-stimulus and under stimulus.

In the case of no external stimulus, particles mainly do the Brownian motion. The results showed the intensities of basic emotions are stable during 20 unit length of time, and the emotions are in the peace state. The intensity values are slightly changing, but not obviously fluctuating (See Fig.1). This is in line with human emotional response without external stimulation.

In the peace state of emotion, virtual agent accepts the external stimulus. For example, the football team, supported by an agent, wins, and the agent's emotion experience is happy. The emotional response of the virtual agent is shown in figure 1. The intensity of happy emotion rise to the higher value 1.88, it means that the virtual agent feels great joy. Then the emotion intensity is decreasing over time, but the virtual agent is still in the joy. At the end, the emotion gradually fell back to the peace emotion. The inner feeling of this virtual agent is pleasant. The system simulated the emotion reactions of different agents with the same external stimulus. The result shows that level of the emotion is determined by an agent's personality. This situation represents a case where the emotion expression of an introvert agent is weaker than those of an extrovert agent. This is entirely consistent with the emotion response of different personality. In non-peaceful state of agent's emotion, the reaction to stimulation will be influenced by the current emotion. For example, the agent is incited by the happy stimulation under the same environment. Results showed that the reaction to the happy stimulation is strong at the time when he is happy, and the reaction is no simply addition (Fig.1).

5 Conclusions

Based on psychic energy of psychology theory, this paper presents a way to build emotion model by using particle system. This emotion model can embody these basic features which are described by Picard [3] in his book *Affective computing*. Emotion model, in the methods, built on the particle system is simple and easy to understand. Moreover, the diversity of particle state fully reflects the part stochastic characteristic of the virtual individual emotion under monolithic certainty. The interactional effect of emotions was fully incarnated through particle movement. The disadvantages of the method are: one is that the emotion model cannot accept indirectly external stimulation. The other is that evaluating the emotion model is difficult because there is no normal assessment standard. Concerning future work, the emotional contagion, as a importance factor influencing on the emotion experience, will be considered in emotion model.

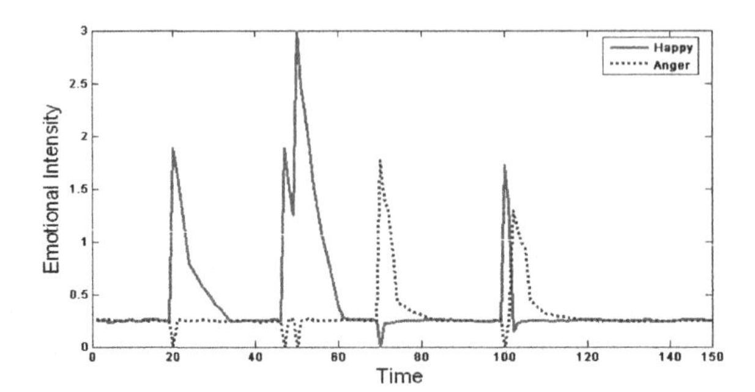

Fig. 1. The emotion levels of an agent with personality (0.5, 0.9).As there is no stimulus about anxiety and relax, the two emotions will not be shown. At the time (t=20s), the agent received happy stimulus (sti-Point=(0,5)). He received the same stimulus at the time (47s and 50s). Then he received an anger stimulus (t=70s, sti-Point= (0,-5)). At the 102s, he received another anger stimulus after he received a happy stimulus (t=100s, sti-Point=(0,5)).

References

1. Maria, K.A., Zitar, R.A.: Emotional agents: A modeling and an application. Information and Software Technology 49(7), 695–716 (2007)
2. El-Nasr, M.S., Yen, J., et al.: FLAME - Fuzzy logic adaptive model of emotions. Autonomous Agents and Multi-Agent Systems 3(3), 219–257 (2000)
3. Picard, R.: Affective Computing. MIT Press, Boston (1997)
4. Bates, J.: The Role of Emotion in Believable Agents. Communications of the ACM 37(7), 122–125 (1992)
5. Ortony, A., Clore, G., Collins, A.: The Cognitive Structure of Emotions. Cambridge University Press, Cambridge (1988)
6. El-Nasr, M.S., Yen, J., et al.: FLAME - Fuzzy logic adaptive model of emotions. Autonomous Agents and Multi-Agent Systems 3(3), 219–257 (2000)
7. Wang, Z.: Artificial emotion. China Machine Press (2009) (in Chinese)
8. Hu, B.-C., Chen, H.-S.: Research of Emotion Model Based on Particle System and Active Field. Mind and Computation 3, 36–44 (2009) (in Chinese)
9. Teng, S.-D., Wang, Z.-L., et al.: Artificial emotion model based on psychology energy idea. Computer Engineering and Applications 3, 1–4 (2007)
10. Turner, J.H.: Human Emotions: A Sociological Theory. Eastern Publish (2009)
11. Zhao, J., Wang, Z., Wang, C.: Study on Emotion Model Building and Virtual Feeling Robot. Computer Engineering 33, 212–215 (2007) (in Chinese)
12. Ren, J., Zhou, L., Luo, J.: Correlation of Zone of Proximal Development with Emotional Changes. Journal of Zhejiang Normal University (social sciences) 35(1), 77–84 (2010)

Hölder Type Inequality and Jensen Type Inequality for Choquet Integral

Xuan Zhao and Qiang Zhang

School of Management and Economics, Beijing Institute of Technology, Beijing, China
zhaoxuan@bit.edu.cn, qiangzhang@bit.edu.cn

Abstract. The integral inequalities play important roles in classic measure theory. With the development of fuzzy measure theory, experts want to seek for the integral inequalities of fuzzy integral. We concern on the inequalities of Choquet integral. In this paper, Hölder type inequality and Jensen type inequality for Choquet integral are presented. As the fuzzy measure are not additive, thus what is the other conditions for integral inequalities are discussed. Besides, examples are given to show that the conditions can't be omitted.

1 Introduction

Since the concept of fuzzy measure and fuzzy integral was introduced in [17], they have been comprehensively investigated [8] [4]. There are interesting properties which have been studied by many authors, including Wang and Klir [9], Pap [13], Grabisch [7] and ,among others. On the other hand, fuzzy integrals, for example Choquet integral [3] and Sugeno integral [17], have been widely used in various fields, such as decision making , artificial intelligence and economics [7], [16], [19] .

Some integral inequalities [6], such as Markov's inequality, Jensen type inequality, Hölder's inequality and Minkowski inequality, play important roles in classic measure space [2]. A natural thought is whether these integral inequalities still hold in fuzzy measure space under the condition of non-additive measure. The study of inequalities for Sugeno integral was developed by Román-Flores, Flores-Franulič, Chalco-Cano, Yao Ouyang, Mesiar, Agahi [15, 5, 14, 12, 1, 11] and so on. All of them enrich the fuzzy measure theory. We focus on the inequalities for Choquet integral.There are hardly any papers concern about inequalities for Choquet integral. R. Wang [18] has done this work, but the Hölder type inequality and Jensen type inequality for Choquet integral are obviously uncorrect. It's easy to find errors in the procedure of the proof and to give counterexamples. Thus the conditions under what the Hölder type inequality and the Jensen type inequality for Choquet integral are discussed.

R.Mesiar and Jun Li [10] gives several inequalities of choquet integral under certain conditions. This paper gives two inequalities of Choquet intergral while from different points of view. This paper is organized as follows. Section 2 provides some basic notations, definitions and propositions. In sections 3, the Hölder type inequality and the Jensen type inequality for Choquet integral are displayed. Section 4 consists of the conclusion and problems for further discussion.

Y. Wang and T. Li (Eds.): Knowledge Engineering and Management, AISC 123, pp. 219–224.
springerlink.com

2 Preliminaries

Let X be a non-empty set, $\mathscr{F}$ be a σ-algebra of X. Throughout this paper, all considered subsets belong to $\mathscr{F}$.

Definition 1. *A set function $\mu : \mathscr{F} \to [0,1]$ is called a fuzzy measure if:*
(1) $\mu(\emptyset) = 0$ and $\mu(X) = 1$;
(2) $A \subseteq B$ implies $\mu(A) \leq \mu(B)$;
(3) $A_n \to A$ implies $\mu(A_n) \to \mu(A)$.
When μ is a fuzzy measure, then $(X, \mathscr{F}, \mu)$ is called a fuzzy measure space. Let $\mathbb{F}$ be the set of all non-negative measurable functions defined on X.

Definition 2. *Suppose $(X, \mathscr{F}, \mu)$ be a fuzzy measure space. Let s be a measurable simple function on X with range $\{a_1, a_2, ..., a_n\}$ where $0 = a_0 \leq a_1 \leq a_2 \leq ... \leq a_n$. The Choquet integral of s with respect to μ is defined as*

$$\int_X s d\mu = \sum_{i=1}^{n} (a_i - a_{i-1}) \cdot \mu(\{x | f(x) \geq a_i\}).$$

For a non-negative measurable function f,

$$\int_X f d\mu = sup \left\{ \int_X s d\mu \,\middle|\, s \text{ is a simple function, } 0 \leq s \leq f \right\}.$$

Proposition 1. *Let f and g be non-negative measurable functions on $(X, \mathscr{F}, \mu)$, A and B be measurable sets, and a be a non-negative real constant. Then,*
(1) $\int_A 1 d\mu = \mu(A)$;
(2) $\int_A f d\mu = \int f \cdot \chi_A d\mu$;
(3) If $f \leq g$ on A, then $\int_A f d\mu \leq \int_A g d\mu$;
(4) If $A \subset B$ then, $\int_A f d\mu \leq \int_B f d\mu$;
(5) $\int_A af d\mu = a \cdot \int_A f d\mu$.

Unlike the Lebesgue integral, the Choquet integral is generally nonlinear with respect to its integrand due to the nonadditivity of μ. That is, we may have

$$\int (f+g) \mathrm{d}\mu \neq \int f \mathrm{d}\mu + \int g \mathrm{d}\mu.$$

Definition 3. *For every pair f and g of measurable functions, f and g are said comonotone if for every pair x_1 and x_2 $(x_1, x_2 \in X)$,*

$$f(x_1) \leq f(x_2) \Longrightarrow g(x_1) \leq g(x_2).$$

Proposition 2. *Let f and g be non-negative measurable functions on $(X, \mathscr{F}, \mu)$, f and g are comonotone. Then for any $A \in \mathscr{F}$,*

$$\int_A (f+g) d\mu = \int_A f d\mu + \int_A g d\mu.$$

Definition 4. *A function $\Phi : [0,\infty] \to [0,\infty]$ is convex if*

$$\Phi\left(\sum_{i=1}^{n} \lambda_i x_i\right) \leq \sum_{i=1}^{n} \lambda_i \Phi(x_i)$$

where $0 \leq \lambda_i \leq 1$ and $\sum_{i=1}^{n} \lambda_i = 1$.

3 Main Results

Theorem 1. *Let $(X, \mathscr{F}, \mu)$ be a fuzzy measure space, $f, g \in \mathbb{F}$, $A \subset X$. If f and g are comonotone, $\alpha + \beta = 1$, $0 < \alpha, \beta < 1$. Then*

$$\int_A f^\alpha g^\beta \, d\mu \leq \left(\int_A f \, d\mu\right)^\alpha \left(\int_A g \, d\mu\right)^\beta.$$

Proof. As $\int f \, \mathrm{d}\mu$ and $\int g \, \mathrm{d}\mu$ are non-negative real numbers, from proposition 1 we know that

$$\frac{\int_A f^\alpha g^\beta \, \mathrm{d}\mu}{(\int_A f \, \mathrm{d}\mu)^\alpha (\int_A g \, \mathrm{d}\mu)^\beta} = \int_A \left(\frac{f}{\int f \, \mathrm{d}\mu}\right)^\alpha \left(\frac{g}{\int g \, \mathrm{d}\mu}\right)^\beta \mathrm{d}\mu.$$

Since the geometric inequality

$$\prod_{i=1}^{n} a_i^{q_i} \leq \sum_{i=1}^{n} a_i q_i,$$

where $q_i > 0$ for all i and $\sum_{i=1}^{n} q_i = 1$. Thus

$$\int_A \left(\frac{f}{\int f \, \mathrm{d}\mu}\right)^\alpha \left(\frac{g}{\int g \, \mathrm{d}\mu}\right)^\beta \mathrm{d}\mu \leq \int_A \left(\frac{\alpha f}{\int f \, \mathrm{d}\mu} + \frac{\beta g}{\int g \, \mathrm{d}\mu}\right) \mathrm{d}\mu.$$

As f and g are comonotone,then

$$\int_A \left(\frac{\alpha f}{\int f \mathrm{d}\mu} + \frac{\beta g}{\int g \mathrm{d}\mu}\right) \mathrm{d}\mu = \int_A \frac{\alpha f}{\int f \mathrm{d}\mu} \mathrm{d}\mu + \int_A \frac{\beta g}{\int g \mathrm{d}\mu} \mathrm{d}\mu = \alpha + \beta = 1.$$

Therefore

$$\frac{\int_A f^\alpha g^\beta \, \mathrm{d}\mu}{(\int_A f \, \mathrm{d}\mu)^\alpha (\int_A g \, \mathrm{d}\mu)^\beta} \leq 1.$$

That is,

$$\int_A f^\alpha g^\beta \, \mathrm{d}\mu \le \left(\int_A f \, \mathrm{d}\mu\right)^\alpha \left(\int_A g \, \mathrm{d}\mu\right)^\beta. \qquad \square$$

Notice that the equation holds if and only if $f(x) = k \cdot g(x) \, (k > 0)$. It is easy to find examples to show that although the comonotone condition may not be necessary, it can't be omitted.

Theorem 2. *Let* $(X, \mathscr{F}, \mu)$ *be a fuzzy measure space and* $f \in \mathbb{F}$. *If* $\Phi : [0, \infty) \to [0, \infty)$ *is a convex non-decreasing function , then*

$$\Phi\left(\int_X f \, d\mu\right) \le \int_X \Phi(f) \, d\mu$$

Proof. Let s be a measurable simple function on X with range $\{a_1, a_2, ..., a_n\}$ where $0 = a_0 \le a_1 \le a_2 \le ... \le a_n$. Denote $A_i = \{x | f(x) \ge a_i\}$, then $\mu(A_1) = \mu(X) = 1$ and $\mu(A_{n+1}) = 0$.

$$\Phi\left(\int_X f \, \mathrm{d}\mu\right) = \Phi\left(\sum_{i=1}^n (a_i - a_{i-1})\mu(A_i)\right)$$
$$= \Phi\left(\sum_{i=1}^n a_i \left(\mu(A_i) - \mu(A_{i+1})\right)\right)$$

For that

$$\sum_{i=1}^n \left(\mu(A_i) - \mu(A_{i+1})\right) = 1$$

and Φ is convex, we have

$$\Phi\left(\sum_{i=1}^n a_i \left(\mu(A_i) - \mu(A_{i+1})\right)\right)$$
$$\le \sum_{i=1}^n \Phi(a_i) \left(\mu(A_i) - \mu(A_{i+1})\right)$$
$$= \sum_{i=1}^n \left(\left(\Phi(a_i) - \Phi(a_{i-1})\right) \mu(A_i)\right).$$

Since Φ is non-decreasing, $\{\Phi(a_0), \Phi(a_1), \Phi(a_2), ..., \Phi(a_n)\}$ is the range of $\Phi(x)$ where $0 = \Phi(a_0) \le \Phi(a_1) \le ... \le \Phi(a_n)$. We get

$$\sum_{i=1}^n \left(\left(\Phi(a_i) - \Phi(a_{i-1})\right) \mu(A_i)\right)$$
$$= \int_X \Phi(s) \, \mathrm{d}\mu.$$

So
$$\Phi\left(\int_X s\,\mathrm{d}\mu\right) \le \int_X \Phi(s)\,\mathrm{d}\mu.$$

For any non-negative function $f \in \mathbb{F}$, Take the supremum of $\{s \le f\}$ on both sides of the inequality, we get

$$\begin{aligned}
&\sup\left\{ \Phi\left(\int_X s\mathrm{d}\mu\right) \middle| s \le f\right\}\\
&= \Phi\left(\sup\left\{\int_X s\mathrm{d}\mu \middle| s \le f\right\}\right)\\
&= \Phi\left(\int_X f\mathrm{d}\mu\right)\\
&\le \sup\left\{\int_X \Phi(s)\mathrm{d}\mu \middle| s \le f\right\}\\
&= \int_X \Phi(f)\,\mathrm{d}\mu.
\end{aligned}$$

That is
$$\Phi\left(\int_X f\,\mathrm{d}\mu\right) \le \int_X \Phi(f)\,\mathrm{d}\mu. \qquad \square$$

Notice that it is easy to find examples to show that the affiliated condition Φ is non-decreasing can't be omitted.

4 Conclusions and Problems for Further Discussion

In this paper, two kinds of integral inequalities are discussed. In the future, there are still two problems to be solved. Firstly, if there exists weaker affiliated conditions or necessary and sufficient conditions for the inequalities. Secondly, what are other inequalities for Choquet integral, such as Hardy type inequality, convolution type inequality, Stolarsky type inequality and so on.

Acknowledgement. This work was supported by the National Natural Science Foundation of China (No. 70471063, 70771010, 70801064).

References

1. Agahi, H., Mesiar, R., Ouyang, Y.: General Minkowski type inequalities for Sugeno integrals. Fuzzy Sets and Systems 161(5), 708–715 (2010)
2. Baǐnov, D., Simeonov, P.: Integral inequalities and applications. Kluwer Academic Publishers, Dordrecht (1992)
3. Choquet, G.: Theory of capacities. Ann. Inst. Fourier 5(131-295), 54 (1953)
4. de Campos, L.M., Jorge, M.: Characterization and comparison of sugeno and choquet integrals. Fuzzy Sets and Systems 52(1), 61–67 (1992)
5. Flores-Franulic, A., Román-Flores, H.: A Chebyshev type inequality for fuzzy integrals. Applied Mathematics and Computation 190(2), 1178–1184 (2007)
6. Littlewood, J.E., Hardy, G.H., Polya, G.: Inequalities. Cambridge University Press (1952)

7. Grabisch, M.: The application of fuzzy integrals in multicriteria decision making. European Journal of Operational Research 89(3), 445–456 (1996)
8. Grabisch, M., Sugeno, M., Murofushi, T.: Fuzzy measures and integrals: theory and applications. Springer-Verlag New York, Inc., Secaucus (2000)
9. Wang, Z.Y., Klir, G.J.: Fuzzy measure theory. Plenum Press, New York (1992)
10. Mesiar, R., Li, J., Pap, E.: The choquet integral as lebesgue integral and related inequalities. Kybernetika 46(6), 1098–1107 (2010)
11. Mesiar, R., Ouyang, Y.: General Chebyshev type inequalities for Sugeno integrals. Fuzzy Sets and Systems 160(1), 58–64 (2009)
12. Ouyang, Y., Mesiar, R.: On the Chebyshev type inequality for seminormed fuzzy integral. Applied Mathematics Letters 22(12), 1810–1815 (2009)
13. Pap, E.: Null-additive Set Functions. Kluwer, Dordrecht (1995)
14. Román-Flores, H., Flores-Franulic, A., Chalco-Cano, Y.: A Jensen type inequality for fuzzy integrals. Information Sciences 177(15), 3192–3201 (2007)
15. Román-Flores, H., Flores-Franulic, A., Chalco-Cano, Y.: The fuzzy integral for monotone functions. Applied Mathematics and Computation 185(1), 492–498 (2007)
16. Romano, C.: Applying copula function to risk management. University of Rome, La Sapienza, Working Paper (2002)
17. Sugeno, M.: Theory of fuzzy integrals and its application. PhD thesis (1974)
18. Wang, R.-S.: Some inequalities and convergence theorems for choquet integrals. Journal of Applied Mathematics and Computing, 1–17 (2009), doi:10.1007/s12190-009-0358-y
19. Xie, X.L., Beni, G.: A validity measure for fuzzy clustering. IEEE Transactions on Pattern Analysis and Machine Intelligence 13(8), 841–847 (2002)

Part IV

Knowledge Engineering and Management

Research on Web Semantic Information Retrieval Technology Based on Ontology*

Hong Zhou and Jun Liu

School of Information Science & Technology, Beijing Wuzi University, 101149, Beijing, China
David_csharp@yahoo.com.cn, liujun@bwu.edu.cn

Abstract. To solve the problems of label clustering, bad theme relevance, etc. of information retrieval especially the semantic retrieval in the context of Chinese environment. The appropriate information retrieval model for the Chinese semantic environment is constructed through introducing ontology and class label mechanism. Based on optimization of the query retrieval submitted by ontology and label to the user and through calculation of the similarity between ontology label and member engine data, the dispatching method for member search engine database is proposed; the method appropriate for extracting this semantic information retrieval model data is proposed through improvement of the traditional STC algorithm. The research on semantic retrieval technology based on ontology and label breaks through the bottleneck of semantic search and information resource management & organization and enhances the scope and quality of the semantic information retrieval.

Keywords: ontology, semantic retrieval, class label, primitive.

1 Introduction

Information retrieval indicates the process of the user's searching for necessary information from various information sets. It contains information storage, organization, representation, inquiry and access, etc..Semantic retrieval indicates the retrieval in combination of information retrieval & artificial intelligence technology (AIT) and natural language technology. It analyzes the retrieval request of information object and searcher from the angle of the semantic comprehension, and it is a retrieval match mechanism based on concept and its relevance. Currently, the main technologies in realizing semantic retrieval include three aspects: natural language processing (NLP), method based on conceptual space as well as method based on ontology, etc. NLP indicates the processing of natural language to enable the computer to understand the content of the natural language. No help will be given to the information retrieval effect by adopting the processes, like simply removing stop words and taking wood root, etc. However, the consumption for processing and storage will be increased if

* Supported by Funding Project for Base Construction of Scientific Research of Beijing Municipal Commission of Education (WYJD200902), Beijing-funded project to develop talents(PYZZ090420001451).

Y. Wang and T. Li (Eds.): Knowledge Engineering and Management, AISC 123, pp. 227–234.
springerlink.com

the processes, like adopting complicated word sense disambiguation and coreference resolution, etc. The semantic retrieval method based on conceptual space indicates applying the conceptual space into the information retrieval and selecting the content related to the concept based on comprehension of the concept connotation by the conceptual space and the concept conveyed by the key words submitted by the user as the retrieval basis to expand the retrieval scope by dint of this. The traditional information retrieval mechanism can not solve the semantic problem under the context of Chinese effectively. However, ontology, as a conceptual model modeling tool that can describe information system at the layer of semantics and knowledge, focused by numerous research personnel at home and abroad since it is proposed and has been widely applied in many fields of the computer, like knowledge engineering, digital library, software reuse, information retrieval and disposal and semantic net of the heterogeneous information in the internet, etc..

2 Clustering Search Technology Based on Ontology

The main advantages of the semantic information retrieval based on ontology are represented in two points: firstly, the system eliminates the concept ambiguity of the key ambiguous words in semantic concept by dint of domain ontology through utilizing the correlation among the key words in the user's query method, and ensures the correctness of the return to the document. Secondly, the system can better comprehend the user's retrieval demand by virtue of domain ontology and perform corresponding reasoning according to the correlation among the key words in the user's retrieval query method to answer the user's questions and tap out the user's real demand.

Definition 2.1 (Ontology): O= (D, W, R), in which D indicates a field; W indicates the clustering of the state of relevant affairs in the field while R indicates the clustering of the conceptual relation in the field space <D & W>. The ontology refers to the description of conceptualization in certain language.

However, at present, in information retrieval technology, the retrieval result information interaction method based on "key word query + user's voluntary browsing" brings about the bottleneck of delivery of user's information demand. To solve this problem, the label is introduced in this paper. The function of the introduction of label refers to the better orientation between the retrieval result and user interaction.

Definitions 2.2 Class tag: Class tag refers to the key terms that indicating the key content of the document but were standardized by ontology techniques, primarily manifested as the search key terms input by users. It possesses three characteristics of subject terms: (1) expressed by letter symbol; (2) reflect the relation model between page data and user search from the perspective of the ontology character; (3) organize structure in the retrieval system according to the ontology model, enabling it to reflect the key content of the document. It composed an assembly concept unit of the retrieved topic. The class tag library, in fact, is a concept table collected with search meaning, which is also a conceptualized key term table for searching and description of various key themes of web page data.

The Information retrieval model based on ontology is shown in Fig 1.The search optimization mainly involves in the optimization of users' search query. The module of tag cluster is responsible for the extraction of original tag data, word segmentation, judgment on key integrity, semantics similarity computing, modularization of tag ontology, and generation of cluster-tree of the clustering search engine. Finally, the corresponding document summary index will be loaded through the class tag, which will be submitted to users in terms of visualized cluster-tree. The following research is mainly about the Semantic Similarity Computation Methods involved in the tag cluster module.

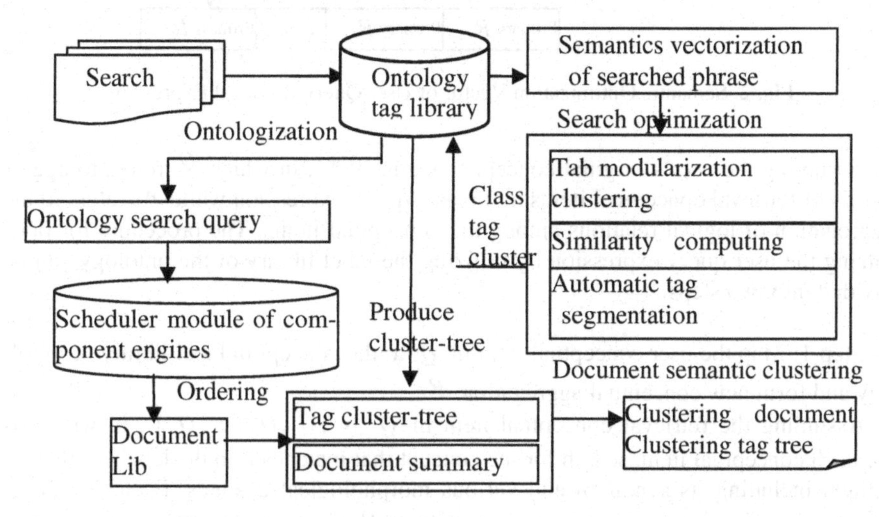

Fig. 1. Information retrieval model based on ontology tags

3 Semantic Query Optimization Based on Ontology and Label

3.1 Query Retrieval Optimization Based on Ontology

As a conceptual model, ontology conveys the universal understanding about the domain knowledge through defining the correlation, axiom and theorem among the shared concepts of the domain. It has very good semantic presentation skill. Therefore, the user's retrieval expression can be optimized through referring to the conceptual relationship in the label library of ontology and the user's retrieval condition can be mapped to the concept and relationship of ontology by utilizing the semantic relationship in the label library of ontology, thus the conformity in user's demand and the machine understanding is realized.

The user's query retrieval expression can ultimately be transformed into the aggregation of a group of concepts and logical relationship (and, or, non-), thus the user retrieval conceptual space is formed. The user query can be expanded into semantic vector. As for the link list composed by the corresponding concept in the ontology, each concept is described by its property. The key words input in the user's query might be the concept of the ontology directly. However, it might be the property description or restriction of certain concept. Therefore, directly represent the user query retrieval expression into the semantic vector of the primary expression of Chinese based on HowNet. Please refer to the Fig. 2.

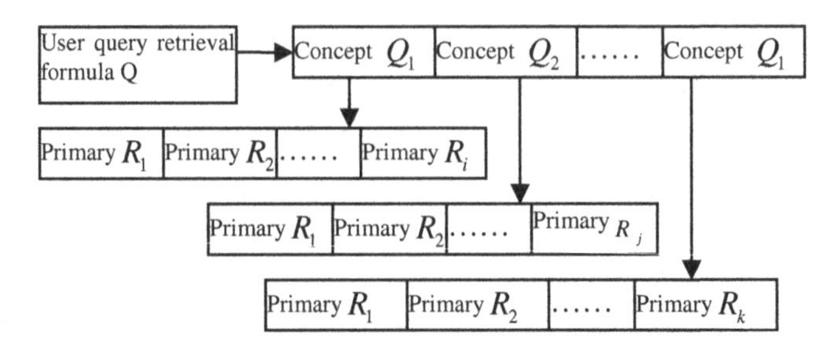

Fig. 2. Semantic Optimization Vector of User Query Retrieval Expression

Assuming the user's original concept space is $\{Q,W\}$, in which Q refers to aggregation of retrieval conceptual items in the user query expression while W refers to the aggregation of logical relations among the conceptual items. The procedure for optimizing the user query expression by utilizing the label library of the ontology can be divided into two steps:

Step 1: Map the user conceptual item in Q to the concept in label library of ontology and form new conceptual aggregation R;

Assuming the retrieval conceptual item in Q is $\{Q_1, Q_2, \cdots, Q_n\}$, in which as for each conceptual item, search for the concept that matches it in the labels of the ontology, including its synonym and various morphological changes. Each successful match will generate a record (Q_i, R_i), in which Q_i is certain retrieval conceptual item in Q while R_i is the concept that Q_i matches it in the labels of the ontology. For one Q_i may have many relevant R_i; therefore, one Q_i might have many records. All Rs form new conceptual aggregation R.

After all user conceptual items in Q are mapped to the concept in the label library of the ontology, then the next step begins.

Step 2: determine the new logical relations among the new concepts by executing logical transformation rule of R according to the semantic relation and original logical relation W to form new conceptual space.

According to the semantic relations among various concepts in aggregation R, transform the former logical relation and form new logical relation among the new concepts. The semantic relations among the concepts include synonymy, hypernymmy/hyponymy, half/full meaning relations and antonym

The transformational rule is mainly realized through application of a suit of logical transformation rules. The logical relation between any two conceptual items R_i in R and R_j mainly includes logical "and" relation, namely $R_i and R_j$; logical "or" relation, namely $R_i or R_j$; logical "non-" relation, namely $not R_j$; provided that there are many logical relation combinations in the concepts of R, the mapping can be realized through combination of aforementioned basic mapping rule. The conformity in user's demand expression and machine's understanding through aforementioned method can enhance the quality of system retrieval effectively.

3.2 Semantic Query Expansion Based on Class Label

The ontology defines the shared concepts of the domain and the relations among the concepts and can expand the semantics of the retrieval condition by utilizing the hierarchical relationship (subclass/superclass relation) and other relations, like synonymy, near-synonymy and inclusion relation.

(1) Synonymous expansion, the phenomenon of "polysemy" extensively exits in the natural language. Substitute the conceptual synonyms in the retrieval space as the retrieval condition. The return result is completely relevant. For instance, "data excavation" can be substituted by "knowledge discovery" in the field of computer and "information retrieval" can be substituted by "information extraction"

(2) Drilling expansion, use more specific subordinate concept (sub-concept) to substitute the user retrieval space concept and define the retrieval scope to a small area.

(3) Generalization expansion, use the superiordinate concept (superclass concept) at higher degree of abstraction to substitute user retrieval space concept and expand the retrieval scope.

From the perspective of the aforementioned process of query expansion, the information retrieval based on ontology is the process of searching for new nodes ceaselessly along the arc from certain node of the figure. Therefore, the heuristic expansion query expression based on ontology is proposed.

Generally, the heuristic expansion expression based on ontology can be defined to the combinatorial calculation of n functions. Its form is: $Q = f_1, f_2, \cdots, f_n$, in which each $f_i(0 \le i \le n)$ represents a search operation $f_i = (\partial)^{\lambda}$;∂ indicates a directed edge e or its reverse edge e^{-1}, which represents the a linking relationship of the label nodes while λ represents the brother's interval length specification.

4 Each Member Engine Dispatching Strategy

The dispatching strategy is to study how the search engine selects appropriate quantity and member search database greatly related to the class label and obtains maximum search benefit by virtue of small resource consumption designated to possibly avoid submitting the query requests to the member database for execution without distinction during the process of query of clustering search.

Definition 4.1: Assuming there is a similarity calculation function, the function calculates the similarity of the each document to the every class label. The similarity can approximately manifest the usefulness of the document. As far as the class label of the automatic clustering, provided that a document d complies with any one of the next condition, then d will be deemed useful to the user query.

(1) Provided the retrieval requests returning to the n documents, then in the similarities of all documents to the L, the similarity of d to L is the among the maximum n similarities.

(2) Provided that the retrieval requests returning to the documents with similarity of class label L larger than certain critical value, then the similarity of d to the class label L will larger than the value.

Definition 4.2: The abstract $R'(l_i, D)$ quoted by the meta search engine database D in the class label L_i is constituted by two parts:

(1) The total actual cites of $D'(l_i)$, $D'(l_i)$;

The quoted document weight included in $D'(l_i)$: $p'(doc \mid D'(l_i))$

Assuming the abstract quoted from the m member search engine databases $D_1, D_2, ..., D_m$ and D_i is $R'(D_i) = \{R'(l_1, D_i), R'(l_2, D_i), ..., R'(l_n, D_i)\}$, in which $R'(l_j, D_i)$ indicates quoted abstract database D_i in the class label l_j.

Then, as for the similarity of the quoted abstract $R'(D_i)$ of the database of the member search, the calculation formula is listed as below:

$sim(L, R'(D_i)) = \sum_{j=1}^{t} lw_{ij} * disl_j$, in which lw_{ij} indicates the weight sum of the document quoted by the class label l_j in ith database, namely $lw_{ij} = \sum p'(doc_j \mid D'_i(l))$, in which $disl_j = \frac{1}{t}\sum_{i=1}^{t} (lr_{ij} - alr_j)^2$, lr_{ij} indicates the proportion of the weights indexed by the label l_j in D_i and all search databases, namely $lr_{ij} = \frac{lw_{ij}}{\sum_{i=1}^{t} lw_{ij}}$, alr_j indicates the mean value of all lr_{ij} s.

During the process of dispatching strategy of the clustering search engine, the relevance of each member search database to the user's interest class label L shall be calculated first. Sequence the member engines according to the relevance and select the several member search engine databases ranked top to provide query service for user.

5 Date Extraction Method Based on STC Algorithm

The traditional STC algorithm refers to the document clustering algorithm of a linear time complication. The main idea is to see the each document as a character string and build suffix tree. The same character strings occurred in the suffix tree are regarded as the fundamental class and then merger the fundamental classes. The characteristics of the suffix tree: (1) there is only root node; (2) the intermediate node shall have at least two child nodes and each edge is identified with the substring, which represents the path of node to the root; (3) there should be not the same identifications at the edge of the same node; (4) the substring of each character string has corresponding suffix node.

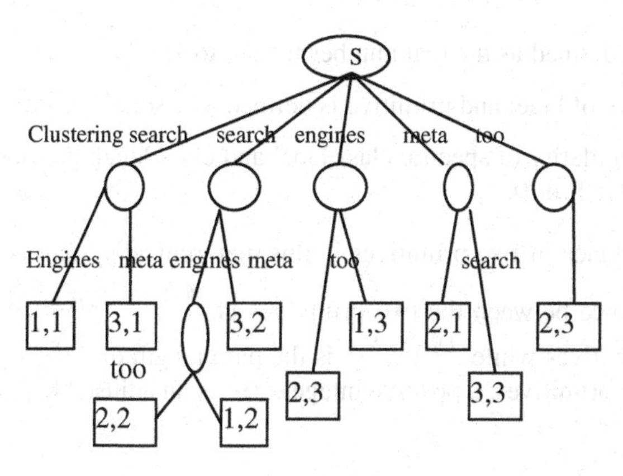

Fig. 3. Suffix Tree of Label Phrase Extraction Result

Traditional STC algorithm is as shown by the Fig. 3. As for the suffix tree formed by three character strings "clustering search engines", "meta search engines too" and "clustering search meta too", the circles in the Fig. indicate the node, each intermediate node indicates the same words showing up in the document, its content is identified at the edge. The first figure in the rectangle indicates the substring belongs to that document while the second figure indicates the which substring among the character strings.

The values of the similarity of the fundamental class labels in STC algorithm are only 0 and 1, as far as T_1, T_2 is concerned, in which the similarity can not the simple 0 and 1. Therefore, the following improvements in STC algorithm of based on ontology are made in this paper:

(1) Similarity calculation in combination of HowNet ontology primitive

By combining the semantic query expansion in 3.2 section, as for the two labels T_1 and T_2, provided that T_1 has n primitive items (concept): $S_{11}, S_{12}, \ldots, S_{1n}$. T_2 has m primitive items (concept): $S_{21}, S_{22}, \ldots, S_{2m}$. Then, the similarity of T_1 and T_2 refers to the maximum value of the similarities of various concepts. That is, it is defined as:

$$Sim(T_1,T_2) = \max_{i=1\ldots n, j=1\ldots m} Sim(S_{1n}, S_{2j})$$

The similarity of T_1 and T_2 is the mean value of the similarities of various concepts, it is defined as: $Sim(T_1,T_2) = \sum_{i=1}^{n}\sum_{j=1}^{m} Sim(S_{1n}, S_{2j}) / mn$

(2) Semantic distance of the added label module

As for the two labels T_1 and T_2 their similarity shall be marked as: $Sim(T_1,T_2)$,

$$Sim(T_1,T_2) = \frac{\alpha}{Dis(T_2,T_2)+\alpha}$$

In which α refers to the adjustable parameter, its meaning is the word distance value when the similarity is 0.5, its word distance is $Dis(T_1,T_2)$. Now a simple transformational relation shall be defined to satisfy the aforementioned condition.

Rule 5.1: Similarity is defined as the real number from 0 to 1;

Rule 5.2: The similarity of label and primitive is defined as a small constant;

Rule 5.3: As for the similarity of specific class label and class label, provided that the two words are same, it is 1, or 0.

Assuming the path distance of two primitives in this hierarchical system is $Dis(S_1,S_2)$ and the semantic distance between the two primitives is $Sim(S_1,S_2)=\frac{\alpha}{Dis(S_1,S_2)+\alpha}$, in which S_1,S_2 indicates two primitives while $Dis(S_1,S_2)$ is the path length of S_1,S_2 at in the hierarchical system of the primitives, a positive integer. α is an adjustable parameter.

6 Conclusions

The semantic retrieval technology based on ontology is mainly studied in this paper. Firstly, the limitation of traditional information retrieval search technology presented, then the new information retrieval model is proposed by combining the ontology and label technology, partial methods and theories in the model are analyzed and studied; the query retrieval expression is decomposed to primitives and mapped to ontology to observe their semantic association in ontology and the optimization of query retrieval expression is realized by utilizing ontology; the dispatching methods for member engine are studied on the basis of meta search technology and by combing the ontology label; the STC algorithm is improved during the process of webpage information retrieval.

References

1. Wang, Z.-H., Zhao, W.: Research on semantic web retrieval model based on ontology and key technologies. Computer Engineering and Design (2011)
2. Shen, Y.-H., Feng, X.-L., Huang, R.-Y.: Search results optimization based on Web clustering algorithm. Journal of Computer Applications (2010)
3. Zeng, H.-J., Chen, Q.-C.H., Ma, W.-Y., Ma, J.: Learning to Cluster Web Search Results. In: Proceedings of the 27th Annual International Conference on Research and Development in Information Retrieval, pp. 210–217 (2004)
4. Kleinberg, J.M.: Authoritative Sources in a Hyperlinked Environment. Journal of ACM (2005)
5. Shen, X.L., Zhou, J.C.: Research on the Data Integration Mechanism Based on Ontology Semantic Mapping. Computer Engineering & Science (2010)
6. Sabou, M.: Learning Web Service Ontologies: an Automatic Extraction Method and its EvaluationIn Ontology Learning. In: Buitelaar, P., Cimiano, P., Magnini, B. (eds.) Ontology Learning and Population. IOS Press (2005)
7. Pedrycz, W., Loia, V., Senatore, S.: P-FCM: A Proximity-based Fuzzy Clustering. Fuzzy Sets and Systems (2004)
8. Coppi, R., D'urso, P.: Three-way Fuzzy Clustering Models for LR Fuzzy Time Trajectories. Computational Statistics & Data Analysis (2003)
9. Zhang, D., Dong, Y.: Semantic, Hierarchical, Online Clustering of Web Search Results. In: Yu, J.X., Lin, X., Lu, H., Zhang, Y. (eds.) APWeb 2004. LNCS, vol. 3007, pp. 69–78. Springer, Heidelberg (2004)

A Knowledge Base Construction Method Based on Cognitive Model

Liwei Tan[1], Shisong Zhu[1,2], and Wenhui Man[1,2]

[1] Xuzhou Air Force College, Xuzhou, 221000, Jiangsu, China
[2] China University of Mining and Technology, Xuzhou, 221116, Jiangsu, China
liweitan000@sina.com, zhushisong@sohu.com, manwenhui2001@163.com

Abstract. In order to solve the problem of the knowledge base construction for a living expert system to adapt the changeable world, a set of method based on the cognitive model include the knowledge acquisition, representation, storage in organization, updating and reasoning conveniently is presented. Simulating the learning procedure of human beings is the core idea of this method from which we can find the ways how to add, delete, amend and use the knowledge in an expert system. Based on the analysis of the common procedure of children's actions during recognizing the world, a cognitive model of concept learning is abstracted. A general concept learning algorithm, a knowledge representation method based on general rules, a logical structure in the forest shape, and a uniform data structure for storage are accordingly presented. Thus, a complete and more scientific management case for the knowledge base of expert system is provided. At last, comparing with some ontology knowledge bases, three different characteristics of this construction method are discussed.

Keywords: knowledge base, construction method, expert system, cognitive mode, knowledge representation.

1 Introduction

The knowledge base is one of the most important factors affecting the performance of expert systems. The quality and quantity of knowledge in an expert system determines its ability to solve problems. Similarly with a human expert, the expert system needs the learning ability to achieve new knowledge, eliminating errors, improving existing knowledge, and to storage them in order, which can create advantage conditions for actual reasoning and application. To construct the knowledge base in scientific can help the expert system maintaining the exuberant vitality.

The concept is a reflection of the nature of things. It can be characteristic the things generally[1].The growth of human knowledge thanks to concept learning methods, although the learned knowledge is not necessarily a reliable, but it is the important way for a man to improve his abilities on recognizing the world. Current research on concept learning is following two different routes [2]: One is based on the engineering method, which sets out from the potential principle(not considering whether the principle is being in the synthesis of the life),attempting to test and confirm a engineering method of concept learning; Another is based on the cognitive model, which tries to

Y. Wang and T. Li (Eds.): Knowledge Engineering and Management, AISC 123, pp. 235–240.
springerlink.com

exploit a computing theory of people's concept learning by analyzing and explaining how the information is being processed during completing their recognition activities.

2 Theory of the Cognitive Model

Simulation is the goal of AI, but also the learning ways and means. 10~24 month old children are the best simulation objects for research on the intelligence growth naturally, who are in an important phase for the intelligence to take off. Because they have the material conditions on the side of physiology and their psychology is also pure. So long as the security sense obtains the guarantee, not have the uncomfortable feelings, they can make the natural responses to acquaintance's simple instruction. Through observed five babies for more than four months, we found that the process of the baby to distinguish things may reappear many times. These are very important for the research on the intelligence's natural increase process.

When a child is interested in a kind of objects at the first time, an approximate impression will be marked in his memory. This impression consists of a few main sensuous features and the concept described with these features appears very imprecise. But with the natural knowledge management, this kind of knowledge will be improved.

If a child meets a new instance and knows he has no concept about it definitely, he may ask the people with more abundant knowledge and then establishes a new concept of the new instance.

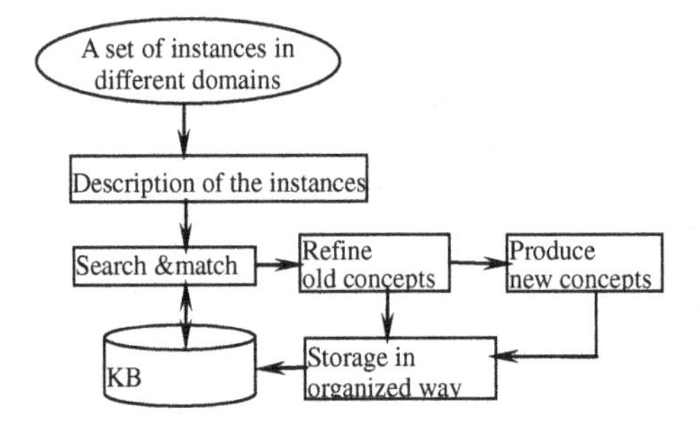

Fig. 1. The cognitive model is the common procedure of children's concept learning

In a word, the procedure of a child to cognize the world is the procedure that during a period many new concepts in different domains are established and many old concepts will be refined from time to time. This procedure is based on existing concepts in his memory, which is in accordance with the "fringe effect" of learning. So we can call this procedure the general concept learning. The model of the learning procedure can be abstracted as Figure1. It is worth noticing that during the procedure of refining the concepts of one kind of instances, there will be some elementary concepts produced, which are called ontology. Ontology is a detailed description of the basic concepts [3]. With the increasing of ontology, the learner's experience will be enriched and the ability of abstract thinking and reasoning will be improved constantly.

3 Knowledge Representation

Knowledge representation is one of the important factors that can influence the knowledge management method and the expert system problem solving performance.

3.1 Requirements

The requirements of knowledge representation in expert system usually include the following sides: (1) Strong ability of expression; (2) Convenient for control, contributing to improving the efficiency of the search and matching; (3) Union structure, easy to expand, modify and consistency check the KB.

3.2 Requirements

Based on the analysis above, an ideal knowledge representation mode is presented, which is called knowledge representation based on generalized rules (KRBGR). Its formal method fully incorporates the benefits from logic, production rules in knowledge representation and reasoning, and has compensated for their weaknesses. Overall, KRBGR likes production rules, has the shape "premise — conclusions". The child conditions of facts and rules' precondition completely uses first-order predicate logic or proposition. The deduction mechanism is the same as the production system, the rule conclusion can be used to further reasoning or the suggestion directly, and it can also be triggered a process command. The description of KRBGR with BNF is shown as follows:

<general rule>:: =<premise>→<conclusion>|<attribute set>→<concept name>

<premise>:: =<single condition>|<compound condition >

<conclusion>:: =<fact>| <operation>

<compounded condition>:: =<single condition> ∧ <single condition>[∧ …]<single condition> ∨ <single condition>[∨ …]

<operation >:: =<single operation>|<compound operation >

<compound operation>:: =<single operation> ∧ <single operation>[∧ …]|<single operation> ∨ <single operation>[∨ …]

<single operation >:: =<operation name>|<predicate name> [(< variable>,…)]

<attribute set>:: =<single attribute>|<compound attribute>

<compound attribute >:: =<single attribute> ∧ <single attribute>[∧ …]

<single attribute>:: =<attribute name>[(<variable>, <attribute value>•]

<concept name>:: =<Instance name>[(<variable >)]

3.3 The Advantages of KRBGR

Application of KRBGR, the knowledge expressed in production rules, framework, semantic networks, relational databases, and other modes can be consistently represented conveniently and modular organization. It will be very useful for large-capacity knowledge base system building undoubtedly.

4 Algorithm of General Concept Learning

4.1 Data Structure

Due to the length of branches is variable, to adopt a uniform data structure for branches is not realistic, but we can use consistent data structure on the leaf node. The data structure of each node not only contains the node content itself, but also including a variety of logical relationships of the nodes in the tree. To store each node in the computer as a record can be easily to recover the logical relationships between the nodes. The data structure description of the node can be shown as follows:

```
struct node
{ int incrs ; // identify of node
  char content[50] ; //content of node
  int  fath ; //father point
  float C_v ; //confidence value
  float T_v ; //threshold and some signs
}
```

4.2 Algorithm

Since the KB adopts a tree-shape structure to organize the concepts and store them, accordingly, the mechanism of searching and match adopts the width-first method, and the description of the instances and the concepts adopts the representation method based on general rules. In order to expatiate on the algorithm more clearly, two definitions are given as follows:

Definition 1. If instance I has n attributes $T_1,T_2,\ldots,T_n$, each one of the attributes has a value V_{Ti}, the name of the instance is N_I, then the description of the instance can be expressed as: $\{T_1(I,V_{T1}), T_2(I,V_{T2}), \ldots, T_n(I, V_{Tn}), N(I)\}$, $i\in N, T_1(I,V_{T1})$ represents the instance.

Definition 2. If concept C has m attributes $A_1, A_2, \ldots, A_n$, each one of the attributes has a value V_{Ai}, the name of the concept is Nc, then the description of the concept can be expressed as: $\{A_1(C, A_{T1}), A_2(C, A_{T2}), \ldots, T_n(I, V_{Tn}), N(C)\}$, $i\in N$, $A_1(C, A_{T1})$ represents the concept.

The description of the algorithm for GMLC is shown as follows:

Step 1: The system accepts a training instance I. If the knowledge base is empty, then storage the nodes sequence directly; else it will get the first attribute T_1 to search for the branch of the concept tree. If it finds a node A_1 has the same name to T_1 and $V_{AI}=V_{TI}$, then it will continue searching for the nodes A_2、A_3、…which match T_2、T_3、…of I along the branch until it finds a leaf node. If $N_I=N_C$ and $n=m$, then the system keeps down the description of the concept, and end.

Step 2: If the system finds a leaf node N_C and $N_C=N_I$, but $n>m$, then it inserts the nodes $T_{m+1}, T_{m+2},\ldots,T_n$ which have not been matched before the leaf node as the new nodes of the branch of the concept tree and ends.

Step 3: If the system finds a leaf node N_C, but $N_C \neq N_I$, then the learner knows that a collision between the two concepts has taken place and returns the branch of the concept to the expert who is required to update the concept. If a node $A_i=T_i$ and $V_{Ai} \neq V_{Ti}$ appears, the system goes to step 5.

Step 4: If the learner finds that the procedure of the match is stopped for the reason n<m, then it l searches for the leaf node N_C of the branch with N_I directly. If the learner finds a leaf node N_C which is similar to N_I, then ends; else he will require the expert to enrich the description of the training instance, which is to add more attributes to the description of the instance, and then goes to step l.

Step 5: If $A_i=T_i$ but $V_{Ai} \neq V_{Ti}$, then the learner will produce a brother node Ai which is the same as T_i, and append the attributes $T_{i+1}, T_{i+2}, \ldots, T_n$ to the node A_i. Let $N_C=N_I$ and end.

Step 6: If the learner can not find a node A_i on the *i*th layer in the branch matching the attribute T_i in the description of the instance in a sequence, then he will move the attribute T_i to the end of the description of the instance, and continue the operation of matching with T_{i+1} until a node A_i is found which matches to T_i. If $V_{AI}=V_{TI}$, then go to step l; else go to step 5. If the attribute node A_i which is similar to T_i can not be found, then a brother node A_i is produce which is the same as T_i and the attributes $T_{i+1}, T_{i+2}, \ldots, T_n$ are appended to the node A_i. Let $N_C=N_I$ and end.

Step 7: If we use an instance without the name NI to test the learner, then the learner will match the attribute nodes with the attributes of the instance along a branch.

(1) If the matching procedure just arrives at a leaf node N_C, then the learner will return N_C to the expert and require him to affirm it. If the concept is right, then the test has succeeded; else go to step 3.

(2) If n>m, the learner will return N_C to the expert and require him to affirm it. If the concept is right, then goes to step 2: else goes to step 3.

(3) If $n<m$, the learner will require the expert to add more attributes to the description of the instance, and then goes to step l.

(4) If a node $A_i=T_i$ but $V_{Ai} \neq V_{Ti}$ appears in the procedure of matching, then the learner will ask the expert the name N_I and goes to step 5.

Step 8: In the learning procedure, if the learner meets some elementary concepts which belong to some kind of objects, he will add them to the component concept base of the domain according to steps l to 6.

5 Discussion and Conclusion

Under the guidance of the cognitive sciences, a set of KB construction methods of expert systems adapting to common fields is presented. Compared with the current representative mass ontology-based KB, such as CYC [4], WordNet [5], NKI [6], there are some significant differences.

1) Different Routes: Ontology-based knowledge acquisition is based on the engineering methods of concept learning (regardless of the mechanism exists in the life organization). The construction method based on cognitive model is a computation

theory of human concept learning through analyzing and interpreting the information processing process when human complete the cognitive activities.

2) Different Goals: Ontology-based KB system is mainly used for knowledge sharing and strived to complete exhaustion. While the KB system construction goal mentioned in this paper is application-oriented directly, such as expert systems, intelligent decision-making, pattern recognition, etc.

3) Different Approaches: Despite the wide variety of types, but the establishment of the basic process of KB ontology-based is similar: At first, based on the idea of ontology, to establish a system structure of the domain knowledge from the domain basic terminologies and relations, class, attribute set, which is convenient for understanding and analyzing and support consistency. After that, under the expert's guidance, the knowledge engineers complete the textual knowledge organization, knowledge acquisition, and then a relatively complete domain knowledge base is established. While the learning procedure of GMCL based on cognitive model is to establish an meta-mechanism of information processing based on the general principles of human knowledge acquisition firstly, and then depending on the problem fields involved and real needs to increase the volume and improve the quality of knowledge gradually and continuously, then improve knowledge architecture, eventually a domain knowledge base satisfying the problem solving is come into being.

Acknowledgments. This work is sponsored by the National Science Foundation Grant such as #50811120111 and #40971275. Thanks CUMT and Xuzhou Air force College for the financial and scientific support.

References

1. Brown, C.: Cognitive Psychology. Cromwell Press, Great Britain (2007)
2. Cai, Z., Xu, G.: Artificial intelligence and applications. TsingHua University Press, Beijing (1996)
3. Kong, F.: Principle of Knowledge Base System. Zhejang University Press, Hangzhou (2000)
4. Lenat, D.: CYC: A Large Scale Investment in Knowledge Infrastructure. Communication of ACM 38(11), 33–38 (1995)
5. Vetulani, Z., Walkowska, J., Obrębski, T., Marciniak, J., Konieczka, P., Rzepecki, P.: An Algorithm for Building Lexical Semantic Network and Its Application to PolNet - Polish WordNet Project. In: Vetulani, Z., Uszkoreit, H. (eds.) LTC 2007. LNCS, vol. 5603, pp. 369–381. Springer, Heidelberg (2009)
6. Sui, Y., Gao, Y., Cao, C.: Ontologies, Frames and Logical Theories in NKI. Journal of Software 16(12), 2046–2053 (2005)

Study on Supplier Selection Model for Supply Chain Collaboration

Xing Xu[1,2], Renwang Li[1], and Xinli Wu[1]

[1] Faculty of Mechanical Engineering & Automation, Zhejiang Sci-Tech University, Hangzhou 310018, China
[2] School of Mechanical & Automotive Engineering, Zhejiang University of Science and Technology, Hangzhou, 310023, China
xuxing3220@163.com

Abstract. Based on current situation that traditional manufacturing enterprises cannot combine the customer demand in supply chain when the supplier is selected, the manufacturing enterprises' model of supplier selection based on improved QFD is proposed through analysis on the index system of supplier selection. Such model is for QFD method the important tool – "House of Quality", which is restructured in content and form, formulates a relation matrix between customer demand and index system, and makes use of AHP method and independent collocation method for supplier selection. The final empirical analysis by examples proves that such model is feasible.

Keywords: improved QFD, supply chain collaboration, supplier selection, index system.

1 Introduction

With the economic globalization and diversification of demand, the lifecycle of product is becoming increasingly shorter and the uncertainty of market is increasingly higher. The supply chain collaboration refers to mutual coordination and endeavor by supply chain enterprises in order to improve the overall competitiveness of supply chain. In the supply chain collaboration, the supplier selection is the key issue[1].

Currently, the research methods regarding supplier selection mainly include ABC cost method, linear planning method, AHP, fuzzy comprehensive evaluation method, neural network method, TOPSIS method, DEA method, principal component analysis method, grey comprehensive evaluation method, integrated application of these methods, and so on[2-5]. The shortcoming of these methods lies in that they do not take customer demand into the supply chain consideration, and it is therefore a must to start from the customer demand so as to build a reasonable, effective model of supplier selection.

Y. Wang and T. Li (Eds.): Knowledge Engineering and Management, AISC 123, pp. 241–246.
springerlink.com

2 Model of Improved QFD

QFD is an important comprehensive planning tool, which converts customer demand into the technical requirement of each link in production while the house of quality is the major tool for such series of conversions, and thus the application of QFD method here mainly means that of the house of quality.

For improvement of core competitiveness of supply chain, the core enterprises always need to outsource some parts, which requires the enterprises to make a choice between different suppliers supplying the parts of the same kind; however, the traditional methods often cannot combine customer demand in the supply chain, enabling the whole evaluation system to lose its target, especially the personalized customer demand requesting modern manufacturing enterprises to respond rapidly, and the customer demand therefore needs to be combined as the supplier is selected. Therefore, the improvement of traditional house of quality must be done.

The supplier selection model based on improved QFD is obtained through tailoring, adding and deepening by the restructured house of quality according to the demand in supplier selection on the basis of traditional house of quality, as shown in Fig.1.

	Evaluation and selection index system	
Customer demand	Relation matrix	Customer demand weight
	Weight of index	
Supplier	Score	Total score

Fig. 1. The model of supplier selection based on improved QFD

In this research, the technical demand is to be replaced by the index system of supplier selection. The elements in the relation matrix is generally marked with four common symbols including ⊙, ○, Δ and × to represent the relation between customer demand and index system of supplier selection: strong correlation, intermediate correlation, week correlation and no correlation. Furthermore, the technical competitiveness evaluation module in traditional house of quality is replaced by evaluation module of supplier.

3 Specific Procedures of Supplier Selection

3.1 Build a Comprehensive Evaluation Index System of Supplier Selection

At first, starting from various demands of customers of core enterprises, classify purchase materials in 80/20 principle (Pareto Principle) taking into consideration cost of purchase materials, supply risk and importance and enthusiasm of supplier and then classify the suppliers[6].

There are various comprehensive evaluation indexes of supplier. This paper comprehensively reflects the traditional indexes such as quality, date of delivery, price and service with combination of quantitative and qualitative principles, along with such indexes as supply flexibility of suppliers, enterprise culture and environmental protection level that embody core competitiveness of enterprises.

The index system is divided into two classes, where the first-class index is divided into quality index, price index, delivery index, enterprise comprehensive index and production capacity index. According to the different emphases to evaluating indexes, the items in first-class index can be future subdivided, as shown in Fig.2.

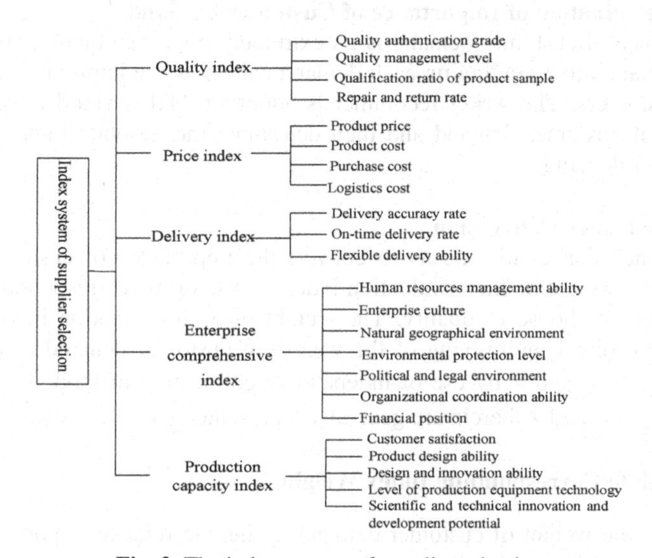

Fig. 2. The index system of supplier selection

3.2 Obtain Customer Demand in Supply Chain

To obtain customer demand is the key in QFD process. The core enterprise in supply chain considers part of suppliers as its "virtual factory" and hands over its parts or components completely to the supplying enterprises for production. With competition and change of market, the customer-directed demand of the core enterprise will unceasingly update and change, and therefore make demands to the suppliers in supply chain. Through investigation, the common table of customer demand is shown in Fig. 3.

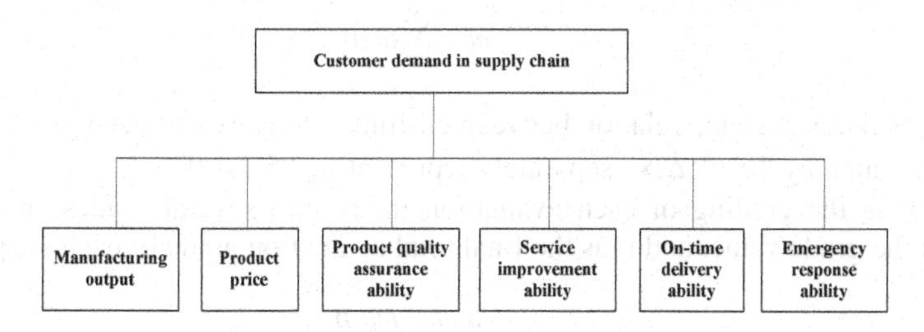

Fig. 3. The Table of Customer Demand

3.3 Build a Model of Supplier Selection Based on Improved QFD

3.3.1 Formulate a Relation Matrix

In accordance with the above established model of supplier evaluation and selection based on improved QFD, formulate the two dimensional table of relation matrix between customer demand in supply chain and evaluation index of supplier, and use separately the symbols of ⊙, ○, Δ and × to represent the relevant relations between each evaluation index and different customers.

3.3.2 Determination of Importance of Customer Demand

The importance of customer demand is an extremely important quantitative index in QFD. The quantitative grading on each demand can show the importance of each demand to customers. The writer recommends adopting AHP method to quantize the importance of customer demand and then determine the absolute importance m_i of each customer demand.

3.3.3 Importance Conversion

The importance conversion means to convert the importance of customer demand (weight) to the weight of each evaluation index by use of the corresponding relations established by the house of quality. The weight of each evaluation index m_j is obtained through direct multiplying of the weight of customer demand and numerical values of ⊙, ○, Δ and × by use of independent collocation method. The numerical values of ⊙, ○, Δ and × therein are generally represented by "⊙:○:Δ:×=5:3:1:0".

3.4 Calculate Corresponding Index Weight

(1) Calculate the weight of customer demand α_i (i.e. the relative importance of each customer demand in supply chain) of i component, where m_i represents the absolute importance of each customer demand and n represents number of item of customer demand.

$$\alpha_i = \frac{m_i}{\sum_{i=1}^{n} m_i} \tag{1}$$

(2) Calculate the weight of each evaluation index m_j through the pertinence relation between each evaluation index and the element of customer demand based on the weight of customer demand α_i, namely:

$$m_j = \sum_{i=1}^{n} a_i \cdot r_{ij} \tag{2}$$

Where r_{ij} is the pertinence relation between customer demand and each evaluation index, represented by "⊙:○:Δ:×" separately representing "5:3:1:0".

(3) Divide the grading of each evaluation index into several grades, and use β_{ij} represent the grade value, and thus the final total evaluation grade of each supplier is:

$$Sum = m_j \cdot \beta_{ij} \tag{3}$$

3.5 Select the Best Supplier

After analysis of the above procedures, keep the total valuation scores of all suppliers in order from high to low, and the enterprise with the highest score is the best supplier. Meanwhile, such evaluation model can also enable the enterprise to make comparison with the model enterprise in higher weight of customer demand to facilitate the enterprise's improvement.

4 Application and Discussion

At present, the automobile industry in China has formed its complete supply chain system of supply – manufacture - sales – service. Taking a certain Automobile Co., Ltd for example, it now needs to evaluate 6 suppliers providing dashboard, which are separately set to be A, B, C, D, E and F.

(1) Use RS theory to normalize customer demand, obtain the importance D_i of each customer demand in Fig. 2, being respectively 7, 44, 21, 21, 5 and 2, and then use the equation (1) to calculate the relevant importance of customer demand, being respectively 0.07, 0.44, 0.21, 0.21, 0.05 and 0.02, as shown in Table 1.

(2) Establish an expert group in special charge of evaluation on suppliers, discuss and determine the degree of correlation r_{ij} between each evaluation index and customer demand, represented by "⊙, ○, △, ×". If the score ⊙ between the evaluation index I1 and customer demand D1 is obtained, then r_{ij}=5; use the equation (2) to get the first evaluation index:

$$m_1 = \sum_{i=1}^{6} a_i \cdot r_{i1} = 5\times0.07 + 3\times0.21 + 1\times0.21 + 5\times0.02 = 1.29 .$$

Similarly, the weights of other 22 evaluation indexes can be obtained.

(3) Divide each index into seven grades, respectively representing very good, good, above average, general, below average, poor, very poor whose corresponding scores are 0.75, 0.50, 0.25, 0, -0.25, -0.50 and -0.75. The expert group gives the score β_{ij} considering the concrete situation of 6 suppliers, e.g. the indexes of A Factory are individually 0.75, 0.25, 0.5, 0.75, 0.25, 0.75, 0.5, 0.25, 0.25, 0.5, -0.75, 0.5, -0.25, 0.5, 0.5, 0.25, -0.5, 0.25, 0.25, -0.25, -0.5, 0.5 and 0.25. By use of the equation (3), $SumA = m_j \cdot \beta_{1j} = 9.3$. Similarly, the scores of the other 22 evaluation indexes can be figured out, and the scores of the other 5 suppliers are individually 12.04, 4.15, 1.69, 6.78 and 7.25. See Table 1 for details.

(4) As seen from the above calculations, the total score of supplier B is the highest, which is 12.04, suggesting that supplier B can better satisfy all demands in supply chain. In the meantime, the highest total score in quality index, price index, delivery index, enterprise comprehensive index and production capacity index shows that such supplier is more advantageous in those aspects.

Table 1. The System of House of Quality for a Certain Automobile Co., Ltd.

Index / Supplier	Quality index				Price index				Delivery index			Enterprise comprehensive index							Production capacity index					Weight of demand
	I1	I2	I3	I4	I5	I6	I7	I8	I9	I10	I11	I12	I13	I14	I15	I16	I17	I18	I19	I20	I21	I22	I23	a_i
D1	⊙		⊙	○		⊙	○	△		⊙	○	○	⊙		○	⊙	△	○		⊙	⊙		△	0.07
D2		⊙	○		⊙		○		⊙		○		○		○				⊙			○		0.44
D3	○			⊙	△	○		⊙		△	△			○			△			○			○	0.21
D4	△	△	⊙		○		⊙				○	⊙			○			⊙		△	○		○	0.21
D5				△					⊙	△			⊙	△			○				△			0.05
D6	⊙		○		⊙	△		○				△				○			○		△		○	0.02
Weight of index m_j	1.29	2.83	2.78	1.41	3.14	1.00	2.58	1.18	2.45	0.61	2.37	1.28	1.92	0.68	2.16	0.37	0.43	1.26	2.26	1.19	1.05	1.33	1.39	Sum
Factory A	0.75	0.25	0.5	0.75	0.25	0.75	0.5	0.25	0.25	0.5	-0.75	0.5	-0.25	0.5	0.5	0.25	-0.5	0.25	0.25	-0.25	-0.5	0.5	0.25	9.3
Factory B	0.75	-0.5	0.75	0.25	0.5	0.5	0.25	0.5	0.5	0.75	0.5	-0.25	0.5	0.75	-0.25	-0.5	0.25	0.5	0.5	0.25	0.25	0.25	0.5	12.04
Factory C	0.75	0.5	0.25	-0.25	-0.25	-0.25	0.25	-0.5	-0.75	0.25	0.25	-0.5	0.75	0.25	0.5	0.25	0.25	-0.5	0.75	0.5	0.25	-0.25	-0.25	4.15
Factory D	0.5	0.75	-0.5	-0.5	-0.5	0.25	-0.25	-0.25	0.25	-0.5	0.25	0.25	0.25	0.5	0.25	-0.25	0.5	0.75	-0.5	0.25	0.75	-0.5	0.25	1.69
Factory E	0.75	0.25	-0.25	0.25	-0.25	0.5	-0.5	0.75	0.75	-0.25	0.5	0.5	-0.5	-0.5	0.75	0.25	0.75	0.25	-0.25	0.75	-0.75	0.75	0.75	6.78
Factory F	0.75	0.25	0	0.5	0.5	-0.25	0.75	0.25	-0.25	0.75	-0.25	0.75	-0.25	-0.25	0.25	0.5	-0.5	-0.25	0.25	-0.25	0.25	0.25	0.5	7.26

Note: 1) I1, I2,…, I23 are their second-class indexes represented in Fig. 2; D1, D2, …, D6 are their indexes represented in Fig. 3.

5 Conclusion

The adoption of the improved model based on QFD to conduct the supply chain collaboration-directed supplier selection overcomes the disadvantage that the customer demand is often ignored by the traditional methods in supplier selection. The model gives full consideration to the importance of customer demand in supply chain collaboration, which enables the core enterprises to address the issue of supplier selection on purpose, reasonable and effectively selects appropriate suppliers in supply chain group and better copes with uncertainty of various customer demands in supply chain.

Acknowledgment. The work is supported by the Natural Science Foundation of Zhejiang Province(Project No.Y6090718, Y107379), and the Science and Technology Project of Zhejiang Province (Project No. 2010C31052, 2011C22070).

References

[1] Li, R., Wang, X., Wang, Y.: Research on Border of Push-Pull Supply Chain Based on Game Theory. Journal of Zhejiang Sci-Tech University 25(5), 625–627 (2008)
[2] Chen, R., Jiang, C.: Study on Strategic Supplier Selection and Evaluation under Supply Chain. Value Engineering (1), 7–10 (2008)
[3] Chen, J.K., Lee, Y.C.: Risk priority evaluated by ANP in failure mode and effects analysis. Quality Tools and Techniques 11(4), 1–6 (2007)
[4] Liu, G., Che, J., Lei, C.: Applications of Fuzzy Comprehensive Evaluation Method and Delphi Method in Support Equipment Evaluation. Journal of Sichuan Ordnance 30(3), 44–45 (2009)
[5] Kumara, M., Vratb, P., Shankarc, R.: A fuzzy goal programming approach for vendor selection problem in a supply chain. Computers & Industrial Engineering 46(3), 69–85 (2004)
[6] Guo, Y.: Evaluation of Capacity Performance in Make-to-order Company Based on Modified QFD. Technoeconomics & Management Research (1), 13–17 (2010)

A Generation Method for Requirement of Domain Ontology Evolution Based on Machine Learning in P2P Network

Jianquan Dong, Mingying Yang, and Guangfeng Wang

School of Computer Engineering and Science
Shanghai University
{jqdong,claireyang,guangfeng}@shu.edu.cn

Abstract. With some critical defects existing in the generation of requirement such as the method to extract concept, property and their relationship from external knowledge sources issues, we propose a novel approach to automatically generate requirement for domain ontology evolution based on machine learning theory in the light of P2P network routing model and storage characteristics of ontology resource. The method takes a comprehensive considering on term frequency, term field, similarity with original ontology and other factors to extract key concept from texts, and then through Naïve Bayes classifier compares those key concept with original ontology and extracts their relationships. We demonstrate that this research can ensure the requirement's reliability and improve the automation and intelligent on the process of requirements generation through simulation experiments and analysis of the results.

Keywords: P2P, similarity, machine learning, ontology evolution.

1 Introduction

In P2P network, the information resources distribute in various independent peers and through the semantic query mechanism users can position the required resources. Ontology is the key to realize the semantic technology, it defines the terms used to describe and express the domain knowledge, and it is a formal specification of a shared conceptualization. Along with the changes of the knowledge in external word, the data consistency between fixed ontology and changing knowledge may be destroyed. As a result, the ontology has not accurately reflected the new state of knowledge [1]. In order to solve the adaptability's problem between ontology and the dynamic changing external knowledge, scholars have proposed ontology evolution. As the foundation and basis for ontology evolution, reliable requirement for ontology evolution can make the ontology evolve more scientifically and reasonably and reflect more practicability [2].

Based on the P2P network, combined with machine learning theory, this paper did a certain studies and experiments on the capture and generation of evolution requirement, and then proposed a method to automatically generate requirement for domain ontology evolution. The method uses the proposed TF-OS algorithm and improved

Y. Wang and T. Li (Eds.): Knowledge Engineering and Management, AISC 123, pp. 247–259.
springerlink.com

BM25F algorithm to extract key concept from Chinese texts, and then takes advantage of the original ontology to train the Naïve Bayes classifier, use the machine learning theory to confirm the relationship between key concept and original ontology. Eventually achieve the goal to automatically generate reliable requirement for domain ontology evolution.

2 Related Works

2.1 P2P Network and Ontology

The integration of P2P (peer-to-peer) equivalence technology and ontologycanmake P2P network support semantic query [3]. The peer in P2P network can use ontology to mark resources and to establish semantic association among peers, and can use ontology reasoning to retrieval implicit knowledge, in turn to improve the efficiency of retrieval and recall ratio in P2P network [4].

2.2 Requirement of Ontology Evolution

According to the different generation methods of the requirement for ontology evolution, it can be divided into manual, semi-automatic and automatic. Aim at the automatic generation of the requirement for ontology evolution, Sun Zhongyi, in reference[5], focus on the research of how to use semantic search and semantic reasoning skills to analyze the environment changes and to generate the evolution requirements in the semantic service composition environment. Zhang Zizhen, in reference[6], proposed to identify the new terminology and the relationship among the ontology elements by taking advantage of the background knowledge of combining multiple data sources (WordNet vocabulary, the online ontology and documents in the Semantic Web). Ouyang, in reference[7], use weighted word frequency algorithms to judge the related degree of the key concepts extracted from pure texts, and combined with ontology search algorithms and the generate rules to automatically generate the evolution requirements.

2.3 Problem to Be Solved in Requirement Generation of Ontology Evolution

There are two key points in the requirement for ontology evolution: one is to extract the concept for evolution, the other is to confirm the relationship and property for concept. The method to extract the concept is closely related to the data sources. While the current studies on automatic generation of requirement are mainly based on regular structured data, the method to extract evolution concept from unstructured data sources needs further study. On the other hand, for the issues such as how to evolve the key concept into ontology and confirm the relationship between key concept and original ontology, there is no effective method. While a lots of third-party mature tools are needed for auxiliary analysis, which needs higher demand on the environment and interface of the system, but poor performance in universality and applicability, meanwhile cannot fully achieve the original intention of automatic.

In the generation of requirement for ontology evolution, in order to enable the process more automatic, and do not lower the reliability of the requirement while reducing human input, at the same time the method has a greater universality and applicability, this paper did a certain studies and experiments on how to automatically generate requirement for domain ontology evolution base on texts.

3 Capture Key Concept for Domain Ontology Evolution

3.1 Domain Ontology

Domain ontology contains correlative knowledge of specific domain, which could provide the relationship between concept definition and the concept and also provide the activities occurred, the main theories, basic principles and so on in specific field[8]. In this paper, domain ontology is defined as follows:

Definition 1: domain ontology can be expressed as follows: $O=\{I,C,P,R\}$.

O is domain ontology, I is information of O, C is a set of concepts or classes, P is a set of properties, R is relation in O.

3.2 Capture Key Concept

By the definition of domain ontology, in order to evolve the existing ontology O, we must first obtain the main subject, candidate concept set C'. Then identify the relationship between C' and the existing concept set C, ultimately, complement with properties to enrich and perfect the concept.

The peers in P2P platform through enrich and update local resources as well as share texts to exchange knowledge. The process of enriching and updating local resources is the process of enriching and updating knowledge. Through preprocess these texts which introduce potential new knowledge to get candidate words and eventually get the key concept of ontology evolution by filtering. In this paper, these texts are defined as 'candidate texts'.

Definition 2: Candidate text collection is defined as: $D_C=\{d_1,d_2,...,d_N\}$.

In order to obtain the candidate concept, we firstly take word segmentation of the N texts in the candidate text set Dc. At the same time, because the concepts of ontology are generally made of nominal words or phrases, in order to filter the non-nominal concepts, pos tagging is needed. Finally, after filtering out these existed words in ontology, the preliminary candidate word set is obtained.

Definition 3: Candidate word set is defined as: $T_C=\{t_1,t_2,..., t_k\}$.

3.2.1 TF-OS Filtering Algorithm

In order to filtering the candidate words, on the basis of referring the classical statistical method TF-IDF this paper propose an algorithm named TF-OS, which takes a comprehensive investigation on the candidate words from both the frequency in the text and the similarity with origin ontology.

TF-IDF(Term Frequency -Inverse Document Frequency)[9] is the most widely used method based on statistics and has various form of formula. The most common form is as follows[10]:

$$W(t_k, d) = tf \cdot idf = \frac{log\ (tf(t_k,d)+1)}{logn} log \frac{N}{df} \quad (1)$$

In order to obtain the concept that has important position in the text and the closerelation with knowledge of the current domain, this paper propose the TF-OS (Term Frequency - Ontology Similarity) method to perspective the candidate words from frequency and similarity with origin ontology, method of calculation is shown as formula (2):

$$w_{TF-OS}(t_k, d) = TF * OS \quad (2)$$

Among them, the TF (Term Frequency) is a calculation of the candidate word's frequency within text, as shown in formula (3).

$$TF = \frac{log\ (tf(t_k,d)+1)}{log\ n} \quad (3)$$

Where $tf(t_k,d)$and n is t_k's frequency and the total number of candidate words in the document d,$t_k \in T_c$, $d \in D_c$. TF show the importance of the word t_k in text d.

This paper argues that, in specific domain ontology with highly recapitulative, the words have a certain similarity in each other. The higher the similarity between candidate word and existed word in ontology, the candidate word is more likely to express the knowledge of this domain.

Definition 4: The words within concept set C and property set P of origin ontology O constitute ontology feature word set together, defined as:$T_O = C \cup P$.

OS (Ontology Similarity) is the similarity value between candidate word and ontology feature word, method of calculation is shown as formula (4).

$$OS = 1 + log\ (1 + m) \quad (4)$$

Where m is the number of word couple whose similarity of candidate word and ontology feature word is bigger than thresholdα_1 (α_1=0.5 in this paper).

This paper use the edit distance method to calculate the name similarity between candidate word and the word in T_O[11]. Edit distance is used to compare the similarity of two string, it is the minimum number of insert, delete and replace operation needed in the conversion from one string to another.

The method of calculation of the name similarity of current candidate word t_k and ontology feature word t_i is shown as formula (5):

$$\mathrm{sim}(t_k, t_i) = \frac{\max(|t_k|,|t_i|)-edit(t_k,t_i)}{\max(|t_k|,|t_i|)} \quad (5)$$

Where T_C and T_O denotes candidate word set and ontology feature word set,$t_k \in T_C$,$t_i \in T_O$.$|t_k|$ and $|t_i|$ is the length of string t_k and t_i respectively. edit(t_k,t_i) denotes the edit distance between t_k and t_i.

Finally, with the comprehensive considering of TF value and OS value, using the formula (2) to calculate the weight value of t_k in text d. Then sort the candidate words by weight and take those words ranked before threshold α_2 (α_2=10 in this paper) into the next selection stage.

3.2.2 Improved BM25F Algorithm

This paper refers multiple weighted filed term frequency function BM25F [12] to make a secondary filtering on the candidate words. The basic ideas of the algorithm is, according to the different appearance field of the word, give each filed different weight to calculate the weighted value $w_{tk}(d,D_C)$ of document d containing candidate word t_k and then calculate the weighted value F_{tk} of $t_k$by$w_{tk}(d,D_C)$, finally set a threshold β, for each candidate word t_k, if the $F_{tk}\geq\beta$, pick t_k out to be a candidate concept. BM25F gains a comprehensive perspective on the candidate word from multi aspect such as frequency, document frequency, important degree of the field, literature-length, the average length of the literature and so on[13].

The formula of BM25F is as follows:

$$wf_{t_k}(\bar{d}, D_c) = \frac{(k_1'+1)tf_{t_k}'}{k_1'\left((1-b)+b\frac{dl'}{avdl'}\right)+tf_{t_k}'}\log\frac{N-df_{t_k}+0.5}{df_{t_k}-0.5} \tag{6}$$

Where$\bar{d}$ is the field weighted version of d,tf_{tk}' denotes the weighted frequency oft_k in$\bar{d}$, dl' is the weighted document length, $avdl'$ is the weighted average document length across the collection, df_{tk} is the document frequency of t_k,N is the total number of document in the collection D_C,$\bar{d} \in D_C$, and k_1' and b are free parameters.

The last item in BM25F formula is a kind of inverse document frequency of t_k in across the collection. That is, the more the document containing t_k, the lower the weight value of the document is. And this is the same in order to give prominence to differentiate when classify the documents. In candidate word filtering with the purpose of obtaining the key concept for ontology evolution, the candidate word needs not only reflect the core knowledge in specific domain, and it's frequency in document collection must have reached a certain extent. In other words, the extracted word must reach a certain level in important degree and extensive extent. Important degree can be measured by term frequency within document, field and other factors, which has been reflected in the original BM25F formula. While the extensive can be measured by document frequency, this paper holds that, on the basis of considering the importance of candidate word, the more the document containing t_k, that is the higher the document frequency, the word is more likely to become key concept.

Therefore, this paper adjusts formula (6) to formula (7) as follows:

$$wf_{t_k}(\bar{d}, D_c) = \frac{(k_1'+1)tf_{t_k}'}{k_1'\left((1-b)+b\frac{dl'}{avdl'}\right)+tf_{t_k}'}\log\frac{N}{N-df_{t_k}+0.5} \tag{7}$$

Where dl' is the number of words within $\bar{d}$ in this paper, k_1' and b are free parameters (k_1' =2 and b=0.75 in this paper) [14].

By contrast formula (6) and formula (7), we can find this paper changes the influence of document frequency to weight. In the original formula, the more documents containing candidate word, the smaller the weigh value is. And in the improved formula (7), the more documents containing candidate word, the grater the weigh value is. Through such adjustment, the extract concept not only has certain importance in local, but also has certain representative in the global scope.

As the unstructured pure text has no clear field segmentation, this paper mainly takes title, abstract (1/5 of the content by length) and main body (other 4/5 of the content) into account. The paper set the coefficients for each fields as: {title, abstract, main body}={3,2,1}.

After taking formula (7) to calculate the weighted value in text of candidate words selected by 3.2.1, we can calculate the word's global weighted value by the formula (8) described as below.

$$F_{t_k} = \sum_{\bar{d} \in D_c} wf_{t_k}(\bar{d}, D_c) \times tf_j \quad (8)$$

Sort the candidate words by F_{tk} and take those words ranked before threshold α_3 (α_3=2 in this paper) to be key concepts. So far won the key concept set C' for ontology evolution.

4 Relationship Extraction

The requirement of ontology evolution, on the basis of having captured key concept, must confirm the relationship between new concept and original ontology as well as its properties. That is the work this chapter going to do.

4.1 Relationship Extraction Based on Machine Learning

With the purpose of evolving the key concept into original ontology accurately, the relationship between key concept and original ontology has to be obtained, namely relationship extraction. In order to realize the automatic generation of requirement for ontology evolution from great quantities of date, this paper introduced Machine Learning method to extract relationship. The essential of Machine Learning is to consider relationship extraction as a classification problem, on the base of hand-tagged corpus to construct a classifier through a specific learning algorithm, and then apply it to the process of relationship judgment within domain corpus [15][16].

In this paper, the relationship extraction based on Machine Learning draw a salutary lesson from the TEXTRUNNER system of Turing center of Washington University. Michele Banko and others proposed the Open Information Extraction method in reference[17] that facilitates domain independent discovery of relations extracted from text and readily scales to the diversity and size of the Web corpus.

4.2 Naïve Bayes Classifier

The whole work of Naïve Bayes classifier can be divided into three stages: preparation stage, classifier training stage, the application stage. Preparation stage needs to confirm features and extract training samples. The main work of training stage is through analysis and statistics on training samples to calculate the $P(y_i)$ of each class and the condition probability of each feature division. Finally, in application stage, this paper uses Bayes theory to calculate the $P(x|y_i)P(y_i)$ of each class and chose the maximum to be the goal class.

4.3 The Application of Naïve Bayes Classifier in Relationship Extraction for Ontology Evolution

4.3.1 Confirm Class and Feature

This paper studies the ontology evolution and supposes that there was a core ontology about a specific domain and the concept, property and instance has been organized together by a certain relationship. And for ontology evolution, most renew and change is occur on concept while the change on relationship is relatively little. Therefore this paper argues that, if we take a kind of relation as a target class, those kinds of relationship of ontology turn out to be the classes in the classifier. This is very different from the information extraction in TEXTRUNNER, of which the classifier is opened to any possible relations.

This paper adopts several basic relationship of ontology as classes for relationship classifier and defined the classes as:Y={subclass_of,sibling_of,property_of}, Y⊂R.

The 'subclass_of' denotes the membership between concepts, the 'sibling_of' denotes the brotherhood between concepts, and the 'property_of' denotes one concept is another's property.

The effect of Naïve Bayes classifier largely depends on whether the relationship vector is with a very good distinguishability besides it expresses the sample accurately. Based on the reference of researches about eigenvector by experts and scholars[18][19][20], for (c_1,c_2), this paper constructs the relationship eigenvector between c_1 and c_2 mainly from the following aspect (1)order of $c_1$andc_2, (2)neighboring words of c_1 and c_2, (3) dependent verb of c_1 and c_2, (4) neighboring words of dependent verb.

Therefore, eigenvector is defined as follows:

X={order,left1,word1,right1,leftv,wordv,rightv,left2,word2,right2}.

Where, 'order' values of 1 when c_1 is on the front of c_2, or 0; 'word1' is the first key word matched in sentence, 'left1' and 'right1' are the word on its left and right respectively; 'wordv' is the dependent verb in sentence, 'leftv' and 'rightv' are the word on its left and right respectively; 'word2' is the first key word matched in sentence, 'left2' and 'right2' are the word on its left and right respectively.

4.3.2 Extract Training Sample

The training sample in Naïve Bayes classifier refers to those already aware of its class.

TEXTRUNNER firstly uses a parser to automatically identify and label a set of trustworthy (and untrustworthy) extractions. The extractions take the form of a tuple and are used as positive (or negative) training sample to train a Naïve Bayes classifier which is then used by the extractor module. Compare with the traditional classifier requires the user to name the target relations and to manually create new extraction rules or hand-tag new training examples, TEXTRUNNER do not need any human input. However because TEXTRUNNER is open relationship extraction without requiring to set relationship class, it can only label whether there is a relationship between entities or not, and it cannot confirm the specific pattern of the relationship, what is far from the demanding for ontology evolution.

For this reason, this paper adopts the following method to gain training sample for the Naïve Bayes classifier.

- Gain the existing concept, property and relationship of the original ontology, and there is a tuple: $(t_i,r_{i,j},t_j)$. Where t_i and $t_j$are words meant to be the concept or property of the ontology, $t_i \in T_O$, $t_j \in T_O$, $r_{i,j} \in Y$. The tuple denotes that there is a relationship$r_{i,j}$ between t_i and t_j.

- Select M texts from the corpus which were used to buildoriginalontology as training corpus. The knowledge in there texts can be well understood by original ontology, relatively, the knowledge of the concept, property and relationship in original ontology has been well reflected in those texts. This paper defines those texts as 'training text'.

Definition 5: Training text collection is defined as: $D_t=\{d_1,d_2,...d_M\}$.

- For each (t_i,t_j), we extract a set of sentence containing t_i, and t_j in D_t, the set is named S. for each (t_i,t_j), these are one or more sentences containing t_i, and t_j, therefore, these is a tuple for each (t_i,t_j) and its sentences as: $(t_i,s_{i,j,q},t_j)$.
- Indicate each $s_{i,j,q}$ as aneigen vector$X_{i,j,q}=\{x_0, x_1, x_2, ...,x_9\}$, and then there is the mapping: each $X_{i,j,q}$ mapping to (t_i,t_j), each (t_i,t_j) mapping to a relationship $r_{i,j}$.

4.3.3 Classifier Training

Classifier training is to calculate and record the frequency of each class in the training samples and the condition probability of each feature division to each class. The inputs of this stage are features and training samples while output is a formed classifier.

The method to calculate prior probability$P(y_i)$ is shown as formula (9).

$$p(y_i) = \frac{N(r_{i,j}=y_i)}{N(y)} \tag{9}$$

Where $N(r_{i,j}=y_i)$ is the number of training sample belong to class y_i and $N(y)$ is the total number of all the training sample.

The method to calculate condition probability$p(x|y_i)$is shown as formula (10).

$$p(X|y_i) = \prod_{k=0}^{9} p(x_k|y_i) \tag{10}$$

Here need to calculate the probability of each feature in specific class, and then calculate the probability that the training sample belongs to this class.

4.3.4 Application Stage

Following classifier training, the experiment samples are put into the classifier to get the relationship.

In order to get the relationship between the key concept and the concept (or property) of original ontology, this paper constructs experiment sample through the following method.

- Construct word pair (c_i,t_j), where $c_i \in C', t_j \in T_O$.
- For each (c_i,t_j), we extract a set of sentence containing c_i and t_j in D_C the set is named S'.
- Indicate each $s_{i,j,q}$ as aneigen vector $X_{i,j,q}=\{x_0, x_1, x_2, ...,x_9\}$, where $s_{i,j,q} \in S'$.

After getting standard experiment samples, we can use the trained Naïve Bayes classifier to confirm the position and relationship of key concept.

According to Bayes' theorem:

$$P(y_i|X) = \frac{P(X|y_i)P(y_i)}{P(X)} \tag{11}$$

Because $P(X)$ is a constant for all kind of class, in order to make $P(y_i|X)$ achieve its maximum value, just need to get the maximum value of $P(X|y_i)P(y_i)$. Therefore, for the unknown sample X, calculate the value of $P(X|y_i)P(y_i)$ for each y_i, then the sample X would be assigned to the class with maximum $P(X|y_i)P(y_i)$.

Using this method, each key concept and each word in original ontology were combined together to be a word pair (c_i,t_j). For each(c_i,t_j)we can extract one or more sentence from candidate texts. So, in the classifier, there will be a double selection of maximum probability.

- The first time, we select which eigenvector has the maximum probability for the specific word pair. This step can confirm the class that the word pair belongs to.
- The second time, we select which word pair has the maximum probability. This step can confirm the key concept's position in the ontology, on the other words, which concept in the original ontology is related with the key concept.

Through such double selection, this paper not only gets the key concept's position in the ontology, but also confirms its relationship with original ontology.

4.4 Process of Auto-generated Requirement of Ontology Evolution

The overall processes of the method for auto-generated requirements of domain ontology evolution based on Machine Learning techniques are divided into three stages: capturing the key concept, classifier preparation, classifier application, the details are described below. The first stage: first step is the preprocessing of the texts, such as word segmentation and pos-tagging, selecting nouns and nominal phrases, as well as filtering out those words having existed in the original ontology. Then, using the proposed TF-OS algorithm to filter the candidate word from two aspects: the term frequency within text and the similarity with the ontology-words. Finally, using the improved BM25F algorithm to make a second selection from various aspects: term frequency within text, global frequency within texts collection, text frequency, weighted fields, text length, the average length of the texts and so on. So far, we get the key concept for ontology evolution. The second stage: Extract training samples and construct Naïve Bayes classifier. This paper takes a full use of the original ontology. Firstly, get the word pairs and the relationship (subclass_of, sibiling_of, property_of) of each pair from the ontology. Then extract the sentences containing these words from training texts and express these sentences as eigenvectors. Finally take these eigenvectors as training samples while each eigenvector mapping to two ontology-words and their relationship. The third stage is classifier application stage, which is to generate the requirement for evolution. Construct a word pair with the key concept and the existing concept (or property) of the original ontology, extract the sentences containing the word pair from candidate texts, express these sentences as eigenvectors, put these eigenvectors into classifier as experiment samples, finally the output is a word pair and its relationship with the maximum probability. This output not only confirms the key concept's position in the ontology, but also confirm its relationship with original ontology.

5 The Experimental Results and Analysis

This paper takes the ontology about naval weapon knowledge domain as original ontology, which was built in the P2P network environment based on semantic query realized by our subject group[21]. This original ontology is built by our group members and the corpus is hundreds of articles about naval weapon from various major websites. After organizing and classifying the professional words in these texts, we use Protege4.0 to construct domain ontology knowledge database. Then we parse this domain ontology by Jena2.4 and stored the results in the SQL Server database. In order to study the base for ontology evolution, the author searched dozens of recent articles with hot topics about naval weapon in 2011 from major websites as experimental corpora; using the proposed method for auto-generated requirements of domain ontology evolution to obtain the standard expression of the requirements, then on the basis of it to adjust and perfect the original ontology as well as to verify the effectiveness of the requirements.

This paper selected 50 articles about naval weapon as a candidate document collection and extracted 2 key concepts complying with the demands of domain ontology.

The top 5 candidate words picked out by TF-IDF and TF-OS respectively were shown in the following Table 1.

Table 1. Compare TF-OS algorithm with TF-IDF algorithm

TF-IDF			TF-OS		
Candidate Word	Frequency	Value	Candidate Word	Frequency	Value
aircraft carrier	22	0.61	aircraft carrier	22	1.59
Ford	18	0.57	aircraft	3	0.64
American	9	0.46	UAV	3	0.64
Nimitz	7	0.40	Electronic ejection	5	0.59
Digital ship	5	0.35	Ford	18	0.57

As we can see from the table, with the considering of name similarity TF-OS method picked out the words which are more closely related with original ontology and filter these words with less related to current domain such as ‘American’.

The top 5 candidate words picked out by original BM25F and improved BM25F respectively were shown in the following Table 2.

Table 2. Compare between different BM25F algorithms

Original BM25F				Improved BM25F			
Candidate word	Global frequency	Text frequency	Value	Candidate word	Global frequency	Text frequency	Value
china	216	12	715.40	aircraft carrier	377	28	784.58
American	105	8	474.18	nuclear submarine	155	23	212.01
France	50	5	303.69	china	216	12	148.15
Brazil	47	5	283.22	CV	76	14	56.18
unit	19	1	227.91	American	105	8	42.91

As we can see from the table, this paper improved the BM25F combined with the purpose of extracting key concept for requirement of ontology evolution. That improvement covered the original algorithm' s insufficiency, which would filter out these words with both high term frequency and document frequency, such as the word 'aircraftcarrier' in the experiment.

This paper sorts those candidate words by their weighted value and takes the top 2 as key concepts, then the experiment extracted key concepts were:

C'={'aircraft carrier', 'nuclear submarine'}.

About 200 articles were selected from the corpus, which been used to build the naval weapon domain ontology, as training document collection to generate training sample for training the NaiveBayes Classifier. Put experiment samples into the classifier and the output were the requirements of ontology evolution shown in the following Table 3.

Table 3. The requirement for ontology evolution

Key concept	To	Relationship	Note
	frigate	sibling _of	'aircraft carrier' is a sibling concept of 'frigate'
Aircraftcarrier`	displacement	property _of	'displacement', 'reconnaissance', and 'patrol' are properties of 'aircraft carrier'
	reconnaissance	property _of	
	patrol	property _of	
	submarine	subclass _of	'nuclear submarine' is a subclass concept of 'submarine'
Nuclearsubmarine	antisubmarine	property _of	'antisubmarine', 'attack' and 'structure' are properties of 'nuclear submarine'
	attack	property _of	
	structure	property _of	

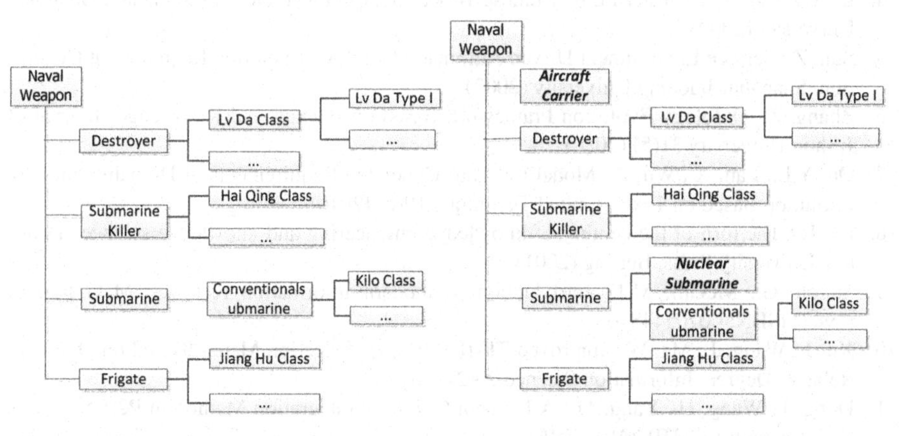

Fig. 1. Original ontology **Fig. 2.** Evolved ontology

6 Conclusions and Future Work

For the issue of the current requirement for domain ontology evolution is mostly manual generated by user, this paper proposed a method for auto-generated requirements of domain ontology evolution based on Machine Learning theory. Experiment shows that the method in this paper can directly extract the key concept from pure text. The extracted candidate key concepts express new things and knowledge, achieve a certain frequency in various texts and closely relate to the current domain knowledge. Moreover, just through analysis of the text and the Naïve Bayes Classifier can we get the exact position and relationship of the key concept in the current ontology as well as numbers of properties matching to the key concept. This method adopts Machine Learning techniques and makes full use of the knowledge structure and relationship model in the existing ontology. Without hand-crafted extraction rules or hand-tagged training examples in the process of relationship extracting, this method fully embodies the automation and intelligent in requirement generation.

Further studies include: more in-depth studies of eigenvector to make it more effective; the update and delete of ontology's concept and property; comprehensive ontology evolution rule; the management of ontology's version after evolution.

Acknowledgments. This work is supported by Shanghai Leading Academic Discipline Project, Project Number: J50103.

References

1. Zhang, Z.: Study of the Ontology Evolution Basec on OWL. Ocean University of China (2007)
2. Ma, W., Du, X.: A study on Domain OntologyEvolution. Library and Information Service 5(6) (2006)
3. Wang, Z., Zhang, D.: Survey of P2P Semantic Search. Computer Science 37(004), 21–26 (2010)
4. Li, L.: The Text Clustering of Chinese Based on Specific Field and Semantic. Shanghai University (2009)
5. Sun, Z.: Service Environment Driven Automatic Ontology Evolution Requirement Generation. Shanghai Jiaotong University (2007)
6. Zhang, Z.: Ontology Evolution Framework Based on Background Knowledge. Journal of JiShou University 31(5) (2010)
7. Ou, Y.L., Lan, X., Wu, Z.: Model for Auto-Generated Requirements of Domain Ontology Evolution Based on Text. Applied Technique 19(6) (2010)
8. Lu, R.: The turn of the century of knowledge engineering and knowledge science. Tsinghua University Press, Beijing (2001)
9. Salton, G., McGill, M.J.: Introduction to modern information retrieval. McGraw-Hill (1983) ISBN: 0070544840
10. Xu, J., Wang, J., Ma, W.: Improved TF-IDF Feature Selection MethodBased on Ontology Relative Degree. Information Science 29(2) (2011)
11. Dong, J., Wang, H., Yang, M.: A Domain Ontology Integration Method in P2P Networks Environment. In: ITIP 2010 (2010)
12. Robertson, S.E., Zaragoza, H., Taylor, M.: Simple BM25 Extension to Multiple Weigthed Fields. In: CIJM 2004 (2004)
13. Robertson, S.E., Zaragoza, H.: The Probabilistic Relevance Framework BM25 and Beyond. Information Retrieval 3(4), 333–389 (2009)
14. Wei, L., Robertson, S.: Field-Weighted XML Document Level Retrieval and Evaluation. Journal of Library Science in China 32(166) (2006)

15. Xu, J., Zhang, Z., Wu, Z.: Review on Techniques of Entity Relation Extraction. New Technology of Library and Information Service 8(168) (2008)
16. Xu, J., Zhang, Z.: The Technical Method Analysis of Typical Relation Extraction System. Digital Library Forum 9(52) (2008)
17. Banko, M., Cafarella, M.J., Soderland, S., et al.: Open Information Extraction from the Web. In: Procs. of IJCAI (2007)
18. Sun, X., Wang, X.: Study on Term Relation Extraction from Domain Text. Computer Science 37(2) (2010)
19. Dong, J., Sun, L.: Automatic Entity Relation Extraction. Journal of Chinese Information Processing 21(4) (2007)
20. Che, W., Liu, T.: Automatic Entity Relation Extraction. Journal of Chinese Information Processing 19(2) (2004)
21. Dong, J.: Chinese. Text Clustering Method Based on Semantics and Special Domain. In: 2009 International Conference on Web Information Systems and Mining, p. 197 (2009)

A Service Chain for Digital Library Based on Cloud Computing

Mengxing Huang and Wencai Du

College of Information Science & Technology, Hainan University, Haikou, 570228, China
huangmx09@gmail.com

Abstract. The main problem of cooperative operation in service chain for digital library (SCDL) is resource sharing and resource unified management under heterogeneous environment. Based on the theory of Cloud Computing and its application, a service chain architecture for digital library based on Cloud Computing is constructed, and an architecture of cloud service platform for SCDL is proposed. Then a method of resource sharing based on publish/subscribe notification mechanisms and concept retrieval models is presented, and a cooperative service architecture in cloud service platform based on resource sharing is proposed. Finally, a sample of content retrieval based on Cloud is demonstrated.

Keywords: Digital library (DL), Service chain for DL (SCDL), Cooperative architecture, Cloud Computing.

1 Introduction

With the development of IT and the popularization of network applications, people obtain their required information and knowledge mainly by means of Internet. A digital library (DL) is an extensible knowledge network system under internet environments and a community service organization which can provide information and knowledge services for people and improve civil education for all-round development. So digital libraries have been emphasized by many countries and developed forcefully since early 1990s. But with the development of search engines, especially of Google which holds several advantages in comparison to digital libraries, digital libraries are faced with the challenge of peripherization [1-2]. Moreover libraries have several advantages over search engines [3]. In order to integrate the advantages of libraries and Google & search engines and provide high efficient, personalized, comprehensive and one-stop-shop information and knowledge services for customers, a supply (or service) chain for digital library (SCDL) has been proposed [4]. But SCDL is composed of the various professional or cross-domain libraries and other enterprises, which can provide users with an integrated multi-disciplinary or cross-domain content services. Then a SCDL is made up of different self-serving organizations that have a lot of difference in the information processing platforms and data storage format, in order to provide users with high-performance and seamless content services, they must work together and share content resources one another, a cooperative working platform need be built where content resources can be dynamically configurated. However, different

Y. Wang and T. Li (Eds.): Knowledge Engineering and Management, AISC 123, pp. 261–266.
springerlink.com

self-serving organizations have lopsided information infrastructure and asymmetrical information, which is becoming the bottlenecks of integration and coordination of SCDL.

Cloud Computing is evolving as a key computing platform for sharing resources that include infrastructures, software, applications, and business processes [5]. Cloud Computing is receiving enormous attention in the industry, mostly due to business-driven promises and expectations, including significantly lower upfront IT costs, a faster time to market, and opportunities for creating new business models and sources of value [6-7]. A cloud service architecture and resource sharing platform for SCDL is constructed making use of Cloud Computing technology.

The rest of the paper is organized as follow. In Section 2, proposes a service chain architecture for DL based on Cloud Computing. Section 3 provides a cloud service framework for SCDL based on resource sharing. In Section 4, a sample of content retrieval based on cloud service is demonstrated. Finally, Section 5 provides the conclusion.

2 A Service Chain Architecture for DL Based on Cloud Computing

According to the requirement of services sharing and cooperation for SCDL and the development trend of Cloud Computing, a service chain architecture for SCDL based on Cloud Computing is presented in Figure 1.

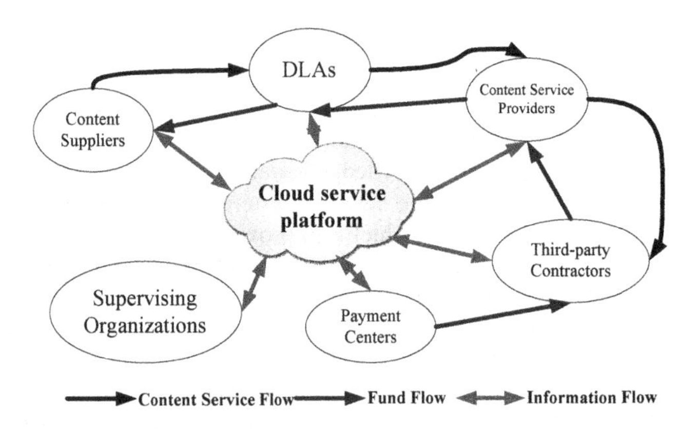

Fig. 1. A service chain architecture for SCDL based on Cloud Computing

1) Content suppliers

Content suppliers are content manufacturers or creators and supply all kinds of media to digital libraries. Content suppliers mainly include copyrighters (authors and copyright owners), publishers, newspaper offices, publishing companies, colleges and universities, academic institutions media importers, professional databases, archive offices and web sites.

2) Digital libraries

DLs are core tier of service chain and in charge of digital content producing, organizing, storing and managing. DLs may be composed of National Digital Library of China, China Academic Digital Library & Information System (CADLIS), National Science Digital Library, National Electronic and Technology Library, party school libraries, military academy libraries, Social Science Library, labor union libraries, locality digital libraries, commercial digital libraries and so on in China.

3) Content service providers

Content service providers lie in down-stream of supply chain, which are service providing and laying out platforms and provide digital content of DL to users by all kinds of channels and according as users' requirement. Content service providers mainly include search engines, personalized and professional portals, digital TVs, broadcasts and mobile communication facilities.

4) Assistant parties

Assistant parties of digital library supply chain mainly include third-party contractors, supervisory institutes and payment centers. Third-party contractors mainly involve the organizations which can provide techniques and funds supporting for digital libraries, such as digitalization service enterprises, rapid printing enterprises, special software providing enterprises and third-party logistics. Supervisory institutes involve copyright bureaus, culture departments and certification authorities. Payment centers involve banks and financial institutions.

5) Cloud service platform

Cloud services platform is to provide each node enterprise a business service platform for synergistic activities which can provide software services, hardware services, platform services, storage services and infrastructure service etc. Figure 2 shows the cloud service platform for SCDL.

In SOA Component Layer, the application of different functional units (called services) are linked to contracts by well-defined interfaces and applying to SOA component model. Cloud services platform uses service-oriented architecture ideas, to provide users with a service interface, service registration, service discovery, service access, service workflow.

Service-oriented Middleware Layer offers support for services sharing and cooperating. The core middleware of Cloud mainly include message oriented middleware (MOM), service aggregation, data mediation service, reliability bulk file transfer (RBFT), service composition, retrieval service, subscription service and so on.

Management Middleware Layer includes user management, task management, resource management and security management.

Cloud computing allows users at any location, using a variety of terminal access to apply services, the application running in the Cloud. Cloud computing can be likened to a huge pool of resources, which make computers and devices distributed on the Internet into a virtual pool of computing resources pool, storage resources pool, network resources pool, data resources pool.

Physical Resource Layer includes computers, memories, network facilities, databases and software.

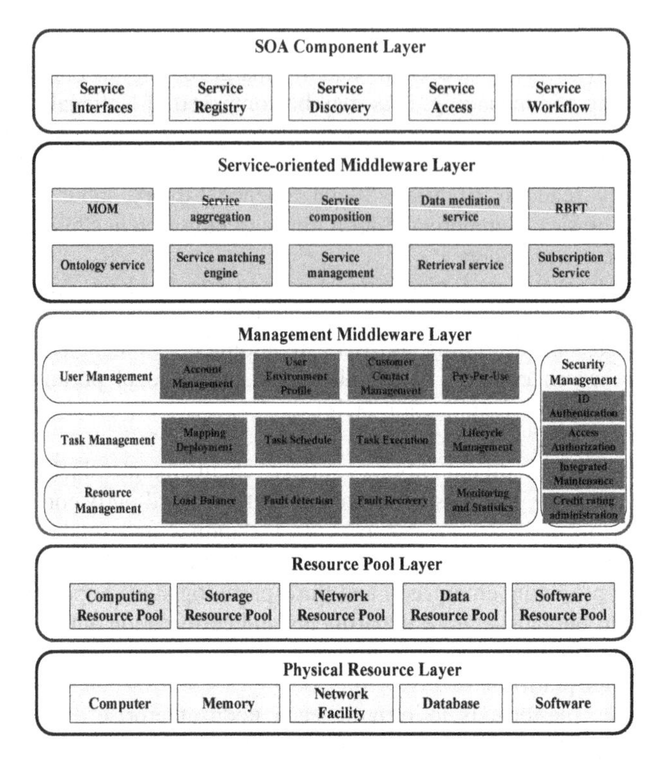

Fig. 2. An architecture of cloud service platform for SCDL

3 A Cooperative Service Framework in Cloud Based on Resource Sharing

In cloud service platform, every digital library will map resources of its autonomous domain to its corresponding virtual resource pool(VP), then cooperation service can be realized by resource sharing in cloud. The matching between user demand and resource characteristics can be realized in different levels of metadata, ontology and semantic.

A resource sharing based on publish/subscribe notification and concept retrieval model is presented, which can realize resources' pushing & pulling service. Figure 3 describe the process of resource sharing in cloud service platform for SCDL. The metadata of resources will be published or mapped to VP, and metadatabase can be encapsulated making use of Web service to unify the operation of resource sharing. When each digital library publishes its resources, VP will automatically produce the messages about resource metadata, and the messages will be broadcasted to overall system by MOM. On the one hand, content resources can be pushed to users according to their subscription information or user characteristics. On the other hand, users' retrieval contents can be supplied according to users' retrieval. Thus the content pushing & pulling service can be realized.

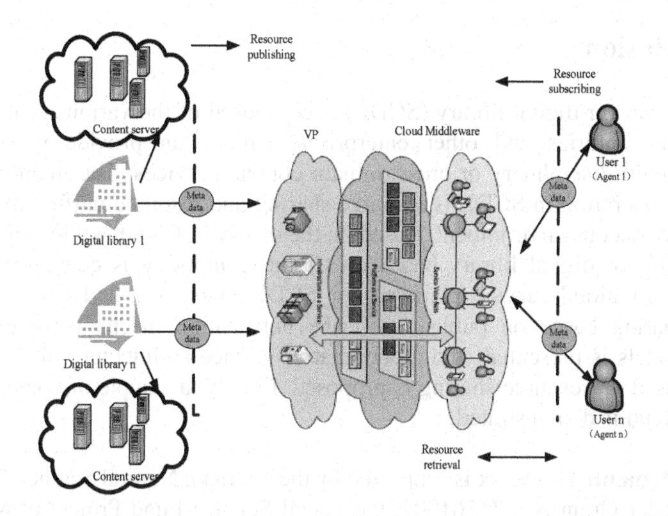

Fig. 3. A cooperative service framework in cloud service platform for SCDL

4 A Sample of Content Retrieval Process

Content retrieval service is one of the major services of digital libraries. In cloud service platform, content retrieval is realized by content retrieval agent and service aggregation middleware. Figure 4 describe the process of content retrieval in cloud service platform.

In that retrieval process, users release a retrieval instruction to content retrieval agent, which includes the metadata information of retrieval scope and contents attributes. The system returns the contents and their link address which match with users' retrieval requests to users. Users do not access content servers which belong to each digital library autonomous domain (including www server, FTP server, streaming Media server, etc.), none but they are interested in or satisfied with the returned contents, they just get the resource through contents link address.

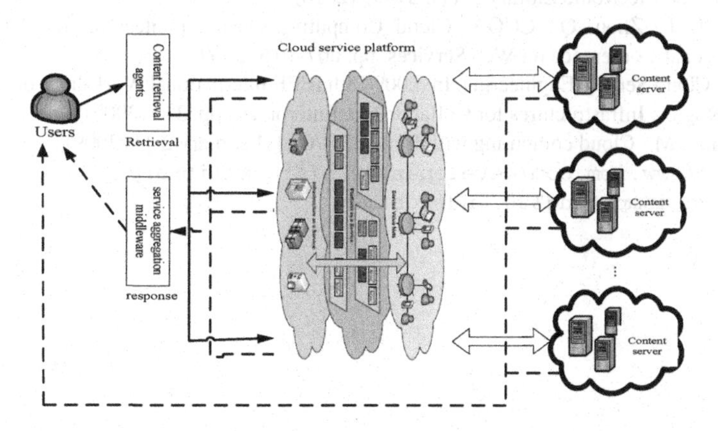

Fig. 4. A content retrieval process

5 Conclusion

A service chain for digital library (SCDL) is composed of the various professional or cross-domain libraries and other enterprises, which can provide users with an integrated multi-disciplinary or cross-domain content services. The main problem of cooperative operation in SCDL is resource sharing and resource unified management under heterogeneous environment. Based on the theory of Cloud and its application , a service chain for digital library based on Cloud Computing is constructed, and an architecture of cloud service platform for SCDL is proposed. Then a method of resource sharing based on publish/subscribe notification mechanisms and concept retrieval models is presented, and a cooperative service architecture in cloud service platform based on resource sharing is proposed. Finally, a sample of content retrieval based on Cloud is demonstrated.

Acknowledgment. This work is supported by the National Natural Science Foundation of China under Grant No. 71161007, the Social Science Fund Project of Ministry of Education under Grant No. 10YJCZH049, the Key Science and Technology Program of Haikou under Grant No. 2010-0067 and the Scientific Research Initiation Fund Project of Hainan University under Grant No. kyqd1042.

References

1. Huang, Z.: A new development phase of digital library: valuation and advices for Google and European's Digital Library Plan. Knowledge of Library and Information Science 5(1), 5–15 (2005)
2. Lopatin, L.: Library digitization projects, issues and guidelines - A survey of the literature. Library Hi Tech. 24(2), 273–289 (2006)
3. Sandusky, R.: Digital library attributes: framing usability research. In: Blandford, A., Buchanan, G. (eds.) Proceedings of the Workshop on Usability of Digital Libraries at JCDL 2002, Portland, OR, July 14-18, pp. 35–38 (2002)
4. Huang, M.-X., Xing, C.-X., Zhang, Y.: Supply Chain Management Model for Digital Library. The Electronic Library 28(1), 29–37 (2010)
5. Zhang, L.-J., Zhou, Q.: CCOA: Cloud Computing Open Architecture. In: 2009 IEEE International Conference on Web Services, pp. 607–616 (2009)
6. Tai, S.: Cloud Service Engineering. In: 2009 18th IEEE International Workshops on Enabling Technologies: Infrastructures for Collaborative Enterprises, pp. 3–4 (2009)
7. Tim Jones, M.: Cloud computing with Linux [EB/OL] (February 2-11, 2009), http://www.ibm.com/developerworks/linux/library/l-cloud-computing%20/index.html

An Enterprise Knowledge Management System (EKMS) Based on Knowledge Evaluation by the Public

Feng Dai, Xinjian Gu, Lusha Zeng, and Yihua Ni

Institute of Manufacturing Engineering, Zhejiang University, Hangzhou 310027, China
{daifeng,xjgu}@zju.edu.cn, lusha1211@yahoo.com.cn,
niyihua@hzcnc.com

Abstract. The decreasing efficiency of knowledge searching and utilization caused by knowledge explosion requires the evaluation of knowledge. An evaluation algorithm based on mass knowledge evaluation was proposed and the evaluation impact brought by evaluation order, evaluation quantity and quality of knowledge was further studied through this algorithm. Grounded in this algorithm, the evaluation capability value of participants and value of knowledge was calculated by common participation action. Different individuals have different evaluation weighing of knowledge and the evaluation capability of an individual is affected by the weighing of the knowledge evaluated. The valuable knowledge can be sorted out relying the self-organization of employees which brings higher utilization efficiency of knowledge, while the knowledge evaluation capability of employees can be sorted which can be used as an index of employee performance appraisal. An enterprise knowledge management system (EKMS) was developed, which proved the feasibility of the proposed algorithm.

Keywords: knowledge evaluation, knowledge management.

1 Introduction

Knowledge explosion results in the decrease of knowledge utilization efficiency. While evaluate and manage the knowledge require the participation of professional technician with large bank of professional knowledge and skilled experience. Hence, it is necessary to study how to encourage all employees, the true users of enterprise knowledge, to participate in this job.

The existing researches in evaluation of internet knowledge and knowledge constructors can be find as follows: An Eigenrumor algorithm[1] ; Another algorithm towards evaluations from shoppers on network commercial city[2]; a similar algorithm[3], the comment and the scores of knowledge evaluation[4] et al..

On the basis of predecessors' work, an integrated evaluation algorithm towards knowledge and capability of participants is proposed and a corresponding enterprise knowledge management system (EKMS) is developed in this paper.

Y. Wang and T. Li (Eds.): Knowledge Engineering and Management, AISC 123, pp. 267–272.
springerlink.com

2 Knowledge and Employees Knowledge Evaluation Capability

2.1 The Method of Knowledge and Employees Knowledge Evaluation

(1) The generalization of knowledge evaluation: when the employees use the knowledge according to their working requirement, the enterprise knowledge platform will automatic keep a record of every operation of employees, such as reading, downloading, recommending, linking, citing and so on. The operations mentioned above are considered to be the evaluation operation towards some knowledge. People with different knowledge background results have different evaluation weightings. For example, old experts and new employees differ a lot in evaluation weightings.

(2) The evaluation of employees' evaluation capability: The employee evaluation capability is obtained by analyzing the effect of the employee evaluation activity. Its effect is: ①To give an appropriate evaluation towards employees' evaluation capability; ②To encourage the employees to evaluate the knowledge with responsibility [5].

(3) The encouraging scheme according to the knowledge contribution: The employees' evaluation of the knowledge should be taken into account. The correct recommendation of knowledge by employees is actually a hidden contribution. The encouragement includes spirit encouragement, material encouragement and evidence for promotion, which is up to the enterprises leaders.

2.2 The Evaluation Model of Knowledge and Employees' Evaluation Capability

While the knowledge is evaluated and sorted by the behavior of enterprise employees, the knowledge evaluation situation is analyzed and sorted to get the employees' knowledge evaluation capability. During the whole process, there are continuous mutual influence between employees' evaluation capability and knowledge value score. The earlier and more precise the employee finds the valuable knowledge, the higher capability he has. And the employees with higher evaluation capability will provide the knowledge with higher score weights. After calculation the relative value score of the knowledge and the relative employees' evaluation capability value are obtained.

The definition of the evaluation of knowledge and employees evaluation capability:

$$r = f(h, e). \tag{1}$$

$$h = g(r, e). \tag{2}$$

r stands for knowledge value score, h stands for the value of employees evaluation capability, e stands for the employees evaluation situation of the knowledge.

2.3 The Evaluation Algorithm of Knowledge and Employees' Evaluation Capability

The rating algorithm of knowledge and employees' evaluation ability synthesized the algorithm of PageRank [6], HITS [7], EigenRumor [1] and the method of linkage analysis. In specific implementation, the actions from participants to information objects is considered as links based on the "Digg" operation which contained reading, downloading, recommending and citing, all of which with different digging weight. Such as one times reading or downloading will be considered as 0.3 digging operation, and one times recommending or citing will be considered as 1 digging operation.

Assume a universe of *n* knowledge objects and *m* participants. One knowledge object can only be evaluated by one participant once. The evaluation matrix $E=\left[e_{ij}\right]$ ($i=m$, $j=n$) will be used to stands for all evaluating knowledge in the universe. When an object *j* receives a digg from participant *i*, $e_{ij}=1$, $e_{ij}=0$ otherwise. Define $\vec{r}$ as a vector that contains the value weight of n knowledge objects, and r_i stands for the value weight of knowledge object *i*. It is considered that the higher the weight, the more valuable the knowledge object. Also define $\vec{h}$ as a vector that contains the evaluation weight of m participants, and h_i stands for the evaluation weight of participant *i*. It is considered that the higher the weight, the better capability the participants have.

The followings are assumptions of integrated evaluation algorithm of knowledge and employees' evaluation ability.

Assumption 1: The higher the evaluating capability of an employee, the greater one's weight in evaluating the knowledge object, and the greater influence one will make to the knowledge. The values of the knowledge object *i* can be calculated as follows.

$$r_i = \sum_{j \in R} h_j. \tag{3}$$

r is all the employees who evaluate the knowledge object *i*. Its vector form is:

$$\vec{r} = E^T \vec{h}. \tag{4}$$

Assumptions 2: If a piece of knowledge is considered as of high value by most participants, those who have evaluated this knowledge will be viewed as employees with high level. The evaluating weight of participant *i* can be calculated as follows:

$$h_i = \sum_{j \in Q} r_j. \tag{5}$$

Q is the knowledge collection dug by participant *i*. Its vector form is as follows:

$$\vec{h} = E\vec{r}. \tag{6}$$

With equation (4) and equation (6), then we can obtain equation (7) and (8):

$$\vec{r} = E^T E\vec{r}. \tag{7}$$

$$\vec{h} = EE^T \vec{h}. \tag{8}$$

$\vec{r}$ and $\vec{h}$ will converge into their positive eigenvector, which can be calculated by the Power-Method [8], and the concrete steps are as follows:

$r_0 = \frac{e}{\|e\|}$;

While $\|r_k - r_{k-1}\| > \varepsilon$ do

$q_k = E^T E r_{k-1}$; $r_k = \frac{q_k}{\|q_k\|}$;

End while.

Where ε is arbitrarily small. $\|\cdot\|$ stands for the normal number of vector. The convergence value of $\vec{h}$ can be calculated by the same method.

In the practical application, in order to guarantee the fairness of evaluating actions, the following situations should be taken into attention.

(1) Problem of random evaluating knowledge. According to the algorithm above, it is entirely possible that some employees may evaluate as more knowledge as possible so as to gain a higher value of evaluating capability. Therefore, in matrix *E*, not only the quantity but also the quality of evaluating should be taken into consideration. According to EigenRumor [2], participant *i* provides *n* knowledge, and its corresponding element in the matrix is $1/\sqrt{n}$. Taking all factors into accounts, for the elements in *E* matrix, $e'_{ij} = \frac{1}{(|E_i|)^{\alpha}}$ Where $|E_i|$ is the total number of knowledge evaluated by participant, and α=0.75 . If α is close to 0, it means "quantity" is more emphasized; If α is close to 1, it means "quality" is more emphasized.

For example, *X*, *Y* and *Z* are knowledge with weight value 0.445, 0.440 and 0.110 respectively. *A*, *B* and *C* are three evaluators. Assuming that *A* recommends *X*, *B* recommends *X* and *Y*, and *C* recommends *X*, *Y* and *Z*. It can be judged that recommendation of *B* is most reasonable. Because *A* leaves out knowledge *Y*, a valuable piece of knowledge, while *C* recommends knowledge *Z*, an unworthy piece of knowledge. Obviously, the weight of knowledge *X* will be further raised than knowledge *Y* and *Z*.

If α=0, the weight values of *A*, *B* and *C* are roughly 0.445, 0.885 and 0.995. In this situation, the quantity of knowledge is given the first priority. The more knowledge participants recommend the higher weight value they will have. It is obviously not reasonable because it is easy for participants to recommend any knowledge without considering whether they are truly valuable; If α=1, the weight values of *A*, *B* and *C* are roughly 0.445, 0.442 and 0.332. In this situation, the quality of knowledge is of great importance while the quantity is overlooked. This method is also not good because it will reduce participants' enthusiasm of recommending knowledge; If α=0.75, the weight values of *A*, *B* and *C* are roughly 0.445, 0.526 and 0.436. Considering both quantity and quality, this situation is the one most close to the real situation.

Furthermore, the recommending of evaluators can be tracked so as to find those who recommend knowledge at will.

(2) Problem of following recommendation. Some evaluators would only recommend the knowledge with uniform conclusion, or those recommended by the authorities. It

helps evaluators safely gain high weight value but fails to facilitate the ordering of knowledge. To avoid this problem, the following methods are proposed:

Assumption 3: The first discover of valuable knowledge contributes more.

Based on assumption 3, evaluators should be given different weight according to their orders of evaluating action. Those who evaluate earlier deserve greater weight. To deal with this problem, the element of matrix E is turned into $e''_{ij} = \frac{1}{(|E_i|)^{\alpha}} \cdot \beta$. Where value of β has negative correlation with evaluating order. That is, the later one evaluates, the smaller β's value would be. β not only rely on the evaluation order, but also rely on the total number of evaluating actions. Moreover, the weight of early evaluating actions decreases more slowly than the later ones. Meanwhile, $\beta \in [0.1,1]$ and $\beta = 0.1$ are used for the last recommending action. If β=0, no one is willing to recommend. According to the above requests, $\beta = 1 - \frac{0.9}{b^{\eta}} \cdot x^{\eta}$., where b is the total number of evaluating actions, x is the evaluating order. $\eta \in [1,+\infty)$, the lower the value of η is, the greater influence evaluating order will make to the final ranking. A high value of η will not result in obvious weakening of the weight of earlier evaluating actions, but may cause sharp reduction of weight of later ones. Therefore, η exerts a great influence to the weight of later recommendations, and in this paper the value of η is 2.5. In addition, the evaluating order of each evaluator can be counted, so as to find the following recommendation problem.

3 Introduction of the Enterprise Knowledge Management System

Based on the evaluation algorithm, an EKMS is developed for an enterprise to support employees' participation of knowledge browsing, downloading, grading and comment. More than 1000 employees participate in the knowledge sharing and evaluation on the system in their daily work. A statistical analysis function mainly provides information of user evaluation capacity and articles on ranking.

Fig.1 shows the top 10 ones ranking list of users' evaluation capacity who can get the enterprise honors and awards in the final of every year. Fig.2 shows the ranking list of corresponding articles to help people to find the truly valuable knowledge in the system.

4 Summaries and Outlook

In this paper, a knowledge evaluation method based on a public knowledge evaluation algorithm and analysis is proposed. Meanwhile, the impact brought by evaluation order, evaluation quantity and quality, and obsolescence of knowledge is studied. Taking advantage of this algorithm, the evaluation capacity value of participants and the value of knowledge is calculated and shown to affect and enhance each other. The purpose is to filter out the most valuable knowledge from massive amounts of knowledge by public wisdom. Based on the above analysis, an enterprise knowledge management system is developed, and the feasibility of proposed algorithm is verified.

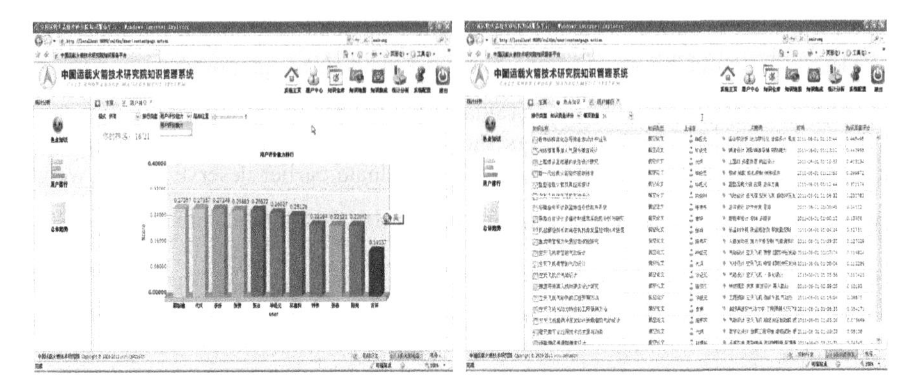

Fig. 1. The ranking of users' rating ability **Fig. 2.** The ranking of knowledge value

The realization of proposed knowledge evaluation algorithm is based on that the knowledge belongs to the same industry or field, and employees are experts in the same industry. Therefore, the next step will study the knowledge and employee evaluation methods when the knowledge and employees belong to different industries and professional fields.

Acknowledgments. The research work is obtained financial support from National Natural Science Foundation (51175463), Zhejiang Province Natural Science Foundation (Y1110414) and Chinese Science and Technology Support Plan Project (2011BAB02B01).

References

1. Fujimura, K., Tanimoto, N.: The EigenRumor Algorithm for Calculating Contributions in Cyberspace Communities. In: Falcone, R., Barber, S.K., Sabater-Mir, J., Singh, M.P. (eds.) Trusting Agents for Trusting Electronic Societies. LNCS (LNAI), vol. 3577, pp. 59–74. Springer, Heidelberg (2005)
2. Fujimura, K.: Reputation Rating System Based on Past Behavior of Evaluators. In: Proceedings of the 4th ACM Conference on Electronic Commerce, pp. 246–247. ACM, San Diego (2003)
3. Chen, M.: Computing and Using Reputations for Internet Ratings. In: Proceedings of the 3rd ACM Conference on Electronic Commerce. ACM, Tampa (2001)
4. Iguchi, M.: Managing Resource and Servant Reputation in P2P Networks. In: Proceedings of the 37th Hawaii International Conference on System Sciences, USA (2004)
5. Tan, J.-R.: Knowledge Engineering Theory Method and Tools in Manufactory Enterprise. Science Press, Beijing
6. Brin, S.: The anatomy of a large-scale hyper-textual Web search engine. Computer Networks and ISDN Systems 30(1-7), 107–117 (1998)
7. Kleinberg, J.: Authoritative sources in hyperlinked environment. Journal of the ACM 46(5) (1999)
8. Wong, D.: Link Analysis: PageRank and Hypertext Induced Topics Search (HITS) [EB/OL], `http://www.cs.helsinki.fi/u/salmenki/sem_s07/linkanalysis.pdf`

Using OWA Operators to Integrate Group Attitudes towards Consensus

Iván Palomares[1], Jun Liu[2], Yang Xu[3], and Luis Martínez[1]

[1] Department of Computer Science, University of Jaén, 23071 Jaén, Spain
{ivanp,luis.martinez}@ujaen.es

[2] School of Computing and Mathematics, University of Ulster, BT37 0QB NewtownAbbey, Northern Ireland, UK
j.liu@ulster.ac.uk

[3] School of Mathematics, Southwest Jiaotong University, Chengdu, 610031 Sichuan, China
xuyang@home.swjtu.edu.cn

Abstract. Nowadays decisions that affect organizations or big amounts of people are normally made by a group of experts, rather than a single decision maker. These decisions would require more than a majority rule to be well-accepted. Consensus Reaching Processes in Group Decision Making Problems attempt to reach a mutual agreement before making a decision. A key issue in these processes is the adequate choice of a consensus measure, according to the group's needs and the context of the specific problem to solve. In this contribution, we introduce the concept of attitude towards consensus in consensus reaching processes, with the aim of integrating it in the consensus measures used throughout the consensus process by means of a novel aggregation operator based on OWA, so-called Attitude-OWA (AOWA).

Keywords: Group Decision Making, Consensus, Attitude, OWA, RIM Quantifiers.

1 Introduction

In Group Decision Making (GDM) problems, which are common in most societies and organizations, two or more decision makers try to achieve a common solution to a problem consisting of two or more alternatives [5]. Traditionally, GDM problems have been solved applying classic approaches such as the majority rule [1]. However, in many contexts it is desirable to achieve an agreement in the group before making the decision, so that all experts agree with the solution, which can be possible conducting a *consensus reaching process* (CRP)[12].

Different consensus approaches have been proposed in the literature, where the notion of *soft consensus* based on fuzzy majority stands out [5]. Also, a variety of consensus models have been proposed by different authors [3,4,8]. However, some crucial aspects in CRPs still require further study, for instance the use of *consensus measures* to determine the level of agreement in the group [2], which usually implies an aggregation process. In this aggregation process, it would be very important reflecting the decision group's *attitude* towards consensus, regarding how to measure consensus as faithfully as possible depending on the experts' prospects and the context of the problem.

Y. Wang and T. Li (Eds.): Knowledge Engineering and Management, AISC 123, pp. 273–282.
springerlink.com

In this paper, we propose integrating the attitude towards consensus of a group participating in a GDM problem. To do this, we present the Attitude-OWA operator (AOWA), that extends OWA operator [13] so that it lets us reflect the desired group's attitude in the aggregation of agreement positions between its members, required to measure consensus. The approach has been put in practice in an automatic consensus support system to solve a problem with a large number of experts.

This contribution is set out as follows: in Section 2 we present some preliminaries related to CRPs and OWA operators based on RIM quantifiers. In Section 3, we present our method to reflect group's attitudes in CRPs with the AOWA operator, and show a practical example where our proposal is applied. Finally, some concluding remarks are given in Section 4.

2 Preliminaries

In this section, we present an overview of CRPs in GDM problems. We then briefly review OWA operators based on RIM quantifiers, which will be the basis for the definition of our proposal.

2.1 Consensus Reaching Processes in GDM

A GDM problem can be defined as a decision situation where a group of *experts* $E = \{e_1,\dots,e_m\}$ $(m > 2)$ tries to achieve a common solution to a problem consisting of a set of *alternatives* or possible solutions, $X = \{x_1,\dots,x_n\}$ $(n > 2)$ [1,5]. Each expert e_i provides his/her opinions over alternatives in X by means of a preference structure. One of the most usual preference structures in GDM problems under uncertainty is the so-called *fuzzy preference relation*. A fuzzy preference relation P_i associated to e_i is characterized by a membership function $\mu_{P_i} : X \times X \to [0,1]$ and represented, given X finite, by a $n \times n$ matrix as follows

$$P_i = \begin{pmatrix} p_i^{11} & \dots & p_i^{1n} \\ \vdots & \ddots & \vdots \\ p_i^{n1} & \dots & p_i^{nn} \end{pmatrix}$$

where each *assessment* $p_i^{lk} = \mu_{P_i}(x_l,x_k)$ represents the degree of preference of alternative x_l over x_k assessed by expert e_i, so that $p_i^{lk} > 0.5$ indicates preference of x_l over x_k, $p_i^{lk} < 0.5$ indicates preference of x_k over x_l and $p_i^{lk} = 0.5$ indicates indifference between x_l and x_k. It is desirable that assessments given by experts are *reciprocal*, i.e. if $p_i^{lk} = x \in [0,1]$ then $p_i^{kl} = 1 - x$.

One of the main drawbacks found in classic GDM rules, such as the majority rule, is the possible disagreement shown by some experts with the decision made, because they might consider that their opinions have not been considered sufficiently. CRPs, where experts discuss and modify their preferences to make them closer each other, were introduced to overcome this problem. These processes aim to achieve a collective agreement in the group before making the decision [12].

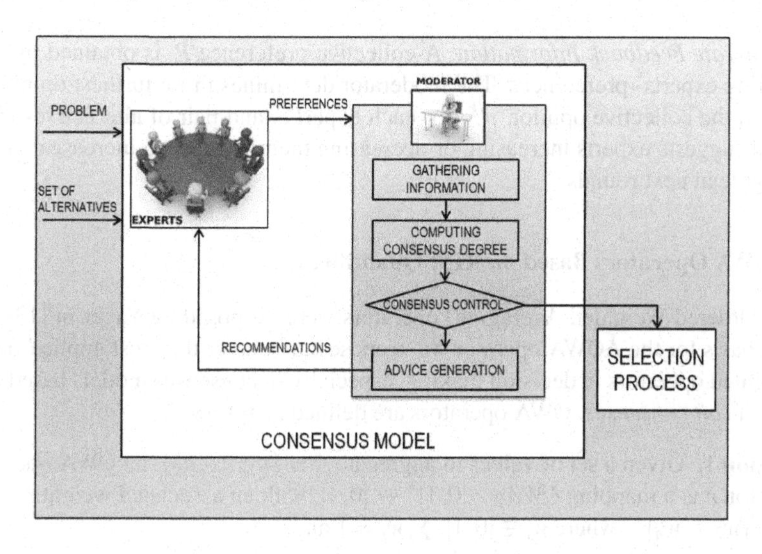

Fig. 1. General CRP scheme in GDM problems

One of the most accepted consensus approaches is the so-called *soft consensus* notion, proposed by Kacprzyk in [5] and based on the concept of fuzzy majority, which establishes that there exists consensus if *most experts participating in a problem agree with the most important alternatives*. This approach can be easily applied by using OWA operators based on linguistic quantifiers to measure consensus [15].

The process to reach a consensus is a dynamic and iterative process, frequently coordinated by a human figure known as *moderator*, responsible for guiding the group in the overall consensus process [7]. Figure 1 shows a general scheme for conducting CRPs. Next, we briefly describe the phases shown in this scheme.

1. *Gather preferences.* Each expert e_i provides moderator a fuzzy preference relation on X, P_i.
2. *Determine Degree of Consensus.* For each pair of experts e_i, e_j $(i < j)$, a similarity value $sm_{ij}^{lk} \in [0,1]$ is computed on each pair of alternatives (x_l, x_k) as [4]

$$sm_{ij}^{lk} = 1 - |p_i^{lk} - p_j^{lk}| \quad (1)$$

   These similarities are aggregated to obtain the consensus degree on each pair of alternatives, $cm^{lk} \in [0,1]$, as [9]

$$cm^{lk} = \phi(sm_{12}^{lk}, \ldots, sm_{1m}^{lk}, sm_{23}^{lk}, \ldots, sm_{2m}^{lk}, \ldots, sm_{(m-1)m}^{lk}) \quad (2)$$

   where the mapping $\phi : [0,1]^{\frac{m(m-1)}{2}} \to [0,1]$ represents the aggregation operator used. We propose using AOWA operator (see Section 3) to reflect the group's attitude in this aggregation. Afterwards, an average operator is used to combine consensus degrees cm^{lk} and obtain an overall consensus degree $cr \in [0,1]$.
3. *Consensus Control*: cr is compared with a consensus threshold μ, previously established. If $cr \geq \mu$, the group moves on to the selection process, otherwise, moderator must move to the feedback generation phase.

4. *Generate Feedback Information*: A collective preference P_c is obtained by aggregating experts' preferences. The moderator determines those furthest opinions p_i^{lk} from the collective opinion p_c^{lk}, for each expert e_i and pair of alternatives (x_l, x_k), and suggests experts increasing or decreasing them in order to increase consensus degree in next rounds.

2.2 OWA Operators Based on RIM Quantifiers

OWA (Ordered Weighted Averaging) operators were proposed by Yager in [13]. They are the basis for the AOWA operator we propose, and one of the most applied families of weighted operators in decision making, especially in consensus models based on the notion of *soft consensus*. OWA operators are defined as follows:

Definition 1. Given a set of values to aggregate $A = \{a_1, \ldots, a_n\}$, an OWA operator of dimension n is a mapping $OWA_W : [0,1]^n \rightarrow [0,1]$, with an associated weighting vector $W = [w_1 w_2 \ldots w_n]^\top$, where $w_i \in [0,1]$, $\sum_i w_i = 1$ and,

$$OWA_W(a_1, \ldots, a_n) = \sum_{j=1}^{n} w_j b_j \tag{3}$$

where b_j is the jth largest of the a_i values. Note that a weight w_i is associated with a particular ordered position, i.e. the ith largest element in A. Therefore a reordering of values is assumed before aggregating.

OWA operators can be classified based on their *optimism degree*, by means of a measure so-called $orness(W) \in [0,1]$, which indicates how close the operator is to maximum (OR) operator.

$$orness(W) = \frac{1}{n-1} \sum_{i=1}^{n} (n-i) w_i \tag{4}$$

Another measure, the *dispersion* or entropy, $Disp(W) \in [0, ln(n)]$, is used to determine the degree to which information in A is really used in aggregation.

$$Disp(W) = -\sum_{i=1}^{n} w_i \ln w_i \tag{5}$$

Several methods have been proposed to compute OWA weights. We consider the use of linguistic quantifiers [15], more specifically RIM (Regular Increasing Monotone) quantifiers, since their implicit semantics become them appropriate to consider the notion of consensus as fuzzy majority.

A RIM quantifier Q is a fuzzy subset of the unit interval, where given a proportion $r \in [0,1]$, $Q(r)$ indicates the extent to which r satisfies the semantics defined in Q [14]. RIM quantifiers present the following properties:

1. $Q(0) = 0$
2. $Q(1) = 1$
3. if $r_1 > r_2$ then $Q(r_1) \geq Q(r_2)$.

For a linear RIM quantifier, its membership function $Q(r)$ is defined upon $\alpha, \beta \in [0,1]$, $\alpha < \beta$ as follows,

$$Q(r) = \begin{cases} 0 & \text{if } r \leq \alpha, \\ \frac{r-\alpha}{\beta-\alpha} & \text{if } \alpha < r \leq \beta, \\ 1 & \text{if } r > \beta. \end{cases} \quad (6)$$

Yager proposed the following method to compute OWA weights based on $Q(r)$ [13]

$$w_i = Q\left(\frac{i}{n}\right) - Q\left(\frac{i-1}{n}\right), i = 1, \ldots, n \quad (7)$$

3 Attitude Integration in Consensus Reaching Processes

Since the aim of this paper is to introduce the concept of group's attitude towards consensus (regarding the aggregation of experts' similarities to measure consensus, as pointed out in Eq. (2)), in this section we present the idea of attitude towards the achievement of consensus and the so-called Attitude-OWA (AOWA) operator used to reflect it in aggregation. Attitude towards consensus must be understood as follows [11]:

- *Optimistic Attitude*: Positions in the group (pairs of experts) where the level of agreement is higher will attain more importance in aggregation, so that when experts modify their opinions, they will require less discussion rounds to achieve the desired consensus.
- *Pessimistic Attitude*: Positions in the group whose level of agreement is lower are given more importance in aggregation, so that experts will require more discussion rounds to achieve the desired consensus.

The choice of an appropriate attitude may depend on the problem to solve and the group's prospects. Figure 2 shows with detail the phase of computing consensus presented in Section 2.1., highlighting the aggregation of similarities where the AOWA operator is used to integrate the group's attitude in the measurement of consensus.

3.1 Attitudinal Parameters and AOWA Operator

Attitude towards consensus is mainly determined by the optimism degree of the group, which will be reflected in the $orness(W)$ measure of its corresponding AOWA operator. Assuming we use a RIM quantifier to compute weights, attitude is also related to the quantifier's shape. Let $Q_{(\alpha,d)}$ be a RIM quantifier, where $d = \beta - \alpha$, $d \in [0,1]$. Considering $orness(W)$, α and d, we define three *attitudinal parameters* which can be used by the group to reflect an attitude towards consensus.

- $orness(W)$ represents the group's attitude, which can be either optimistic, pessimistic or indifferent, depending on its value to be higher, lower or equal to 0.5, respectively. The higher $orness(W)$, the more importance is given to higher similarities in aggregation.
- α determines whether higher or lower similarity values are given non-null weight when aggregating (assuming they are ranked in decreasing order). The lower α, the higher ranked values are considered.

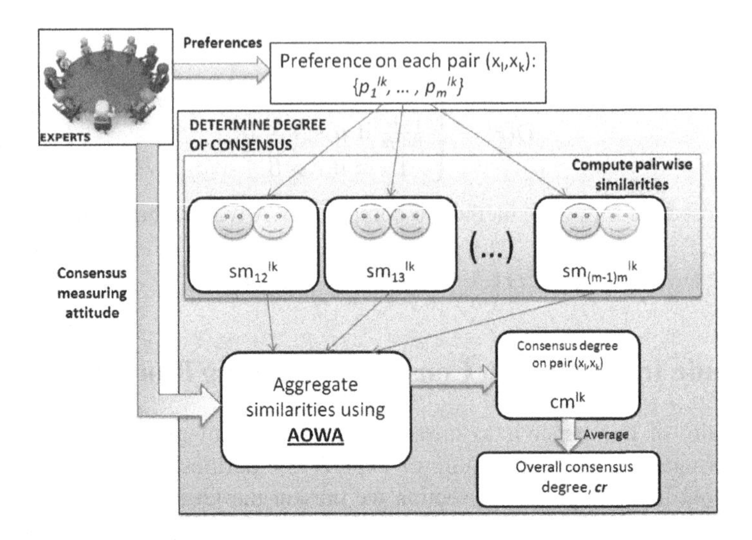

Fig. 2. Procedure to measure consensus based on group's attitude

- d indicates the amount of similarity values which are given non-null weight when aggregating, and has a relation with the dispersion of the corresponding operator. The higher d, the more values are considered and the higher dispersion.

Due to the existing relation between these parameters, it is only necessary that the group uses two of them to define the attitude they want to reflect, therefore we focus on considering $orness(W)$ and d as the two input attitudinal parameters used by the group. Next, we define a class of OWA operator so-called AOWA for reflecting attitudes towards consensus.

Definition 2. An AOWA operator of dimension n on $A = \{a_1, \ldots, a_n\}$ is an OWA operator based on two attitudinal parameters ϑ, φ given by a decision group to indicate how to measure consensus between their members,

$$AOWA_W(A, \vartheta, \varphi) = \sum_{j=1}^{n} w_j b_j \tag{8}$$

where b_j is the jth largest of values in A, $\vartheta, \varphi \in [0,1]$ are two attitudinal parameters and weights W are computed using RIM quantifier $Q_{(\alpha,d)}$. As aforementioned, we consider $\vartheta = orness(W)$ and $\varphi = d$ as the two attitudinal parameters used to define an AOWA operator.

Given a RIM quantifier with a membership function as shown in Eq. (6), when the number of elements (pairs of experts) to aggregate n is large enough, it is possible to compute the optimism degree ϑ of an AOWA operator based on a quantifier $Q_{(\alpha,\varphi)}$ as follows [6,14]

$$\vartheta = 1 - \alpha - \frac{\varphi}{2} \tag{9}$$

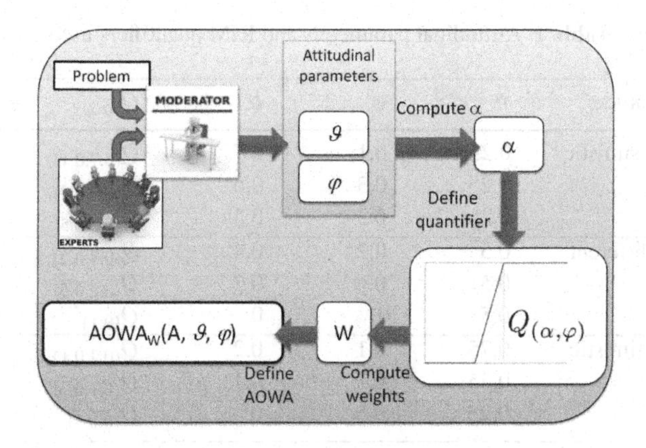

Fig. 3. Process to determine the AOWA operator used to measure consensus based on ϑ and φ

therefore, we conclude that given ϑ, φ it is possible to compute the value of α, required to define the RIM quantifier associated to the AOWA operator as

$$\alpha = 1 - \vartheta - \frac{\varphi}{2} \tag{10}$$

Relations between attitudinal parameters make necessary defining some restrictions in their values, to ensure the definition of a valid quantifier upon them. A RIM quantifier $Q_{(\alpha,\varphi)}$ with $\alpha, \varphi \in [0,1]$ is valid only if $\alpha + \varphi \leq 1$. In order to guarantee this condition, the following restrictions are imposed to the group when they provide an input parameter's value, depending on the value given for the other one.

- Given φ, the value of ϑ provided by the decision group must fulfill,

$$\frac{\varphi}{2} \leq \vartheta \leq 1 - \frac{\varphi}{2} \tag{11}$$

- Given ϑ, the value of φ provided by the decision group must fulfill,

$$\varphi \leq 1 - |2\vartheta - 1| \tag{12}$$

If these restrictions are considered by the group when expressing their desired attitude towards consensus, a valid definition of AOWA operator will be guaranteed. Figure 3 shows the process to integrate the attitude towards consensus, which results in the definition of the associated AOWA operator. This process should be conducted before beginning the discussion process. The moderator is responsible for gathering and introducing attitudinal parameters, considering both the context and characteristics of the decision problem to solve, and the experts' individual desired attitudes.

3.2 Example of Performance

Once presented AOWA operator to integrate attitudes towards consensus, we illustrate its performance on a multi-agent based consensus support system [9,10], which has let

Table 1. Attitudinal parameters and RIM quantifiers used

Attitude	ϑ	φ	α	$Q_{(\alpha,\varphi)}$
Pessimistic	0.25	0.1	0.7	$Q_{(0.7,0.1)}$
	0.25	0.3	0.6	$Q_{(0.6,0.3)}$
	0.25	0.5	0.5	$Q_{(0.5,0.5)}$
Indifferent	0.5	0.2	0.4	$Q_{(0.4,0.2)}$
	0.5	0.6	0.2	$Q_{(0.2,0.6)}$
	0.5	1	0	$Q_{(0,1)}$
Optimistic	0.75	0.1	0.2	$Q_{(0.2,0.1)}$
	0.75	0.3	0.1	$Q_{(0.1,0.3)}$
	0.75	0.5	0	$Q_{(0,0.5)}$

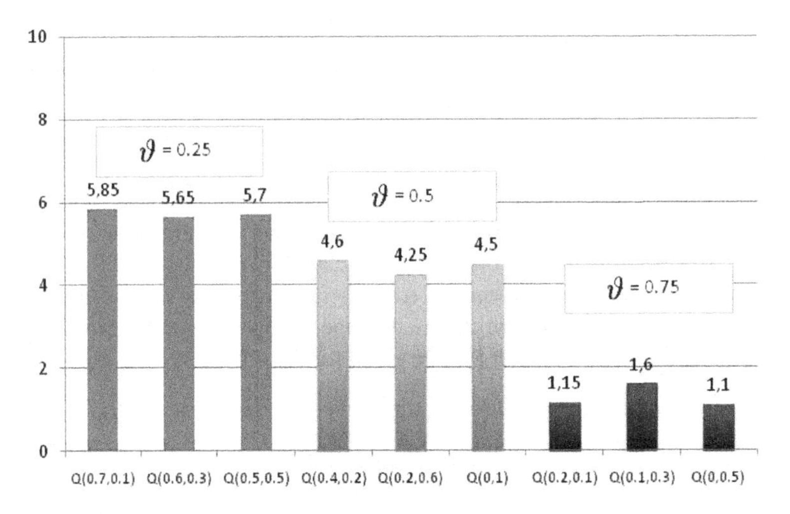

Fig. 4. Average number of required rounds of discussion for RIM quantifier-based AOWA operators with different attitudinal parameters given by ϑ and φ

us simulate the resolution through consensus of a GDM problem under uncertainty, using AOWA operators based on different attitudes, and considering a large number of experts in the group. We assume the hypothesis that using different attitudes directly affects the process performance, so that optimism favors a greater convergence towards consensus with a lower number of rounds, whereas pessimism favors a lower convergence, so that more discussion rounds will be required.

The problem simulated consists of 4 alternatives $X = \{x_1, x_2, x_3, x_4\}$, 50 experts, $E = \{e_1, \ldots, e_{50}\}$ and a consensus threshold $\mu = 0.85$. Different attitudes have been defined (including optimistic, pessimistic and indifferent attitudes) and summarized in Table 1, with the corresponding RIM quantifier obtained. Experiments consisted in running 20 simulations for each attitude.

Results are shown in Figure 4, indicating the convergence to consensus as the average number of rounds required to exceed μ for each AOWA operator. These results confirm our hypothesis, regardless of the amount of agreement positions considered, φ. As a result, if the group's priority is achieving a consensus quickly (no matter the highest agreement positions are rather considered), they would adopt an optimistic attitude. On the contrary, if they consider that the problem requires more discussion and/or they want to ensure that even the most disagreing experts finally reach an agreement, they would rather adopt a pessimistic attitude.

4 Concluding Remarks

In this contribution, we have presented the AOWA operator for expressing group's attitudes in consensus reaching processes. Basing on the definition of attitudinal parameters, we have proposed a method where a decision group can easily reflect the attitude they consider towards the measurement of consensus in the group. A simulation has been conducted on a consensus support system to show both the importance of integrating these attitudes in the process and the effect of using different attitudes in the process performance. Future works focus on integrating this proposal in a linguistic background, as well as extending it to a wider range of different linguistic quantifiers to compute AOWA weights.

Acknowledgements. This work is partially supported by the Research Project TIN-2009-08286, P08-TIC-3548 and FEDER funds.

References

1. Butler, C., Rothstein, A.: On Conflict and Consensus: A Handbook on Formal Consensus Decision Making, Takoma Park (2006)
2. Cabrerizo, F.J., Alonso, S., Pérez, I.J., Herrera-Viedma, E.: On Consensus Measures in Fuzzy Group Decision Making. In: Torra, V., Narukawa, Y. (eds.) MDAI 2008. LNCS (LNAI), vol. 5285, pp. 86–97. Springer, Heidelberg (2008)
3. Herrera, F., Herrera-Viedma, E., Verdegay, J.: A model of consensus in group decision making under linguistic assessments. Fuzzy Sets and Systems 78(1), 73–87 (1996)
4. Herrera-Viedma, E., Martínez, L., Mata, F., Chiclana, F.: A consensus support system model for group decision making problems with multigranular linguistic preference relations. IEEE Transactions on Fuzzy Systems 13(5), 644–658 (2005)
5. Kacprzyk, J.: Group decision making with a fuzzy linguistic majority. Fuzzy Sets and Systems 18(2), 105–118 (1986)
6. Liu, X., Han, S.: Orness and parameterized RIM quantifier aggregation with OWA operators: A summary. International Journal of Approximate Reasoning 48(1), 77–97 (2008)
7. Martínez, L., Montero, J.: Challenges for improving consensus reaching process in collective decisions. New Mathematics and Natural Computation 3(2), 203–217 (2007)
8. Mata, F., Martínez, L., Herrera-Viedma, E.: An adaptive consensus support model for group decision-making problems in a multigranular fuzzy linguistic context. IEEE Transactions on Fuzzy Systems 17(2), 279–290 (2009)

9. Mata, F., Sánchez, P.J., Palomares, I., Quesada, F.J., Martínez, L.: COMAS: A Consensus Multi-Agent based System. In: Proceedings of the 10th International Conference on Intelligent Systems Design and Applications (ISDA), Cairo, Egypt), pp. 457–462 (2010)
10. Palomares, I., Sánchez, P.J., Quesada, F.J., Mata, F., Martínez, L.: COMAS: A multi-agent system for performing consensus processes. In: Abraham, A., et al. (eds.) Advances in Intelligent and Soft Computing, pp. 125–132. Springer, Heidelberg (2011)
11. Palomares, I., Liu, J., Martínez, L.: Modeling Attitudes towards Consensus in Group Decision Making by means of OWA Operators. Technical Report 4-2011, Escuela Politécnica Superior, Universidad de Jaén (2011), `http://sinbad2.ujaen.es/sinbad2/?q=node/158`
12. Saint, S., Lawson, J.R.: Rules for Reaching Consensus. A Modern Approach to Decision Making. Jossey-Bass (1994)
13. Yager, R.R.: On ordered weighted averaging aggregation operators in multi-criteria decision making. IEEE Transactions on Systems, Man and Cybernetics, Part B: Cybernetics 18(1), 183–190 (1988)
14. Yager, R.R.: Quantifier guided aggregation using OWA operators. International Journal of Intelligent Systems 11(1), 49–73 (1996)
15. Zadeh, L.A.: A computational approach to fuzzy quantifiers in natural languages. Computing and Mathematics with Applications 9, 149–184 (1983)

Criteria-Based Approximate Matching of Large-Scale Ontologies

Shuai Liang[1,2], Qiangyi Luo[3], Guangfei Xu[2], Wenhua Huang[1], and Yi Zhang[1]

[1] Institute of Command Automation, PLA University of Science & Technology, Nanjing, 210007, China
[2] The 63892nd unit of the PLA, Luoyang, 471003, China
[3] Institute of Electronic Equipment System Engineering Corporation, Beijing, 100141, China
liangshuai12@163.com

Abstract. Large ontology matching problem brings a new challenge to the state of ontology matching technology. In this paper we present a criteria-Based approximate matching approach to solve the problem. The main principle of our approach is based on fully explore semantic information hidden in ontology structures, either physically or logically, and tightly coupled element-level and structure-level features to obtain matching results. Through joining the quantitative structural information for the matching process can significantly decrease the execution time with pretty good quality.

Keywords: ontology matching, ontology modularization, approximate ontology matching, combinatorial optimization.

1 Introduction

Ontology matching is a critical way of establishing interoperability between different related ontologies. Many ontologies have reached a size that overburdens matching procedures. Most of the existing ontology matching approaches are not applicable for large-scale ontology. These methods are always considering the element and structural characteristics separately and loosely coupled two level similarities to obtain the final mapping result [1]. In this paper, we aim at fully exploring semantic information hidden in ontology structures by a set of quantitative criteria, and tightly coupled the element-level and structure-level features to obtain matching results.

Our approach can be divided into four steps: firstly, analyzes the network characteristics of large-scale ontology; secondly, propose a set of semantic and theoretical criterions to measure the semantic information of nodes and edges inherently hidden in ontology topological structures; thirdly, model the Ontology corresponding Undirected weighted hierarchical network to electrical systems for ontology modularization; fourthly, propose an approximate method on ontology matching approach which combines information on ontology structures and element similarity. Use convex relaxation algorithm to solve quadratic programming.

Y. Wang and T. Li (Eds.): Knowledge Engineering and Management, AISC 123, pp. 283–288.
springerlink.com

2 Network Characteristics Analysis

Well-defined ontology can be viewed as concept network [2]. By mapping ontology concepts to the nodes of network, and the relationships to the edges of network, we can create the corresponding network [3].

There are two steps to transform ontology to networks: First, adding functional nodes, OWL graph expressed functional relationship as edges between concept nodes. In many applications, functional relationship plays a great role in data processing, such as ER diagram in DB. To fully use the functional relationship, it is needed to convert functional relationship to function node; Second, edges in OWL graph always has direction, while actually each edge consists of two inverse relations, so it is needed to replace directed edges with weighted undirected edges.

As the result, the corresponding concept network can be viewed as a weighted undirected hierarchical network with a variety of types of nodes and edges. There are two types of nodes: concept nodes and function nodes, and two types of edges: hierarchical relationship and non-hierarchical relationship.

3 Ontology Structure Quantitative Criteria

Node Quantitative Criteria

Assumption 1. In ontology, the nodes closer to the middle-level has stronger semantic description ability, can represent more details of ontology.

Define mid(c) as the node-level coefficient, measure the proximity of node's level closer to the middle-level:

$$mid(c) = 1 - \left| \frac{D(c) - \frac{H(c)}{2}}{\frac{H(c)}{2}} \right| \quad (1)$$

H(c) represents the longest distance from the root node to the leaf nodes of all branches pass c. D(c) represents the shortest path from the root to concept c. mid (c) $\in [0,1]$, the more closer to the middle level of ontology, the greater its value.

Assumption 2. In ontology, node distribution density is different in different location [4]. The higher of node's global density and local density, the more important role it plays.

The formula of node global density Gden (c) is:

$$Gden(c,O) = \frac{\sum_{i=1}^{3} w_i |S(c)_i|}{\max(\left\{ \forall N \in O \rightarrow \sum_{i=1}^{3} w_i |S(N)_i| \right\})} \quad (2)$$

Here $S(c) = (S(c)_1, S(c)_2, S(c)_3)$ = (direct-subclasses[c], direct-superclasses[c], Relations[c]), respectively represent the number of node c's direct subclasses, direct superclass and functional relations. w_i judging the different role of different types of link in density calculation.

The local density criterion favours the densest concept in a local area, for being potentially the most important for this particular part of the ontology. It is computed using the formula below, where by "nearest concepts" to c, we refer to the set which includes sub- and super-classes reachable through a path of maximum length k in the hierarchy from c, as well as c itself. The formula of Node local density Lden (c) is:

$$Lden(c) = \frac{Gden(c,O)}{\max(\{\forall N \in nearest_k(c) \rightarrow wGD(c,N)\})} \quad (3)$$

Here $wGD(c,N) = (1-(ratio_D * dis(c,N))) * Gden(N,O)$, $nearest_k(c)$ is all the nodes within the shortest path from c whose length less than k. $ratio_D$ defined as the ratio of distance effects. dis(c, N) is the length of shortest path from c to N.

Finally, the overall node density den(c) ∈ [0,1] can be computed as weighted sum of global density and local density, each of these sub-measures being associated with a particular weight:

$$den(c) = w_G * Gden(c) + w_L * Lden(c) \quad (4)$$

The Local Centrality, lc(c), of a concept C is a measure between 0 and 1, which indicates how 'central' C is in the structure of the ontology. lc(c) can be derived from combining the node's depth (level coefficient) and density.

$$lc(c) = \frac{w_h * mid(c)}{\max_{1 \le j \le |O|}\{mid(j)\}} + \frac{w_d * den(c)}{\max_{1 \le j \le |O|}\{den(j)\}} \quad (5)$$

Assumption 3. In ontology, for the specified number of nodes collection, the higher level of coverage, the more accurate description of the overall ontology.

We define Coverage(S) as the measure of the level of coverage of a set of concepts S in a given ontology. Specifically, coverage({c_1,…,c_n}) is computed using the following formula (with |O| being the size of the ontology O given as the number of concepts included in O):

$$\operatorname{cov}erage(\{c_1,...,c_n\}) = \frac{|\{\operatorname{cov}(c_1) \cup \operatorname{cov}(c_2)... \cup \operatorname{cov}(c_n)\}|}{|O|} \quad (6)$$

Here cov(c) represent the set of concepts covered by concept c, which is computed on the basis of its number of all subclasses, all superclasses and functional relations.

In the limited set of nodes, single node on the contribution of the set coverage can be represented by the node coverage set difference between coverage.

$$con(c_i,\{c_1,...,c_n\}) = \frac{|cov(c_i) - (cov(c_i) \cap (\cup_{1 \le j \le n \wedge j \ne i} cov(c_j)))|}{|\cup_{1 \le k \le n} cov(c_k)|} \quad (7)$$

Comprehensive Local Centrality and coverage of the node, we can judge node significance all-round by:

$$sig(c_i,\{c_1,...,c_n\}) = w_c * \frac{con(c_i,\{c_1,...,c_n\})}{\max con(\{c_1,...,c_n\})} + w_{LC} * lc(c_i) \quad (8)$$

Edge Quantitative Criteria

The weight of edge is dependent on the link types, edge depth and density. We define hierarchy coefficientαt ∈ [0,1] as a link type factor applied to a sequence of links of type t, t=0, 1, 2 respectively represent hyper/hyponym, hol/meronym and functional relations. The most strongly related concepts are the hyper/hyponym case, we assign a higher weight $\alpha_0 = 1$. For the link type of hol/meronym, we assign an intermediate weight, $\alpha_l = 0.9$. Similarly, we assign a lower weight for the function relation, $\alpha_l = 0.8$.

The weight of hierarchical relation is relevant with potential difference between hierarchy levels. As the common superclass of two nodes deeper, the potential difference is getting small, the structural proximity of two nodes more closely. We define the level of level coefficient $\beta(c_i, c_j)$:

$$\beta(c_i, c_j) = \frac{2dep(c_{ij})}{dep(c_i) + dep(c_j)} \tag{9}$$

Here dep(c_{ij}) represents for the nearest common superclass of two nodes c_i, c_j.

In order to quantitative of edges in the ontology, we expand Sussna[5] semantic-based edge weight calculation method. We proposed a type-level weight measurement method, according to the endpoints with the kind of relationship types of local density and the edge of the depth to calculate the semantic distance between two endpoints:

$$dis(c_i, c_j) = (1 - \beta(c_i, c_j)) * \frac{w(c_i \rightarrow_t c_j) + w(c_i \rightarrow_t c_j)}{2\max(dep(c_i), dep(c_j))} \tag{10}$$

$$w(X \rightarrow_t Y) = 1 - \frac{1 - \alpha_t}{n_t(X)} \tag{11}$$

Where $\rightarrow_t$ is a relation of type t, $\rightarrow_{t'}$ is its inverse. nt(X) is the number of relations of type t leaving node X.

4 Ontology Modular Partitioning

As the semantic distance between two nodes in network can be determined by the sum of all including edges' shortest path weight. Edge of the network can be seen as a circuit. The weight of edge can be seen as the resistance. The network can be viewed as an electric circuit with current flowing through each edge. By solving Kirchhoff equations we can obtain the voltage value of each node. As edges are sparser between two clusters, the local resistivity should be larger compared to the local resistivity within the two clusters. The voltage drops primarily at the junction between clusters. According nodes' voltage value the network can be divided into clusters, and we are able to judge which cluster it belongs to.

The Kirchhoff equations show that the total current flowing into a node should sum up to zero. That is, the voltage of a node is the average of its neighbours.

$$V_C = \frac{1}{n} \sum_{i=1}^{n} V_{D_i} \tag{12}$$

According to weighted network each edge's conductivity proportional to its weight:

$$R_{ij} = w_{ij}^{-1} \tag{13}$$

Randomly choosing two core nodes from core concept set in the network as poles. Attaching the battery between the poles, so that they have fixed voltages, say 1 and 0.

Considering the core node set selection rules, the chosen poles lie in different clusters. Following Eq. (12), the Kirchhoff equations of a n-node circuit can be written as:

$$V_i = \frac{1}{k_i}\sum_{(i,j)\in E} V_j = \frac{1}{k_i}\sum_{j\in G} V_j a_{ij} = \frac{1}{k_i}\sum_{j=3}^{n} V_j a_{ij} + \frac{1}{k_i} a_{i1} \tag{14}$$

where k_i is the degree of node i and a_{ij} is the adjacency matrix of the network. Define:

$$V = \begin{pmatrix} V_3 \\ \vdots \\ V_n \end{pmatrix}, B = \begin{pmatrix} \frac{a_{33}}{k_3} \cdots \frac{a_{3n}}{k_3} \\ \vdots \quad\quad \vdots \\ \frac{a_{n3}}{k_n} \cdots \frac{a_{nn}}{k_n} \end{pmatrix}, C = \begin{pmatrix} \frac{a_{31}}{k_3} \\ \vdots \\ \frac{a_{n1}}{k_n} \end{pmatrix} \tag{15}$$

the Kirchhoff equations can be further put into a matrix form:

$$V = BV + C \tag{16}$$

As edges are sparser between two clusters, the local resistivity should be larger compared to the resistivity within the two clusters. We suggest placing the threshold at the largest voltage gap near the middle. Note that the global largest gap often appears at the two ends of the voltage spectrum, but it does not make sense to cut there at all, which would divide the graph into two extremely asymmetrical clusters, one of which has only one or two nodes. We simply cut at the right largest gap in the middle of the voltage interval.

$$V_p = \max_{v_i<v_j \wedge v_i,v_j\in(0.5\pm\alpha)\wedge\neg\exists v_k, v_i<v_k<v_j} V_i + 1/2(V_j - V_i) \tag{17}$$

We define α as a tolerance to describe the range of allowed division. A tolerance 0.2 means we only select the largest voltage gap of the range of 0.3 to 0.7. First we sort the voltage values. Then we find the largest gap among 0.3 to 0.7 and cut there. The sort can be done in O(V) time by using a standard linear time sort. This method can divide the network into almost two equal-sized clusters.

The steps of ontology modular partitioning are:

Input: ontology O corresponding undirected weighted network G (V, E, w)
Output: Ontology modular set M (T_1, T_2, ...)
(1) Obtain ontology core node S based on the core nodes selection algorithm;
(2) Arbitrarily select two core nodes c_i, c_j from node set T, and set their value as $V_i = 1$, $V_j = 0$;
(3) According Kirchhoff equation to calculate the remaining nodes voltage V_k ($k \neq i, j$);
(4) Computing the largest voltage gap in the division range. Cut the notes to two disjoint clusters of nodes, T_1 and T_2;
(5) Respectively determine the number of core nodes in T_1, T_2:
(a) If there is only one core node in T_1 (or T_2), remove this set from T, and join this set to the collection of modules M;
(b) If there are more than one core nodes in T_1 (or T_2), then go to step 2;
(6) When the node in T is empty, and then ends iteration and output the set of cluster M.

5 Approximate Ontology Matching

We solve ontology matching problem as a Labelled Weighted Graph Matching Problem [6]. It permits the enforcement of two-way constraints on the correspondence.

The proposed structure-based approximate ontology matching method has four steps:

1. Extract core concept set from nodes of the two ontologies. According to label information, using terminology-based similarity analysis methods to obtain the core nodes matching matrix A_C;
2. Calculate the weight of different types of edges to obtain weighted adjacency matrix A_G, A_H corresponding to ontologies;
3. Define approximate matching matrix P, combined with A_C and A_G, A_H, propose objective function;
4. Use combinatorial optimization algorithm to compute the value of P.

6 Conclusion

This method establishes a series of structure quantization criteria, which measure the semantic information inherently hidden in ontology structures. Results have showed that the structural characteristics of ontology have great influence on ontology matching and will significantly improve the accuracy of matching.

References

[1] Hu, W., Qu, Y., Cheng, G.: Matching large ontologies: A divide-and-conquer approach. Data & Knowledge Engineering 67(1), 140–160 (2008)

[2] Ashburner, M., et al.: Gene ontology: tool for the unification of biology. the gene ontology consortium. Nature Genet. 25(1), 25–29 (2000)

[3] Newman, M.E.J.: The structure and function of complex networks. SIAM Review 45, 167–256 (2003)

[4] Peroni, S., Motta, E., d'Aquin, M.: Identifying Key Concepts in an Ontology, through the Integration of Cognitive Principles with Statistical and Topological Measures. In: Domingue, J., Anutariya, C. (eds.) ASWC 2008. LNCS, vol. 5367, pp. 242–256. Springer, Heidelberg (2008)

[5] Sussna, M.: Word sense disambiguation for free-text indexing using a massive semantic network. In: Proceedings of the Second International Conference on Information and Knowledge Management, pp. 67–74 (1993)

[6] Zaslavskiy, M., Bach, F., Vert, J.: Path following algorithm for the graph matching problem. CoRR (2008)

Route Troubleshooting Expert System Based on Integrated Reasoning

Hui-fen Dong[1], Chu Cui[1], and Guang Li[2]

[1] Civil Aviation University of China, Aeronautical Automation College, Tianjin, 300300
[2] Tianjin Maintenance Base of Air China, Tianjin, 300300
newdong@163.com, none_smxc@126.com, liguang@airchina.com

Abstract. The single reasoning mechanism is adopted in most of the expert system designed for fault diagnosis, the diagnosis results are single and insufficiency. In order to handle the problems which are more and more complex and improve the efficiency and accuracy, the integrated reasoning mechanism which is suitable for the route troubleshooting is designed in the paper. How to combine the route troubleshooting and the expert system by the design of integrated reasoner, the establishment of the integrated knowledge database, and the selection of the reasoning strategy and the realization of the soft ware are introduced in this paper. The route troubleshooting and expert system are combined well, the design and develop of the soft ware are completed, a diagnosis results with high accuracy, wide coverage and strong reliability can be obtained in this system, data can be conversed among three databases, then the diagnosis efficiency is improved.

Keywords: expert system, route troubleshooting, integrated reasoner, integrated knowledge database.

1 Introduction

Expert system is one of the artificial intelligences which developed fast and has been paid much more attention. The study in this field from abroad is earlier and the relatively mature software system is from Boeing and Airbus which is a collection of maintenance, management and prevention designed for theirs own aircraft. The software system is being employed in some domestic airlines. However, on one hand, the high cost prevent some small-scale Regional Aviation to adopt the software, on the other hand, with today' s increasing prosperity of large aircraft project, to develop a soft ware with proprietary intellectual property rights to break the European technical blockade in this field is extremely urgent. At present, domestic expert system concerning aircraft troubleshooting is mainly from the major Air University, for example Nanjing University of Aeronautics and Astronautics, Xi'an Polytechnic University and so on[1]. However, their expert system is based mainly on a single reasoning mechanism whose results are single and insufficiency.

Route maintenance includes preflight check, transit check and after flight check[2]. Three tasks should be done within each check: routine check, maintenance, and troubleshooting. However, the third task is the key and relatively complicated whose

Y. Wang and T. Li (Eds.): Knowledge Engineering and Management, AISC 123, pp. 289–296.
springerlink.com

efficiency will have a direct influence on the safety and punctuality of the flight. Based on the former theories, combined with the characteristics of route troubleshooting and with a collection of cases of air bleed system troubleshooting of B737-600 from a certain airline as well as some relative manuals, a route troubleshooting expert system is designed and developed in this paper, the system' s core is that integrated with three reasoning mechanism to enhance the efficiency and accuracy of aircraft maintenance according to origin of fault information.

2 The Configuration Design of the Integrated Reasoning Expert System

The structure of integrated expert system includes: integrated reasoner, integrated knowledge database, knowledge acquisitor, user interface, and explanation mechanism[3]. The diagram is as follows:

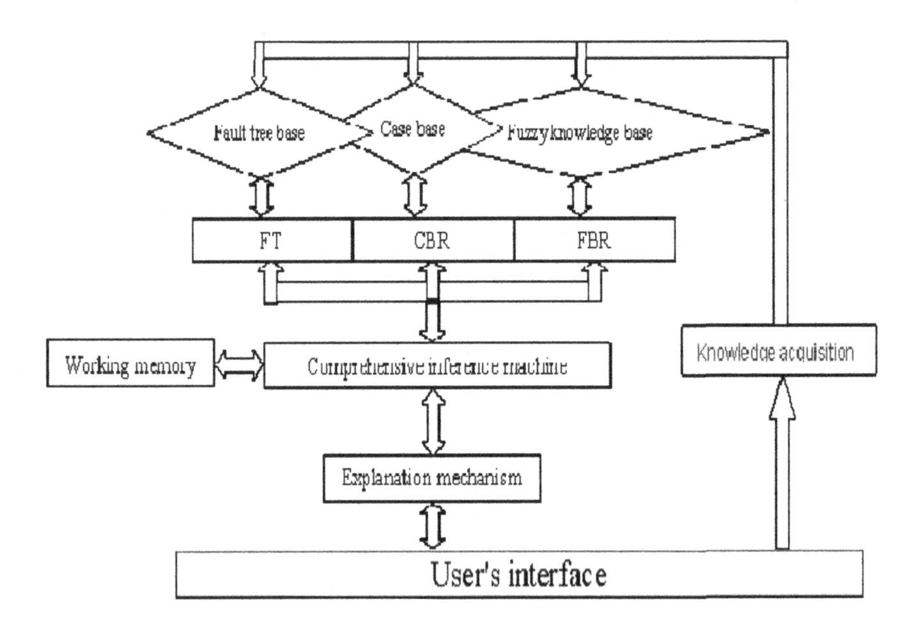

Fig. 1. The structure diagram of integrated reasoning expert system

3 The Establish of the Reasoning Mechanism

3.1 Case-Based Reasoning(CBR)

Case-based reasoning include case representation, case base index, case storage, case retrieval, case reuse, case changes and case studies, and several major aspects. Case representation, is to taken the characteristics and the relationship of the characteristics, and input the system[4]. Based on the features of the route troubleshooting, this article uses the frame to describe cases, the content of the case representation and a specific example are shown in the table 1.

Table 1. Frame description of a specific example

The property of fault description	Symptom or code of the fault	When the plane climb to 11539 meters, the right engine bleed trip off light on.
	Fault type	Parts / Accessories Function failure
	Fault time	Air cruise
The property of solution	Measure number	XXX
	Measure program	Check associate piping by AMM36, replace the 390F 、 450F sensors and the pre-cooler control value, test, and the right engine bleed air pressure normal.
	Replaceable parts	390F sensor , 450Fsensor, pre-cooler control value
The property of Auxiliary information	Aircraft type	B737-600
	ATA chapter	36
	Reference material	AMM36
	Remarks	390F sensor is a mistaken replaceable part.

Specific, this system will classify all of the cases by ATA chapter, and then classify the cases in each chapter according to the "fault description" given in the FIM for the second time. Case retrieval uses a combination of retrieval mode, first, it retrieves according to the character selected by the user, this retrieval is ripe SQL operation, easy, and can effectively narrow the search range[5]; then retrieve the cases which are meet the condition for the similar search, this search use the character matching technique, then we can get one or a few cases which are most relevant with the current issue.

The retrieved cases are ordered by the comprehensive calculated matching degree, the calculation of the matching degree consists of two parts: first, the senior maintenance staff will give a initial value for every character weight when the system initialization; second, the correlation calculated by using character matching technique according to the unique attributes input by user. The system set a successful rate of matching "a/b" for each case, where the "a" is a number of successful matches, "b" is a number of retrieved, the system will automatically make appropriate changes for the two numbers after each retrieval to reflect the occurrence probability of the case, and reflect it in the future reasoning results. This work will play a decisive role in the order of reasoning results to guide the crew to select cases. And the experts in related fields can enrich the case base according to their experience.

3.2 Fault Tree-Based Reasoning (FT)

Combining the features of route troubleshooting, all possible manual data in the process of route troubleshooting which includes AMM、FIM、TSM、WDM、MM etc are collected in this paper. Then a binary tree is extracted from those data and stored directly in the system in a way of standard ATA100. The diagram can provide the necessary circuit layout for certain node on the fault tree and connect the Fault Segregate Program to the basic event in which way the time for consulting manuals will be saved, then the efficiency will be improved. The following binary tree is based on the "High Duct Pressure" in Chapter 36.

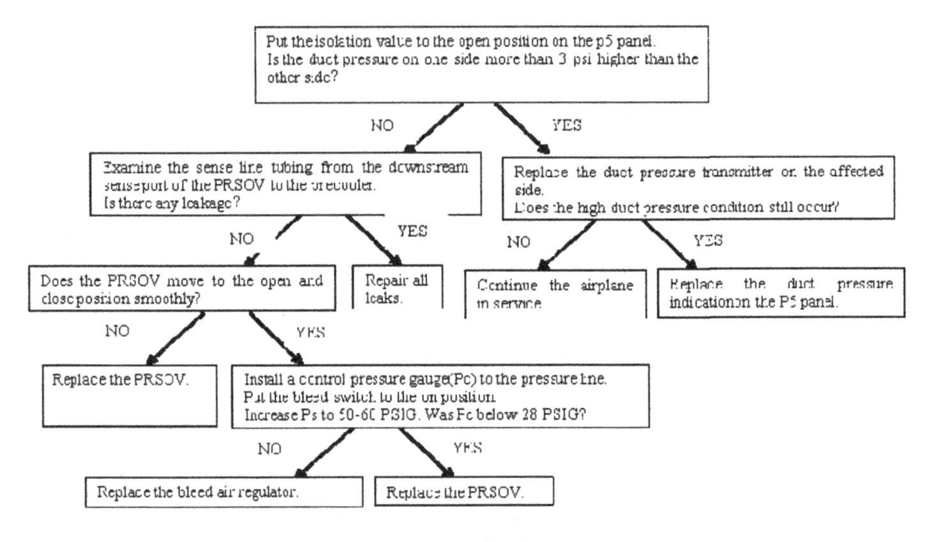

Fig. 2. Example for fault tree

3.3 Fuzzy-Based Reasoning, FBR

Fuzzy-Based Reasoning has to be employed whose basic ideology is to connect the fault and symptom by using fuzzy relation matrix R, then fault will be drawn from the fuzzy relation equation. The description is as follows:

Let's suppose X is input which represent fault reasons, and Y is output which represent fault symptom quantification assembles[6].

$$X = (x_1, x_2, \cdots, x_m) \tag{1}$$

$$Y = (y_1, y_2, \cdots, y_m) \tag{2}$$

X and Y both are fuzzy collections ($x_1, y_1 \in [0,1]$). The membership grade of fault symptom $r_{ij} \in [0,1]$ is called strength of ties between x_1 and y_1 . Thus, unit fault assemble K is drawn:

$$r_k = (r_{k1}, r_{k2}, \cdots, r_{kn}) \tag{3}$$

It is a fuzzy sub collection on the fault symptom collection R, in this way, a general fuzzy estimation matrix is extracted from M fault estimation collection[7].

$$R = \begin{bmatrix} r_{11} & r_{12} & \cdots & r_{1n} \\ r_{21} & r_{22} & \cdots & r_{2n} \\ \vdots & \vdots & & \vdots \\ r_{n1} & r_{n2} & \cdots & r_{nn} \end{bmatrix} \tag{4}$$

The odds of the diagnosis' result rests on the reliability of the matrix above, and the initial value of the matrix can be decided by the combination of experience and statistics from experts which can be perfected gradually through the systematic learning mechanism of the experts in the real-time application. Finally, the principle of maximum degree of membership can determine the source of fault.

4 To Integrate the Reasoning Mechanism

The reasoning strategy of the integrated reasoner is that in the very beginning of the diagnoses, the system will let the user to input the source of the fault information and a best reasoning mechanism will be selected by the system for the user. Generally, the case-based reasoning which is simple operation and effective will be chosen. However, if there is no matching case, a next operation should be operated according to the source and amount of the information. Fault Tree can be chosen where the source of the information is reliable and adequate, while Fuzzy-Based Reasoning is necessary if the information is fuzzy and inadequate. The system will support a multi-reasoning mechanism when the fault information is sufficient; the result will be compared voluntarily by the system to drawn out a comprehensive result in this situation. The process of reasoning is shown in the following diagram:

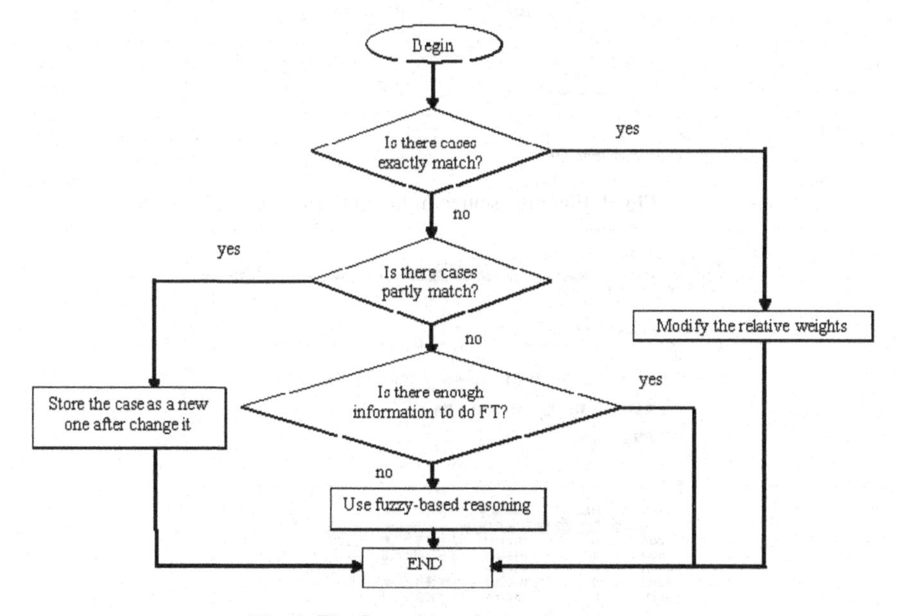

Fig. 3. The flow chart of reasoning strategy

5 To Realize the Route Troubleshooting Expert System

This expert system is composed by the following two parts:

The first part: SOL Server 2000 will be adopted to be the ground database server to establish case database, fault tree database and fussy regulation database.

The second part: Visual C will be adopted to develop a sub-system for the expert system which can function as a nozzle, an explanation mechanism and an inference engine for user-user or man -machine communication. The pasteurized interface provided by the sub-system can be employed by the user to diagnose the fault. After the reasoning has been finished, fault reason will be confirmed; solutions will be found in the database and provided by the system. Certain man-machine interfaces are as follows:

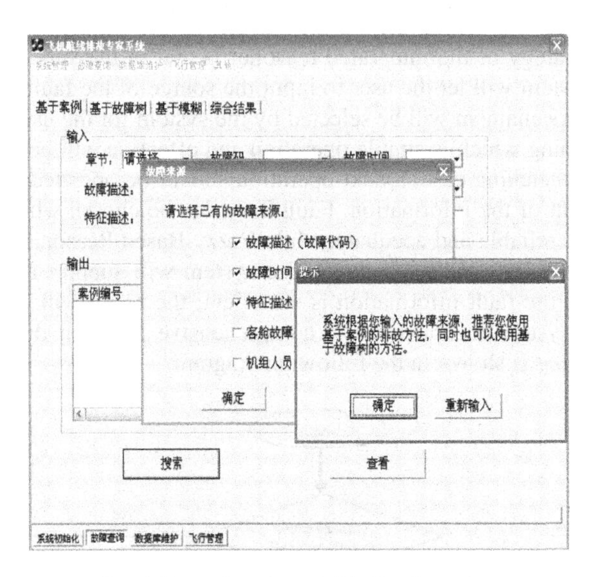

Fig. 4. Enter the source of fault information

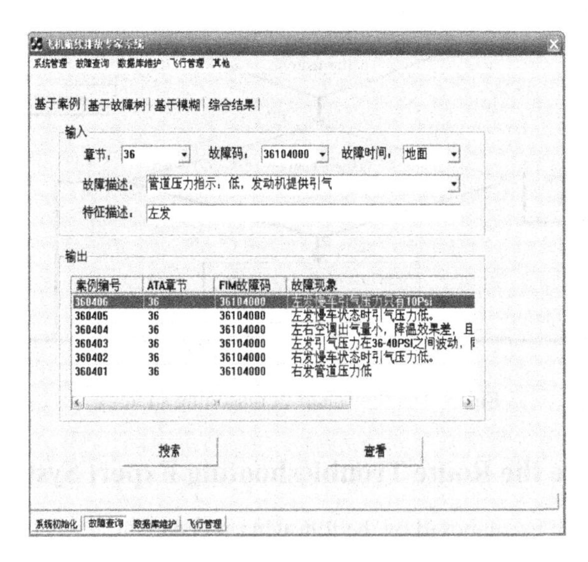

Fig. 5. Case-based troubleshooting

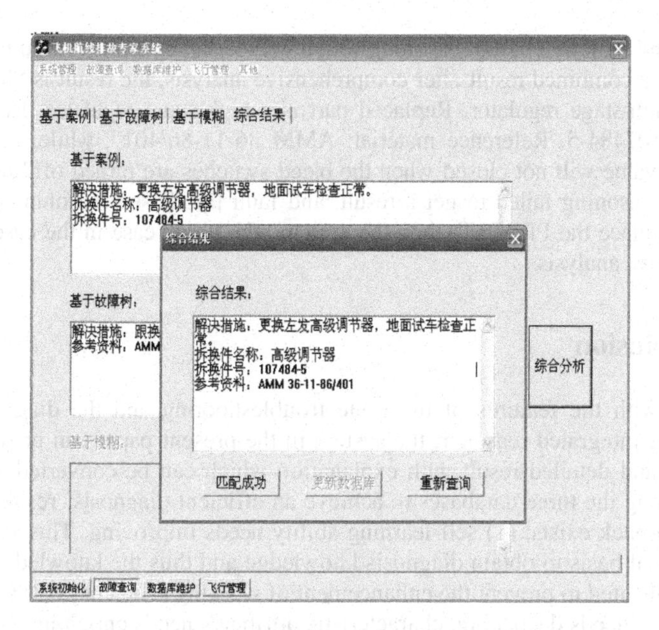

Fig. 6. Comprehensive analysis

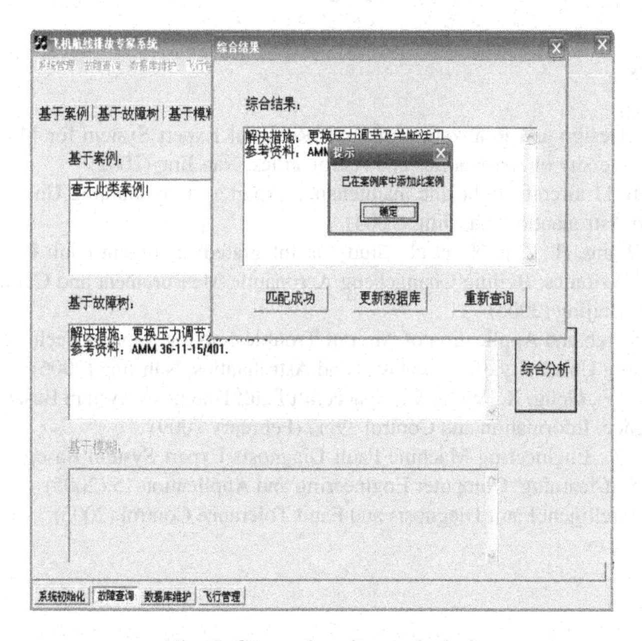

Fig. 7. Comprehensive analysis 2

Figure 4 is the interface which is used to input the fault information source. Figure 5 is case-based reasoning interface, it take the case that "low duct pressure, the engine was the bleed source" for example, and give the results. Figure 6 and figure 7 is comprehensive analysis interface, figure 6 take the case in figure 5 for example, there is case-based reasoning and fault tree based reasoning which get the same results in figure 6, then the system give a combined result after comprehensive analysis, the result is "Measure: replace the high-stage regulator. Replaced part is a high-stage regulator. Replaced part number is 107484-5. Reference material: AMM 36-11-86/401", while, figure 7 take "The bleed value will not closed when the bleed switches are turned off" for example, case-based reasoning failed to get a result, and fault tree based reasoning get a result which is "replace the PRSOV", then the system add a new case in the case base after comprehensive analysis.

6 Conclusion

Combined with the features of the route troubleshooting and the diagnosis' result made by the integrated reasoner, the system in the present paper can provide a comprehensive and detailed result with explanation which can be converted and supplemented among the three databases to achieve an efficient diagnosis' result. However, certain drawback exists: (1) self-learning ability needs improving. This system relies on the manual basis to obtain diagnosis knowledge and thus the knowledge maintaining is complicated to prevent the enhancement of working efficiency. (2) Case reasoning technique needs deepening, characteristic attributes needs enriching and the rationality of weight distributes needs verifying.

References

1. Zhang, X.: Design and Realization of Fault Removal Expert System for Modern Airliner. Nanjing University of Aeronautics and Astronautics, Nan Jing (2002)
2. Qiu, Y.: MD11 aircraft flight line maintenance expert system. Nanjing University of Aeronautics and Astronautics, Nan Jing (2004)
3. Yang, Z., Wang, H., Zhu, Y., et al.: Study on Integrated Intelligent Fault Diagnosis Expert System for Avionics. Beijing Changcheng Aeronautic Measurement and Control Technology Co, Ltd, Beijing (2006)
4. Li, Y.: Research and Application of Aircraft Troubleshooting Support Technology Based on CBR. Nanjing University of Aeronautics and Astronautics, Nan Jing (2006)
5. Xie, L., Liu, F., Gong, X., Wang, Z.: Spacecraft Fault Diagnosis System Based on the Hybrid Intelligence. Information and Control 39(1) (February 2009)
6. Li, X., Hu, Z.: Engineering Machine Fault Diagnosis Expert System Based on Fuzzy Inference and Self-learning. Computer Engineering and Application 15 (2005)
7. Wang, Z.: Intelligent Fault Diagnosis and Fault Tolerance Control (2005)

Design and Implementation of Ontology-Based Search Algorithm for Internal Website

Wenyuan Tao, Haitao Zhang, Wenhuan Lu, and Shengbei Wang

School of Computer Software
Tianjin University
Weijin Road, Nankai District, Tianjin, China
taowenyuan@tju.edu.cn, 353815060@qq.com, wenhuan@tju.edu.cn, 466209046@qq.com

Abstract. Searching records in internal website is an important way for people to obtain information, however, accurate and satisfying results cannot be acquired easily with conventional methods. In order to improve the effectiveness of search algorithm, in this paper we adopt ontology technology to the searching process and build up an ontology model for searching. Finally, some corresponding tests were carried out in order to valid the performance of the algorithm.

Keywords: Ontology model, Searching algorithm, Results classification.

1 Introduction

The term "ontology" comes from philosophy, which was used to study the inherent characteristics of things in philosophical problems. With the in-depth study of "ontology", "ontology" has been introduced to computer in recent years, such as artificial intelligence, computer languages, information retrieval and so on.

With the rapid development of information technology and the popularity of internet, the function of searching for records in internal website has been used more and more common. This paper existed mainly to design a searching algorithm that is more efficient and accurate. As to achieve this aim, we establish a model of record's elements based on "ontology" in which the relationships between elements are built up. The establishment of ontology model provides many conveniences during the process of searching by excluding some unnecessary searching paths, especially when the elements of the ontology model increase gradually, the classified approximation glossary can be easily obtained under the limitation of relationship between elements. According to "ontology", we design a search algorithm without using a professional glossary as a basis for dividing words.

2 Establishment of Ontology Model

2.1 Introductions of the Relationship in Ontology Model

In some commonly used algorithm, the searching of records is always limited to match the elements of record to obtain a weight of similarity between the search

Y. Wang and T. Li (Eds.): Knowledge Engineering and Management, AISC 123, pp. 297–302.
springerlink.com

keywords and the content of the record. However, since the relationship between elements is not considered, this algorithm usually results in a great waste of time and an incomplete result inevitably. In order to improve the searching efficiency and accuracy, we can establish a model of the elements in records according to "ontology" and define the relationship between elements. In this model, we can express any element with a two-tuples, which we can see in formula (1), where *R* is the relationship between the element and other elements and *B* is related information of the element:

Here we can see the first part of two-tuples *R*, in order to get the shortest search path and reflect the relationship between element and element, we divided the relationship *R* into two types: parent-child relationship and parallel relationship, in formula (2) *f* presents the parent-child relationship and *h* presents the parallel relationship. Then is *B*, from formula (3) , we know *B* is constituted by four items, named as *n*, *t*, *f*, *c*, where n is the number of the element; t represents the element data; *f* indicates the appearance frequency of the element; *c* shows that the record which the element belongs to.

$$\text{Element} <R, B> . \quad (1)$$

$$R=\{ f, h\} . \quad (2)$$

$$B=\{n, t, f, c\} . \quad (3)$$

2.2 The Structure of Ontology Model

In the relation *R*, if two elements have parent-child relationship, the child-node will inherit the basic properties from the parent-node, apart from this, the child-node also has its own properties, that is to say, the properties of the child-node are much clearer and refined than its parent-node, so the two nodes exist in different layer of the model. Different from this, in parallel relationship, some properties of the two elements are same, but the rest of the properties are different. These parallel-nodes exist in the same layer. All the relationship we talked above can be illustrated in Fig. 1:

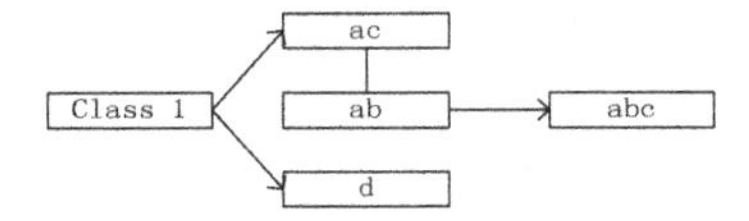

Fig. 1. This shows a figure of the searching process based on the ontology model

Once the relationship between elements has been established, we can use the relationship to increase the speed of search of the desired record. If the searching content at the same level, we can access to the parallel node through the parallel relationship, and then make a further search without matching with the other child-nodes. If the content we search is included in a major class or it is sub-problem of it, we can find the record by parent-child relationship directly, without matching with other extraneous elements. This approach can not only reduce the marching time but also improve the searching efficiency and accuracy significantly. The ontology model we constructed is shown Fig. 2:

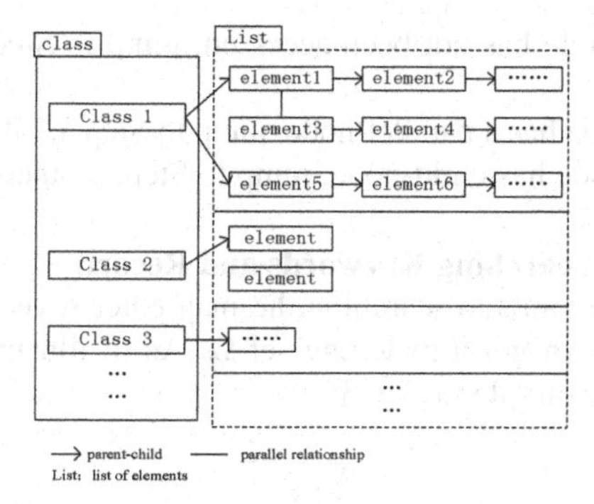

Fig. 2. This shows a figure of ontology model

3 Design and Implementation of Search Algorithm

According to the theoretic we obtained in previous section, the classified approximation glossary can be easily obtained under the limitations of relationship between elements when the elements in the ontology model increases and the two kind relationships: parallel relationship and parent-child relationship existed in the model, we can design our search algorithm and make the search process conveniently.

3.1 Principle of the Searching Algorithm

Matching Rules

In a process of search, we should consider the relationship between the search keywords and the records as well as the relationship between record and record influenced by search keywords. The relationship between the searching keywords and one record can be viewed as a partial matching, if we simply use this relationship to represent the searching results, we may get a disorganized result from the overall perspective. In order to avoid the appearance of this case, the relationship between record and record influenced by search keywords should be incorporated to rectify the existed result. Thus, the search result will be more reasonable and meet the user's search requests much better by providing them with an ordered result as well as a simple classification of it.

Once the user has chosen a search category and entered one keyword, the keyword will start its searching process and match layer by layer from the first child-node of the category, the matching rules are as follows:

Step 1. If the length of matching is zero, jump to the next child-node of the category;

Step 2. If the length of matching is bigger than zero, save the length and jump to Step 6; otherwise jump to Step 7;

Step 3.Starting from the current node and move to its child-node to make a matching ,then turn to Step 2;

Step 4. Starting from the current node and return to its parent-node, then jump to Step 6;

Step 5. Starting from the current node and move to its parallel-node to make a matching, then turn to Step 2;

Step 6. If the current node has not been accessed, jump to Step 8, otherwise jump to Step 7;
Step 7. If the current node has a parallel-node, jump to Step 5, otherwise jump to Step 4;
Step 8. If the current node has child-node, jump to Step 3, otherwise jump to Step 7;

Relationship between Searching Keywords and Record

Usually, when the user wants to searching, he may enter several searching keywords, for each keyword, we can get a matching set Ω. As to illustrate the search process, consider the following example in Fig. 3:

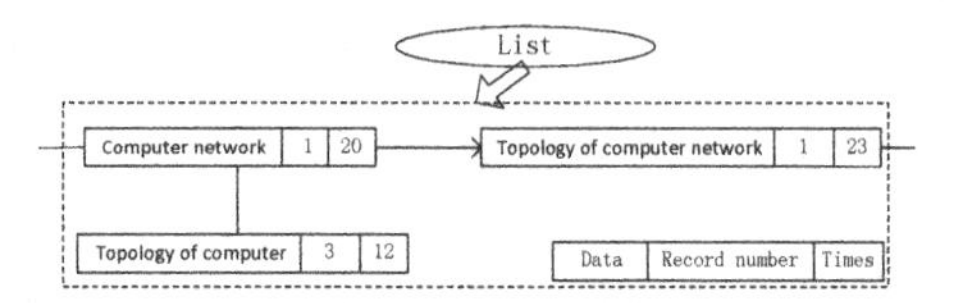

Fig. 3. There is a set of elements which is made up of three elements, where "data" is the element's data, "record number" stands for which record the element belongs to, "times" represents appearance number of the element in the record

If the user entered these two search keywords: "Computer" and "Topology", according to the matching rules introduced before, we will get the following matching results: the matching set of "Computer" belongs to record 1 is $\{1/2, 1/4, \cdots\}$, and the matching set of "Topology" belongs to record 1 is $\{0, \cdots\}$.

From the matching sets that belong to the same record, we should select the set that match the record with the biggest degree, then we can get a matching vector λ, the item number of the vector is equal to the number of search keywords.

In this step, we will get the matching probability of the search keyword with all the elements in the record by calculating total probability, which can be combined to the other matching vector V. Now, we can use the similarity value W of the two vectors to represent the relationship between search keywords and record.

Relationship between Record and Record Influenced by the Searching Keywords

If we take all the elements in the same category as a whole, we can get the appearance probability P of each element which could reflect the weight of the element in the whole category. However, for each record, we can use λ to get the maximum matching length of the search keyword and its corresponding the record's elements, thus, we can get an expectation θ of the entered searching keywords with each record by means of expectation formula using each element's overall appearance probability P and the maximum matching length x, θ is the average correlation degree of the searching keywords from the global perspective.

If several records have been searched, and the expectations of them have not much difference from each other, these records should be regarded as same category, otherwise, they should be divided in different category. θ is the value of expectation that reflects the relationship between record and record influenced by the searching keywords as we said before.

In the actual searching process, in order to expand the distance between different categories, we can use the following formula (4) to make an adjustment. So the records we obtained will have an obvious classification as expected.In the last, we can use formula (5) to get the final matching weight of record.

$$\theta = e^{\theta} . \tag{4}$$

$$\Theta = \theta \cdot W . \tag{5}$$

3.2 Realization of Search Algorithm

In this section, the realization procedures of our proposed algorithm will be description rigorously.

Calculate the matching probability g_k of the number K search keyword with all elements of record by means of total probability formula (6):

$$g_k = \sum_{i=1}^{n} \Omega_i \eta_i \tag{6}$$

Here are the means of the symbols: n is the total number of element in one record, Ω_i stands for the value of the number i item in Ω and η_i represents the appearance probability of element i in one record. Since the search keywords may be more than one, we can get the matching vector V.

Then, the similarity value W, the expectation θ and final matching weight Θ can be followed to get by the formula (7), formula (8) and formula (5):

$$W = \frac{\langle V, \lambda \rangle}{|V||\lambda|} \tag{7}$$

$$\theta = \sum_{i=1}^{n} P_i \cdot x_i \Rightarrow \theta = e^{\theta} \tag{8}$$

4 Algorithm Testing and Discussion

In order to verify the efficient and accuracy of our algorithm, we take a small test on the following example in Table 1, there are six articles, the words' total number of each articles shown in the second column, then the third column is keywords and keyword's appearance frequency of the article.

When we enter "apple" and "computer" as search keywords and apply the proposed algorithm to search relative articles, the matching weight $\Theta$of six articles are illustrated in the fourth column as 1.01719, 0.116278, 0.582098, 1.02299, 0.25867, 0. It's easy to find that: the number 1, number 3 and number 4 articles' weight $\Theta$are biggest than others, we know these three articles are more relevant to the search keywords, this is because the article 1 maybe mainly about the computer, the article 4 is about apple as a kind of fruit and the article 3 is about "Apple Computer".

Table 1. Test data

Number j	Total number	Keyword and frequency		Θ_j
1	1321	CPU design	4	1.01719
		China heart CPU	5	
		Computer	3	
2	3122	Algorithm design	10	0.116278
		Computer algorithm structure	8	
		Algorithm optimization	12	
3	3811	Apple CPU	20	0.582098
		Apple motherboard	7	
		Apple computer	20	
4	7467	Apple	4	1.02299
		Apple nutrition	10	
		Eating apple	8	
5	8003	C++	20	0.25867
		Development language	9	
		Computer language	6	
6	4001	Artistic content	10	0
		Body art	12	
		Performance art	8	

5 Conclusions

In this paper, we introduce "ontology" to the searching algorithm for internal website based on the ontology model, which is validated by a number of records, and the result reflects the relationship between the search keywords and the records as well as the relationship between records and records influenced by search keywords. Because the model of the algorithm is very simple and the process of searching does not need a professional glossary as a basis for dividing words therefore the algorithm can be used in software development for saving costs.

References

1. Berners Lee, T., Hendler, J., Lassila, O.: The Semantic Web. Scientific American 284, 34–43 (2001)
2. Lowe, D.G.: Distinctive image features from scale-invariant keypoints. International Journal of Computer Vision 60, 91–110 (2004)
3. Guo, X., Liu, W., Qian, M., Zhang, Z.: Information retrieval model based on ontologies. Journal of Yunnan University 25, 324–327 (2003)
4. Uschold, M., Gruninger, M.: Ontologies principles methods and application. Knowledge Engineering Review 11, 93–155 (1998)
5. Break, D.: Lost in cyberspace. New Scientist 154, 12–13 (1997)
6. Wu, F., Zhuang, Y.: Cross Media Analysis and Retrieval on the Web: Theory and Algorithm. Journal of Computer Aided Design & Computer Graphics 22, 1–9 (2010)

Knowledge-Based Model Representation and Optimization of Micro Device

Zheng Liu[1] and Hua Chen[2]

[1] School of Mechatronic Engineering, Xi'an Technological University, Xi'an China
[2] Xi'an Technological University, Xi'an China
zheng.liumail@gmail.com,
chenhua126@163.com

Abstract. Feature modeling technology makes micro device design in a more intuitive and efficient way. To support the feature-based information organization and rule-based features mapping procedure, a design framework and key enabling techniques are presented. Firstly, on the basis of hierarchical feature structure, the architecture for the model construction and optimization is constructed. Secondly, in accordance with the ontological representation, the knowledge of different features is hierarchically organized. Then, the feature mapping relationship is built with the rule-based reasoning process. By ontology-based features representing framework and reasoning procedure, the constraint features and associate rules are connected with other features. Finally, the design model and manufacturing model are optimized simultaneously.

Keywords: Micro device, Knowledge representation, Feature modeling, Feature reasoning.

1 Introduction

The traditional mask-beginning design flow is unintuitive and inefficient to micro device designers. To improve efficiency, the top-down design methodology proceeds from system to device level [1]. Feature modeling is one of the key technologies, which support the more reasonable "function-to-shape-to-mask" design flow [2]. This approach relieves the designers from the fussy considering of fabricating issues at early design stage. However, concerning the more and more complex structure of the micro devices, the requirement of a reasonable and extensible knowledge organizing form comes forth. So problem-solving of feature system construction and rule-based mapping technology are the key factors to accomplish the methodology.

Ontology was used to organize knowledge in design environments for many aspects, such as feature representation [3,4], multidisciplinary design optimization processes [5], and CAD modeling [6]. Besides the above Ontology applications, the "forward" and "inverse" design flow problems were also studied as key issues [7]. In order to utilize the advantages in the traditional CAD system, feature recognition method was suggested [8]. In addition, an approach accomplished the mask creation by investigating the vertical topology of the model [9].

Y. Wang and T. Li (Eds.): Knowledge Engineering and Management, AISC 123, pp. 303–310.
springerlink.com

To sum up, efforts have been made to get the manufacturing information from the original design model. However, the original geometric model has hardly good manufacturability. Knowledge-based reasoning technology provides a bridge to connect the design model and the manufacturing model. Because of the different natures between the feature information and the knowledge representation, how to build architecture to interface the feature modeling and the knowledge-based reasoning process is the key problem.

This paper is intended to address the key issues for optimizing the manufacturing model with knowledge-based reasoning method. These issues include the knowledge representation, the rule-base reasoning technology for feature mapping and model optimization in cooperative way.

2 The Knowledge-Based Designing Framework

The three level modeling framework is constructed as shown in Fig. 1. The process features and constraint features are essential for the model optimizing process. The system level modeling focuses on the function and behavior, while the process level modeling works at manufacturability. With features mapping procedure, these different features are connected to present a reasonable design flow.

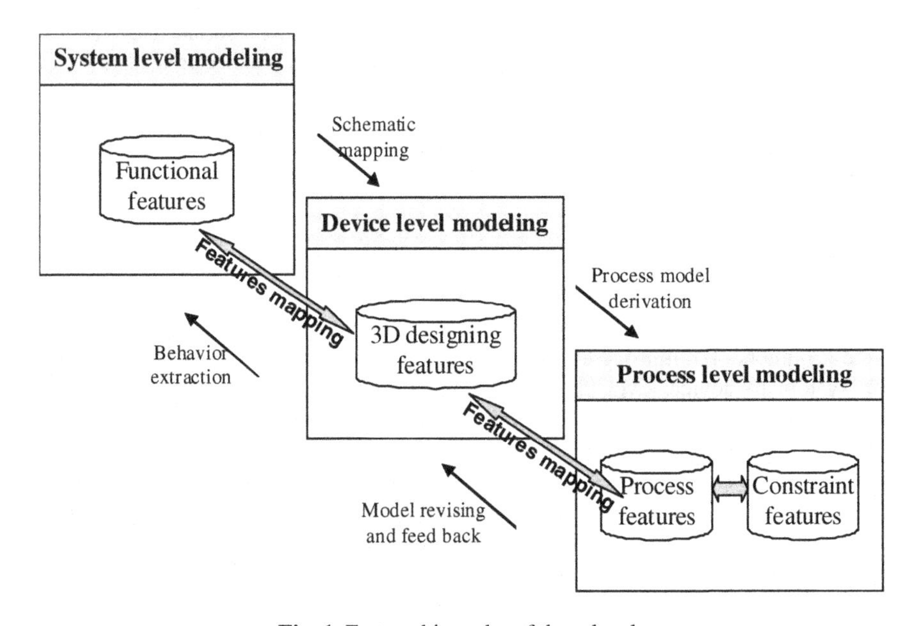

Fig. 1. Feature hierarchy of three levels

The knowledge-based designing framework is shown in Fig. 2. There are three ways to build the model. Functional features mapping is the normal way to construct model, which begins with the simulation components in functional features library. For the similar devices, the template library is presented to support template-based design procedure. For those anomalous parts, direct geometric element constructing method is recommended.

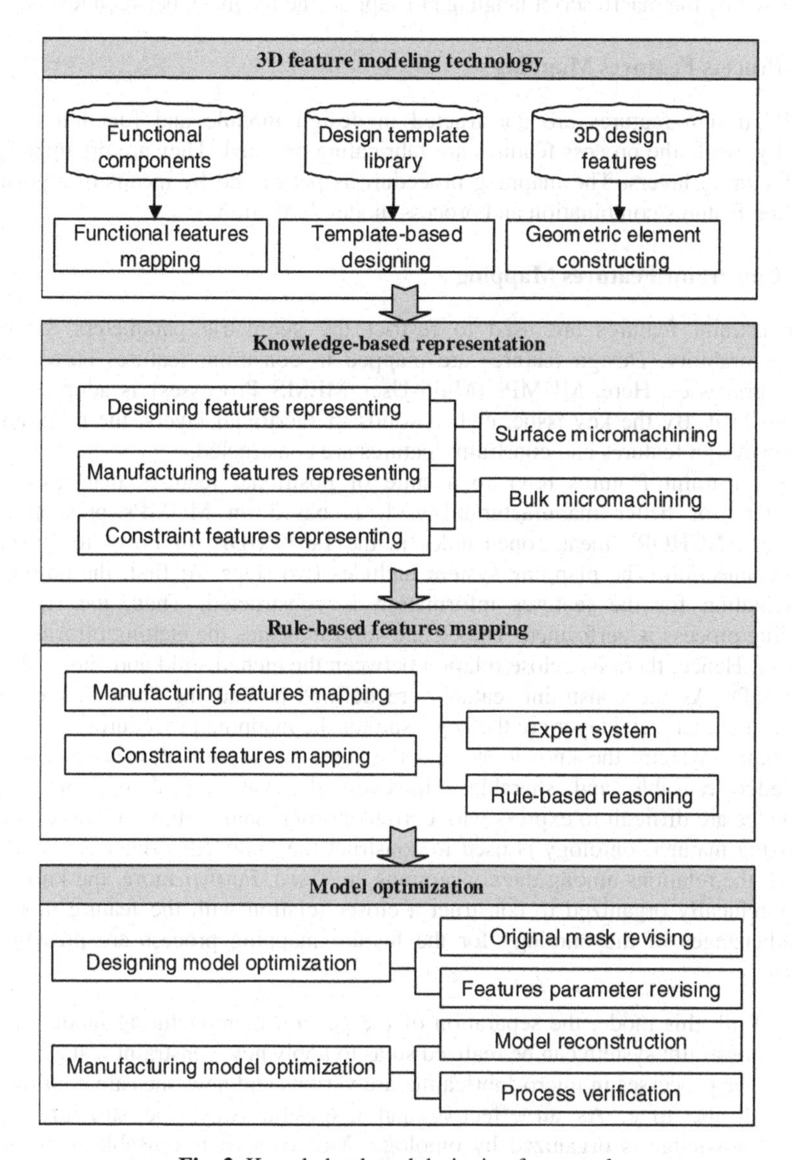

Fig. 2. Knowledge-based designing framework

2.1 Functional Features Mapping

In the system level, lumped bond graph is used to construct system dynamical simulation models to represent the functional requirements. The functional features library includes many physical simulation components. By the conversion from the predefined physical parameters to the geometric and material parameters, the functional features are mapping to the 3D design features. The mapping process is formalized by the macro script language to support the feedback between levels.

2.2 Process Features Mapping

The 3D design features are constructed in design module and function oriented normally, while the process features are fabricating oriented. They are organized with manufacturing layers. The mapping procedure is performed by means of algorithms including features combination and process model derivation.

2.3 Constraint Features Mapping

The constraint features are used to restrict the geometric parameters for better manufacturability. Design features are mapped to constraint features based on the mature processes. Here, MUMPs (Multi-User MEMS Processes) is adopted as the standard [10]. By the key issue, etched solids in sacrificial layers, the relationships between design features and constraint features are constructed.

The constraint features refer to a suite of positional features that restrict the parameters for better manufacturability. It is based on MUMPs presently. For example, ANCHOR1 means open holes in the 1st OXIDE for Poly1 to Nitride or Poly0 connection. The mapping system includes two steps. At first, the ontological representation for the features information is constructed. Then, the rule-based reasoning process is performed. The etched solid indicates the etching information of the layer. Hence, there is a close relation between the etched solid and "hole" defined in MUMPs. As the constraint features are defined around the "holes", the etched solids of the sacrificial layer are the key issue for the mapping procedure.

In expert system, the knowledge and the rules are separated so as to make the knowledge reusable and sharable. However, the hierarchical relations among knowledge are difficult to express with current storing manner. For a more reasonable organizing manner, ontology is used to construct the knowledge database. With this method, the relations among the conceptions are fixed. Further more, the knowledge is hierarchically organized to construct a closer relation with the feature modeling. The advantages of the ontology for the feature mapping process are presented as follows:

- With this mode, the separation of the general manufacturing model and the constraint system can be realized so as to apply new constraint features.
- The processes in micro-fabricating are various and new ones are coming forth all the time. As an effective and extensible way, the structure of the knowledge is organized by ontology. Moreover, it is reusable to avoid the iterative analysis of domain knowledge.
- The ontology provides a hierarchical form to describe the domain knowledge. It is more comfortable to construct the feature modelling information for the rule-based reasoning process.

To construct an ontology-based knowledge base, all the features information is arranged with ontological habit. The main factors of ontology are class, slot, instance and axiom. Protégé is adopted as the ontology tool, which is developed by Stanford Medical Informatics. The expert system utilizes Jess (Java Expert System Shell) as the rule engine. JessTab (a plugin used in Protégé) is adopted as the interface between them to implement the transmission of knowledge. The organization of features is illustrated in Fig. 3.

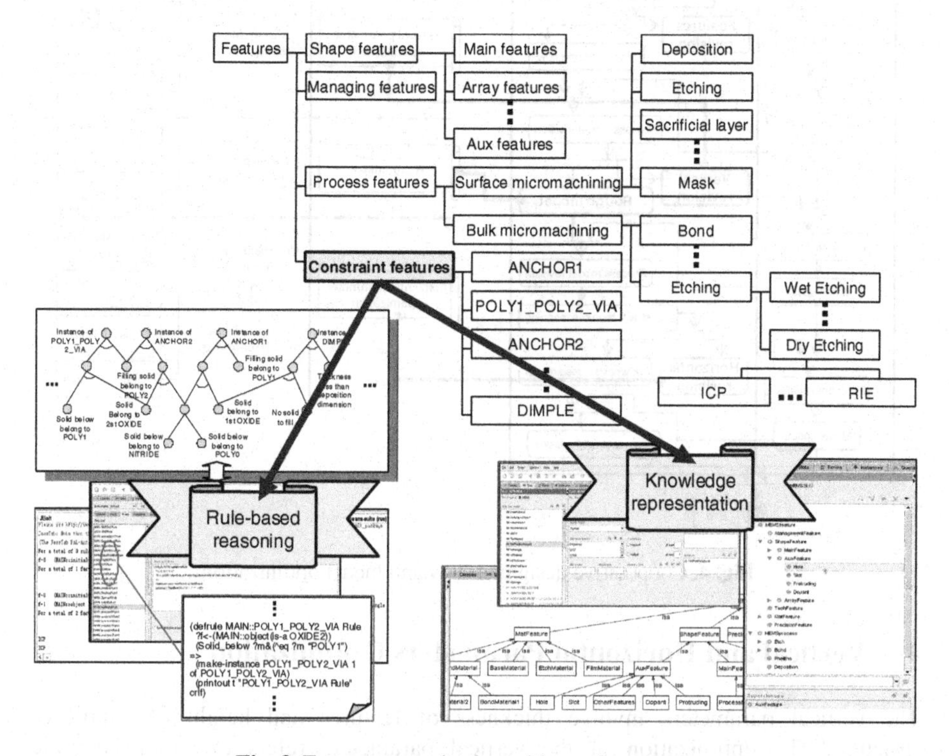

Fig. 3. Features organization with ontology manner

3 Cooperative Design Flow and Parametric Optimization

From the perspective of activities, Fig. 4 illustrates the cooperative design flow. In the activity diagram, the device designer and process designer are responsible for the activities accordingly in swim-lanes, which carry out simultaneously or in turn. The key optimal activities are marked. With these steps, both vertical and horizontal parameters are optimized accordingly. Fig. 4 also illustrates the category of optimization in different stages. Firstly, to complete the vertical parameter optimization, the feature thickness and height of cantilever structure is adjusted by means of features combination method. Then, the generation of rough process model provides the rough mask that is the original reference for the subsequent model revision. Finally, with the constraint features, the horizontal parameters are optimized to perform the model revision.

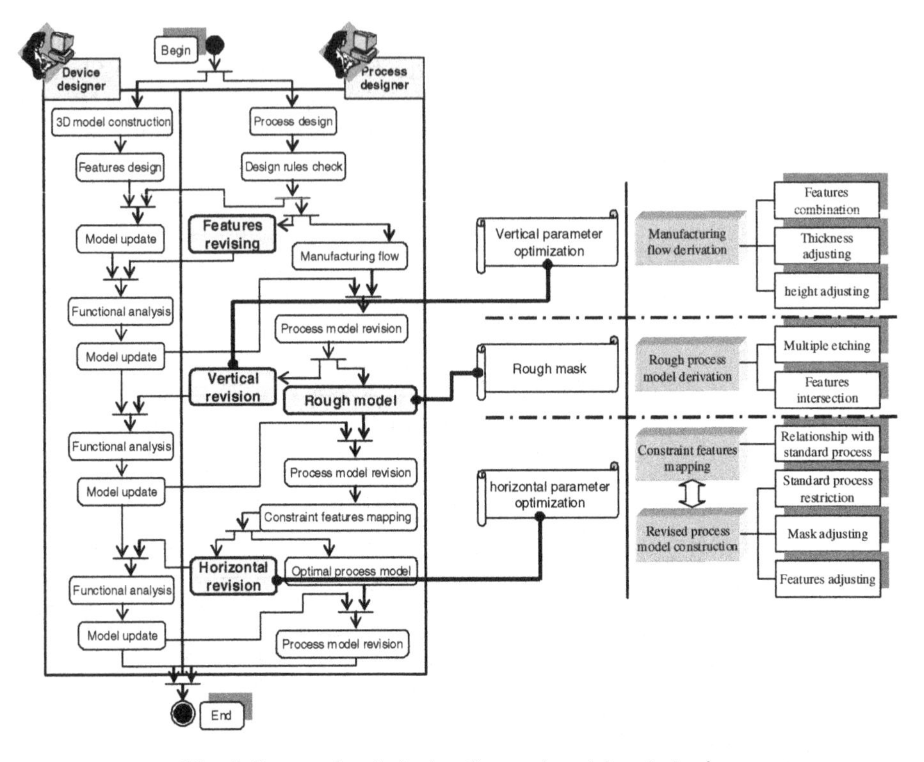

Fig. 4. Cooperative designing flow and model optimization

4 Vertical and Horizontal Parameters Optimization

The vertical parameters involve thickness of features and height of cantilever structure. The optimization of the vertical parameters refers to simplifying the fabricating process by adjusting the parameters. The horizontal parameters mainly refer to the mask and relevant dimensions. With the constraint features, the rough mask is revised to improve the manufacturability.

4.1 Features Combination

The features combination algorithm deals with organizing the design features with layer architecture to generate fabricating flow, which is the basis of optimization. The directed graph is adopted to perform the combination in accordance with principles. With features combination, the "loops" in the graph are detected to correct the design defect, which imply fabricating confliction. The minimal number of layers is calculated, which is restricted in standard processes commonly. Firstly, the thickness of features is optimized. The features combined into one layer are fabricated with one deposition process. To avoid the superfluous etching step, the thicknesses of these features are expected to be uniform. Secondly, the height of cantilever structure is optimized. The cantilever structure indicates that the material under it is removed. In fact, it is the sacrificial layer. The height of cantilever structure determines the deposition of sacrificial layer. Therefore, for the features in same group, these parameters are expected to be uniform to simplify the sacrificial layer fabrication.

4.2 Connecting with Constraint Features

The rough process model gives us an approximate fabricating solution for the original design model, which is foundational for optimization. The constraint features mapping is a rule-base reasoning procedure on the basis of feature definition and location analysis of solids. Without layer information, it is difficult to directly map the designing features to the constraint features. The rough model provides a middle link. It constructs the relationship between the designing features and fabricating layers. By analyzing the positional relations among solids, the constraint features are fixed. For example, if three conditions are met, an etched solid is an instance of constraint feature, ANCHOR1.

- Condition 1, the location of the etched solid is on the 1st sacrificial layer.
- Condition 2, the solid close under it belongs to NITRIDE or POLY0.
- Condition 3, the solid that takes the place of it, when etched, belongs to POLY1.

Firstly, the standardized processes database is constructed, in which some representative processes are classified. It provides designers a reference to revise the planning result. Then, based on the knowledge of ontological presentation and the experience in micro-manufacturing, rules are defined with the CLIPS language form. They are imported into the Jess system. These constraint features and associated rules provide the evaluation criteria of manufacturability.

4.3 Process Model Revising

Firstly, the etched solids of sacrificial layer are picked up and connected with constraint features. The projection pattern of the etched solid is compared with the rough mask. The rough mask is revised based on the constraint rules. Secondly, the models of structural layer are reconstructed based on the revised mask. The optimized structural layer model is obtained by the Boolean operation of subtraction between the deposition model and the new etched solids. Then, feedback is provided to transmit the revised model to the design part to confirm whether these changes are allowed. The influence mainly involve function (e.g., force, motion, etc) and performance (e.g., rigidity, fatigue life, etc). Finally, the sacrificial layer is reconstructed based on the optimized structural layers close above and below it.

5 Conclusion

In this paper, method for incorporating the knowledge-based representation and reasoning process into the feature modeling of micro device is proposed. With this method, the manufacturing model is derived from the 3D design model together with the optimizing process to improve manufacturability. The knowledge-based reasoning capabilities can enhance the knowledge organization and features mapping process of micro device. In this way, by updating the Ontology-based knowledge database, the reasoning system can be expanded and become more powerful to provide better manufacturability for feature modeling method.

Acknowledgments. This work was financially supported by the Science and Technology Development Plan Foundation of Shaanxi Province (No. 2011K07-11), Scientific Research Program Funded by Shaanxi Provincial Education Department (Program No. 11JK0864), President Fund of Xi'an Technological University (No. XAGDXJJ1007) and Shaanxi Major Subject Construction Project.

References

1. McCorquodale, M.S., Gebara, F.H., Kraver, K.L., Marsman, E.D., Senger, R.M., Brown, R.B.: A Top-Down Microsystems Design Methodology and Associated Challenges. In: Proceedings of the Design, Automation and Test in Europe Conference and Exhibition, vol. 1, pp. 292–296. IEEE Computer Society, Los Alamitos (2003)
2. Gao, F., Hong, Y.S., Sarma, R.: Feature Model For Surface Micro-machined MEMS. In: Proceedings of ASME Design Engineering Technical Conferences, Chicago, USA, vol. 1, pp. 149–158 (2003)
3. Dartigues, C., Ghodous, P., Gruninger, M., Pallez, D., Sriram, R.: CAD/CAPP Integration using Feature Ontology. Concurrent Engineering Research and Applications 15, 237–249 (2007)
4. Chen, G., Ma, Y.S., Thimm, G., Tang, S.H.: Knowledge-Based Reasoning in a Unified Feature Modeling Scheme. Computer-Aided Design & Applications 2, 173–182 (2005)
5. Maier, F., Mayer, W., Stumptner, M., Muehlenfeld, A.: Ontology-based Process Modelling for Design Optimisation Support. In: Proceedings of the 3rd International Conference on Design Computing and Cognition, pp. 513–532 (2008)
6. Andersen, O.A., Vasilakis, G.: Building an Ontology of CAD model information. Geometric Modelling, Numerical Simulation, and Optimization, 11–40 (2007)
7. Ananthakrishnan, V., Sarma, R., Ananthasuresh, G.K.: Systematic Mask Synthesis for Surface Micromachined Microelectromechanical Systems. Journal of Micromechanics and Microengineering 13, 927–941 (2003)
8. Li, J., Gao, S., Liu, Y.: Feature-based Process Layer Modeling for Surface Micro-machined MEMS. Journal of Micromechanics and Microengineering 15, 620–635 (2005)
9. Schiek, R., Schmidt, R.: Automated Surface Micro-machining Mask Creation from a 3D Model. Microsystem Technologies 12, 204–207 (2006)
10. Carter, J., et al.: MUMPs Design Handbook. Revision 11.0. MEMSCAP Inc., Durham (2005)

Modeling Supply Chain Network Design Problem with Joint Service Level Constraint

Guoqing Yang, Yankui Liu*, and Kai Yang

College of Mathematics & Computer Science, Hebei University
Baoding 071002, Hebei, China
ygqfq100@gmail.com, yliu@hbu.edu.cn, yangk09@sina.com

Abstract. This paper studies a supply chain network design problem with stochastic parameters. A Value-at-Risk (VaR) based stochastic supply chain network design (VaR-SSCND) problem is built, in which both the transportation costs and customer demand are assumed to be random variables. The objective of the problem is to minimize the allowable invested capital. For general discrete distributions, the proposed problem is equivalent to a deterministic mixed-integer programming problem. So, we can employ conventional optimization algorithms such as branch-and-bound method to solve the deterministic programming problem. Finally, one numerical example is presented to demonstrate the validity of the proposed model and the effectiveness of the solution method.

Keywords: Supply chain network design, Joint service level, Stochastic programming, Mixed-integer programming.

1 Introduction

A supply chain is a network of suppliers, manufacturing plants, warehouses, and distribution channels organized to acquire raw materials, convert these raw materials to finished products, and distribute these products to customers. Since the pioneering work of Geoffrion and Graves [1] on multi-commodity distribution network design, the supply chain problem was investigated by a number of researchers (see, Cohen and Lee [2], Aikens [3] and Alonso-Ayuso et al. [4]).

In supply chain network design systems, there are many forms of uncertainties which can affect the network configurations designing processes such as cost information, capacity data and customer demand. Considering these uncertainties will result in more realistic supply chain models. The significance of uncertainty has prompted a number of researchers to address stochastic parameters in supply chain problem. For example, Sabri and Beamon [5] incorporated production, delivery, and demand uncertainty in their model; Cheung and Powell [6] considered the distribution problem with uncertain demands as a two-stage stochastic program and incorporated customer demand as the parameter; Georgiadis et al. [7] considered a mathematical formulation for the problem of designing supply chain networks comprising multiproduct production facilities with shared

* Corresponding author.

Y. Wang and T. Li (Eds.): Knowledge Engineering and Management, AISC 123, pp. 311–318.
springerlink.com

production resources, warehouses, distribution centers and customer zones and operating under time varying demand uncertainty. The interested readers may refer to Owen and Daskin [8] for recent developments about the design of supply chain networks under uncertainty.

The purpose of this paper is to develop a new class supply chain network design problem with VaR objective, in which the transportation costs and customer demand are stochastic parameters with known joint distribution. In contrast to the expected value model [9], the VaR criteria is a powerful strategy for modeling stochastic phenomena in decision systems. The main idea underlying it is to optimize the critical value of the total cost objective with certain confidence level subject to some constraints. Two key difficulties in solving this VaR based stochastic supply chain network design problem are in evaluating the critical value in the objective and checking the joint service level constraints. Note that discrete distributions arise frequently in applications, which may also available through experience distribution or approximating continuous probability distribution. In the case when the stochastic parameters are general discrete distributions, we can turn the VaR objective and the service level constraints into their equivalent deterministic forms. Then the original uncertain supply chain network design problem is equivalent to a deterministic mixed-integer programming problem which can be solved by standard optimization solvers. At the end of this paper, we perform a number of numerical experiments to illustrate the new modeling ideas and the efficiency in the proposed method.

The rest of this paper is organized as follows. In Section 2, we develop a new modeling approach to supply chain network design problem and highlight its difficulties. In Section 3, we discuss the equivalent deterministic forms for the VaR objective and the joint service level constraints of the supply chain model. The solution method to the equivalent problem with the standard optimization solver LINGO and the numerical experiments are discussed in Section 4. Finally, Section 5 gives our conclusions.

2 Problem Formulation

In this section, we formulate a VaR-based VaR-SSCND problem with joint service level constraint. We introduce the notation for this VaR-SSCND model. Let I, J, L, K and M be the sets of suppliers, plants, warehouses, products and customers respectively; cm_j and cw_l denote the cost of building the plant j and the warehouse l; cq_i^I is the unit cost of raw material from supplier i and cq_{jk}^J is the unit production cost of product k in plant j; s_i denotes the capacity of raw material for supplier i; h_l is the storage capacity in warehouse l; a_j is the product capacity in plant j; r_{jk}^P is the processing requirement for per-unit product k in plant j; r_{lk}^L is the processing requirement for per-unit product k in warehouse l; r_k is the raw material required for per-unit product k; β is the confidence level of the Value-at-Risk; γ is the service level. For the transportation cost, cp_{ij} is the unit one from supplier i to plant j for raw material, cp'_{jlk} is the unit one between plant j and warehouse l for product k and cp''_{lmk} denotes the unit one from warehouse l to customer m for product k; d_{mk} is the demand of customer m for product k; x_{ij} is the amount of raw material transported from supplier i to plant j; y_{jlk} denotes the amount of product k transported from plant j to warehouse l; z_{lmk} is the amount of product k transported from warehouse l to customer m; $u_j = \begin{cases} 1, \text{ if plant } j \text{ is built;} \\ 0, \text{ otherwise.} \end{cases}$ $w_l = \begin{cases} 1, \text{ if warehouse } l \text{ is built;} \\ 0, \text{ otherwise.} \end{cases}$

Using the notation above, we present the following VaR-SSCND, model with joint service level constraints:

$$\min C$$

$$\text{s.t.} \Pr\{\sum_{j\in J} cm_j u_j + \sum_{l\in L} cw_l w_l + \sum_{i\in I}\sum_{j\in J} cp_{ij}x_{ij} + \sum_{k\in K}\sum_{l\in L}\sum_{m\in M} cp''_{lmk} z_{lmk} +$$
$$\sum_{k\in K}\sum_{j\in J}\sum_{l\in L} cp'_{jlk} y_{jlk} + \sum_{i\in I} cq^I_i \sum_{j\in J} x_{ij} + \sum_{k\in K}\sum_{j\in J} cq^J_{jk} \sum_{l\in L} y_{jlk} \leq C\} \geq \beta \quad (1)$$

$$\Pr\{\sum_{l\in L} z_{lmk} \geq d_{mk}, \forall m \in M, \forall k \in K\} \geq \gamma \quad (2)$$

$$\sum_{i\in I} x_{ij} = \sum_{k\in K} r_k \sum_{l\in L} y_{jlk}, \forall j \in J \quad (3)$$

$$\sum_{j\in J} y_{jlk} = \sum_{m\in M} z_{lmk}, \forall l \in L, \forall k \in K \quad (4)$$

$$\sum_{j\in J} x_{ij} \leq s_i, \forall i \in I \quad (5)$$

$$\sum_{k\in K} r^P_{jk} \sum_{l\in L} y_{jlk} \leq a_j u_j, \forall j \in J \quad (6)$$

$$\sum_{k\in K} r^L_{lk} \sum_{j\in J} y_{jlk} \leq h_l w_l, \forall l \in L \quad (7)$$

$$x_{ij} \geq 0, y_{jlk} \geq 0, z_{lmk} \geq 0, u_j, w_l \in \{0,1\}, \forall i,j,k,l,m. \quad (8)$$

The objective of the VaR-SSCND model is to minimize the sum of total cost for a given confidence level $\beta \in (0,1)$. The constraint (2) imposes the amount of product k from warehouses to customer m should meet the customer's demand with a given service level γ. The constraints (3) ensure that the raw material to the plant j should meet the need in plant, while the constraints (4) ensure the amount of product k from warehouse l to customers should be equal to the one from plants to the warehouse l. Constraints (5) require raw material r from supplier i should be less than the supply capacity. Constraints (6) and (7) enforce all product produced by plant j should be less than its production capacity and from plants to warehouse l should be less than the warehouse's capacity. Finally, constraints (8) ensure non-negativity of the flow variables corresponding to origins, destinations and commodities, also enforce the binary nature of the configuration decisions.

It is easy to see that the VaR-SSCND model is a type of mixed-integer stochastic programming. It clearly belongs to the class of probabilistic constraint programming problems. The traditional solution methods require conversion of probabilistic constraints to their respective deterministic equivalents. As we know, this conversion is usually hard to perform and only successfully for special case. We will discuss the equivalent formulation of VaR-SSCND model in the the case when random parameters are characterized by discrete distributions.

3 Deterministic Equivalent Programming Problems

3.1 Handling VaR Cost Function

We consider the case when the random transportation cost parameters are general discrete random variables. For the sake of simplicity of presentation, we denote $cp = (cp_{11}, \ldots, cp_{IJ}, cp'_{111}, \ldots, cp'_{JLK}, cp''_{111}, \ldots, cp''_{LMK})$, which is a discrete random vector with the following probability distribution $\begin{pmatrix} \hat{cp}^1 & \cdots & \hat{cp}^N \\ p_1^1 & \cdots & p_1^N \end{pmatrix}$, where $p_1^n > 0$, $n = 1, 2, \ldots, N$, and $\sum_{n=1}^N p_1^n = 1$, $\hat{cp}^n$=$(\hat{cp}^n_{11}, \ldots, \hat{cp}^n_{IJ}, \hat{cp'}^n_{111}, \ldots, \hat{cp'}^n_{JLK}, \hat{cp''}^n_{111}, \ldots, \hat{cp''}^n_{LMK})$ is the nth scenario.

In this case, we consider the VaR cost function (1). We introduce a "big enough" constant M and a vector B of binary variables whose components B_n, $n = 1, 2, \ldots, N$, the corresponding constraint has to be satisfied and 1 otherwise. Then, the objective can be turned into the following equivalent form:

$$\begin{cases} \min C \\ \text{s.t.} \sum_{j\in J} cm_j u_j + \sum_{l\in L} cw_l w_l + \sum_{k\in K}\sum_{j\in J}\sum_{l\in L} \hat{cp'}^n_{jlk} y_{jlk} + \sum_{k\in K}\sum_{l\in L}\sum_{m\in M} \hat{cp''}^n_{lmk} z_{lmk} \\ \quad \sum_{i\in I}\sum_{j\in J} \hat{cp}^n_{ij} x_{ij} + \sum_{i\in I} cq_i^I \sum_{j\in J} x_{ij} + \sum_{k\in K}\sum_{j\in J} cq^J_{jk} \sum_{l\in L} y_{jlk} - MB_n \le C, \\ \qquad\qquad n = 1, 2, \ldots, N \\ \quad \sum_{n=1}^N p_1^n B_n \le 1-\beta, B_n \in \{0,1\}, n = 1, 2, \ldots, N. \end{cases} \tag{9}$$

where $\sum_{n=1}^N p_1^n B_n \le 1-\beta$, $B_n \in \{0,1\}$, $n = 1, 2, \ldots, N$, define a binary knapsack constraint ensuring that violation of stochastic service level constraints is limited to $(1-\beta)$.

3.2 Service Level Constraint

Only the right-hand-side vector is stochastic in constraint (2). In this case the probability function has the form: $F(z)$=$\Pr(z \ge d)$. The constraint (2) is obviously equivalent to $F(z) \le \gamma$. Now, we assume that demand vector $d = (d_{11}, \cdots, d_{MK})$ has a finite discrete distribution and corresponding the probability distribution $\begin{pmatrix} \hat{d}^1 & \cdots & \hat{d}^{N'} \\ p_2^1 & \cdots & p_2^{N'} \end{pmatrix}$, where $\sum_{n=1}^N p_2^n = 1$, $\hat{d}^{n'}$=$(\hat{d}^{n'}_{11}, \ldots, \hat{d}^{n'}_{MK})$ is the n'th scenario.

In VaR-SSCND model, we may restrict ourselves to those function values $F(\hat{d}^{n'})$ for which we have $F(\hat{d}^{n'}) \ge \gamma$, provided that it is not difficult to select these values or at least a somewhat large category of the probability distribution function values which surely contains those which are greater or equal to γ.

In addition, we introduce a vector T of binary variables whose components $T_{n'}$, $n'=1,2,\cdots,N'$. Then, we can express the service level constraints (2) as

$$\begin{cases} \sum\limits_{l\in L} z_{lmk} \geq \sum\limits_{n'=1}^{N'} \hat{d}_{mk}^{n'} T_{n'} \quad \forall m \in M, \forall k \in K \\ \sum\limits_{n'=1}^{N'} F(\hat{d}^{n'}) T_{n'} \geq \gamma \\ \sum\limits_{n'=1}^{N'} T_{n'} = 1, \quad T_{n'} \in \{0,1\}, \quad n' = 1,2,\ldots,N'. \end{cases} \tag{10}$$

3.3 Equivalent Mixed-Integer Programming

Substituting (1) and (2) into (9) and (10), the VaR-SSCND model can be turned into the following equivalent mixed-integer programming model.

$$\begin{cases} \min C \\ \text{s.t.} \\ \quad \sum\limits_{j\in J} cm_j u_j + \sum\limits_{l\in L} cw_l w_l + \sum\limits_{i\in I}\sum\limits_{j\in J} \hat{cp}_{ij}^n x_{ij} + \sum\limits_{k\in K}\sum\limits_{l\in L}\sum\limits_{m\in M} \hat{cp''}_{lmk}^{n} z_{lmk} + \\ \quad \sum\limits_{k\in K}\sum\limits_{j\in J}\sum\limits_{l\in L} \hat{cp'}_{jlk}^{n} y_{jlk} + \sum\limits_{i\in I} cq_i^I \sum\limits_{j\in J} x_{ij} + \sum\limits_{k\in K}\sum\limits_{j\in J} cq_{jk}^J \sum\limits_{l\in L} y_{jlk} - MB_n \leq C, \\ \qquad\qquad n = 1,2,\ldots,N \\ \quad \sum\limits_{n=1}^{N} p_1^n B_n \leq 1-\beta \\ \quad \sum\limits_{l\in L} z_{lmk} \geq \sum\limits_{n'=1}^{N'} \hat{d}_{mk}^{n'} T_{n'}, \forall m \in M, \forall k \in K \\ \quad \sum\limits_{n'=1}^{N'} T_{n'} = 1 \\ \quad \sum\limits_{n'=1}^{N'} F(\hat{d}^{n'}) T_{n'} \geq \gamma \\ \quad (3),(4),(5),(6) \text{ and } (7) \\ \quad B_n \in \{0,1\}, T_{n'} \in \{0,1\}, n = 1,2,\ldots,N, n' = 1,2,\ldots,N' \\ \quad x_{ij} \geq 0, y_{jlk} \geq 0, z_{lmk} \geq 0, u_j, w_l \in \{0,1\}, \forall i,j,k,l,m. \end{cases} \tag{11}$$

The crisp equivalent model (11) is a mixed-integer programming problem with binary variables. It belongs to a class of NP-hard problems. Because of its many variables and constraints, equivalent problem (11) would be difficult to solve for large instances. As such, we focus our research on the solution approaches, which will be described in the next section.

4 Solution Methods and Numerical Experiment

4.1 Solution Methods

One possibility for solving the equivalent problem (11) is of course to use the branch and bound method. Branch and bound method is a basic workhorse technique for solving integer and discrete programming problems. It is based on the observation that the enumeration of integer solutions has a tree structure. The main idea of the branch and bound method is to avoid growing the whole tree as much as possible. And pruning is very important for the method since it is precisely what prevents the search tree form growing too much [10].

An efficient approach is the preprocessing of the scenario set. It is worthwhile noting that on the basis of the comparison of the value $F(\hat{d}^{n'})$, $n'=1,2,\ldots,N'$, and the service level γ, the value that some variables $T_{n'}$ will take in the optimal solution can be established a priori. In particular, if $F(\hat{d}^{n'})<\gamma$, then $T_{n'}$ can be set to 0. Thus, the constraints (10) corresponding to such scenarios are accordingly reduced. The preprocessing may be useful in the reduction of the search tree in the branch and bound method.

The advantage of the proposed solution approach is its simplicity of implementation if one has a standard mixed integer programming software. The task becomes really easy if one can use, moreover, a modeling tool such as LINGO. The structure of the constraints in the problem makes the use of modeling language particularly appropriate. This yields a rather efficient solution method for this kind of problem.

4.2 Numerical Experiments

In this section we present an application example about stochastic supply chain network design problem. We use I=3, J=3, L=4, M=5 and K=2. The fixed parameters are set as follows. The plant cost cm_i is chosen randomly from the interval [5000, 6000] for every plant $i \in I$. For each warehouse $j \in J$, a fixed cost cw_j is chosen randomly in the interval [3000, 4000]. The specific capabilities a_j, h_l and s_i are chosen randomly from the interval [1000, 15000]. The requirement parameters r^P_{jk}, r^L_{lk} and r_k are chosen randomly in the interval [1, 3].

Now we consider the random parameters. The discrete distributions are used for these uncertain parameters. The discrete distributions structure is determined as follows. We choose the scenarios of transportation cost cp_{ij}, cp'_{jlk} and cp''_{lmk} from the interval [5, 30] according to uniform distribution, and the probability of the scenario is $1/N$. The scenarios of customer's demand d_{mk} are chosen in the interval [70, 200]. The probability of the $\hat{d}^{n'}$ is $1/N$. Finally, we obtain the discrete distributions of cp and d.

All of the numerical experiments were executed on a personal computer (Intel Pentium(R) Dual-Core E5700 3.00GHZ CPU and RAM 2.00GB), using the Microsoft Windows 7 operating system. All mixed-integer programs were solved to optimality using the LINGO 8.0.

Table 1 reports the computational results when the parameter β and γ take different values. From the computational results, it is obvious that the optimal objective value varies while the β and γ change between 0 and 1.

Table 1. Computational results for different parameters

N	N'	γ	β	Optimal objective	CPU(sec)
30	100	0.75	0.80	149985.4	33
			0.90	156242.3	35
		0.80	0.80	150687.7	38
			0.90	156995.6	29
		0.90	0.80	152849.6	28
			0.90	159319.2	34
	300	0.75	0.80	149632.5	84
			0.90	155706.1	35
		0.80	0.80	150482.4	32
			0.90	156609.4	31
		0.90	0.80	152704.6	38
			0.90	158974.6	18
40	100	0.75	0.80	153682.9	174
			0.90	159165.2	39
		0.80	0.80	154408.8	410
			0.90	159904.3	56
		0.90	0.80	156654.2	166
			0.90	162201.3	47
	300	0.75	0.80	153459.8	105
			0.90	158713.9	40
		0.80	0.80	154351.6	337
			0.90	159623.5	43
		0.90	0.80	156692.9	102
			0.90	162015.5	55
50	100	0.75	0.80	154424.0	788
			0.90	159374.8	59
		0.80	0.80	155146.1	627
			0.90	160124.9	159
		0.90	0.80	157361.5	621
			0.90	162440.7	115
	300	0.75	0.80	154143.6	746
			0.90	158822.1	160
		0.80	0.80	155024.8	553
			0.90	159745.2	119
		0.90	0.80	157343.4	655
			0.90	162171.7	147
55	100	0.75	0.80	154244.4	1793
			0.90	158798.9	622
		0.80	0.80	154960.3	3830
			0.90	159557.0	546
		0.90	0.80	157171.9	2532
			0.90	161905.5	907
	300	0.75	0.80	154007.2	4797
			0.90	158531.4	347
		0.80	0.80	154883.7	2029
			0.90	159449.5	362
		0.90	0.80	157180.6	2328
			0.90	161886.7	157

5 Conclusions

In this paper we have considered the VaR based stochastic supply chain network design problem with random transportation costs and random customer demand. To the best of our knowledge, such issues have not been studied in the literature. For general discrete distributions, the problem can be formulated as an equivalent mixed-integer programming model by introducing auxiliary binary variables. To solve the equivalent model, a branch and bound method is proposed, in which a preprocessing producer was utilized to reduce the branch of the search tree. The proposed method scheme was implemented by LINGO solver.

From computational point of view, it may be infeasible to employ LINGO solver for quite large instances of the supply chain problem. Therefore, other intelligent algorithms such as a combination of genetic algorithm and neural network should be further developed, which will be addressed in our future research.

Acknowledgments. This work is supported partially by the Natural Science Foundation of Hebei Province (A2011201007), and National Natural Science Foundation of China (No.60974134).

References

1. Geoffrion, A.M., Graves, G.W.: Multi-commodity distribution system design by benders decomposition. Manage. Sci. 20, 822–844 (1974)
2. Cohen, M.A., Lee, H.L.: Resource deployment analysis of global manufacturing and distribution networks. J. Manuf. Oper. Manage. 2, 81–104 (1989)
3. Aikens, C.H.: Facility location models for distribution planning. Eur. J. Oper. Res. 22, 263–279 (1985)
4. Alonso-Ayuso, A., Escudero, L.F., Garin, A., Ortuno, M.T., Perez, G.: An approach for strategic supply chain planning under uncertainty based on stochastic 0-1 programming. J. Global Optim. 26, 97–124 (2003)
5. Sabri, E.H., Beamon, B.M.: A multi-objective approach to simultaneous strategic and operational planning in supply chain design. Omega 28, 581–598 (2000)
6. Cheung, R.K.M., Powell, W.B.: Models and algorithms for distribution problems with uncertain demands. Transport. Sci. 30, 43–59 (1996)
7. Georgiadis, M.C., Tsiakis, P., Longinidis, P., Sofioglou, M.K.: Optimal design of supply chain networks under uncertain transient demand variations. Omega 39, 254–272 (2011)
8. Owen, S.H., Daskin, M.S.: Strategic facility location: A review. Eur. J. Oper. Res. 111, 423–447 (1998)
9. Santoso, T., Ahmed, S., Goetschalckx, M., Shapiro, A.: A stochastic programming approach for supply chain network design under uncertainty. Eur. J. Oper. Res. 167, 96–115 (2005)
10. Atamtürk, A., Savelsbergh, M.W.P.: Integer-Programming Software Systems. Ann. Oper. Res. 140, 67–124 (2005)

An Ontology-Based Task-Driven Knowledge Reuse Framework for Product Design Process

Jun Guo[1], Xijuan Liu[2], and Yinglin Wang[1]

[1] Department of Computer Science and Engineering, Shanghai Jiao Tong University, Shanghai 200240, China
ylwang@sjtu.edu.cn
[2] School of Mechanical Engineering, Shanghai Dianji University, Shanghai 200240
xijuan.liu@yahoo.com

Abstract. This paper puts forward a task-driven knowledge reuse framework for complex product design process. Domain ontology is used in the framework, which is fundamental for customization of the systems developed under this framework. The task representation model, as part of the domain ontology, is discussed in this paper; then a mechanism of sharing knowledge among instances of the same task class or between tasks which have some special relationships is discussed. The algorithm for generating knowledge learning schedule is also given based on the task relationship and the dependency of knowledge items. Through this framework, the task instances containing historical experiences of input and output can be stored and retrieved seamlessly during the execution of tasks without manual efforts.

Keywords: Product Design Process, Task-driven, Ontology, Knowledge Reuse.

1 Introduction

The design of complex products, such as aircraft, is a challenging task [1]. Because product design is a knowledge-intensive activity, and it's also very difficult to coordinate the work of a lot of teams in the design process. Another reason is because of the limited time and budget. So, an effective Product Design Process is crucial for the success of certain product. It can help the project manager to organize the design activities, coordinate between different teams, and manage the whole progress of the design, and so on.

Product design processes are quite different from other processes, for the reason that in design process there are too many uncertainties. Activities may be iterated many times. Lots of parts needed to be redesigned because of the failure of the initial design [2]. So, it would be very helpful if the designers can refer to historical design activities quickly and efficiently.

Currently, there are two major model of design process: process-based model and task-based model. There are many process-based models for design process [3], in which the design process is divided into four phases: clarification of the task, conceptual design, embodiment design and detail design [1]. However, process-based

Y. Wang and T. Li (Eds.): Knowledge Engineering and Management, AISC 123, pp. 319–327.
springerlink.com

model is not very adaptable. In task-based design, the process is represented as many tasks. Each task requires inputs and generates outputs. And tasks are often related to knowledge library. Task modeling is at the heart of the task-based design [4]. Task modeling is a complex process, driven by the need to find the most effective definition of products. Eppinger et. al. have put forward a model-based method for organizing tasks in design process using design structure matrix (DSM) [5]. They use the DSM to capture both the sequence of and the technical relationships among the many design tasks to be performed. And this allows the process to reduce iteration and increase parallel tasks. Petri nets are also used to represent the tasks.

Besides of that, design processes need a large amount of knowledge. There are also many researches about knowledge management systems, e.g., Yinglin Wang et. al. proposed an ontology-based framework for building adaptable KM systems, and a model for integrating the KM process and the business process [6]. Helin Wen proposed a context-based knowledge supply mechanism for product design process [7]. He models the context of design activities, and then builds a mechanism for active knowledge supply, knowledge requirements mining, and knowledge evaluation. However, some practical issues, such as how to use the task classification ontology to facilitate the delivery of knowledge, were not fully addressed in the previous researches.

This paper proposes a task-based knowledge reuse framework for product design process based on our previous work [6][7]. The remainder of the paper is organized as follows. Section 2 describes the task modeling method and task categorization method. Section 3 puts forward the entire framework. Section 4 describes the implementation of the key modules in the framework. Section 5 gives a case study. Section 6 concludes the paper.

2 Task Modeling and Task Categorization

2.1 Task Categorization

In task-based design process, task categorization is the first step we have to do. It would be helpful to manage the design process if we can divide the design task into a lot of design units [8]. Each unit is logical constrained and depended on other units. On one hand, there are some time sequence limits on these units; because the design process can be divided into several major parts and each part depended on the former part. On the other hand, these units can be divided into several major levels, and task units in certain level are more detailed than the units in former level. So, first, we use a top-down method to divide the design task into several levels of detailed task units, and then we add the timing constraints on these task units. Fig 1 is an example of the task categorization of the design of a trainer aircraft wing.

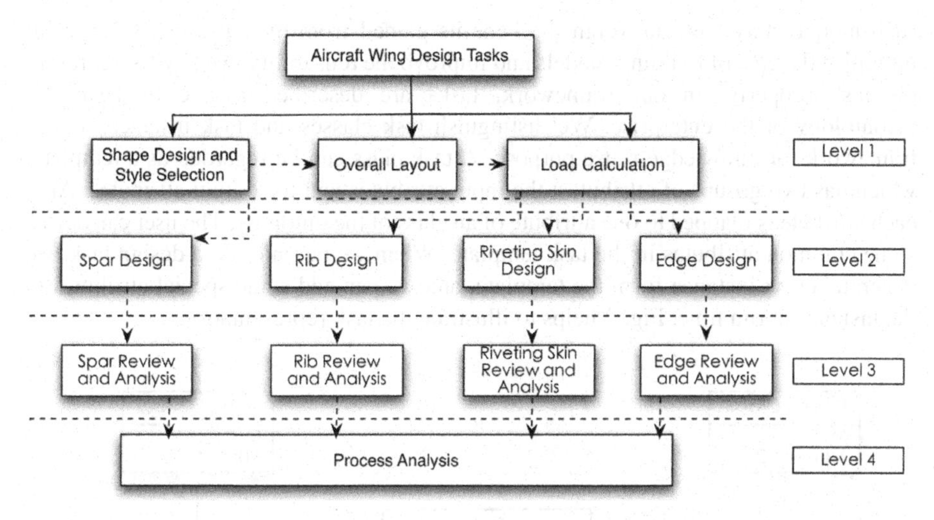

Fig. 1. Task categorization of the design of a trainer aircraft wing

2.2 Ontology Models Used in This Paper

Ontology is a conceptualization which is an abstract, simplified view of the world that we wish to represent for some purpose [6]. In this paper, we use the ontology models defined by Wang [6].

Definition: An ontology model includes five different parts: set of classes, set of enumeration types, set of relationships between classes, set of relationships between objects, set of axioms. It is as follows.

OntologyModel = (S_class, S_Enum, S_RelC, S_RelO, S_Axiom)

In this definition, *S_RelC* is the set of relations between classes; *S_RelO* is the set of relationships among objects except properties of the classes [6]. For example:

Class *Rib Review and Analysis* can be the subclass of class Review and *Analysis*.

We can also define a relationship, called *occur_after*, between two design class objects. Such as,

occur_after(DesignObject o1, Designobject o2) iff o1.time > o2.time

Another useful relationship we can define is *knowledge_dependency relationship.*

x, y∈knowledgeSet, dependentOn(x, y) iff y should be learned before x can be learned

This relationship can be used by the knowledge push model to push knowledge items to the users.

2.3 Task Representation

Generally, the task generates the output values from input parameters. And task-based models are generally inflexible and project specific. In addition, the cost of building such models may well outweigh the benefits gained from their use [1]. In order to minimize the cost of building models and improve the reusability, we have to represent the task properly. In our framework, tasks are described as the fundermental theontology of the enterprise. We distinguish task classes and task instances in the framework for knowledge reuse purpose. A task class can be regarded as a template, which has two groups of attributes: the input attributes and the output attributes. And, each attribute is mapped to one attribute of an class of the ontology. The user can define some common attributes in the task template. When we begin to do a design task, we generate a task instance form the template, and we can add some special attributes to the instance at run time. Fig. 2 helps to illustrate the task representation.

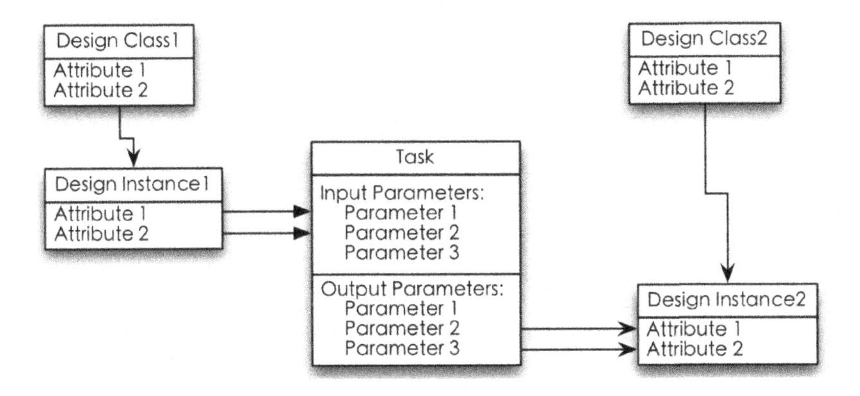

Fig. 2. Task representation illustration

In this way, we can reuse the task template and the classes in the ontology. Customers can also customize the task model, and the cost of building a model is minimized. Another advantage is we can easily find the similar task instances we have done before by comparing the values of input parameters of two instances, which is a special way of using case based reasoning seamlessly integrated in the design process. Here a case will be automatically stored while the task instance is completed during the process, and moreover, similar historical cases will be retrieved just in the time the designer login the system and try to do the task. Hence, designers do not need to intentionally store and retrieve the knowledge items during the task execution phase, thus greatly reduce the burden of designers.

Besides, task instances has type (class), which means that instances of the same task class inherently share some common knowledge. Hence, the relevant knowledge items should be linked previously to the task class and then be pushed to the task instances for reuse.

Moreover, knowledge belongs to different task classes may be shared among tasks based on their relationships, which can be used in the task driven knowledge reuse model. Task classification is one of the relationships, i.e., a task class may be a special case of another task class, e.g., the task "aircraft wing structure design" is a special kind of "mechanical structure design" task, hence knowledge used for "mechanical structure design" should be the basis and can be used in "aircraft wing structure design" task.

Hence, we can use the following rule to provide knowledge items:

$\forall$A, B $\in$ taskClassSet, $\forall$x$\in$ knowledgeSet, subclassOf(b, a) $\wedge$ revelantKnowledeOf(x, a)

$\Rightarrow$ revelantKnowledeOf(x,b)

The dependency relationship between knowledge items may also be used in the knowledge reuse process of product design. We have the following rule:

$\forall$a$\in$ taskClassSet, $\forall$x,y$\in$ knowledgeSet, revelantKnowledeOf(x, a) $\wedge$ dependentOn(x, y)

$\Rightarrow$ revelantKnowledeOf(y, a)

Based on the above relationships in an ontology, we can build a knowledge learning schedule for each designer before the task is executed, assuming that the classes of tasks the designer are going to do is already known. Through learning earlier before the task execution time, the executing time of the tasks will be reduced. Thus, the overall product design time will also be reduced. The following is the algorithm that can be used to generate the learning schedule for product designers.

Algorithm: Learning_Schedule_generation (tasksToBeDone, Ontology):

For each task (S)a $\in$ tasksToBeDone,

1) Find all the tasks which are superclasses of a:

   C={b| $\forall$b $\in$ taskClassSet, $\forall$a $\in$ tasksToBeDone, superClassOf(b, a)}

2) Find all the knowledge items that are relevant to the tasks of C,

   K={x| $\forall$x$\in$ knowledgeSet, $\forall$a$\in$ C, revelantKnowledeOf(x, a) }

3) Use topology ranking rank knowledge items in K, via the dependency relationships in K;

4) The ranking of knowledge items in K is the learning schedule the person should follow;

Note: some post-processing of the schedule may be required through eliminating some already learned knowledge items from the schedule.

3 The Framework

There are four levels in this framework, which are illustrated in Fig. 3.

Level 1: User interface level interacts with the customers, and provides the access to the design processes.

Level 2: Process level organizes the tasks in *Level 3*, and provides the chance to modify the task instances at runtime.

Level 3: Task level generates the task instances from templates, and allows the customers to customize the task instances.

Level 4: Ontology Library provides a lot of domain ontologies, which can be used to store knowledge or can be used to customize tasks.

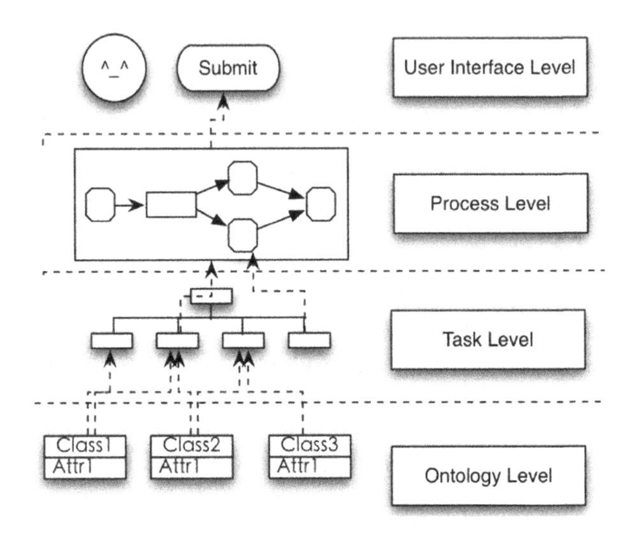

Fig. 3. The structure of framework

4 Implementations

The implementation of *level 1* and *level 2* is based on our previous work [6]. Task modeling and representation have been discussed in section 2. Here, we will discuss two modules: task instance library module and knowledge supply module.

4.1 Task Instance Library Module

This module is used to store historical task instances, and help users manage finished tasks. In addition, the module can push related instances to the user when the user is doing certain task, and these instances are sorted by similarity of the values of the input parameters in the task. So the user can refer to these useful instances easily. The workflow of this module is described in Fig 4.

Step 1, a task instance is generated when an activity began.

Step 2, the task instance is stored in the Task Instance Library when the activity is finished.

Step 3, the Extract Engine extracts similar instances from the library, according to the similarity of the input parameters.

Step 4 &5, these extracted similar instances are displayed on the activity pages, and sorted by the similarity.

4.2 Knowledge Supply Module

The knowledge supply module is quite the same as the module describe by H. Wen [7] except for two exceptions. One is that the context inference module is integrated at task level, and each task template defines a context. The second exception is that knowledge items can be pushed to a designer just when the tasks assigned to the designer are already known but they are still not executed, in this situations, the algorithm we have discussed in section 2.3 can be used. The Modified knowledge supply engine is in Fig 4. The detailed implementation is presented at H. Wen[7].

Step 1, a task instance is generated when an activity began.

Step 2, knowledge requirement is extracted from the task instance, according to the ontology classes under the task instance and the relationships defined.

Step 3&4, knowledge supply module gets knowledge requirements, and retrieve relevantknowledge items from knowledge base and knowledge requirement base.

Step 5, the retrieved knowledge items are pushed to the activity page and designer can refer to the releveant knowledge items quickly.

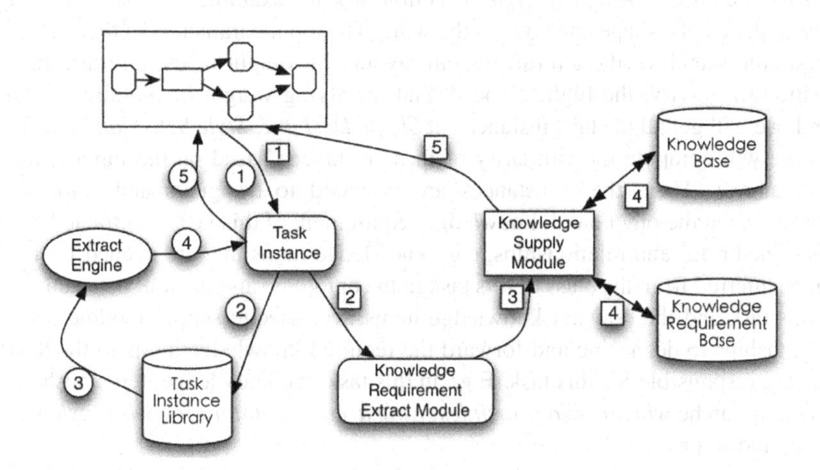

Fig. 4. Task instance library and knowledge push model illustration

5 A Case Study

A system based the above idea has been developed and used in the product design process in a mechanical company. The process and the task template are shown in Fig 5. In the design process, similar instances from the task history instance library are pushed to designers, and sorted by similarity. Besides, other related knowledge is also pushed to the activity page to help designers to make a decision.

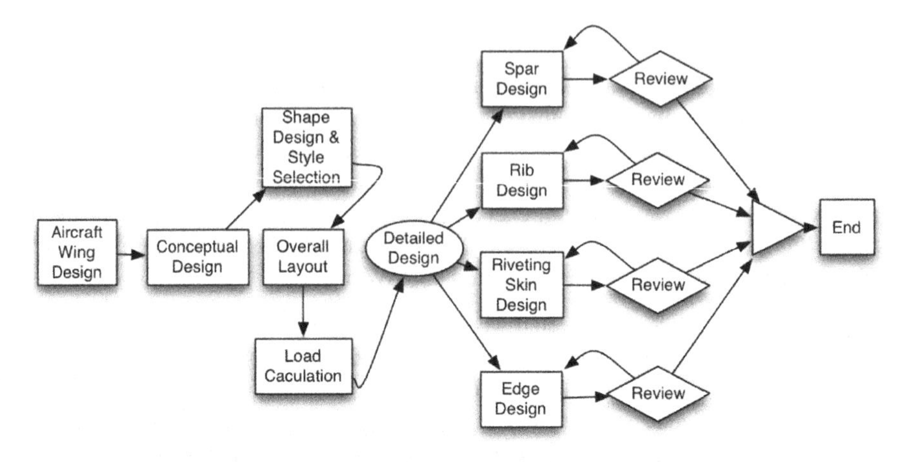

Fig. 5. Design process of an aircraft wing

Take the *Shape Design & Style Selection* task for example. In this task, designers have to decide the shape and style of the wing. The input parameters of this task contain the simple sketch of the aircraft overall layout, the weight of the aircraft, the center position of gravity, the highest speed, and the flying weight of the aircraft. On one hand, we will get all the task instances of *Shape Design & Style Selection* in the library, and we will compute the similarity of these instances based on the input parameters listed above. Then, these instances are provided to designers and sorted by the similarity. On the other hand, knowledge requirement of this task is extracted based on predefined rules and relationships, e.g., knowledge items that are needed for this task can be inferred from the class of this task in the ontology, and the link between the class of this task and the relevant knowledge items. Knowledge supply module uses these relationships to determine and forward the required knowledge items to the designers who are responsible for this task. E.g., in this task, the knowledge items pushed to the designers can be *wing design principles*, *wing area calculator tool*, *wing style selection guide*, and so on.

At last, the designers can evaluate these design instances and some knowledge items, which will help the system to improve the accuracy and performance.

6 Conclusions

This paper proposed a task-driven knowledge reuse framework and its implementation for complex product design process based on domain ontology. The four levels of this framework, the modeling of task templates and the algorithms for knowledge pushing and learning schedule generation have been introduced. Compared with other task-based frameworks, this one can be easily customized. By building task template and storing design instances at each task activity, historical design instances can be pushed to designers and reused efficiently in the design process. And this framework provides a reference model for incorporating knowledge management in complex product design process.

Acknowledgment. Thank the National High Technology Research and Development Program of China (Grant No. 2009AA04Z106) and the Natural Science Funds of China (Grant No. 60773088) for financially supporting this research.

References

1. Clarkson, P.J., Hamilton, J.R.: Signposting: a parameter-driven task-based model of the design process. Research in Engineering Design 12(1), 18–38 (2000)
2. Eckert, C., Clarkson, P.: Planning development processes for complex products. Research in Engineering Design 21(3), 153–171 (2010)
3. Paul, G., Beitz, W., Feldhusen, J., Grote, K.H.: Engineering Design, A Systematic Approach. Springer, London (2007)
4. Dittmar, A., Forbrig, P.: Task-based design revisited. In: Proceedings of the 1st ACM SIGCHI Symposium on Engineering Interactive Computing Systems, pp. 111–116. ACM, Pittsburgh (2009)
5. Eppinger, S.D., Whitney, D.E., et al.: A model-based method for organizing tasks in product development. Research in Engineering Design 6(1), 1–13 (1994)
6. Wang, Y., Guo, J., Hu, T., Wang, J.: An Ontology-Based Framework for Building Adaptable Knowledge Management Systems. In: Zhang, Z., Siekmann, J.H. (eds.) KSEM 2007. LNCS (LNAI), vol. 4798, pp. 655–660. Springer, Heidelberg (2007)
7. Helin, W., Yinglin, W., Jianmei, G.: A Context-Based Knowledge Supply Mechanism for Product Design Process. In: Proceedings of 2010 International Conference on Progress in Informatics and Computing, Shanghai, China, pp. 356–359 (2010)
8. Xie, Q., Mao, Z.: Aircraft product design process model based on Petri net. Engineering Journal of Wuhan University 39(1), 131–136 (2006)

Part V

Data Mining, NLP and Information Retrieval

A LDA-Based Approach to Lyric Emotion Regression

Deshun Yang, Xiaoou Chen, and Yongkai Zhao

Institute of Computer Science & Technology, Peking University,
100871, Beijing, China
{yangdeshun,chenxiaoou,zhaoyongkai}@icst.pku.edu.cn

Abstract. Lyrics can be used to predict the emotions of songs, and if combined with methods based on audio, better predictions can be achieved. In this paper, we present a new approach to lyric emotion regression. We first build a Latent Dirichlet Allocation (LDA) model from a large corpus of unlabeled lyrics. Based on the model, we can infer the latent topic probabilities of lyrics. Based on the latent topic probabilities of labeled lyrics, we devise a scheme for training and integrating emotion regression models, in which separate models are trained for latent topics and the outputs of those models are combined to get the final regression result. Experimental results show that this scheme can effectively improve the emotion regression accuracy.

Keywords: Latent Dirichlet Allocation, lyric emotion regression.

1 Introduction

The explosion of digital music calls for new ways of music retrieval. Organizing and searching music according to its emotional content is such a new way. In supporting emotion-based retrieval, automatic music emotion recognition plays an important role, in that it can act as a means of automatic music emotion annotation.

Due to the emotion representation adopted, music emotion recognition can take either the form of emotion classification where categorical emotion models are adopted, or the form of emotion regression where emotion is scaled and measured by a continuum of two or more real-valued dimensions.

Compared to other related tasks of music concept recognition, such as genre recognition, emotion recognition is still in its early stage, though it is attracting more and more attention from the research community. As an indicator, MIREX run Audio Music Mood Classification contest for the first time in 2007. Since then, the contest is run each year and the performance of the best algorithm improves year on year, but is still far behind that of the recognition algorithms for other music concepts.

Until recently, most of the research work on music emotion recognition is based on music audio analysis, where only audio data is used to predict music emotions. However, there appears a few research papers on music emotion recognition through lyrics. In fact, for a song, lyric is an essential component, and contains information useful for predicting the emotion of the song. In addition, studies show that lyrics are complementary to audio in predicting song emotions.

Y. Wang and T. Li (Eds.): Knowledge Engineering and Management, AISC 123, pp. 331–340.
springerlink.com

Lyric emotion recognition is mostly carried out as an supervised machine learning task, where training lyrics are labeled with emotions and represented by vectors of features and a learning model, e.g., SVM, is employed on the training examples. N-gram features, having been well explored in other natural language processing tasks, are naturally introduced to the task of lyric emotion recognition. In addition to n-gram features, semantic model based features have also been tried for lyric emotion recognition.

LDA (Latent Dirichlet Allocation) [1] was proposed by David M. Blei in 2003. It is a generative graphical model that can be used to model and discover the underlying topic structures of any kind of discrete data. LDA has exhibited superiority on latent topic modeling of text data in the research works of recent years. Ralf Krestel et al [2] introduced a method based on LDA for recommending tags of resources. István Bíró et al [3] applied an extension of LDA for web spam classification, in which topics are propagated along links in such a way that the linked document directly influences the words in the linking document.

In this paper, we try to exploit both LDA model based features and n-gram features to build a better-performing lyric emotion regression model. For this purpose, we propose an emotion regressor training and integration scheme. First of all, we build a LDA model from a large corpus of unlabelled lyrics. The model learns a range of readily meaningful topics relating to emotion. Then, based on the model, we infer the latent topic probabilities of labeled lyrics to be used as training examples. According to the topic probabilities of the lyrics, we distribute them among subsets of examples, with each subset dedicated to a unique topic. We then train a regressor for each topic on its subset of training lyrics which are represented by n-gram features. To compute the emotion of a given lyric, we first call the individual regressors with the lyric's n-gram features and then combine the multiple results into a final value according to the lyric's LDA topic probabilities.

The rest of the paper is organized as follows: Section 2 will review some related works. Section 3 gives the emotion model we adopt. Section 4 describes in detail our scheme. Section 5 explains the details of experiments we did, including datasets and features. Section 6 shows evaluation experiments for proposed method. Finally, Section 7 gives conclusions and future work directions for this research.

2 Related Works

In the beginning, most research works of this field focus on acoustic analysis, finding new features or new classification models to improve the accuracy. Liu et al [4] presented a mood recognition system with a fuzzy classifier. In this system, all the music file were translated into MIDI form. Some music-related features, like tempo, loudness, timbre etc were extracted, by which they build a fuzzy classifier to recognize the mood of music. Lie Lu et al [5] presented a framework for automatic mood detection from audio data. The framework has the advantage of emphasizing suitable features for different tasks. Three feature sets, including intensity, timbre, and rhythm are extracted for classification.

At the same time, there appears a group of researchers who begin to notice the importance of lyrics. Some researchers make an effort to construct effective emotion lexicon and use it to compute the emotion of a piece of lyric [6,7]. Others try to use machine learning methods to solve this problem.

Hu et al [8] made a detailed comparative analysis between audio and lyrics. Among the 18 mood categories, lyric features outperformed audio spectral features in seven categories. Only in one category, audio outperformed all lyric features.

Xia et al [9] proposed a model called sentiment vector space model (s-VSM) to represent song lyric documents which uses only sentiment related words. Their experiments prove that the s-VSM model outperforms the VSM model in the lyric-based song sentiment classification task.

Y. Yang et al [10] exploited both audio and lyrics for song emotion classification. They used bi-gram bag-of-words and Probabilistic Latent Sentiment Analysis(PLSA) to extract lyric features. The results show that the inclusion of lyric features significantly enhances the classification accuracy of valence.

3 Emotion Model

Roughly speaking, emotion models can be classified into categorical models and dimensional models.

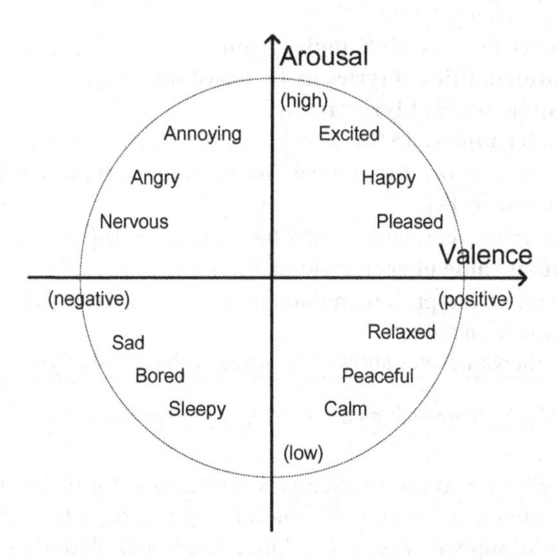

Fig. 1. Thayer's arousal-valence emotion plane

In a categorical model, a number of distinct emotion classes are identified, each of which is described by a group of related adjectives or nouns. Hevner's Adjective Circle [11] is a typical categorical model. Dimensional models reduce emotion to two or more dimensions of continuous values. Thayer's model[12] (see Fig. 1) is a dimensional model. It consists of two dimensions: Arousal and Valence. The Arousal dimension (the horizontal axis in Fig. 1) reflects the degree of stress and Valence (the vertical axis) the degree of energy.

Dimensional models are thought to be superior to categorical models in that they can represent the fine details of emotion and therefore don't have the inherent ambiguities between emotion categories in categorical models. When using a dimensional model, human annotators label a song in a continuous multi-dimensional emotion space freely without being confined to predefined emotion tags.

In the presented work, we adopt Thayer's two dimensional emotion model and we focus only on the problem of *Valence* value regression in this paper.

4 Regression Model Training and Integration Process

The process is shown in Fig. 2.

Input:
D_u: a set of unlabeled Chinese song lyrics;
D_l: a set of Chinese song lyrics which have been labeled with emotion values;
K: the number of latent topics in the LDA model built;
$\overline{P}$: topic probability threshold for example selection;
$D_l[k]$: subset of labeled lyrics selected for topic k;
$\widehat{R}$: a basic regression algorithm;
Build a LDA model M_{LDA} **with** K **topics from** D_u
Infer the topic probabilities of lyrics in D_l **based on** M_{LDA}
Extract uni-gram features of lyrics in D_l
Train regressors for topics, do for $k = 1,2,\dots,K$
1) For each lyric in D_l, if its probability to generate topic k is larger than $\overline{P}$, then put it into $D_l[k]$;
2) Call $\widehat{R}$ to train a regression model for topic k on $D_l[k]$;

Predict the emotion value of each lyric in D_l
1) Call $\widehat{R}$ with each topic's regression model and the lyric's uni-gram features to get an emotion value;
2) Combine the values got above into a final value for the lyric;

Fig. 2. Regression model training and integration process

The process works on two sets of examples; one is used for fitting the LDA model, and the other for training and testing the emotion regressors. A basic regression algorithm, such as SVM(Support Vector Machine) regression algorithm, is used in the process.

First of all, We fit a LDA model from a corpus of unlabeled lyrics and, with the model, infer the topic probability distributions of labeled lyrics. Instead of using the topic probabilities to represent lyrics in the training instances, we use them to base a framework for training and integrating emotion regression models. We assume that there exist more explicit relationships between the latent topics of the LDA model and the two dimensional emotion space of lyrics. And, lyrics which have higher probabilities to generate the same topic express emotions in similar ways, depending on the topic. So, for each latent topic, we choose those labeled lyrics whose probability to generate the topic is greater than a threshold to compose a training set, and train a regression model on the set of selected training lyrics.

To predict the emotion value of a given lyric, all the regressors are called and provided with the lyric's uni-gram features. Then the outputs of the regressors are integrated to get the final value. We devised and experimented with two alternative mechanisms for combining the multiple values to get the final emotion value.

1) ***Mechanism A*** (MA)
For a lyric, we choose the topic on which the lyric has the largest generation probability and call the topic's regressor with the lyric's uni-gram features to get the emotion value. This mechanism is based on the assumption that a lyric is only talk about one topic, on which the lyric has the largest generation probability. Of cause this assumption is sometimes not true, because there exit lyrics which generate multiple topics with almost the same high probabilities. So we propose the other mechanism.
2) ***Mechanism B*** (MB)
For a lyric, we first choose those topics on which the lyric has a probability higher than a pre-set threshold. Then we call the regressors corresponding to the topics, and provide them the lyric's uni-gram features. To get the final emotion value, we adopt a weighted-voting method to integrate the multiple values returned by the regressors. For a lyric, the weight of a regressor is computed as in formula (1):

$$w_l[\hat{r}_k] = \frac{p(z_k|l)}{\sum_{j:p(z_j|l)\geq\bar{p}} p(z_j|l)} . \quad (1)$$

where $w_l[\hat{r}_k]$ denotes the weight of regressor k for lyric l. Regressor k corresponds to topic k. $p(z_k|l)$ denotes the probability of lyric l generating topic k. This mechanism may select multiple topics that the lyric most probably talks, not only the single most probable topic. It doesn't have the weakness of the first one.

Our scheme for regression model integration differs from the Adaboost-style methods in that different lyrics have different topic probability distributions and therefore, different weights for the regressors. That is to say, the weights for the regressors vary with lyrics whose emotion value needs to be figured out.

It needs to be pointed out that the LDA model data consists of two parts: document-topic probability distribution and topic-word probability distribution. The document-topic distribution is usually taken as features of documents. In our work, we only use the document-topic distribution of the LDA model.

5 Experiments

5.1 Data Sets

We employed two sets of lyric examples. One is for building the LDA model and the other for training the emotion regression model. We downloaded about 35,000 song lyrics from Internet, covering most of the contemporary Chinese popular songs, and use them to build the LDA model.

The training set we use to train the emotion regression models is the same as that used in [7]. We had downloaded 981 Chinese songs(including both the waveform data and lyric text data) from www.koook.com, which was then one of the most popular Chinese music websites on Internet. The songs were labeled manually with VA(Valence and Arousal) values. The values of VA emotion dimensions were confined in the range of [-4.0,+40]. Seven people took part in the annotation work, and everyone labeled all of the songs. From the original 981 songs, we selected only those songs which had been labeled with similar VA values by at least six people. At last, we got 500 songs labeled with VA values and they will be used to train the emotion regression models. More than 270 artists appear in our final collection and the genres of the songs in the collection include pop, rock & roll and rap.

For Chinese lyric texts, word segmentation need to be done before any other processing. We do word segmentation for the lyrics by a Chinese NLP tool.

5.2 Lyric Features

For training a regression model with a basic regression algorithm, such as SVM (Support Vector Machine) regression algorithm, we need to represent lyrics by feature vectors. In the field of NLP(Natural Language Processing), n-gram based bag-of-words (BOW) feature model is a commonly used model. In this paper, we focus on the issue of integration of regression models based on the LDA topic probabilities of lyrics, so we simply use uni-gram features to represent lyrics.

The original distinct uni-grams in the lyric corpus count up to tens of thousands. A feature space of this high dimensionality is not feasible for computational processing. To reduce feature dimensions, we divided the real value range of V dimension into four segments as shown in Table 1, and then apply the chi-square feature selection technique. At last we get a feature space of 2000 dimensions.

Table 1. Four segments of the value of V dimension

	$V< -2$	$-2 \leq V < 0$	$0 \leq V < 2$	$V \geq 2$
Class	1	2	3	4

Where n-grams are used as features, there are two commonly used methods of measuring the n-gram weight, tf*idf and Boolean value. Here, tf represents term frequency and idf represents inverse of document frequency. A Boolean value of 1 means that the corresponding n-gram appears in the document, and a value of 0 means that the n-gram doesn't appear. Previous experiments have shown that Boolean values are better than tf*idf, so we use Boolean values in our experiments.

5.3 LDA Model Parameter Estimation Method

It is intractable to compute the parameters of a LDA model exactly. Therefore, all the three parameter estimation methods currently used are approximate solutions. Variational Bayes was proposed by David M. Blei [1], which introduced variational inference method into standard Expectation Maximization method. Expectation Propagation was proposed by T. Minka [13], which is another approximate solution, also based on variational inference. Gibbs Sampling was proposed by Tomas L. Griffiths [14] which is a approximate method theoretically different from the former two. It is based on the Markov chain Monte Carlo (MCMC) and is a special case of Metropolis-Hasting method. Thomas L. Griffiths and Mark Steyvers compared these three methods and made the conclusion that Gibbs sampling method gives the best result. So, in our experiment we use Gibbs Sampling to estimate the LDA model parameters.

6 Evaluation Results

In our experiments, we use Pearson Correlation Coefficients to measure the performance of the regression models.

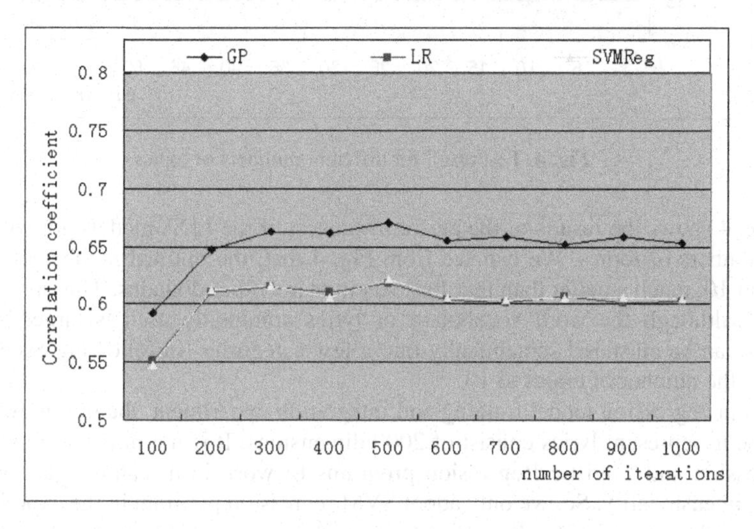

Fig. 3. Test result for different numbers of iterations

The number of iterations the LDA model fitting program runs and the number of latent topics the model contains may affect the performance of the model. We determine these two numbers through experiments. With a series of different values for the two numbers, We build LDA models and infer the topic probabilities of the lyrics based on each model respectively. Then, representing lyrics with their topic probabilities, and employing three basic regression algorithms, SVM (Support Vector Machine) regression, LR (Logistic Regression) and GP (Gaussian Regression), we train emotion regression models and test their performances. We then choose the number of topics and the number of iterations with which the best-performing LDA model has been built.

Fig. 3 shows the results of the performance test of the LDA models got with different numbers of iterations and a fixed number of topics, 15. From Fig. 3, we can see that when the number of iterations exceeds 300, the performance of the LDA model becomes stable. So, in the following experiments, we set the number of iteration to 500.

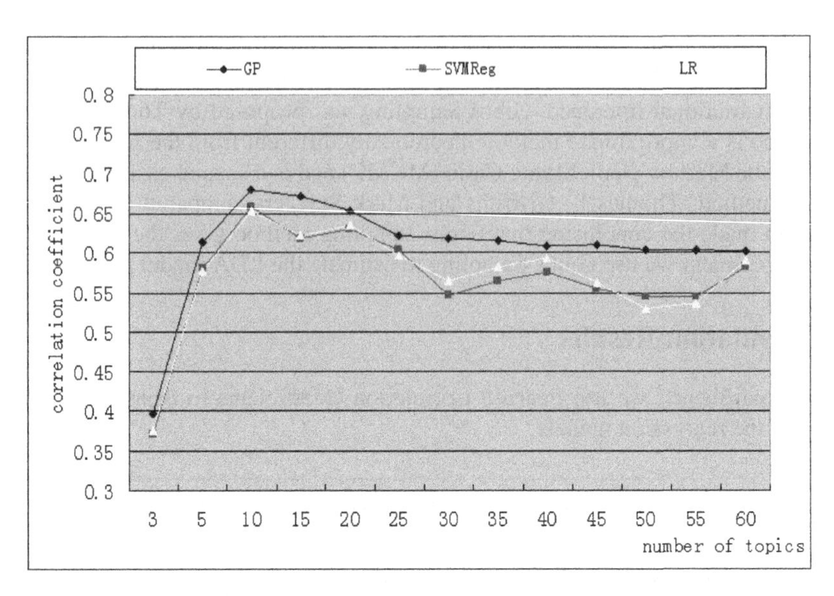

Fig. 4. Test result for different numbers of topics

Fig. 4 shows the results of the performance test of the LDA models got with different numbers of topics. We can see from Fig. 4 that, the optimal number of topics is around 10, much smaller than that for texts in many other domains. This shows that in lyrics, although the word vocabulary of lyrics commonly used is large, but these words can be clustered semantically into a few categories. In following experiments, we set the number of topics to 10.

In our regression model training and integration experiment, the uni-gram features we use to represent lyrics consist of 2000 dimensions. It is not practical for Gaussian Regression and Logistic Regression programs to work in a feature space with this high dimensionality. So, we only adopt SVM regression program in the experiment.

The value of parameter $\overline{P}$ in Fig. 2 affects the selection of training examples. So it has influences on the performance of regression models trained. We do experiments with the value set to 0.03, 0.05, 0.1, 0.15, 0.2, 0.25, and 0.3, respectively and find that at first, with the increase of the value, the performance gets better, which is in our expectation. But when the value becomes greater than 0.15, the performance starts to get worse. The reason for the worse regression models may be that the training set gets too small. In fact, if the value is too large, few lyrics will be chosen as training examples.

The last parameter to be set is $\overline{p}$ in formula (1). We set its value to 0.1, which is the average probability a lyric generates each topic (1/10).

To get a reliable evaluation result, we took 5-fold cross-validation on the set of labeled lyrics. The final result is shown in table 2. The baseline SVM regression model has been trained on the set of all labeled lyrics. It can be seen from the table that the two model-integration mechanisms both outperform the baseline SVM model. MA raises the correlation coefficient by 3.3% and MB increases it by 4.1%.

Table 2. Final results

	SVM Model (baseline)	Topic SVM Models-MA	Topic SVM Models-MB
CF	0.731	0.764	0.772

We can get the conclusion that, for a specific lyric to be recognized, if we select only those examples that are similar to the test example to train a regression model, the resulted model will be better than the model trained on all examples. In other words, an example which talks about topics totally different from that talked about by the test example is considered to be noise for training the regression model, and we should remove it from the training set.

But we can also see that the accuracy does not improve much. This may be caused by the small size of the training set. After the screening process, the set of selected training examples for each topic are all much smaller than the original training set. So, the regression models may be under-trained. If the size of the training set is too small, the hyper plane of the SVM model can not reflect the true information of example distribution.

Mechanism MB is better than mechanism MA. It proves our idea mentioned before. Some lyrics talk about mainly a single theme from the beginning to the end, expressing a happy feeling, for example. For these songs, MA mechanism can give a much better prediction. But there are a number of songs that express multiple themes at the same time.

Overall, the regression model training and integration scheme which uses LDA model based information is effective to improve the final regression accuracy.

7 Conclusion and Future Work

In this paper, we investigated a lyric emotion regression model training and integration scheme which is based on the LDA model based information about lyrics. We train a separate SVM regression model for each latent topic and integrate these regression models based on the latent topic probabilities of lyrics. The experimental results show that this method can effectively improve the regression accuracy.

In the future, we will try LDA models which include both bi-grams and uni-grams. By considering bi-gram patterns of words, we believe the resulted LDA models will be more semantically expressive. In addition, we will do experiments on a larger training set of about 2,000 Chinese songs with emotion annotations, to see if better-performing regression models can be obtained.

Acknowledgments. This work was supported by Beijing Natural Science Foundation(Multimodal Chinese song emotion recognition) and National Development and Reform Commission High-tech Program of China under Grant No. [2010]3044.

References

1. Blei, D.M., Ng, A.Y., Jordan, M.I.: Latent Dirichlet Allocation. In: NIPS 2002 & JMLR 2003
2. Krestel, R.: Latent Dirichlet allocation for tag recommendation. In: Proceedings of the Third ACM Conference on Recommender Systems, pp. 61–68 (2009)
3. Bíró, I., Siklósi, D., Szabó, J., Benczúr, A.A.: Linked latent Dirichlet allocation in web spam filtering. In: Proceedings of the 5th International Workshop on Adversarial Information Retrieval on the Web, pp. 37–40 (2009)
4. Liu, D., Zhang, N.Y., Zhu, H.C.: Form and mood recognition of Johann Strauss's waltz centos. Chin. J. Electron. 12(4), 587–593 (2003)
5. Lu, L., Liu, D., Zhang, H.-J.: Automatic Mood Detection and Tracking of Music Audio Signals. In: IEEE Transactions on Audio, Speech and Language Processing (2006)
6. Meyers, O.C.: A mood-based music classification and exploration system. Master's thesis, Massachusetts Institute of Technology (2007)
7. Hu, Y., Chen, X., Yang, D.: Lyric-based song emotion detection with affective lexicon and fuzzy clustering method. In: Proc. of the International Society for Music Information Conference, Kobe, Japan (2009)
8. Hu, X., Downie, J.S.: When lyrics outperform audio for music mood classification: a feature analysis. In: Proceedings of 11th International Society for Music Information Retrieval Conference (2010)
9. Xia, X.Y., Wang, L., Wong, K., Xu, M.: Sentiment vector space model for lyric-based song sentiment classification. In: Proc. of the Association for Computational Linguistics, ACL 2008, pp. 133–136. Columbus, Ohio, U.S.A (2008)
10. Yang, Y.-H., Lin, Y.-C., Cheng, H.-T., Liao, I.-B., Ho, Y.-C., Chen, H.H.: Toward Multimodal Music Emotion Classification. In: Huang, Y.-M.R., Xu, C., Cheng, K.-S., Yang, J.-F.K., Swamy, M.N.S., Li, S., Ding, J.-W. (eds.) PCM 2008. LNCS, vol. 5353, pp. 70–79. Springer, Heidelberg (2008)
11. Hevner, K.: Experimental studies of the elements of expression in music. American Journal of Psychology 48(2), 246–268 (1936)
12. Thayer, R.E.: The Biopsychology of Mood and Arousal. Oxford Univ. Press, New York (1989)
13. Minka, T.: Expectation propagation for approximate Bayesian inference. In: Proc. 17th Conf. on Uncertainty in Artificial Intelligence (2001)
14. Griffiths, T.L., Steyvers, M.: Finding Scientific Topics. PNAS (101), 5228–5235 (2004)

An Online Fastest-Path Recommender System

Yun Xun and Guangtao Xue

Department of Computer Science and Engineering, Shanghai Jiao Tong University
Shanghai 200240, China
{xunyun_sj,xue-gt}@sjtu.edu.cn

Abstract. This paper presents an online traffic system to recommend taxi drivers the fastest-path of picking passengers up. Several systems have been studied to find and recommend the shortest-paths on distance in mobile scenarios. However, in practical traffics, we discover that the shortest-path is usually not the fastest-path due to congestion. Especially for the taxi drivers, the fastest-path to pick up passengers is the best choice. Analyzing a real trace data including about 2000 taxis in a 22 square kilometers area in 7 days in Shanghai. Then we design a practical recommendation system to process the fastest-path selection. Experimental results show that our online system can quickly recommend the almost exact fastest-paths to taxi drivers for picking up passengers in real traces.

Keywords: Mobile recommender system, Knowledge Discovery, Fastest-path.

1 Introduction

As the development of sensor, wireless as well as communication facilities [6] such as GPS, Wi-Fi and RFID, large amount of location data traces are easy to be collected. In several fields, intelligent systems are developed to recommend optimal choice after mining these traces. It is so called recommender system. In recent years, recommender systems are increasingly advocated in both leisure and business applications [3][4][8][9]. E.g., K. Cheverst [1] developed a system to guide tourists the following routes according to their visited parts, current positions and the scenery conditions.

In this paper, we focus on developing an online fastest path recommender system for taxi drivers. The closest work [2] introduced a system to recommend taxi drivers a shortest-path to pick-up passengers. We argue and that the shortest-path on distance metric is usually not the fastest-path in time metric in a dynamic and complex traffic environment, which is also verified by mining real taxis traces in Shanghai. Hence, the key function of our system is to help the empty taxi to pick up passengers as soon as possible, i.e., improving the business performance of taxis and reducing the waiting time of passengers.

We simulate this recommender system in a real trace data including about 2000 taxis in a 22 square kilometers area in 7 days in Shanghai. The performance evaluation shows when more than 200 passengers distribute in this 22 km^2 area, the travel time is equal to the minimum time for a taxi to pick up passengers if this taxi drives along the fastest-path recommended by our system.

Y. Wang and T. Li (Eds.): Knowledge Engineering and Management, AISC 123, pp. 341–348.
springerlink.com

2 Problem Formulation

2.1 Hot Terminals of Pick-Up

With our traces data, we introduce the hot terminal method. The traces [5] recorded 4000 taxis driving in Shanghai. They reported the information tuple <*id, position, state, time*> per second, where <*state*> uses '1' or '0' to present a taxi with or without passengers. A pick-up point is denoted when <*state*> changes from 0 to 1. We set the area of interest to be limited in a 22 square kilometers area in the center of city since there are many taxi requests in urban area. After all the pick-up points are extracted in this area, *K*-MEANS algorithm is adopted to cluster these points according to their geographical locations. The pick-up points are denoted as $\{x_1, x_2, \dots, x_i, \dots\}$ in the map, where each point is presented by its longitude and latitude. The aim of *K*-MEANS is to divide the all pick-up points into *K*=*n* hot terminals sets $H = \{H_1, H_2, \dots, H_n\}$, so as to minimize the within terminal sum of squares. Fig. 1 shows the hot terminals situation when setting $n = 6$.

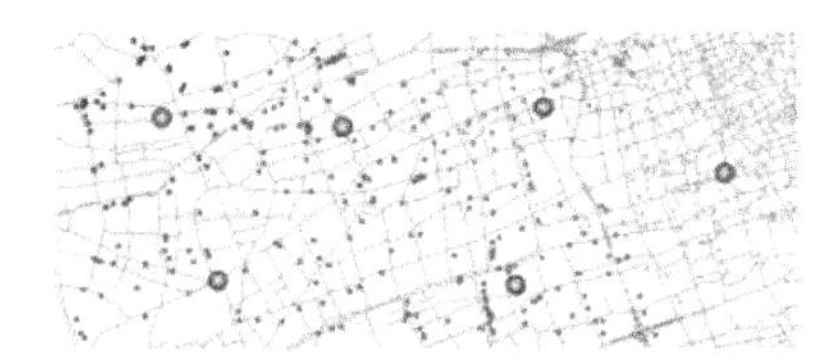

Fig. 1. Distribution of hot terminals (6 black points) in rebuilt map based on real pick-up (colored points belong to its closest hot terminals) data

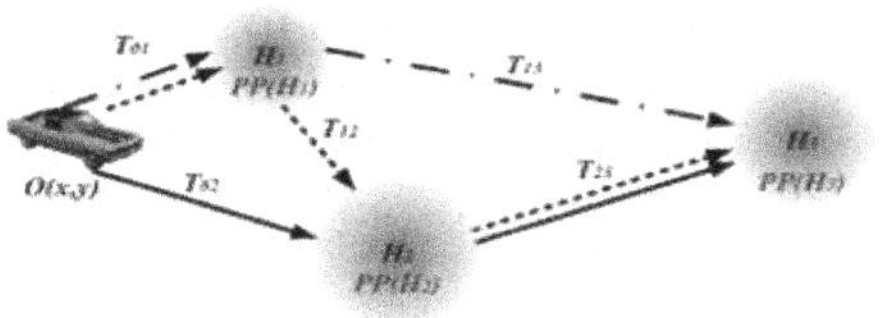

Fig. 2. An illustration example that a taxi can select the fastest-path in several potential paths from its current position to hot terminal H_3

The result of *n* sets is defined *n* hot terminals with all pick-up points. Consequently, the center of a hot terminal can be served as the destination of a fastest-path, where has a high pick-up probability.

2.2 Pick-Up Probability of Hot Terminal

Before we prepare to recommend the destination of the fastest-path, the metric of pick-up probability (PP) for these *k* hot terminals needs to be considered.. A minimal round area is used to cover a hot terminal within almost all pick-up points belong to this terminal. From the trace data, let N_e as the number of empty cars enter the round area, and let N_p as the number of pick-up events happen in this round area. Thus, $PP = \frac{N_p}{N_e}$ is the pick-up probability for each terminal.

2.3 Problem Statement

As mentioned above, the preparations including the positions of hot terminals $H = \{H_1, H_2, \dots, H_n\}$ with pick-up probability $PP(H_i)$, rebuilt map *G*, road velocity *V* and fastest-path *FP* definition are ready. Afterward the recommendation is a path traversing a sequence of hot terminals combination with minimum time to pick-up.

In addition, let $\vec{R} = \{\overrightarrow{R_1}, \overrightarrow{R_2}, \ldots \overrightarrow{R_m}\}$ be the set of all the possible directed sequences (a potential driving path traversing several hot terminals) generated from H. Next, let $h_{\overrightarrow{R_i}}$ be the number of hops of path $\vec{R}_i (1 \le i \le m)$, where $1 \le h_{\vec{R}_i} \le n$. Finally, for any directed sequence, let $PP_{\vec{R}_i}$ be a set including all the probability of all hot terminals containing in $\vec{R}_i$. E.g., if a path $\overrightarrow{R_1}$ traverses H_2 and H_6, $PP_{\overrightarrow{R_1}} = \{PP(H_2), PP(H_6)\}$.

Thereby, the *Fastest-path Along Sequential Terminals* (FAST) selection problem can be formulated as following table. The function $f_T(\cdot)$ is to calculate the potential travel time before pick up passengers. The fastest-path $\overrightarrow{FP}$ is one of the path $\vec{R}_i$ that has the minimal T after computation of $f_T(\cdot)$.

Table 1. FAST Problem Formulation

FAST problem
Given: A set of hot terminals H with $\lvert H \rvert = n$, a probability set $\{PP(H_1), PP(H_2), \ldots, PP(H_n)\}$. A directed sequence set $\vec{R}$ with $\lvert \vec{R} \rvert = m$, map G, road velocity V, and the current location $L(x, y)$ of a taxi.
Objective: Select a fastest-path $\overrightarrow{FP}$ ($\overrightarrow{FP} \in \vec{R}$), which minimizes the driving time T. $\overrightarrow{FP} = \text{argmin}\, f_T(O(x,y), \vec{R}_i, PP_{\vec{R}_i}, V, G)$

2.4 Fast Problem with Constraint

A straightforward solution to find the fastest-path in FAST problem is the brute-force method for searching all possible sequences in set $\vec{R}$. If we assume that the cost for computing $f_T(\cdot)$ once is 1, the complexity of searching the optimal $\overrightarrow{FP}$ in FAST problem by is $O(n!)$. Apparently, the complexity for searching the optimal solution for FAST problem by brute-force method is unacceptable. Thus, from a practical view, we simplify the FAST problem by setting a constraint on the number of hops in a recommended path $h_{\vec{R}_i}$. i.e., the number of hot terminals in a recommendation path is a given constant $h_{\vec{R}_i} = s$. Let $\vec{R}_i^s$ denote a path selection with s hops. We can extend the FAST problem with the hops constraint as follow:

Table 2. FAST-HC Problem Formulation

FAST-HC problem
Objective: Select a fastest-path $\overrightarrow{FP}$ ($\overrightarrow{FP} \in \vec{R}$)under the same condition of FAST problem, which minimizes the driving time T with a condition of hops s. $\overrightarrow{FP} = \text{argmin}\, f_T(O(x,y), \vec{R}_i^s, PP_{\vec{R}_i}, V, G)$

Since the number of hop is known, the quantity of all possible directed sequence from H is $\binom{n}{s} s!$. So the computational complexity is $O(\binom{n}{s} s!) = O(n^s)$, which is acceptable in practical computation due to s is a constant. In this paper, we focus on the FAST-HC problem.

3 Online Recommender System

In this section, we design an online recommender system for searching the fastest-path. First, we derive the function $f_T(\cdot)$. Second, a pruning method is developed. Third, we propose the process of this mobile recommender system.

3.1 Potential Travel Time Function

We design a potential travel time (PTT) function to describe $f_T(\cdot)$ in detail, which is used to compute the travel time of a given path. Fig. 2 illustrates three directed sequence from $O(x,y)$ to H_3: the first driving path $\overrightarrow{R_1}$ is $O(x,y) \to H_2 \to H_3$, the second driving path $\overrightarrow{R_2}$ is $O(x,y) \to H_1 \to H_2 \to H_3$, and the last path $\overrightarrow{R_3}$ is $O(x,y) \to H_1 \to H_3$, where $h_{\overrightarrow{R_1}} = 2$, $h_{\overrightarrow{R_2}} = 3$, and $h_{\overrightarrow{R_3}} = 2$. PTT function computes the potential time cost in these three paths, and then the system selects the path that has the minimal value as the fastest-path.

When a taxi moves along a path $\vec{R}_i^s$, it may pick up passengers at each hot terminals H_i with a probability of $PP(H_i)$. In Fig.2, if a taxi select the path $O(x,y) \to H_2 \to H_3$, its pick-up probability in H_2 is $PP(H_2)$. Moreover, its pick-up probability in H_3 is $\overline{PP(H_2)}PP(H_3)$, which is a conditional probability that the pick-up event does not happen in in H_2 but in H_3, where $\overline{PP(H_1)} = 1 - PP(H_1)$.

The travel time from one hot terminal H_i to another H_j is denoted as T_{ij}. Since the any possible length between H_i and H_j can be got from map and the velocity of each road V is obtained as the same method in [7], the travel time T_{ij} is the one who has the minimal time in all possible paths from H_i to H_j by their lengths dividing the V of all included roads. As shown in Fig. 2, T_{12} presents the travel time from H_1 to H_2, which can be treat as the weight of edge between H_1 and H_2.

It is possible that a taxi takes no passengers after moving through the recommended path. Since the FAST-HC problem has limited the number of hops, the travel time beyond the last hot terminal is assumed to be T_∞ for all candidate paths. In the sample path $\overrightarrow{R_1}$: $O(x,y) \to H_2 \to H_3$ of Fig.2, the distribution of the travel time before next pick-up can be represented by two vectors: $T_{\vec{R}_1^2} =< T_{02}, (T_{02} + T_{23}), T_\infty >$ and $PP_{\vec{R}_1^2} =< PP(H_2), \overline{PP(H_2)}PP(H_3), \overline{PP(H_2)PP(H_3)} >$. Thus, the PTT function f_T is defined as:

$$f_T = T_{\vec{R}_i^s} \cdot PP_{\vec{R}_i^s} , \tag{1}$$

where $\cdot$ is the dot product of two vectors. In order to note easily, we consider using a vector *TP* for presenting the travel time and pick-up probability of a selected path, where a *TP* has $2s$ elements. For instance, the path $\overrightarrow{R_1}$ in Fig.2, $s = 2$, it *TP* is $< T_{02}, PP(H_2), T_{23}, PP(H_3) >$,who has 4 elements.

Lemma 1: The potential travel time $f_T(TP)$ is monotonically increasing with the increasing of any element.

Proof 1: First, $f_T(TP)$ is a polynomial formula. Second, the value of each element is positive. Third, the coefficient of each element is positive. It is easy to get that the derivation of $f_T(TP)$ is always positive. Thus, the increasing monotone property of $f_T(TP)$ is proved.

3.2 Dominated Pruning Method

As proof in Section 3, the complexity is unacceptable if the brute-force method is adopted to compute the f_T for all the hot terminals sequence with size s. However, from Lemma 4.1, we know that the PTT function strictly monotonically increasing with each element in TP. Thereby, plenty of path sequences can be pruned to reduce the complexity according to the monotone property.

Suppose there are two paths $\vec{R}$ and $\vec{R}'$, who have the same source point, the same destination point, as well as the same number of hops s. Representing them as two $2s$ elements vectors TP and TP'. Iff $\exists 1 \le m \le 2s$, $TP_{\mathrm{m}} < TP'_{\mathrm{m}}$ and $\forall 1 \le m \le 2s$, $TP_{\mathrm{m}} \le TP'_{\mathrm{m}}$. Hence, we consider that the path $\vec{R}$ dominates the path $\vec{R}'$, which means $\vec{R}'$ can be pruned from the candidate paths. It is so called domination pruning method.

For example, we compare two paths $\overrightarrow{R_1}$ and $\overrightarrow{R_3}$ in Fig. 2. The corresponding vectors are $\langle T_{02}, PP(H_2), T_{23}, PP(H_3)\rangle$ and $\langle T_{01}, PP(H_1), T_{13}, PP(H_3)\rangle$ for TP_1 and TP_3 respectively. If $\overrightarrow{R_1}$ dominates $\overrightarrow{R_3}$ after computing, $\overrightarrow{R_3}$ is pruned.

3.3 Recommendation System

We develop a system to recommend the fastest-path to taxi drivers to help them pick up passengers as soon as possible. This system operates in a client/server pattern. When a taxi wants to know where to drive, it sends a request to the server through wireless communication. The server computes the fastest-path and delivers it to the client quickly.

The distributed method on every taxi to search the fastest-path is feasible according to our analysis. However, we argue that the client/server pattern is more efficient due to the resource problem. Since the traffic situation is time varying, lots of data need to be gathered and computed for real-time path recommendation. If every taxi collects all others' data, it costs many computation and storage resource, in addition, the wireless bandwidth. Nevertheless, a server such as cloud computing could provide strong computation and storage resource. And the current infrastructure such as cellular network base station or wireless AP could help the server to gather all clients' data.

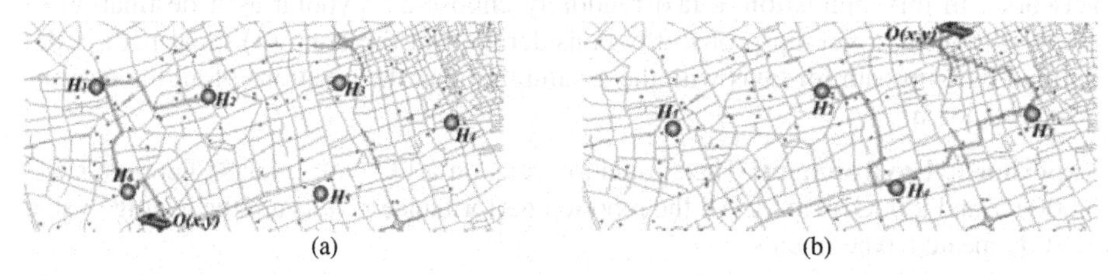

(a) (b)

Fig. 3. A taxi drive along the fastest-path recommendation and picks up passengers quickly in different parameter setting. (a) n=6, s=3; (b) n=4, s=3

Table 3. Pseudo Code of Fastest-Path Recommendation

```
Algorithm FastestPathRecommendation(H,PP,T,s,G, O(x,y))
Input:
        H:  the set of hot terminals generated by K-MEANS
        PP:  the probability set of hot terminals
        T: the set of travel time between any two hot terminals
        s: the number constraint of hops.
        G: the map graph
        O(x,y): current position of  a taxi.
Output:
        FP→ : the fastest-path including a list of road sequence.
Online operation:
1. WHILE hear the request from any taxi DO
2. Enumerate all path candidates with hops of s from the position O(x,y) of
   this taxi to hot terminals in H.
3. Use domination pruning algorithm to prune the computational complexity of
   path candidates.
4. Calculate the f_T() of each path candidates
5. Select the path with min(f_T) as the FP→.
6. Send FP→ to the taxi
```

After selecting the client/server pattern for recommendation system design, the preparation of operation on server part is to mine the collected historical data for map rebuilding, hot terminals division with a given value of *n*, pick-up probability computation of every hot terminal, travel time computation through road velocity, and set the value of parameter *s*. The online operation on the server part is shown in Table 3.

4 Performance Evaluation

4.1 Experimental Setting

Path Comparison: four kinds of recommended paths are compared in this simulation. (1) The fastest-path is computed as the method in Section 4; (2) the shortest-path is the path with the shortest length from $O(x,y)$ to hot terminals, which is got according to [2]; (3) Random waypoint is widely adopted mobility model in vehicle networks. In this simulation, a taxi randomly choose a waypoint as it destination to move without the pick-up probability consideration in the area; (4) Ideal path is the optimal pick-up path for taxis with the assumption that the positions of all the passengers are known.

Experiment Environment: The simulation runs on a PC with Intel(R) Pentium CPU 3.00GHz and 2.00GB RAM. All the reported performance results are the average value of 100 repeated experiments.

4.2 Performance Comparison

It is an illustration to show the effect of our fastest-path recommender system. When an empty taxi just enters the area of interest, it sends its position and the request to our system and gets the response of a fastest-path shown in Fig. 3. If this taxi drives along this recommended path, it could pick up passengers quickly. Even this taxi miss the first passenger, it will meet many other passengers along the recommended path.

In Fig. 4, the travel time is defined as the time that a taxi travels from its source position $O(x,y)$ to the position that it meets the first passenger. The performance is simulated when the number of hot terminals is set $n = 6$ and the number of hops is constrained by $s = 3$. It is obvious when the number of passengers is increasing in the 22 km^2 area, the travel time of all kinds of paths are decreasing. The reason is that the higher density of passengers leads to the higher pick-up probability. We can find that the performance of fastest-path is always better than shortest-path and random waypoint path. The important observation in Fig. 4 is when there are more than 200 passengers in the 22 km^2 area, the travel time of fastest-path is the same of the ideal path, which verifies the efficiency of this novel recommender system.

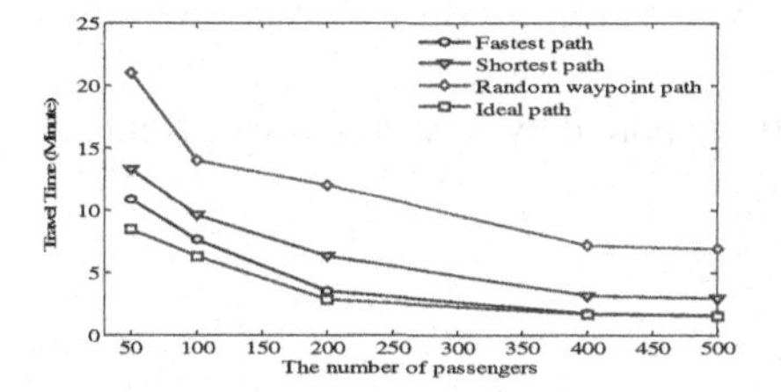

Fig. 4. Travel time comparison among 4 recommended paths varying with the passenger density, when *n=6, s=3*

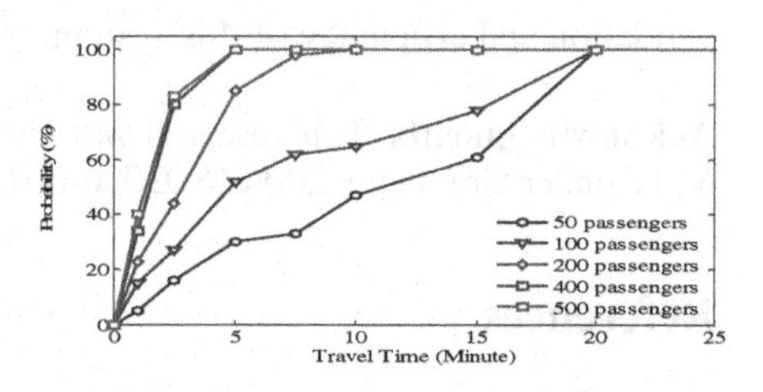

Fig. 5. Travel time CDF with different passenger density when using the fastest-path for picking-up, when *n=6, s=3*

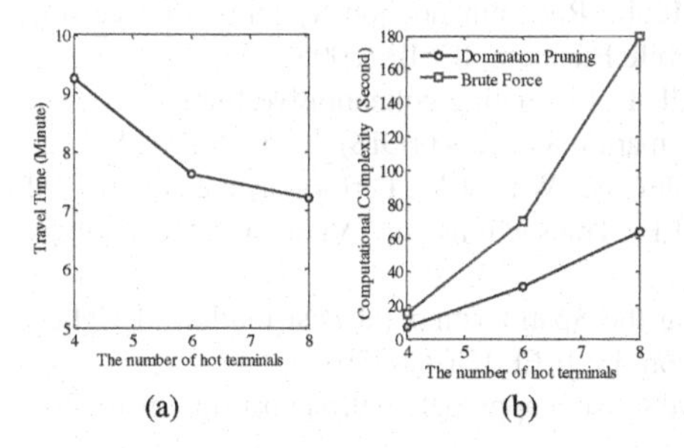

Fig. 6. The travel time of picking-up and the computational complexity affected by the different number of hot terminals *n,* when using the fastest-path recommendation with 100 passengers in the area and *s=3*

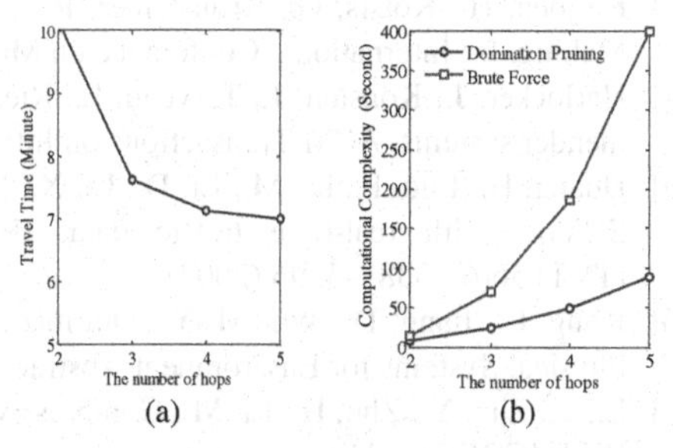

Fig. 7. The travel time of picking-up and the computational complexity affected by the different number of hops *s,* when using the fastest-path recommendation with 100 passengers in the area and *n=6*

Fig. 5 is a CDF varying with travel time under fastest-path recommendation. Then, Fig. 6 and Fig. 7 provide the impact of two parameters *n* and *s* on the travel time performance and computational complexity on the server.

Fig. 6 exhibits the performance results when the number of hot terminals is varying from 4 to 8. In Fig. 7, we display the performance curves when the number of hops is varying from 2 to 5.

5 Conclusion

In this paper, we propose and develop a fastest-path recommendation system that can help a taxi to pick up passengers as soon as possible. Through data mining, we divide the hot terminals in the area of interest and design the method to compute the pick-up probability. Then, we propose a potential travel time function for evaluating the optimal fastest-path. Finally, the performance of this system is simulated based on a real trace data including about 2000 taxis in a 22 square kilometers area in 7 days in Shanghai. The recommended fastest-path shows the travel time is nearly equal to the minimum one when more than 200 passengers are in this area, which verifies the validation and efficiency of this system.

Acknowledgments. This research was partially supported by Key Program of National MIIT under Grants no. 2009ZX03006-001.

References

[1] Cheverst, K., Davies, N., Mitchell, K., Friday, A.: Experience of developing and deploying a context-aware tourist guide: the GUIDE project. In: ACM MOBICOM (2000)
[2] Ge, Y., Xiong, H., Tuzhilin, A., Xiao, K.: An energy-efficient mobile recommender system. In: ACM KDD (2010)
[3] Heijden, H., Kotsis, G., Kronsteiner, R.: Mobile Recommendation Systems for Decision Making. In: International Conference on Mobile Business, ICMB (2005)
[4] Herlocker, J., Konstan, J., Terveen, L., Riedl, J.: Evaluating collaborative filtering recommender systems. ACM Transactions on Information Systems (TOIS) 22, 5–53 (2004)
[5] Huang, H., Luo, P., Li, M., Li, D., Li, X., Shu, W., Wu, M.Y.: Performance evaluation of SUVnet with real-time traffic data. IEEE Transactions on Vehicular Technology (TVT) 56(6), 3381–3396 (2007)
[6] Kong, L., Jiang, D., Wu, M.-Y.: Optimizing the Spatio-Temporal Distribution of Cyber-Physical Systems for Environment Abstraction. In: IEEE ICDCS (2010)
[7] Li, Z., Zhu, Y., Zhu, H., Li, M.: Compressive sensing approach to urban traffic sensing. In: IEEE ICDCS (2011)
[8] Resnick, P., Varian, H.R.: Recommender systems. Communications of the ACM 40, 56–58 (1997)
[9] Ricci, F.: Mobile recommender systems. International Journal of Information Technology and Tourism 12(3) (2010)

A Computer Dynamic Forensic Technique Based on Outlier Detection

Bin Huang, Wenfang Li, Deli Chen, and Junjie Chen

Electronic Information Engineering Department
Putian University
Putian, China
hb0219@126.com

Abstract. This essay firstly introduces a survey of computer dynamic forensics, and then proposes a computer dynamic forensics system model. Aiming at solving the problems that the computer dynamic forensics faces in the data analysis stage, we apply the Outlier Detection to maguuiimous data analysis in the computer dynamic forensics. We apply this technique on KDDCUP99 data set and get satisfactory results.

Keywords: Outlier, Data Mining, computer dynamic forensic.

Introduction

With the rapid development of the Internet, the gross volume of computer and Internet crimes is rising continuously. Therefore, the study of Internet security has become imminently important. Currently, the study has been mainly concentrated on defensive security, the core technologies of which include: firewall, intrusion detection, security loopholes scanning, virtual private network. However, with the means of Internet crimes keeping upgraded, it has no longer been effective to rely solely on Internet security technologies to combat them. Eventually, people need to resort to the force of the whole society and criminal laws to deal with them, making the computer forensic become necessary. The development of computer dynamic forensic is a symbol of the maturity of Internet defensive security theory. The computer forensic is the process of confirmation, protection, extraction, and filing of the electronic evidence, existing in computers and related peripheral equipment, that is accepted by the criminal courts, reliable and convincing enough. Just like traditional evidence, electronic must also be truthful, accurate, integrated, conforming to laws and regulations, and acceptable by criminal courts. According to different timings of the occurrence of the computer forensic, it is categorized into an ex post facto static forensic and beforehand dynamic forensic. An ex post facto static forensic refers to the applying of various technological means to analyze and extract evidence after the computer has been intruded. The success of it highly depends on the scene the computer criminals have left behind. It deals with the computer crimes passively and is not very reliable. The upgrade of Internet crimes technologies has made this forensic inadequate. The solution is to carry out the real-time dynamic forensic [1].

Y. Wang and T. Li (Eds.): Knowledge Engineering and Management, AISC 123, pp. 349–356.
springerlink.com

1 A Computer Dynamic Forensic Technique Based on Data Mining

1.1 The Definition of Dynamic Forensic

The computer dynamic forensic is the process of integrating the forensic into firewall and intrusion detection to obtain and analyze the real-time data of all possible computer crimes, intelligently analyzing the attempts of intruders, taking measures to cut off links or to lure the intruders to take further actions, so long as the system security is intact, in order to obtain maximum data, and thereafter appraising, saving, and submitting the evidence [2,3].

1.2 The Use of Data Mining in Dynamic Forensic

Data mining is the process of specifically applying of data analysis to extract the maximum hidden information from data that contains a great amount of redundant information, in order to provide the base of forming correct judgment. As it is characterized by having high automation, data mining has been frequently used in the field of intrusion detection, which is closely related with the field of computer forensic, to intelligently process massive data of security appraising, for the purpose of abstracting feature models that are convenient to determine and compare data[4].

In the field of computer crimes and forensic, data mining has been gaining increasing attention of researchers because its traits fitly solve the problems faced by computer forensic.

In China, there are few reports of researches of the application of data mining in computer forensic, as data mining is mainly directed at the field of intrusion detection. However, with the further development of the study, researchers are beginning to cast their eyes upon data mining. And as the need of dynamic forensic in computer forensic is brought forward, the study of intrusion detection and the study of computer forensic are highly complementary and closely relevant. The application of data mining in computer crimes and forensic would inevitably become a new focus of researchers [5].

The system of multi-intelligent surrogate dynamic forensic based on data mining consists of subsystems of data obtaining, data mining, data analyzing, evidence appraising, evidence saving, and evidence submitting. Each module is constructed by intelligent surrogate technique. The communication and cooperation among module surrogates and within each module surrogate is supported by intelligent surrogate accessing technology to accomplish the process of dynamic forensic. The model of the system is illustrated in Figure 1.

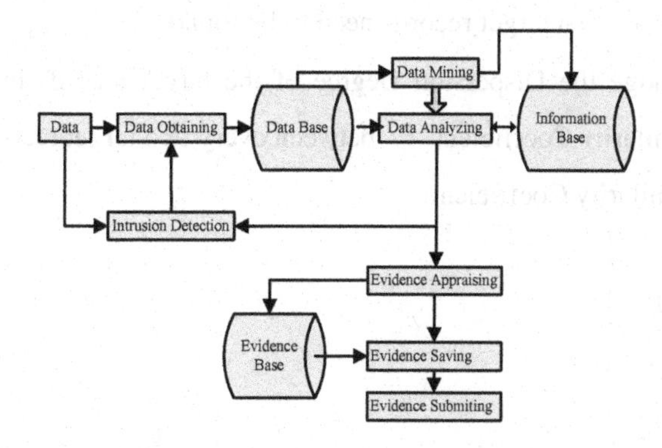

Fig. 1. A Computer Dynamic Forensic Model based on Data Mining

There are many ways of data mining, some of which frequently used in dynamic forensic are: clustering analysis, connection rules, sequence analysis, and outlier mining. The main purpose of adopting outlier analysis is to find out some user's behavior patterns anomaly by analyzing the outlier phenomenon in the data. While these "outlier" and "anomaly" behavior patterns are usually ignored in connection rules and sequence analysis or are considered as "tiny" pattern that could be abandoned, they are "interesting" and "useful" information in computer forensic because they indicate the irregularity of the user's behavior, which usually carries with it some purpose, which is worthwhile for detectives to pay attention to. Therefore, in the module of data mining, we introduce the outlier mining, presenting a computer dynamic forensic technique based on outlier detection.

2 Computer Forensic Applying the Outlier Mining Based on Similar Coefficient Sum

2.1 The Outlier Mining Algorithm Based on Similar Coefficient Sum

The Outlier Mining based on Similar Coefficient Sum is described as follows[6]:

Given domain $X = \{x_1, x_2, \cdots, x_n\}$ as the target record of detection and each target record has m indexes, i.e. $x_i = \{x_{i1}, x_{i2}, \cdots x_{i3}\}$ $i = (1,2,\cdots n)$,the data matrix is denoted as:

$$X = \begin{bmatrix} x_{11}x_{12} \cdots x_{1m} \\ x_{21}x_{22} \cdots x_{2m} \\ \cdots\cdots\cdots\cdots\cdots \\ x_{n1}x_{n2} \cdots x_{nm} \end{bmatrix} \quad (1)$$

Now the Outliers set in n target records need to be found.

In order to judge the Dispersion Degree of the target records in X, we need to calculate the Similarity Coefficient r_{ij} between every two target records, thus to form the matrix of Similarity Coefficient:

$$R = \begin{bmatrix} r_{11} r_{12} \cdots r_{1n} \\ r_{21} r_{22} \cdots r_{2n} \\ \cdots\cdots\cdots\cdots \\ r_{n1} r_{n2} \cdots r_{nn} \end{bmatrix} \quad (2)$$

in which $r_{ij} = 1 - \sqrt{\frac{1}{n}\sum_{k=1}^{m}(x_{ik} - x_{jk})^2}$ (3)

Let $P_i = \sum_{j=1}^{n} r_{ij}$ (4)

p_i is the sum of the ith row of the matrix of Similarity Coefficient. The smaller p_i is, the farther the target record in the row i is from other records, and therefore becoming as the candidate of Outliers Set.

$$\lambda_i = \frac{p_{\max} - p_i}{p_{\max}} \times 100\% \quad (5)$$

λ denotes the threshold. Members conforming to $\lambda_i \geq \lambda$ form the Outliers Set.

2.2 Forensic Based on Similar Coefficient Sum

In the forensic module, we use outlier detection to obtain anomaly data, which is crucial to the success of computer forensic and may become the important breaking point in the cracking of computer crimes cases.

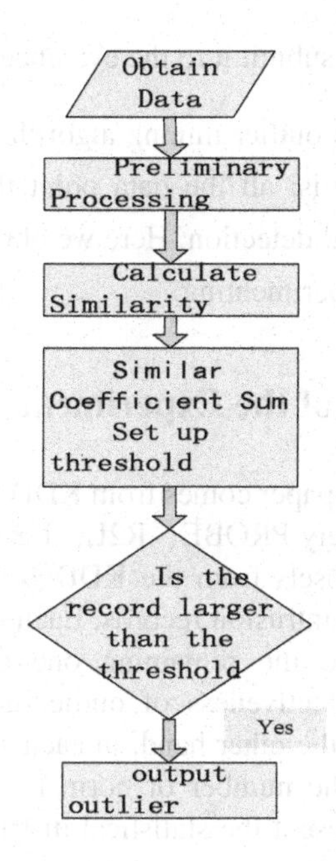

Fig. 2. Process of Computer Dynamic Forensic based on Outlier Mining

We will apply the outlier mining algorithm based on similar coefficient sum to dynamic forensic, as illustrated in Figure 2. The detailed processes are as follows:

1) Input data set: given the matrix with n rows and m columns denotes each record set of n intrusion detections of initial network connections has m characteristic properties.
2) Preliminary process of data: Standardize data set to make the value of each characteristic property ranges from 0 to 1 and let the initial value of the incidence matrix be 0.
3) Calculate similarity degree: calculate the similarity coefficient r_{ij} between every two records of n network connections, thus form the matrix of similarity coefficient.
4) Calculate Similar Coefficient Sum: get p_i, the sum of the ith row of the matrix of similarity coefficient. The smaller p_i is, the farther the record in the row i is from other records, and therefore becoming as the candidate of outliers set. Let the threshold be $\lambda_i = \frac{p_{\max} - p_i}{p_{\max}} \times 100\%$, where λ is the threshold, and members conforming to $\lambda_i \geq \lambda$ form the outlier set.

5) Output the outliers set and submit it to the evidence base.

There is a parameter λ in the outlier mining algorithm based on similar coefficient sum. The definition of outlier is: all the data point that is larger than λ, which is usually determined by empirical detection. Here we obtain a proper λ in the Internet connection data set through experimenting.

3 Result and Analysis of the Experiment

The data set we adopted in this paper comes from KDD Cup 1999[7], which includes 4 major types of intrusions, namely PROBE、R2L、U2R and DoS. During the process of analyzing, we choose 5 subsets from the KDD99 data set. Each subset contains 10000 normal records and 200 intrusion records. Each of the first 4 subsets contains a distinctive type intrusion, and the remaining one contains mixed intrusion. By arranging in this way, the effectiveness of outlier detection of different types of intrusions could be tested. On the other hand, in each subset, the number of intrusion records are far smaller than the number of normal connection records, so that the experiment can truthfully represent the statistical distribution of proper behavior and anomaly behavior on the Internet.

Not every property in the KDD set contributes to the result of detection. Srinivas Mukkamala and other people have used SUM in experiment to point out that there are 13 most important properties in the KDD Cup 1999 data set[8]: duration、src_bytes、dst_bytes、urgent、count、srv_count、same_srv_rate、dst_host_count、dst_host_srv_count、dst_host_same_srv_rate、dst_host_same_src_port_rate、protocol_type and service. We choose these 13 properties, standardize the 11 numerical properties, execute code mapping on character properties,i.e. service and protocol_type and then apply Principal Components Analysis on the latter two to generate a new data set[9], and finally the results of forensic of this new set by applying the outlier mining algorithm based on similar coefficient sum are as follows:

Table 1. Result of Computer Dynamic Forensic Experiment

Intrusion Type / Para. Range	DOS		R2L		U2R		PROBE		Data Set	
	DR	FPR	DR	FPR	DR	FPR	DR	FPR	DR	FPR
λ=8	100%	3.57%	54.1%	4.20%	38.1%	5.52%	100%	4.18%	10000	200
λ=9	100%	2.14%	54.1%	2.46%	28.6%	4.49%	100%	2.34%	10000	200
λ=10	100%	1.02%	50%	1.23%	19.1%	3.58%	100%	1.63%	10000	200
λ=11	0	0	37.5%	0.20%	14.3%	2.04%	95%	0	10000	200
λ=12	0	0	37.5%	0	4.76%	1.43%	0	0	10000	200

From the result of the experiment we can see that when λ is 10, the detection rates and fault detection rates of the first four types of intrusions are most desirable, while those of the fifth type of data sample, the mixed intrusion data set are 90% and 1.63% respectively.

From the above experiment results it can be found that the forensic based on outlier mining accomplishes considerably high success rate when applied to DOS and PROBE intrusions, but accomplishes undesirable results when applied to R2L and U2R intrusions. This is basically consistent with what we have predicted. By analyzing we find that there are mainly two reasons: (1)Unlike DOS and PROBING which launch many links to some host computers, R2L and U2R intrusions, on the contrary, are embedded into packages, which normally only exist in one link, making it impossible for it's traits to be adequately portrayed and rendering us to resort to other more effective ways. (2) The types of Probe and DoS intrusions are relatively more stable. On the other hand, the types of R2L and U2R intrusions are various, and even some types of intrusions are disguised as legitimate user account□ making their traits more homogeneous to the normal data packages and therefore difficult to be detected by forensic algorithm. In the KDD99 contest□ the champion approach's detection rates of the connection data of these two types intrusions are 13.2% and 8.4% respectively. In the light of this, there is a certain degree of superiority in the approach presented in this essay.

4 Summary

This essay has analyzed the computer dynamic forensic based on data mining by applying the outlier detection based on similarity coefficient sum into it. The process of the forensic has been presented. The emulational experiment of the forensic based on outlier detection has been carried out within the KDD99 experiment data set to prove that the success rate of this approach is relatively desirable, especially when applied to DoS and PROBE intrusions. The results of the experiments have shown that outlier detection could be used to accomplish computer forensic, and when the anomaly data is far smaller than the proper data, the results are more satisfactory, which is encouraging as the statistical distribution of anomaly behaviors and proper behaviors of the Internet data is basically in accordance with the conditions required by the carrying out of outlier mining under most circumstances.

Acknowledgment. This work was supported by the Pu Tian Technology ProjectsNO.2011G04 (2) and NO.2009G26.

References

[1] Wang, L., Qian, H.-L.: Computer Forensics and Its Future Trend. Journal of Software 9, 1635–1644 (2003)

[2] Guo, J.-C.: Application and Research of Computer Forensics Technology. Lanzhou University (2007)

[3] Jiang, Z.-Y.: The Research of Dynamic Network Intrusion Forensics Based on Multi-Agent. Jiangnan University (2006)

[4] Han, J., Kamber, M.: Data Mining Concepts and Techniques. Morgan Kaufmann Publishers (August 2000) ISBN 1-55860-489-8H

[5] Huang, L.: The Application of Data Mining Theory in Security Audit Analysis. Microcomputer Information (27) (2007)

[6] Jiang, L.-M.: Clustering Algorithm to Check Outlier Based on Similar Coefficient Sum. Computer Engineering 7, 183–185 (2003)

[7] KDD99 Cup dataset [DB/OL] (1999), `http://kdd.ics.uci.edu/databases/dcup99/kddcup99.html`

[8] Mukkamala, S., Janoski, G., Sung, A.H.: Intrusion detection using support vector machines and neural networks. In: Proc. of the IEEE Int'l Joint Conf. on Neural Networks, pp. 1702–1707 (2002)

[9] Yang, W., Tianshu, H., Guangyu, D.: An Anomaly Detection Method Based on Clustering and Principal Component Analysis. Computer Engineering and Applications (21) (2006)

Sentence Compression with Natural Language Generation

Peng Li and Yinglin Wang*

Department of Computer Science and Engineering,
Shanghai Jiao Tong University
{lipeng,ylwang}@sjtu.edu.cn

Abstract. We present a novel unsupervised method for sentence compression which relies on a Stanford Typed Dependencies to extract information items, then generates compressed sentences via Natural Language Generation(NLG) engine. An automatic evaluation shows that our method obtains better results. We demonstrate that the choice of the parser affects the performance of the system.

Keywords: Sentence Compression, Natural Language Generation.

1 Introduction

In recent years, there has been much interest in the task of redundancy information detection in text. In this paper, we study the task of sentence compression which is to transform long complex sentence to multiple grammatical correct short sentences. There are many applications which would benefit from a robust compression system, such as summarize news events, compression for mobile devices with a limited screen size. In summarization scenario, sentence compression are a common way to deal with redundancy.

In recent years, there has been so many works[3,1] to sentence compression. Many explicitly rely on a language model to produce grammatical output. Others use hand-crafted rules to prune tree of the sentence.

In this paper we present a novel unsupervised English grammatical approach (See Table 1) to sentence compression which is motivated by the belief that the natural language generation engine can work well in sentence compression. In particular, given a information item which contribute important information to the content of the sentence, we want to rebuild the sentence. Instead of a tree pruning approach, hand-crafted rules may not cover all cases, our approach can generate meaning sentences, the main requirement being that there are a dependency parser. We test our approach on English data sets and obtain better results.

2 Information Item Extraction

In [2], they define information items as subject-verb-object triple. The question is how to find verbs. Instead of defining some pruning rules, we use English grammatical

* Corresponding author.

Y. Wang and T. Li (Eds.): Knowledge Engineering and Management, AISC 123, pp. 357–363.
springerlink.com

Table 1. An example of sentence compression

Original Sentence: The Cypriot airliner that crashed in Greece may have suffered a sudden loss of cabin pressure at high altitude, causing temperatures and oxygen levels to plummet and leaving everyone aboard suffocating and freezing to death, experts said Monday. **Information Items:**

1. airliner – crash – null
2. airliner – suffer – loss
3. loss – cause – null
4. loss – leave – null

Generated Sentences:

1. A Cypriot airliner crashed.
2. A Cypriot airliner may have suffered a sudden loss of cabin pressure at high altitude.
3. A sudden loss of cabin pressure at high altitude caused temperatures and oxygen levels to plummet.
4. A sudden loss of cabin pressure at high altitude left everyone aboard suffocating and freezing to death.

Original Sentence: At least 25 bears died in the greater Yellowstone area last year, including eight breedingage females killed by people. **Information Items:**

1. bear – die – null
2. person – kill – female

Generated Sentences:

1. 25 bears died.
2. Some people killed eight breeding-age females.

relations defined by Stanford Typed Dependency Manual[1]. We first recognize possible information items(see Algorithm 1), then filter information items(see Algorithm 2).

3 Sentence Generation

We generate a new sentence for each information item. To help other people to follow this approach, we give more implementation details as a complement of the work originally proposed by [2]. Our procedure can simplify the generation process and make it easier to understand the whole generation process. Algorithm 3 shows the procedure that generate compressed sentences. The key subroutine is generate Noun Phrase (See Algorithm 4) and generate Verb Phrase(See Algorithm 5)

[1] `nlp.stanford.edu/software/dependencies_manual.pdf`

Algorithm 1. Recognize Information Items

```
EnglishGrammaticalStructure eg
SET predicates
Collection typedDependencyCollection
for all td in typedDependencyCollection do
   TreeGraphNode gov = td.gov()
   GrammaticalRelation gr = td.reln()
   if gr = "nsubj" or gr = "dobj or gr = "xcomp" or gr = "agent" then
      predicates.add(gov)
   end if
end for
List tmpItems
Set subjects, objects
for all n in predicates do
   subjects = getSubjects(eg,n)
   if subjects.size() == 0 then
      continue
   end if
   for all s in subjects do
      objects = getObjects(eg,s)
      if objects.size()! = 0 then
         for all o in objects do
            item = new InformationItem(s,n,o)
         end for
      else
         item = new InformationItem(s,n,null)
      end if
      tmpItems.add(item)
   end for
end for
return tmpItems
```

Algorithm 2. Filter Information Items

```
Require: Run Algorithm 1 get tmpItems
  List filteredItems, objs
  TreeGraphNode subj, obj
  for all item in tmpItems do
    subj = item.getSubject()
    obj = item.getObject()
    if !predicates.contains(subj) and obj! = null then
      objs.addobj
      items.add(item)
    end if
  end for
  for all item in tmpItems do
    TreeGraphNode s = item.getSubject(), p, o
    InformationItem newItem
    if predicates.contains(s) then
      for all TreeGraphNode obj in objs do
        p = item.getPredicate()
        o = item.getObject()
        newItem = new InformationItem(obj, p, o)
        items.add(newItem)
      end for
    end if
  end for
  return filteredItems
```

Algorithm 3. Generate compressed sentence

```
Require: build DependencyGraph
  SPhraseSpec newSent
  NPPhraseSpec subjectNp
  VPPhraseSpec vp
  TreeGraphNode s ,p
  for all item in filteredItems do
    s = item.getSubject()
    subjectNp = generateNP(graph, s)
    newSent.setSubject(subjectNp)
    p = item.getPredicate()
    vp = generateVP(graph, p)
    newSent.setVerbPhrase(vp)
  end for
```

Algorithm 4. Generate Noun Phrase

```
NPPhraseSpec np, tmpNp;
PPPhraseSpec pp;
Stack stack
stack.add(head)
while !stack.isEmpty() do
   children = grpah.adj(head);
   for all td in children do
      GrammaticalRelation gr = td.reln()
      if gr = "prep" then
         pp = generatePrepP(td)
         np.setPostModifier(pp)
      else if gr = "nn" or gr = "conj" then
         tmpNp = generateNP(graph, td.dep())
         np.setPostModifier(tmpNp)
      else if gr = "det" or gr = "num" or gr = "amod" then
         np.setPostModifier(td.dep())
      else
         continue
      end if
   end for
end while
```

Algorithm 5. Generate Verb Phrase

```
VPPhraseSpec vp
NPPhraseSpec dirobjNp, indirObjNp
vp.sertVerb(verb)
if object! = null then
   dirobjNp = generateNP(graph, object)
   vp.setObject(dirobjNp)
   children = grpah.adj(verb);
   for all td in children do
      GrammaticalRelation gr = td.reln()
      if gr = "iobj" then
         indirObjNp = generateNP(graph, td.dep())
         vp.IndirectObject(indirObjNp)
         break
      end if
   end for
else
   for all td in children do
      GrammaticalRelation gr = td.reln()
      if gr = "ccomp" then
         vp.setPostModifier(complement)
         break
      end if
   end for
end if
```

Table 2. Average results on the English corpus

	F–measure	compr.rate
Our Method (Stanford Parser)	53.6%	70.4%
Our Method (RASP)	50.2%	67.8%
GOLD	–	72.1%

4 Evaluation

4.1 Data

We use English Compression Corpus, it is a document-based compression corpus from the British National Corpus and American News Text Corpus which consists of 82 news stories[2].

4.2 Quality of Compression

We evaluate the results automatically as well as with human subjects. To assess the performance of the method on the English data, we calculate the F–measure on grammatical relations. Following [4], we calculate average precision and recall as the amount of grammatical relations shared between the output of our system and the gold standard variant divided over the total number of relations in the output and in the human-generated compression respectively. According to [1], this measure reliably correlates with human judgements. We also calculate the compression rate for the gold standard ignoring punctuation. Table 2 give the evaluation results.

5 Conclusions

We presented a new compression method which compresses dependency trees with Natural Language Generation engine. The method is unsupervised and can be easily adapted to languages other than English. It does not require elaborated hand-crafted rules to decide which arguments can be pruned. We demonstrated that the performance of the system depends on the parser. The results indicate that the approach is an alternative to the hand-crafted ruled-based approaches.

Acknowledgments. This work was supported by the National Natural Science Foundation of China (NSFC No. 60773088), the National High-tech R&D Program of China (863 Program No. 2009AA04Z106), and the Key Program of Basic Research of Shanghai Municipal S&T Commission (No. 08JC1411700).

[2] `http://homepages.inf.ed.ac.uk/s0460084/data`

References

1. Clarke, J., Lapata, M.: Models for sentence compression: A comparison across domains, training requirements and evaluation measures. In: Proceedings of the 21st International Conference on Computational Linguistics and the 44th Annual Meeting of the Association for Computational Linguistics, pp. 377–384. Association for Computational Linguistics (2006)
2. Genest, P.E., Lapalme, G.: Text generation for abstractive summarization. In: Proceedings of the Second Text Analysis Conference, National Institute of Standards and Technology, Gaithersburg (2010)
3. Jing, H.: Cut-and-Paste Text Summarization. Ph.D. thesis, Computer Science Department, Columbia University, New York, N.Y (2001)
4. Riezler, S., King, T.H., Crouch, R., Zaenen, A.: Statistical sentence condensation using ambiguity packing and stochastic disambiguation methods for lexical-functional grammar. In: Proceedings of the 2003 Conference of the North American Chapter of the Association for Computational Linguistics on Human Language Technology, vol. 1, pp. 118–125. Association for Computational Linguistics (2003)

Feature and Sample Weighted Support Vector Machine

Qiongsheng Zhang, Dong Liu, Zhidong Fan, Ying Lee, and Zhuojun Li

College of Computer and Communication, China University of Petroleum,
Qingdao, 266555, Shandong, China
287605365@qq.com

Abstract. In this paper, we analyzed the shortcoming of Feature Weighted SVM and Sample Weighted SVM, then a new SVM approach is proposed based on the comprehensive feature and sample weighted. This method estimates the relative importance (weight) of each feature by discernibility matrix. It utilizes the weights for computing the inner product in kernel functions. In this way the computing of kernel function can avoid being dominated by trivial relevant or irrelevant features. Then we estimate the weight of each training samples by the feature weight and similarity between samples, in order to reduce the influence of non-critical samples and noise data on the SVM learning and improve the noise immunity. Experimental results show that comparing with simply considering the importance of feature or sample, the proposed method can more effectively improve the classification accuracy of SVM.

Keywords: Support Vector Machine, Feature weighting, Sample weighting.

1 Introduction

The traditional classification algorithm of SVM assume that each features of the samples vector is equivalent for classification and each samples vector is equivalent for classification. It ignores the different influence of different features and different samples on the classification. In practical applications, this assumption may not established, at this time the classification accuracy will be bad.

To solve this problem, many researchers have proposed feature weighted SVM and sample weighted SVM. However, these methods consider the importance of feature or sample singly, not both of them. In the actual problem, considering importance of feature or sample unilaterally usually can not obtain accurate classification results.

According to this problem, we propose improved feature and sample weighted support vector machine, which combines feature and sample weighted support vector machine. This method estimates the relative weight of each feature by discernibility matrix. Then we estimate the weight of each samples through feature weight and similarity between samples. Finally we take the feature weight and sample weight into the SVM and get the feature and sample weighted classification model. The experimental results show that weighting method of considering the importance of feature or the sample unilaterally may not improve classification accuracy, and the method proposed in this paper can improve the SVM classification accuracy.

Y. Wang and T. Li (Eds.): Knowledge Engineering and Management, AISC 123, pp. 365–371.
springerlink.com

2 The Method of Feature Weighted

In the features of the sample vector, some features of strongly correlated with the classification, some features of weakly correlated with the classification, or uncorrelated. If we do not consider the different importance of different features for the classification, kernel function may be dominated by weakly related or not related to features. It will reduce the performance of SVM. In this paper, we measure the importance of different features for the classification by discernibility matrix in the theory of rough sets. We assign greater weight to the more important feature for the classification. The weight will be used in the calculation of kernel function.

2.1 The Introduction of the Theory of Rough Sets and Discernibility Matrix

The following is the definitions that need to be used in the theory of rough sets:

Definition 1: $\forall$ P, Q$\in$R, U is domain. R is all collections of features of U, posQ (P) =$\cup$Q (X), Q (X) is lower approximation set of the knowledge Q. Card (posQ (P)) represent the number of elements which belong to knowledge P according to Q in U. Card(posQ(P))/card(U) is defined the dependent degree of knowledge P relative to Q.

Definition 2: Discernibility matrix M = {m_{ij} } is defined as below

$$m_{ij} = \begin{cases} a\in C: f(x_i,a)\neq f(x_j,a), while\ , f(x_i,D)\neq f(x_j,D) \\ \text{null}\ , Otherwise \end{cases}$$

Here C is condition features set, D is category label. Category label also be called the decision feature of sample. The following is an example for discernibility matrix. Table 1 represents that there are six elements in domain, each element has four condition features called a, c, d, e. Category label is o. The feature value of each element is shown in table 1. According to definition 2, we get discernibility matrix in table 2:

Table 1. Value of elements

U	a	c	d	e	o
x1	0	1	1	1	1
x2	1	1	0	1	0
x3	1	0	0	1	1
x4	1	0	0	1	0
x5	1	0	0	0	0
x6	1	1	0	1	1

Table 2. Discernibility matrix

	x1	x2	x3	x4	x5	x6
x1	null	a,d	null	a,c,d	a,c,d,e	null
x2	a,d	null	c	null	null	null
x3	null	c	null	null	e	null
x4	a,c,d	null	null	null	null	c
x5	a,c,d,e	null	e	null	null	c,e
x6	null	null	null	c	c,e	null

2.2 Improved Discernibility Matrix

Due to the scale of discernibility matrix square grow with the number of sample increases, and it needs to ergodic discernibility matrix more than once in the subsequent operation. If matrix scale will be excessively large, the ergodic operation should be inefficient. And it will consume a lot of memory resources. According to the characteristic that discernibility matrix is symmetrical and the diagonal is empty, we improve it as is shown in table 3:

Table 3. Improved discernibility matrix

	m_{12}	m_{14}	m_{15}	m_{23}	m_{35}	m_{46}	m_{56}
a	1	1	1	0	0	0	0
c	0	1	1	1	0	1	1
d	1	1	1	0	0	0	0
e	0	0	1	0	1	0	1

In table 3, we only store the elements that its value is not null. Such as column m_{12}, the value of row a and row d is 1 and the value of row c and row e is 0. It means x_1 and x_2 belong to different categories, x_1 minus x_2 equals 1 in the features a and d. Compared with the original method, this method is at least 50% reduction in space complexity. It is better adapted to handle large-scale data problems.

2.3 The Calculation of Features Weight Based on Discernibility Matrix

Measuring importance of the feature can be considered from following factors: expert experience, correlation degree of feature for classification, the number of samples distinguished by the feature and the difference degree between samples in same feature.

Expert experience which expresses known knowledge is presented by experts.

Correlation degree of feature for classification can be measured by the dependent degree of knowledge between the feature and the decision-making feature (category label).We can calculate it by definition 1. The bigger is correlation degree of feature, the greater is the correlation between feature and category label.

The samples distinguished by the condition feature can be obtained by counting times that this feature value in the discernibility matrix is not 0. The more samples the condition feature distinguished, the more important is the feature. The same feature which has different discernibility degree in some pairs of samples will also have different effects on the classification. Therefore, we must consider the discernibility degree between samples in same feature as well as the sample number distinguished.

Suppose discernibility matrix has m column and n row. That is to say, it has n condition features. We also suppose that the weight of the feature i is W_i, the weight of expert experience is Ex_i, dependent degree of knowledge is K_i, the weight of feature i obtained from column j is F_{ij}, the difference degree of feature i between sample x and y is $Diff_i(x, y)$. Now we need an algorithm, which consider each of the factors. In table 3, x_1 and x_2 are distinguished by feature a and d. Therefore, feature a and d will totally get 1 unit weight. Feature a will get large part of the 1 unit weight, if the four factors of feature a are more important than feature d. We multiply Ex_i, F_{ij} and $Diff_i(x, y)$, the result is the standard of weight gained. Algorithm is as follows:

(1) Giving the expert weight of each condition feature by the experts, constructing discernibility matrix and computing the dependent degree of knowledge of each condition feature according to the above-mentioned method.

(2) In the discernibility matrix, each element divided by the maximum element of the row. The result is the difference degree of feature.

(3) Calculate the weight of each feature in each column by the formula 5 as follows:

$$F_{ij} = Ex_i * K_i * Diff_i(x,y) / \sum_{i=1}^{n} Ex_i * K_i * Diff_i(x,y) \tag{1}$$

(4) We can get the weight of the feature i by the formula 6 as follows:

$$W_i = \sum_{j=1}^{m} F_{ij} \tag{2}$$

2.4 The Method of Feature Weighting

According to the above-mentioned method, we get the weight value of each feature, then we construct feature weighting matrix P as follows:

$$P = \begin{bmatrix} w_1 & & & \\ & w_2 & & \\ & & \ddots & \\ & & & w_n \end{bmatrix}$$

P is a n rank opposite angles array, $P_{ii}=W_i$ represents the weight of feature i.

Definition 3: Making the k is the kernel function defined in the $X \times X$, $X \subseteq Rn$. P is the n rank feature weighting matrix. The number of dimensions of input space is n. The feature weighting kernel function k_p is defined as follows:

$$k_P(x_i, x_j) = k(x_i^T P, x_j^T P)$$

According to the definition 3, before computing the kernel function, the two samples will respectively multiply the feature weighting matrix, which carry out the weighting to the feature. In this way the computing of kernel function can avoid being dominated by trivial relevant or irrelevant features.

3 The Method of Sample Weighted

When we predict category of a sample, the different training samples have different importance for classification. In general, the more similar is between training sample and the test sample which will be classified, the more important the training sample is to the classification. In addition, when the SVM applied to some practical problems, data errors are inevitable. SVM is very sensitive to the noise data and isolated points of the training sample. We need to reduce the interference of the noise samples and isolated points. To solve this problem, we give a greater weight to the training samples with high similarity to the test sample, and give a smaller weight to the training samples with low similarity to the test sample in the classification.

Here we added a membership function in the conventional SVM sample. All samples are fuzzed by membership function. Fuzzy training set is expressed as follows:

$$D = \{(x_1,y_1, u_1),\ldots(x_l, y_l, u_l)\},\quad x\in Rn,\ 1\le i\le l\quad y\in\{+1,\ -1\}\quad 0\le u_i\le 1$$

Here u_i membership function of sample x_i, each sample has its u_i. For this training set, through Lagrange multiplier and kernel function, the problem that find the best separating hyperplane of SVM can be transformed into the dual problem as follows:

$$\max W(\alpha) = \sum_{i=1}^{l} \alpha_i - \tfrac{1}{2}\sum_{i=1}^{l} \sum_{j=1}^{i} \alpha_i Q(i, j)\alpha_j$$
$$s.t. \sum_{i=1}^{l} y_i\alpha_i = 0, 0 \le \alpha_i \le u_i * C, i = 1,\ldots, l \tag{3}$$

Punish parameter C is a constant, and u_i will be fuzzed C. u_i*C will set different punish parameter for each sample. It can be drawn from the formula 3 that the greater u_i*C is, the smaller the possibility of misclassification of sample x_i will be. To the important samples, like support vector, we set a greater value for its u_i. So the important samples for classification will be classified more exactly. To the noise data and isolated points, we set a smaller value for its u_i and u_i*C also will be small. It will greatly reduce the influence of noise data and isolated points for classification. By this way, we can reduce the interference of noise data and isolated points for classification, while the normal sample will not be affected in classification.

Membership function is the key to sample weighting algorithm. It must be able to objectively reflect the importance of each sample. Here membership function is the similarity between test sample and the training samples, calculated as follows:

(1) Normalize the training samples and the test sample which will be classified.

(2) The test sample vector minus one of the training samples vector, and calculate absolute value. The result is called the difference vector for the training sample vector.

(3) The difference vector multiply feature weighting matrix, and we stored it in the temporary vector. Then we sum each dimension of temporary vector, the result is called different degree.

(4) Repeat the step 2 and 3, until we get each different degree of training samples. Each different degree was normalized. The similarity between test sample and the training sample is 1 minus different degree which has been normalized.

4 Experimental Results and Analysis

4.1 Experiment Design

Experimental data was taken from the Iris, Wine and part of Ecoli database in the UCI website. Take three categories of data from each database for testing. Training samples account for ten percent, testing samples account for ninety percent, then make the cross-validation. The experiments used the radial basis kernel function. The parameters were set up as follows: C=1.0, gamma = 2.0. As the SVM solve the problem of two-class, three categories of data from each database were tested in pairs. As for each database, take class 1 and class 2, cass1 and cass3, class 2 and class 3 for the training samples to train the SVM classifier. In this four conditions that is weighting the sample and feature, only weighting the sample, only weighting the feature and without any weighting, to statistic the accuracy of the classification results.

4.2 Experiment Results

Table 4. Weighting the sample and feature

Database	Class1 and 2	Class1 and 3	Class2 and 3
Ecoli	94.44%	95.42%	97.41%
Iris	100%	100%	100%
Wine	97.72%	100%	97.19%

Table 5. Only weighting the sample

Database	Class1 and 2	Class1 and 3	Class2 and 3
Ecoli	94%	95%	97.41%
Iris	100%	100%	90%
Wine	94.01%	100%	94.39%

Table 6. Only weighting the feature

Database	Class1 and 2	Class1 and 3	Class2 and 3
Ecoli	52.45%	64.64%	59.48%
Iris	100%	100%	86.66%
Wine	54.7%	55.2%	58.87%

Table 7. Without any weighting

Database	Class1 and 2	Class1 and 3	Class2 and 3
Ecoli	94.14%	95%	96.55%
Iris	97.77%	98.88%	80%
Wine	54.7%	55.2%	58.87%

The result can be drawn from the Ecoli database that if only unilaterally consider weighting feature or weighting sample, the accuracy of classification may not be does improved. Compare with table 6 and table 7, in the condition of only weighting the feature, the classification accuracy of Ecoli has a great fall. Compare with table 5 and table 7, in the condition of only weighting the sample, the classification accuracy of Ecoli is that one is increase, one is lower and the other is unchanging. It means that the impact of sample weighted to the classification accuracy is uncertain. Compare with the Ecoli classification results from table 3 and table 7, if considering the feature and the sample weighted, the classification accuracy was definitely improved.

According to the description of UCI, class 1 of Iris is linear separable to class 2 and class 3, class 2 and class3 are linear inseparable. Compare with the classification results of the Iris database from table 4 to table 7, the conclusion can be drawn that sample weighted make a large improvement in the accuracy of classification for the case of linear inseparable, feature weighted is relatively small.

The result can be drawn from the Wine database that only sample weighted could make a large improvement to the classification accuracy, only feature weighted has no effect on the classification accuracy .Take feature weighted and sample weighted into account has higher accuracy compared with the single-weighted method.

The comprehensive conclusion can be drawn that only considering the feature weighted or the sample weighted will not necessarily improve the classification accuracy. In our method we take feature and sample weighted into account. It can obtain much higher accuracy than unweighted or single-weighted method.

5 Summary

In this paper, based on the traditional SVM algorithm, at first we designed two calculation methods to measure the importance of features and samples, then combined with the importance of the features and the importance of samples, we proposed a new support vector machine algorithm to consider both the sample weighted and the feature weighted. Finally, we prove the feasibility of this algorithm through the experiments. Machine learning algorithms in practical applications often need to handle the multi-class problems. In the future work, we will study how to make the improvement strategies applied to multi-class problem in practice.

References

1. Vapnik: The Nature of Statistical Learning Theory. Springer, NewYork (1995)
2. Pawlak Rough Sets: Theoretical Aspects of Reasoning about Data. Kluwer Academic Publishers, Boston (1991)
3. Lin, C.F., Wang, S.D.: Fuzzy support vector machines. IEEE Trans. on Neural Networks 13(2), 464–471 (2002)
4. Wang, T.-H., Tian, S.-F., Huang, H.-K.: Feature Weighted Support Vector Machine. Journal of Electronics & Information Technology (March 2009)
5. Chang, C., Lin, C.: LIBSVM: a library for support vector
6. Yang, S.-L., Ni, Z.-W.: Machine learning and intelligent decision support system. Science Press (2004)
7. Hu, J., Wang, G.-Y.: Uncertainty Measure Rule Sets of Rough Sets Pattern Recognition and Artificial Intelligence Vol. 10 (2010)
8. Baojian, Yiming, Fengjun: Text classification based on fuzzy support vector machine. Journal of Liaoning Technical University, Natural Science 10 (2010)

Research on Dictionary Construction in Automatic Correcting

Xueli Ren and Yubiao Dai

Department of Computer Science and EngineeringQujing Normal University
Qujing 655000, China
oliveleave@126.com, abiaodai@163.com

Abstract. It is a key technology in online examination system that automatic correcting. Semantic similarity is the main way to solve the auto-correcting, but the exact calculation depends on the words similarity of the student's response and accurate answers. They are the important reasons to decrease the accuracy of word segmentation that identify the new words (Glossary) and segmentation ambiguity problem. A new method to construct dictionary having glossary is proposed that the new words are identified by PAT array and ambiguities are eliminated by association rule mining. The accuracy of segmentation may achieve 95% and have 5% of increase in automatic correcting of computer organization.

Keywords: automatic correcting, PAT array, Support, Confidence.

1 Introduction

The rapid development of computer technology and the deepening of the reform of education and teaching, online testing with preventing subjective factors effectively improves the efficiency and reliability of tests and achieves "separation of teaching and testing, so it is more and more popular for an increasing number of teachers and students, which is important to the improve teaching effectiveness and teaching quality. Automatic correcting is not only a key technology in online examination system, but also a research focus in natural language processing. Currently, the technology to correct automatically of objective questions has been quite mature, but not an examination system is well correct subjective questions automatically because of answers' features and complexity. It is a difficult problem in online examination system that correct subjective questions automatically as many problems need to solve in technology which involve artificial intelligence, pattern recognition, natural language processing and so on. Many methods to correct subjective problems automatically have been researched. Automatic correcting of text entry in computer foundation examination and filing in the programming blank is researched using string matching by Wang Han et al [1], it may correct blank filled of C programming automatically. Score of subjective questions based on Chinese word segmentation is proposed by Zhenbo Zhou who research on automatic interpretation of the outlined problems in limited areas [2]. An algorithm to score subjective questions automatically is designed by Aiguo Meng who introduces the concept of one-way close-degree based on close

Y. Wang and T. Li (Eds.): Knowledge Engineering and Management, AISC 123, pp. 373–378.
springerlink.com

degree in fuzzy mathematics. Sentences similarity is involved in automatic scoring of subjective questions by Sidan Gao who compute sentences similarity using dynamic programming algorithm. Many breakthroughs are made in correcting subjective questions automatically by these methods, but the implementation of these methods are based on word segmentation of the standard answers and the answers, and methods to segment words are based on the general dictionary. Because it has a strong professional in the automatic correcting and the recognition rate of professional words is not high for general vocabulary dictionary, these affected the accuracy of correcting to a certain extent. An effective way to solve this problem is a dictionary constructed including specialized vocabularies which is studied in this paper.

2 PAT Array and Ambiguity Elimination

2.1 PAT Array

PAT-array called suffix array is a compact data structure to handle arbitrary-length strings and performs much powerful on-line string search operations. It has less space because it's supported by PAT-tree. Several important definitions are given to understand easily [3-5].

Definition 1. S' is a string in the character set $\sum$, in which $ is only unique terminator and less than all of the characters. Let LS=S'$ as a new string which is constructed through adding $ to then end of a string S' and called left suffix array. If |S| is the length of string and LS[i] is a character lied in position i, then Lsuffi=S[i]S[i+1]...S[|S|] is left suffix array of S that start at position i. For example, if S'=" aedfhg ", then LS=" aedfhg $" is a new string by adding $, LS1=" edfhg $" is the left suffix of S that start at position 1.By the method RS= $S' is a new string which is constructed through adding $ to the initial of a string S' and called right suffix array. If |S| is the length of string and RS[i] is a character lied in position i, then Rsuffi=S[i]S[i+1]...S[|S|] is right suffix array of S that start at position i. For example, if S'=" aedfhg ", then RS="$ aedfhg " is a new string by adding $, LS1="hfdea$" is the right suffix of S that start at position 1.

Definition 2. Longest Common Prefix(also known as LCP) The Suffix array Left-index[0..n] (Right-index[0..n]) is an array of indexes of LSi (RSi),where LSLeft-index[i] < LSLeft-index[j] (RSRight-index[i] <RSRight-index[j]), i<j, in lexicological order. Let LLCP[i] (RLCP[i]), i=0..n-1, as the length of Longest common Prefix (LCP) between two adjacent suffix strings, LSLeft-index[i] and LSLeftindex[i+1] (RSRight-index[i] and RSRight-index[i+1]). These arrays on both sides are assistant data structures for speeding string search.

With the string "upcfpsopcf" as an example□ it shows a simple Suffix array sorted on left and right context, coupled with the LCP arrays respectively. We apply the sort-algorithm proposed by Manber, which takes O(nlogn) in worst cases performance to construct the Suffix arrays, and sort all the suffix strings in lexicological order.

The conception of suffix array is explained using the string “vqdgqtpqdg”.

Table 1. Cconstruct suffix array

0	1	2	3	4	5	6	7	8	9	10	11
$	v	q	d	g	q	t	p	q	d	g	$

Table 2. Sorting suffix array by the initial character

	0	1	2	3	4	5	6	7	8	9	10	11
left	0	11	3	9	10	4	7	8	2	5	6	1
right	0	11	9	3	4	10	7	5	2	8	6	1

Table 3. Computing the Longest Common Prefix

	0	1	2	3	4	5	6	7	8	9	10	11
LLCP	0	0	2	0	1	0	0	3	1	0	0	/
RLCP	0	0	2	0	3	0	0	1	1	1	0	/

2.2 Ambiguity Elimination

Ambiguity elimination seriously affected accuracy of the Chinese word segmentation, and an important reason that affects accuracy of the Chinese word segmentation is the ambiguities identified. Combination ambiguity and cross ambiguity are two common ambiguities.

Definition 3. Combination ambiguity if AJ ∈ W and AJB ∈ W in AJB then there is combination ambiguity in AJB in that w is word-list.

Definition 4. Cross ambiguity if AJ ∈ W and JB ∈ W in AJB then there is cross ambiguity in AJB in that w is word-list.

The method to mine association rules is used to eliminate ambiguity in the paper.

Definition 5. Word support called sup (w) is the times that the word appears in the text.

Definition 6. If sup(w1) is the word w1 support , sup(w2) is the word w2 support and sup (w) is the word w support which link w2 to w1, then conf (w1/w) = (sup(w1) -sup (w))/sup(w1) is w1confidence to w. conf(w2/w) can be computed using the same principle.

Support and confidence of word are computed, and then ambiguity is eliminated by the following strategies:
Strategy 1: if conf(w1/w)< α, then the probability that w is the true word is larger than w1, so w1 is filtered out from candidate list. In that α=0.2
Strategy 2: if conf(w1/w)> β, then the probability that w1 is the true word is larger than w, so w is filtered out from candidate list. In that β=0.8
Strategy 3: if conf(w1/w)> α and conf(w1/w)< β, then the probability that w1 is the true word is larger than w, so w is filtered out from candidate list. In that β=0.8, then w1 and w are considered as the true words.

The strategies that located the front also apply to w2.

3 Word Segmentation Based on Statistics with PAT Array

3.1 The Method to Segment Words

"word + suffix" is recognized as a word in the dictionary-based method with forward maximum matching, for example, the word "duan yu's" in the novel 《tian long ba bu》 is a candidate word but not a true word, in fact "duan yu" is a true word. In the same, "prefix+word" is recognized a word in the dictionary-based method with reverse maximum matching, such as "with duan yu". To eliminate these two kinds of mistake and improve the accuracy of word segmentation, firstly, the two side suffix arrays are constructed for the input texts; secondly, the suffix arrays are sorted and get the candidate words; finally, the candidate words are compared and given the true words by mining association rules. The process of word segmentation with suffix arrays is shown in figure 1.

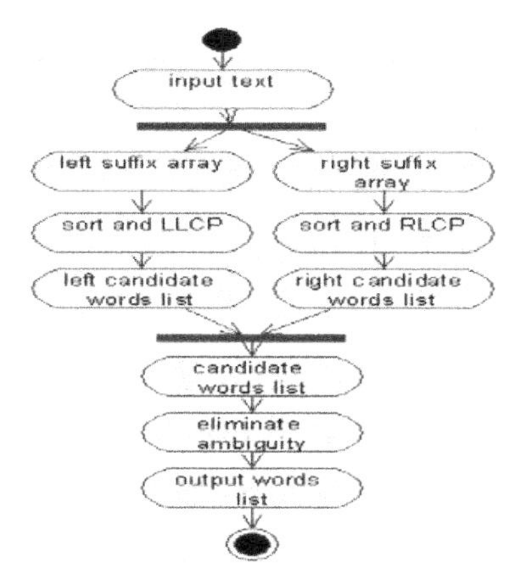

Fig. 1. The process of word segmentation with suffix arrays

Pretreatment. The contents which do not make sense are transformed into "/", such as punctuation.

Construct positive and negative suffix arrays. In this phase, Suffix arrays of the input texts are constructed on both left and right sides. We call them Left-index and Right-index respectively;

Sort suffix arrays and compute LCP. In this phase, firstly, suffix arrays are sorted on both left and right sides context for each occurrence of Chinese character. Then the Longest Common Prefix is computed on both positive and negative suffix arrays.

Extract frequent n-gram candidate terms. In this phase, firstly we extract n-grams, appearing more than one time in different contexts according to Left-index and Right-index of source texts, into Left-list and Right-list respectively. Then, we combine Left-list with Right-list, and extract n-grams which appear in both of them as candidates (C-list, for short);

Eliminate ambiguity. Candidates in C-list are filtered with 2.2.

3.2 Analysis of Algorithm

If the size of input texts is n and there averagely are 10 words in a sentence, then the input texts include n/10 sentences. Suffix array of a sentence is constructed using 110 Bytes; suffix array of the input texts is constructed using 11n Bytes. Capacity of memory in the method is 11n*2=22n Bytes that is 22 sizes of the input texts. Changli Zhang proposes a method of word segmentation without a dictionary which constructs suffix array of the input texts with the size of 1000 using 50 sizes of them.

4 Special Dictionary Constructions

Word segmentation not only needs correct and effective but also recognize new words in automatic correcting, so word segmentation based on statistics is used to improve new words recognized. Accuracy is affected by ambiguity of texts, so the method to mine association rules is used to eliminate ambiguity. Special dictionary is constructed in figure 2. The process of word segmentation is as following: firstly, the input texts are segmented using reverse maximum matching by the dictionary, and they are segmented using statistic based on PAT array to identify new words which are not in the dictionary; then ambiguity is eliminated in C-list; finally, new words recognized are added to the dictionary. The way to store a dictionary in [4] is used to improve the efficiency of segmentation.

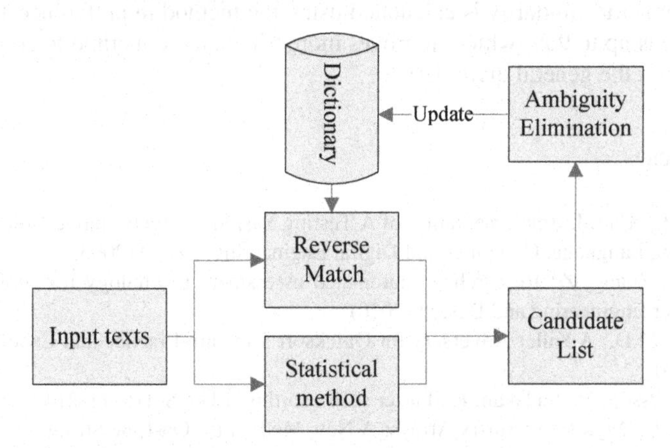

Fig. 2. The process of dictionary construction

5 Experiments

1000 documents are collected which relate to computer, news, sports, healthy and technology as test documents, and accuracy is index. Accuracy of every document is above 96% from Figure 3 which plots accuracy of word segmentation(Y axis) versus test documents which have 5 Categories(X axis).

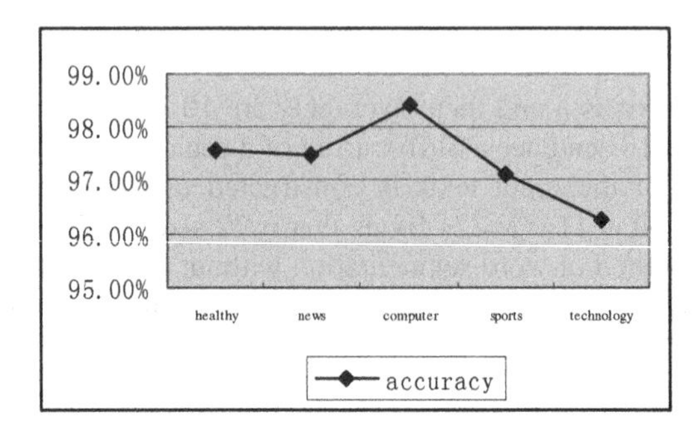

Fig. 3. Results of word segmentation

6 Conclusions

Two problems are solved in the process of dictionary construction. Firstly, new words are recognized using statistical word segmentation based on two sides suffix arrays; secondly, ambiguity is eliminated using support and conference in association rule mining. Accuracy of word segmentation is above 96% from experiments and dictionary where glossaries are stored is updated using new words that are recognized. The questions for the principle of computer organization are corrected using the dictionary, and semantic similarity is calculated using the method in preference 10, the consistency rate is up to 90% which improves more 5% than the method to compute similarity based on the general dictionary.

References

1. Wang, H., Xiao, J.: Implementation of A Testing System on Web - based Course of C Programming Language. Computer and Digital Engineering, 37- 39 (2003)
2. Tian, T., Zhang, Z.: Research on automated assessment technology for subjective tests. Computer Engineering and Design (2010)
3. Mcilroy, M.D.: A Killer Adversary for Quicksort Software-Practice and Experience 29, 5–12 (1999)
4. Jesper Larsson, N., Sadakan, K.: Faster suffix sorting. LU-CS-TR, 99–214 (1999)
5. Manber, U., Myers, G.: Suffix Arrays: A New Method for On-Line String Searches. SIAM Journal on Computing, 10–11 (1993)
6. Fang, G., Yu, H.: Web Translation Mining Based on Suffix Arrays. Journal of Chinese Language and Computing, 6–7 (2007)
7. Lee, F.C.: PAT-Tree-Based Keyword Extraction for Chinese Information Retrieval (2010), `http://citeseerx.ist.psu.edu/viewdoc/download`
8. Ma, Y.- C., Song, H.- T.: Research of Chinese word segmentation based on the Web. Computer Application, 134- 136 (2004)
9. Zhang, C.: An automatic and dictionary-free Chinese word segmentation method based on suffix array. Journal of Jilin University: Science Edition (2004)
10. Zhang, X.: Similar Sentence Search Algorithm in Automated Assessment System of Subjective Test Based on Chinese. Journal of Nan Jing Normal University (2007)

TEM Research Based on Bayesian Network and Neural Network

Fu-li Bai*, Hai-feng Cao, and Tong Li

Civil Aviation Management Institute of China
(Grant from) The National Natural Science Funds (60979006)
whitewelfare@hotmail.com

Abstract. This paper established the structural database using the method of Threat and Error Management (TEM), and obtained the characteristics of three types of data which are classified by threat, error and unexpected situation. Based on TEM frame, Neural Network is applied to add data, and Bayesian Network is applied to study the correlation among threat, error and unexpected situation. Here comes the conclusion: 1) Through applying Bayesian Evaluation to 625 selected samples, we found that specific unexpected situations have high correlation with some specific threat and error, Bayesian Evaluation reveals that when some unexpected situation happen, the high correlated threat and error to the unexpected situation. For instance, the threat and error that have highest correlation to in air unexpected situation are: Procedure Internal threat and communication error among Air Traffic Control and aircrew. 2) The high occurrence of some threat and error don't necessarily have high correlation with some high unexpected situation; high probability of some threat and error don't necessarily lead to unexpected situation. Research achievements provide data analysis skill to Air Traffic Control on safety management control, who could make corresponding prevention measure.

Keywords: Air Traffic Control, Human Error, TEM, Neural Network, Bayesian Evaluation, Correlation.

0 Introduction

Civil Aviation safety management is changing from traditional experience management into active prevention risk control [1]. In the field of civil aviation, one classified research method about human error data can be based on Threat, Errors and unexpected situation risk Management model (TEM), which was first proposed by the International Civil Aviation Organization (ICAO) [2] and was further development by the International Air Transport Association (IATA) later, who has linked the TEM method with the existing accident and incident sigh database. How to quantitatively research a large number of safety data information and to achieve the reliability of different types data source and correlation to each other, thereby to obtain a more real

* Corresponding author.

Y. Wang and T. Li (Eds.): Knowledge Engineering and Management, AISC 123, pp. 379–388.
springerlink.com

risk warning mechanism based data information is an important work on the human error [3] research field in the Civil Aviation management. At present, main factors limiting human error research to obtain more practical application are unreal operating situation, not to carry on quantitative analysis and the lack of unified structured database to support.

Researchers in the field of civil aviation safety management paid more attention to data applied in safety management now. Quantitative research was gradually carried out on data set of civil aviation human error, and correlation between the different data information was studied in-depth. For example, Lv Ren Li[4] has studied the TEM distribution characteristics of three types safety information of China Civil Aviation; DODD B. [5] has used Bayesian to estimative process various safety data and how to use these data to realize risk management ,which was published on the international civil aviation magazine; Anthony [6] and others used the United States forced report data to research incident relationship between the ground running, airspace operation and Air traffic control (ATC) ; Luxhj [7] has established the aviation safety management models to the low probability factors but with serious consequences, and then has established complex system risk management concentrated model for the human, machine and environment in repair situation[8]; Milan [9] has built safety risk assessment model by using mathematical methods in Poisson process.

In terms of research method, Event Chain Theory, Fault Tree Analysis, Failure Mode and Effect Analysis, Event Tree Analysis and Leeson Model were widely used into the aviation accident reasons in recent years. But these methods can not objectively and accurately find out the weak links of management system, don't mention about improvement.

In 1986, Pearl [11] has proposed Bayesian network (BN). BN is an effective method using system uncertainty and probability inference in analysis of complex systems. It uses local conditional dependence relationship in model to reasoning two-ways uncertainty, which was widely used in prediction, classification, causal analysis and diagnostic analysis. Zhou Wei Jie [12] has planned one kind civil aviation safety analysis method based on the BN, which fully utilized expert information and safety accident data to find the weak link of system.

This paper attempts to improve the TEM and ATC human error structured database, to quantitative research about human error risk model based on the actual operation of ATC, and then to identify error probability of particular error type in certain situations. The result provided a system method of human error risk quantification for the ATC operational units. BN can provide us a quantitative research method of human error probability, its greatest advantage is a distribution conversion from prior distribution to posterior. It is a inductive reasoning network statistical method from the existing sample to derivation rule. Neural network (NN) is a method using nonlinear mapping ideal and parallel processing, and with its own structure, it can express implicit function code of input and output correlation knowledge. Using the analytical solution ability of artificial NN in nonlinear problems and expressing nonlinear mapping relation of causal network can implement BN data.

This paper presents an ATC threat and error analysis method based on combining with BN and NN. Supplementing real safety event data and achieving higher precision data added by the NN, applying system inherent logic to safety analysis and finding the weak link of system by the BN, this paper provides some methods and tools for the study of human factors and controls.

Using BN to quantitative research intrinsic correlation of a variety of information, the paper can more fully reflect the safety risk state of ATC system. For example, an air accident state can be linked with various types' threats, which distinguish the effects of air accident status from the various threats. Not only is advantageous to understand all kind of air accident state threat categories, also to promote the application of ATC threat, error and accidental state data source.

1 ATC Threat, Error and Unexpected Situation Data Statistics

In accordance with certain rules, TEM divides several situations affected or may affect the normal traffic operation. Threaten is also known as risk or hazard, means the all external source or situations those are out of the operated personnel control, increase operational complexity, improve the operation difficulty, easy to cause injury and loss. Errors are usually divided into failure and violations. Failure is the inherent limitations caused by involuntary behavior, refers to the inappropriate behavior deviated from the intention or plan of operational person. Unexpected situation refers to the system has deviated from the expected normal status into a dangerous condition and has produced the actual deviation. Unexpected condition is caused by improper management to threaten or error, its risks is more than threat and error and lead to more obvious hazards. The appearance of unexpected condition indicates that system is losing safety degree. It is more urgent to management accidental situation than to threat and error and it is a last chance to avoid incidents or accidents.

TEM data analysis frame gets all safety hazards, threats, cognitive or operation errors into an accident prevention structure, it hierarchy defines the threat which does not have a direct consequences, errors of minor consequences as well as the Undesired States prior to accident. TEM structures the controller's Countermeasures, we can use TEM structure to accurately define any kind of error status and control countermeasure, it can structured and quantitative any uncorrelated, fragmented voluntary reporting or events so that we can control all kinds of safety threats and errors or states of emergency probably exists in the operation and also to summarize measures will be taken.

This paper statistics and classifies 625 ATC threat, error, Unexpected Situation data and information. In Table 1, A_i represents threat (i = 1, ... , 15), B_i represents error (i = 1, ... , 11), C_j represents Unexpected Situation (j = 1, 2, 3).

Table 1. Threat, Error, Unexpected Situation Data and Information

Type of Threat		Data	Type of Error		Data	Type of Unepected Situation	Data
Internal threats of ATSP	A_1 equipment	28	Device process-ion error	B_1 The use of radar	16	C_1 ground	5
	A_2 location factors	1		B_2 automation	3	C_2 air	61
	A_3 program	88		B_3 radio / intercom	1	C_3 landing	3
	A_4 Other controllers	45		B_4 flight process of single	16		
External threats of ATSP	A_5 Airport Planning	2	Program error	B_5 shift process	13		
	A_6 navigation equipment	1		B_6 information	49		
	A_7 airspace structure / design	15		B_7 file	3		
	A_8 adjacent units	72		B_8 check list	1		
				B_9separation standard	9		
Threats in air	A_9 pilots	41	Communication error	B_{10} controller to generator	31		
	A_{10} aircraft performance	6		B_{11} controllers to controllers	21		
	A_{11} R / T communication	16					
	A_{12} air traffic	51					
Environmental threats	A_{13} weather	19					
	A_{14} geographical environment	2					
	A_{15} interference	6					

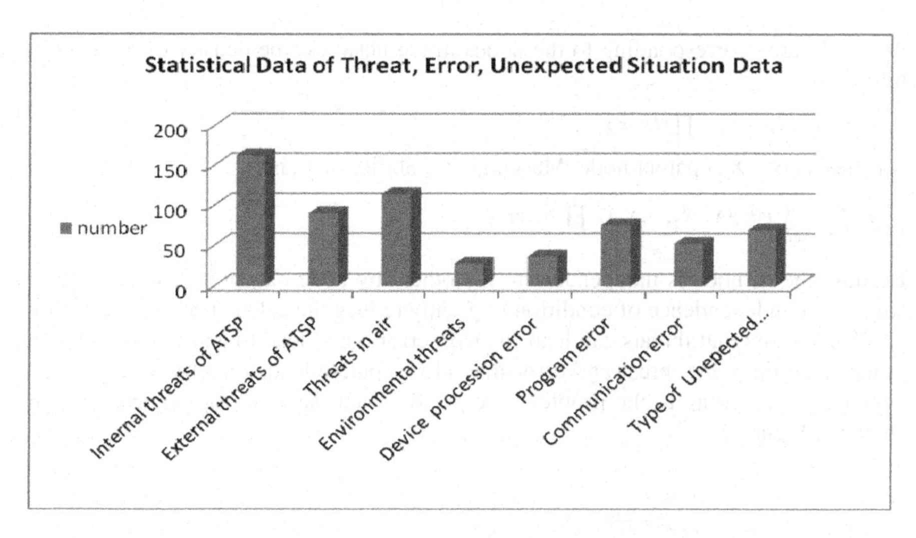

Fig. 1. Statistical data of Threat, Error, Unexpected Situation Data

Based on analyzing the TEM statistical data, this paper quantitative the probability of error and threat happened at the same time, and further to study the correlations between these data by using BN to estimate respectively on threat and unexpected situations, errors and unexpected situations. Taking relationships between the threat and an air accident state for an example, we can get the probability that the pilot is exposed to some threats and accident and that a threat occurs accident through the known data. To avoid the air accident, it is particularly important to study the relationship between the appearance of an air accident and the threat. The quantitative study on the prevention for the threat highly correlated air accident states and on the establishment of alarm mechanism is totally in accordance with risk management thought promoted in aviation safety management.

2 BN Topology Structure

Bayesian network is also known as Bayesian belief network, it is an directed acyclic graph (DAG) probability model.

It consists of model structure and parameters, in which the nodes of a directed graph is corresponding to the variables in the model, directed edges represents conditional dependencies, related parameter is the conditional probability tables (CPT) specified by each variable. In the network, no inputting arc node is called the root node and inputting arc node is called child node, for root node, we must determine a priori probability. According to the Bayesian formula, the conditional probability is definite as below:

$$P(A|B) = \frac{P(B|A)P(A)}{P(B)} \tag{1}$$

P(B) is the prior probability, P (A|B) is the posterior probability. For sets of variables, $U = (X_1, X_2, \cdots, X_n)$, the BN network is a graphical joint probability distribution,

$X_1, X_2, \cdots, X_n$ are corresponding to the nodes in the network, the density of joint probability is

$$P(U) = (X_1, X_2, \cdots, X_n) = \prod_{i-1}^{n} P(X_i | \pi_i) \tag{2}$$

π_i is the set of X_i,s parent node. Marginal probability of X_i is

$$P(X_i) = \sum_{exceptX_i} P(U)(X_1, X_2, \cdots, X_n) = \prod_{i-1}^{n} P(X_i | \pi_i) \tag{3}$$

Because the X_i node is independent to the other any node except its parent node set, we can use independence of condition to greatly reduce the calculation of probability.

We all know that threats can lead to error, error could lead to an air accident state, at the same time, the threat and error may also separate leads directly to an air accident state. According to the problem, we get Bayesian network topological structure shown in Figure 2.

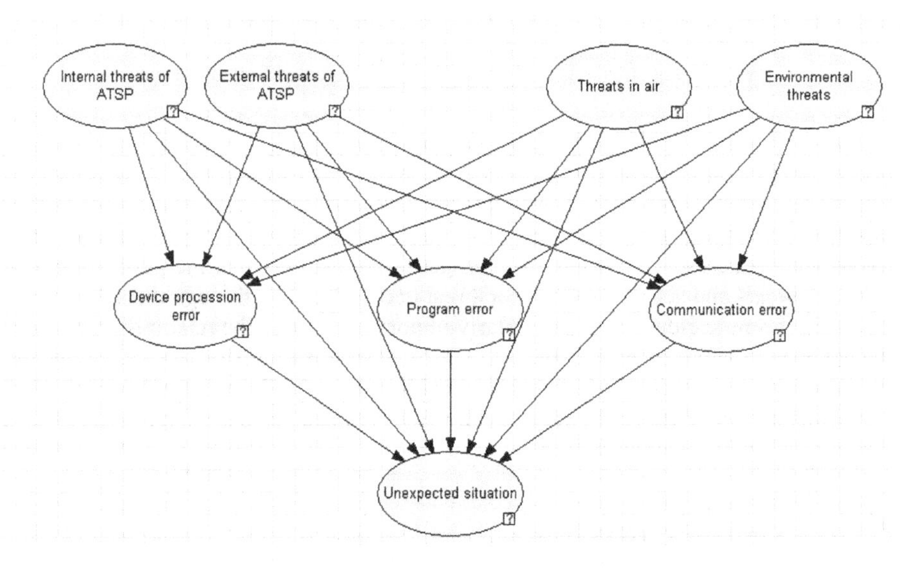

Fig. 2. The ATC Human Factors in BN Topology Structure

In Figure 2, error layers have three variables, which are Device procession error, program error and communication error. Each variable is with 4 different reason variables. Reason variables and outcome variables were two values of state variables. So the number of combination state in error layer is 2^5=32, but each outcome variables in the actual error layer lacks 4 variables, due to incomplete data all has 1 case, actual incomplete data is 2. The air accident state variables have 7 reason variables and its reason variables and outcome variables were two values of state variables. So the number of combination state in air accident state variables is 2^8=256. But the actual probability data only have 52, the data is incomplete so this study uses BP neural network to train and to add the unknown data.

3 Completing Data by Using BP Network

The Back Propagation Network (BP Network) consists of input layer, hidden layer, and output layer.

Because the three layer BP network can approximate any continuous function in any accuracy, research and application model is provided with a hidden layer, Sigmoid as a neuronal excitation function was used in BP networks which use a certain threshold characteristics of continuously function, this study used a linear transfer function. The learning process of BP network consists of two components: the forward propagation and propagation. For forward propagation, the input information is from the input layer and is disposed from hidden layer to the output layer; neurons' condition of each layer only affects the neurons in the next layer. If the output is unwanted in the output layer, then the reverse transmission will be used. Error signal will be return along the original neuronal connections path. In the return process, the weights and thresholds of neuronal connection will be modified one by one. The error signal is within the allowable range so that finally it makes the actual network output more closer to the expected output through constantly modification.

This paper aims at incomplete data of device procession error, program error, communication error, air accident state to establish the BP nerve network. the reason dates of variables is as the input samples of neural network for training. For device procession error, program error, communication error, the reason variables are four threat statuses, for air accident state, the reason variables are four threats statuses and three errors. Output sample for each variable are device procession error, program error, communication error, air accident state itself.

Table 2. Fitting and Precision of Network to Samples

Date Simulation	Precision
Program error	78.26%
Communication error	80.80%
Device procession error	71.73%
Air accident state	74.78%
Average precision	76.39%

As shown in table 2, all the fitting precision of network to sample are higher than 70%, the average is above 76%. The prior probabilities are shown in table 3. After putting the simulation data into a Bayesian network (Figure 1), the posterior probabilities are shown in table 4.

Table 3. The Prior Probabilities

Threat and error		Internal threats of ATSP		External threats of ATSP		Threats in air		Environmental threats	
Logic value		F	T	F	T	F	T	F	T
Unexpected situation	F	0.576	0.424	0.507	0.493	0.4	0.6	0.472	0.528
	T	0.547	0.453	0.542	0.458	0.48	0.52	0.48	0.52

Table 3. The Prior Probabilities (continue)

Threat and error		Device procession error		Program error		Communication error	
Logic value		F	T	F	T	F	T
Unexpected situation	F	0.504	0.496	0.533	0.467	0.408	0.592
	T	0.613	0.387	0.6	0.4	0.385	0.615

Table 4. The Posterior Probabilities

Threat and error		Internal threats of ATSP		External threats of ATSP		Threats in air		Environmental threats	
Logic value		F	T	F	T	F	T	F	T
Unexpected situation	F	0.196	0.804	0.434	0.566	0.399	0.601	0.833	0.167
	T	0.244	0.756	0.46	0.54	0.269	0.731	0.849	0.151

Table 4. The Posterior Probabilities (continue)

Threat and error		Device procession error		Program error		Communication error	
Logic value		F	T	F	T	F	T
Unexpected situation	F	0.721	0.279	0.687	0.313	0.781	0.219
	T	0.817	0.183	0.704	0.296	0.614	0.386

The results show the severity based on the posterior probability and the threat of the mistake is different from that based on prior probability, computing is meaningful. Posterior probability is the probability of threat and the error in the case of accident. So a posterior probability can be regarded as the importance degree to judge and the error and the threat factors be occured.

The internal threats is the highest of the posterior probability, and it shows that internal threats is the most sensitive causes in the air accident state. So reducing internal threats can dramatically reduce accident state.

In addition, the internal state threat, air threat, an external threat posterior probability is greater than 50% when the air accident occurs. They should be keys as the control factors. This shows that the corresponding control procedures and countermeasures formulated by the unit operation to avoid errors and various kinds of threats and avoid accident play the most important role.

4 Conclusion and Prospect

Based on neural network and the Bayesian network model, this paper discussed the relationship between the air accident state and all kinds of threats and errors, built the air accident risk of state estimation model. the model provided the prior probability of threat and errors and also the calculation method of the posterior probability of an unexpected condition to all kinds of threats and errors, to make up for the human error data collection is difficult to end all possible combinations of insufficient. Through the comparison of occurrence frequency to the threat and error to corresponding probability in the estimation of Bayesian, this paper point out the rank correlation of all kinds of threats and errors, further from the point of view of the quantitative it cleared the type of threat and error need to be controlled in the daily management.

This paper did not discuss association rules of threats, error and the air accident state, and this kind of accident chain research should pay more subsequent attention into it. We can use data mining algorithm method, focus on accident case of accidental, research objectively the association rules of all kinds of threats and errors and unexpected situation.

References

1. International Civil Aviation Organization. Line operations safety audit, DOC 9803-AN/761, International Civil Aviation Organization, Montreal (2002)
2. International Civil Aviation Organization. Threat and error management in air traffic control(circular 314-AN/178). International Civil Aviation Organization, Montreal (2008)
3. Kathleen, L., McFadden, E.R.: Towell: Aviation human factors: a framework for the new millennium. Journal of Air Transport Management 5, 177–184 (1999)
4. Lv, R.-L., Cao, H.-F., Zhou, Y., Bai, F.-L.: A comparative study of three sources for safety information in civil aviation. Journal of Safety and Environment 10(3), 170–174 (2010)
5. Dodd, B.: Safety data integration can lead to better understanding of risk. International Civil Aviation Organization Journal (2), 21–22 (2007)
6. Anthony, M.P., Douglas, A.W., Scott, S.: Air traffic control related accidents and incidents: a human factors analysis. In: The 11th international symposium on aviation psychology Columbus, The Ohio State University, OH (2001)
7. Luxhøj, J.T., Coit, D.W.: Modeling Low Probability/High Consequence Events:An Aviation Safety Risk Model. In: Reliability and Maintainability Symposium RAMS 2006. Annual, January 23-26, pp. 215–221 (2006)
8. Luxhøj, J.T.: Risk Analysis of Human performance in aviation maintenance. In: 16th Human factors in Aviation Maintenance Symposium, April 2-4 (2002)
9. Janic, M.: An assessment of risk and safety in civil aviation. Journal of Air Transport Management (6), 43–50 (2000)
10. Horowitz, B.M., Santos, J.R.: Runway safety at airports: A systematic approach for implementing ultra-safe options. Journal of Air Transport Management 15, 357–362 (2009)
11. Pearl, J.: Probabilistic reasoning in intelligence systems. Networks of Plausible Inference. Morgan Kaufmann, San Francisco (1988)
12. Zhou, W.-J., Wang, H.-W., Zhao, F.: Study on Bayesian Networks based Aviation Safety Analysis. Aeronautical Computing Technique 1, 45–47 (2011)

Morphological Analysis Based Part-of-Speech Tagging for Uyghur Speech Synthesis

Guljamal Mamateli[1], Askar Rozi[2], Gulnar Ali[1], and Askar Hamdulla[1]

[1] Institute of Information Science and Engineering of Xinjiang University, 830046 Urumqi, China
[2] Institute of Mathematics and System Science of Xinjiang University, 830046 Urumqi, China
guljamal123@gmail.com, {askhar,gulnar,askar}@xju.edu.cn

Abstract. Accuracy of part-of-speech tagging is critical to downstream sub-tasks in front-end text analysis model of text-to-speech System. Uyghuris an agglutinative language in which numbers of words are formed by suffixes attaching to a stem (or root). Owing to there are unlimited new formed and derived syntactic words in Uyghur, Sizes of part-of-speech tagging set were big and out-of-vocabulary words often occurred in conventional Uyghur part-of-speech tagging method which directly trained and predicted the part-of-speech of word. To address this problem, this paper proposes the idea that trains the part-of-speech of stem and predicts the part-of-speech of word mainly by stem. Bi-gram language model is used to segment the stem and affix boundary of word, hidden markov model is used to train and predict part-of-speech of stem. In the end, rule adjusting method is used to adjust the changed part-of-speech of word when suffix attaching to a stem. Experimental result shows that proposed method obviously reduces the part-of-speech tagging error rate comparing to conventional part-of-speech tagging method.

Keywords: Uyghur-language, part-of-speech tagging, bi-gram language model, hidden markov model.

1 Introduction

One of the fundamental building blocks of text processing for text-to-speech system is the assignment of a part-of-speech tag to each input word. Part-of-speech tags augment the information contained within words by explicitly indicating some of the structure inherent in language. Their accuracy is therefore critical to downstream sub-tasks, including chunking and semantic role labeling. This in turn affects proper placement of prosodic markers such as accent types and phrase boundaries, which greatly influences how natural synthetic speech sounds. It is therefore important to make sure part-of-speech tagging works well in the context of text-to-speech synthesis.

Part-of-speech tagging assigns a correct part-of-speech tag to input words according to the contextual information. There are three commonly used part-of-speech tagging methods such as rule based method[1], data-driven method[2] and hybrid method of combining rule and data-driven methods[3].

Y. Wang and T. Li (Eds.): Knowledge Engineering and Management, AISC 123, pp. 389–396.
springerlink.com

Historically, part-of-speech tagging has predominantly been based on manually specified rules. With the growing availability of natural language training resources in recent years, mainstream tagging has increasingly involved some form of data driven statistical processing. As is well known, the size and pertinence of the training data is critical to the quality of the resulting models. There is, unfortunately, an inherent trade-off between size and pertinence. So as to reap the benefits of both sides, some research efforts are done to combine data- driven and rule-based methods.

Throughout the previous work, hidden markov model and viterbi searching algorithm are widely applied and it almost becomes a standard model and algorithm ofpart-of-speech tagging method. Sizes of training data were big and numbers of out-of-vocabulary words often occurred in conventional hidden markov model based Uyghur part-of-speech tagging system [4] which directly trained and predicted the part-of-speech of words, because of a large number of derived syntactic words are formed by suffixes attaching to a stem in Uyghur-language. Therefore, this paper proposes the idea that trains the part-of-speech of stem and predicts the part-of-speech of words mainly by stem. Bi-gram method is used to segment the stem and affixes of word, hidden markov model is used to train and predict the part-of-speech of stem, and statistical rule based adjusting method is integrated to adjust the changed part-of-speech of words when suffix attaching to a stem.

The remainder of the paper is organized as follows. Section 2 presents the morphological and grammatical structure of Uyghur-language related to our work. Section 3 describes the part-of-speech tagging method proposed in the paper. Experimental results and discussion are given in section 4. Section 5 gives the conclusion and future work.

2 Morphological Structure of Uyghur-Language

Uyghur-language belongs to the Turkish language family of the Altaic language system and agglutinative language in structural grammar. Uyghur text is written from right to left in Arabic scripts with some modifications. Figure 1 shows the morphological structure of Uyghur sentence “she cares for us”.

Uyghur sentence consists of words, which are separated by blanks or other punctuations. Morphological structure in figure 1 has four levels which are phoneme, syllable, word and sentence from the bottom up.

There are 32 phonemes corresponding to 32letters in Uyghur-language, in which 8 vowel and 24 constants. Uyghur syllable structure is “[C]+V+[C]”,where V stands for vowel, there is only one vowel in one syllable, vowel is the center of the syllable, and C stands for constant, there can be 0 to 3 constants in one syllable.

Morpheme structure of Uyghur word is “Prefix+Stem+Suffix1+Suffix2+…”.Words are formed by affixes attaching to the stem (or root).Root is the minimum significant unit which can’t be segmented, A root (or stem) is attached in the rear by zero to many (longest is about 10 suffixes or more) suffixes. Suffixes that make semantic changes to a root are derivational suffixes. Suffixes that make syntactic changes to a root are inflectional suffixes. A root linked with the derivational suffixes becomes a stem. So the root set is included in the stem set. Sometimes the words “stem” and “root” are used without distinguishing. Figure 2 shows the morphological structure of Uyghur word “teachers”.

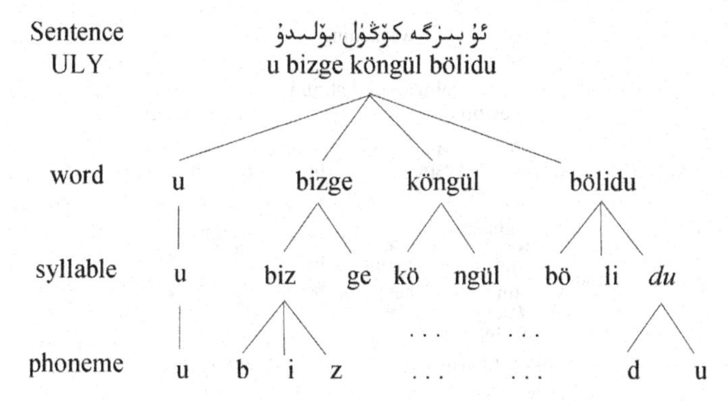

Fig. 1. Morphological structure of Uyghur sentence

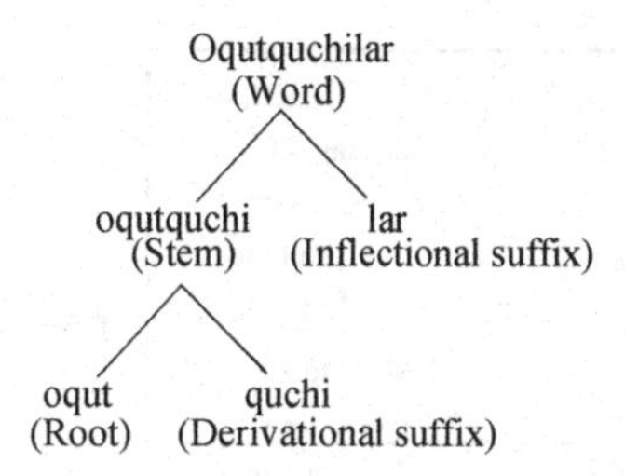

Fig. 2. Morphological structure of Uyghur word

There are some language phenomenon that occurs when suffix attaching to a stem, which are weakening, deletion, insertion, phonetic harmony and ambiguity. They are returned back to the original stem-suffix form when segmenting morpheme units. Figure 3 shows this natural language phenomenon of Uyghur-language.

3 Part-of-Speech Tagging Method

Owing to there are unlimited new formed and derived syntactic words in Uyghur, Sizes of part-of-speech tagging set were big and out-of-vocabulary words often occurred in conventional Uyghur part-of-speech tagging method, it affected the performance of tagging method to a great extent. Part-of-speech of stem basically decides the part-of-speech of word in context. Therefore, this paper uses the method that trains and model the part-of-speech of stem and predict the part-of-speech of word mainly by stem, stem and affix are segmented through bi-gram model, then part-of-speech of stem is predicted through the hidden markov model, in the end, part-of-speech of the word is adjusted by using the rule based method according to the suffixes. Figure 4 shows the framework of part-of-speech tagging method proposed in the paper, dashed line is the training part.

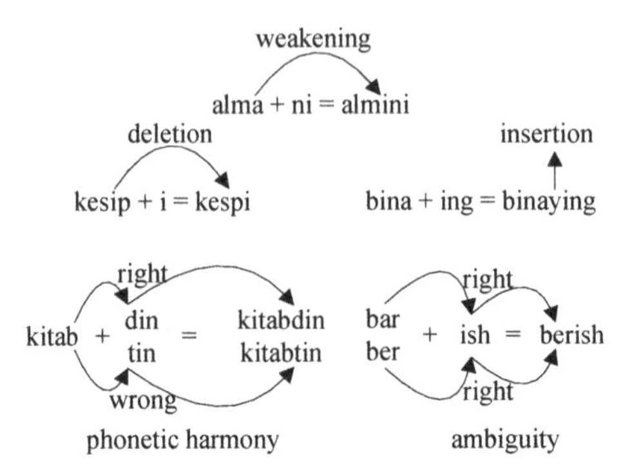

Fig. 3. Natural language phenomenon in Uyghur-language

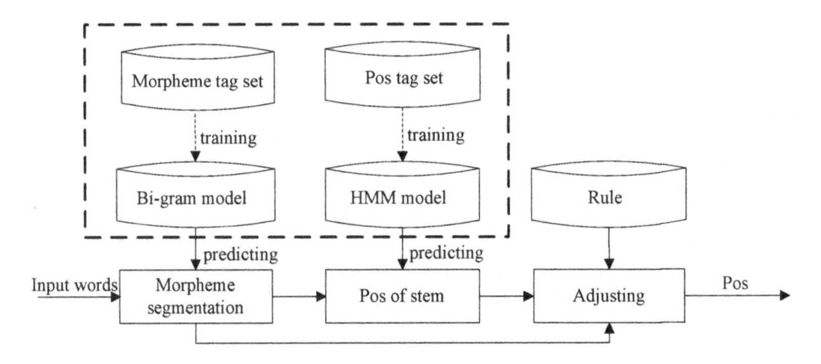

Fig. 4. Framework of part-of-speech tagging method

3.1 Morpheme Segmentation Based on Bi-gram Method

For an input word, all the possible segmentation results are extracted in reference for both stem and suffix, and their probabilities are computed to get the best result.

Generally, an intra-word bi-gram method based on the following probabilities is used, and the identification of stem-suffix boundary is the most important part in segmentation.

$$\begin{cases} P(Stem, FirstSuffix) \\ P(Stem)\sum_{i=1} P(suffix_i|Stem) \text{ for smoothing} \end{cases} \quad (1)$$

At first, a word is split into two parts, a stem and a combined suffix, and several possible stem-suffix pairs are obtained.

Prefix+ Stem+ Combined Suffix

Prefix+ Stem+Suffixl +Suffix2 +···

Then, the combined suffix is segmented into singular-suffixes, because each combined suffix (word endings or stem endings in some papers) may have several different singular-suffix segmentations.

For insertion in figure 3, we add the inserted phoneme to the subsequent suffix, and form a new surface form of the same suffix type. For deletion, because it happens in the stem only, a list of deleted stems are learned from the training corpus.

3.2 Part-of-Speech Tagging Based on Hidden Markov Model

In this part, we use hidden markov model to build the part-of-speech model of stem and predict the part-of-speech of stem which is delivered from the upper bi-gram model.

Suppose $W = w_1 w_2 \cdots w_n$ isstem sequence, $C = c_1 c_2 \cdots c_n$ ispart-of-speech sequence corresponding to$W$, we choose the part-of-speech sequences $C^*$with the maximum $P(W,C)$as the final result of part-of-speech tag combinations, then

$$C^* = \underset{C}{\operatorname{argmax}} P(W,C) = \underset{C}{\operatorname{argmax}} P(W|C)P(C) \tag{2}$$

C* can be expressed approximately by using first-order hidden markov model.

$$C^* = \underset{C}{\operatorname{argmax}} \prod_{i=1}^{n} p(w_i|c_i)p(c_i|c_{i-1}) \tag{3}$$

To facilitate the calculation, the negative logarithm is used, then

$$C^* = \underset{C}{\operatorname{argmax}} \sum_{i=1}^{n} [-\ln p(w_i|c_i) - \ln p(c_i|c_{i-1})] \tag{4}$$

Out-of-vocabulary words are predicted by the contextual information. Taking into account the text analysis model idiosyncrasies of text-to-speech system, this paper uses the 39part-of-speech tags, includes 14first class part-of-speech tag and 25 second class part-of-speech tag (mainly the branches of first class part-of-speech tag).

3.3 Part-of-Speech Adjusting Based on Rule

Training and predicting the part-of-speech of words mainly by the stem not only reduces the out-of-vocabulary words, but also reduces the size of training corpus. Part-of-speech of stem basically decides the part-of-speech of word in the context. For example, if the training corpus includes a word "kitab" which only consists of single stem and not includes it's derived syntactic words "kitablar, kitabni, kitabdin…",the conventional hidden markov based part-of-speech tagging method regards these derived words as out-of-vocabulary words. In our work, we can get the stem "kitab" of these derived words through the bi-gram based morpheme segmentation method and their part-of-speech are same as their stem "kitab" which already exists in the corpus.

But, not all the part-of-speech of derived syntactic words are same as their stem, since some suffixes which are first attached to a stem change the part-of-speech when forming a new derived syntactic words. For example, the suffixes "sh" and "ghuchi" change the part-of-speech from "verb"(of stem) to "gerund" and "noun"(of words) respectively when attaching to a stem "oqu" which is the stem of both "oqush" and "oqughuchilarning" in the sentence "oqushbashlashmurasimighaoqughuchilarning-hammisiqatnishidu".

There are some suffixes which change the part-of-speech when forming new derived words by attaching to stem in Uyghur. For example, we can also decide the part-of-speech of words will be "noun" when suffix "ghuchi" attaching to a stem. In this work we collected the 33 suffixes which have a part-of-speech changing function. According to the suffixes information from bi-gram model, we detect the part-of-speech changing problem of stems which are tagged from hidden markov model. Suppose$S = s_1 s_2 \cdots s_{33}$is suffix sequence, then

$$C(\text{word}) = \begin{cases} C'_{suffix} & \text{if suffix } s_i \text{ exists in S} \\ C_{stem} & \text{otherwise} \end{cases} \quad (5)$$

C'_{suffix} is a part-of-speech which is changed in accordance with the first attached suffix. if the stem is attached with suffix which has part-of-speech changing function, we changes the part-of-speech of word, if they are not, we regard the $C(\text{word})$ same as C_{stem} and output the part-of-speech information directly from the hidden markov model based on stem.

4 Experiments and Analysis

4.1 Experimental Data

When segmenting stem and affix boundary of words, we adopt the morpheme segmentation method used in paper [5], 10025 sentences are manually segmented, 9025 sentences are trained, and the results on the test of 1000 sentences get to 97%. Word coverage is 86.85%and morpheme coverage is 98.44%.

In the part-of-speech tagging module, words in the 13,324 sentences are tagged manually with part-of-speech and segmented into morpheme units, length of each sentence approximately 10~25 words. 12,324 sentences are used for training, including 70559 words, 9071 stem, 296 suffixes and 7 prefixes. Experimental results are made on the test of 1000 sentences including 9424 words.

4.2 Experimental Results and Analysis

Figure 5 shows the error rate of three different part-of-speech tagging methods. First one is the error rate of conventional hidden markov model based part-of-speech tagging method which trains and predicts the part-of-speech of words, second one is the error rate of part-of-speech tagging method which uses the bi-gram to segment the morpheme, then uses the hidden markov model to predict part-of-speech of stem, last one is the error rate of the method after adding the rule based part-of-speech adjusting method. It can be seen from figure 5 that the proposed method of the paper corresponds to a relative reduction of 51.5% compared to conventional part-of-speech tagging method.

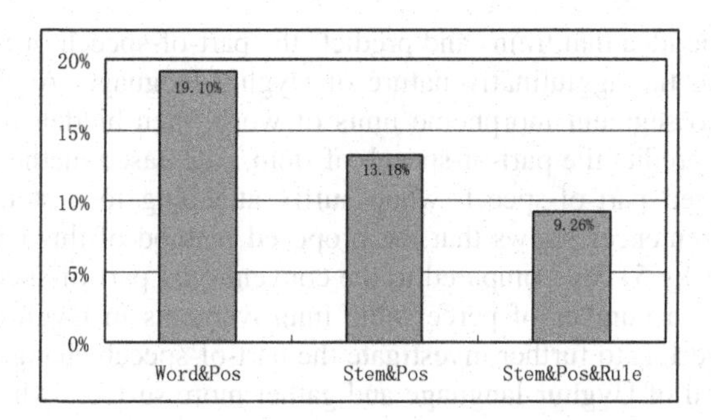

Fig. 5. Error rate of three different part-of-speech tagging method

Table 1 list the error rate of open and closed test of conventional hidden markov model based part-of-speech tagging method and the proposed method for first and second class part-of-speech (POS) tag.

The reason result in the part-of-speech tagging error rate mainly comes from segmenting error and unmatched stem, suffix and contextual information. Figure 6 shows the proportion of these errors in the tagging error rate.

Table 1. Error rate of different method

		First classPOS	Second class POS
Open test	Word&POS	18.23%	19.10%
	Stem&POS	7.97%	9.26%
Closed test	Word&POS	1.39%	2.63%
	Stem&POS	1.32%	2.56%

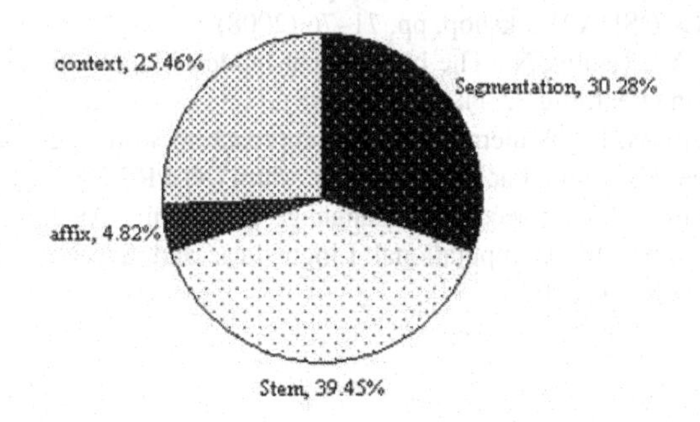

Fig. 6. Proportion of different errors in the tagging error rate

5 Conclusions

To address the problem such as big training corpus and high occurrence of out-of-vocabulary words in the conventional Uyghur part-of-speech tagging method, this paper proposes the idea that trains and predicts the part-of-speech of words mainly by stem according to the agglutinativenature of Uyghur-language. At first the bi-gram method is used to segment morpheme units of word, then hidden markov model is used to train and predict the part-of-speech of stem, rule based method is used for adjusting the changed part-of-speech when suffix attaching to a stem. Experimental resolution 1000 sentences shows that the proposed method of this paper reduces the tagging error rate by 51.5% compared to the conventional part-of-speech tagging method and leads to a number of perceivable improvements in Uyghur text-to-speech quality. Future work is to further investigate the part-of-speech changing phenomenon from stem to word of Uyghur-language and gather more suffixes which have part-of-speech changing function.

Acknowledgments. This work was supported by National Natural Science Foundation of China under the grant of 61065005 and 61062008.

References

1. Brill, E., Magerman, D.: Marcus Met al. Deducing linguistic structure from the statistics of large corpus. In: Proceedings of the DARPA Speech and Natural Language Workshop, pp. 275–282. Addison Wesley Longman Limited, Hidden Valley (1990)
2. Merialdo, B.: Tagging English text with a probabilistic model. Computational Linguistics 20(2), 1–29 (1994)
3. Sun, M., Bellegarda, J.R.: Improved Pos Tagging For Text–To–Speech Synthesis. In: ICASSP 2011, pp. 5384–5387 (2011)
4. Mamateli, G.: Research and Implementation of key Technologies in UTTS Based on two-level Speech Units and Prosodic Parameters.Master thesis of Xinjiang University (2009)
5. Ablimit, M., Neubig, G., Mimura, M., Mori, S., Kawahara, T., Hamdulla, A.: Uyghur Morpheme-based Language Models and ASR. In: Proc. ICSP 2010, pp. 581–584 (2010)
6. Ablimit, M., Eli, M., Kawahara, T.: Partly supervised Uyghur morpheme segmentation. In: Proc.Oriental-COCOSDA Workshop, pp. 71–76 (2008)
7. Fine, S., Singer, Y., Tishby, N.: The hierarchical hidden Markov model: Analysis and applications. Machine Learning 32(1), 41–62 (1998)
8. Peng, F., Schuurmans, D.: A hierarchical EM approach to word segmentation. In: The 6th Natural Language Processing Pacific Rim Symposium (NLPRS 2001), Tokyo (2001)
9. Jurafsky, D., Martin, J.H.: Speech and Language processing: An Introduction to Natural Language Processing. In: Computational Linguistics, and Speech Recognition, vol. 12, Posts & Telecom Press (2010)

Where Scene Happened in Text on Relevant Measures

Hanjing Li[1] and Ruizhi Dong[2]

[1] Specail Eduaction School, Beijing Union University,
100075, Beijing ,China
tjthanjing@buu.edu.cn

[2] MOE-MS Key Laboratory of Natural Language Processing and Speech,
Harbin Institute of Technology, No. 92, West Dazhi Street, NanGang,
Harbin, 150001, China
rzdong@mtlab.hit.edu.cn

Abstract. This paper proposes a method to identify scenes in texts with relevant measures, namely to infer the locations where actions take place without any explicit description in texts. Firstly, 48 scene categories are classified by the sememe hypernym and concept similarity on HowNet. Secondly, the hierarchical scene information tuples generated on action tuples and scenes from corpus are extracted by calculating the relevancy for each tuple, and then the knowledge base is generated. Finally, the constructed knowledge base is used to identify scenes in novel texts, and the experimental results on 1-best and voting shows the effective of the method.

Keywords: Dunning's likelihood ratio, text-to-scene, scene identification, scene category.

1 Introduction

There has been a great deal of work on inferring semantic information from text corpora over the past decades [1,2]. In real life, people use computers to draw 3D renderings from the plot in the text. In this situation, we need to know the locations of these circumstances occur. When it refers to Text-to-Scene (TTS)[3,4], we care more about the invisible scenes in natural language. The paper makes a statistical method to infer the explicitly scenes of actions.

The paper gets the associations on four statistics method of Dunning's likelihood ratio test, Cosine coefficient, $\chi2$ test, mutual information MI [5] so to compare the performance for scene identification. The experiments proved that Dunning's likelihood ratios test is not only applicable to the English Scene Identification, but also equally applicable to Chinese.

2 Preparing Work

At present, researchers often use their own terminology approved to describe their work. The notion "scene identification" means inferring the scene category which is

Y. Wang and T. Li (Eds.): Knowledge Engineering and Management, AISC 123, pp. 397–402.
springerlink.com

not explicitly described in the text[4]. Scene category refers to the natural or artificial environments where the action occurs, and is denoted by "SCENE", such as "天空(sky)". The action tuples are composed of the verbs and the nouns, such as (升起(rise), 太阳(sun)), in which the verb is recorded as "V", the noun "N", and the action tuples VN (V, N). Hierarchical scene information tuples, for example((升起(rise) 太阳(sun)), 天空(sky)), are recorded as VNS (VN, SCENE).

The candidates of category nouns are clustered on HowNet[6] sememe hypernym and concept relevance in the following algorithms.

Initialization. scene categories sememe sets: Se={waters|水, land|陆, sky|空}; scene categories seed values: S_0={水(water)}; S_1={陆(land)}, S_2={空(sky)}, S_3={ }; information words: X = $\{x_i\}$, 1<= i<= n;

Construct the set of the scene categories seeds.

```
for k= 0 to 2
  if (the hypernym of some  record on Sk is in the set
      Se)   then put the record in set RSk;
  else (there's no hypernym of records on Sk is in the
       set Se) then put all of the records on Sk in RSk;
```

Construct the sets of the candidates records.

```
for i = 1 to n
  if  (the hypernym of some  record on xi is in the set
       Se ) then put the record in set Ri;
  else (there is no hypernym of records on xi is in the
       set Se) then put all of the records on xi in Ri;
```

Cluster information words.

```
for i = 1 to n {
   for k =0 to 2 {
```

$$sim(x_i, S_k) = \max \sum_{j=1}^{m} \sum_{h=1}^{l} \text{Concept\_Similarity}(R_{ij}, RS_{kh}),$$

$$m = | R_i |, l = | RS_k |;$$

```
   }
   for l= 0 to 2 {
```

if $\forall j \ sim(x_i, S_j) = \max_{l=0}^{2} sim(x_i, S_l) \ j = 0,1,2$ then put x_i in S_3;

```
    else put xi in Sj;     } }
```

3 Construction of the Knowledge Base

3.1 Extraction of Various Elements of VNS

The following algorithms can realize extraction of various elements of VNS on initializations of action tuples VN set noted as ActionPri, hierarchical scene information tuples VNS set BasePri, scene categories SCENE set SetScene = {a|a∈Scene categories}, and context set of extracting VN and VNS Setcon = {c | c∈context}.

```
Add  V ( V∈CON ) to the set of VERB;
Add  N ( N∈CON ) to the set of NOUN;
Add Scene (Scene∈CON∩Scene∈SetScene) to the set of SCENE;
for ∀V∈VERB, and ∀N∈NOUN, add the tuple VN (V, N) to the set of ACTION and ActionPri;
for ∀VN∈ACTION, and ∀Scene∈SCENE, constitute VNS(VN,Scene), add it to BasePri;
If ∃c∈Setcon, and c has not been accessed, then go to (b); otherwise, end the process.
```

3.2 Statistical Results of VNS

The details are as follows:

```
Count the frequencies of various elements (VN, frequency C2) in ActionPri;
Count the frequencies of various elements (SCENE, frequency C1) in SetScene;
Count the frequencies of various elements (VNS, frequency C12) in BasePri;
for the tuple (VNS, frequency C12) in step (c)
    find the corresponding (VN, frequency of C2) and (Scene, frequency C1) from (a), (b), constitute a new tuple (VN, C2, Scene, C1, C12); end.
```

A part of statistical results are shown in Table 1.

3.3 Generate VNS Knowledge Base

We use Cosine coefficient, $\chi 2$ test, mutual information MI, Dunning likelihood ratios test to get VNS knowledge bases[7,8]. On Dunning's likelihood ratios test also called LLR (Log-Likelihood Ratio), we compute association between each tuple and the target term on Section 3.2. It directs the correlation between the actions and the scenes category. Every row in Table 1 is one part of VNS knowledge bases.

Table 1. Some actions associated with scenes by Dunning's likelihood ratios formula

Action tuples	C2	scene	C1	C12	LLR
(清算,金融) (liquidate, finance)	63	银行 (bank)	4028	63	954.134
(装卸,集装箱) (load, container)	11	港口 (port)	211	11	231.887
(演出,戏曲) (perform, opera)	10	舞台 (stage)	482	5	83.2777

4 Experimental Results

4.1 Prepare Work

People's Daily (1998.01-1998.06), the Chinese encyclopedia and children's tales are combined into a training database of a total of 20,071,674 words. The open testing database is 113 sentences including both verbs and nouns from 16 articles extracted from Beijing Normal University pupil corpus (www.mclass.cn).

The 113 sentences are divided in average into three set of set A. set B and set C. Set A as the baseline is used for random identification, set B for system identification, and set C for manual identification. Random means that one is assigned one class from 48 categories randomly. System identification has two methods. One is 1-best of getting the scene category corresponding to the maximum correlation from the knowledge base. The other is the voting of indentifing the maximum correlation after sum of all the correlation values of one category with various action tuples. Manual identification is selecting one from 48 categories by hand. In addition, we obtain the golden standard by hand.

If O denotes the multi-set of the output scene categories and I denotes the multi-set of the golden standard, then precision p, recall r, and their harmonic mean $F1$ can be computed as:

$$p=\frac{|O\cap I|}{|O|} \quad r=\frac{|O\cap I|}{|I|} \quad \mathrm{F1}=\frac{2\times p\times r}{p+r}$$

Two sample t-test to show the significances for the differences between the baseline and the system, the system and the hand on 45 subjects. The number of error identification the answers give is on the following formula:

$$m=\frac{\sum_{i}^{n} \textit{the}\ \text{number of the}\textit{sentence i}\ \text{is considered to be wrong}}{\textit{the}\ \text{total of sentences}\ n}$$

4.2 Results

The experiment is separated into two groups. The first group is combining the different thresholds and various contexts to study every statistics method performance. The results on four statistics method with 1-best method under best thresholds are shown in Table 2. The second group is used to take the best statistics method under the best context performances by 1-best and voting methods, and the results under the best F1 are shown in Figure 1. Proof of statistical significance is shown in Table 3.

The first group is divided into three situations on three contexts: comma(Comma), period(Sen) and fixed-window (Regu), the size of which is set to 81 words with 40 words on each side of the target word.

From Table 2, fixed-window system on voting performs best. Although comma or sentences can express a complete meaning, the their sparseness is more significant than the fixed-window. In fixed-window system, the meaning of the sentence is split, however, the scenes between two adjacent sentences are relevant in a certain range from a comprehensive point of view. Therefore it can achieve a compromise between semantic information and the data sparseness.

Table 2. Results from four system on voting method with various contexts under best thresholds

systems	contexts	thresholds	Precision	Recall	F1
	Comma	60%	47%	26%	34%
χ2	Sen	70%	52%	47%	49%
	Regu	20%	54%	47%	50%
	Comma	80%	42%	26%	33%
Dunning	Sen	70%	54%	47%	50%
	Regu	30%	62%	60%	61%
	Comma	10%	78%	23%	36%
MI	Sen	100%	48%	43%	46%
	Regu	90%	52%	50%	51%
	Comma	20%	67%	26%	38%
Cosine	Sen	70%	58%	50%	54%
	Regu	30%	57%	53%	55%

From Figure 1, the scene identification on voting is better than 1-best and the Dunning system gets the best results. F1 can reach 61% on voting. The Dunning system on the voting method should be used.

From Table 3, the results from the system identification are better than random identification, and worse than hand identification. This result shows that inferring the location by the relevance of action and scene with LLR on voting achieves positive results and the results can meet the statistical significance.

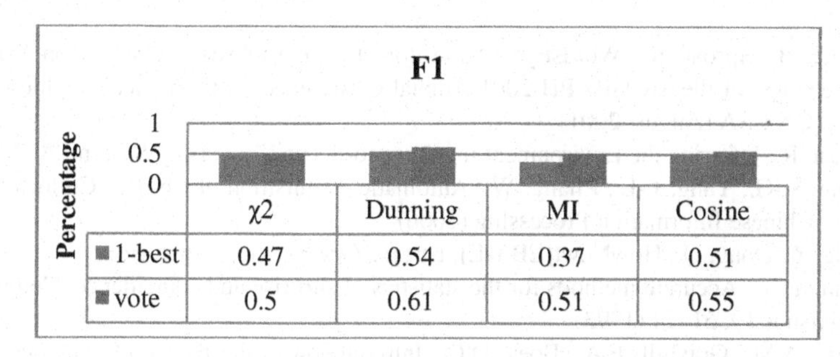

Fig. 1. The results of the the four systems with fixed-window on 1-best and voting methods under the best F1

Table 3. The human evaluation of the scene identification

	ERR	Real ERR
Baseline	39.4**(5.7)	37**(6.6)
System	17.4(17.7)	16.9(17.9)
Hand	8.1**(9.1)	4.2**(3.4)

Where: "Baseline" denotes the randomly tagged 30 sentences; "System" the sentences tagged by the Dunning system on the voting method; and "Hand" the human-judged sentences. Shown are the means and (standard deviations) for 45 subjects for: total errors ERR (= all error categories including "any"); total real errors (= error categories not counting "any"); Differences between the baseline and the system, and the system and the hand are significant at $p<0.01$ on a two sample t-test where indicated with a "**"on the mean for baseline.

4.3 Conclusions and Future Work

In this paper, Cosine coefficient, mutual information MI, $\chi 2$ test, and Dunning's Likelihood Ratios test are conducted in Chinese scene identification automatically. The results shows significantly that system scene identification is better than the random. Dunning's likelihood ratios test on voting is most effective, and F1 of 61% with a fixed-window of 81 words is gotten, but also which is far from the manual annotation. One reason is on the existence of POS tagging errors. Reducing the impact of the noise from the knowledge base is a feasible method. The future work is to filter the knowledge base with collocation[9]. In addition, semantic expansion for action tuples and scenes will be used to resolve data sparseness.

Acknowledgments. Thanks for the supports from National Natural Science Foundation of China under Grant No. 60803094 and Research Project in Beijing Union University.

References

1. Davidov, D., Rappoport, A.: Translation and extension of concepts across languages. In: Proceedings of the 12th Conference of the European Chapter of the ACL, pp. 175–183 (2009)
2. Roxana, G., Nakov, P., Nastase, V., Szpakowicz, S., Turney, P., Yuret, D.: Classification of semantic relations between nominals. Language Resources and Evaluation 43(2), 105–121 (2009)
3. Coyne, B., Sproat, R.: WordsEye: An Automatic Text-to-Scene Conversion System. In: Proceedings of the SIGGRAPH 2001 Annual Conference on Computer Graphics, Los Angeles, CA USA (August 2001)
4. Sproat, R.: Inferring the Environment in a Text-to-Scene Conversion System. ACM (2001)
5. Wang, S.-G., Yang, J.-L., Zhang, W.: Automatic Acquisition of Chinese Collocation. Journal of Chinese Information Processing (2006)
6. Dong, Z., Dong, Q.: HowNewt[EB/OL], `http://www.keenage.com`
7. Dunning, T.: Accurate methods for the statistics of surprise and coincidence. Computational Linguistics 19, 61–74 (1993)
8. Mood, A.M., Graybill, F.A., Boes, D.C.: Introduction to the theory of statistics, 3rd edn. McGraw-Hill, New York (1974)
9. Yang, S.: Machine Learning for Collocation Identification. In: 2003 IEEE International Conference on Natural Language Processing and Knowledge Engieering (NLP-KE 2003), pp. 315–320 (2003)

Chinese Zero Anaphor Detection: Rule-Based Approach

Kaiwei Qin, Fang Kong*, Peifeng Li, and Qiaoming Zhu

School of Computer Science and Technology, Soochow University, Jiangsu 215006, China;
Key Lab of Computer Information Processing Technology of Jiangsu Province,
Suzhou, Jiangsu, 215006, China

Abstract. A rule-based approach for Chinese zero anaphor detection is proposed. Given a parse tree, the smallest IP sub-tree covering the current predicate is captured. Based on this IP sub-tree, some rules are proposed for detecting whether a Chinese zero anaphor exists. This paper also systematically evaluates the rule-based method on OntoNotes corpus. Using golden parse tree, our method achieves 82.45 in F-measure. And the F-measure is 63.84 using automatic parser. The experiment results show that our method is very effective on Chinese zero anaphor detection.

Keywords: Natural Language Processing, Chinese Zero Anaphor, Parse Tree, Rule-based Approach.

1 Introduction

Zero anaphors generally refer to noun phrases that can be understood in the preceding utterances and do not need to be specified in a discourse. In English, definite noun phrases and overt pronouns are frequently employed as referring expressions, which refer to preceding entities. In comparison with English, Chinese is a pro-drop language. Kim (2000) compared the use of overt subjects in English and Chinese. He found that overt subjects occupy over 96% in English, while this percentage drops to only 64% in Chinese. This indicates the prevalence of zero anaphors in Chinese and the necessity of zero anaphora resolution in Chinese anaphora resolution.

For resolving zero anaphors, we have to detect them first. Referring to the linguistic studies, especial the work of Liao (1992), the omission of noun phrases is heavily related to the verbs in sentences and only the elements governed by the verbs can be omitted. In this paper, we adopt this notion and propose a rule-based approach to detect zero anaphors considering the lexical and syntactical knowledge.

2 Related Work

Although zero anaphors are prevalent in many languages, such as Chinese, Japanese and Spanish, there only have a few works on zero anaphora resolution. In comparison

* Corresponding author.

Y. Wang and T. Li (Eds.): Knowledge Engineering and Management, AISC 123, pp. 403–407.
springerlink.com

with zero anaphora resolution, the works on zero anaphor detection are much fewer. Representative work includes: Yeh and Chen (2004) [2]proposed an approach for Chinese zero anaphora resolution based on the centering theory. They used a set of hand-engineered rules to perform Chinese zero anaphor detection and gained 85% in accuracy. But the corpus is private. Yang and Xue (2010) [3]proposed a scheme for recovering empty categories. Their system not only used a set of rules to do this task, but also applied a machine learning-based method by using a set of lexical features and syntactic features to recovery empty categories. They mainly detect and recover all the empty elements from syntactic perspective, not consider zero anaphora resolution, which has a close relationship with semantic information. Kong and Zhou (2010) [4]proposed a unified framework for zero anaphora resolution. They divided zero anaphora resolution into three sub-tasks: zero anaphor detection, anaphoricity determination and antecedent identification. Their system used a tree kernel-based method and found that tree kernel-based method were better appropriate for Chinese zero anaphora resolution.

3 Rule-Based Chinese Zero Anaphor Detection

In this section, after introducing the Onto Note corpus, we will give out the rule-based Chinese zero anaphor detection method.

3.1 Corpus

The OntoNotes project aims to annotate a large corpus comprising various genres of text (news, conversational telephone speech, weblogs, use net, broadcast, talk shows) in three languages (English, Chinese, and Arabic) with structural information (syntax and predicate argument structure) and shallow semantics (word sense linked to an ontology and coreference). In this paper, we use part of Chinese corpus, including 100 documents of Xinhua news. In this corpus, empty elements have been annotated. There are six types of empty elements. All these empty elements are annotated from syntactic perspective and cover all the zero anaphors. The distribution of these empty elements is shown in Table 1. From Table 1, we can find that:

1) '*T*' type and 'OP' type account for most of the empty elements. Observing the corpus and considering the generative grammar proposed by Chomsky, we find that '*T*' and 'OP' generally appear in pairs. Due to most of the pairs of '*T' and 'OP' have nothing with zero anaphors, we do not consider them.
2) Most zero anaphors play as subject or object roles in Chinese sentences. In table 1, we also list the distribution of empty elements from grammatical roles. We can find that empty elements acting as subject account for 53.92%, and the empty elements acting as object account for 9.33%.
3) Among the empty elements excluding the 'OP' type, the elements playing as subject or object roles account for about 91.2%. Obviously, the performance of zero anaphors detection heavily depends on the detection of them.

Table 1. Distribution of the six types of empty elements of the corpus

Types	*T*	*pro*	*PRO*	*RNR*	*OP*	Others	Tota l
Subjects Numbers	430	442	397	0	0	2	1271
Objects Numbers	191	4	0	25	0	0	220
Others	121	0	2	19	722	2	866
Total	742	446	399	44	722	4	2357

3.2 Rule-Based Chinese Zero Anaphor Detection

Considering the distribution of empty elements, similar to Wu and Liang (2006) [1], Yeh and Yang (2004)[2], we propose 4 rules to detect zero anaphors. Due to the close relationship between the empty elements and grammatical roles, we present the clause as a 3-tuple.

T=[S,P,O]

In this 3-tuple, S is a list of nouns whose grammatical role is the subject of a clause, P is a list of verbs or a preposition whose grammatical role is the predicate of a clause, O is a list of nouns whose grammatical role is the object of a clause.

Following are the rules for zero anaphor detection:

1) Triple1(S,P,O) → np(S), vtp(P), np(O).

2) Triple2(S,P,none) → np(S), vip(P).

3) Triple3(S,P,O) → np(S), prep(P), np(O).

4) Triple4(S,none,none) →np(s).

In order to reduce the influence of parser errors, we apply the above rules on clause level. That is to say, we only consider part of parse tree (sub-tree) when conduct the zero anaphor detection. The algorithm detecting zero anaphors is:

1) Divide the complete parse tree into a number of 'IP' sub-trees.
2) For every 'IP' sub-tree, traverse the sub-tree hierarchically and find the first 'VP' node. Then find the left sibling node of the 'VP'. If the left sibling node bears the subject grammatical role, we think, there's no zero anaphor in subject position. Otherwise we think there is a zero anaphor in the front of the 'VP' node and it is a subject empty.
3) Determine whether the "VP" node is transitive or not. If the verb is transitive, traverse the "VP" sub-tree to find out whether the right child plays the object role. If there is a object, no zero anaphor exists in object position. Otherwise there is a zero anaphor behind the "VP" node. If the verb is intransitive, we do not need to determine the object role.

4 Experiments and Discussion

4.1 Evaluation Metrics

We use precision, recall and F-measure as our evaluation metrics for Chinese zero anaphor detection similar to Zhao and Ng (2007)[5]. Precision is defined as the number of correctly identified zero anaphora divided by the total number of ZAs that our system produced. Recall is defined as the number of correctly identified ZAs divided by the total number of the ZAs labels in the ontonote_3.0 gold standard data. F-measure is defined as the geometric mean of precision and recall.

4.2 Results Using Golden Parse Tree and Auto Parse Tree

Table 2 gives the performance of zero anaphor detection using golden parse tree, which achieves 91.37%, 75.11% and 82.45 in precision, recall and F-measure, respectively. High precision shows the effect of our rules, but the low recall will influence the performance of zero anaphora resolution. In order to find the reason leading to low recall, we evaluate the recall of Chinese zero anaphor detection over different empty types. Table 3 shows the results.

Table 2. Performance of zero anaphor detection

Golden parse tree			Auto parse tree		
P%	R%	F	P%	R%	F
91.37	75.11	82.45	75.84	54.98	63.84

Table 3. Recall of Chinese zero anaphor detection over different empty categories

Golden parse tree				Auto parse tree			
Type	Total	Correct	Recall(%)	Type	Total	Correct	Recall(%)
T	742	422	56.87	*T*	742	336	45.28
pro	446	434	97.31	*pro*	446	253	56.73
PRO	399	370	92.73	*PRO*	399	309	77.44
RNR	44	0	0	*RNR*	44	0	0
Other	4	2	50	Other	4	1	25

From Table 3 we can see that: the recall of "*pro*" category is about 97.31 and the recall of "*PRO*" category is 92.73. And the recall of "*T*" category is only 56.87. Because "*T*" accounts for the majority of all the empty elements, the overall recall is only 75.11%. Fortunely, most of "*T*" empty elements are non-anaphoric. There will be little influence on zero anaphora resolution.

In order to study the influence of parse errors on Chinese zero anaphor detection, we evaluate our method on auto parse trees. In this paper, we use the Stanford parser as tools to generate the auto parse trees. In comparison with the performance using golden parse tree, the precision, recall and F-measure lowered about 15.53%, 20.13% and 18.61% respectively. For different empty categories, the performance of "*pro*" lowered most. Because the number of "*pro*" account for the majority of Chinese zero anaphors, the performance of Chinese zero anaphora resolution will reduce much using automatic parse tree comparing with using golden parse tree.

5 Conclusion and Future Work

This paper proposes a rule-based method for Chinese zero anaphor detection. Given a parse tree, the smallest IP sub-tree covering the current predicate is captured. Based on this IP sub-tree, some rules are proposed for detecting whether a Chinese zero anaphor exists. This paper also systematically evaluates our method on Onto Notes corpus. The experiment results show that our method is very effective on Chinese zero anaphor detection. In comparison with other previous related work, the performance of our system is better than them using both golden parse tree and automatic parse tree.

Our study also find that the performance of Chinese zero anaphor detection heavily depend on the performance of syntactic parser. In the future work, we will explore joint learning between syntactic parser and Chinese zero anaphor detection.

References

[1] Wu, D.-S., Liang, T.: A Case-Based Reasoning Approach to Zero Anaphora Resolution in Chinese Texts. In: Matsumoto, Y., Sproat, R.W., Wong, K.-F., Zhang, M. (eds.) ICCPOL 2006. LNCS (LNAI), vol. 4285, pp. 520–531. Springer, Heidelberg (2006)

[2] Yeh, C.-L., Chen, Y.-C.: Zero anaphora resolution in Chinese with shallow parsing. Journal of Chinese Language and Computing (2004)

[3] Yang, Y., Xue, N.: Chasing the ghost:recovering empty categories in the Chinese Treebank. In: Colin 2010, pp. 1382–1390 (2010)

[4] Kong, F., Zhou, G.: A Tree Kernel-based Unified Framework for Chinese Zero Anaphora Resolution. In: Proceedings of the 2010 Conference on Empirical Methods in Natural Language Processing, pp. 882–891 (2010)

[5] Zhao, S., Ng, H.T.: Identification and Resolution of Chinese Zero Pronouns:A Machine Learning Approach. In: Proceedings of the 2007 Joint Conference on Empirical Methods in Natural Language Processing and Computational Natural Language Learning, pp. 541–550 (2007)

Enhancing Concept Based Modeling Approach for Blog Classification

Ramesh Kumar Ayyasamy[1], Saadat M. Alhashmi[1], Siew Eu-Gene[2], and Bashar Tahayna[1]

[1] School of Information Technology, [2] School of Business
Monash University, Malaysia
{ramesh.kumar,alhashmi,siew.eu-gene,bashar.tahayna}@monash.edu

Abstract. Blogs are user generated content discusses on various topics. For the past 10 years, the social web content is growing in a fast pace and research projects are finding ways to channelize these information using text classification techniques. Existing classification technique follows only boolean (or crisp) logic. This paper extends our previous work with a framework where fuzzy clustering is optimized with fuzzy similarity to perform blog classification. The knowledge base-Wikipedia, a widely accepted by the research community was used for our feature selection and classification. Our experimental result proves that proposed framework significantly improves the precision and recall in classifying blogs.

Keywords: Blog classification, Wikipedia, Fuzzy clustering, Fuzzy similarity.

1 Introduction

Blog classification plays a very important role in many information management and retrieval tasks. It refers to the task of assigning blogs one or more pre-defined categories. Blog classification is cumbersome than the normal web classification, because blog posts are frequently updated and the bloggers disseminate information and present their ideas on various topics. According to Blogpulse[1] statistics, there are 169.56 million identified blogs, 96,488 new blogs in last 24 hours and 502,525 blogposts are indexed in last 24 hours. This proves that the weblogs are growing at a rapid rate, which is a rich source of information requiring an efficient and effective automatic classification and categorization techniques. Text clustering received a significant attention in recent years in the area of machine learning and text mining applications such as webpages and blogs. The purpose of text clustering is to create vector space model [1]. The entire text collection could be represented as *text by document* matrix. Clustering of blogposts enables automatic categorization and facilitates certain types of blog search. In any clustering method, if two documents are similar to each other, it has to embed in a suitable similarity space. In contrast with *boolean logic*, where binary sets have two-valued logic, fuzzy logic variables may have a truth value that ranges in degree between 0 and 1. Fuzzy logic values are known to be many-valued logic [2]. The difference between fuzzy clustering and

[1] http://www.blogpulse.com

Y. Wang and T. Li (Eds.): Knowledge Engineering and Management, AISC 123, pp. 409–416.
springerlink.com

regular clustering is that in the former case each data point belongs to more than one cluster are fuzzified in accordance with certain membership functions.

Although fuzzy C-means clustering was not widely researched on blog classification, there is a considerable amount of literature [5- 7] carried out on document clustering. Mendes and Sacks [5] analyzed various clustering methods to discover relevant document relationship. Their experiments with various test documents [5] have proven that fuzzy C-means clustering performs better than the hard K-means algorithm. Miyamoto [6] developed a method using fuzzy clustering for fuzzy multi-sets. This work used two dissimilarity measures for computing cluster centers. Saraçoglu et al. [7] used fuzzy clustering to find similar documents. This work is similar to ours, but their clustering is done using training documents. Our work is automatic and does not depend on training documents. Widyantoro and Yen [8] adopted a fuzzy similarity approach originated from Rocchio's algorithm for text classification. This research work [8] used only the fuzzy term-category relation to predict the category of documents.

Despite widespread adoption on clustering tasks, only few studies have investigated using Wikipedia as a knowledge base for document clustering [10 - 13]. Gabrilovich et al., [10] applied structural knowledge repository-Wikipedia as feature generation technique. Their work confirmed that background knowledge based, features generated from Wikipedia can help text categorization. In our previous work [11] we followed two-value logic (or *boolean logic*), and utilized Wikipedia to index and classify text documents. Huang et al. [12] clustered documents using Wikipedia knowledge and utilized each Wikipedia article as a concept. This work extracted related terms such as synonyms, hyponyms and associated terms. Hu et al. [13] used Wikipedia concept feature for document clustering. In our work, we use concepts which are derived from Wikipedia to classify blogposts.

In this paper, we developed a framework where fuzzy clustering is optimized with fuzzy similarity for blog classification. The knowledge base-Wikipedia, a widely accepted by the research community was used for our classification. The reminder of the paper is organized as follows: Section 2 describes our proposed framework, and in Section 3, we evaluate the proposed framework with real data set and discuss experimental results. Finally, we conclude our paper in Section 4.

2 Our Proposed Framework

We present our fuzzy clustering framework (Figure 1) that has different stages in classifying blogposts using fuzzy clustering and fuzzy similarity. Our contribution comes in three folds: we use Wikipedia to find the membership value of categories. Through blogpost, we find the membership value of concepts using *n-gram based concept extraction*. We optimized fuzzy c-means clustering with fuzzy similarity to do blog classification.

2.1 n-gram Based Concept Extraction

We use relatedness measurement method that computes the relatedness among words based on co-occurrence analysis. As a simple case, we map each word to a concept and the combination of these two or more concepts/words can create a new concept (compound concept). For example, assume A is a word mapped to a concept Con_1, and

B is a word mapped to a concept Con_2. Then there is a probability that the combination of A&B can produce a new concept: Let us consider the following example,

A = *Yellow*, B = *River:* A: *Color* concept, B: *Nature* concept.
So, A+B→*Yellow River* → *Geographical place in China*: *Place* concept
and B+A→ *River Yellow*: no Concept.

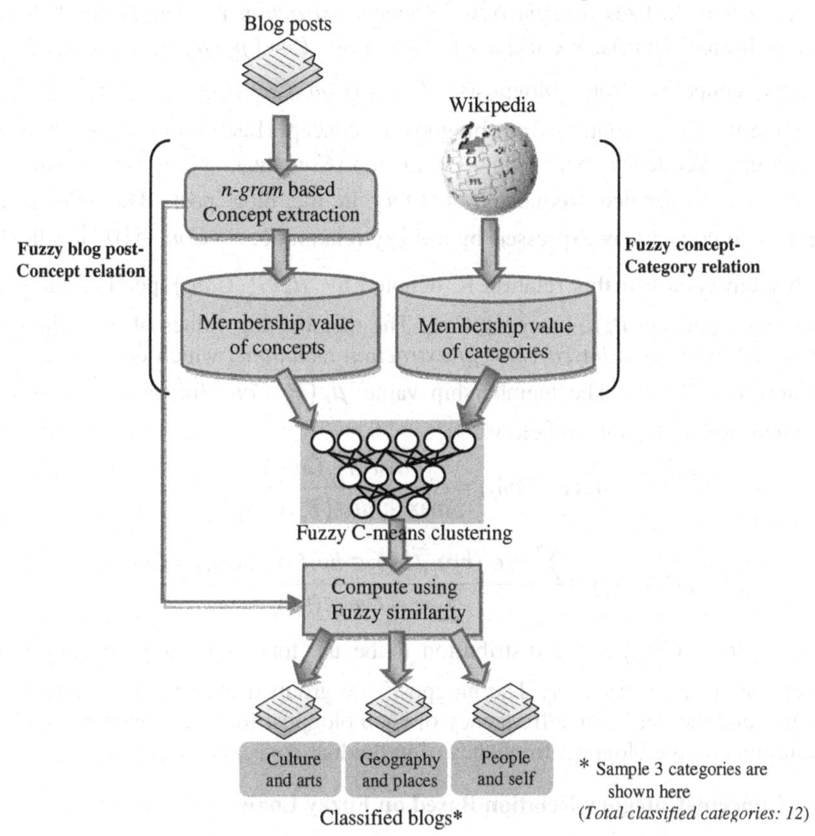

Fig. 1. Our Proposed Framework

Since the order of A and B can give different concept or no concept at all, we neglect the reordering. For effective classification, we mine existing knowledge bases, which can provide a conceptual corpus. Wikipedia is one of the best online Knowledge Base to provide an individual English article with more than 3 million concepts. With that said, an important feature of Wikipedia is that on each wiki page, there is a URI, which unambiguously represent the concept. Our concept-based representation method was guided by the research carried out in concept vectorization and categorization [11]. Two words co-occur if they appear in an article within some distance of each other. Typically, the distance is a window of *k* words. The window limits the co-occurrence analysis area of an arbitrary range. The reason of setting a window is due to the number of co-occurring combinations that becomes too huge when the number of words in an article is large. Qualitatively, the fact that two words often occur close to each other is more likely to be significant, compared to the fact that they occur in the same article, even more so when the number of words in an article is huge. The *n-gram based concept extraction* algorithm was briefly discussed in Ayyasamy et al.'s work [11].

2.2 Blogpost-Concept Relation Based on Fuzzy Logic

Our framework utilizes *n-gram based concept extraction* to identify the belonging concepts from blogposts. We use a set of blogpost $P = \{ p_1, p_2, p_3 \ldots . p_n \}$ and set of concepts collected from blogposts, $Con = \{Con_1, Con_2, Con_3, \ldots .. Con_m\}, m \in \mathbb{Z}^+$ to determine the fuzzy relation from blogpost to concept. Each blogpost is represented by concept frequency pairs $bp = \{(Con_1, w_1), (Con_2, w_2), \ldots (Con_m, w_m)\}$ where w_m denotes the occurrence frequency of Con_m in the blog post. The relevance of blogposts to concepts is expressed by a fuzzy relation $R : P \times Con \rightarrow [0,1]$, where the membership values of this relation R, denoted by $\mu_R(P_n, Con_i)$ specifies the degree of relevance concept P_n to category Con_i. The membership values of this relation are determined by *n-gram based concept extraction technique*, which consists of a blog post and its concepts. The membership value $\mu_R(P_n, Con_i)$ for each concept (1) is computed from (2) is shown below:

$$\mu_R(P_n, Con_i) = \frac{dist(P_n, Con_i)}{\max_{m,k} dist(P_m, Con_k)} \tag{1}$$

$$dist(P_n, Con_i) = \frac{\sum w_i \in (bp)_l, (bp)_l \in bp, Con((bp)_l) = Con_i w_i}{\sum w_i \in (bp)_l, (bp)_l \in bpw_i} \tag{2}$$

Where $dist(P_n, Con_i)$ is the distribution to be the total number of occurrences of concepts P_n in category Con_i. The blogposts are grouped together according to their concepts and the occurrence frequency of each blogpost for each concept is collected by summing up the blogpost frequency of individual concepts.

2.3 Concept-Category Relation Based on Fuzzy Logic

We use a set of concepts $Con = \{Con_1, Con_2, Con_3, \ldots .. Con_m\}, m \in \mathbb{Z}^+$ and Wikipedia category set $C = \{c_1, c_2, c_3 \ldots . c_{12}\}$ to determine the fuzzy relation from concept to category. Each category is represented by concept-frequency pairs $cp = \{(Con_1, w_1), (Con_2, w_2), \ldots (Con_m, w_m)\}$ where w_m denotes the occurrence frequency of Con_m in the category. The relevance of concepts to categories is expressed by a fuzzy relation $R : Con \times C \rightarrow [0,1]$, where the membership values of this relation R, denoted by $\mu_R(Con_i, c_j)$ specifies the degree of relevance concept Con_i to category c_j. The membership value $\mu_R(Con_i, c_j)$ for each category (3) is computed from (4) is shown below:

$$\mu_R(Con_i, c_j) = \frac{dist(Con_i, c_j)}{\max_{k,l} dist(Con_k, c_l)} \tag{3}$$

$$dist\left(Con_i, c_j\right) = \frac{\sum w_i \in (cp)_k, (cp)_k \in cp, c(cp_k) = c_j w_i}{\sum w_i \in cp_k, cp_k \in cpw_i} \tag{4}$$

Where $dist\left(Con_i, c_j\right)$ is the distribution to be the total number of occurrences of concepts Con_i in category c_j. The concepts are grouped together according to their categories and the occurrence frequency of each concept for each category is collected by summing up the concept frequency of individual category.

2.4 Fuzzy C-Means Clustering

For fuzzy C-means clustering, membership value of concepts (1) and membership value of categories (3) are involved. We utilize Fuzzy C-means clustering (FCMC), which was developed by Dunn [3] and improved by Bezdek [4]. FCMC allows one piece of data to belong to one or more clusters with different membership degrees (between 0 and 1) and fuzzy boundaries between clusters. Let blogpost $P = \{p_1, p_2, p_3 .....p_n\}$ be a finite dataset and U be the fuzzy matrix: where each $p_i = \{p_{i1}, p_{i2}, p_{i3} .....p_{if}\}$ has f features, such that $M_{fcn} = \{U \in R^{c\times n} : u_{ik} \in [0,1]$. To classify P into c $(2 \le c \le n)$ categories, let V be $V = (v_1, v_2, v_3 .....v_c)^{CON}$ the cluster centers or prototype matrix, $V \in R^{c\times f}$. To find the optimization of fuzzy clustering, a fuzzy matrix U is selected, where $U \in M_{fcn}$, we equate as:

$$M_{fcn} = \left\{ U \in R^{c\times n} : u_{ik} \in [0,1]; \sum_{i=1}^{c} u_{ik} = 1; 0 < \sum_{k=1}^{n} u_{ik} < n \right\} \tag{5}$$

FCMC executes iteratively these following steps until a satisfactory objective is reached. The objective here is defined as reducing the total error to a given threshold value ε or stopping after a certain number of iterations are completed. To minimize the objective function J_m to calculate U and V [4],

$$J_m(U,V) = \sum_{k=1}^{n}\sum_{i=1}^{c} (u_{ki})^m d_{ki}^2 = \sum_{k=1}^{n}\sum_{i=1}^{c} (u_{ki})^m \left\| P_k - v_i \right\|^2 \tag{6}$$

where m is the fuzzifier, $m>1$, which controls the *fuzziness* of the method. u_{ki} is the membership degree of P_k in the cluster i, P_k is the kth of d-dimensional measured data, v_i is the d-dimension center of the cluster, and ||*|| is any norm expressing the similarity between any measured data and center. Both the parameters need to be specified beforehand. $d_{ki}^2 = \left\| P_k - v_i \right\|^2$ is the square Euclidean distance between data object P_k to cluster center v_i .Thus the fuzzy clustering is carried out by minimizing the objective function (4) with the update of membership u_{ki} (7) and the cluster centers v_i (8) as,

$$u_{ki} = \left[\sum_{l=1}^{c} \left[\frac{\left\| P_k - v_i \right\|}{\left\| P_k - v_l \right\|} \right]^{\frac{2}{m-1}} \right]^{-1} \tag{7}$$

$$\text{Where } v_i = \frac{\sum_{k=1}^{n} u_{ki}^{m}.P_k}{\sum_{k=1}^{n} u_{ki}^{m}} \tag{8}$$

The choice of the above appropriate objective function (5) is the point to the success of the cluster analysis. The iteration (6) terminates when

$$\text{MAX}_{ij}\{|u_{ki}^{(l+1)} - u_{ki}^{(l)}|\} < \Phi \tag{9}$$

Where Φ is between 0 and 1. Thus for a finite set of blogposts, the fuzzy C-means clustering returns a list of cluster centers $V = (v_1, v_2, v_3.....v_c)^{CON}$ and a fuzzy matrix u_{ki}.

2.5 Fuzzy Similarity

The aim of using fuzzy similarity is to identify the category of the blogpost based on fuzziness. One can suggest that, fuzzy C-means clustering returns the cluster centers, which is nothing but identified category of the blogpost. In our framework we would like to improve the fuzzy C-means clustering result, by finding the fuzzy similarity with the returned list of *n-gram based concept extraction.*

In this work [9], they highlighted the six fuzzy similarity measures using fuzzy conjunction and disjunction operators such as *Einstein, Hamacher, Bounded Difference, Algebraic, Drastic and Min-Max.* Their empirical result on [9] showed that similarity measure using *Einstein* operators achieved the best performance in classifying blogs than the other measures. We utilize Einstein similarity measure for classifying blogs by measuring the similarity between the returned list of *n-gram based concept extraction* and the fuzzy C-means cluster center.

3 Experiments

This section describes the experiment that test the efficiency of our proposed framework. We carried out experiments using part of TREC BLOGs08[2] dataset. TREC dataset is a well-known dataset in blog mining research area. This dataset consists of the crawl of Feeds, associated Permalink, blog homepage, and blogposts and blog comments. This dataset is a combination of both English and Non-English blogs. We used blogpost extraction program named *BlogTEX*[3] to extracts blog posts from TREC Blog dataset. During preprocessing, we removed HTML tags and used only English blog posts for our experiment. To test our approach, we collected around 41,178 blog posts and assigned based on the found blog concepts related to each category. Table 1 shows the *CategoryTree* structure used in our experiment. We downloaded the Wikipedia database dumps on 15 February 2010 and extracted 3,207,879 Wikipedia articles. After pre-processing and filtering we used 3,101,144 article titles and are organized into 145,990 subcategories.

---

[2] http://ir.dcs.gla.ac.uk/test_collections/blogs08info.html
[3] http://sourceforge.net/projects/blogtex

Table 1. The *CategoryTree* Structure used in this experiment

Categorical Index	# of documents	# of concepts
General Reference	878	3758
Culture and Arts	8580	18245
Geography and Places	4558	23220
Health and Fitness	4082	21409
History and Events	2471	12746
Mathematics and Logic	213	2561
Natural and Physical sciences	1560	8530
People and Self	6279	10125
Philosophy and Thinking	2345	7591
Religion and Belief	2345	8315
Society and Social sciences	4657	13170
Tech and Applied sciences	3210	26310

We ran our experiment as mentioned in our framework and fuzzy similarity based on fuzzy C-means method, and classified blogposts based on 5 mentioned categories. To evaluate the proposed framework, two popular evaluation measures, *precision* and *recall* are used. *Precision measures* the percentage of *"categories found and correct"* divided by the *"total categories found"*. *Recall* measures the percentage of *"categories found and correct*" divided by the *"total categories correct"*. We evaluated the classification performance among the top 10, 20, and 50 of the classified blogs respectively. From our dataset, most of the blogposts are categorized on *Culture and Arts*, and *Geography and Places*, and very few posts are categorized on *Mathematics and Logics*.

Table 2. Evaluation Result

Categorical Index	Precision	Recall
General Reference	0.810	0.764
Culture and Arts	0.792	0.663
Geography and Places	0.841	0.735
Health and Fitness	0.928	0.762
History and Events	0.787	0.683
Mathematics and Logics	0.741	0.769
Natural and Physical sciences	0.910	0.785
People and Self	0.850	0.694
Philosophy and Thinking	0.962	0.815
Religion and Belief	0.827	0.897
Society and Social sciences	0.942	0.895
Tech and Applied sciences	0.903	0.823
Average	**0.8577**	**0.7654**

From Table 2 we can note that *Mathematics and Logics* and *Religion and Belief* have less precision and more recall than the other four categories. Our framework produced better precision of 85.77% and recall of 76.54% on classifying blogs.

4 Conclusions

In this paper, a framework in optimizing fuzzy clustering with concept based modeling approach for blog classification is presented. We demonstrated the framework's effectiveness by measuring *precision* and *recall* through real blog dataset. We conclude that using of Wikipedia knowledge base combined with fuzzy clustering and fuzzy similarity measures produce better results than the traditional clustering techniques. In our future work, we plan to improve our experiment using the full TREC Blogs08 dataset and combining the blog features such as blog tags to measure the clustering performance and its scalability.

References

1. Salton, G., Buckley, C.: Term-weighting approaches in automatic text retrieval. J.Information Processing & Management 24, 513–523 (1988)
2. Zadeh, L.A.: Fuzzy Sets, Information and Control 8, 338–353 (1965)
3. Dunn, J.C.: A fuzzy relative of the ISODATA process and its use in detecting compact well-separated clusters. J. of Cybernetics 3(1), 32–57 (1973)
4. Bezdek, J.C.: Pattern Recognition with Fuzzy Objective Function Algorithms. Kluwer Academic Publishers, Norwell (1981)
5. Mendes, M.E.S., Sacks, L.: Evaluating fuzzy clustering for relevance-based access. In: IEEE International Conference on Fuzzy Systems, pp. 648–653 (2003)
6. Miyamoto, S.: Fuzzy multisets and fuzzy clustering of documents. In: 10th IEEE International Conference on Fuzzy Systems, pp. 1191–1194 (2001)
7. Saraçoglu, R., Tütüncü, K., Allahverdi, N.: A fuzzy clustering approach for finding similar documents using a novel similarity measure. Expert Systems with Applications 33(3), 600–605 (2007)
8. Widyantoro, D.H., Yen, J.: A Fuzzy Similarity Approach in Text Classification Task. In: IEEE International Conference on Fuzzy Systems, pp. 653–658 (2000)
9. Ayyasamy, R.K., Tahayna, B., Alhashmi, S., Eu-gene, S.: Concept Based Modeling Approach for Blog Classification using Fuzzy Similarity. In: 8th IEEE International Conference on Fuzzy Systems and Knowledge Discovery, pp. 1007–1011 (2011)
10. Gabrilovich, E., Markovitch, S.: Overcoming the brittleness bottleneck using Wikipedia: Enhancing text categorization with encyclopedic knowledge. In: AAAI, Park (2006)
11. Ayyasamy, R.K., Tahayna, B., Alhashmi, S., Eu-gene, S., Egerton, S.: Mining Wikipedia Knowledge to improve Document Indexing and Classification. In: 10th Int. Conf. on Information Science, Signal Processing and their Applications, pp. 806–809 (2010)
12. Huang, A., Milne, D., Frank, E., Witten, I.H.: Clustering Documents Using a Wikipedia-Based Concept Representation. In: Theeramunkong, T., Kijsirikul, B., Cercone, N., Ho, T.-B. (eds.) PAKDD 2009. LNCS, vol. 5476, pp. 628–636. Springer, Heidelberg (2009)
13. Hu, J., Fang, L., Cao, Y., Hua-Jun Zeng, H., Li, H.: Enhancing Text Clustering by Leveraging Wikipedia Semantics. In: ACM SIGIR, pp. 179–186 (2008)

A Distributed Clustering with Intelligent Multi Agents System for Materialized Views Selection

Hamid Necir and Habiba Drias

Research Laboratory in Artificial Intelligence (LRIA) Department of Computer Science, Faculty of Electrical and Computer Science, USTHB, University Science and Technology Houari Boumediene, El Alia BP 32, Bab Ezzouar, Algiers, Algeria
ncrhmd@yahoo.fr, hdrias@usthb.dz

Abstract. Materialized views are the most common approach that can provide optimal performance in processing time, especially for *OLAP* queries known for their great complexity. Due to the large computation and storage limitation, materialization of all possible views is not possible. Therefore, the key issue is to choose an optimal set of views to materialize. However, this task is a very hard, especially in the data warehouses context, where a trade-of-between performance and view storage cost must be taken into account when deciding which views should be materialized. Addressing this problem, we propose a new approach with two main phases. The first involves pruning the search space to reduce the number of views candidates. In this order, we use a distributed clustering approach using multi agents system that can significantly reduces the complexity of the selection process The second phase uses also a multi agent's architecture to capture the relationships between views candidates to select the final set of materialized views. This set minimizes the query processing cost and satisfy the storage constraint. We validate our proposed approach using an experimental evaluation.

Keywords: Clustering, data warehouse, materialized view, multi agents, OLAP.

1 Introduction

Data warehouses are usually modeled by a star schema represented by a single table, called fact table, and a set of dimensional tables. Each dimension table is joined to the fact table via foreign keys, but the dimension tables are not joined to each other [22]. In this context, *OLAP* queries are complex and can take many hours or even days to run, if executed directly on the raw data [15], [16]. The performances of queries will be greatly enhanced by storing some intermediate results of these queries processing in the data warehouse. These stored results are called materialized views [29], [33].

Therefore, some queries can be efficiently processed, because they require only access to materialized views and do not need to the original data. We have also the possibility to index materialized views.

Y. Wang and T. Li (Eds.): Knowledge Engineering and Management, AISC 123, pp. 417–426.
springerlink.com

Several studies and experiments have shown that the materialized views are one of the most judicious options to improve *OLAP* queries performances [19], [29], [5], [13], [40], [33], [1]. Several commercial database-management systems (DBMS) use materialized views to speed up query evaluation [9]. Nevertheless, these views require additional storage space and entails maintenance overhead when refreshing the data warehouse [3]. The storage limitation prohibits the materialization of all possible views [30]. Therefore, the selection of an optimal set of materialized views is a hard problem [16], [7], [15], [20], especially in the data warehouses context [37] where a trade-off between reducing storage costs and improvement query performance should be determined.

Addressing this problem, many methods and algorithms have been published in the literature [7], [19], [17], [41], [42], [16], [24], [40], [4], [12]. Unfortunately, to the best of our knowledge, no studies have been proposed for considering a distributed method for materialized view selection.

In this order, the main idea of our work consists in adopting a distributed intelligence approach [8], and to take the advantages of the multi-agent approach [35], [10], [14], [39]; like: parallelism, robustness, scalability [36], as a mean of coping with the complexity of this problem in dense databases. In order to reflect a geographically compartmentalized business in which each region is relatively autonomous, we adopt a distributed clustering using a COordination Agent (*COA*) to manage each of the regions. Each *COA* is in charge of managing a set of Cluster Agents (CAs). Each CA manages a set of predicates that is very similar for each functional segment (sales department, the shipping department, the finance department...). The centralized clustering is a computationally intensive process [26] that limits its use a relatively small set of data. Overeinder et al., [27] have shown also that clustering with a decentralized agent method produces a better clustering than the centralized algorithms. We believe that a similar observation holds for the materialized view selection problem.

Our approach uses two main phases to select an optimal set of materialized views. The first prunes the search space to reduce the number of views candidates. In this pruning phase, we adapt *K-means* algorithm [25], [2] that allows grouping common sub expressions within a set of queries that can be resolved by one single materialized view. In order to reduce the complexity of the selection process, we proceed in the following way: (1) for all functional segments of one region of business, *COA* creates a set of initial clusters, using an incremental clustering approach [18]. This method improves the performance of the k-means algorithm versus the dependency on initial parameters. (2) For all regions of business, we improve these partial solutions by an exchange of predicates between the coordination agents until no further exchange action is possible. Once the pruning phase is processed, we combine these partial solutions using a merge agent to produce a global selection. This set of materialized views minimizes the query processing cost and satisfy the storage constraint.

The rest of the paper is organized as follows. Section 2 presents our formalization of the problem of selecting materialized views. Section 3 presents our approach for generating materialized views. Section 4 validates our work using experimental studies. Finally, a conclusion and research perspectives are presented in section 4.

2 Formulation of Materialized View Selection Problem

Materialized view selection problem is formalized as follows: given a source of data (in our case, a data warehouse modeled using star schemas with a set of dimension tables and a fact table), a workload of queries and a set of constraints on materialized views (e.g., space available for storing the materialized views). The goal of this problem consists in automatically selecting an appropriate set of views would that satisfy the constraints and reduce the evaluation costs of the queries.

3 Our Approach

Our new approach, consider the selecting materialized views problem in data warehouse environments using three main phases: (1) pre-processing of queries (2) generation of views candidates and (3) selection of a final configuration. Fig. 1 shows the set of agents used in our approach.

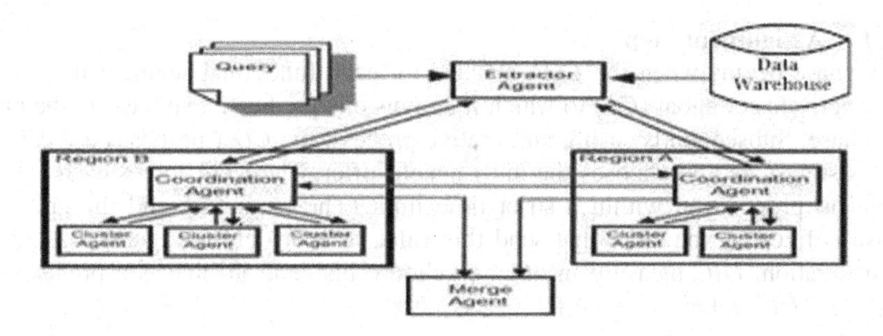

Fig. 1. The set of agents used in our approach

3.1 Pre-processing of Queries

To implement this step, we use an extractor agent that is in charge of extraction of all predicates that are referenced in the clause *WHERE* from a given set of queries workload.

Recall that most prevalent star queries typically include restrictions on multiple dimension tables that are used for selecting specific facts. These join constraints are defined between the fact table and one or many dimension tables on key attributes [21]. In this order, the predicates are divided into two main categories.

1. Restriction predicate: This predicate is defined by: $A\theta v$, where A represents a descriptive attributes of dimensional table, $\theta \in \{>, =, <, \leq, \geq\}$ and v is a constant with the numeral or alphabetic value.

2. Join predicate: This predicate is defined by: $A = B$, where A and B represent respectively a two key attributes.

Once the extraction phase is processed, extractor agent builds a binary matrix called predicates usage matrix. In this matrix each row and column represents one query and its predicates. A value of this matrix is given by:

$$\begin{cases} 1 \text{ if query } q_i \text{ uses the predicate } p_i \\ 0 \text{ otherwise} \end{cases}$$

The extractor agent uses also specific metadata that describe informations about queries, tables and attributes in order to transmit adapted predicates to each *COA*.

3.2 Generation of Candidates' Views

This phase prunes the search space of materialized view selection problem by an adapted distributed approach. Since we are interested by materialized view selection in relational data warehouse modeled by star schema, in this phase, we use only the restriction predicates to generate candidates clusters. In this order, we proceed in the following way:

3.2.1 Assignment Step

This phase begins when the *COA* creates, for each functional segment in its region, the single cluster agent (*CA*) to which it assigns one predicate extracted in the preceding phase. Subsequently, using an iterative process, the *COA* proposes a predicate to the existing *CA*s and chooses the most benefit offer. The *COA* forces each *CA* to respond to proposition within a strict time limit (The beginning and the end of this phase is fixed by the *COA* that send this value to every cluster agent). *CA* uses, for each iteration, *DIC* measure in order to choose his centroid that is a predicate with minimum *DIC* value.

$$DIC(o_i) = \frac{\sum_{j=1}^{j=k} Dist(o_j, o_i)}{k}. \quad (1)$$

Where o_i and k represent respectively an object and the number of predicate of the cluster.

$Dist(o_i, o_j)$ evaluates the distance between to predicate o_i and o_j. This measure is computed using the Manhattan measure:

$$Dist(o_i, o_j) = \sum_{k=1}^{m} |M(o_i, q_k) - M(o_j, q_k)|. \quad (2)$$

Where m defines a number of queries that contain the object o_i or o_j and $M(o_i, q_k)$ represents a cell of object usage matrix.

To evaluate each proposition, the cluster agent computes the distance between this predicate o_i and his centroid c_j using $Dist(o_i, c_j)$.

In order to reduce the effect of outliers, a cluster agent considers only the set of predicates with distance values less or equal than a given minimum distance, called *MinDist*. When a value of *MinDist* is selected, it can vary between the value that obtain the largest number of clusters to the value obtaining the smallest. Each cluster agent sends its respond with a personal message. We have two cases: (1) each cluster agent sends "accept_proposal" message if the distance value between this predicate and his centroid provides more advantages by considering the new predicate. This message includes this distance value. (2) each cluster agent sends "refuse_proposal" message if the distance value between this predicate and his centroid does not provide any advantage by considering the new predicate. The *COA* assigns the predicate to the cluster agent with superior offer that means minimizing his distance. In the case or no stake was registered, *COA* creates a new cluster agent and also assigns this predicate. This state comes to an end if *COA* has no predicates to propose. In this case, each *CA* sends the set of his clusters to *COA*.

In order to take into account the limitation of the communication range of each agent, thus limiting the amount of information that can be exchanged, we have assumed that when a new agent joins the system, it will only communicate with the *COA*.

3.2.2 Cooperation Step

In this step, we consider all environments of agents. There is no central controller that directs the behavior of the agents, meaning that cooperation collective emerges [31] because for a Coordination Agent the predicates exchange can achieve not only its own goals, but also the goals of agents other than itself. Therefore, cooperation is beneficial to both parties. In this step, *COA* uses two roles (announce_exchange, propose_exchange), each with different specific capabilities. The role may change, but it can only play one role at a time.

The cooperation phase begins when a *COA* that sends an "announce_exchange" message to announce a tender for predicate exchange. In this step, we consider a random policy in the choice of first agent who made the announcement. We consider that this model is sufficient to capture the collective behavior of large population agents. Indeed, [38] showed that greater efficiency can be attained by introducing random assignment to the initial preference of agents. The new offer is communicated for all existing *COA*s and each *COA* is allowed to make one bid for each announcement it receives, and the bids of the other parties are not revealed to it. The exchanges are directed in the sense that an announcement is not sent to all other agents [28]. After the reception of the "announce" message, the *COA* evaluates this offer using *Dist* measure. It sends a reply with a personal message in a strict time limit (The beginning and the end of this period is fixed by the *COA* that sends this value for every other *COA*s). We have two cases:

1. Each *COA* sends "accept_proposal" messages if it accepts the proposal. The "accept_proposal" message is materialized by a predicate that symbolizes the estimation to the request of the exchange.
2. Each *COA* sends "refuse_proposal" message if this proposition does not improves these clusters. In this case, the proposition (predicate) is not incorporated into any cluster.

The exchange is accepted with the most superior received offer and if no stake was registered, the exchange will be rejected (no cooperation is possible). This state comes to an end with a message of confirmation.

3.3 Building of Candidates' Views

Since we are interested by materialized views, we use the context matrix for adding to the set of generating clusters by the last step, the join predicates that always appear with them. Finally, we use the *DBMS* optimizer to automatically choose the most efficient execution of queries (implicitly rewriting query).

3.4 Final Selection of Views

Building all resulting views from the last step is not feasible in general because of some limitations as the storage constraint. Therefore, we must select from the set of proposed views a subset which is the most appropriate to the workload.

To achieve this task, a merge agent improves partial solutions using two steps. In the first one, merge agent rejects all clusters (view candidates) selected by preceding phase that don't respect the constraint of space disk *S*. After this operation, merge agent captures dependency between clusters by detecting common sub-expressions within clusters using adapted Harinarayan et al., [19] approach. Merge Agent uses also a mathematical cost model which is an adaptation of Aouiche's model [4] to evaluate each candidate's view. In the second, merge agent uses configuration having views which provides more advantages while executing query by considering the storage constraint, and then improve iteratively the initial configuration until no further reduction in total query processing cost and no violation of the storage is bound.

4 Experimental Evaluations

To validate our approach of materialized view's selection we use *JADE* (Java Agent Development Framework) as the tool to implement the multi-agent infrastructure.

We then conducted a set of experiments using a star schema generated by the tool Apb.exe. The star schema of data has one fact table *Activars* (23 780 000 tuples), four dimension tables: *Prodlevel* (10 000 tuples), *Custlevel* (1000 tuples), *Chanlevel* (9 tuples), *Timelevel* (24 tuples). The proposed benchmark includes also 200 *OLAP* queries. The experiments were conducted according to three scenarios: (1) Evaluation of our approach by executing the 200 queries without considering storage constraint, (2) evaluation of our approach by considering the storage constraint and (3) measurement of CPU bound of our approach.

4.1 Evaluation without Storage Constraint

Figure2 shows how our approach reduces the cost of executing the 200 queries. The main result is that our approach gives better performance for almost all values of *MinDist*. The performance of our approach deteriorates for *MinDist* value becomes high (which reduce the number of generated candidates' views). We notice that for values of *MinDist* exceed 16.5, our approach ceases to generating new candidates' views. Moreover, we also observe that the maximal gain is reached for *MinDist* values equal to 14.5.

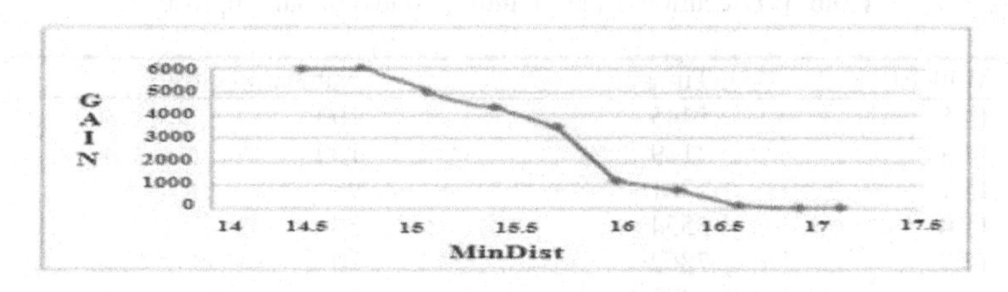

Fig. 2. Behavior of our approach without Storage

4.2 Evaluation with Storage Constraint

All clusters generated by the pruning phase represent some of candidates' views. These candidates cannot be selected in practice, considering the different constraints of storage space. We execute our approach for selecting views by considering various storage values with a fixed value of *MinDist* equal to 14.5. This value allows the generation of a large number of candidates views. Figure 3 shows that as we increase the storage space occupation, that the performance also increases. This is predictable because we select more materialized views when storage space is large. For used scenario, a significant reduction in total query processing cost is reached for a space occupation greater than 64%. On the other hand, for the percentage values of storage space ranging between 45% and 64%, we have a similar performance.

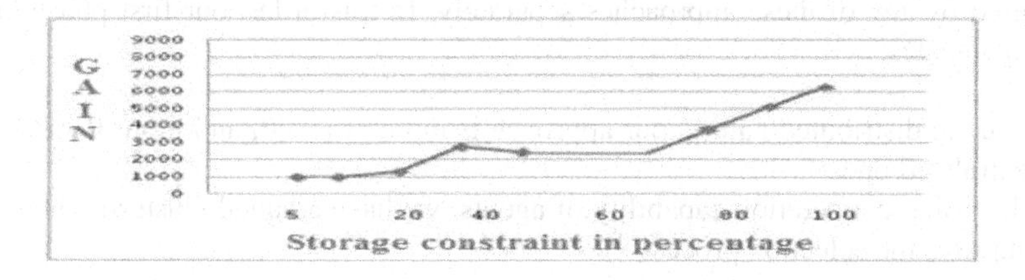

Fig. 3. Behavior of our approach depending on storage space

4.3 Evaluation of Our Approach Based on CPU Bound

We have conducted another bunch of experiments about the processing time of our algorithm (in microseconds) according to a varying value of *MinDist*.

Table 1 shows that approach requires the smallest time for almost all *MinDist* values that we consider. Processing time required for our approach decreases for large values of *MinDist*. These results are foreseeable because big values of *MinDist* decrease the number of selected clusters.

Table 1. Execution Time (in milliseconds) of our Approach

MinDist	Time	Number of cluster
14,5	5445	100
14.8	4119	100
15.1	2761	93
15,4	5554	87
15,7	7862	70
16	1435	24
16,3	1779	16
16,6	312	3
16,9	0.5	1
17,1	0.5	1

5 Conclusions

In this work, we have considered the problem of selection materialized views in data warehouse context. In order to deal with this complex challenge, we have used two main phases: (1) generation of views candidates and (2) selection of a final configuration of materialized views.

In first phase, we have proposed a new approach that takes the advantages of the integration of agents and data mining in order to correctly prune search space in selection process. This integration has the potential to result in advantages that cannot be delivered by any of these approaches separately. In this order, our first phase brings two advantages:

1. Due to distributed nature, our approach is more efficient, and more flexible than the centralized ones.
2. By using cooperation capability of agents, we have adapted clustering algorithm that improve the selection process.

In the second phase, we have also proposed agent architecture to select a set of materialized views that minimize the total query response time cost under the storage constraint. However, it will be interesting to extend our approach by considering other optimization techniques, like index, fragmentation, etc. Another possible extension is to consider other clustering algorithms like Fuzzy C-means, etc.

In order to make our approach usable in real-world applications, we also plan to extend this work to the problem of selecting indexes and materialized views for the efficient sharing of the available storage space because these data structures are often used together. Furthermore, this work may also be extended to dynamic change of materialized views when significant changes in the workload and table sizes occur.

References

1. Agrawal, R., Aggarwal, C., Prasad, V.V.: Depth first generation of long patterns. In: 7th International Conference on Knowledge Discovery and Data Mining, pp. 108–118 (2000)
2. Anderberg, M.R.: Cluster analysis for applications. Academic Press (1973)

3. Aouiche, K., Darmont, J.: Data mining based materialized view and index selection in data warehouses. Journal of Intelligent Information Systems 33(1) (2009)
4. Aouiche, K., Jouve, P.-E., Darmont, J.: Clustering-based materialized view selection in data warehouses. In: Manolopoulos, Y., Pokorný, J., Sellis, T.K. (eds.) ADBIS 2006. LNCS, vol. 4152, pp. 81–95. Springer, Heidelberg (2006)
5. Baralis, E., Paraboschi, S., Teniente, E.: Materialized view selection in a multidimensional database. In: Very Large Data Base, pp. 156–165 (1997)
6. Baril, X., Bellahsene, Z.: Selection of materialized views: a cost-based approach. In: Proceedings of the 15th International Conference on Advanced Information Systems Engineering, pp. 665–680 (2003)
7. Bellatreche, L.: Utilisation des vues matérialisées, des index et de la fragmentation dans la conception logique et physique d'un entrepôt de données. Phd thesis. Clermont-ferrant University. France (2000)
8. Gasser, L., Bond, A.H.: Readings in distributed artificial intelligence. Morgan Kaufmann Publishers, San Mateo (1988)
9. Cao, L.: Data mining and multi-agent integration, 1st edn., pp. 77–92. Springer, Heidelberg (2009)
10. Chaib-draa, B., Jarras, I., Moulin, B.: Systèmes multi-agents: principes généraux et applications. In: Principes et architectures des systèmes multi-agents. Collection IC2, hermes science publication (2002)
11. Chan, G.K.Y., Li, Q., Feng, L.: Design and selection of materialized views in a data warehousing environment: a case study. In: Proceedings of the 2nd ACM International Workshop on Data Warehousing and OLAP, pp. 42–47 (1999)
12. Derakhshan, R., Stantic, B., Korn, O., Dehne, F.: Parallel simulated annealing for materialized view selection in data warhousing environments. In: Proceedings of the International Conference on Algorithms and Architectures for Parallel Processing, pp. 121–132 (2008)
13. Grumbach, S., Rafanelli, M., Tininini, L.: On the equivalence and rewriting of aggregate queries. Acta informatica 40(8), 529–584 (2004)
14. Guessoum, Z., et al.: Monitoring and organizational-level adaptation of multi agent systems. In: Proceedings of the Third International Joint Conference on Autonomous Agents and Multi Agent Systems, pp. 514–521 (2004)
15. Gupta, H.: Selection of views to materialize in a data warehouse. In: Afrati, F.N., Kolaitis, P.G. (eds.) ICDT 1997. LNCS, vol. 1186, pp. 98–112. Springer, Heidelberg (1996)
16. Gupta, H., Mumick, S.: Selection of views to materialize under a maintenance cost constraint. In: Proceeding of 7th International Conference on Extended Database Theory, pp. 453–470 (1999)
17. Gupta, H., Mumick, S.: Selection of views to materialize in a data warehouse. IEEE Transactions on Knowledge and Data Engineering 17(11), 24–43 (2005)
18. Hammouda, K., Kamel, M.: Incremental document clustering using cluster similarity histograms. In: The 2003 IEEE/WIC International Conference on Web Intelligence, pp. 597–601 (2003)
19. Harinarayan, V., Rajaraman, A., Ullman, J.D.: Implementing data cubes efficiently. In: Proceedings of International Conference of ACM SIGMOD, pp. 205–216 (1996)
20. Horng, J.T., Chang, Y.J., Liu, B.J.: Applying evolutionary algorithm to materialized view selection in data warehouse. In: Soft Computing, pp. 574–581 (2003)
21. Karayannidis, N., Tsois, A., Sellis, T., Pieringer, R., Markl, V., Ramsak, F., Fenk, R., Elhardt, K., Bayer, R.: Processing star queries on hierarchically clustered fact tables. In: International Conference on Very Large Data Bases, VLDB, pp. 730–741 (2002)

22. Kimball, R., Ross, M.: The data warehouse toolkit: the complete guide to dimensional modeling, 2nd edn. John Wiley & Sons (2002)
23. Kotidis, Y., Roussopoulos, N.: Dynamat: A dynamic view management system for data warehouses. In: Proc of the ACM SIGMOD, pp. 371–382 (1999)
24. Lee, M., Hammer, J.: Speeding up materialized view selection in data warehouses using a randomized algorithm. International Journal of Cooperative Information System 10(3), 327–353 (2001)
25. MacQueen, J.B.: Some methods for classification and analysis of multivariate observations. In: Proceedings of 5th Berkeley Symposium on Mathematical Statistics and Probability, pp. 281–297. University of California press, Berkeley (1967)
26. McCallum, A., Nigam, K., Ungar, L.: Efficient clustering of high dimensional data sets with application to reference matching. In: Sixth ACM SIGKDD International Conference on Knowledge Discovery and Data Mining, pp. 169–178 (2000)
27. Overeinder, B., van Steen, M., Brazier, F.: Group formation among peer-to-peer agents: Learning group characteristics. In: Proceedings of the Second International Joint Conference on Autonomous Agents and Multi-Agent Systems, pp. 789–796 (2003)
28. Parunak, H.V.D.: Manufacturing experience with the contract net, pp. 285–310. Morgan Kaufmann (1987)
29. Rizzi, S., Saltarelli, E.: View materialization vs indexing: balancing space constraints in data warehouse design. In: The 15th International Conference on Advanced Information Systems Engineering, pp. 502–519 (2003)
30. Shukla, A., Deshpande, P., Naughton, J.: Materialized view selection of multidimensional datasets. In: The 24th International Conference on Very Large Data Bases, pp. 488–499 (1998)
31. Steels, L.: Cooperation between distributed agents through self organization Decentralized AI. In: Demazeau, Y., Müller, J.-P. (eds.), pp. 175–196. Elsevier Science Publishers B.V (1990)
32. Steinbrunn, M., Moerkotte, J., Kemper, A.: Heuristic and randomized optimization for the join ordering problem. International Journal on Very Large Database, 191–208 (1997)
33. Valduriez, P.: Join indices. ACM Transaction on Database Systes 12(2), 218–246 (1987)
34. Valluri, S.R., Vadapalli, S., Karlapalem, K.: View relevance driven materialized view selection in data warehousing environment. In: The 13th Australasian Database Technologies, pp. 187–196 (2002)
35. Wang, A., Conradi, R., Liu, C.: A multi-agent architecture for cooperative software engineering. In: Twelfth International Conference of Software Engineering and Knowledge Engineering (2000)
36. Weiss, G., Sen, S. (eds.): Adaption and Learning in Multi-Agent Systems. LNCS, vol. 1042. Springer, Heidelberg (1996)
37. Widom, J.: Research problems in data warehouse. In: 4th International Conference on Information, Knowledge and Managment, pp. 25–30 (1995)
38. Wong, K.Y.M., Lim, S.W., Luo, P.: Diversity and adaptation in large population games. International Journal Mod Phys 18, 222–243 (2004)
39. Wooldridge, M.: An introduction to multi agent systems. John Wiley & Sons, Chichester (2002)
40. Yang, J., Karlapalem, K., Li, Q.: Algorithm for materialized view design in data warehousing environment. In: 23rd International Conference on Very Large Data Bases, pp. 136–145 (1997)
41. Zhang, C., Yang, J.: Genetic algorithm for materialized view selection in data warehouse environments. In: Mohania, M., Tjoa, A.M. (eds.) DaWaK 1999. LNCS, vol. 1676, pp. 116–125. Springer, Heidelberg (1999)
42. Zhang, C., Yao, X., Yang, J.: An evolutionary approach to materialized view selection in a data warehouse environment. IEEE Transactions on Systems, Man, and Cybernetics - Part C: Applications and Reviews 31(3), 282–294 (2001)

One Method for On-Line News Event Detection Based on the News Factors Modeling

Hui Zhang[1] and Guo-hui Li[1,2]

[1] Department of Engineering, School of Information System and Manegement, National University of Defense Technology, Changsha, 410073, China
[2] Science and Technology Foundation of State Key Laboratory of Information System Engineering, National University of Defense Technology, Changsha 410073, China
zhanghui_nudt09@yahoo.com.cn, guohli@nudt.edu.cn

Abstract. On-line news event detection is detecting the first news report of a news event from various news sources in real time. Related to on-line news event detection, in this article, the author firstly introduces a news representation method for the news factors modeling based on the time, locations, characters (or organization), contents, and so on, and deducing a method related to the features of different types of news factors to calculate the weight of those news factors. Considering the insufficient of the traditional detection algorithms, then the author presents the algorithm of Micro-clusters-based on-line news event detection with Window-Adding and conducts an experiment based on news data which is collected in reality. The author achieved a satisfied experimental result verifying the validity of the proposed method.

Keywords: News factors modeling, News event detection, Hierarchical clustering algorithm with agglomeration.

1 Introduction

Nowadays, Internet technology is developing rapidly. On-line news is gradually surpassing the traditional mediums such as newspapers and TV and becoming the first-hand access for people to receive news and information. Even many traditional mediums also have established websites to report news and supply video-news and download services. Internet is a vast amount of information database. As one of the most primary information carriers on Internet, on-line news reports thousands of news events happening anywhere in the world day after day, including political, economic, military, culture, religion, conflict, and other aspects.

How to search the latest news events from the vast amount of news data on Internet is the problem that on-line news event detection, ONED [1] going to solve. In this article, according to news factors, the author establishes news reports modeling. Aiming at the insufficient of the traditional detection algorithms, the author introduces the algorithm of Micro-clusters-based on-line news event detection with Window-Adding.

Y. Wang and T. Li (Eds.): Knowledge Engineering and Management, AISC 123, pp. 427–434.
springerlink.com

2 The News Report Modeling Based on News Factors

A complete news report generally includes the following four elements: time (When), locations (Where), characters (organization) (Who), and contents (What), named the four Whs. These elements are the key to understand news event. The traditional detection methods usually combine the four elements to form a simple vector to make the similarity comparison between news reports. However, because jumping the key role to classify events through the time, locations, characters (organization), and other news elements, those traditional methods can not solve some special circumstances efficiently. Such as, the same or similar contents (consistent theme) which belong to different news events.

2.1 News Reports Modeling

In the article, for each news elements, the author assigned a semantic class, which is a set of words having the same type meaning. Location semantic class contains all place names related to a news report; time semantic class contains all time nouns related to a news report; characters (organization) semantic class contains all person's and organization names related to a news report; content semantic class contains other news report elements other than time, locations, characters (organization).

According to news four elements (When, Where, Who, WHAT, 4 Whs), the author creates modeling of one news report responding to different semantic classes. This modeling method is to express each news elements as a sub-vector, in other words, the time sub-vector $T = \{t_1, t_2, \dots, t_n\}$, to location sub-vector $L = \{l_1, l_2, \dots, l_n\}$, characters (organization) sub-vector $P = \{p_1, p_2, \dots, P_n\}$, content sub-vector $C = \{c_1, c_2, \dots, c_n\}$. Therefore, a news report can be expressed as $S = \{T, L, P, C\}$, which each elements in S is a real sub-vector. The time, locations, characters (organization) sub-vector elements are extracted from news report by named entity relation extraction technology (NEE) [2] . Sometimes, the same time, locations, characters (organization) may appear in the same report for several times and on different places; thus, the degree of importance that the four elements reflects in reports will be various. The following formula [3] is for calculating weight for time, locations, character (organization), and other name entity features.

$$\omega(f, d) = \log\left[(1 + \frac{C_d}{N}) \times \frac{1 + (N - L_d)}{\sum_{j=1}^{N} j}\right] \quad (1)$$

C_d is the frequency how the feature f appears in the report d. N is the sum of times that each entity appears in the name entity which belongs to the feature f. L_d is the first place where the feature f appears in d.

2.2 The Calculation Method of the Featured Weight of Content Sub-vector

To calculate the featured weight of content sub-vector is an important part of the modeling based on news elements. In the article, the traditional $tf * idf$ method to calculate weight is applied [4]. The method works by using a single news report as the basic statistics unit to calculate featured weight and reflects the importance degree that features appear in reports.

The similarity of each of sub-vectors is measure by new events detection Hellinger [5] methods, in other words, the sum of square root of the product of the corresponding featured weight. Hellinger methods express as following:

$$sim(d_1, d_2) = \sum_{1 \le i \le |d_1 \cap d_2|} \sqrt{\omega(f_i, d_1) \times \omega(f_i, d_2)} \quad (2)$$

3 The Algorithm of On-Line News Event Detection

The traditional on-line news event detection employs the statistics principle-based text representation form. The most commonly used represetation method is vector space model. The similarity between events and reports is accordingly calculated by cosine included angle formula [6] and Hullinger distance calculation formula [5]. Detection algorithms commonly use agglomerative method [1], single-Pass method [7] and increment *k*-mean method [7], etc. Carefully analysing traditional method, it can be seen that the traditional methods have a lot of drawbacks: 1) the biggest flaw of the method is not able to efficiently distinguish different events having the same topic; 2) the noise informations in event space interfere new event detection; 3) it increases the time comsumed for the traditional similarity comparisons strategy in on-line detection; especially, when data flow rate is large.

According to the deficiency of the traditional algorithms, in this article, the author applys the algorithm of Micro-clusters-based on-line news event detection with Window-Adding (MONEDW). In the window, the system can adopt condensedly hierarchical clustering algorithm to conduct micro-clustering, generating micro clusters, and use a earliest report representation of time stamp for micro cluster. By setting a higher similarities threshold value, the similarity of micro cluster internal report is extremely high. After the completion of window-interior micro-clustering, each micro cluster will make similarity comparisons to ready-detected event to determine whether the micro clusters describe a new events.

Using Window-Adding strategy in the algorithm, the size of a window is defined by time. In this article, the time interval for periodical collection news from web is defined to be the size of a window. The purpose is to avoid the case where intervals that cover fixed numbers of news at different time points may vary dramatically. Such as, setting the window size in hours. For example, setting one hour for the window size means adding all collected news reports in an hour to the window.

This on-line news event detection algorithm include two sub-algorithms: Algorithm 1 is micro clustering within window algorithm; Algorithm 2 is the similarity comparisons algorithm between micro clusters and events. Window micro clustering choose condensedly hierarchical agglomerative clustering algorithm. Hierarchical agglomerative algorithms find the clusters by initially assigning each object to its own cluster and then repeatedly merging pairs of clusters until a certain stopping criterion is met.

Algorithm 1: Micro Clustering within Window Algorithm

Input: News reports set $D=\{d_1,d_2, \cdots, d_n\}$ collected from window, threshold θ;

Output: Micro clusters;

Step 1: Start by assigning each item of D to a cluster, so that D have n clusters, each cluster containing just one item. Let the distances (similarities) between the clusters the same as the distances (similarities) between the items they contain;

Step 2: Find the closest (most similar) pair of clusters and merge them into a single cluster as long as the similarity between pair of clusters above the θ;

Step 3: Compute similarities between the new cluster and each of the old clusters;

Step 4: Repeat steps 2 and 3 until all similarities not above θ.

Algorithm 2: The similarity comparisons algorithm between micro clusters and events

Input: micro clusters from algorithm 1, ready-detected event model (featured weight vector) saved in memory;

Output: whether micro clusters describe new event;

Step 1: Select a micro cluster in order (choose the news reports represented the micro cluster);

Step 2: Make comparisons between the sub vectors of time, locations, characters of micro cluster and corresponding sub-vectors of ready-detected events. if each similarity of sub-vectors is greater than the corresponding threshold values, the micro clusters belongs to the event, adding the micro cluster to the event, and updating the event. Go to the first step otherwise going to the third step;

Step 3: If all the similarities of the sub-vectors of locations and characters are greater than the corresponding threshold values, the micro cluster belongs to the event, adding the micro cluster to the event, and updating the event. Go to the first step otherwise going to the fourth step;

Step 4: If all the similarities of the sub-vectors of time and characters are greater than the corresponding threshold values, the micro cluster belongs to the event, adding the micro cluster to the event, and updating the event. Go to the first step otherwise going to the fifth step;

Step 5: Comprehensively calculate the similarities of four sub-vectors, and weighting sum the result. Finally, make comparison to threshold values. If the similarity is greater than threshold value and the similarities of time and location are greater than the corresponding threshold value, the micro cluster belongs to the event, adding the micro cluster to the event, and updating the event. Go to the first step; Otherwise, the cluster is a new event. The news report which is on behalf of the micro cluster is the first report of the event, creating a new event mark and the representation model of the new event, saving to local, going to the step 1.

4 Experimental Results and Evaluation

4.1 Experiment Data

This section presents our experiments performed on news report texts which come from CNTV's XinWenLianBO. All the news report texts are collected from March

2006 to July 2008, including a total of 25776 news stories, each news unit is treated as a text. In our experiments, we select a total of 4214 news stories, which belong to 18 event, each news story associated with a reporting category being a member of event. As the news programs of XinWenLianBo report only a variety of important news, so the number of news reports in each category is uneven distribution, the number of various types of news stories used in this study range from 26 to 828. We randomly select the 30% samples of the total number of each category as being training samples, the remaining 70% samples as being testing samples.

Topic detection and tracking(TDT) is often used Recall Rate(R), the Precision Rate (P), the Miss Rate(P_M), the False_alarm Rate(P_F), Normalized System Cost(C_N) to evaluation the performance of news event detection [7]. This study refer to the performance evaluation criterion of TDT, using the same evaluation system.

4.2 The Result of Experiment

First of all, for the sake of obtaining the performance difference between our method and traditional method, we compare this news event detection method with the traditional method which use single vector of features splited from news story for new event detecton. There are two tables which display the result of comparison, table 1 shows the performance of news event detection based on multi-semantics class, table 2 shows the performance of news event detection based on single vector of features.

Table 1. The Performance of News Event Detection based on Multi- semantics Class

ID	Event Name	R(%)	P(%)	P_M(%)	P_F(%)	C_N
1	Fire	97.50	98.73	2.50	0.02	0.0262
2	Flood	96.72	96.72	3.28	0.05	0.0351
3	Earthquake	87.08	83.61	12.92	2.98	0.2754
4	Storm	97.17	90.35	2.83	0.27	0.0414
5	Olympics	89.13	88.60	10.87	2.81	0.2462
6	Avian Influenza	100.00	96.3	0.00	0.02	0.0012
7	Taiwan Problem	96.46	97.32	3.54	0.07	0.0390
8	Korea Nuclear	97.37	97.37	2.63	0.02	0.0275
9	United Nations	99.44	98.90	0.56	0.05	0.0080
10	America	89.81	89.32	10.19	1.01	0.1516
11	Russia	88.80	86.47	11.20	0.91	0.1566
12	Japan	89.86	86.71	10.14	0.96	0.1487
13	Iran	97.54	94.82	2.46	0.33	0.0406
14	Iraq	95.77	93.15	4.23	0.25	0.0543
15	Terrorists Attacks	96.47	95.35	3.53	0.10	0.0400
16	Countryside	88.96	86.27	11.04	2.55	0.2353
17	Oil Price	99.12	99.12	0.88	0.02	0.0100
18	Football	96.55	100.00	3.45	0.00	0.0345
Total		94.65	93.28	5.35	0.69	0.0873

Table 2. The Performance of News Event Detection based on Single Vector of Features

ID	EventName	$R(\%)$	$P(\%)$	$P_M(\%)$	$P_F(\%)$	C_N
1	Fire	85.00	91.89	15.00	0.15	0.1571
2	Flood	80.33	87.50	19.67	0.17	0.2050
3	Earthquake	86.12	82.44	13.88	3.21	0.2959
4	Storm	87.74	85.32	12.26	0.39	0.1417
5	Olympics	88.65	88.22	11.35	2.89	0.2553
6	Avian Influenza	88.46	82.14	11.54	0.12	0.1212
7	Taiwan Problem	84.07	87.16	15.93	0.34	0.1760
8	Korea Nuclear	84.21	82.05	15.79	0.17	0.1661
9	United Nations	88.33	89.83	11.67	0.45	0.1385
10	America	85.95	89.40	14.05	0.96	0.1876
11	Russia	87.64	85.34	12.36	0.99	0.1719
12	Japan	88.77	85.37	11.23	1.07	0.1646
13	Iran	89.34	88.26	10.66	0.73	0.1424
14	Iraq	88.73	89.36	11.27	0.37	0.1307
15	Terrorists Attacks	84.71	86.75	15.29	0.27	0.1660
16	Countryside	87.71	85.45	12.29	2.69	0.2546
17	Oil Price	87.72	90.09	12.28	0.27	0.1360
18	Football	89.66	83.87	10.34	0.12	0.1093
Total		86.84	86.69	13.16	0.85	0.1733

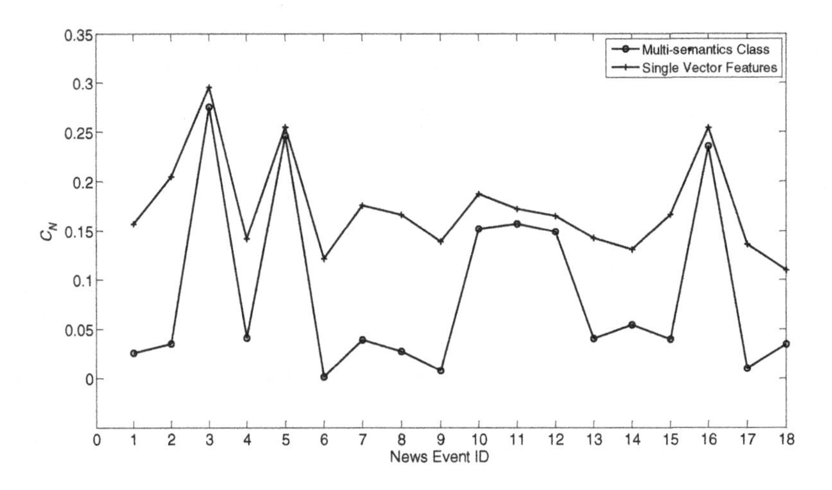

Fig. 1. The Normalized System Cost(C_N) Comparison with Different Text Representation Model

From the experimental results can be found that the news event detection algorithm based on news elements of this study obtains the average Recall Rate of 94.65%, the average Precision Rate of 93.28%, it is 7.81% and 6.59% separately higher than the traditional algorithm which is based on single vector of features. Two kinds of detection algorithms for events, there are some performance criterion including the Miss Rate, the False_alarm Rate and the Normalized System Cost being relatively high for the Earthquake, the Olympics, the Taiwan Problem, the United States, the Russia, the Japan, the Countryside, it is because that these types of events cover a wide range and topics are more dispersed. Figure 1 shows that the Normalized System Cost(C_N) Comparison with Different Text Representation Model.

To further verify the validity of the proposed method, we made a set of comparative experiments, the methods used for comparison are reported by literature [8], including C3 method, CMU method and Dragon method, their performance criteria are relatively better. The results of comparison shown in Figure 2. Figure 2 shows that the Normalized System Cost is the lowest for the proposed news event detection algorithm in this study, the comprehensive performance is the best in the four methods. The performance of Dragon method is the worst in the four methods, the Normalized System Cost reach about 28%, the CMU method in regard to the C3 method and the Dragon method performed better. The algorithm presented in this paper significantly improve the system detection performance.

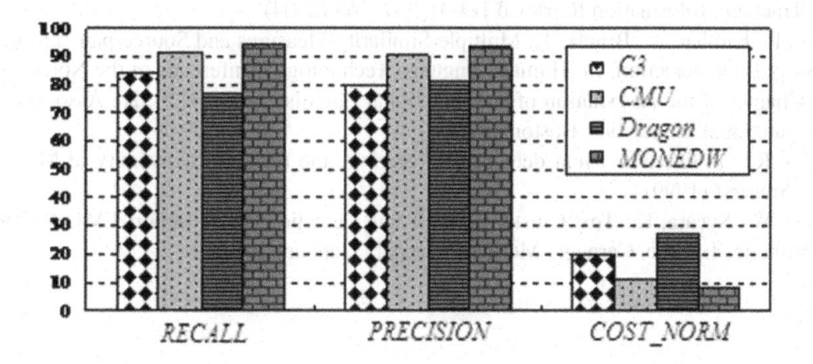

Fig. 2. The Performance Comparison with Different Methods for News Event Detection

5 Conclusion

This paper, being against the lack of traditional event detection algorithm for news and combining with news features, proposed a new algorithm, which is called micro-clusters-based on-line news event detection with window-adding(MONEDW) and make use of news elements for modeling news document. From the experimental results, the algorithm avoied the mutual interference between different events, which subject to the same topic, and promoted the efficiency and accuracy of online news event detection, is a feasible way for news event detection.

Acknowledgments. Thanks to Lian Lin, Chen Jun for discussions on News Event Detection, to Liu Fengzeng for his spell check, and to anonymous reviewers for their comments.

References

1. Allan, J., Papka, R., Lavrenko, V.: On-line new event detection and tracking. In: Proceedings of the SIGIR 1998: 21st Annual International ACM SIGIR Conference on Research and Development in Information Retrieval, pp. 37–45. ACM Press, New York (1998)
2. Zhang, H.-P., Liu, Q., Yu, H.-K., Cheng, X.-Q., Bai, S.: Chinese Name Entity Recognition Using Role Mode. Special Issue Word Formation and Chinese Language Processing of the International Journal of Computational Linguistics and Chinese Language Processing 8(2), 29–602 (2003)
3. Fu, Y., Zhou, M.-Q., Wang, X.-S., Luan, H.: On-line Event Detection from Web News Stream. ICPCA 2010 5th International Conference on Pervasive Computing and Applications, 105–110 (2010)
4. Sun, J.: The Topic Tracking Research with Document Title. Beijing City College, Beijing (2006)
5. Makkonen, J., Ahonen-Myka, H., Salmenkivi, M.: Simple Semmantics in Topic Detection and Tracking. Information Retrieval 7(3-4), 347–368 (2004)
6. Chen, F., Farahat, A., Brants, T.: Multiple Similarity Measures and Source-pair Information in Story Link Detection. In: Human Language Technology Conference of the North American Chapter of the Association of Computational Linguistics, pp. 313–320. Association for Computational Linguistics, Boston (2004)
7. Papka, R.: On-line new event detection, clustering, and tracking. University of Massachusetts Amherst (1999)
8. Seo, Y.W., Sycara, K.: Text Clustering for Topic Detection. Tech. Report CMU-RI-TR-04-03, Robotics Institue, Carnegie Mellon University (January 2004)

Automatic Recognition of Chinese Unknown Word for Single-Character and Affix Models

Xin Jiang[1], Ling Wang[2], Yanjiao Cao[1], and Zhao Lu[1,*]

[1] Department of Computer Science and Technology, East China Normal University 200241, Shanghai, China
[2] Shanghai Interactive TV CO., LTD. Shanghai 200072, China
{xjiang,yjcao}@ica.stc.sh.cn, zlu@cs.ecnu.edu.cn

Abstract. This paper presents a novel method to recognize Chinese unknown word from short texts corpus, which is based our observation of both single-character and affix models of Chinese unknown word. In our approach, we collect some news titles of a news site and view these titles as short texts. There are three steps in our approach: First, all collected news titles are segmented with ICTCLAS, and statistics of potential unknown words are conducted. Second, all potential unknown words are classified into either single-character model or affix model based on structures of unknown word. Some filtration methods are used to filter garbage strings. Finally, unknown word is extracted according to the frequencies of word. We have got the excellent precision and the recalling rates, especially for the single-character model. The experiment results show that our approach is simple yet effective.

Keywords: Chinese unknown word, single-character model, affix model.

1 Introduction

As we all know that word segmentation is the key to Chinese information processing. Previous researches have made great progresses in word segmentation, but cases with unknown word are not satisfied, and any existing lexicon in Lexical analyzer is limited. What's more, it is difficult to cover all words in real texts or speeches. With the development of information technology, there are more and more Chinese unknown words emerging on the Internet, and they are difficult to be identified correctly for several reasons: (1) The meanings or usages of existing word can be changed; (2) There is neither a general role of the composition of unknown word, nor an unique regular pattern to distinguish these unknown words; (3) Unknown word which seldom appears in the corpus is extremely difficult to identify.

According to the definitions in linguistics, Chinese unknown word is not only a word that has not been recorded in popular dictionaries, but also an existed word that possess new meanings or new usages. In natural language processing, unknown word mainly refers to the word that has not been registered in the dictionary of Lexical analyzer, which mainly includes word with new morphology, named entities. Owning to the fuzzy definition of Chinese word, it is difficult to define a Chinese word explicitly.

---

* Corresponding author.

Y. Wang and T. Li (Eds.): Knowledge Engineering and Management, AISC 123, pp. 435–444.
springerlink.com

The Chinese unknown word discussed in this study refers to a word which is not recorded in the dictionaries of ICTCLAS [1], that is developed by Institute of Computing Technology Chinese Academy of Sciences. After segmenting, a potential unknown word may be mixed with some single Chinese characters and always assembled by some single Chinese characters. It is a good idea to identify Chinese unknown word by combining some single characters with their adjacent words. It is discovered that a Chinese unknown word is always formed with two to four characters, and it is rare more than five. A Chinese unknown word is always formed as follows [2]:

1. Two-character new word, denoted as NW11 (two single-character word, "1+1");
2. Three-character new word, denoted as NW111 (three single-character word, "1+1+1"), NW12 (a single character followed with a bi-character word, "1+2"), NW21 (a bi-character word followed by a single character, "2+1");
3. Four-character new word, such as NW1111 (four single-character word, "1+1+1+1"), NW22 (a bi-character word followed by a bi-character word, "2+2"), NW211 (a bi-character word followed by two single character, "2+1+1"), NW121 (a single character followed by a bi-character word followed with a single character, "1+2+1"), NW112 (two single character followed by a bi-character word, "1+1+2"), NW13 (a single character followed by a tri-character word, "1+3"), NW31 (a tri-character word followed by a single character, "3+1");

In general, NW11, NW22, NW21 and NW12 cover more than 89% of all unknown Chinese words, they are 53%, 2%, 31% and 3%, respectively, and the others cover less than 11% [2]. In this study, we focus on unknown word of two surface patterns which account for more than 92% of Chinese unknown word, they are single-character model and affix model. The former model refers to NW11, NW111 and NW1111, the latter one refers to NW12, NW13, NW21 and NW31.

There are many approaches in Chinese unknown word recognition. These methods can be classified into three categories: (1) Rule-based methods, such as two methods suggested in [4, 5], they have advantages of high accuracy and strong pertinence. However it is difficult to define and evolve rules. The rules are always domain-specific which result in bad portability and flexibility. (2) Statistical methods, such as the method based on a SVM classifier in [2, 9] focus on two surface patterns, NW11 and NW21, the model of conditional random field in [6, 8] and the method based on Independent Word Probability (IWP) [7]. These methods need training through large-scale corpus and produce sparse data owing to few countable structure laws, which will lead to low accuracy rate. (3) Hybrid methods, which combine the two kinds of method. The author focus on bi-gram, tri-gram, quad-gram word and other common model, garbage string dictionary is used by self-learning method to filter [11].

Although these practicable methods achieve reasonable precision or recalling rates in some special cases, they have inherent deficiencies: (1) It is difficult to define and evolve the rules. And the rules are always domain-specific which result in bad portability and flexibility; (2) For statistical method, it needs training through large-scale corpus and owing to few countable structure laws sparse data will be produced; those finally lead to a low accuracy rate. Furthermore it's time-consuming, expensive and inflexible. What's more many approaches focus on the unknown word on specific areas or specific categories of unknown word. But the categories of unknown word are diverse and the amount of such word is huge. With the rapid development of the Internet, this situation is becoming more and more serious.

This paper presents a novel approach considering features of unknown word, they are single-character model and affix model. The remainder of the paper is structured as follows: Section 2 details our method; Section 3 shows experiments and evaluations; Section 4, concludes this paper and future work.

2 Our Approach

2.1 Single-Character Model and Affix Model

In this section, several definitions of the single-character model and the affix string model are given.

Definition 1 Single-character string. A Chinese word containing two or more two single Chinese characters is called a single-character string.

For example: *上海/ns 公布/v 购买/v 经/n 适/n 房/n 细则/n*. Here, "*经适房*", "*经适*", "*适房*" are all single-character strings.

Definition 2 Parental string and substring. For three single-character strings, *A, B* and *C,* if *A*=*B*+*C* and both lengths of *B* and *C* are large than 1 or equal to 1, then *A* is viewed as the parental string of *B* and *C*, *B* and *C* are viewed as two substrings of *A*.

Definition 3 Potential Unknown Word (PUW) and Longest Potential Unknown Word (LPUW). Given a string *T*, $T = \{X_1X_2 \dots X_i \dots X_n\}$, $(1 \le i \le n)$, and each X_i in the string *T* is a single Chinese character. The string *NW*, $NW(i.j) = \{X_jX_{j+1} \dots X_k\}$, $(1 \le j, k \le n)$, is viewed as a potential unknown word if the string *NW* meets:

1. *NW* is a substring of the string *T*.
2. The length of the string *NW* is equal to 2 or larger than 2.

and more, the string *NW* is viewed as a longest potential unknown word if it meets:

1. $j = 0$ or X_{j-1} is not a single Chinese character,
2. $k = n$ or X_{k+1} is not a single Chinese character.

In above example, the string *NW* "*经适房*" is the longest potential unknown word, and the string *NW* "*经适*" and "*经适房*" are all potential unknown words.

For unknown words of the affix string model, they can be classified two forms, the postfix string such as NW21 and NW31, the prefix string such as NW12 and NW13.

Definition 4 Affix string model. Given a string T, that is $T = \{X_1X_2\}$, X_1 is an existing word that consists of two or three single Chinese characters, and X_2 is a single Chinese character. Then the sting *T* is viewed as a postfix string. If X_1 is a single Chinese character and X_2 is an existing word, the string *T* is viewed as a prefix string.

For NW21, "2+1", a postfix string example: ***国土/n 部/q*** *拟/v 全国/n 推广/vn 土地/n 问/v 责/ng 连/d 坐/v 制/v*

For NW31, "3+1", a postfix string: *中医/n 治疗/vn* ***扁桃体/n 炎/n*** *常用/a 的/ude1 四/n 种/q 方法/n*

For NW12, "1+2", a prefix string: *广东/ns 廉江/ns 公安局/n* **副/b** *局长/n 迁/v 豪/ag 宅/ng 收/v 红包/n*.

2.2 The High-Level Structure of Our Approach

For Chinese unknown word of the two modes, the single character model and the affix mode, our idea is to construct a candidate word list based on statistics of a large-scale corpus of short text. We collected the news titles from Sina.com, and these titles are parsed as a corpus and segmented by ICTLACS. Based on our observations about the structure features of Chinese unknown word, our aim is to identify Chinese unknown word of the single-character model and the affix model.

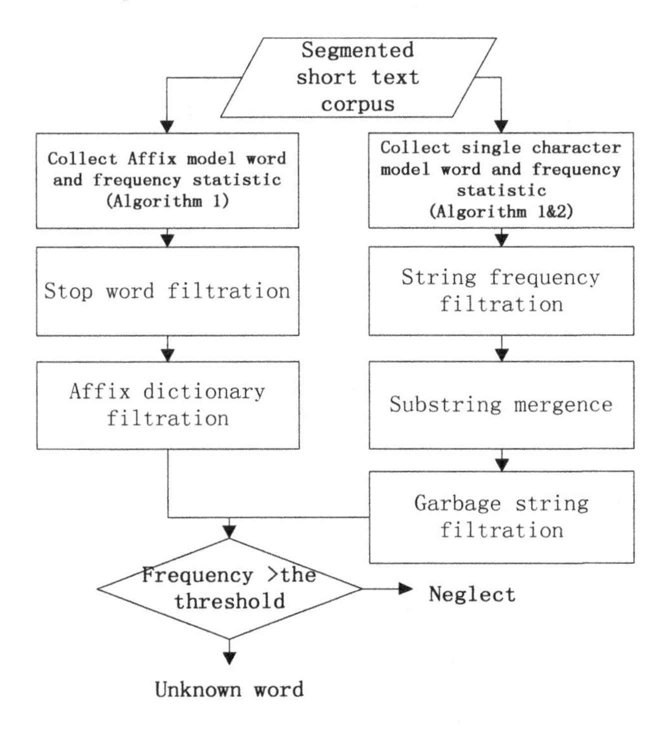

Fig. 1. The high-level structure of our approach

As shown in Fig.1, all potential unknown words of the two models are extracted from the corpus and their frequencies are also collected. All unknown words are added into a set of potential unknown word. For the single-character model, we adopt the string frequency filtration algorithm and the substring mergence algorithm to filter a word which is low frequency or is a redundant string. And we do the garbage string filtration. For the affix model, the stop word filtration algorithm and the affix dictionary filtration algorithm are adopted to filter a potential unknown word which is based on stop words and an affix dictionary respectively. In the end, we confirm an unknown word according to its frequency.

2.3 Potential Unknown Word Recognition

We introduce the PUW recognition algorithm to extract all potential unknown words. We pay our attention on the single-character model and the affix model: (1) For the Single-character model, the longest potential unknown word is extracted and its frequency is collected. All substrings of the longest potential unknown word are extracted as a potential unknown word, and their frequencies are collected also. (2) For the Affix model, postfix strings such as NW21 and NW31, and prefix strings, NW12 and NW13, are extracted and their frequencies are collected.

Two algorithms of potential unknown word recognition: (1) The Potential Unknown word Detection (PUD) algorithm is used to detect longest potential unknown words for the single-character model and potential unknown words for the affix model. All frequencies are collected; (2) The Sliding Window Algorithm (SW) is used to get all substrings of longest potential unknown words as potential unknown words for the single-character model, and get their frequencies

Here are some important variables used in the algorithm and their meanings: *SSTC* is the set of segmented short text corpus; *a* is one of segmented short text in *SSTC*, and *a*={*w[0]w[1]*... *w[k]*},*w* is a segmented fragment of *a*; *N(w)* is the frequency of *w*; *length(w)* is the length of *w*.

Algorithm 1: The Potential Unknown word Detection Algorithm (PUD)

```
Input:  set SSTC
Output:  A set slpuw of  LPUW, and a set spuw of  the affix model PUW
1.  for each a_i in SSTC
2.        get  a_i ={w[0]w[1]... w[k]}
3.        for each w[j]w[j+1] in a_i
4.             if just one length of w[j] and w[j+1] is equal to 1
5.                  if  w[j]w[j+1]  not in spuw
6.                       add to spuw
7.                  else
8.                       N(w[j]w[j+1])++
9.             end if
10.       end for
11.       Set temp to null
12.       for each w[j] in a_i
13.            if length(w[j])==1
14.                 w[j] appended to temp
15.            end if
16.            else if length(temp)>1
17.                if temp not in slpuw
18.                     add to slpuw, temp set to null
19.                else
20.                     the N(temp)++, temp set to null
21.            end else
22.       end for
23. end for
```

The main steps of the PUD algorithm are: Get a short text from the segmented corpus, traverse the segmented fragments, extract all longest potential unknown words and all potential unknown words of the affix model, and their frequencies are recorded. After using the PUD algorithm, all longest potential unknown words are recorded in the set *slpuw*. Then the Sliding Window (SW) algorithm is used to detect all substrings of the longest potential unknown word, their frequencies are recorded.

Algorithm 2: The Sliding Window Algorithm (SW)

```
Input:  The set slpuw of LPUW
Output: A set subset of substring
1.  for each c_k in slpuw
2.      let s=c_k, j=2, substring is null
3.      for ( ; j< length(s); j++)
4.          for( i=0; i+j-1<length(s); i++)
5.              substring=s.sub(i,i+j) //substring of s the index is between i and
    i+j;
6.              if substring not in subset
7.                  added to subset; N(substring)=N(s);
8.              else
9.                  N(substring)++
10.         end for
11.     end for
12. end for
```

2.4 Filtration of PUW

If a new word appears in the network, its frequency may be high in a certain time. But in the set of our collecting potential unknown word, some strings will be neglected either for their low frequencies or their massive redundant information. For above two kinds of situations, the string frequency filtration and the substring merging are used to filter single-character string.

Filtration of Single-Character Model

Word is the smallest linguistic unit which could be used independently. Once a new word or expression is accepted by people, they would be used repeatedly. Based on this observation, we mainly study these single-character strings repeatedly used in a corpus. That is to say, strings whose frequencies are larger than a threshold in the potential unknown word set would be our objects.

A potential unknown word of the single-character model is not only a longest potential unknown word but also substrings whose length is at least 2. These substrings are parts of the longest potential unknown word. Such as, if "□□□" is a longest potential unknown word, both "□□" and "□□" are potential unknown words, and the substring merging is used to filter potential unknown word [10]:

If $N\{C_iC_{i+1}\dots C_{i+j}\} = N\{C_{i+1}C_{i+2}\dots C_{i+j+1}\} = N\{C_iC_{i+1}\dots C_{i+j}C_{i+j+1}\}$, then delete $N\{C_iC_{i+1}\dots C_{i+j}\}$ and $N\{C_{i+1}C_{i+2}\dots C_{i+j+1}\}$

If $N\{C_iC_{i+1}\dots C_{i+j}\} > N\{C_iC_{i+1}\dots C_{i+j}C_{i+j+1}\}$ or $\{C_{i+1}C_{i+2}\dots C_{i+j+1}\} >$ $N\{C_iC_{i+1}\dots C_{i+j}C_{i+j+1}\}$, then delete $N\{C_iC_{i+1}\dots C_{i+j}C_{i+j+1}\}$.

Here, $C_iC_{i+1}\dots C_{i+j}$ represents a potential unknown word, and C_i represents a single Chinese character, $N\{C_iC_{i+1}\dots C_{i+j}\}$ represents the frequency of $C_iC_{i+1}\dots C_{i+j}$.

However, for single-character model, there are no significant rules to follow. To any combination of single Chinese characters, if the meaning of the combination is clear and widely used, it can be considered as a Chinese unknown word. Garbage strings have obvious linguistic features also. So the problem of Chinese unknown word recognition can be turned into garbage string filtering from the set of potential unknown words. The garbage strings of single-character model mainly refer to noise strings generated by prepositions, adverbs and other functional words.

In this study, we extract news titles as a corpus to extract unknown word. It is obviously that there are a lot of Arabic numbers and some special characters such as Roman alphabets or English letters whose frequencies are considerably high. Strings with these characters could not be viewed as words, such as *"20年"* (20 years), *"32级"* (32 level) and *"300余"* (a little more than 300). Strings with these characters will affect recognition accuracy, so they must to be neglected.

Filtration of Affix Model

Potential unknown word of the affix model is consisted of a Chinese word and a single Chinese character. It means that a single Chinese character is suffixed or prefixed to a word. For the prefix form, NW12 and NW13 are two types to be mainly identified. They always appear with some prefixes such as *"副, 近, 新, 原, 创, 反, 亚, 非..."* . For these prefixes, the corpus is trained to get them and add them into a prefix dictionary. The training steps are:

1. For a well segmented corpus, look for all potential unknown words which are in the form 1+2 or 1+3.
2. Figure out every first word that appears in this mode.
3. Count out a number of the highest *N* words and add them to the prefix word dictionary.

As well, for the postfix form, NW21 and NW31 are two types to be mainly identified. Word with this kind of form always possesses obvious linguistic features. The last character can always generate a large number of three-character words and a few four-character words. We also call them postfix, such as *"门, 热, 控, 秀, 局, 案, 部, 者..."* .. For these postfixes, a training corpus is adopted to extract them and add them into the postfix dictionary. The training steps are shown as follows:

1. For a well segmented corpus, look for all potential unknown words which are in two forms of 2+1 and 3+1.
2. Figure out every last word that appears in this mode.
3. Count out the highest *N* words and add them to the postfix word dictionary.

Stop characters are firstly used to filter potential unknown word during the process of affix model filtration. Here, stop characters are ones which often appear in news titles but cannot be viewed as a part of a word with other words, such as "*被*", "*致*", "*在*", "*将*", "*为*", "*称*" .... Therefore, potential unknown word with stop characters as their postfix or prefix would be neglected.

In addition, further filtration for potential unknown word is conducted based on composition of the affix model. If the prefix of a potential unknown word with the form of NW12 or NW13 is the character that is in the postfix dictionary, the potential unknown word must be neglected. For example, the "*案*" in "*案抓获*" (a potential unknown word) always exists in the postfix dictionary. However, it appears as a prefix here, so "*案抓获*" would be neglected from the set of unknown word. Similarly, if the last character of a potential unknown word with the form of NW21 or NW31 is the one that is in the prefix dictionary, the potential unknown word must be neglected.

3 Experiment Evaluations

In this paper, the news titles are taken from sina.com as a short text corpus, because the news titles are always in real times and we can get the latest information from news titles. In addition the words in news titles are always simple and formal; it will be performance better in the experience, especially in the result of filtration. In our experience 100,000 news titles are selected randomly. The news titles are in two years period, from April, 2009 to April, 2011, and we collected in April, 2011.The titles are segmented, potential unknown words and their frequencies are collected.

In this study, two experiments are conducted to recognize Chinese unknown word, one is for the single-character model and the other is for the affix model. Both two experiments choose an optimal threshold mentioned in Fig.1. The programs of two experiments are developed by JAVA and MySql.

The performance of two experiences is measured by precision, recall and F of unknown words identification, which are defined as follows:

Precision= (number of correct identification)/(total number of identification made)*100% (1)
Recall= (number of correct identification)/(total number of identification made+number of flitted ones)*100% (2)
F= (2*Precision*Recall)/(Precision+Recall)*100% (3)

3.1 Experiments for Single-Character Model

In the experiment, we extract domestic news titles from Sina.com with the size of 1,000,000. They are used to identify unknown word after segmented by ICTCLAS. After a word of single-character model is identified, its frequency filtration and substring merging are used to filter single-character string, and we do the garbage string filtration. Fig.2 shows the PRF values when frequencies of all words of the single-character model are higher than a specific threshold to determine an optimal threshold.

All unknown words of the single-character model are extracted, these extracted words are with either forms of NW11, NW111 and NW1111. As shown in Fig.2, the best performance when the threshold is set to 75. In this experiment, there are 112,154 words of the single-character model extracted in total. The number of low frequency words that their frequencies are below 20 is 111,310, and the result in Fig.2 is based on the words that their frequencies are larger than 20.

3.2 Experiments for Affix Model

We use the same dataset to extract words of the Affix model. Stop characters and the prefix (or the postfix dictionary) are used to filter garbage strings from all potential unknown words. And their frequencies are viewed as the threshold to identify unknown word. Fig.3 shows the PRF when the frequencies of unknown words of the affix model are more than a threshold.

As shown in Fig.3, the best performance is that the frequency of unknown word is larger than 100. However, the number of identified unknown word is little in that case. So on the whole, the threshold should be set to 75. In this experiment, there are 145,900 words of the affix model extracted in all. 144,872 words can be viewed as the low frequency words, that is to say, their frequencies are lower than 20, and the result in Fig.3 is based on the left 1,028 words that their frequencies are larger than 20.

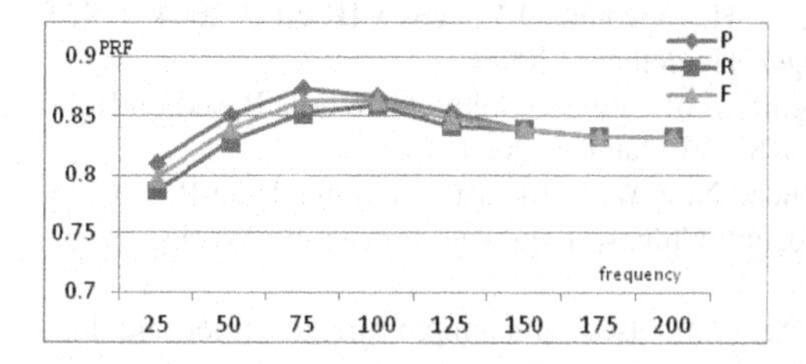

Fig. 2. The curve graph of PRF of the single-character model

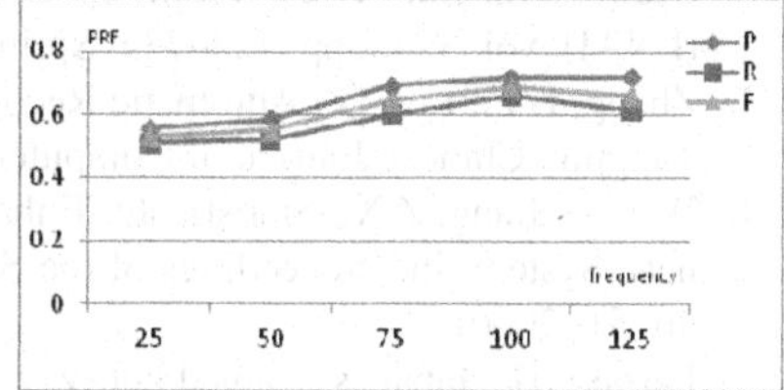

Fig. 3. The curve graph of PRF of the affix model

4 Conclusions and Future Work

Nowadays, there are many researches on Chinese unknown word recognition. However, most of them are not suitable or flexible to all types of words. The unknown word identification method discussed in this study is based on a large-scale corpus of short texts. In our approach, we collect news titles of sina.com and view these titles as short texts corpus. The single-character model and the affix model are two models we focus on. Potential unknown words are extracted and their frequencies are collected. The different methods are used to filter junk strings of the two models. Finally unknown word is extracted based on their frequencies.

Compared with other existing methods, this method has some advantages. First, the method gives the detailed forms and characteristics of unknown word, and we divide unknown word into the single-character model or the affix model. Different method is used to identify unknown word of two models to improve accuracy. In addition, the filtration method based on stop words is used in the affix model, and it performs a good effect. Second, in the experiments we get the latest news titles as the short text corpus, we can get the latest unknown word timely. And the word in news title is always formal. We can also get the unknown words however Li, H.Q [2] and Qin,

H.W [9] cannot, for example, NW111, NW13 and NW31. The experiment results show that our approach is effective.

This study focus on two models of Chinese unknown words, the approach cannot detect unknown words which are not belonging to the two modes, such as NW22. The experiment in this study does not take some real-time news titles, and our approach does not extract the latest high frequency of unknown words. For future work, we will try to design independent modules for other models of unknown words. And there is a need to expand the corpus to identify more words. Moreover, it is necessary to find some efficient methods used in filtration of junk strings.

Acknowledgments. This work is supported by an Opening Project of Shanghai Key Laboratory of Integrate Administration Technologies for Information Security (No. AGK2010004).

References

1. http://ictclas.org/ (September 2011)
2. Li, H., Huang, C.-N., Gao, J., Fan, X.-z.: The Use of SVM for Chinese New Word Identification. In: Su, K.-Y., Tsujii, J., Lee, J.-H., Kwong, O.Y. (eds.) IJCNLP 2004. LNCS (LNAI), vol. 3248, pp. 723–732. Springer, Heidelberg (2005)
3. Zhang, H.P., Liu, Q.: Automatic Recognition of Chinese Unknown Words Based on Roles Tagging. Chinese Journal of Computers, 85–91 (January 2004)
4. Wu, A., Jiang, Z.X.: Statistically-Enhanced New Word Identification in a Rule-Based Chinese System. In: Proceedings of the Second Chinese Language Processing Workshop, pp. 46–51 (2000)
5. Isozaki, H.: Japanese named entity recognition based on a simple rule generator and decision tree learning. In: Proceedings of the 39th Annual Meeting on Association for Computational Linguistics, pp. 314–321 (2001)
6. Chen, K.J., Ma, W.Y.: Unknown word extraction for Chinese documents. In: The 19th International Conference on Computational Linguistics, pp. 169–175 (2002)
7. Meng, Y., Yu, H., Nishino, F.: Chinese new word identification based on character parsing model. In: Proceedings of First International Joint Conference on Natural Language Proceeding Sanya, Hainan Island, China, pp. 489–496 (2004)
8. Xu, Y.S., Wang, X., Tang, B.Z.: Chinese Unknown Word Recognition using improved Conditional Random Fields. In: Eighth International Conference on Intelligent Systems Design and Applications, pp. 363–367 (2008)
9. Qin, H.W., Bu, F.L.: Research on a Feature of Chinese New word Identification. Computer Engineering (2004)
10. Lv, H.L.: Chinese New Word Identification Based on Large-scale Corpus. Dalian University of Technology (2008)
11. Cui, S.Q., Liu, Q., Meng, Y., Yu, H., Nishino, F.: New Word Detection Based on Large-Scale Corpus. Journal of Computer Research and Development (2006)
12. Ding, J.L., Ci, X., Huang, J.X.: Approach of Internet New Word Identification Based on Immune Genetic Algorithm. Computer Science. 240–245 (Janruary 2011)
13. Zhang, Y., Sun, M., Zhang, Y.: Chinese New Word Detection from Query Logs. In: Cao, L., Zhong, J., Feng, Y. (eds.) ADMA 2010, Part II. LNCS, vol. 6441, pp. 233–243. Springer, Heidelberg (2010)
14. Zhu, Q., Cheng, X.Y., Gao, Z.J.: The Recognition Method of Unknown Chinese Words in Fragments Based On Mutual Information. Journal of Convergence Information Technology (2010)

Social Media Communication Model Research Bases on Sina-weibo

Ming Wu, Jun Guo, Chuang Zhang, and Jianjun Xie

Pattern Recognition and Intelligent System Lab
Beijing University of Posts and Telecommunications, Beijing China
wuming@sina.com, zhangchuang@bupt.edu.cn

Abstract. The popularity of microblog brings new characters to information diffusion in social networks. Facing new challenges of understanding information propagation in microblog, the framework of information producing and receiving was proposed. A general model named competing-window is also presented based on human behavior. The detailed composition of the model and its basal mathematical description are also given. In addition, a parameter called information lost as a supplement to measure dynamics of information diffusion. Meanwhile, the further application of our model to information processing and propagating was pointed out. All those work is based on the studies on human dynamics. Finally, to verify applicability, the model was applied to empirical data crawled from Sina-weibo. The interesting patterns extracted from empirical data indicate that microblog in deed is fundamentally characterized by human dynamics.

Keywords: communication model, social media, human behavior, competing window, microblog, Sina-weibo.

1 Introduction

Sina-weibo is the most popular microblog service in China. After its launch on July 2009, Sina-weibo users have increased rapidly. They are currently estimated as two thousand million users worldwide. Like twitter, Sina-weibo is an online social network used by millions of people, but which only shares Chinese information around the world to remain socially connected to their friends, family members and co-workers through their computers and mobile handset.

Microblog has many new characteristics. In traditional communicating ways, such as letter, email, phone calls and short messages, users are almost equal in message publishing and replying. Thus communications are mainly peer-to-peer (P2P). Since it would take efforts to deliver messages, each message has its own destination. For example, you would surely not text all your friends what you have visited through short messages. Instead, you may publish a tweet (the common form of messages in microblog) wishing some of your friends would read it. As to the receiving end, letter or email readers may often be expected to reply in some time. However, tweets readers may choose to ignore any message and reply nothing at all. User relations in

Y. Wang and T. Li (Eds.): Knowledge Engineering and Management, AISC 123, pp. 445–454.
springerlink.com

microblog are mostly asymmetric, where a user can have many people following without a need for reciprocity. As the impact of asymmetric relations, microblog is a broadcast communication medium where information dissemination is in large scale involving multi-node interactions. Users have full autonomy to decide or choose how to behave rather than being forced to act. The instantly updated contents are pushed to related users, which advanced the ease of information publishing and disseminating. User relations mainly are asymmetric, namely Asymmetric Follow.

Previous studies of social networks paid intensive attention to structure-based research [1] [2] in network evolution [3] [4], information diffusion [5] [6] and data mining [7] [8]. The studies on communications, e.g. letter and email, short message, mobile phone calls, web information access, blog posting and other social networks, have shown that human activities have the characteristics of non-Poisson distribution (mostly Power Law distribution) with heavy tails. However, microblog has not been covered until recent researches on Twitter, Facebook, etc., including network property analysis [9], prediction of information diffusion [10] [11] and spam detection [12]. However, behaviors of communication in microblog haven't been covered yet.

In this paper, after review previous researches on human dynamics, we model basic user behaviors including tweets publishing, browsing, replying and retweeting. By introducing interest-driven hypothesis, we explain the process of broadcast communication in microblog, which provides a possible explanation to the origin of heavy-tailed Power Law distribution in collective communicating behaviors. Finally, as verifications to the model, empirical statistics are presented.

2 Related Studies

Traditionally, human actions are modeled as Poisson process [13] [14], where events independently occur at a constant rate. Thus the time interval of two consecutive events obeys negative exponential distribution $P(\tau) = \lambda e^{-\lambda\tau}$. Recent studies have shown that human activities are non-Poisson in various fields [1]–[8], where human activities are characterized by bursts of rapidly occurring events separated by long periods of inactivity. Interevent time obeys heavy-tailed power law distribution [15]. Several models were proposed to explain the origin of bursts and heavy tails in human dynamics. Priority-queuing model shows the burst nature of human activities is a consequence of a decision-based queuing process [1] [16] [17]. When individuals execute tasks based on some perceived priority, the timing of the tasks will be heavy tailed. Most tasks being rapidly executed, whereas a few experience very long waiting times. In contrast, priority blind execution is well approximated by uniform interevent statistics. Further development of this model introduced limitations and variations to the queue length [16].

Interest adjustment model [18] [19] was based on interest variation. Facing a new thing, people often show strong interests. As time goes by and repetition of actions, the interests would gradually descend and finally disappear. Activities would then stop. But after a period of idleness, interests would recover and drive activities again. The automatic adjustment mechanism of interests will produce heavy tails of human behaviors in interevent time distribution. Other models, such as Poisson processes modulated by circadian and weekly cycles [20] [21], preferential linking [18] and memory-based activity adjustment [22] also give possible explanations to bursts and heavy tails in human activities. Previous studies have mainly focused on separated individuals and the communication is P2P model. However, study of human dynamics in Web2.0 instant broadcast medium, such as microblog, is still insufficient.

3 Communication Models of Social Media

3.1 The Framework of Microblog Communication Model

Basically, there are two kinds of relations: *unidirectional* and *bidirectional* relation. See Fig 1. If user *A* follows user *B*, user *A* is called the follower of *B*. Here the word "follow" means user *A* subscribes to user *B* and *A* will receive tweets from *B*.

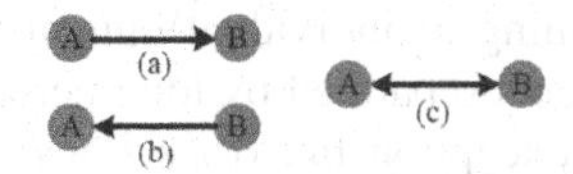

Fig. 1. User relations in microblog: (a) unidirectional relation. User *A* is following user *B* and *B* is followed by *A*. *A* will receive tweets from *B*, but the inverse is not. (b) *A* is followed by *B*. (c) bidirectional relation. *A* and *B* are following each other and receive tweets from each other.

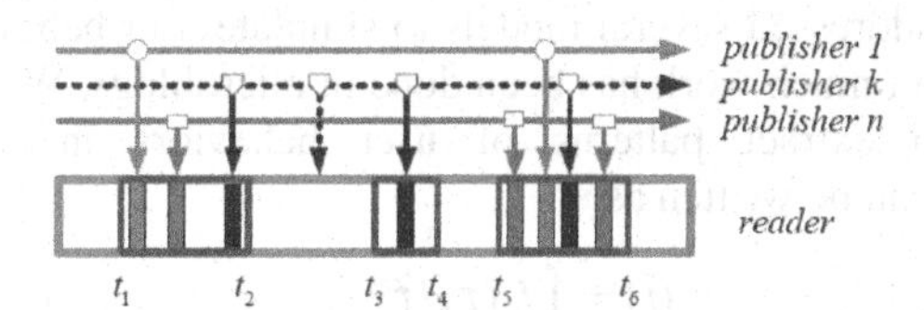

Fig. 2. Visualization of Competing-window model. The specific reader has *n* followers labeled as *publisher 1*, *publisher 2* to *publisher n*. Through the time line, each publisher independently publishes tweets which are instantly pushed to the reader (denoted by down arrows). In microblogs, all tweets are received but wheather to be read completely denpends on the readers. The time period of reading tweets is called time window (denoted by (t_1, t_2), (t_3, t_4) and (t_5, t_6)).

Note that publisher and reader relation is relative. We focus on unidirectional relation when studying information diffusion and on bidirectional relation when comments, replies and retweets are concerned.

Now we can isolate one specific reader and all of his or her friends to get a clear observation of information producing and receiving, which is also the micro node of information dissemination. The whole process is visualized by Fig 2.

From the reader's perspective, individual publishers form relations of competing without even noticing that themselves. The general picture of competing process can be literally described as: information produced by publishers, stretching out on the time line like a stream, is crowding into the reader's limited processing time periods, namely *time windows*.

From the microscopic view, the whole process of microblog can be generally divided into four stages in our model: information publishing, receiving, processing and propagating. Next we will give fundamental mathematical definition and description of those stages and define information lost in microblog.

3.2 The Models for the Different Stages in Microblog Communication

Stage A: Information Publishing

The production of information in microblog is extensively broad participation involving nearly every user of it, which represents one of the remarkable differences between microblog and traditional social media, e.g. blog. We can observe it in two dimensions.

One is from the distribution of time intervals between two successive messages. Barabasi [23] and Vazquez's [16] works pointed out that in email, mobile communications and web browsing, timing of individual human actions are characterized by bursts of rapidly occurring events separated by long periods of inactivity. Tweet publishing and browsing are no exception but obey *Power Law* [15] distribution with heavy tails. $C = e^c$, α is called the *exponent*.

$$P(x) = Cx^{-\alpha} \tag{1}$$

The other dimension is from the information density distribution of time, which is as individual as the person but will show statistical stability as a whole. Though previous studies have addressed several models to simulate user behaviors in online network or web site , not much work has been done in microblogs. We suggest using statistical analysis to extract patterns of user behaviors in microblogs whose mathematical form can be written as

$$F_u(t) = \int_{\Delta t} f_u(\tau) d\tau \tag{2}$$

where $F_u(t)$ is the *information entropy* during time period Δt and $f_u(\tau)$ is its density of time.

Stage B: Information Receiving

Microblogs adopt mechanism of pushing friends' messages (or say tweets) to their followers automatically. Since the amount of friends of one specific reader may range from zero to hundreds, thousands or even more, and his or her friends post tweets in a particular way which obeys power law distribution, the time interval of received tweets will form a new distribution. Theoretically, when a reader has enough friends to satisfying the assumption of mathematical derivation of a typical Poisson distribution, we can prove the quantity of messages (denotes as k) arrived in time duration t is Poisson process

$$p_k(t) = \frac{(\lambda t)^k}{k!} e^{-\lambda t} \tag{3}$$

Let τ be the successive time interval of received messages, then the distribution of τ can be derived from (3). We have

$$p\{T < t\} = 1 - p\{T \geq t\} = 1 - p_0(t) = 1 - e^{-\lambda t} \quad (4)$$

where $t \geq 0$. Thus probability density function (PDF) would be negative exponential distribution

$$f_i(\tau) = \lambda e^{-\lambda\tau} \quad (5)$$

Stage C: Information Processing

Email or mobile phone message readers almost always read the emails or messages from their friends even though they don't need to reply. But in social networks like microblog network, reader may ignore the message, which, unlike email or phone message reading will not upset his or her friends. Such phenomenon is characterized by human dynamics. microblog is user centric communication, which means reader is no longer a mere and passive information container or processer. Instead, not only publishers can freely choose to post messages, but readers also can freely choose whether to read the messages. People can enjoy this kind of freedom only after the emergence of new social network forms.

It is relatively easier to model or do statistical analysis on publisher's behaviors than understand reader's, which means precisely deciding whether messages or which message being read is rather difficult. Traditional webpage browsing can be modeled as Random Walk [24]. However, in microblog, there is neither web links directing user to the next webpage, nor web logs indicating when and how the page is read. Besides, message reading is interests-related and affected by the strength of relationship between publisher and reader.

Here we address two possible methods as solutions. One is doing surveys among microblog users, from which empirical model can be created. Another one is building a support vector (denoted as V_s) and a weight vector (denoted as V_w) based on reading habits and relation strength, which can be applied to predicting the reading and reposting behavior of a reader. Each dimension of the vector is an initially normalized parameter representing one factor that attributes to the probability of message reading. Then what we need to do is dynamically adjusting V_w according to the algorithms we take. The algorithms can be borrowed from related subjects, e.g. neural network, pattern recognition and machine learning, and revised if needed. The ultimate result would be presented as formula (6), where R is the predicting factor.

$$R = V_s \cdot V_w^T \quad (6)$$

Stage D: Information Propagating

Information diffusion in blogs and microblogs has been studied in some aspects. In blogosphere, dynamics of information propagation in environments of low-overhead personal publishing is studied in both macro and micro scope. But it's far from enough to fully understand information diffusion in microblog, especially when facing heavy information lost. Investigation on Twitter [10] also gives empirical conclusion on the speed, scale and range of topic diffusion. But those studies only deliver an overall prospect of topic diffusion and encounter difficulties when answering what topic will be propagated.

By adopting methods much like those mentioned in Stage C, we can build a support vector model of information diffusion, which would further explain when and what information propagates by introducing analysis of human dynamics.

Stage E: Information Lost

We can define *information density* and *processing ability*, of which the distribution of time is $f_i(t)$ and $f_p(t)$ respectively. Information density and processing ability refers to the entropy density of information received and processed in per unit time. Comparing $f_i(t)$ and $f_p(t)$, we notice that *information lost*, which is defined as the entropy of information being ignored by the reader, exists if the integration of $f_i(t)$ is greater than that of $f_p(t)$. Thus information lost during time period $\Delta\tau$, denoted as L, can be calculated as follows

$$L = \begin{cases} \int_{\Delta\tau} f_i(t) - f_p(t)dt & , \int_{\Delta\tau} f_i(t)dt > \int_{\Delta\tau} f_p(t)dt \\ 0 & , otherwise \end{cases} \tag{7}$$

Note that sometimes it could be difficult to decide the entropy of information, so for simplification we assume that each message has the same entropy $H(X_i)$, where X_i is the i_{th} message being received and get

$$LM = \begin{cases} (M-N)\cdot H(X_i) & , M > N \\ 0 & , M \le N \end{cases} \tag{8}$$

where M and N are the count of messages being received and processed by the reader, respectively.

4 Empirical

4.1 Data Source

In our study, we choose Sina-weibo as main empirical data source, which now is the largest microblog community in China with over 200 million users. A standard data set of about 2,000 typical users and 750,000 tweets has been built. Tweets from 3/1/2011 to 4/20/2011 were crawled. Due to privacy and limits of the Sina-weibo API, we can't obtain users' all tweets or whole follower list.

4.2 Individual Information Source

The interevent time of consecutive tweets obeys heavy-tailed Power Law distribution. We adopted the power law fitting methods. The mean exponent α is 1.8, see Fig 3.

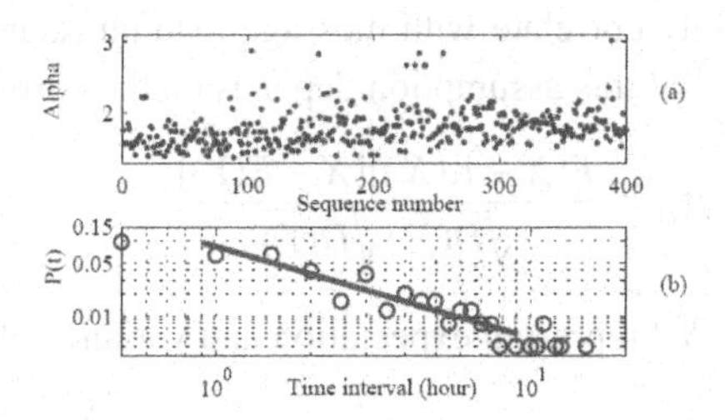

Fig. 3. User behavior of tweets publishing, fitted by Power Law. (a) value of power law exponent α. The mean value is α =1.78 with standard deviation σ =0.25. The goodness of fit, namely *p-value* is p = 0.52, σ =0.25. Note that if p is large (close to 1), the difference between the empirical data and the model can be attributed to statistical fluctuations alone; if it is small, the model is not a plausible fit. (b) plots a user's fitting (ID: 1463172125) with α =1.79 and p = 0.81. Heavy tails widely exist in time interval distributions. Overall tests show power law fit is applicable (p > 0.5 with support of 38.0% and p > 0.1 with support of 94.1%).

To understand the dynamics of individual publisher more comprehensively, we perform observations on message density through time line. The results indicate that no one curve can fit most distributions. Reversely, user behavior of information producing differs from one to another.

4.3 Information Receiving Patterns

Time interval of consecutive messages at the receiving end is our first concern. We have modeled this process as negative exponential distribution theoretically, which in fact matches empirical results pretty well. See Fig 4.

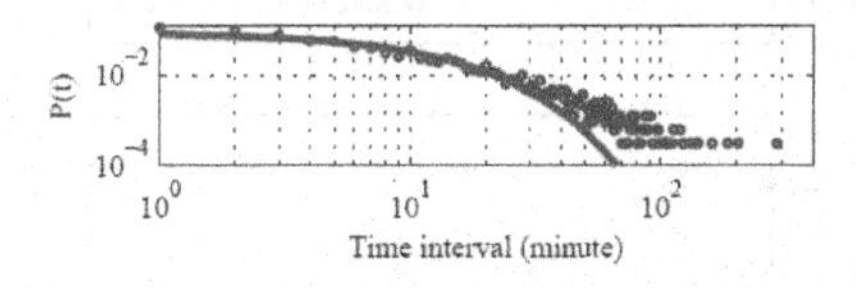

Fig. 4. Time interval distribution of consecutive tweets at individual follower. Tweets received in one month time duration were analyzed. Fitted by $P(t) = ae^{bt}$, axis x and y both are in logarithm. The figure shows a fit for a follower (ID: 1912337980, follows 50 publishers, 3,255 tweets received and counted.) where a = 0.089, b = -0.102 and $R^2 = 0.905$.

4.4 Information Processing and Diffusion in One Node

Firstly, we try to understand when one specific reader process messages received. The time window distribution of readers is shown in Fig 5. We draw the time window distributions indirectly using the data of posting and reposting time. Since there is no direct evidence of user online, we assume that posting and reposting (denoted as random variable X) closely correlate with message reading (denoted as random variable Y). The confidence of this assumption depends on the correlation coefficient

$$\rho_{XY} = \frac{E[X - E(X)][X - E(Y)]}{\sqrt{D(X)}\sqrt{D(Y)}} \quad (9)$$

where $E(X)$ and $D(X)$ represent expectation and variance of X respectively.

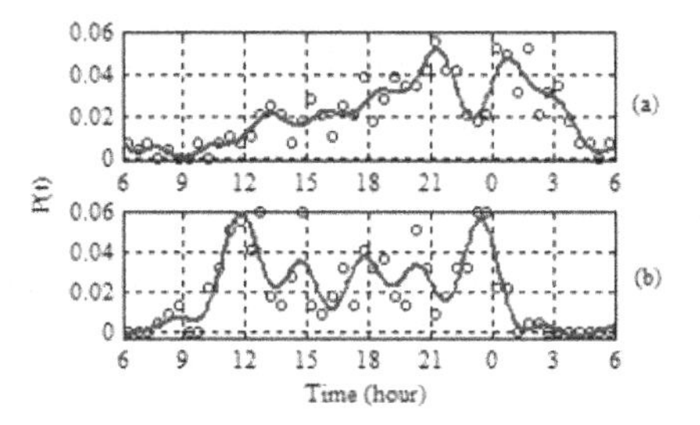

Fig. 5. Time window of readers: (a) time window of Yao Chen. (b) time window of Xiao S.We assume a day begins at 6:00 which corresponds to the time table of Chinese people. We found that most users' time window has two peaks indicating availabily of inforamtion processing.

Till now, we are no able to give a detailed description of information propagating through one node, or reader more precisely, which involves in-depth study of human dynamics. However, based on our data, we can determine the distribution of reposting time, illustrated in Fig 6.

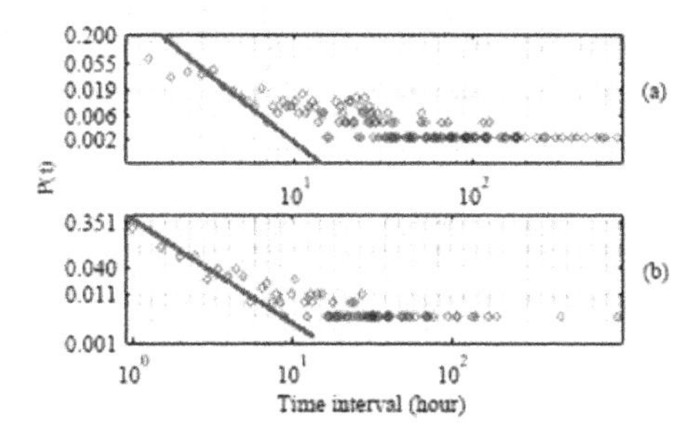

Fig. 6. The distribution of time interval between the original message being posted and being reposted by the reader. We performed power law fit on the data: (a) reader Yao Chen, where $p = 0.88$, $\alpha = 2.78$, fit area is $t < 7.0$ hours. (b) reader Xiao S, where $p = 0.56$, $\alpha = 2.17$, fit area is $t < 12.5$ hours. In the fit area, reposting time interval well matches the power law, and outside distribution tends to be the heavy tail of power law distribution.

The rudimentary conclusion of information diffusion in one node would be as follows: reposting is directly characterized by human dynamics, or more specifically, bursts of rapidly occurring events separated by long periods of inactivity [23]. In application, the "fit area" would be the golden time of information propagating through one node.

4.5 Information Lost Predictions

Here we assume the amount of information being processed is proportional to the time when the reader is online. Due to the diversity in time window distribution of different users, we can only deal with specific reader using statistical methods.

Now we apply our model to the prediction of information lost of user Yao Chen. We have already extracted the information receiving distribution (see Fig 5) and time window distribution (see Fig 6). We define three parameters: m, the average receiving messages per minute; v, the messages can be read by reader per minute time; $\Delta t_n = t_{n2} - t_{n1}$, the n_{th} online duration. Thus formula (8) can be written as

$$LM = \sum_n \{m \int_{t_{n1}}^{t_{n2}} f_i(t)dt - v\Delta t_n \int_{t_{n1}}^{t_{n2}} f_r(t)dt\} \tag{10}$$

5 Conclusions

Our work aims at building a framework to explain interactions between users and information diffusion between publishers and readers. Thus we have introduced the competing window model, which provides the fundamental framework to answer questions about information producing, receiving, processing and propagating. The model applies to microscopic perspective, which is also the foundation of macro network and information diffusion.

To verify this framework, we applied it to real social network, Sina-weibo. The empirical data provides detailed observation of microblog and realistic evidence of human dynamics and proves the feasibility and robustness of our general framework.

A framework can't solve every concrete problem. This is our rudimentary work and further studies are needed to better understand human dynamics involved information diffusion.

References

1. Barabasi, A.-L., Albert, R.: Emergence of Scaling in Random Networks. Science 286(5439), 509–512 (1999)
2. Fu, F., Liu, L.: Empirical Analysis of Online Social Networks in the Age of Web2.0. Physica A (2007)
3. Gross, T., Blasius, B.: Adaptive Coevolutionary Networks- A Review. Journal of the Royal Society – Interface 5, 259–271 (2008)

4. Bringmann, B., Berlingerio, M., Bonchi, F., Gionis, A.: Learning and Predicting the Evolution of Social Networks. IEEE Intelligent Systems 25(4), 26–35 (2010)
5. Xu, B., Liu, L.: Information diffusion through online social networks. In: 2010 IEEE International Conference on Emergency Management and Management Sciences (ICEMMS), August 8-10, pp. 53–56 (2010)
6. Yang, J., Leskovec, J.: Modeling Information Diffusion in Implicit Networks. In: 2010 IEEE 10th International Conference on Data Mining (ICDM), December 13-17, pp. 599–608 (2010)
7. Bird, C., Gourley, A., Devanbud, P., et al.: Mining email social networks. In: Proceedings of the 2006 International Workshop on Mining Software Repositories, Shanghai, China, pp. 137–143 (2006)
8. Yassine, M., Hajj, H.: A Framework for Emotion Mining from Text in Online Social Networks. In: 2010 IEEE International Conference on Data Mining Workshops (ICDMW), December 13, pp. 1136–1142 (2010)
9. Teutle, A.R.M.: Twitter: Network properties analysis. In: 2010 20th International Conference on Electronics, Communications and Computer (CONIELECOMP), February 22-24, pp. 180–186 (2010)
10. Yang, J., Counts, S.: Predicting the Speed, Scale, and Range of Information Diffusion in Twitter. In: Proc. ICWSM (2010)
11. Boyd, D., Golder, S., Lotan, G.: Tweet, Tweet, Retweet: Conversational Aspects of Retweeting on Twitter. In: 2010 43rd Hawaii International Conference on System Sciences (HICSS), January 5-8, pp. 1–10 (2010)
12. Wang, A.H.: Don't follow me- Spam detection in Twitter. In: Proceedings of the International Conference on Security and Cryptography, SECRYPT (2010)
13. Haight, F.A.: Handbook of the Poisson distribution. Wiley, New York (1967)
14. Reynolds, P.: Call center staffing. The call Center School Press, Lebanon (2003)
15. Newman, M.E.J.: Power laws, Pareto distributions and Zipf's law. Contemporary Physics 46(5), 323–351 (2005)
16. Vazquez, A., Oliveira, J.G., DezsöGoh, K.I., Kondor, I., Barabasi, A.L.: Modeling bursts and heavy tails in human dynamics. Phys. Rev. E 73, 36127 (2006)
17. Gabrielli, A., Caldarelli, G.: Invasion percolation and critical transient in the Barabasi model of human dynamics. Phys. Rev. Lett. 98, 208701 (2007)
18. Goncalves, B., Ramasco, J.: Human dynamics revealed through Web analytics. Phys. Rev. E 78, 26123 (2008)
19. Han, X.P., Zhou, T., Wang, B.H.: Modeling human dynamics with adaptive interest. New. J. Phys. 7, 73010–73017 (2008)
20. Malmgen, R.D., Stouffer, D.B., Motter, A.E., Amaral, L.A.N.: A Poissonian explanation for heavy tails in e-mail communication. Proc. Natl. Acad. Sci. USA 105, 18153–18158 (2008)
21. Malmgen, R.D., et al.: On universality in human correspondence activity. Science 325, 1696–1700 (2009)
22. V´azquez, A.: Impact of memory on human dynamics. Physica A 373, 747–752 (2007)
23. Barabási, A.L.: The origin of bursts and heavy tails in human dynamics. Nature 435, 207–211 (2005)
24. Pearson, K.A.R.L.: The Problem of the Random Walk. Nature 72(1867), 342–342 (1905)

A Simple and Direct Privacy-Preserving Classification Scheme

Rongrong Jiang[1] and Tieming Chen[2]

[1] School of Information and Engineering, Zhejiang TV and Radio University, Hangzhou, 310012, China
[2] College of Computer Sci. and Tech., Zhejiang University of Technology, Hangzhou, 310023, China
jiangrr_zjtvu@163.com, tmchen@zjut.edu.cn

Abstract. An inconsistency-based feature reduction method is firstly proposed in this paper, based on which a simple and direct rule extraction method for data classification is addressed. Because the proposed classification method depends directly on data inconsistency without leaving any value contents of datasets, it could be utilized to build a novel privacy-preserving scheme for isomorphic distributed data. In this paper, a simple privacy-preserving classification model is proposed based on the inconsistency-based feature reduction and its direct rule extraction method. Experimental results on benchmark datasets from UCI show that our method is both in good correctness performance and model efficiency.

Keywords: Date inconsistency, Data classification, Feature reduction, Rule extraction, Privacy preserving.

1 Introduction

Research and application of knowledge discovery have been booming over the past decade[1], where data mining, the core technique of knowledge discovery, has employed a number of practical methods, such as clustering, association rule, Bayesian method, decision tree, neural network, support vector machine (SVM) and so on. However, data mining is just the last job of data processing as during the whole knowledge discovery. Yet another crucial part of data mining is the preliminary job, say data acquisition, preprocessing, data cleaning, etc[2]. Feature reduction[3]is the most key problem in data processing. A good feature reduction method not only can effectively simplify the data modeling, but also improve the mining efficiency without compromising the model accuracy. Generally, the difficulty of the feature reduction for knowledge discovery depends on the way of interaction between feature reduction and data mining, since different models employ different feature selection methods[4,5]. Meanwhile, with rapid spread of applications of information technology, problems of personal privacy and information security are concerned. The data privacy protection of data mining is therefore becoming a very important issue. Privacy protection in data mining was firstly proposed in paper [6], following which the research on privacy protecting data mining is rapidly developed and various new

Y. Wang and T. Li (Eds.): Knowledge Engineering and Management, AISC 123, pp. 455–464.
springerlink.com

methods are emerging. At current, many privacy-preserving data mining techniques are employed, such as random perturbation technique[7], K-anonymity technique[8], data encryption[9].

A general privacy-preserving data mining framework for distributed database is proposed in paper[10]. It employs the randomness technique to transform data in order to achieve the purpose of privacy-preserving, but it noticeably scarifies the predictive accuracy. Paper [11] presents a novel method through the single-attribute random matrix to disrupt data and multi-attribute joint matrix to rebuild the original dataset, but this method is only applied to decision tree classification method, leaving a large cost of computation. As to effectively protect privacy and reduce the loss of information on anonymous data, paper [12] proposed the entropy classification method applied to K-anonymity model, but it did not make an in-depth analysis on the connection attack. Using encrpytion tachniques, paper [13] achieved privacy protection classification mining method without disclosing the precise data information, but the algorithm is limited to two entities, not supporting the multi-party model. Then, a RSA cryptosystem-based method in paper [14] is proposed to solute secure multi-program multi-data to scheduling problem, but the efficiency of this model will be sharply frustrated with the increase of data set.

This paper will propose a fast privacy-preserving data classification method based on discretization technique. In order to utilize data discretization for each group, the Chi-merge technique is employed in data processing. Then the inconsistency rate based contingency table is used to direct the feature reduction, as well as to generate the classification rules. The proposed data classification method can also be applied to distributed homogeneous data classification with privacy preserving. The rest of paper is organized as follows. The proposed algorithm is described into detail in section 2. The application model and experimental results for privacy-preserving distributed data classification is proposed in section 3. A brief conclusion on our methodology is drawn at final.

2 Direct Feature Reduction

2.1 Inconsistency Calculation

In order to utilize inconsistency-based data classification for feature reduction and rule extraction, continuous features must be effectively discretized. There are a number of discretization techniques in the literature, for example Chi-merge technique which is a supervised discretization technique based on Chi-square (x^2) test of independence. It begins by placing each observed data value into its own interval and proceeds by using the x^2 test to determine when adjacent intervals should be merged. The extent of the merging process is controlled by the use of a x^2 threshold α. This method is commonly used in the field of data mining. However, it has been reported that for certain data a very high threshold must be set to avoid creating too many final intervals. Another drawback is that the merging process could be tedious because there are too many initial intervals.

The improved version of Chi-merge has two innovations. The first improvement is the determination of initial intervals based on changes in class label, which generally results in far fewer initial intervals. The second improvement is the merging process. Instead of using a fixed threshold α, we keep merging the two intervals with the smallest x^2 until the terminal condition is met. The termination condition could be a certain number of intervals, x^2 Minimum Entropy optimization principles, or a combination of both. The improved Chi-merge not only reduces the process of merger, but also decides the number of final intervals, or deserves the best number of intervals based on Minimum Entropy principles. We take the feature Pedal Length of IRIS problem as an example to illustrate the proposed discretization method.

We set a certain number of intervals as the terminal condition, for example three. So the continuous values of feature Pedal Length are discretized into three intervals small, medium and large. The initial rough partition based on the change of output yields a result of (1.0,3.0,4.6,4.9,5.0,5.1,5.2,6.9) where each value within the parenthesis is a cutoff point. It represents intervals [1.0,3.0), [3.0,4.6), …, [5.1,5.2), and [5.2,6.9]. The x_2 values for each pair of adjacent intervals are calculated from left to right and they are 87.0,3.074,3.192,2.225,0.30 and 4.354 respectively. The minimum is 0.30 and thus its corresponding pair of intervals, [5.0,5.1) and [5.1,5.2), are merged. Then the domain partition becomes (1.0,3.0,4.6,4.9, 5.0,5.2,6.9) and the corresponding contingency table is updated. This process continues until the final result is reached, which is (1.0,3.0,4.9,6.9). As a result, the continuous values in the regions of [1.0,3.0),[3.0,4.9) and [4.9,6.9] can be respectively labeled as Small, Medium and Large and the corresponding contingency table is updated as shown in Table 1.

Table 1. Chi-Merge Demonstration - After Merge for Pedal Length

Interval	Setona	Versicolor	Virginica
Small [0.1,3.0)	50	0	0
Medium [3.0,4.9)	0	46	3
Large [4.9,6.9]	0	4	47

Inconsistency rate is a popular measure that accurately characterizes the class separability of a dataset with discrete features. Without loss of generality, assume a data set has d features $F_1,F_2 \dots F_d$ with discrete values and a class label L that has c categories. There are n data instances. For each data instance j, let x_{ji} denote the value of feature Fi (j = 1, 2, …, n; i = 1, 2, …d) and y_j denote the value of class label L. Thus, data instance j is denoted $[X_j, y_j]$, where $X_j^* = [x_{j1}, x_{j2}, x_{j3} \dots x_{jd}]$.

Assume there are p distinct patterns of x, denoted as X_k^* (k = 1, 2, …, p). Note that because F_i is discrete usually $p << n$. A contingency table can be created with p rows and c columns where f_{kl} denotes the number of instances with pattern X_k^* that belong to class l (l = 1, 2, …, c). The inconsistency rate of the dataset with feature set $S = \{F_1, F_2, \dots, F_d\}$ is calculated as

$$\varepsilon_s = \frac{\sum_{k=1}^{p}(\sum_{l=1}^{c} f_{kl} - \max_{l}\{f_{kl}\})}{n}$$

For the data set IRIS, $\{F_1,F_2,F_3,F_4\}$={Sepal Length, Sepal Width, Pedal Length, Pedal Width}. According to Table 3, the contingency table can be produced by distinct patterns X_1^*, X_2^*, X_3^* of feature pedal length shown in Table 2.

Table 2. Contingency table for feature Pedal Length

	L=1(Setona)	L=2(Versicolor)	L=3(Virginica)
X_1* (Small)	50	0	0
X_2* (Medium)	0	46	3
X_3* (Large)	0	4	47

The inconsistency rate of the dataset with feature pedal length is calculated as

$$\varepsilon_{\{F_3\}} = \frac{\sum_{k=1}^{3}(\sum_{l=1}^{3} f_{kl} - \max_{l}\{f_{kl}\})}{n} = \frac{7}{150} = 0.047 \;.$$

2.2 Feature Reduction Scenario

Let Z be a subset of S ($Z \subset S$) and ε_Z be the inconsistency rate of the reduced data set (only features belonging to Z are retained). The task of feature reduction is to find Z with the smallest number of elements where ε_Z is equal or very close to ε_S. Because S totally has $d!$ non-empty subsets, it is time-consuming and impractical to calculate all possible ε_Z when d is large. Therefore, greedy search algorithms such as sequential forward search (SFS) or sequential backward search (SBS) are commonly used to find the optimization subset. Note that if we require that ε_Z equals ε_S then Z is called a reduced set of S in rough set theory. However, ε_Z usually decreases with the increasing features of Z and it is hard to directly find Z where ε_Z is nearly equals to ε_S. We propose a new feature reduction method using SFS based on the minimum inconsistency rate described as follows.

1. Suppose data set D has d features, denoted as $F_1,F_2,\ldots,F_d$.
2. Initialize the optimal feature subset as empty, namely S={ }.
3. While S<> $\{F_1,F_2,\ldots,F_d\}$ do

3.1 Calculate the inconsistency rate ε_S of data set D with feature subset which is composed up by feature set S and the each remaining feature (only features which are element of set $\{F_1,F_2,\ldots,F_d\}$ out of S).

3.2 Select the feature Fi which produces the smallest ε_S , and update the optimal feature subset S to $\{S, F_i\}$.

4. According to the sequential forward search, an inconsistency rate table can be created for each feature subset with feature number from small to large.

5. Select the feature subset Z with feature number as few as possible. If ε_Z equals to or is very close to ε_S , or $\varepsilon_{Z'} / \varepsilon_Z$ (Z' is the adjacent feature subset)is the smallest within all items from all adjacent feature subsets, such Z is identified as the optimal feature subset.

The well-known IRIS dataset is used as an example to illustrate the feature reduction algorithm. Known from the above, the inconsistency rate of feature Pedal Length is 4.67%. The same calculation is conducted for the other three variables Pedal Width, Sepal Width, and Sepal Length and the results are 4%, 28.67% and 41.33% respectively. Obviously the first iteration selects Pedal Width into the optimal subset. The next iteration is to choose the optimal feature from the remaining three to join Pedal With, producing patterns {Pedal Width, Pedal Length},{Pedal Width, Sepal Width},{Pedal Width, Sepal Length}. We choose {Pedal Width, Pedal Length} as the optimal subset with the smallest inconsistency rate 0.040 in this iteration. The process continues until all variables are included. The result for the IRIS dataset is shown in Table 3. Observe that iteration 3 does not produce better inconsistency rate than iteration 2. Therefore those tow features Pedal Length and Pedal Width are selected finally.

Table 3. Result of forward sequential search-based feature reduction for IRIS dataset

No.	Sepal Length	Sepal Width	Pedal Length	Pedal Width	Inconsistency rate
1				*X*	0.040
2			*X*	*X*	0.033
3	*X*		*X*	*X*	0.033
4	*X*	*X*	*X*	*X*	0.027

3 Simple Data Classification

3.1 Rule Extraction

Let *d'* be the number of elements in the selected subset *Z* (usually *d'*<<*d*). We re-label the elements in *Z*, so $Z=\{F_1,F_2,\ldots,F_{d'}\}$. Following previous notations, let *p'* be the number of distinct patterns of Xi^* (i=1,2,…,p', generally $p' << p$) in the reduced dataset. We now have a new contingency table shown as Table 4.

We can directly extract p' rules as shown in Table 7 and the No.k rule extracted can be expressed as below.

IF F_1 is x_{k1} AND F_2 is x_{k2}…AND F_d is $x_{kd'}$THEN

L=1 with probability $$P_{k1} = \frac{f_{k1}}{\sum_{l=1}^{c} f_{kl}}$$

…

L=C with probability $$P_{kc} = \frac{f_{kc}}{\sum_{l=1}^{c} f_{kl}}$$

Rule k (k=1,2,…,p') with weight $W_k = \frac{\sum_{l=1}^{c} f_{kl}}{n}$.

Table 4. Classification rule table

Rule	Antecedent	Consequent				Weight
		L=1	L=2	…	L=c	
1	$X_1^*=\{X_{11}, X_{12},…, X_{1d`}\}$	P_{11}	P_{12}	…	P_{1c}	W_1
2	$X_2^*=\{X_{21}, X_{22},…, X_{2d`}\}$	P_{21}	P_{22}	…	P_{2c}	W_2
..	..	…	…	…	…	…
$p`$	$X_{p`}^*=\{X_{p`1}, X_{p`2},…, X_{p`d`}\}$	$P_{p`1}$	$P_{p`2}$	…	$P_{p`c}$	$W_{p`}$

Back to dataset IRIS to illustrate the power of proposed algorithm. Under features Pedal Length and Pedal Width selected, the corresponding contingency table can be computed. According to the rule extraction method aforementioned, rules as shown in Table 5 can be easily extracted from the contingency table.

Table 5. Extracted rules for Pedal Length and Pedal Width of IRIS dataset

Rule No	Pedal Length	Pedal Width	Setona	Versicolor	Virginica	Weight
1	Small	Small	100%	0%	0%	0.33
2	Medium	Medium	0%	98%	2%	0.31
3	Medium	Large	0%	33%	67%	0.02
4	Large	Medium	0%	50%	50%	0.05
5	Large	Large	0%	0%	100%	0.29

When extracted rules are used for prediction, for a new data point, the predict model is divided into two kinds of discussion. The first case is the new data point completely matches those features of the antecedent. The instance then is labeled as the class with the highest probability in the corresponding consequent shown in Table 8. There is another case that an instance has missing values. If those features with missing values are ignored, multiple matched antecedents may be identified. A weighted-average aggregation of all corresponding consequents is then used to produce the final prediction. Without loss of generality, assume that $\{x_1^*, x_2^*, \ldots, x_m^*\}(m<<p')$ are matched antecedents for a new instance with missing values. The estimated probabilities on each class are L=1 with $P_1 = \frac{P_{11}W_1 + P_{21}W_2 + \cdots + P_{m1}W_m}{W_1 + W_2 + \cdots + W_m}$, L=2 with $P_2 = \frac{P_{12}W_1 + P_{22}W_2 + \cdots + P_{m2}W_m}{W_1 + W_2 + \cdots + W_m}$, L=C with $P_c = \frac{P_{1c}W_1 + P_{2c}W_2 + \cdots + P_{mc}W_m}{W_1 + W_2 + \cdots + W_m}$.

For example, for a new sample IRIS plant with Pedal Length of Medium and Pedal Width of Medium, Rule 2 in Table 5 will be triggered and class Versicolor will be directly assigned. If the Pedal Width information is missing, Rule 2 and Rule 3 will be activated and the estimated probabilities on each class are $P_1 = \frac{0\% * 0.31 + 0\% * 0.02}{0.31 + 0.02} = 0\%$ for Setona, $P_2 = \frac{98\% * 0.31 + 33\% * 0.02}{0.31 + 0.02} = 94\%$ for Versicolor, and $P_3 = \frac{2\% * 0.31 + 67\% * 0.02}{0.31 + 0.02} = 6\%$ for Virginica. According to the maximum probability, class Versicolor will be assigned.

3.2 Privacy-Preserving Classification for Distributed Isomorphic Data

Distributed data mining is a hot topic of current research, how to improve the efficiency of the distributed data mining and how to protect privacy of the distributed data become a focus part of the study. In this section, a new type of feature reduction and data classification method will be applied to distributed data, which quickly and safely achieves classification.

Let H denotes the mentioned discretization technique Chi-Merge, M denotes the frequency calculation method based on feature division pattern, F denotes the forward-continuous search method based on inconsistent rate and C denotes the quick data classification method based on feature reduction. Suppose there are t distributed users $U_1, U_2, \ldots, U_t$ which have a homogeneous data set $D_1, D_2, \ldots, D_t$ respectively. Each data set contains the same d features $F_1, F_2, \ldots, F_d$. Assume the data classification processing center G possesses F and C, while the distributed user Ui (i=1,2,…,t) only approach H and M.

The distributed data classification method is based on the interaction between distributed users and data classification processing center, which is described as follows:

1. Suppose the feature j of data set has the same number of final discrete interval, thus we will get discrete interval $I^{ij} = \{I_1^i, I_2^i, \ldots, I_{N_i^j}^i\}$ (i=1,2,…,t; j=1,2,…,d;) after

Ui implementting *H*. *Ui* sends the discrete interval set $I^i = \{I^{i1}, I^{i2}, ..., I^{id}\}$ (i=1,2,...,t) to *G*.

2. Suppose $I_1^1 < I_1^2 < ... < I_1^t < I_2^1 < I_2^2 < ... < I_2^t < ... < I_{N_1^j}^1 < I_{N_2^j}^2 < ... < I_{N_t^j}^t$, for the feature *j*, *G* re-arranges values $\{I_1^i, I_2^i, ..., I_{N_i^j}^i\}$ of set I^{ij} (i=1,2,…,t; j=1,2,…,d) by size and finds the interval set $I^G = \{I^{G1}, I^{G2}, ..., I^{Gd}\}$; then *G* sends the interval set to *Ui* (i=1,2,…,t) Where $I^{Gj} = \{I_1^1, I_1^2, ..., I_1^t, I_2^1, I_2^2, ..., I_2^t, ..., I_{N_j^1}^1, I_{N_j^2}^2, ..., I_{N_j^t}^t\}$ (j=1,2,…,d).
3. For the feature *j*, *U*i carries out *M* and gets frequency table T_j^i (i=1,2,…,t; j=1,2,…,d) with corresponding interval set I^{Gj}. Then *U*i sends T_j^i to *G*..
4. The frequency table of the data set with feature *j* is calculated as $T_j = \sum_{i=1}^{t} T_j^i$ and then *H* is executed by *G* to get the final discrete interval I_j (j=1,2,…,d).
5. Initialize the optimal feature subset as empty *S*={ }, the candidate feature subset as *S'*={{F_1},{F_2},…, {F_d}} and *d'*=*d*.
6. while *S* <>{F_1,F_2,…, F_d} do

   6.1 *G* sends the feature subset *S'* being selected to *U*i(i=1,2,…,t)

   6.2 For the feature subset of *S'*, *U*i carries out *M* and gets frequency table T_j^i (i=1,2,…,t; j=1,2,…,d') with corresponding patterns. Then *U*i sends T_j^i to *G*..

   6.3 The frequency table with the patterns corresponded the feature subset of *S'* is calculated as $T_j = \sum_{i=1}^{t} T_j^i$ and then *F* is executed by *G* to select the optimal feature subset *Z* (*Z*⊂*S'*); update *d'* to (*d'*-1) and *S* to (*S*+*Z*). Note that $\overline{Z} = \{F_{i_1}, F_{i_2}, ..., F_{i_{d'}}\}$.Thus $S' = \{S + \{F_{i_1}\}, S + \{F_{i_2}\}, ..., S + \{F_{i_{d'}}\}\}$.
7. According to *S* and the rate of inconsistency rate, *G* determines the optimal feature subset and completes *C*. Now the data classification is successfully achieved.

3.3 Experiment Results

To inspect the effect of distributed data classification model, we randomly and equally divide the data set Indian Diabetes into 2,3,4,5 groups respectively and completes four times experiment with distributed data classification. Each experiment involving different number of users selects the same optimal features of plas and mass and extracts classification rules. In the 10-fold cross-validation experiment environment, the classification accuracy is described as shown in Table 6.

Table 6. Performance comparion on between centralized data set and distribute data set of Indian Diabetes

Data Set	Accuracy
centralized dataset (con)	75.84%
two-party distributed dataset (dis2)	73.80%
three-party distributed dataset (dis3)	75.76%
four-party distributed dataset (dis4)	76.41%
five-party distributed dataset (dis5)	74.24%

Furthermore, Figure 1 shows that the ROC curves of different distributed datasets are basically with no difference on the term of its areas.

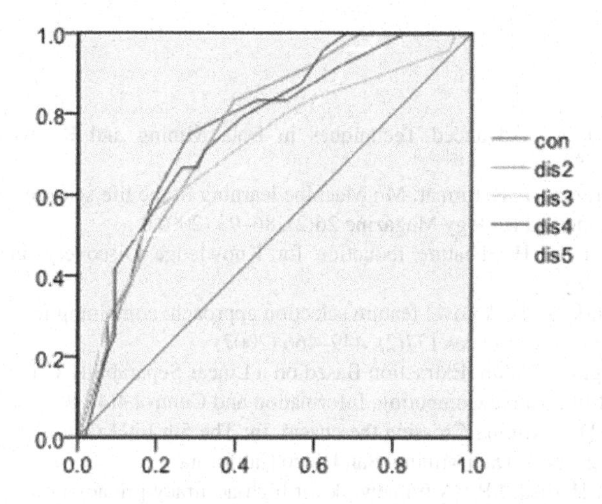

Fig. 1. Different ROC curves for different distributed datasets

The fast classification method presented in this paper is also applied to the distributed homogeneous dataset. The method consists of a processing center to complete data classification and the classification result is basically stable. Without having to know the precise data of those various distributed data sets, the processing center only requires a number of statistical information exchange with those distributed users to complete the automatic classification rule extraction. The information contains discrete interval points and the frequency table. Not only does it hold high classification efficiency, but also supports function of privacy-preserving on distributed data.

4 Conclusion

This paper presents a simple statistical method for inconsistency rate. It can directly and fast determine the optimal feature subset with the smallest inconsistency rate. The discovered knowledge is presented in easy-to-understand rules, which are intuitively derived from contingency tables summarized from data. Contingency table is calculated with the patterns of the selected optimal feature subset. The efficient and fast method for feature reduction and data classification has a distinct advantage in performance, which is demonstrated through those numerous experiments. For the distributed homogeneous data set, this paper finally builds a privacy-preserving data classification model with the proposed method. The method consists of a processing center to complete data classification. Without having to know the precise data of those various distributed data sets, the processing center is only required to exchange mass statistical information of discrete interval points and the frequency table with those distributed users to complete the automatic rule extraction. The classification result of the distributed homogeneous dataset is basically consistent with that of the original dataset.

Acknowledgments. This work was partially supported by Zhejiang Science and Technology Program of China under Grant No. 2010C31126 and 2011C21046.

References

1. Pal, N., Jain, L.: Advanced Techniques in Data Mining and Knowledge Discovery. Springer (2005)
2. Cios, K., Kurgan, L., Reformat, M.: Machine learning in the life sciences. IEEE Engineering in Medicine and Biology Magazine 26(2), 86–93 (2007)
3. Liu, H., Motoda, H.: Feature reduction for Knowledge Discovery and Data Mining. Springer (1998)
4. Uncu, O., Turksen, I.: A novel feature selection approach: combining feature wrappers and filters. Information Sciences 177(2), 449–466 (2007)
5. Xu, Y., Song, F.: Feature Extraction Based on a Linear Separability Criterion. International Journal of Innovative Computing, Information and Control 4(4), 857–865 (2008)
6. Rakesh, A.: Data Mining:Crossing the chasm. In: The 5th Int'l Conf. Knowledge Discovery in Databases and Data Mining, San Diego,California,
7. Agrawal, S., Haritsa, J.R.: A framework for high-accuracy privacy-preserving Mining. In: 21st International Conference on Data Engineering(UCDE), pp. 193–204. IEEE Computer Society, Tokyo (2005)
8. Friedman, A., Wolff, R., Schuster, A.: Providing k -anonymity in data mining. The VLDB Journal 17(4), 789–804 (2008)
9. Pinkas, B.: Cryptographic techniques for privacy-preserving data mining. ACM SIGKDD Explorations Newsletter 4(2), 12–19 (2002)
10. Agrawal, S., Haritsa, J.R.: A framework for high-accuracy privacy-preserving Mining. In: 21st International Conference on Data Engineering(UCDE), pp. 193–204. IEEE Computer Society, Tokyo (2005)
11. Lindell, Y., Pinkas, B.: Privacy Preserving Data Mining. In: Bellare, M. (ed.) CRYPTO 2000. LNCS, vol. 1880, pp. 36–54. Springer, Heidelberg (2000)
12. Kisilevich, S., Elovici, Y., Shapira, B., Rokach, L.: kACTUS 2: Privacy Preserving in Classification Tasks Using k-Anonymity. In: Gal, C.S., Kantor, P.B., Lesk, M.E. (eds.) ISIPS 2008. LNCS, vol. 5661, pp. 63–81. Springer, Heidelberg (2009)
13. Han, J.-M., Cen, T.-T., Yu, H.-Q.: Research in Microaggregation Algorithms for k-Anonymization. Acta Electronica Sinica 36(10) (2008)
14. Mei, Q., Luo, S.-S., Liu, W., Chen, P.: A Solution of Secure Multi-Party Multi-Data Ranking Problem Based On RSA Encryption Scheme. Acta Electronica Sinica 37(5) (2009)

Guess-Answer: Protecting Location Privacy in Location Based Services

Yingjie Wu, Sisi Zhong, and Xiaodong Wang

Dept. of Computer Science, Fuzhou University,
Fuzhou 350108, China
{yjwu,wangxd}@fzu.edu.cn,
zhong-sisi@163.com

Abstract. When mobile users retrieve interested location information through location based service (LBS), they should always provide their accurate location, which may leak their location privacy. In order to prevent privacy leakage when service provider is not reliable, some solutions based on two ties architecture are proposed. However, these solutions are either hard to implement or may lead to huge communication cost. In this paper, we present a novel technique, called Guess-Answer. Through the interaction between user and server, the location information within a region that satisfies the minimum privacy requirement of the user would be delivered. Through proper analysis and experimental study, we demonstrate that Guess-Answer is an efficient and effective way to achieve personalized privacy.

Keywords: Location Privacy, Location-based services, spatial databases, GIS.

1 Introduction

Recently, with the development of wireless network communication technologies, more and more people use location based services (LBS), that is, a kind of service that response various requests of mobile device users according to the location information they provide. When mobile users search points of interest (POI) from location based service databases, they always should provide their accuracy location, which may leak users' location privacy.

Various solutions have been proposed to protect location privacy [1-5]. The architecture of existing solutions can be grouped roughly into two categories: (1) three ties, which should employ a trust third-party as an anonymizer [1, 2, 3] and (2) two ties, which is based on client-server framework, doesn't need an anonymizer [4, 5]. For three ties [1, 2, 3], the anonymizer should be a trust third-party, and all techniques based on this architecture at least have one of the following drawbacks: (1) the anonymizer may become a performance bottleneck, especially when a huge number of users request the services; (2) the anonymizer may be vulnerable to malicious attack; (3) some users may not trust the anonymizer.

Y. Wang and T. Li (Eds.): Knowledge Engineering and Management, AISC 123, pp. 465–474.
springerlink.com

As a result, two ties architecture becomes more and more popular. ExactNN [5] is a famous technique which adopts two ties architecture. ExactNN is based on theoretical work on Private Information Retrieval[9]. By ExactNN, we could privately retrieve the exact nearest neighbor from server. However, ExactNN may suffer Man-In-the-Middle Attack (MIMT) where attackers intrude into the communication between server and users to inject false information and intercept the data transferred between them. Furthermore, except NN, ExactNN can not process any other type query, such as R-Range query. Moreover, due to an expensive PIR protocol, ExactNN will inevitably lead to huge computational and communication cost, even for a small POI databases.

In this paper, we propose a novel two ties framework, called Guess-Answer, which allows a user to generate cloaking area without revealing users' accurate locations for KNN query. Guess-Answer is based on client-server framework and does not need a trusted third party. In the Guess-Answer solution, a user can specify his/her personalized privacy requirement by given a threshold A_{min} before issuing his/her request. Here, A_{min} means that the final region issuing from server should not be smaller than A_{min}.

When a user submits a location based service request, for example, a KNN query, the server guesses a region where the user may locate in. After the user receives the server-guessed region, he/she answers a message about the region according his/her actual location, for example, "north of the region" or "east-west of the region" or "inside the region". According to the user's answer, the server can guess a more accurate region and deliver it to the user. This guess-answer process will repeat until the server can not generate a better region that is larger than the threshold A_{min}. In the guess-answer process, the server only holds the user's vague location information (e.g., "south-west of the region") and can not obtain user's exact position information. Even in the final step, the server would just know that the user is within a region no smaller than A_{min}.

The contributions of this paper are as follows. (1) We introduce a novel two ties solution, Guess-Answer, which allows users to enjoy location based services without revealing users' accuracy locations by specifying their privacy requirement. (2) We provide experimental evidence to show that Guess-Answer is efficient in terms of time and scalable in terms of supporting large maps, while providing high-quality query results without revealing users' exact location information.

2 Algorithms

In this section, we introduce the detail of Guess-Answer (GA) algorithm. The aim of GA is to protect users' location privacy. In reality, both location based service provider and communication between users and servers are probably unsafe. In fact, there exist base stations, which are operated by trust agents, e.g. communications service providers, between users and location services providers. In this paper, we eliminate base stations and assume that the map is a rectangle.

Figure 1 show the architecture of Guess Answer (GA) algorithm. The GA algorithm consists of five steps: Firstly, users submit a location based service request. Then the GA interaction protocol between servers and users works. After generating NN query candidate set, servers forward query candidate set to users. In the final step, users filter accurate query result. When a user submits a location based service request, the server and the user communicates under GA interaction protocol to generate a cloaked region. Using the cloaked region, the server generates NN query candidate set through the intersections of Voronoi diagram and cloaked region, and then sends the candidate set to the user. When the user receives the candidate set, he/she filters the wrong results from candidate set and gets the accurate query result. In section 2.1, we discusses the GA interaction protocol in detail, while the server guess algorithms are shown in section 2.2, and section 2.3 focuses on how to generates NN query candidate.

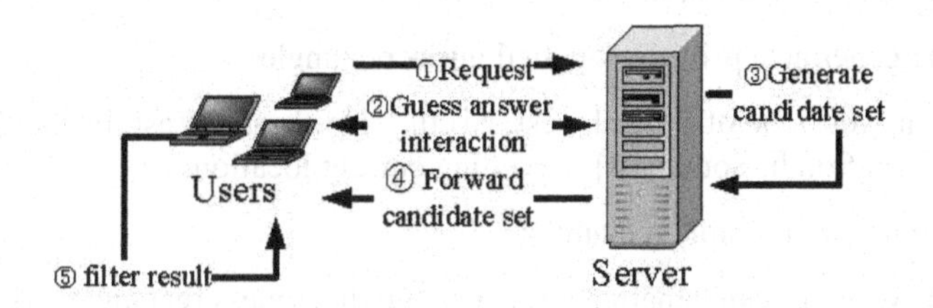

Fig. 1. The Architecture of Guess Answer Algorithm

2.1 Guess Answer Interaction Protocol

When a user submits a location based service request, she/he should specify a minimum privacy requirement, A_{min}, which is the measure of area. It means that the user wish the cloaked region (CR) should satisfy following three conditions. Firstly, CR must contains the user's exact location. Secondly, the area of CR is not smaller than A_{min}. In the third, nobody knows where the user is in CR. Consequently, users who need stronger privacy requirements could specify a larger A_{min}. However, larger A_{min} always means larger service cost.

Possible Rectangle. The rectangle where the user may locate in. Initially, possible rectangle is set to the whole map since the server could get nothing about the user's accurate location information when the server receives the user's request at the first time.

Guess Rectangle. A rectangle that the server guesses the user may locate in. Obviously, guess rectangle should be included in the possible rectangle.

Privacy-Satisfied Guess Rectangle. A guess rectangle (GR) is privacy satisfied when the areas of all sub-regions that divide from possible rectangle should be not smaller than A_{min}.

After the user receives the privacy-satisfied guess rectangle, s/he answers a message according to her/his current location, such as "north of the rectangle", "southwest of the rectangle", "inside the rectangle", et al. According to the user's answer message, the server could get more information about the user's actual location and could reduce the possible rectangle. In the next step, the server will generate another privacy-satisfied guess rectangle. The new-generated region would be delivered to user again, and the user would answer the server again too. The guess-answer procedure will repeat until the possible rectangle can not be further improved according to the user's privacy requirement A_{min}. As a result, the possible rectangle is the cloaked region. The pseudo-code of GA Interaction Protocol is shown in Fig.2.

Guess Answer Interaction Protocol

Input: A_{min}, user specify privacy requirement

M, map.

1. **Server:** set possible rectangle to M.
2. **Server:** generate a privacy-satisfied guess rectangle
3. **User:** answer one of { Northwest, North, Northeast, West, Inside, East, Southwest, South, Southeast} according current location
4. **Server:** update possible rectangle.
5. **Server:** if server can generate a privacy-satisfied guess rectangle, go to step 2, else go to step 6.
6. Terminate. Return possible rectangle as cloak region for query

---

Fig. 2. Protocol for Guess Answer Interaction

For example, in Fig.3, a user locates in the grid labeled U and the user's privacy requirement is 4 grids. At first, the server does not know in which grid the user locates, and sets possible rectangle to [A0, E7]. Then the server generates a privacy-satisfied guess rectangle R_1 [D4, E5] and assumes that the user is within the rectangle. When the user receives R_1 [D4, E5] from the server, he/she answers "north-west" of the rectangle. According to this information, the server would know that the user should locate in the rectangle [A0, C3], i.e. the possible rectangle now is [A0, C3]. Then, in the next step, the server guesses another privacy-satisfied guess rectangle R_2[C0, C3], which is included in the rectangle [A0, C3], and the user answers "west" of the rectangle. Thus, the server update possible rectangle to [A0, B3], which is more accurate than the rectangle [A0, C3]. In third time, the server guesses rectangle R_3 [A0, B1], the user answers "south" of the rectangle. Currently, the possible rectangle is [A2, B3], which is the smallest rectangle that satisfy the user's privacy requirement. Consequently, [A2, B3] is the cloaked region. In the whole guess-answer interaction process, the area of possible rectangle is not small than A_{min}, and the server could only know the possible rectangle that the user locate, without any other accuracy information. Therefore, guess answer interaction protocol can generate cloaked region without leaking any users' location privacy.

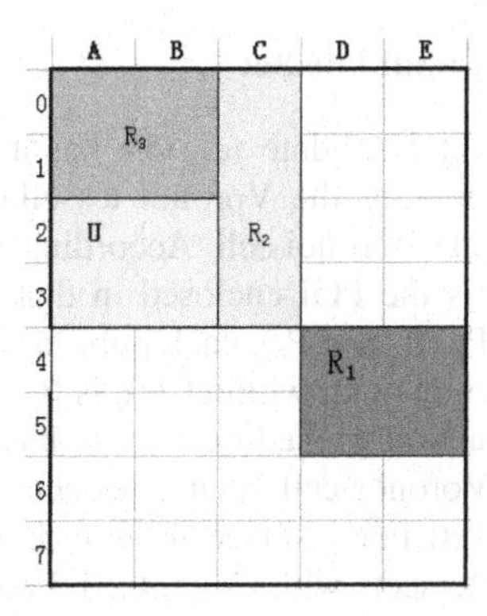

Fig. 3. An Example of Guess Answer Process

2.2 Server Guess Algorithms

For each time a privacy-satisfied guess rectangle is generated, it divides the possible rectangle into two or more sub-rectangles. For example, the guess rectangle in Fig. 4 divides the current region into nine sub-regions. In order to meet users' privacy requirements, the areas of all those nine sub-regions should not be smaller than A_{min}..

In fact, there are two guess approaches.

(1) Static guess. Static guess is based on a pre-defined pattern such as dividing the current rectangle into nine equal sub-rectangles and using the center partition as guess rectangle.

(2)Random guess. Randomly generating a privacy-satisfied guess rectangle, including the position and width of the rectangle.

When possible rectangle is too small to use static guess or random guess, the rectangle can be divided into quad or binary until the area of possible rectangle is smaller than twice of A_{min}.

Though static guess approach is simple and fast, it suffers from a number of weaknesses. For example, it will disclose map information, which is valuable for LBS, to clients. In contrast to the static guess approach that generates guess using a pre-divide structure, the random guess approach would not disclose any server side information.

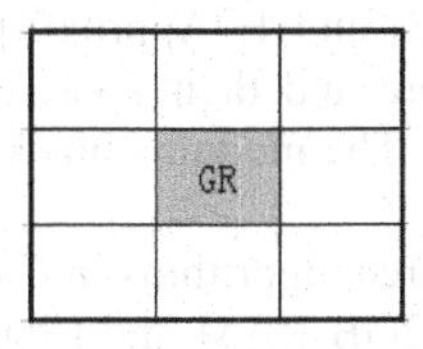

Fig. 4. An Example of Server Guess Algorithm

2.3 Generate NN Query Candidate Set

In order to generate NN query candidate set, GA has a preprocessing phase. In the preprocessing phase, GA computes the Voronoi tessellation [8] of the set of POIs, where each POI contains in one Voronoi cell. According the definition, the NN of any point within a Voronoi cell is the POI enclosed in that cell. For example, Figure 5 contains five points (P_1, P_2, P_3, P_4 and P_5), each point has one corresponding Voronoi cell. For example, the NN of any point within $Cell_3$ is P_3.

After guess answer interaction procedure, server has a cloaked region, CR. For each CR, it determines all Voronoi cells that intersect with it, and adds the corresponding POIs to candidate set. For instance, Figure 5 contains three cloak regions, R1, R2 and R3. For R1, it intersects with Cell5, Cell 3 and Cell4. Hence, R1's candidate set is {P5, P3, P4}. Similarly, R2's candidate set is {P1} since R2 is contained in Cell1. R3 contains Cell3 and intersects with Cell5, Cell4, Cell 2 and Cell 1, therefore its candidate set is {P1, P2, P3 P4, P5}.

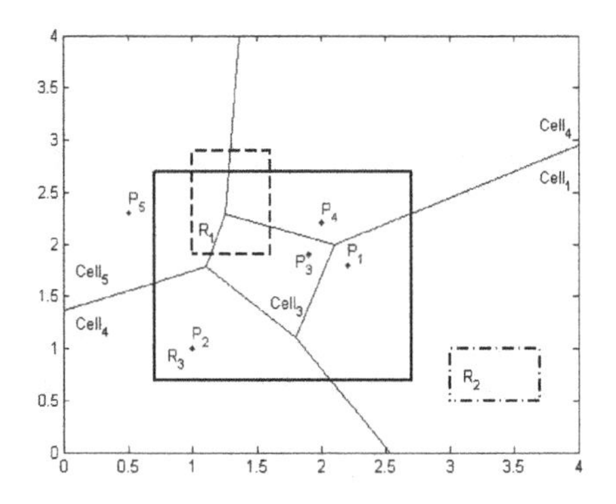

Fig. 5. NN query based on Voronoi diagram

3 Experiments

3.1 Experiment Settings

Experiments are designed to study the performance of the following algorithms: Guess-Answer (GA), ExactNN[3] and 1DApprox[3]. Both ExactNN and 1DApporx are based on two-ties architecture and their approaches of computing the NN of a query point are similar with GA. The modulus bits k is set to 256, which is the smallest parameter in [3].

We implemented the above three algorithms in C++ language on a machine with a 2.70 GHz dual-core Intel CPU, 2GB RAM and Linux OS. We tested the algorithms using both real (Sequoia[7], 65K POIs in California) and synthetic (uniform and Gaussian) datasets. Table 1 gives a summary of the parameters and settings used in experiments, the bold means default value. In each experiment, we randomly generate uniformly distribute 120 queries on the whole spaces and use the average value as the performance metrics. We use the following performance metrics:

(1) Communication cost, in numbers of TCP/IP packets. Since a 2D data point takes 8 bytes, a packet has a 40 bytes header, and the typical value of a Maximum Transmission Unit over network is 576bytes, so we set the packet capacity to (576-40)/8=67.
(2) CPU Time of the server.

Table 1. Experiment parameters and settings

Parameter	**Values**
Dataset size (K)	10, 25, 50, **75**, 100
Average Privacy Requirement (K * m^2)	5 10 50 100 **500** 1000

3.2 Experiment Results

Effect of Dataset Size

Figure 6(a) shows the server CPU time of NN queries as a function of dataset size. For all algorithms and datasets, the average server CPU time increases as the dataset size increases. However, GA performs much better as compared with its counterparts, since both ExactNN and 1DApprox are based on PIR, which may need a lot of computation and cost CPU time. For skew dataset, ExactNN and 1DApprox may be necessary to use a finer grid than uniform dataset, which results in CPU time increasing. In contrast, GA is made up of five parts, all of those parts are not sensitive to the skew, and thus the server CPU time is almost similar between skew dataset and uniform dataset.

Figure 6(b) shows the communication cost of NN queries as a function of dataset size. For ExactNN and 1DApprox, the results show that the average communication cost is almost linear to the dataset size increases on uniform data and skew data. On the other hand, the curves under GA are almost flat, regardless of the change of dataset size. Although the size of candidate set increases as the increasing of dataset size, the increase is too small to affect the communication cost. Furthermore, GA performs much better as compared with its counterparts, since both ExactNN and 1DApprox need send grid divide information to client. For skew dataset, ExactNN and 1DApprox may be necessary to use a finer grid than uniform dataset, which also resulting in increase of communication cost. On the contrary, GA is made up of five parts, all of those compositions are not sensitive to the skew dataset, and thus the effectiveness of data distribution is limited to our GA solution. According the following figures, the server communication cost of skew and uniform data is nearly the same for GA.

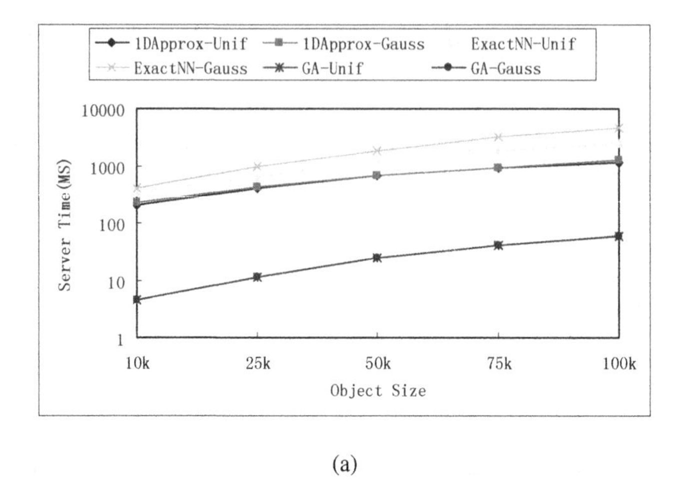

(a)

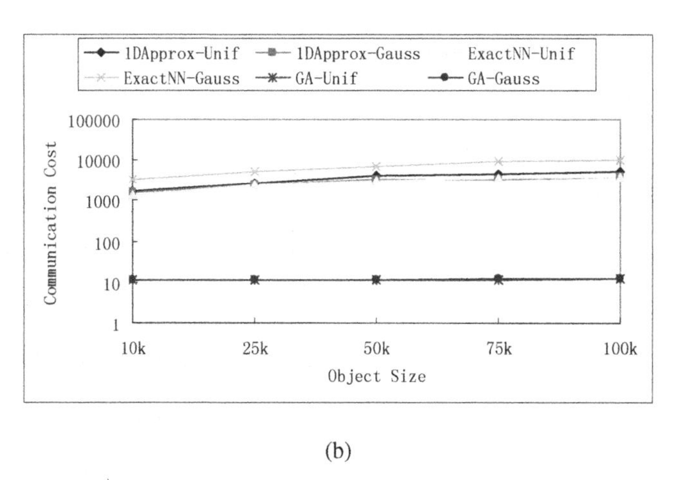

(b)

Fig. 6. Effect of Dataset Size

Effect of Privacy Requirement

Figure 7(a) plots the server CPU time of NN queries, by varying the privacy requirements. For server CPU time, the curves under all datasets are almost flat, regardless of the change of privacy requirements. Server CPU time on real dataset Sequoia is lower than uniform and gauss datasets, since the dataset size of Sequoia is smaller than uniform and gauss datasets.

Figure 7(b) shows communication cost of NN queries, as a function of the privacy requirement. Interestingly, the communication cost of GA first decreases, and then, increases. This is attributed to the guess-answer times and number of disclosed POIs. For weak privacy requirement, GA should communicate with users under guess-answer interaction protocol more times, leading to many TCP/IP packets delivering. For strong privacy requirement, more candidate POIs are disclosed and the candidate set size is enlarged, while increasing the number of communication packets.

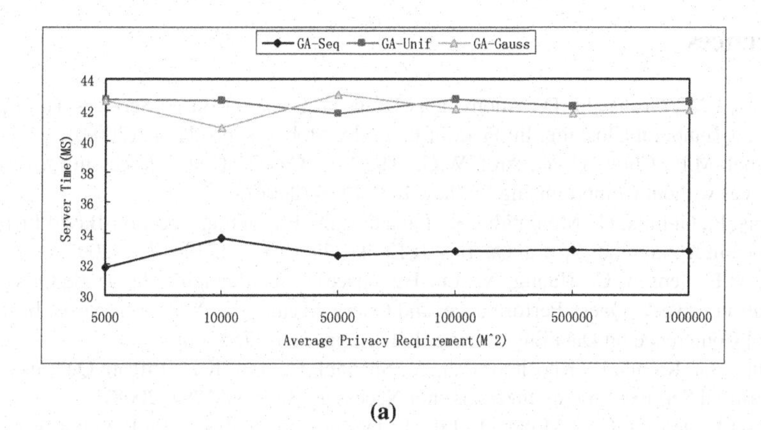

(a)

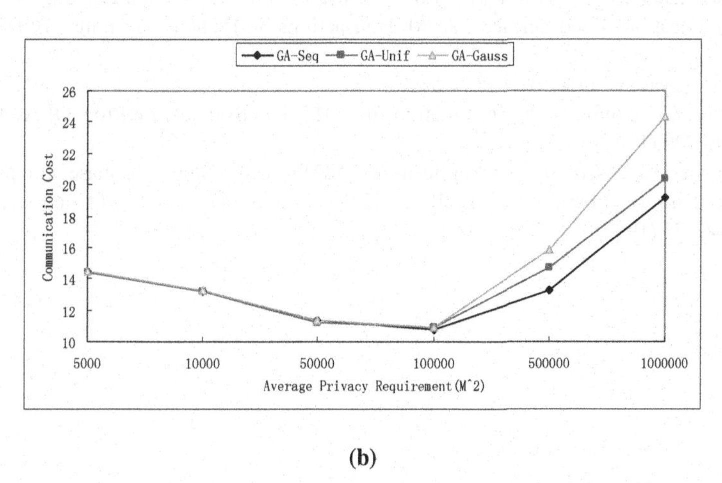

(b)

Fig. 7. Effect of privacy requirement

4 Conclusions

In this paper, we present a novel two-ties architecture, Guess-Answer, for preserving users' location privacy. Compared with previous work, Guess-Answer is simpler, lower server-side loads and more secure (i.e., personal privacy requirement specify, does not require Trust Third-Party). In the future work, we will consider integrating Guess-Answer with k-anonymity to provide stronger privacy protection in Location Based Services.

Acknowledgments. The research is supported by the Natural Science Foundations of Fujian Province under Grant No. 2009J01295 and No. 2010101330, and by the Foundation of Fujian Education Department under Grant No. JA09004. The authors would like to thank Dr. G.Ghinita for kindly sharing the implementation of the ExactNN[5] and 1DApprox [5].

References

1. Gruteser, M., Grunwald, D.: Anonymous Usage of Location-Based Services Through Spatial and Temporal Cloaking. In: Proc. of USENIX MobiSys, pp. 31–42 (2003)
2. Mokbel, M.F., Chow, C.-Y., Aref, W.G.: The New Casper: Query Processing for Location Services without Compromising Privacy. In: VLDB (2006)
3. Kalnis, P., Ghinita, G., Mouratidis, K., Papadias, D.: Preventing Location-Based Identity Inference in Anonymous Spatial Queries. IEEE TKDE 19(12), 1719–1733 (2007)
4. Yiu, M.L., Jensen, C., Huang, X., Lu, H.: SpaceTwist: Managing the Trade-Offs Among Location Privacy, Query Performance, and Query Accuracy in Mobile Services. In: International Conference on Data Engineering (ICDE), pp. 366–375 (2008)
5. Ghinita, G., Kalnis, P., Khoshgozaran, A., Shahabi, C., Tan, K.L.: Private Queries in Location Based Services: Anonymizers are not Necessary. In: SIGMOD (2008)
6. Yiu, M.L., Jensen, C.S., Moller, J., Lu, H.: Design and Analysis of a Ranking Approach to Private Location-Based Services. ACM Transactions on Database Systems (TODS) (to appear)
7. `http://www.rtreeportal.org/`
8. Fagin, R.: Combining Fuzzy Information from Multiple Systems. In: Proc. Of ACM PODS, pp. 216–226 (1996)
9. Kushilevitz, E., Ostrovsky, R.: Replication is NOT needed: Single database, computationally-private information retrieval. In: IEEE Symposium on Foundations of Computer Science, pp. 364–373 (1997)

Clustering Blogs Using Document Context Similarity and Spectral Graph Partitioning

Ramesh Kumar Ayyasamy[1], Saadat M. Alhashmi[1], Siew Eu-Gene[2], and Bashar Tahayna[1]

[1] School of Information Technology, [2]School of Business
Monash University, Malaysia
{ramesh.kumar,alhashmi,siew.eu-gene,bashar.tahayna}@monash.edu

Abstract. Semantic-based document clustering has been a challenging problem over the past few years and its execution depends on modeling the underlying content and its similarity metrics. Existing metrics evaluate pair wise *text similarity* based on text content, which is referred as *content similarity*. The performances of these measures are based on co-occurrences, and ignore the semantics among words. Although, several research works have been carried out to solve this problem, we propose a novel similarity measure by exploiting external knowledge base-Wikipedia to enhance document clustering task. Wikipedia articles and the main categories were used to predict and affiliate them to their semantic concepts. In this measure, we incorporate context similarity by constructing a vector with each dimension representing contents similarity between a document and other documents in the collection. Experimental result conducted on TREC blog dataset confirms that the use of context similarity measure, can improve the precision of document clustering significantly.

Keywords: Blog Clustering, Bipartite graph, Similarity Measures, Wikipedia.

1 Introduction

Clustering is an unsupervised learning technique that organizes similar members into groups or clusters and dissimilar members into other clusters. Clustering helps to discover the patterns and correlation in large data sets [2]. Document clustering received a significant attention in the last few years in the area of machine learning and text mining applications, such as webpages and blogs [4]. The entire text collection could be represented as *term by document* matrix, where each term is used as features for representing documents. Most traditional clustering systems are based on *bag-of-word* (BOW) representation and disregards semantic information and word order [7]. This BOW representation is restricted in calculating only term frequency in a document. Blogs - a new form of webpages, which consists of group of blogposts and each blogpost could be written on various topics, and BOW assumption cannot perform well [18]. There is no suitable way to identify and categorize blogposts; hence, it is an open problem for research [1]. Clustering similar posts based on different categories helps the user identify or search the particular topic of interest. Recently, blogs became the subject of research as the information

Y. Wang and T. Li (Eds.): Knowledge Engineering and Management, AISC 123, pp. 475–486.
springerlink.com

content is large and diverse, and creates a need for automated organization of blog pages. Traditional clustering cannot find the hidden relationship between heterogeneous objects. Many clustering algorithms have been proposed in the various contexts of literature. Berkhin [3] investigated the applications of clustering algorithms for data mining. In most of the works, partitioned clustering algorithm [4] is suited for large data set clustering due to their low computational requirements. Baker et al. [11] used the distributional clustering method, which clusters together those terms that tend to indicate the presence of the same category, or group of categories. Xu et al. [10] presents a simple application using *Non-negative Matrix Factorization* (NMF) for document clustering. As *K-means* clustering algorithm cannot separate clusters that are non-linearly separable in input space, Dhillon et al. [13] used Jensen-Shanon divergence to cluster words in K-means fashion in text clustering. Despite widespread adoption on clustering tasks, only few studies have investigated using Wikipedia as a knowledge base for document clustering [5 - 7]. Gabrilovich [5] have applied structural knowledge repository-Wikipedia as feature generation technique. Their work confirmed that background knowledge based, features generated from Wikipedia can help text categorization. Huang et al. [6] have clustered documents using Wikipedia knowledge and utilized each Wikipedia article as a concept. This work extracted related terms such as synonyms, hyponyms and associated terms. Hu et al. [7] have used Wikipedia concept feature for document clustering. Development in data mining applications have demanded for clustering highly inter-related heterogeneous objects, such as *documents* and *terms* in a text corpus. Clustering each type of objects independently might not work well, since one type of objects can be defined by other type of objects. This paved the way for many researchers to co-cluster two or more heterogeneous data. The authors in [17], [22], expanded the traditional clustering algorithms and proposed bipartite spectral graph partitioning algorithm to cluster documents and terms simultaneously. Similar bipartite techniques were applied in medicine [8], image [20] and video processing [21]. Gao et al. [9] have suggested consistent bipartite graph co-partitioning (CBGC) by treating the tripartite graph as two-bipartite graphs. This work proved that consistent partitioning provides the optimal solution using positive Semi-Definite Programming (SDP). To co-cluster triplet data [9], terms, category and documents (Web pages) were used. This work utilizes, terms from web pages and does clustering based on it. Like bipartite graph partitioning, it has limitations that the clusters from different types of objects must have one-to-one associations and it fails to consider the surrounding text and context information. On the contrary, each blogpost consists of various features and a framework is needed beyond the using of terms representing documents. With the above motivation in mind, in this paper we constructed a tripartite spectral graph clustering, using *concepts*, *document*, and *document contexts* to cluster similar topical blogposts based on Wikipedia categorical index.

The outline of this paper is structured as follows: Section 2 defines, the different terms used in our paper. Section 3 explains our CSSGP framework which uses similarity based clustering on *content similarity*, *context similarity* and our *Concept frequency measure*. Section 4 describes the experiments and results in detail. Finally, section 5 concludes with the summary of the work and future research.

2 Definition of Terms

In this paper, we use these following terms:

- *"Document (D)"* refers to single blogpost from the collection. Let χ be the data set, where $\chi = \{D_1, D_2, ..., D_n\}, n \in \mathbb{Z}^+$
- *"Term(T)"* refers to a meaningful word from a particular blogpost. Let D be the Document which consists of set of terms, where $D = \{T_1, T_2, ..., T_m\}, m \in \mathbb{Z}^+$
- *"Document context(Dc)"* refers to document co-occurrence information and are constructed against the set of documents.
- *"Concept(C)"* refers to a Wikipedia article title. Let φ be the set of concepts, where $\varphi = \{C_1, C_2, ..., C_i\}, i \in \mathbb{Z}^+$
- *"Categories"* refers to Wikipedia's 12 main categories, where $\mathrm{T} = \{\varphi_1, \varphi_2, ..., \varphi_{12}\}$
- *"T-D-Cat"* refers to tripartite clustering of *Terms*, *Documents* and *Categories*
- *"C-D-Dc"* refers to tripartite clustering of *Concepts*, *Documents* and *Document Contexts*
- *"T-D"* refers to bipartite clustering of *Terms*, and *Documents.*

3 Clustering Blogs Using Context Similarity and Spectral Graph Partitioning Framework- CSSGP

We present our blog clustering framework using Context similarity and Spectral graph partitioning to deal with the above observations (See Figure 1).To illustrate the CSSGP framework effectiveness, we have conducted an extensive experiment aiming at document clustering.

Our contribution comes in two folds: First by using Wikipedia, we perform an explicit semantic based topic analysis by converting terms to concepts. We calculate surrounding text similarity using *conf.idf.* Second we used Content Similarity and surrounding text similarity to calculate Document Context Similarity. We used *document context, concepts,* and *document*s to produce a tripartite spectral graph, which helps for efficient document clustering. There is a general notion that the combination of terms and the BOW approach hold the greatest promise in text clustering. The fundamental challenges in bridging the gap between the syntactic and semantic elements is the design of distance function that measure the perceptual similarity between text features. To evaluate the content-based text similarity, we utilize the term weighting function, which is one of the most widely used metrics due to its simplicity and effectiveness [15]. Adding to this document-based text similarity (in section 3.1), we compute the concept-based similarities that two documents have in common (i.e., *the surrounding text*). Thus, we further compute this kind of similarity, which we denote as context similarity.

3.1 Document Context Similarity

Content Similarity Measure.
When two documents contain similar topics, there are higher chances for existence of many common terms. After pre-processing, we represent the stemmed words as

vectors for each document in a vector space model. The term weighting function *tf-idf* [15] been defined as;

$$tf.idf(t_k, D_j) = \#(t_k, D_j).\log\frac{|T_r|}{\#T_r(t_k)} \quad (1)$$

Where t_k denotes k'th terms, D_j denotes j'th documents, $\#(t_k, D_j)$ denotes the number of times t_k occurs in D_j, and $\#T_r(t_k)$ denotes the number of documents T_r in which t_k occurs at least once (also known as the document frequency of term t_k).

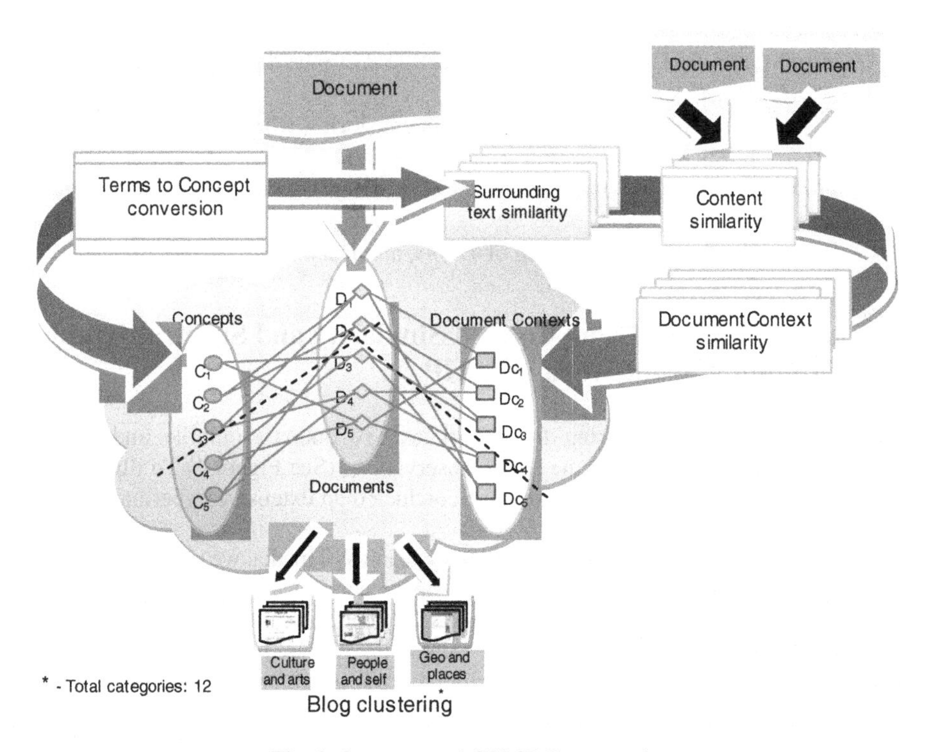

Fig. 1. Our proposed CSSGP Framework

Given this representation, the cosine similarity measure of the document pairs is represented as:

$$sim^{\cos ine}_{content}(D_p, D_k) = \frac{D_p.D_k}{|D_p||D_k|} \quad (2)$$

Where d_p and d_k are p'th and k'th document respectively.

Surrounding Text Similarity Measure (*Terms* to *Concepts* Conversion)

Our proposed method is a relatedness measurement method that computes the relatedness (i.e. among words) based on co-occurrence analysis. As a simple case, we map each word to a *concept*; the combination of these concepts/words can create a new *concept* (compound concept). Let us assume A is a word mapped to a *concept* C_1, B is a word mapped to a *concept* C_2, and ω is the set of extracted terms from blogpost.

Then there is a chance that the combination of A & B can produce a new concept (see Figure 2). Consider the following example,

$$\$s = \text{"President of the United States"}$$
$$X = \text{"President"}, Y = \text{"United"}, Z = \text{"States"}, \omega = \{X, Y, Z\}$$

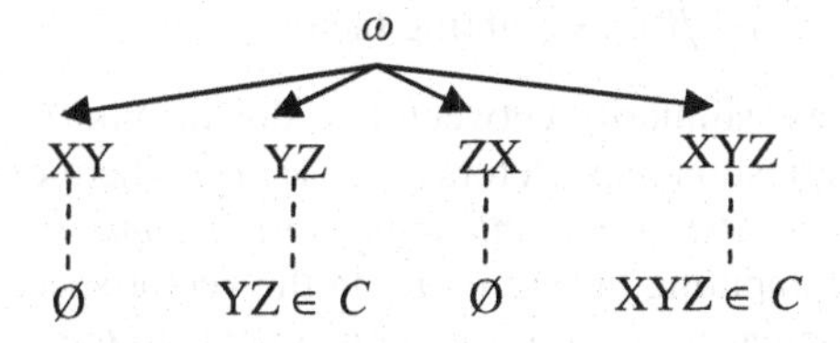

Fig. 2. Related measurement method

Where *C* is the set of concepts such that YZ is mapped to *Country* concept and XYZ is mapped to *President* Concept. Since the order of X, Y and Z can give different concept or no concept at all, we neglect the reordering. For effective clustering, we mine existing knowledge bases, which can provide a conceptual corpus. Wikipedia is one of the best online Knowledge Base to provide an individual English article with more than 3 million concepts. With that said, an important feature of Wikipedia is that on each wiki page, there is a URI which unambiguously represent the concept. Two words co-occur if they appear in an article within some distance of each other. Typically, the distance is a window of *k* words. The window limits the co-occurrence analysis area of an arbitrary range. The reason of setting a window is due to the number of co-occurring combinations that becomes too huge when the number of words in an article is large. Qualitatively, the fact that two words often occur, close to each other is more likely to be significant, compared to the fact that they occur in the same article, even more so when the number of words in an article is huge. The *n-gram based concept extraction* algorithm was briefly discussed in Ayyasamy et al.'s work [14]. We transform the surrounding text from *terms* into a weighted vector of *concepts* to represent in a vector space. The concept weights are computed with concept weighting scheme *conf.idf* [19].

$$conf \cdot idf(con,T) = conf(con,T).idf(con), \tag{3}$$

$$idf(con) = \log \frac{N_t}{pf(con)} \tag{4}$$

Where $conf(con,T)$ denotes the number of times a concept *C* occurs in the surrounding text *T*, *pf(con)* denotes the number of *documents* contains the concept *con,* and N_t is the number of documents in the collection. Our vector space model consists, the set of all vectors representing distinct documents. We need a similarity measure that compares two vectors and returns a numeric value that shows the level of similarity between them. There exists several similarity measures and *cosine metric* is the one among commonly employed similarity metric for text mining.

Document Context Similarity Measure

Document context consists of documents co-occurrence information and are constructed against the set of documents. The *document context* is a feature vector with each element representing the content similarity of the document and each document in the collection. In a given set of *n* documents $\chi = \{D_1, D_2, ..., D_n\}, n \in \mathbb{Z}^+$, the context vector of a document D_p is an *n-dimensional* vector $f(D_p) =< f_1(D_p), f_2(D_p).....f_n(D_p) >$, where each element is equal to the corresponding content similarity as follows:

$$f_k(D_p) = \alpha.Sim^{\cos ine}_{content}(D_p, D_k) + \beta.Sim^{\cos ine}_{concept}(D_p, D_k), 1 \le k \le n \quad \{\alpha, \beta \text{ are weighting factors}\} \tag{5}$$

We obtain the context similarity between two documents D_p and D_q by computing the similarity of the corresponding vectors $f(D_p)$ and $f(D_q)$. There exist various measures such as *Dice coefficient, Overlap Coefficient, Cosine Similarity, and Jaccard Similarity Coefficient* for computing the similarity in the vector space model [16]. We use *Cosine similarity*, a popular measure for text documents to measure our external cluster validation. For example, the context similarity using *Cosine Similarity* between two documents D_p and D_q defined as follows (6):

$$sim^{\cos ine}_{context(D_p, D_q)} = \frac{\sum_k f_k(D_p) \times f_k(D_q)}{\left(\sum_k f_k(D_p)^2 \times \sum_k f_k(D_q)^2 \right)^{\frac{1}{2}}} \tag{6}$$

3.2 Spectral Graph Clustering

Clustering is an unsupervised learning method, which can be formulated by partitioning the graph such that the edges between different groups have very lower weights, and the edges within the same group have higher weights. The importance of spectral graph clustering can be found on Luxburg's work [12]. The algorithm for bipartite spectral graph clustering (Figure 3) was briefly discussed in Dhillon's work [17].

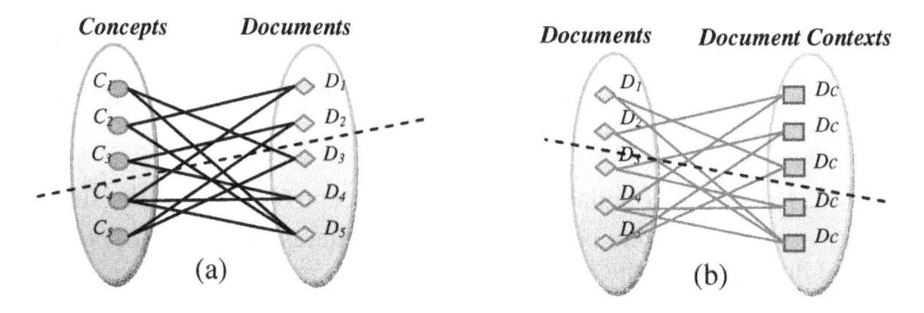

Fig. 3. The Bipartite Graph of (a) *C-D* , (b) *D-Dc* Clustering

Tripartite Spectral Graph Partitioning

In order to benefit from both the *concepts* and the *document contexts* for text clustering, we use tripartite graph as shown in Figure 4 to model the relations among the *concepts*, *documents*, and *document contexts.* In this tripartite graph (Figure 4), circle symbol represents the *concepts* $C = \{C_1, C_2, C_3, ...C_i\}, i \in \mathbb{Z}^+$, diamond symbol represents the *documents* $D = \{D_1, D_2, D_3, ...D_j\}, j \in \mathbb{Z}^+$ and the rectangle symbol represents the *document contexts* $Dc = \{Dc_1, Dc_2, Dc_3, ...Dc_k\}, k \in \mathbb{Z}^+$ respectively. An undirected tripartite graph can be represented as quadruplet $G = \{C, D, Dc, E\}$, where E is the set of edges connecting vertices $\{\{C_i, D_j, Dc_k\} \mid C_i \in C, D_j \in D, Dc_k \in Dc\}$. An edge $\{C_i, D_j\}$ exists if *concepts* C_i occurs in *document* D_j, where the edges are undirected which denotes there are no edges between *concepts* or between *documents*.

The edge-weight between C_i and D_j equals the value of C_i in D_j, while the edge-weight between D_j and Dc_k equals the frequency of Dc_k in the surrounding text of document k. Consider i x j x k as *concept*-by-*document*-by-*document context* matrix,

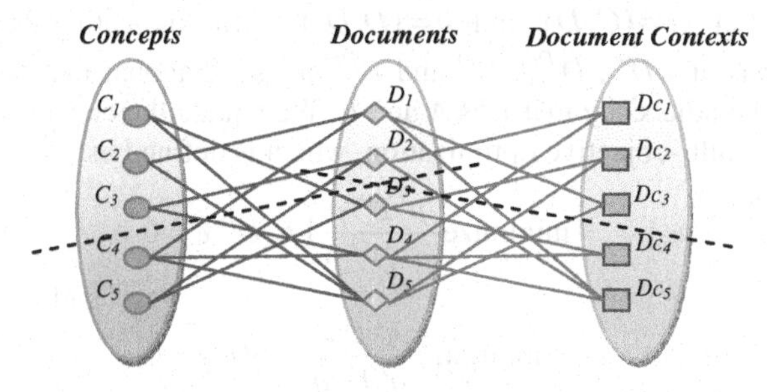

Fig. 4. Tripartite Graph of Concepts, Documents, and Document Contexts(C-D-Dc)

where A and B represent feature weight matrices for *Concepts-Documents* and *Documents-Document Contexts*, in which the A_{ij} equals the edge-weight E_{ij} and B_{jk} equals the edge-weight E_{jk} respectively. The adjacency matrix M of the tripartite graph could be written as:

$$M = \begin{array}{c} \\ C \\ D \\ Dc \end{array} \begin{array}{c} \begin{array}{ccc} C & D & Dc \end{array} \\ \begin{pmatrix} 0 & A & 0 \\ A^T & 0 & B \\ 0 & B^T & 0 \end{pmatrix} \end{array} \tag{7}$$

where the vertices been ordered in such a way that the first i vertices index the *concepts* while the next j vertices index the *documents* and the last k vertices index the *document contexts*. Every entry in A and B represents the importance of particular *concepts* to *document* and *document contexts* respectively. To co-cluster C, D, and Dc simultaneously, it seems common to partition the graph in Figure 3(b) by solving the generalized eigenvalue problem corresponding to the adjacency matrix. In parallel, we found that this solution does not work, as it seems. Originally, if we move the vertices of *concepts* in Figure 3(b) to the side of the vertices of *document contexts*, then the drawn tripartite graph would turn to be a partite graph. As our work is related on (*documents*, *concepts*, and *document contexts*) bipartite graph and the loss incurred on cutting an edge between a *document* and a *concept*, which contributes the loss function similar to the loss of cutting an edge between a *document* and *document context*. In addition, these two kinds of edges are heterogeneous and incomparable. To deal with such problems, Gao et al. [9] have treated tripartite graph as two-bipartite graphs, where each graph share the central nodes. Therefore, we convert the actual problem by combining the pair-wise problems over these two bipartite graphs. During our implementation of bipartite graph (as in Figure 3), will be generated from the tripartite. In the mean time, if we transfer bipartite spectral graph partitioning [17] on Figure 3(a) and Figure 3(b) individually, that provides us with a maximum probability that the partitioning schemes for documents are different in the two solutions (i.e., *not locally optimal*). To solve this optimization problem, this work [9] has suggested a *consistent bipartite graph co-partitioning* (CBGC), which provides an optimization algorithm using positive semi-definite programming. Our work only focuses on bi-partitioning the *concepts*, *documents* and the *document contexts* into two groups simultaneously. To achieve this clustering, we allow $\{C, D, Dc\}$ to perform as the column vectors of i, j, k dimensions for *concept, document* and *document context* respectively. Let $p=(C,D)^T$ and $q=(D,Dc)^T$ as the indicating vectors for the two bipartite graphs and $D^{(c)}$, $D^{(Dc)}$, $L^{(c)}$ and $L^{(Dc)}$ as the diagonal matrices and Laplacian matrices for the adjacency matrices A and B. We equate the consistent co-partitioning problem with multi-objective optimization, which is defined as:

$$\underset{p\neq 0}{\text{minimize}} \frac{p^T L^{Dc} p}{p^T D^{Dc} p} \mid p^T D^{Dc} e = 0, \tag{8}$$

and

$$\underset{q\neq 0}{\text{minimize}} \frac{q^T L^{C} q}{q^T D^{C} q} \mid q^T D^{C} e = 0, \tag{9}$$

where e is the column vector, which is equal to 1. To solve the multi-objective optimization problem, this work [17] provides a complete detail on the ways of using the positive semi-definite programming as an efficient solution to the optimization problem. We utilize their algorithm to solve the co-clustering of *concepts*, *document* and *context* (*C-D-Dc*) and compared with *terms*, *documents* and *category* (*T-D-Cat*). Partitioning this graph lets us achieve blog clustering by integrating information from *concepts* and *document context* synchronously.

4 Experimental Evaluation

4.1 TREC Dataset

In this section, to evaluate our CSSGP framework, we used part of TREC BLOGs08[1] dataset. TREC dataset is a well-known dataset in blog mining research area. This dataset consists of the crawl of Feeds, associated Permalink, blog homepage, and blogposts and blog comments. The BLOGs08 corpus is a sample of blogosphere crawled over from 14 January 2008 to 10 February 2009. The crawled feeds are mostly from *BlogSpot, LiveJournal, Xanga,* and *MSN Spaces*. The blog permalinks and homepages are encoded using HTML. This dataset is a combination of both English and Non-English blogs.

4.2 Experiment Dataset Preparation

As our research focus is only on blog clustering and not on language identifier, we filtered the blogs by English language. In order to extract the blogposts from blog

[1] `http://ir.dcs.gla.ac.uk/test_collections/blogs08info.html`

documents, the HTML source code of blog pages should be parsed (which includes removal of HTML tags, scripts, etc.), and to be converted into plain text. We used blog-post extraction program named *BlogTEX*[2] to extracts blog posts from TREC Blog dataset. We utilized the list of stop words in our program to identify the English blogs. During preprocessing, we removed HTML tags and used only blogposts for our experiment. Furthermore, as the dataset is huge and our motive is to test our tripartite clustering approach, we collected around 41,178 blogposts and assigned based on each category. Table 1 below shows the *CategoryTree* structure of Wikipedia and the cluster sizes. Our expected clustering result should be the same as Table 1. We built, *Document context* (*Dc)* by constructing a vector with each dimension representing the content similarity between the blogpost, and any blogpost in the dataset. We consider leftover words as textual representation of blogposts. The entire document collections represented as *word* × *document matrix*, where rows corresponds to *words* and columns to *documents*. We carried out concept indexing, using *conf.idf*, 1,55,980 *concepts* and term indexing by using *tf.idf* [15], 2,83,126 *terms* is reserved for our experiment.

Table 1. The Main *CategoryTree* Structure of Wikipedia and the Cluster Size

Categorical Index	Sub-Categories	# of documents	# of concepts	Cluster
General Reference	Almanacs • Atlases ... more	878	3758	C_1
Culture and Arts	Baseball • Basketball •......	8580	18245	C_2
Geography and Places	Cities • Continents • ...more	4558	23220	C_3
Health and Fitness	Health promotion • health ,...	4082	21409	C_4
History and Events	Africa • Asia• ..... more	2471	12746	C_5
Mathematics and Logic	Algebra • Analysis ..... more	213	2561	C_6
Natural and Physical sciences	Astronomy • Chemistry....	1560	8530	C_7
People and Self	Children • Political people•...	6279	10125	C_8
Philosophy and Thinking	Theories • Aesthetics•....	2345	7591	C_9
Religion and Belief	Allah • Bible • Buddha•....	2345	8315	C_{10}
Society and Social sciences	Business • Family • .... More	4657	13170	C_{11}
Tech and Applied sciences	Bioengineering • Chemical ...	3210	26310	C_{12}

4.3 Evaluation Results

Our CSSGP framework have been evaluated by comparing clustering output with true class labels (Wikipedia categories). There are number of comparison metrics and we use cluster *purity* and *entropy*. The clustering metric, *purity* measures the extend to which each cluster contained documents from primarily one class. The second clustering metric, *entropy* measures the various classes of documents that are distributed within each cluster. The value for the relative *entropies* and *purities* metrics are shown in Table 2. The higher purity $C_2$and lower entropy C_4 values are

[2] `http://sourceforge.net/projects/blogtex`

highlighted in the above Table 2. As most of the documents have discussed on *culture and arts*, *health and fitness*, we got higher purity and lower entropy during clustering.

Table 2. Results from Clustering evaluating metrics *Purity* and *Entropy*

	C_1	C_2	C_3	C_4	C_5	C_6	C_7	C_8	C_9	C_{10}	C_{11}	C_{12}
Purity	0.843	**0.881**	0.835	0.859	0.455	0.668	0.856	0.753	0.817	0.795	0.720	0.813
Entropy	0.285	0.259	0.223	**0.210**	0.675	0.419	0.265	0.299	0.238	0.329	0.368	0.233

4.4 Performance Comparison

In this section, we compared the clustering performance for all 12 possible categories with existing clustering technique. As blog document clustering research is in its nascent stage, there is no current benchmark to prove our work. To test the performance of our tripartite spectral graph clustering (*C-D-Dc*), we compared with the existing technique, *T-D* bipartite spectral graph clustering, and *T-D-Cat* tripartite spectral graph clustering. *T-D* bipartite graph clustering [18] consists of raw terms collected from the documents. *T-D-Cat* tripartite spectral graph clustering [9] consists of raw terms, documents and the Wikipedia categories. We plot two-dimensional graph to compare performance between *C-D-Dc* tripartite spectral graph clustering accuracy and *T-D* bipartite spectral graph clustering accuracy in Figure 5.

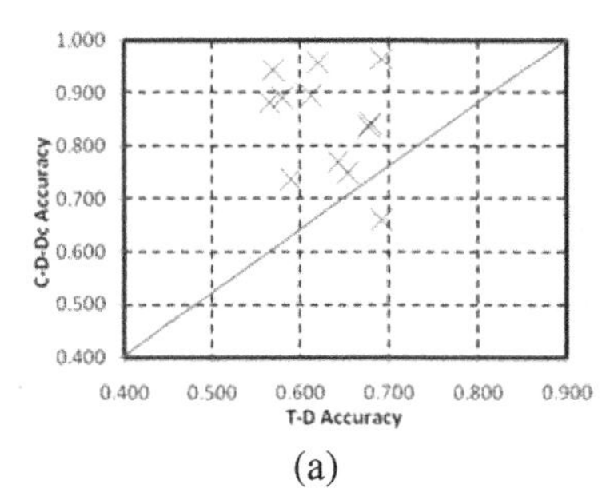

(a)

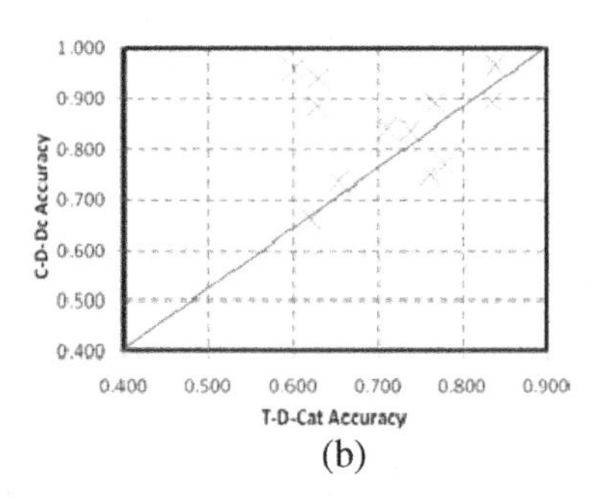

(b)

Fig. 5. Performance Comparison between (a) ***C-D-Dc*** vs. ***T-D***, (b) ***C-D-Dc*** vs. ***T-D-Cat*** clustering

C-D-Dc Accuracy is plotted in the X-axis and *T-D* Accuracy is plotted in the Y-axis. Each dot in the graph represents a possible category pair. We see that most of the dots are in the *C-D-Dc* clustering accuracy side, which indicates that *C-D-Dc* clustering performs better than the *T-D* clustering. This graph interprets that tripartite spectral graph clustering using *document context* (*Dc*) data gives better clustering. As similar to the above comparison, we plot two-dimensional graph to compare tripartite spectral graph clustering performance between *C-D-Dc* and *T-D-Cat* clustering accuracy in Figure 5. *C-D-Dc* Accuracy is plotted in the X-axis and *T-D-Cat* Accuracy is plotted in the Y-axis. Each dot in the graph represents a possible category pair. We see that most of the dots are in the *C-D-Dc* clustering accuracy side, which indicates that *C-D-Dc* clustering performs better than the *T-D-Cat* clustering. This graph interprets that tripartite spectral graph clustering using *document context* (*Dc*) data gives better clustering. To summarize this performance comparison, *C-D-Dc* tripartite spectral graph clustering provides better performance than the other two clustering *T-D*, and *T-D-Cat* clustering.

4.5 Average Performance

To measure the average performance, we used tripartite graph clustering results using *C-D-Dc* with the *T-D-Cat* and bipartite graph clustering using *T-D*. The average performances between each clustering and all other 12 categories are shown in Table 3. We highlighted the exceeding values. Here *Terms and Documents (T-D) clustering* follows conventional bipartite spectral graph clustering. Our three clustering results are compared with the expected result, which is mentioned in Table 3.

Table 3. The Average Performance Result among *T-D, T-D-Cat* and *C-D-Dc* clustering

Category Name	*T-D*	*T-D-Cat*	*C-D-Dc*
General Reference	0.590	0.655	0.736
Culture and Arts	0.622	0.660	0.692
Geography and Places	0.581	0.768	**0.890**
Health and Fitness	0.692	**0.840**	**0.966**
History and Events	0.655	0.761	0.749
Mathematics and Logic	0.681	0.711	**0.841**
Natural and Physical sciences	0.565	0.629	**0.883**
People and Self	0.678	0.738	**0.836**
Philosophy and Thinking	0.621	0.601	**0.957**
Religion and Belief	0.643	0.750	0.768
Society and Social sciences	0.570	0.630	**0.942**
Tech and Applied sciences	0.613	**0.837**	**0.895**
Average performance	**0.626**	**0.715**	**0.846**

Out of 12-main categories in Table 3, 8-categories-*Health and Fitness, Philosophy and Thinking, Society and Social sciences, Tech and Applied sciences, Geography and Places, Natural and Physical sciences, Mathematics and Logic* and *People and Self* shown better performance on *C-D-Dc* graph clustering. The *T-D-Cat* clustering shown good performance on categories such as *Health and Fitness,* and *Tech and Applied sciences.* This implies that our proposed method has 84% success rate and achieved better performance, compared to the other two clustering, *T-D*, and *T-D-Cat.*

5 Conclusion and Future Work

In this paper, we developed a CSSGP framework that represents concepts, documents and document contexts as a tripartite graph structure and demonstrated the effectiveness of document context as a significant feature that improves the document clustering results. The primary contributions of this paper are the following: First by using Wikipedia, we perform an explicit semantic based topic analysis by converting terms to concepts. We calculated the surrounding text similarity using *conf.idf*. Second we used Content Similarity and surrounding text similarity to calculate Document Context Similarity. We used *document context, concepts,* and *documents* to produce a tripartite spectral graph, which helps for efficient document clustering. As shown in Table 3, our experiment results indicate that our framework enhances cluster quality than the traditional document clustering approaches. In addition, our proposed solution blends well with real examples. In the future, we plan to expand our experiment using the full

TREC Blogs08 dataset to measure the clustering performance and its scalability. We plan to include other blog features such as comments and tags, as one of the data for tripartite graph clustering and measure its performance for document clustering.

References

1. Ounis, I., Macdonald, C., Soboroff, I.: On the TREC BlogTrack. In: ICWSM, USA (2008)
2. Halkidi, M., Batistakis, Y., Vazirgiannis, M.: On Clustering Validation Techniques. J. of Intelligent Information System (2001)
3. Berkhin, P.: Survey of clustering data mining techniques. Accrue Software Inc., Technical report (2002)
4. Xu, R., Wunsch II, D.: Survey of clustering algorithms. IEEE Trans. Neural Netw. 16(3), 645–678 (2005)
5. Gabrilovich, E., Markovitch, S.: Overcoming the brittleness bottleneck using Wikipedia: Enhancing text categorization with encyclopedic knowledge. In: AAAI (2006)
6. Huang, A., Milne, D., Frank, E., Witten, I.: Clustering documents using a Wikipedia-based concept representation. In: PAKDD, pp. 628–636 (2009)
7. Hu, J., Fang, L., Cao, Y., Hua-Jun Zeng, H., Li, H.: Enhancing Text Clustering by Leveraging Wikipedia Semantics. In: ACM SIGIR, pp. 179–186 (2008)
8. Yoo, I., Hu, X., Song, I.Y.: Integration of semantic-based bipartite graph representation and mutual refinement strategy for biomedical literature clustering. In: KDD (2006)
9. Gao, B., Liu, T., Zheng, X., Cheng, Q., Ma, W.: Consistent Bipartite Graph Co-Partitioning for Star-Structured High-Order Heterogeneous Data Co-Clustering. In: SIGKDD (2005)
10. Xu, W., Liu, X.: Gong. Y.: Document clustering based on nonnegative matrix factorization. In: SIGIR 2003, pp. 267–273 (2003)
11. Baker, L., McCallum, A.: Distributional Clustering of Words for Text Classification. In: ACM SIGIR, pp. 96–103 (1998)
12. von Luxburg, U.: A tutorial on Spectral Clustering. In: MPI-Technical Reports No.149. Tubingen: Max Planck Institute for Biological Cybernetics
13. Dhillon, I., Guan, Y., Kulis, B.: Kernel k-Means, Spectral Clustering and Normalized Cuts. In: KDD, pp. 551–556 (2004)
14. Ayyasamy, R.K., Tahayna, B., Alhashmi, S.M., Siew, E., Egerton, S.: Mining Wikipedia knowledge to improve document indexing and classification. In: 10th International Conference on Information Science, Signal Processing and their Applications, ISSPA 2010, pp. 806–809 (2010)
15. Salton, G., Buckley, C.: Term-weighting approaches in automatic text retrieval. J. Information Processing & Management 24, 513–523 (1988)
16. Strehl, A., Ghosh, J., Mooney, R.: Impact of similarity measures on web-page clustering. In: AAAI Workshop on AI for Web Search, pp. 58–64 (2000)
17. Dhillon, I.: Co-clustering documents and words using bipartite spectral graph partitioning. In: ACM SIGKDD, pp. 269–274 (2001)
18. Sun, A., Suryanto, M.A., Liu, Y.: Blog Classification Using Tags: An Empirical Study. In: Goh, D.H.-L., Cao, T.H., Sølvberg, I.T., Rasmussen, E. (eds.) ICADL 2007. LNCS, vol. 4822, pp. 307–316. Springer, Heidelberg (2007)
19. Tahayna, B., Ayyasamy, R.K., Alhashmi, S.M., Siew, E.: A Novel Weighting Scheme for Efficient Document Indexing and Classification. In: 4th International Symposium on Information Technology, ITSIM 2010, pp. 783–788 (2010)
20. Rui, X., Li, M., Li, Z., Ma, W.Y., Yu, N.: Bipartite graph reinforcement model for web image annotation. In: Multimedia 2007 (2007)
21. Zhang, D.Q., Lin, C.Y., Chang, S.F., Smith, J.R.: Semantic Video Clustering Across Sources Using Bipartitie Spectral Clustering. In: ICME (2004)
22. Zha, H., Ding, C., Gu, M.: Bipartite graph partitioning and data clustering. In: CIKM (2001)

Enhancing Automatic Blog Classification Using Concept-Category Vectorization

Ramesh Kumar Ayyasamy[1], Saadat M. Alhashmi[1], Siew Eu-Gene[2], and Bashar Tahayna[1]

[1]School of Information Technology, [2]School of Business
Monash University, Malaysia
{ramesh.kumar,alhashmi,siew.eu-gene,bashar.tahayna}@monash.edu

Abstract. Blogging has gained popularity in recent years. Blog, a user generated content is a rich source of information and many research are conducted in finding ways to classify blogs. In this paper, we present the solution for automatic blog classification through our new framework using Wikipedia's category system. Our framework consists of two stages: The first stage is to find the meaningful terms from blogposts to a unique concept as well as disambiguate the terms belonging to more than one concept. The second stage is to determine the categories to which these found concepts appertain. Our *Wikipedia based blog classification* framework categorizes blog into topic based content for blog directories to perform future browsing and retrieval. Experimental results confirm that proposed framework categorizes blogposts effectively and efficiently.

Keywords: Blog classification, Wikipedia, Weighting Scheme.

1 Introduction

The blogosphere is expanding in an unrivalled speed till this day. As per *BlogPulse*[1] statistics, there are 170 million identified blogs, 954,136 new blogs in last 24 hours and 1,010,813 blogposts are indexed in last 24 hours. Blogs are user-generated content, where blogger writes any topic of his/her interests and topics could be diverse. These diverse and dynamic natures of blogs make blog classification much more challenging task than the traditional text classification. Blog classification could provide a structure for organizing blog pages for efficient indexing and classification. Several text classification approaches have been developed to leverage this rich source of information. Notable approaches are mostly drawn from machine learning techniques such as SVM [1], ANN [2], Naïve-Bayes [3], Bayesian Network [4] and K-NN classifiers [5]. Over the years, the research on text classification has become more mature [18], and these techniques are applied only for webpages and other text documents. Traditional classification algorithms [1, 2, 3, 4, 5] follows *bag-of-Word* approach and accounts only for the term frequency in the documents and ignores semantic relationship between key words. To resolve this problem, is to enrich document representation with background knowledge represented by Ontology.

[1] http://www.blogpulse.com

Y. Wang and T. Li (Eds.): Knowledge Engineering and Management, AISC 123, pp. 487–497.
springerlink.com

Research has been done to exploit general ontologies such as WordNet [6, 7] and domain specific ontology [8, 9] for content based categorization of large corpora of documents. However, they all have limited coverage and are not regularly updated. To solve this, ontology terms could be enriched, and then it has its own drawback. While enriching original content with ontology terms may cause information loss.

To address the above issues, recent research [10, 11, 12, 13, 14] explored the usage of knowledge base derived from the Internet, such as Wikipedia. Wikipedia is based on wiki, where articles are registered and uploaded, and links are built in real time. Each wiki articles describes a single topic. It covers wide concepts and new domains. This wide coverage attracted the researchers' attention to treat Wikipedia as a knowledge base. In Wikipedia, the association relation between a concept and a category is communicated by a link, called *category link*. The category and the category link express its own direction towards its belonging concepts. The category system is edited and maintained by Wikipedia users as well as articles.

In our work, we first build the *concept by category* matrix using Wikipedia, which explicitly derives concept relationships. We then propose a framework which consists of two stages: first stage is to find the meaningful terms from blogposts to a unique concept as well as, disambiguate the terms belonging to more than one concept. The second stage is to determine the categories to which these found concepts appertain.

We summarize our contribution as follows:

- Unlike traditional classification techniques, our framework does automatic blog classification and does not use any manual training data to classify blogs.
- As Wikipedia refines the categories into narrow subcategories, our framework combines the use of vast number of these organized human knowledge.

The reminder of the paper is organized as follows: Section 2 defines the key terms used in our paper. Section 3 reviews the existing works on blog classification and Wikipedia. Section 4 describes our proposed framework, including n-gram based concept extraction, mapping terms to concepts, mapping concepts to categories and automatic blog classification. In Section 5, we evaluate the proposed framework with real data set and discuss experimental results. Finally, we conclude our paper in Section 6.

2 Definition of Terms

In this paper, we use these following terms:

- *"Blogpost (B)"* refers to single blogpost from the collection. Let χ be the blog data set, where $\chi = \{B_1, B_2, \ldots\ldots, B_n\}, n \in \mathbb{Z}^+$
- *"Term(T)"* refers to a meaningful word from a particular blogpost. Let D be the blogpost which consists of set of terms, where $B = \{T_1, T_2, \ldots\ldots., T_m\}, m \in \mathbb{Z}^+$
- *"Concept(C)"* refers to a Wikipedia article title. We treat each article title as a concept.
- *"Subcategories"* refers to Wikipedia's subcategories. Let *SC* consists of set of concepts, where $SC = \{C_1, C_2, \ldots., C_i\}, i \in \mathbb{Z}^+$
- *"Categories"* refers to Wikipedia's 12 main categories, where each category *Ct* consists of set of subcategories, $Ct = \{SC_1, SC_2, \ldots., SC_j\}\, j \in \mathbb{Z}^+$ and main parental categories (*Pcat*) where $Pcat = \{Ct_1, Ct_2, \ldots..Ct_{12}\}$

3 Related Work

Recently there has been a huge interest in utilizing Wikipedia to enhance the text mining tasks such as text/web mining. To deal with Ontology coverage, research [10, 11, 12, 13, 14], explored the usage of knowledge base-Wikipedia derived from the Internet. Wikipedia has been used and demonstrated by researchers to improve the performance of text classifiers. Schonhofen et al. [10] exploited the titles and categories of Wikipedia articles to characterize documents. This method does not prove the relation between the input terms. In addition, it requires a lot of pre-processing of the Wikipedia articles themselves. In our work, we conceptualize that using this huge knowledge base-Wikipedia, the accuracy problem deriving from Natural Language Processing can be avoided. Wang et al.[11] and Syed et al.[12] constructed a thesaurus of concepts from Wikipedia and demonstrated its efficacy in enhancing previous approaches for text classification. Gabrilovich et al.[13] applied structural knowledge repository-Wikipedia as feature generation technique. Their work confirmed that background knowledge based features generated from Wikipedia can help text categorization. Shirakawa et al.[14] proved category system in Wikipedia is not in a tree structure but a network structure.

As blog classification is in the early stages of the research, there is no suitable way to identify and categorize blogposts. Hence it is an open problem for research [17]. A very recent work on blog classification is presented in [16, 19, 20, 21]. Qu et al. [19] proposed an approach to the automatic classification of blogs into four genres: personal diary, news, political, and sports. Using *tf.idf* [22] document representation and Naïve Bayes classification, this work [19] achieved the accuracy of 84%. The authors [16] compared the effectiveness of using tags against titles and descriptions. They trained a support vector machine on several categories and implemented a tag expansion algorithm to better classify blogs. Bayoudh et al. [20] have used the K-Nearest Neighbor (K-NN) algorithm for blog classification. This classifier was built by using Part-Of-Speech (POS) tagger knowledge. This work empirically states that, nouns are relevant for the determination of the documents meaning and topics. Elgersma [21] addressed the task of separating personal from non-personal blogs and achieved upto 90% classified scores.

4 Wikipedia Based Blog Classification Framework

This section introduces our framework (Figure 2) leveraging Wikipedia concept and category information to improve blog classification. We explain our proposed framework by defining *n-gram based concept extraction*, mapping terms to concepts, mapping concepts to categories and automatic blog classification respectively.

4.1 n-gram Based Concept Extraction

Our proposed method is a relatedness measurement method that computes the relatedness (among words) based on co-occurrence analysis. As a simple case, we map each word to a *concept*; the combination of these concepts/words can create a new *concept* (compound concept). Let us assume A is a word mapped to a *concept* C_1, B is a word mapped to a *concept* C_2, and ω is the set of extracted terms from blogpost.

Then there is a chance that the combination of A & B can produce a new concept (see Figure 1). Consider the following example,

$$\$s = "President\ of\ the\ United\ States"$$

$$X = "President",\ Y = "United",\ Z = "States",\ \omega = \{X, Y, Z\}$$

$$\omega = \begin{cases} L^* - (\overline{\beta}); \omega \in C \\ else \quad \phi \end{cases}$$

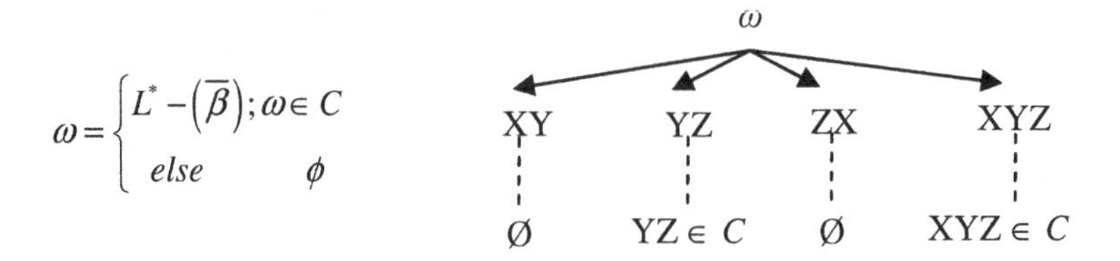

Fig. 1. Related measurement method

Where L^* is all possible combination of concepts and $\overline{\beta}$ is set of non-related terms, such as *XY* and *ZX*. *C* is the set of concepts such that *YZ* is mapped to *Country* concept and *XYZ* is mapped to *President* concept. Since the order of *X*, *Y* and *Z* can give different concept or no concept at all, we neglect the reordering. For effective classification, we mine existing knowledge bases, which can provide a conceptual corpus.

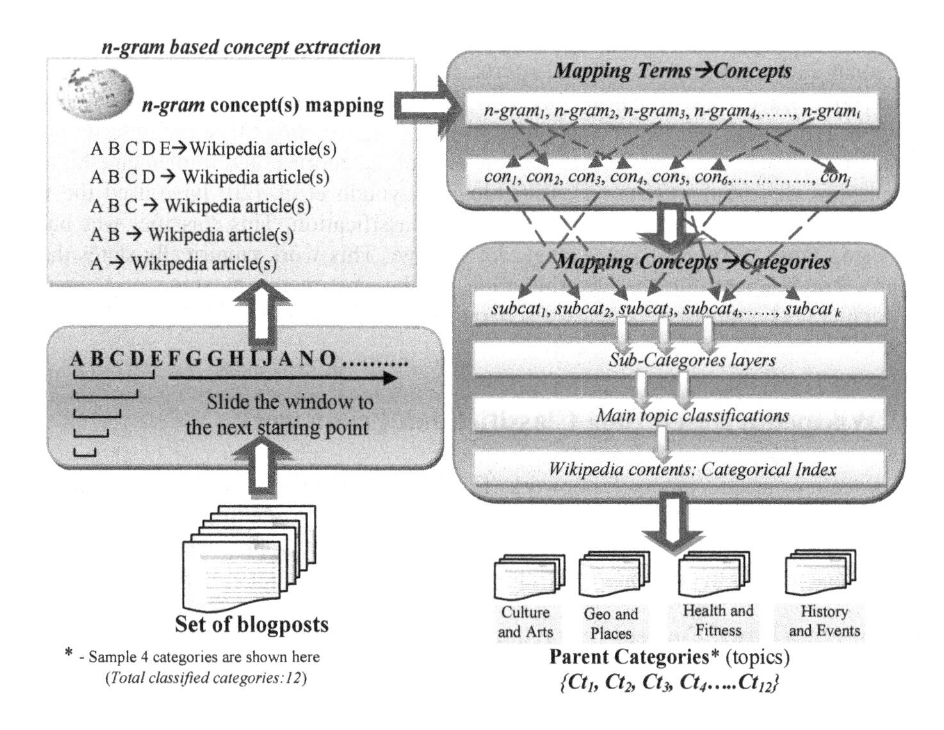

Fig. 2. Wikipedia based blog classification Framework

Wikipedia is one of the best online knowledge base, to provide an individual English article with more than 3 million concepts. With that said, an important feature of Wikipedia is that on each wiki page, there is a URI, which unambiguously represent the concept. Two words co-occur if they appear in an article within some distance of each other. Typically, the distance is a window of k words. The window limits the co-occurrence analysis area of an arbitrary range. The reason of setting a window is due to the number of co-occurring combinations that becomes too huge when the number of words in an article is large. Qualitatively, the fact that two words often occur close to each other is more likely to be significant, compared to the fact that they occur in the same article, even more so when the number of words in an article is huge. As explained above, our concrete process flow algorithm (Algorithm 1) using *n-gram* model (n = 3) and Wikipedia is shown:

Algorithm 1. *n-gram based concept extraction*

Input:
w is the *WindowSize* , and $max = n = 3$
Output:
The concepts W_i
Description:
1: Set Scanning $WindowSize\ w = n$, $MaxnGramsize = max$, $i = 1$
2: Parse a given blogpost text at position $p{=}i$ until $p{=}EOF$
3: Extract word-sequence and set $W = w_i w_{i+1} \ldots\ldots w_{i+MaxnGramsize}$
4: **If** W exists as a Wikipedia page (concept)
5: Retrieve W , set $p = i + MaxnGramsize$
6: Insert W into W_i
7: Repeat from *Step 3*
8: **else**
9: Set $MaxnGramsize = MaxnGramsize - 1$ & Update $W = w_i w_{i+1} \ldots\ldots w_{i+MaxnGramsize}$
10:**If** $Size(W) = 0$, then set $p = p + 1$; **Goto** *Step 3*
11:**Goto** *Step 4*

4.2 Mapping Terms→Concepts

To a certain degree, few *n-gram extracted terms* can map more than one *concept* in Wikipedia. This is called *concept disambiguation*. For example,

$$\$s = "CIA"$$

$$S = \{SC_1, SC_2, \ldots\ldots SC_n\}; SC_i \in one\ or\ more\ Ct, i \geq 1$$

$$\$s \leq_m S_i ; i \in \mathbb{Z}^+$$

where "CIA" can be mapped to several concepts that correspond to different subcategories like *Educational Institutes*, *Organizations*, *Airports*, and *Technology*. To address this issue, we use the disambiguation based on the context, i.e., each ambiguous concept in the text will be corresponding to a category, based on the scheme described in the following subsection. Then, for each ambiguous concept, we make a voting based on the majority from the "*Category histogram*" (Figure 3) . A *category histogram* is a distribution of the main categories and their number of appearances in a blogpost.

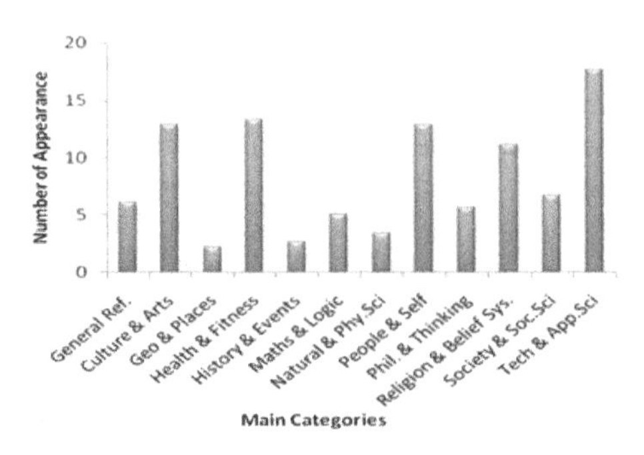

Fig. 3. Category Histogram

For example, if A is an ambiguous concept mapped to a set $SC = \{C_1, C_2, ..., C_i\}, i \in \mathbb{Z}^+$, then, we find the relevant main category (see Table 1 for the list of main categories) of each SC_j. Next we select the major category that has the highest frequency in the given blogpost. By doing this, we insist on selecting the major context category as the most relevant topic. We measure the influence of each individual post on the overall classification. To that end, we use our concept based weighting scheme [15]. In a blog, a post corresponds to concepts and category(s) and a common category clearly represent semantic associations between concepts. Therefore, extracting associations between concepts is achieved by extracting important categories in the blog by using *conf.idf* (equation 1). The importance of each concept in a blogpost (B) can be defined as follows:

$$Conf \cdot idf(C_k, B_j) = Conf(C_k + C_s, B_j).\log\frac{|N|}{N(C_k)} \tag{1}$$

Where $Conf(C_k, B_j)$ denotes the number of times a concept C_k occurs in a blogpost B_j, and $N(t_k)$ denotes the number of blogposts N in which C_k occurs at least once.

4.3 Mapping Concepts→Categories

Since Wikipedia has a well-structured category system [11], nearly all concepts belong to more than one sub-category, and nearly all sub-categories belong to categories each other form a category mapping. Shirakawa et al. [14] have proposed concept vectorization methods. A vector value between two nodes on a graph, consisting of category network in Wikipedia, is defined based on the number of paths and the length of each path. It is true that the length of the hopcount (Figure 4) gives an intuitive measure on the belonging degree between a concept and a category. However, this is a very bias measure. Without generalization, a concept can be mapped to huge number of categories. In other words, concepts that appear in a blogpost are totally context dependent.

As mentioned in the below example (Figure 4), "Barack Obama" belongs to several categories, namely: *Society and Social Sciences*, *People and Self*, *Geography and Places*, and *General Reference*. For an educational blogpost discussing about "Columbia University", it is possible to find the previous concept several times (since Obama is an alumnus); therefore, it should not be classified into, or tagged as a "political" blogpost. As we can see from the Figure 4, the work proposed by [14] will assign a localized stronger belonging degree ("Barack Obama", "*Society and Social Sciences*") and ignores the underlying context.

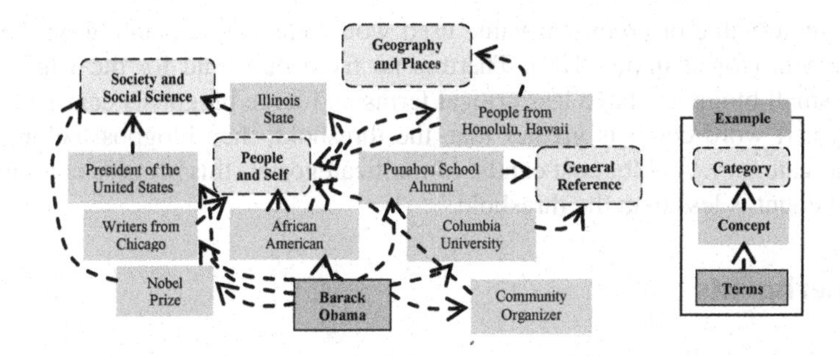

Fig. 4. An example: Wikipedia Sub-Category. A concept can belong to more than one Category.

We constructed *Concept* × *Category* matrix based on frequency of concepts that occur in a collection of categories, such that each row represents a *Category* and each column represents a *Concept.*

4.4 Automatic Blog Classification

Obviously, bloggers are free to post anything on their blogs. For example, a blog that extremely concerned about *entertainment*, and *music* can also have a post on a political issue. However, this blog must not be categorized as *politics*. Based on this observation, we measure the influence of each individual post on the overall classification. The frequency of the category is calculated based on the frequency of the belonging concept. If more than one concept is mapped to a blogpost, then the sum of the frequencies of these concepts is the category frequency.

Algorithm 2. Automatic blog classification

Input:
Blogpost *B*
Output:
Blogposts to Categories
Description:
1: W_i = {Ct, *Wc*} // W_i =*n-gram based concept extraction(B), Wc is the word count*
2: Selectionsort (W_i)
3: post_length : = {*B*} //total number of words in blogpost *B*
4: Threshold = get (Empirical Threshold value) / 100 * post_length //
5: $\forall Ct \in W_i$
6: **If** *Wc* ≥ Threshold
7: $B \rightarrow Ct$
8: **Else**
9: Continue
10: **End** ∀
11:**End**

We explained our automatic blog classification through Algorithm 2 above. For a given blogpost W_i consists *of* the categories of the concept found, and the word count for each category. *Selection sort* is used to sort the W_i on descending order using word count (*Wc*). Threshold is calculated (Line 4) based on dividing the empirical threshold value by 100 and multiplying the blogpost length (post_length). Empirical threshold value was calculated using the number of lines for each individual document.

The threshold value or empirical value used would change, depending on the size of the document (for example: 100). There is no fixed threshold for the whole corpus, because small blogposts have less critical terms and large blogposts can fit into many categories. If word count is greater than the threshold, then blogpost belongs to the particular category. As blogpost can be of multicategories, this process continues until the word count is less than the threshold.

5 Experiments

This section describes the experiment that test the efficiency of our proposed framework. We carried out experiments using part of TREC BLOGs08 dataset. We compared our framework that uses *conf.idf* with traditional *tf.idf* weighting scheme.

5.1 TREC Dataset

TREC BLOGs08[2] dataset is a well-known dataset in blog mining research area. This dataset consists of the crawl of Feeds, associated Permalink, blog homepage, and blogposts and blog comments. This is a sample of blogosphere crawled over from 14 January 2008 to 10 February 2009. The crawled feeds are mostly from *BlogSpot, LiveJournal, Xanga,* and *MSN Spaces*. The blog permalinks and homepages are encoded using HTML. This dataset is a combination of both English and Non-English blogs.

5.2 Wikipedia Data

We downloaded the Wikipedia database dumps[3] on 15 February 2010 and extracted 3,207,879 Wikipedia articles. After pre-processing and filtering (as illustrated in Figure 5), we used 3,101,144 article titles and are organized into 145,990 subcategories.

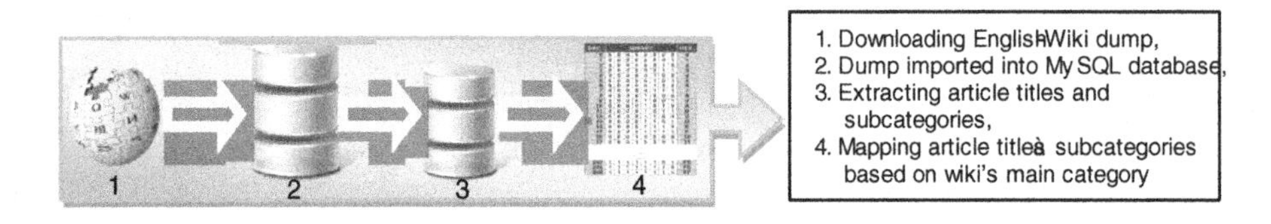

Fig. 5. Stages in creating the knowledge base

5.3 Experimental Dataset Preparation

As our research focus is only on blog classification and not on language identifier, we filtered the blogs by English language. In order to extract the blogposts from blog documents, the HTML source code of blog pages should be parsed (which includes removal of HTML tags, scripts, etc.), and to be converted into plain text. We used blogpost extraction program named *BlogTEX*[4] to extracts blog posts from TREC Blog

[2] `http://ir.dcs.gla.ac.uk/test_collections/blogs08info.html`
[3] `http://download.wikipedia.org`
[4] `http://sourceforge.net/projects/blogtex`

dataset. We utilized the list of stop words in our program to identify the English blogs. Furthermore, as our dataset is huge and our motive is to test our framework, we collected around 41,178 blogposts for our blog classification. We treated blog title and blog content as blogposts in our experiment.

5.4 Experimental Results

Based on our concept weighting scheme (*conf.idf*) total of 155,980 *concepts* are retrieved by using our framework. Table 1 lists the results of our experiment that shows the *Categorical index* of Wikipedia, the number of blogposts classified, and number of concepts retrieved.

Table 1. Experimental Result

Categorical Index	Our framework Classification using *conf.idf*		Categorical Index	Our framework Classification using *conf.idf*	
	# of blogposts	# of concepts		# of blogposts	# of concepts
General Reference	878	3758	*Natural and Physical sciences*	1560	8530
Culture and Arts	8580	18245	*People and Self*	6279	10125
Geography and Places	4558	23220	*Philosophy and Thinking*	2345	7591
Health and Fitness	4082	21409	*Religion and Belief*	2345	8315
History and Events	2471	12746	*Society and Social sciences*	4657	13170
Mathematics and Logic	213	2561	*Tech and Applied sciences*	3210	26310

TREC BLOGs08 dataset was crawled during US election. During classification, we noticed that certain posts are classified on multi categories. Blogposts which is classified under *People and Self* category is also in *Society and Social sciences*. For example, a blogpost which discussed about Obama, (*People and Self*) has also discussed about the Democratic Party (*Society and Social sciences*). Majority of posts are classified under *Culture and Arts*, *Society and Social sciences*, *People and Self*, and *Tech and Applied sciences*. From our dataset very few blogposts are classified under *Mathematics and Logic* category.

5.5 Framework Comparison Based on Weighting Scheme

Feature selection and feature extraction plays an important role in identifying meaningful terms or concepts for blog classification. In our framework, we used *conf.idf* during feature selection and feature extraction process. Traditional text classification uses *tf.idf* weighting scheme and follows *bag-of-word* approach. Majority of research works in the area of text classification used Support Vector Machine (SVM). We compared our framework which uses *conf.idf* with the traditional SVM framework using *tf.idf* weighting scheme. We utilized the same set of 41,178 blogposts and partitioned into two sets: two-third blogs were used for training and the rest one-third for testing. The experiment was conducted based on the same 12 categories (Table 2), using *SVM*^{light} package[5]. We performed traditional pre-processing methods such as infectional stemming, stop word removal, and conversion to lower case for blogposts.

[5] `http://svmlight.joachims.org`

5.6 Performance Evaluation

The performance of blog classification can be measured in several ways. We use the standard definition of *precision,* and *recall* as performance measure to evaluate our framework with *tf.idf* based blog classification. *Precision* measures the percentage of *"categories found and correct"* divided by the *"total categories found"*. Recall measures the percantage of *"categories found and correct"* divided by the *"total categories correct"*. We evaluated the classification performance by collecting random 250 blogposts from the classified output respectively (shown in Table 2 below).

Table 2. Performance evaluation based on random 250 classified blogposts

Categorical Index	Our classification framework using *conf.idf*		SVM Classification framework using *tf.idf*	
	Precision	**Recall**	**Precision**	**Recall**
General Reference	0.787	0.791	0.719	0.621
Culture and Arts	0.761	0.700	0.655	0.548
Geography and Places	0.778	0.671	0.726	0.615
Health and Fitness	0.744	0.735	0.691	0.723
History and Events	0.732	0.732	0.663	0.640
Mathematics and Logic	0.764	0.738	0.730	0.598
Natural and Physical sciences	0.792	0.653	0.623	0.531
People and Self	0.847	0.729	0.715	0.692
Philosophy and Thinking	0.790	0.611	0.671	0.580
Religion and Belief	0.750	0.708	0.698	0.659
Society and Social sciences	0.806	0.742	0.728	0.620
Tech and Applied sciences	0.828	0.674	0.725	0.653
Average	**0.782**	**0.707**	**0.695**	**0.623**

Table 2 shows that our framework classification using *conf.idf* performs better than the SVM classification framework using *tf.idf* weighting scheme. For example, blogposts which discusses about "Iraq war", as per *conf.idf* was classified under *History and Events* (sub: Modern History) and *Society and Social sciences* (sub: Politics). When traditional classification *tf.idf* scheme is used, "Iraq war" blogpost was wrongly classified under *Geography and places* (sub: places). Our framework produced better precision (78%) and recall (70.7%) than the SVM classification framework.

6 Conclusion

In this paper, we presented an automatic classification of blogs using the *n-gram* technique to extract the possible concepts from the underlying posts. Wikipedia was used as an external knowledge base to map each n-gram concept to its corresponding categories. We conducted extensive experiments to measure the efficiency of our proposed framework. The experimental results proves that proposed system distinguishes blogs that belong to more than one category and has a better performance and success than the traditional SVM classification approaches.

References

1. Joachims, T.: Text categorization with Support Vector Machines: Learning with many relevant features. In: Nédellec, C., Rouveirol, C. (eds.) ECML 1998. LNCS, vol. 1398, pp. 137–142. Springer, Heidelberg (1998)
2. Ng, H.T., Goh, W.B., Low, K.L.: Feature selection, perceptron learning, and a usability case study for text categorization. In: Proc. of ACM SIGIR, pp. 67–73 (1997)
3. McCallum, A., Nigam, K.: A Comparison of Event Models for Naïve Bayes Text Classification. In: AAAI 1998 Workshop on Learning for Text Categorization (1998)
4. Friedman, N., Geiger, D., Goldszmidt, M.: Bayesian Network Classifiers. Machine Learning 29, 131–163 (1997)
5. Yang, Y.: Expert network: Effective and efficient learning from human decisions in text categorization and retrieval. In: Proc. of ACM SIGIR, pp. 13–22 (1994)
6. Hotho, A., Staab, S., Stumme, G.: WordNet improves text document clustering. In: Proc. of ACM SIGIR (2003)
7. Budanitsky, A., Hirst, G.: Evaluating wordnet-based measures of lexical semantic relatedness. Computational Linguistics 32(1), 13–47 (2006)
8. Bloehdorn, S., Hotho, A.: Boosting for text classification with semantic features. In: Proc. of the MSW 2004 Workshop at the 10th ACM SIGKDD, pp. 70–87 (2004)
9. Jing, L., Ng, M.K., Huang, J.Z.: Knowledge-based vector space model for text clustering. KAIS 25, 35–55 (2009)
10. Schonhofen, P.: Identifying Document Topics Using the Wikipedia Category Network. In: Proc. of the IEEE/WIC/ACM International Conference on Web Intelligence, pp. 456–462 (2006)
11. Wang, P., Hu, J., Zeng, H.J., Chen, Z.: Using Wikipedia knowledge to improve text classification. KAIS 19(3), 265–281 (2009)
12. Syed, Z., Finin, T., Joshi, A.: Wikipedia as an Ontology for Describing Documents. In: Proc. of the AAAI International Conference on Weblogs and Social Media (2008)
13. Gabrilovich, E., Markovitch, S.: Overcoming the brittleness bottleneck using Wikipedia: Enhancing text categorization with encyclopedic knowledge. In: AAAI (2006)
14. Shirakawa, M., Nakayama, K., Hara, T., Nishio, S.: Concept vector extraction from Wikipedia category network. In: Proc. of the ICUIMC, pp. 71–79 (2009)
15. Tahayna, B., Ayyasamy, R.K., Alhashmi, S.M., Siew, E.: A Novel Weighting Scheme for Efficient Document Indexing and Classification. In: 4th International Symposium on Information Technology, pp. 783–788 (2010)
16. Sun, A., Suryanto, M.A., Liu, Y.: Blog Classification Using Tags: An Empirical Study. In: Goh, D.H.-L., Cao, T.H., Sølvberg, I.T., Rasmussen, E. (eds.) ICADL 2007. LNCS, vol. 4822, pp. 307–316. Springer, Heidelberg (2007)
17. Ounis, I., Macdonald, C., Soboroff, I.: On the TREC BlogTrack. In: ICWSM (2008)
18. Mahinovs, A., Tiwari, A.: Text classification method review. Decision Engineering Report Series, pp. 1-13 (2007)
19. Qu, H., Pietra, A.L., Poon, S.: Automated Blog Classification: Challenges and Pitfalls. Computational Approaches to Analyzing Weblogs, pp. 184–186 (2006)
20. Bayoudh, I., Béchet, N., Roche, M.: Blog classification: Adding linguistic knowledge to improve the k-nn algorithm. In: Intelligent Information Processing, pp. 68–77 (2008)
21. Elgersma, E.: Personal vs non-personal blogs. In: Proc. of ACM SIGIR, pp. 723–724 (2008)
22. Salton, G., Buckley, C.: Term-weighting Approaches in Automatic Text Retrieval. Information Processing and Management 24(5), 513–523 (1988)

A Naïve Automatic MT Evaluation Method without Reference Translations

Junjie Jiang, Jinan Xu, and Youfang Lin

School of Computer and Information Technology, Beijing Jiaotong University
Beijing, 100044, China
{jiangjunjie,jaxu,yflin}@bjtu.edu.cn

Abstract. Traditional automatic machine translation (MT) evaluation methods adopt the idea that calculates the similarity between machine translation output and human reference translations. However, in terms of the needs of many users, it is a key research issues to propose an evaluation method without references. As described in this paper, we propose a novel automatic MT evaluation method without human reference translations. Firstly, calculate average *n-grams* probability of source sentence with source language models, and similarly, calculate average *n-grams* probability of machine-translated sentence with target language models, finally, use the relative error of two average *n-grams* probabilities to mark machine-translated sentence. The experimental results show that our method can achieve high correlations with a few automatic MT evaluation metrics. The main contribution of this paper is that users can get MT evaluation reliability in the absence of reference translations, which greatly improving the utility of MT evaluation metrics.

Keywords: machine translation evaluation, automatic evaluation, without reference translations.

1 Introduction

There are some disadvantages of human evaluation, such as time consuming and high-cost. For this reason automatic evaluation methods were developed. Most of them are based on calculating the similarity between machine translation output and human reference translations. Automatic evaluation metrics can be divided into three categories according to calculation methods and resources. The first way is based on the number of matching word sequences between machine translation output and human reference translations, such as BLEU (Papineni et al., 2002), NIST (Doddington, 2002), etc. The second way employs the edit-distance between both translations such as WER (Nißen et al., 2000), PER (Tillmann et al., 1997), TER (Snover et al., 2006), etc. The third way usually addresses the sentence structure using syntactical tools, for instance, METEOR (Banerjee and Lavie, 2005), PosBLEU (Popović and Ney, 2007), etc, which improving the lexical level by the synonym dictionary or stemming technique. Nevertheless, those methods depend strongly on the quality of the syntactic analytical tools.

Y. Wang and T. Li (Eds.): Knowledge Engineering and Management, AISC 123, pp. 499–508.
springerlink.com

By providing immediate feedback on the effectiveness of various techniques, automatic evaluation has guided machine translation research and facilitated techniques rapid advances in the state-of-the-art. However, for the most common users of MT, it is illogical to judge the quality of machine translation output with human reference translations. On the other hand, with the rapid development of MT, the number of MT users is increasing quickly. How to judge the quality of machine translation for users without human reference translations is emerged as an urgent and challengeable issue.

In this paper, we propose a novel automatic MT evaluation method without reference translations for resolving this kind of problem. The basic idea of our proposed approach consists of three parts: Firstly, calculate average *n-grams* probability of source sentence with source language models, and similarly, calculate average *n-grams* probability of machine-translated sentence with target language models, finally, use the relative error of two average *n-grams* probabilities to mark machine-translated sentence. The experimental results show that our method can achieve high correlations with a few MT evaluation metrics, such as BLEU, NIST, etc.

The remainder of this paper is organized as follows: Section 2 reviews the related researches on automatic MT evaluation in the absence of human reference translations. In Section 3, we describe our proposed approach based on average *n-grams* probability. In Section 4, we present our experimental results on different data sets. And conclusions and future works are drawn in the Section 5.

2 Related Works

The need of common users requires an automatic MT evaluation metric without human reference translations. In this section we briefly present some relational research literatures.

Michael Gamon et al. (2005) proposed a method to measure the system performance, which are based on language model and SVM classifier. This method can identify highly dysfluent and ill-formed sentences. Andrew Mutton et al. (2007) developed an automatic evaluation metric to estimate fluency alone, by examining the use of parser outputs as metrics. Joshua S.Albrecht et al. (2007) developed the metrics which using regression learning and are based on a set of weaker indicators of fluency and adequacy. Reinhard Rapp (2009) advocated that the method of BLEU has some drawbacks, for example, it does not perform well at the sentence-level. He suggested conducting automatic MT evaluation by determining the orthographic similarity between back-translation and the original source text. This metric attained an improved performance at the sentence-level.

Previous works show that the research of automatic MT evaluation without reference translations is a challenging topic. Therefore, we analyze the role of *n-grams* probability and find that it can contribute to the automatic evaluation of translation quality. In our study, we firstly calculate average *n-grams* probabilities of source and machine-translated sentences, and then we calculate the relative error of two average *n-grams* probabilities to judge the quality of machine-translated sentence.

3 Our Proposed Approach

The procedure of our proposed approach is described in Figure 1.

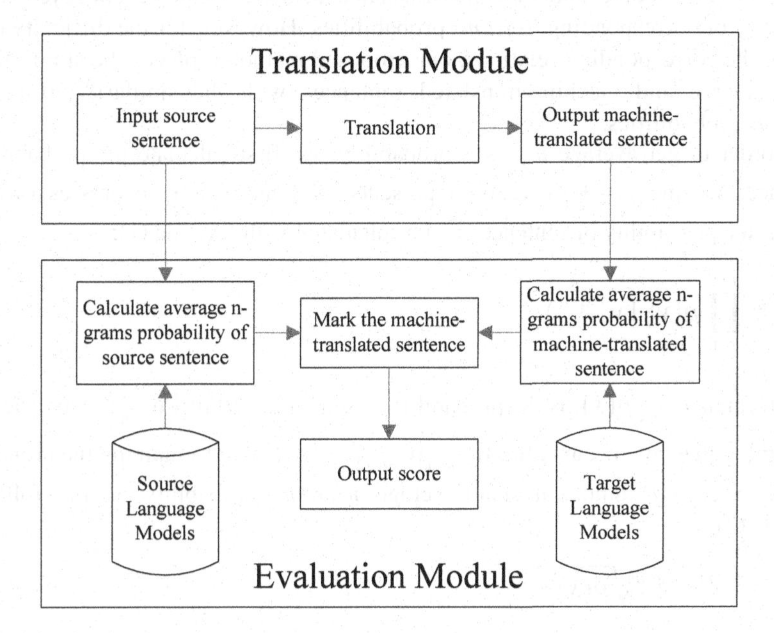

Fig. 1. Outline of Our Proposed Approach

As shown in Figure 1, the outline of our proposed approach consists of translation module and evaluation module. Translation module involves three basic steps: input source sentence, translate source sentence and output machine-translated sentence. Evaluation module consists of several steps including the following: calculate average *n-grams* probability of source sentence with source language models, calculate average *n-grams* probability of machine-translated sentence with target language models, use the two probabilities to mark machine-translated sentence and output the score.

3.1 Ultimate Principle

Many automatic MT evaluation methods are based on the comparison of a machine translation with one or more reference translations produced by humans. The comparison is done by calculating the similarity between the two translations. Similarly, in this paper, we calculate the similarity between source and machine-translated sentences to measure translation performance. Concretely, we use the relative error of average *n-grams* probabilities in source and machine-translated sentences to judge the quality of machine-translated sentence.

Translation of a certain source sentence can be seen as an interpretation in another language. In theories, there are certain *n-grams* mapping relations between source sentence and its correct translation. Well-translated *n-grams* in source sentence convey similar information with its corresponding *n-grams* in machine-translated sentence. From the perspective of information theory, the corresponding *n-grams* have similar amount of information and probabilities. Therefore, the similarity between source and machine-translated sentences can be conveyed by the similarity of corresponding *n-grams* probabilities. However, for the difficulty of detecting the corresponding relations, our proposed method replaces the similarity between source and machine-translated sentences with the similarity of average *n-grams* probabilities.

In order to get average *n-grams* probability, we first calculate the probability of sentence. Assume $s = w_1 w_2 \ldots w_l$ represents one sentence, w_i expresses a word or phrase, the probability of sentence can be calculated with formula (1):

$$p(s) = \prod_{i=1}^{l+1} p(w_i \mid w_{i-n+1}^{i-1}) \quad . \tag{1}$$

In formula (1), $p(s)$ is the probability of sentence, $p(w_i \mid w_{i-n+1}^{i-1})$ indicates the probability that w_i occurs after $w_{i-n+1} w_{i-n+2} \ldots w_{i-1}$. After obtaining the probability of sentence, we could calculate average *n-grams* probability as the following formula (2):

$$\overline{p(s)} = \sqrt[Count(n-gram)]{p(s)} \quad . \tag{2}$$

In formula (2), $\overline{p(s)}$ represents average *n-grams* probability of sentence s, $p(s)$ is the probability of sentence s, $Count(n-gram)$ is the number of *n-grams* in sentence, *n* is the value used when calculating sentence probability with formula (1).

According to our proposed method, after calculating average *n-grams* probabilities of source and machine-translated sentence, we use the relative error of two average *n-grams* probabilities to mark machine-translated sentence. The formula (3) is defined as:

$$score = \left| \frac{\overline{p_t} - \overline{p_s}}{\overline{p_s}} \right| \quad . \tag{3}$$

Formula (3) represents the score of machine-translated sentence, $\overline{p_t}$ is average *n-grams* probability of machine-translated sentence, and $\overline{p_s}$ is average *n-grams* probability of source sentence. The lower the score is, the higher performance machine-translated sentence has.

3.2 Sentence Length Penalty

According to our proposed method, we use the relative error of average *n-grams* probabilities to mark machine-translation sentence. Unfortunately, some "scheming" machine-translated sentence will get a high score by translating only those certainly correct substrings. Consider the following situation:

Source sentence: 党指挥枪是党的行动指南。

Candidate 1: It is a guide

Candidate 2: It is a guide to action that ensures that the military will forever heed Party commands.

With our baseline method, candidate 1 will get a higher score than candidate 2. It is clearly wrong. Obviously, it is necessary to take the length of sentence into account.

We noticed the fact that longer sentences in one language tend to be translated into longer sentences in the other language, and that shorter sentences tend to be translated into shorter sentences. As was shown in the sentence alignment literature (Church, K.W.1993), the sentence length ratio is a very good indication of the alignment of a sentence pair for different languages. In our method, for machine-translated sentence t, if $\frac{|t|}{|s|} \in [minLenRatio, maxLenRatio]$, there is no need to penalize them again. The length ratio is assumed to be a Gaussian distribution. The mean μ and variance σ^2 are calculated from the parallel training corpus. In our case, [*minLenRatio*, *maxLenRatio*] =[$\mu-\sigma$, $\mu+\sigma$]. Therefore, the brevity penalty *BP* is calculated by using following formula (4):

$$BP = \begin{cases} e^{(1-\frac{|s|\cdot\mu}{|t|})} & |t| < |s| \cdot minLenRatio \\ 1 & |s| \cdot minLenRatio \le |t| \le |s| \cdot maxLenRatio \\ e^{(1-\frac{|t|}{|s|\cdot\mu})} & |t| > |s| \cdot maxLenRatio \end{cases} . \quad (4)$$

In formula (4), *BP* represents the brevity penalty, ranging from 0 to 1, $|t|$ represents the sentence length of machine-translated sentence t, and $|s|$ is the length of source sentence s. Modified average *n-grams* probability of machine-translated sentence will be calculated according to formula (5):

$$\overline{p_t}' = \begin{cases} \overline{p_t} / BP & \overline{p_t} \ge \overline{p_s} \\ \overline{p_t} \cdot BP & \overline{p_t} < \overline{p_s} \end{cases} . \quad (5)$$

In formula (5), $\overline{p_t}'$ represents average *n-grams* probability of machine-translated sentence after penalizing, $\overline{p_t}$ expresses average *n-grams* probability before penalizing, and $\overline{p_s}$ represents average *n-grams* probabilities of source sentence. The final score will be calculated with formula (6).

$$score = \left| \frac{\overline{p_t}' - \overline{p_s}}{\overline{p_s}} \right| . \tag{6}$$

3.3 Document-Level Score

Our basic unit of evaluation is the sentence. However, we usually need to evaluate MT systems on a corpus of entire documents. The score of a document that consist of multiple sentences can be calculated by the relative error of average *n-grams* probabilities in entire text as formula (7):

$$score(text) = \left| \frac{\sum \overline{p_t}' - \sum \overline{p_s}}{\sum \overline{p_s}} \right| . \tag{7}$$

4 Experiments

4.1 Experimental Data

In order to verify the validity of our proposed method, we did some experiments based on the parallel corpus of French-English and German-English, which are from European Parliament Proceedings Parallel Corpus[1].

We randomly selected 2,000 sentences as open test set (T3) from the parallel corpus, the remaining as training data. Then we use the training data to build *trigram* language model, and randomly select two groups of 2,000 sentences as closed test set (T1 and T2), as the following Table 1:

Table 1. Experimental Data

		French-English	German-English
Training Data		1,825,077 sentences	1,739,154 sentences
Test Data	T1(closed)	2,000 sentences	2,000 sentences
	T2(closed)	2,000 sentences	2,000 sentences
	T3(open)	2,000 sentences	2,000 sentences

[1] http://www.statmt.org/europarl

4.2 Sentence Length Ratio Model

The sentence length can be defined in bytes and in words. In our case, we applied words based length ratio model only. The means and variance of the length ratios are calculated from parallel corpus in table 1. The statistics of the sentence length ratio model are shown in Table 2:

Table 2. Sentence length ratio statistics

	French-English	German-English
Mean μ	0.9239	1.0731
Var σ^2	0.0442	0.0533

Therefore, for French-English translation, μ = 0.9239, σ^2 = 0.0442, [*minLenRatio*, *maxLenRatio*] = [0.7137, 1.1341]. For German-English translation, μ =1.0731, σ^2 =0.0533, [*minLenRatio*, *maxLenRatio*] = [0.8589, 1.3045].

4.3 Experimental Procedure

The test sets were translated using the online Bing MT-system[2] and online Google MT-system[3].

The machine translation outputs are evaluated by two methods based on our baseline method. In the case of method 1, the probability of sentence is calculated based on *trigram model.* in the cased of method 2, the probability is calculated based on *trigram model* interpolated with *class-based trigram model.* After obtaining the probabilities of source and machine-translated sentences, we calculate average *n-grams* (*n*=3) probabilities of two sentences as formula (2), and penalize average *n-grams* probability of machine-translated sentence as formula (4) and (5). Finally, we calculate the score of document according to formula (7).

4.4 Experimental Results

As described in this paper, we performed experiments based on method 1 and 2. Additionally, we compared our experiments results with a few automatic MT evaluation methods, such as BLEU, NIST, etc.

Table 3 shows the results of experiments on French-English corpus. Table 4 shows the results of experiments on German-English corpus. T1+S1 means the machine-translated sentence set which is translated from test set T1 based on system S1. Similarly, T1+S2 means the machine-translated sentence set which is translated form test set T1 based on system S2, etc. The Italic type represents the better translation. According to our proposed method, the lower the score is, the higher performance machine translation outputs have.

[2] http://www.microsofttranslator.com/

[3] http://translate.google.cn/

Table 3. Experimental Results on French-English Corpus

	T1+S1	T1+S2	T2+S1	T2+S2	T3+S1	T3+S2
Method1	*0.3882*	0.4254	*0.3956*	0.4052	*0.2721*	0.3223
Method2	*0.3150*	0.3688	*0.3239*	0.3412	*0.2134*	0.2860
BLEU	*0.3506*	0.3173	*0.3408*	0.3176	*0.3064*	0.2789
NIST	*8.2050*	8.0268	*8.0536*	7.9966	*7.6305*	7.4436
Meteor	*0.6639*	0.6468	*0.6584*	0.6459	*0.6272*	0.6105
TER	*0.4852*	0.4888	0.4949	*0.4900*	*0.5309*	0.5322

Table 4. Experimental Results on German-English Corpus

	T1+S1	T1+S2	T2+S1	T2+S2	T3+S1	T3+S2
Method1	0.1069	*0.0398*	0.1210	*0.0888*	*0.1177*	0.1844
Method2	*0.1983*	0.2595	*0.1631*	0.1768	*0.2568*	0.2996
BLEU	*0.2985*	0.2953	*0.3004*	0.2948	*0.2733*	0.2667
NIST	7.5710	*7.6270*	7.6151	*7.6570*	7.1820	*7.1861*
Meteor	0.6313	*0.6324*	*0.6348*	0.6343	0.5960	*0.5980*
TER	0.5461	*0.5342*	0.5452	*0.5330*	0.5814	*0.5694*

4.5 Discussion

As shown in Table 3, our proposed approach can achieve high correlations with BLEU, NIST, etc. As shown in Table 4, method 2 is more robustness than method 1. The reason is that sentence probability based on *n-gram* model interpolated with *class-based n-gram* model can be calculated more accurately than *n-gram* model.

The results of BLEU and NIST metrics are quite opposite on the German-English Experiments. The reason is that, compared with BLEU, NIST revise the scoring methods of common units and brevity penalty. According to the NIST researchers' experimental results, NIST is more robust and reliability than BLEU. Especially for adequacy evaluation, NIST correlates more highly with human evaluation than BLEU.

As mentioned above, our method can generate the acceptable evaluation results under certain conditions. Besides, the scale of language models will directly affect the precision of sentence probability, and then, indirectly affect the final evaluation results. However, depending on the specific area, building a large monolingual corpus is easier than parallel corpus. Therefore, it is not difficult to improve the precision of our proposed approach with large scale monolingual language model.

5 Conclusion and Future Works

This paper describes an automatic MT evaluation metric without reference translations based on average *n-grams* probability. Our proposed method also takes sentence length penalty into account. The experimental results show that our proposed method can judge the quality of machine translation output effectively, and achieve high correlations with many automatic MT evaluation methods, such as BLEU, NIST, etc. We also notice that:

- The efficiency of our proposed method requires the large scale language models. It's easy to lead to data sparseness problem and affect the final evaluation results based on the small scale language models.
- Using *n-gram* model interpolated *class-based n-gram* model can appropriately alleviate the data sparseness problem.
- The score on document-level should be regarded as an average value. Integrated with the score on sentence-level, we can eliminate the poor-translated sentence.
- Our proposed approach works well for similar familial languages. The main reason is that the corresponding relations between two similar familial language are certain and simple.

The advantage of our proposed method is that it can perform simply and fast in the absence of human reference translations. MT systems usually contain the source and target language models.

The main contribution of this research consists of two parts: i) for researchers, our method can help get rid of reference translations and minimize the person's efforts. ii) For the most common users of MT, our method can help them get the reliability of machine translation output in the absence of human reference translations. Therefore, our method has a good prospect of application.

In the future, we will take syntactic structure and semantic information into account, and try to use the large scale language model to improve the reliably of our method. Besides, we will test our method based on different languages further to validate the effectiveness and stability.

Acknowledgements. This work was supported by the Fundamental Research Funds for the Central Universities (2009JBM027) (K11JB00210) and by the RenCai Funding of Beijing Jiaotong University (2011RC034).

References

1. Mutton, A., Dras, M., Wan, S., Dale, R.: GLEU: Automatic Evaluation of Sentence-Level Fluency. In: Proceedings of the 45th Annual Meeting of the Association of Computational Linguistics, pp. 344–351 (2007)
2. Lin, C.-Y., Och, F.J.: Automatic Evaluation of Machine Translation Quality Using Longest Common Subsequence and Skip-Bigram Statistics. In: Proceedings of ACL 2004, pp. 606–613 (2004)
3. Coughlin, D.: Correlating Automated and Human Assessments of Machine Translation Quality. In: Proceedings of MT Summit IX, pp. 63–70 (2003)
4. George, D.: Automatic Evaluation of Machine Translation Quality Using N-gram Co-Occurrence Statistics. In: Proceedings of the 2nd International Conference on Human Language Technology, pp. 138–145 (2002)
5. Leusch, G., Ueffing, N., Ney, H.: A Novel String-to-String Distance Measure with Applications to Machine Translation. In: Proceedings of MT Summit IX, New Orleans, USA, pp. 240–247 (2003)

6. Hiroshi, E.-Y., Kenji, A.: Automatic Evaluation of Machine Translation based on Recursive Acquisition of an Intuitive Common Parts Continuum. In: Proceedings of MT Summit XII, pp. 151–158 (2007)
7. Hiroshi, E.-Y., Kenji, A.: Automatic Evaluation Method for Machine Translation using Noun-Phrase Chunking. In: Proceedings of the 48th Annual Meeting of the Association for Computational Linguistics, pp. 108–117 (2010)
8. Joshua, S.: Albrecht and Rebecca Hwa: Regression for Sentence-Level MT Evaluation with Pseudo References. In: Proceedings of the 45th Annual Meeting of the Association of Computational Linguistics, pp. 296–303 (2007)
9. Church, K.W.: Char_align: A Program for Aligning Parallel Texts at the Character Level. In: Proceedings of ACL 1993, Columbus, OH (1993)
10. Papineni, K., Roukos, S., Ward, T., Zhu, W.-J.: BLEU: a Method for Automatic Evaluation of Machine Translation. In: Proceedings of 40th Annual Meeting of the Association for Computational Linguistics, pp. 311–318 (2002)
11. Popović, M., Ney, H.: Syntax-oriented evaluation measures for machine translation output. In: Proceedings of the 4th EACL Workshop on Statistical Machine Translation, pp. 29–32 (2009)
12. Snover, M., Dorr, B., Schwartz, R., Micciulla, L., Makhoul, J.: A Study of Translation Edit Rate with Targeted Human Annotation. In: Proceedings of the 7th Conference of AMTA 2006, pp. 223–231 (2006)
13. Gamon, M., Aue, A., Smets, M.: Sentence-level MT evaluation without reference translations: Beyond language modeling. European Association for Machine Translation, (EAMT 2005) (May 2005)
14. Rapp, R.: The Back-translation Score: Automatic MT Evaluation at the Sentence Level without Reference Translations. In: Proceedings of the ACL-IJCNLP 2009 Conference Short Papers, pp. 133–136 (2009)
15. Banerjee, S., Lavie, A.: METEOR: An Automatic Metric for MT Evaluation with Improved Correlation with Human Judgments. In: Proceedings of ACL 2005 Workshop on Intrinsic and Extrinsic Evaluation Measures for Machine Translation and/or Summarization, pp. 65–72 (2005)
16. Nieβen, S., Och, F.J., Leusch, G., Ney, H.: An Evaluation Tool for Machine Translation: Fast Evaluation for MT Research. In: Proceedings of the 2nd International Conference on Language Resources and Evaluation (2000)
17. Christoph, T., Vogel, S., Ney, H., Zubiaga, A., Sawaf, H.: Accelerated DP based Search for Statistical Translation. In: Proceedings of European Conference on Speech Communication and Technology (1997)

Combining Naive Bayes and Tri-gram Language Model for Spam Filtering

Xi Ma[1], Yao Shen[1], Junbo Chen[2], and Guirong Xue[2]

[1] Department of Computer Science and Engineering
Shanghai Jiao Tong University
Shanghai 200240, China
maxisjtu@gmail.com, shen_yao@cs.sjtu.edu.cn
[2] Alibaba Cloud Computing
Hangzhou 310012, Zhejiang China
{junbo.chenjb,grxue}@alibaba-inc.com

Abstract. The increasing volume of bulk unsolicited emails (also known as spam) brings huge damage to email service providers and inconvenience to individual users. Among the approaches to stop spam, Naive Bayes filter is very popular. In this paper, we propose the standard Naive Bayes combining with a *tri-gram* language model, namely TGNB model to filter spam emails. The TGNB model solves the problem of strong independence assumption of standard Naive Bayes model. Our experiment results on three public datasets indicate that the TGNB model can achieve higher *spam recall* and lower *false positive*, and even achieve better performance than support vector machine method which is state-of-the-art on all the three datasets.

Keywords: Naive Bayes, tri-gram, email anti-spam, machine learning, statistical approach.

1 Introduction

With the developing of Internet technology, emails have become a common and important communication medium for almost every person. While since the first junk email written in program and sent by Canter and Sigel, the situation of bulk unsolicited email as known as spam has not changed better but worse. According to [1] the fraction of all emails that can be considered spam is higher than 90%. Spam causes much misuse of bandwidth, storage space and computational resource [2]; spam takes up email users much time to check the additional emails, and causes loss of work productivity; also, spam causes legal problems by advertising pornography, phishing site[3]. The total worldwide spam cost reached almost 200 billion US dollars in 2007 with roughly 50 billion junk mails daily according to Radicati Group.

There are mainly two directions to filter spam, one is based on the behavior of users and ip [4] [5]; another is based on the content of emails [6] [7]. According to the study by Siponen and Stucke about the use of different kinds of anti-spam tools and techniques in companies, filters, black lists, restricting the disclosure of email addresses, and white lists were seen as the most effective anti-spam techniques in order of effectiveness [2]. In respond to the keyword filtering, spammers try to bypass the filter by

Y. Wang and T. Li (Eds.): Knowledge Engineering and Management, AISC 123, pp. 509–520.
springerlink.com

misspelling of keywords or adding symbols between letters. As for the black lists and white lists, spammers use spoofed legitimate email addresses or new email addresses to send out spam. Among the proposed methods, much interest has focused on machine learning techniques. They include rule learning [8], Naive Bayes [9][10], support vector machine [11], decision tree [12] , K-nearest neighbor(K-NN) [13]. Many studies have shown that many anti-spam system use Naive Bayes filtering and the main reason is that the approaches based on Naive Bayes have the advantage of being very fast and are fit for continuous on-line tasks. However, the main problem of Naive Bayes is its strong independence assumption between its attributes, relaxing the assumption could allow for better text classification [14], and it has been shown in practice [4] [15].

The objective of spam filtering is not totally the same as text classification's: spam email does not have restrict definition but bulk unsolicited mails; misclassifying a legitimate mail is generally much worse than misclassifying a spam [16]. Users can tolerate a few spam emails in their inbox but hardly tolerate their legitimate emails being classifying as spam.

[15] proposed *Tree Augmented Naive* Bayes classifier(TAN) which allows for a tree-structured dependence among observed variables. However, this model has rarely been used in text classification application. Recently,[17] proposed the *Chain Augmented Naive Bayes* (CAN) model, which combines Naive Bayes and n-Gram language model for text classification. It lies between pure Naive Bayes and TAN Bayes model, and restricts dependencies among observed variables to form a Markov chain instead of tree. In this paper, we find that the sensitive words that are usually related with spam often appear continuously in spam but not in legitimate email. For example, the words "she wants it" are more likely to appear in pornography or drug advertising, and using traditional Naive Bayes cannot find it easily, while combining Naive Bayes and n-gram language model are more likely to identify it with spam. For this we make the following contributions:

1. We suggest combining Naive Bayes and n-Gram language model for spam filtering which is similar to the work of [17].
2. We try to combine Naive Bayes and four different(for n) n-Gram language models for spam filtering on three different public datasets. The experimental results demonstrate that combining with the tri-gram language model is the first choice, in this paper, we call the model as TGNB model.
3. We propose flexible threshold for spam filtering to reduce possible *false positive* and possible *false negative* which help spam filtering get almost perfect results.
4. Our spam filtering method could run much faster than most other machine learning method and is fit for continuous on-line jobs.

The rest of this paper is organized as follows: in Section 2, we present common performance measures and the Naive Bayes filtering; n-Gram language model is presented in Section 3; We combine Naive Bayes and n-Gram language model for spam filtering in Section 4; in Section 5, the experimental datasets and results are presented, and the results demonstrate TGNB model achieves the best performance; in Section 6, we give some conclusions and some effective improvement methods in future work.

2 The Naive Bayes Filtering

2.1 Performance Measures

At the beginning, we give some important indices for spam filtering applications. Out of all spam emails, let a be the number of emails classified correctly as spam and the remaining b emails classified as legitimate. In the same way, out of all the legitimate emails, let c be the number of emails misclassified as spam and the remaining d emails classified correctly as legitimate. Then:

$$Accuracy = \frac{a+d}{a+b+c+d} \quad (1)$$

$$Spam\ recall = \frac{a}{a+b}\ Spam\ precision = \frac{a}{a+c} \quad (2)$$

$$Ham\ recall = \frac{d}{c+d}\ Ham\ precision = \frac{d}{d+b} \quad (3)$$

False positive: A result that misclassifies a legitimate email as spam email.

False negative: A result that misclassifies a spam email as legitimate email.

Generally *False positive* error that misclassifies a legitimate email as spam is considered more serious than a *False negative*.

2.2 Naive Bayes Filtering

In our paper, we mainly focus on the body textual content and subject of emails for Naive Bayes filtering. Spam filtering is to classify an email E to be one of a set $|C|$ pre-defined categories $C = \{c_{spam}, c_{ham}\}$, where c_{spam} and c_{ham} stand for spam and legitimate email separately. We use the random variables E and C to denote the email and category values respectively. Naive Bayes filtering is based on a simple application of *Bayes' rule*[17]:

$$P(c|e) = \frac{P(c) \times P(e|c)}{P(e)} \quad (4)$$

Using the equation, Bayes's rule can get a posterior probability by computing a likelihood and a prior probability. In spam filtering, an email e contains a vector of K grams $e = (v_1, v_2, \ldots, v_K)$, where the grams are also called features or attributes in many papers. Since the space of possible email $e = (v_1, v_2, \ldots, v_K)$ is vast, computing $P(c|e)$ is not usually easy. To simplify this computation, Naive Bayes model makes an assumption that all of the gram values, v_j, are independent given the classifying label, c. This strong independent assumption greatly simplifies the computation by reducing Eq.(4)to:

$$P(c|e) = P(c) \times \frac{\prod_{j=1}^{K} P(v_j|c)}{P(e)} \quad (5)$$

Maximum a posterior(MAP) filtering can be constructed based on Eq.(5) by seeking the optimal classifying which maximizes the posterior $P(c|e)$:

$$c^* = max_{c \in C}\{P(c|e)\} \tag{6}$$

$$c^* = max_{c \in C}\left\{P(c) \times \frac{\prod_{j=1}^{K} P(v_j|c)}{P(e)}\right\} \tag{7}$$

$$c^* = max_{c \in C}\left\{P(c) \times \prod_{j=1}^{K} P(v_j|c)\right\} \tag{8}$$

The *P(e)* in Eq.(7) is constant for every category *c*, thus the Eq.(7) going to Eq.(8) is valid. The MAP filter (Eq.(6)) is optimal in sense of minimizing *zero-one* loss (misclassification error). If the independence assumption holds, the filter based on Eq.(8) is also optimal [18]. The prior distribution *P(c)* can also be used to incorporate additional assumption, for example, when uniform distribution is used as the prior, the Map filter becomes equivalent to the maximum likelihood classifier:

$$c^* = max_{c \in C}\left\{\prod_{j=1}^{K} P(v_j|c)\right\} \tag{9}$$

Eq.(9) is called the *maximum likelihood* Naive Bayes classifier, and there are several variants of Naive Bayes for spam filtering [10]. According to [10], the multinomial Naive Bayes has the best result, both in *spam recall* and *spam precision*, therefore we will only consider the multinomial Naive Bayes model in this paper. Fig.1 [17] presents a graphical representation of standard Naive Bayes classifier, showing that each attribute node is independent of the other attributes given the class label C.

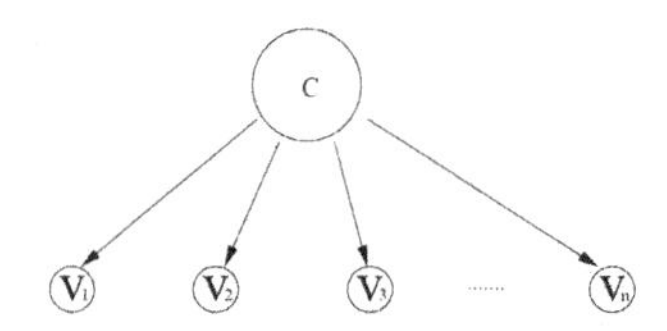

Fig. 1. Standard Naive Bayes classifier

The parameters of a multinomial Naive Bayes filter are given by

$$\Theta = \{\theta_j^c = P(v_j|c)\}$$

. The likelihood of a given set of emails E^c for a given category *c* is given by

$$P(E^c|\Theta) = \frac{N^c!}{\prod_j N_j^c!}\prod_j (\theta_j^c)^{N_j^c} \tag{10}$$

where N_j^c is the frequency attribute j occurring in E^c and $N^c = \Sigma_j N_j^c$.

$$\frac{P(E^{c_{spam}}|\Theta)}{P(E^{c_{ham}}|\Theta)} > T \tag{11}$$

where T is the threshold which classifies an email as spam or legitimate. A maximum likelihood estimate yields the parameter estimates

$$\theta_j^c = \frac{N_j^c}{N^c} \tag{12}$$

Eq.(12) puts zero probability on attribute values that do not actually occur in E^c(i.e., N_j^c=0). When we encounter a new email with attribute values having not been in the training corpus (i.e. OOV values)E^c, we choose to classify this email as legitimate to reduce possible *false positive*, and this rarely happens in fact.Also, several smoothing techniques have been developed in statistical language modeling to solve the problem, and we will discuss later.

3 The Markov n-Gram Language Model

Statistical language models have been widely used in many other application areas such as information retrieval,speech recognition, machine translation [19]. The goal of language modeling is to get a probability of word sequences and helps to find which category the words belong to. Given a test corpus, a word sequence $w_1w_2 \dots w_T$ composes the corpus. The quality of a language model is measured by the empirical perplexity on this corpus

$$Perplexity = \sqrt[T]{\prod_{i=1}^{T} \frac{1}{P(w_i|w_1 \dots w_{i-1})}}$$

The smaller perplexity means the language model works better. In n-gram language model, we can write the probability of any word sequence as

$$P(w_1w_2 \dots w_T) = \prod_{i=1}^{T} P(w_i|w_1 \dots w_{i-1}) \tag{13}$$

An n-gram language model assumes that only the precious n-1 words are relevant to predicting P($w_i \mid w_1 \dots w_{i-1}$); in other words, here assumes the Markov n-gram independence assumption

$$P(w_i|w_1 \dots w_{i-1}) = P(w_i|w_{i-n+1} \cdots w_{i-1})$$

The maximum likelihood estimate of n-gram probabilities from a corpus can be given by the observed frequency:

$$P(w_i|w_{i-n+1} \cdots w_{i-1}) = \frac{Num(w_{i-n+1} \cdots w_i)}{Num(w_{i-n+1} \cdots w_{i-1})} \tag{14}$$

where Num(.) is the number of occurrences of a specified gram in training data. Unfortunately, using grams of length n could result in huge model data which contains the probability of W^n entities, where W is the size of gram vocabulary. Meanwhile, due to the heavy tailed nature of language(i.e. Zipf's law), encountering novel n-garms that were never witnessed during training is common. Here, we use one standard approach with some sort of back-off estimator to smooth the probability estimates, and this could handle the problems of sparse data and potential missing n-grams.

$$P(w_i|w_{i-n+1}\dots w_{i-1}) = \begin{cases} \hat{P}(w_i|w_{i-n+1}\dots w_{i-1}), \\ \qquad if\ Num(w_{i-n+1}\dots w_{i-1}) > 0 \\ \beta(w_{i-n+1}\dots w_{i-1}) \times P(w_i|w_{i-n+2}\dots w_{i-1}), \\ \qquad otherwise \end{cases} \tag{15}$$

where

$$\hat{P}(w_i|w_{i-n+1}\dots w_{i-1}) = \frac{discount * Num(w_{i-n+1}\dots w_i)}{Num(w_{i-n+1}\dots w_{i-1})} \tag{16}$$

is the discounted value, and β =($w_{i-n+1}\dots w_{i-1}$) is the normalization constant which is calculated as

$$\beta = (w_{i-n+1}\dots w_{i-1}) = \frac{1-\sum_{x\in(w_{i-n+1}\dots w_{i-1})}\hat{P}(x|w_{i-n+1}\dots w_{i-1})}{1-\sum_{x\in(w_{i-n+1}\dots w_{i-1})}\hat{P}(x|w_{i-n+2}\dots w_{i-1})} \tag{17}$$

There are several different smoothing approaches for computing discounted probability in (16): Written-Bell smoothing, Good-Turing smoothing, absolute smoothing and linear smoothing [20]. In [17], absolute and linear smoothing performed best. However, in our experiment, Witten-Bell smoothing gets the best result, and we will use it on all the datasets.

Witten-Bell smoothing Witten-Bell smoothing reserves probability mass for OOV values, and the discounted probability is calculated as

$$\hat{P} = \frac{Num(w_{i-n+1}\dots w_i)}{Num(w_{i-n+1}\dots w_{i-1}) + W}$$

where W is the number of distinct words that follow $w_{i-n+1}\dots w_{i-1}$ in the training data. We now show how to combine Naive Bayes model and n-gram language model for spam filtering.

4 Combining Naive Bayes and n-Gram Language Model for Spam Filtering

Spam filtering attempts to identify attributes which distinguish spam and legitimate emails. Up to now each attribute has corresponded to an uni-gram in standard Naive Bayes, and this has too strong independence assumption between the uni-grams [14]. Thus, combining Naive Bayes and n-gram language model can weaken the independence assumption, which could help in spam filtering. In this case

$$c^* = max_{c\in C}\{P(c|e)\} = max_{c\in C}\{P(e|c)P(c)\} \tag{18}$$

$$= max_{c\in C}\{P(e|c)\} \tag{19}$$

$$= max_{c\in C}\left\{\prod_{i=1}^{T} P(w_i|w_{i-n+1}\dots w_{i-1}, c)\right\} \tag{20}$$

where in the step from Eq.(18) to Eq.(19), a uniform distribution is used as the prior over different categories, and the step form Eq.(19) to Eq.(20) makes a Markov n-gram independence assumption. An n-gram language model can be used in spam filtering as Eq.(20) generates a probability value for a given email according to the email's text content.

$$\frac{\prod_{i=1}^{T} P(w_i|w_{i-n+1}\ldots w_{i-1}, c_{spam})}{\prod_{i=1}^{T} P(w_i|w_{i-n+1}\ldots w_{i-1}, c_{ham})} > T \tag{21}$$

Thus, we train two separate language models for the two categories (spam and legitimate mails), and calculate c^*s on both models. The threshold T in Eq.(21) is the threshold which can be used to classify an email as spam or legitimate. However, in practice, we use the logarithmic value ratio instead

$$\frac{log(\prod_{i=1}^{T} P(w_i|w_{i-n+1}\ldots w_{i-1}, c_{spam}))}{log(\prod_{i=1}^{T} P(w_i|w_{i-n+1}\ldots w_{i-1}, c_{ham}))} > T \tag{22}$$

is equal to

$$\frac{\sum_{i=1}^{T} log(P(w_i|w_{i-n+1}\ldots w_{i-1}, c_{spam}))}{\sum_{i=1}^{T} log(P(w_i|w_{i-n+1}\ldots w_{i-1}, c_{ham}))} > T \tag{23}$$

The following two figures Fig.2 [17] and Fig.3 present a graphical representation of Naive Bayes classifier combined with bi-gram and tri-gram language model. The two figures demonstrate that the standard Naive Bayes classifier and n-gram augmented Naive Bayes classifier differ: The bi-gram and tri-gram introduce an additional Markov chain dependence between two adjacent and three sequential attributes separately, which help weaken the strong independence assumption between uni-grams. Thus, we could combine Naive Bayes with tetra-gram and n-gram language model.

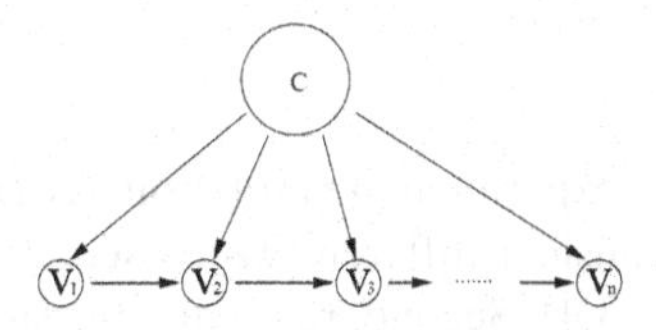

Fig. 2. Naive Bayes classifier combined with bi-gram language model

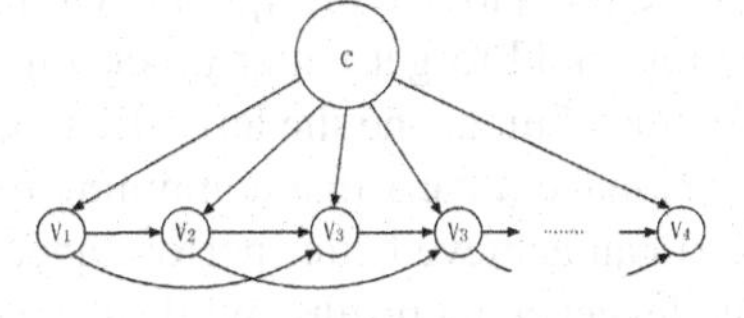

Fig. 3. Naive Bayes classifier combined with tri-gram language model

We now experimentally evaluate the performance of Naive Bayes spam filtering combined with different n-gram language model on three public datasets.

5 Experimental Evaluation

5.1 Datasets

The Naive Bayes combined with n-gram language model performance is evaluated on three widely used datasets namely LingSpam, EnronSpam(both available at `http://www.iit.demokritos.gr/skel/i-config`) and SpamAssassin (available at `http://www.spamassassin.org/publiccorpus`). Table 1 summarizes the basic properties of each dataset, including the number of spam, the number of ham and the spam ratio. In each dataset, duplicate spam emails have been removed. If an email contains html tags, the html tags are also removed. We only consider the subject and body of each email in experiment.

Table 1. Summary of datasets in experiment

Dataset	No. of spam	No. of ham	Spam Rate
LingSpam	481	2412	16.6%
Enron1	1500	3672	29.0%
SpamAssassin	1896	4150	31.3%

5.2 Performance Measures

The performance measures to evaluate the Naive Bayes spam filtering combined with n-gram language model include: *Accuracy, Spam Recall, Spam Precision, Ham Recall, Ham Precision*, and how to calculate the measures have been presented in the precious part.

5.3 Experimental Result

Next we give the details of experimental results on different datasets with different parameters. First, we will consider different words sequences of length one, two and three (i.e. n-gram for n = 1,2,3,4). Second, removing the stopwords can help to enlarge the difference between spam and legitimate, and we will show the influence of removing stopwords. Thirdly, due to the threshold T in Eq.(21) determines whether an email is spam, we will try flexible threshold to get better precision. In experiment, we select randomly 90% of total data for training, and the left 10% to test the performance.

The following three tables make it clear that combining Naive Bayes spam filtering and n-gram language model can achieve better performance. Comparatively, tri-gram and tetra-gram language model get better results. While using tetra-gramed Naive Bayes spam filtering, the size of tri-gram trained model data becomes 25-51 times larger than the uni-gram model and it costs much memory space (more than 1.5GB) which easily causes OutOfMemeoryError, and its performance does not have apparent improvement. Thus, combining Naive Bayes and tri-gram language model is the first choice.

Table 2. Different n-gram on LingSpam

Performance	LingSpam			
	uni-gram	bi-gram	tri-gram	tetra
Accuracy	97.6%	98.6%	99.7%	98.3%
Spam Recall	93.9%	93.9%	100%	91.8%
Spam Precision	92.0%	97.9%	98.0%	97.8%
Ham Recall	98.3%	99.6%	99.6%	99.6%
Ham Precision	98.8%	98.8%	100%	98.4%

Table 3. Different n-gram on Enron1

Performance	Enron1			
	uni-gram	bi-gram	tri-gram	tetra
Accuracy	94.4%	97.1%	98.1%	98.5%
Spam Recall	80.7%	90.0%	93.3%	95.3%
Spam Precision	100%	100%	100%	99.3%
Ham Recall	100%	100%	100%	99.7%
Ham Precision	92.7%	96.1%	97.4%	98.4%

Table 4. Different n-gram on SpamAssassin

Performance	SpamAssassin			
	uni-gram	bi-gram	tri-gram	tetra
Accuracy	96.0%	96.7%	97.1%	97.0%
Spam Recall	90.0%	90.5%	92.1%	97.4%
Spam Precision	97.2%	98.9%	98.9%	93.4%
Ham Recall	98.8%	99.5%	99.5%	96.9%
Ham Precision	95.6%	95.8%	96.7%	98.8%

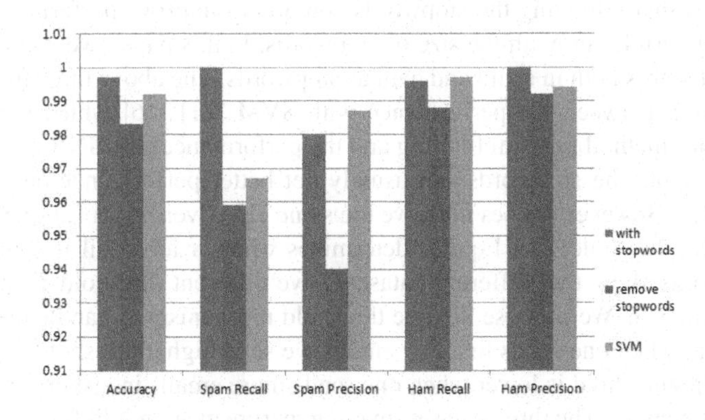

Fig. 4. TGNBs vs. SVMs on LingSpam

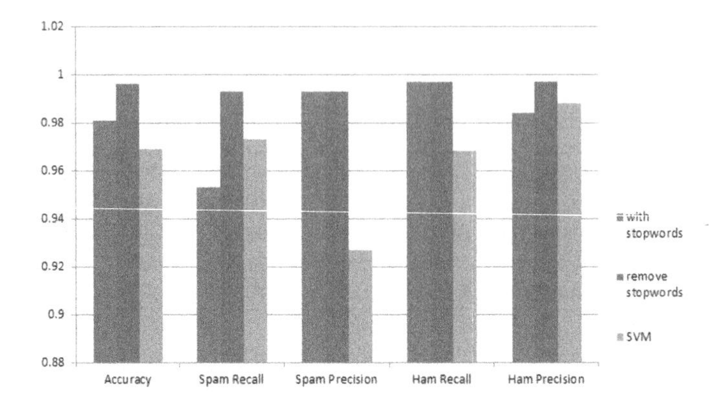

Fig. 5. TGNBs vs. SVMs on Enron1

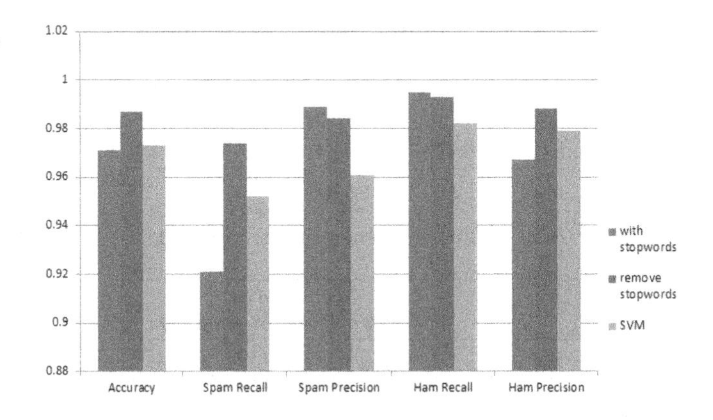

Fig. 6. TGNBs vs. SVMs on SpamAssassin

[16] shows that removing the stopwords could help improve performance to some extent, and it's irrelevant with the size of stopwords. In this paper, we choose the most 200 frequent words both in spam and ham as stopwords. The above three figures shows the comparison between the performance with SVM [21] [16] which constitute the state-of-the-art method in spam filtering and the performance of TGNB model. It turns out that removing the stopwords can usually get better performance on Enron1 and SpamAssassin. However, it does not have the same effectiveness on LingSpam.

Finally, the threshold T in Eq.(23) determines whether an email is spam. The experiment results show that different datasets have different threshold T values to get better performance. We propose flexible threshold mechanism to handle the *false positive* problem: when one sends first few emails we set a higher threshold to filter spam email, and the threshold is lower when one sends more emails in a short time which is likely spam behavior. The three datasets in our experiment are not fit for testing flexible threshold, however, it's an effective method in practical spam filtering implement.

6 Conclusions and Future Work

We discussed and evaluated experimentally the Naive Bayes combined with n-gram language model in spam filtering. Our investigation included Naive Bayes combined with four different n-gram language model for spam filtering, and the help of removing the stopwords for enlarging the probability difference between spam and legitimate emails.The three datasets that we used in experiment are publicly available, and the ham and spam messages are freely mixed in different proportions. Experiment results demonstrated that TGNB model could weaken the independence assumption in standard Naive Bayes filtering, and got better performance than the state-of-the-art SVMs method in spam filtering.Finally,. We are currently focusing on the calculation of user reputation, which helps to determine its email is legitimate or spam, and the users include email address, email domain and ip. Meanwhile, we are trying several methods to filter image spams in practical email service, where the text-based filtering cannot help.

References

1. Kreibich, C., Kanich, C., Levchenko, K., Enright, B., Voelker, G.M., Paxson, V., Savage, S.: Spamcraft: An inside look at spam campaign orchestration. In: Proceedings of the 2nd USENIX Conference on Large-Scale Exploits and Emergent Threats: Botnets, Spyware, Worms, and More, pp. 4–4. USENIX Association (2009)
2. Siponen, M., Stucke, C.: Effective anti-spam strategies in companies: An international study (2006)
3. Moustakas, E., Ranganathan, C., Duquenoy, P.: Combating spam through legislation: a comparative analysis of us and european approaches. In: Proceedings of the Second Conference on Email and Anti-Spam, Citeseer (2005), `http://www.ceas.cc`
4. Li, F., Hsieh, M.H.: An empirical study of clustering behavior of spammers and group-based anti-spam strategies. In: CEAS 2006: Proceedings of the 3rd Conference on Email and Anti-Spam (2006)
5. Wu, C.H.: Behavior-based spam detection using a hybrid method of rule-based techniques and neural networks. Expert Systems with Applications 36(3), 4321–4330 (2009)
6. Blanzieri, E., Bryl, A.: A survey of learning-based techniques of email spam filtering. Artificial Intelligence Review 29(1), 63–92 (2008)
7. Yu, B., Xu, Z.: A comparative study for content-based dynamic spam classification using four machine learning algorithms. Knowledge-Based Systems 21(4), 355–362 (2008)
8. Cohen, W.W.: Learning rules that classify e-mail. In: AAAI Spring Symposium on Machine Learning in Information Access, vol. 18, p. 25 (1996)
9. Sahami, M., Dumais, S., Heckerman, D., Horvitz, E.: A bayesian approach to filtering junk e-mail. In: Learning for Text Categorization: Papers from the 1998 workshop, vol. 62, p. 28 (1998)
10. Metsis, V., Androutsopoulos, I., Paliouras, G.: Spam filtering with naive bayes-which naive bayes. In: Third Conference on Email and Anti-Spam (CEAS), Citeseer, vol. 17, pp. 28–69 (2006)
11. Islam, M.R., Chowdhury, M.U., Zhou, W.: An innovative spam filtering model based on support vector machine (2005)
12. Carreras, X., Marquez, L.: Boosting trees for anti-spam email filtering. Arxiv preprint cs/0109015 (2001)

13. Zhang, L., Zhu, J., Yao, T.: An evaluation of statistical spam filtering techniques. ACM Transactions on Asian Language Information Processing (TALIP) 3(4), 243–269 (2004)
14. Lewis, D.: Naive (bayes) at Forty: The Independence Assumption in Information Retrieval. In: Nédellec, C., Rouveirol, C. (eds.) ECML 1998. LNCS, vol. 1398, pp. 4–15. Springer, Heidelberg (1998)
15. Friedman, N., Geiger, D., Goldszmidt, M.: Bayesian network classifiers. Machine learning 29(2), 131–163 (1997)
16. Hovold, J.: Naive bayes spam filtering using word-position-based attributes. In: Proceedings of the 2nd Conference on Email and Anti-Spam (CEAS 2005), Citeseer (2005)
17. Peng, F., Schuurmans, D.: Combining Naive Bayes and n-Gram Language Models for Text Classification. In: Sebastiani, F. (ed.) ECIR 2003. LNCS, vol. 2633, pp. 335–350. Springer, Heidelberg (2003)
18. Duda, R.O., Hart, P.E.: Pattern classification and scene analysis, vol. 1. A Wiley-Interscience Publication, Wiley, New York (1973)
19. Liu, X., Croft, W.B.: Cluster-based retrieval using language models. In: Proceedings of the 27th Annual International ACM SIGIR Conference on Research and Development in Information Retrieval, pp. 186–193. ACM (2004)
20. Chen, S.F., Goodman, J.: An empirical study of smoothing techniques for language modeling. In: Proceedings of the 34th Annual Meeting on Association for Computational Linguistics, pp. 310–318 (1996)
21. Burges, C.J.C.: A tutorial on support vector machines for pattern recognition. Data mining and knowledge discovery 2(2), 121–167 (1998)

Rare Query Expansion via Wikipedia for Sponsored Search

Zhuoran Xu, Xiangzhi Wang, and Yong Yu

Department of Computer Application Technology, Shanghai Jiao Tong University, Shanghai 200240, China
{zhuoranxu,xzwang,yyu}@apex.sjtu.edu.cn

Abstract. Sponsored Search has evolved as the delivery of relevant, targeted text advertisements for Web queries. To match the most relevant advertisements for queries, query expansion algorithms were deeply researched during previous works. While most of current state-of-the-art algorithms appeal to Web search results as external resources to expand queries, we propose a novel approach based on Wikipedia for query augmentation against rare queries in sponsored search. By retrieving the top-*k* relevant articles in Wikipedia with Web query, we can extract more representative information and form a new ad query for the web query. With the new ad query, more relevant advertisements can be identified. To verify the effectiveness of our *wiki-based* query expansion methodology, we design a set of experiments and the results turn out that our approach is very effective for rare queries in sponsored search.

Keywords: Query Expansion, Sponsored Search, Pseudo-relevance Feedback, Wikipedia.

1 Introduction

Matching advertisements is more challenging compared with document retrieval due to the shorter lengths of ads and the sparsity of the bid phrase distribution over ads. Moreover, users intentionally choose terms to formulate query to retrieve the most relevant Web pages rather than relevant advertisements. To resolve the problem, techniques broadly used in Web search such as query expansion can be in consideration. Researchers studied possible ways to enhance queries with additional information such as electronic dictionaries and relevance feedback techniques that make use of a few top-scoring search results. In [6], Broder *et al.* firstly applied Pseudo-relevance Feedback (PRF) to sponsored search. They processed a large number of popular queries offline by augmenting them with features extracted from Web search results and store the results in a look-up table. The expanded query then can be used as a new ad query to retrieve relevant advertisements in ads database during online phase when corresponding web query issued.

As we described in Section 4.1, however, distribution of Web queries follows the power law [11]. To expand rare queries which accounts for most portion of

Y. Wang and T. Li (Eds.): Knowledge Engineering and Management, AISC 123, pp. 521–530.
springerlink.com

the volume of Web queries, Broder *et al.* in [2] built an inverted index with the expanded query vectors and expanded rare queries by retrieving the most related expanded queries from the index. Even though the performance of this approach is verified as effective in their experiments, we believe the effectiveness of rare query expansion relies on the quality of top-k relevant documents retrieved by its related popular queries. Its performance will be reduced terribly once the search results are irrelevant. In other words, we can say the side-effect of this methodology makes queries inter-dependent.

In this paper, we propose a novel methodology to expand rare queries leveraging the largest human knowledge base, Wikipedia. As a multilingual, free-content encyclopedia, Wikipedia has 75,000 regular editing contributors. More than 3 million entries covering different domains and concepts make it a perfect external knowledge corpus for query expansion. Instead of expanding query by search results of common queries offline, we build an inverted index with Wikipedia articles and extract more representative information to form a new ad query. The online query expansion is composed by the top-k ranked wiki articles of the input query and a second retrieve on ads space is performed.

The contributions of our work are as follows:

- We propose a novel approach of query expansion of rare queries for sponsored search, which base on the world's most powerful human knowledge base.
- With bag of words model and Wikipedia data source, this approach is very general and can be applied to any search tasks facing to vocabulary missing problem.
- Extensive experiments with a real data set show the effectiveness and feasibility of our approach.

The remainder of this paper is organized as follows. Section 2 provides a brief description of related work. The detail of our wiki-based query expansion methodology is presented in Section 3. In Section 4, we describe evaluation methodology, experiments and results. Conclusion and future work are finally illustrated in Section 5.

2 Related Work

Pseudo-relevance Feedback (PRF) was proved as an effective method for improving retrieval performance and widely researched [7] . With this approach, queries are executed offline and the top-k relevant documents are obtained as query expansion resources for future queries in real-time. Broder *et al.* applied this approach in sponsored search [6]. Besides of bag-of-words model, Broder *et al.* proposed augmenting common queries with additional knowledge rich features. Applying PRF on the rare queries, Broder *et al.* built an expanded query representation by leveraging offline processing done for related popular queries [2]. The limitation of PRF related approach is that it is based on the assumption that the search results are relevant which can not be guaranteed. What's more, applying PRF to rare query expansion makes queries dependent.

Other external resources are exploited to enrich information in queries. Xue *et al.* proposed to obtain relevance of queries by mining click-through data[12]. Term relationships and sophisticated inferential relationships extracted with Information Flow are used to expand query model instead of document model [13]. Term based query expansion methods are also researched. Atsuyoshi *et al.* paid attention to users' short-term interests and studied corresponding customization techniques [9].

For sponsored search, Zhang *et al.* and Neto *et al.* proposed that the advertisement vocabulary is often very limited while matching with user queries [1].Jin *et al.* found out sensitive content in the advertisement by its classification [10]. Yih *et al.* introduced their solution to the above vocabulary problem by extracting phrases from the Web pages and matching them to the bid phrases of the ads [8].

3 Methodology

In this section, we present details of our wiki-based approach for matching advertisements to rare queries. To start with, we present the definition of problem our methodology to resolve first. Then a high-level overview of the algorithm framework is described. After that we give the details about how an advertisement query is generated based on Wikipedia. In the end, we give out the score method in advertisement matching.

3.1 Problem Definition

When a user query is issued, the search engine is expected to retrieve top-k relevant web pages to users. In the meanwhile, advertisers expect their sites can be populated and displayed if the query is relevant with their advertisements. Here, our task is to identify advertisements which are most relevant to rare queries. In turn three main sub problems need to be resolved:

- Extract external representative resource with the original query.
- Generate new ad query.
- Match relevant ads with the new ad query.

3.2 Approach Overview

Figure 1(a) gives a high-level overview of our algorithm framework. During offline phase, the inverted index of Wikipedia articles is built. To accelerate the retrieval process in ads space with expanded query, we also built inverted index for advertisements. In real-time, when a rare query is issued in the system, we retrieve the top-k relative articles from wiki index. We use these search results to augment the web query and combine them as a new version of query which then is evaluated against the index of advertisements. Vector space-based model is used to formulate scoring formula to rank the ads list. The rest of this section will give more detail description for: how to retrieve the top-k relative wiki articles; how to generate expanded queries with these articles; scoring formula used to calculate relevance of advertisements.

3.3 Top-k Wiki Articles Retrieval

Neto *et al.* pointed out the vocabulary impedance problem in text-based advertisement matching [1], in form of the frequent mismatching between the vocabulary of Web pages and the vocabulary of advertisements. As a result, the naive method to match the Web query against the advertisements do not work well. To address this problem, we pull Wikipedia in to help making an expansion of the original Web query. We first index the whole documents in Wikipedia and make a top-k retrieval using the Web query. To calculate the similarity between wiki articles and query, we use TFIDF scheme to define term weighting function on wiki documents:

$$w(t, W) = (1 + \log tf(t, W)) \cdot idf(t) \tag{1}$$

where $tf(t, W)$ stands for the t's term frequency in the wiki article W. Terms from the original query are weighted as follows:

$$w(t, Q) = (1 + \log tf(t, Q)) \cdot idf(t) \tag{2}$$

where $tf(t, Q)$ is the number of times t occurs in Web query Q. As [2], we define $idf(t)$ is the inverse document frequency of t in the advertisement database:

$$idf(t) = \log \frac{N}{N_t + 1} \tag{3}$$

with N is the total number of ads and N_t is the number of ads containing term t.

With term weight in wiki articles and query definition, the similarity score between query and wiki articles can be calculated as:

$$sim(Q, W) = \sum_{t \in T(Q) \bigcap T(W)} w(t, Q) \cdot w(t, W) \tag{4}$$

where $T(Q)$ represents the term set of query Q while $T(W)$ is the term set of wiki article of W.

With the similarity score $sim(Q, W)$, the most relative wiki articles can be selected from the inverted index of Wikipedia.

3.4 Ad Query Generating

After we get top-k most relevant wiki articles for the web query, we expect to generate a new ad query with combination of these articles and the original Web query. Actually we just need to calculate the term weights of the expanded query. We improve Rocchio's query expansion [3] and weight the terms in the ad query as follows:

$$w(t, AQ(Q)) = (1 - \lambda) \cdot w(t, Q) + \lambda \sum_{W \in W(Q)} \frac{sim(Q, W) \cdot w(t, W)}{Score(Q)} \tag{5}$$

where $AQ(Q)$ is the new ad query to generate, $w(t, AQ(Q))$ is the weight of term t in ad query $AQ(Q)$, λ is a free parameter that allows us to tune the weight of wiki articles impacting the new term weight, $w(t, Q)$ is the t's weight in the original query Q expressed in Equation 2, $W(Q)$ is document set of the top-k wiki articles retrieved by web query Q, $sim(Q, W)$ is the similarity score between wiki article W and query Q we defined in Equation 4, $w(t, W)$ is the weight of term t in wiki document W using in Equation 1, $Score(Q)$ is the sum similarity score between each wiki article W and query Q in the top-k wiki articles, which used to normalize the similarity score $sim(Q, W)$.

3.5 Advertisement Matching

This part will describe last task of matching relevant advertisements. As modeling above, we first define term weight for each term in advertisement. Since our approach works within the vector space model, we employ BM25 weighting method[4] for term on ads:

$$w(t, A) = \sum_{f \in F(A)} \frac{(k+1) \cdot wtf_f(t, A) \cdot idf(t)}{k \cdot \left((1-b) + b \cdot \frac{|A|}{|A|_{avg}}\right) + wtf_f(t, A)} \tag{6}$$

where $F(A)$ is the fields set of advertisement including title, description, display URL and bid phrase. $|A|$ is the length of the advertisement, while $|A|_{avg}$ is the average length of the advertisements, and $wtf_f(t, A)$ is described as follows:

$$wtf_f(t, A) = w(f) \cdot tf(t, f) \tag{7}$$

where $w(f)$ denotes the weight of field f.

Taking structure of advertisement into consideration, we apply the title matching boosting method in [2]. Thus we have the following boost factor:

$$prox(Q, A) = \frac{\sqrt{\sum_{u \in f_T(A)} w(t, Q) \cdot w(t, A)}}{\sqrt{\sum_{t \in T(Q)} w(t, Q)^2}} \tag{8}$$

where $f_T(A)$ are the terms extracted from the ad A's title, and $T(Q)$ is the term set of original web query Q.

With similar concepts drawn from [2], we can finally define the similarity between an ad query and an advertisement in the vector space model. An improved similarity function is given out as follows:

$$sim(Ad(Q), A) = \sum_{t \in W(Q)} w(t, Ad(Q)) \cdot w(t, A) \tag{9}$$

where $W(Q)$ is the top-k retrieved wiki articles of original query Q. With the title match boost function and the similarity score, the scoring function of a Web query and an advertisement is:

$$S(Q, A) = sim(Ad(Q), A) \cdot (1 + prox(Q, A)). \tag{10}$$

As a first step of exploiting knowledge from Wikipedia to enrich rare query for ad search, we adopt bag of words model and map query, advertisement and wiki articles into vector-based space. However, it does not mean only term feature can be applied with our methodology. Wikipedia is a multi-dimension knowledge corpus with organized categories and finely structured link graphs, which can be potential information for query expansion. From the formulation of our methodology, it can be seen that such features can easily added to the scoring model, which makes our approach very extensible.

4 Experiments

We have implemented our methodology of query expansion based on Wikipedia to retrieve relevant advertisements for rare queries. In this section, we give a detailed description of our data set, evaluation metrics and corresponding experimental results.

4.1 Data Set

The data of Wikipedia used in our experiments can be freely downloaded from `http://download.wikipedia.org`, which was released on November, 2009. For query log, we used MSN Search Spring 2006 Query logs, which contain more than 14 million queries during May 1-31 2006. Figure 1(b) shows that queries follow power law distribution [11] exactly, which indicates the expansion task of rare query is critical for sponsored search. After cleaning the data by only keeping the frequency less then 3 and well-formatted English queries, we had a total of 6.5 million distinct queries. To verify the effectiveness of our solution, we sampled 200 rare queries randomly from the data set. With the 14 million queries, we crawled 68k distinct advertisements from MSN Live Search.

4.2 Evaluation Methodology

We adopted two widely accepted evaluation metrics: Normalized Discount Cumulative Gain(NDCG) and Presicion@k as two different evaluation dimension for our system. To avoid evaluation bias caused in dreary label task, each query-ad pair is classified into one of 5 different relevant degrees defined in Table 1. Annotators are asked to extract the main topic and sub topics for each query and advertisement. Then each query-ad pair can be classified according to the portion of overlap of topics as defined in Table 1. The column of Gain Value is used to calculate NDCG and we treat each query-ad as irrelevant with *Somewhat Attractive* or *Somewhat Attractive* to calculate P@k, vise versa. Three annotators with good English language skills labeled the relevance of top ten advertisements for each rare query. As most commercial search engines display at most 3 relevant advertisements for each Web query on the result page, we only discuss P@1~3 and NDCG@1~3 in the rest of this section.

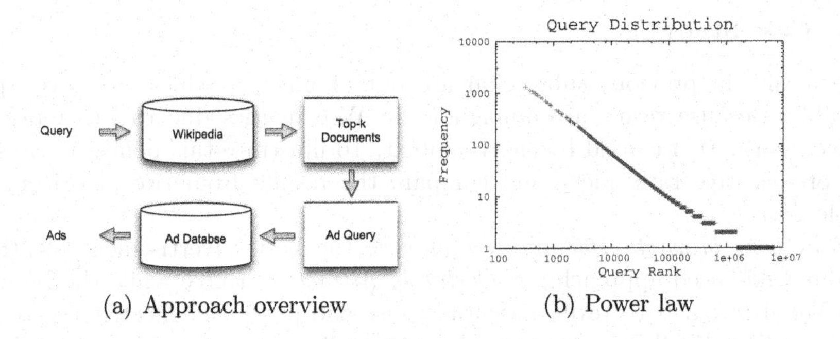

(a) Approach overview (b) Power law

Fig. 1. Our approach overview and the power law distribution of our Queries

Table 1. Advertisement relevance criterion

Relevant Degree	Gain Value	Criterion
Perfect	10	Both main topic and all sub-topics are matched
Certainly Attractive	7	Main topic and most of sub-topics are matched
Probably Attractive	5	Main topic and few of sub-topics are matched
Somewhat Attractive	2	Only few of sub-subjects are matched
Not Attractive	0	Nor main topic neither sub-topic are matched

4.3 Experimental Results

The baseline approach we choose to compare the effectiveness of our methodology is searching the ads database with original query without any query rewriting. We believe the baseline chosen is reasonable as for rare queries, little can be done using existing on-line query expansion techniques. Even though Broder's method proved features extracted from related augmented queries composed during offline with common queries brought great relevance improvements in [2], our method is totally differentiate from that in nature.

λ in Equation 5 reflects the weight of Wikipedia in the combined bag-of-words model. A group of experiments with increased value of λ from 0 to 1 are executed. Actually, the baseline method can also be viewed as a wiki-based query expansion method with parameter λ set to 0. From Figure 2(a), we can see the precision of ad matching changing according to the changes of Wikipedia's impact in calculation of similarity between advertisements and rare queries. Comparing with baseline method($\lambda = 0.0$), it shows that our wiki-based query expansion algorithm consistently and significantly improves P@1, P@2 and P@3. With λ set to 0.2, we get the best performance for P@1 which exceeds baseline 12%, while λ=0.8 beats against baseline with improvement 12% and 7% for P@2 and P@3 respectively. It is worth notice that the improvement of precision does not increase with λ monotonously. We confer this that noisy information is also extracted during the augmentation process. Future analysis will be engaged on this phenomenon. Figure 2(b) also illustrate that wiki-based query expansion out performs the baseline method, which indicates the ranking of top-k advertisements returned by our approach is better.

4.4 Case Study

The results in previous subsection are out of unexpectedness but even predictable. Because terms are ambiguous in Web queries due to less representative information caused by short context. To illustrate this point, we present a representative rare query and compare the results from two algorithms in Table 2(a).

For query *bravo tv 100 funniest movies*, the best advertisement identified by our wiki-based approach got *Certainly Attractive* degree while the baseline got *Not Attractive*. To reveal essential cause and prove the effectiveness of our method theoretically, we explored the intermediate result described in 3. From Table 2(b), we can see term *bravo* in the original query is given a high query weight $w(t, Q)$ because of its low idf in ad database according to Equation 2. As we mentioned before, the baseline method can be seen as a wiki-based algorithm with λ set to 0. So Equation 5 can be also applied to baseline. As indicated by Equation 10, the ad covering this term obtains a higher score while other terms' weight can not balance this bias. Contrarily in our approach, the top-k wiki articles related to this query are taken into consideration. New ad query weight $w(t, Ad(Q))$ in Equation 5 of term *tv* and *movies* got reinforced under the impact of corresponding wiki weight $w(t, W)$ as indicated by Equation. Finally reinforcement of query weight came from wiki articles balanced the bias caused by *bravo* effectively and then a salient improvements were obtained by the wiki-based methodology. From this example, we conclude that our approach can balance the bias caused by some rare term in ad database with the help of Wikipedia, and in turn extract the topic of the original query, which finally result in more relevant matching advertisements.

5 Conclusion and Future Work

This paper proposes a novel methodology in sponsored search to augment rare queries based on Wikipedia as an external corpus. Our approach differs from previous works on queries expansion in that we leverage Wikipedia rather than

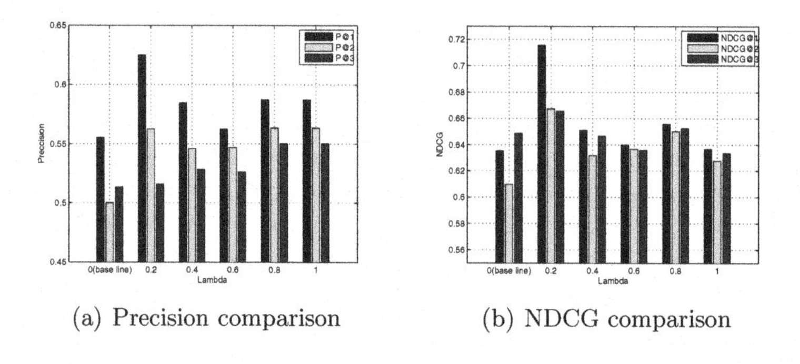

(a) Precision comparison (b) NDCG comparison

Fig. 2. Precision comparison and NDCG comparison of our approach and baseline

Table 2. This table makes a detailed result description of the query *bravo tv 100 funniest movies* by our approach and baseline. The first table simply displays the most relevant ad results with different methods, and the second one reveals more detailed weights, which can make a fine interpretation for the reason why our approach is more effective.

(a) Most Relevant Ad Results

	Best Ad Title	Best Ad Description
Baseline	Men's Propet **Bravo**	Excellent customer service. Secure. 90 day guarantee & free shipping
Our Approach	MAGIX **Movies** on CD & DVD	Burn your videos on disc - Watch them on **TV**

(b) More detailed Weights Value

Term	$w(t, Q)$	$w(t, W)$	w(t, Ad(Q))
bravo	9.89	13.7	10.66
tv	5.50	22.9	8.98
100	4.42	27.98	9.13
funniest	10.59	31.87	14.84
movies	7.64	41.38	14.39

search results as information complement to augment queries, which makes query expansions independent with each other. By retrieving the top-k most related documents from the inverted index of Wikipedia articles, we extracted more representative terms to complement rare queries. The baseline method in our experiment retrieves advertisements with original queries directly without any expansion. We believe it is reasonable as there is few query expansion techniques can be applied during online phase for rare queries. Our experimental results show that the wiki-based query expansion algorithm outperforms the baseline both in term of NDCG and precision. Finally, a query is presented as an example to explain why the augmentation with Wikipedia works effective for query expansion.

Our work on wiki-based query expansion is the first step in augmenting rare queries based on Wikipedia. More features can be extracted such as Wiki-Graph and Wiki Categories. In the future, we would like to combine these features with currently bag-of-words model. Moreover, the effectiveness of this approach can also be verified on common queries.

References

1. Neto, B.R., Cristo, M., Golgher, P.B., de Moura, E.S.: Impedance coupling in content-targeted advertising. In: SIGIR 2005: Proceedings of the 28th Annual International ACM SIGIR Conference on Research and Development in Information Retrieval (2005)

2. Broder, A., Ciccolo, P., Gabrilovich, E., Josifovski, V., Metzler, D., Riedel, L., Yuan, J.: Online expansion of rare queries for sponsored search. In: WWW 2009: Proceedings of the 18th International Conference on World Wide Web (2009)
3. Rocchio, J.: Relevance Feedback in Information Retrieval. The SMART Retrieval System (1971)
4. Robertson, S.E., Walker, S., Jones, S., Hancock-Beaulieu, M.M., Gatford, M.: Okapi at TREC-3 (1994)
5. Broder, A.Z., Fontoura, M., Gabrilovich, E., Joshi, A., Josifovski, V., Zhang, T.: Robust classification of rare queries using web knowledge. In: SIGIR 2007: Proceedings of the 30th Annual International ACM SIGIR Conference on Research and Development in Information Retrieval (2007)
6. Broder, A.Z., Ciccolo, P., Fontoura, M., Gabrilovich, E., Josifovski, V., Riedel, L.: Search advertising using web relevance feedback. In: CIKM 2008: Proceeding of the 17th ACM Conference on Information and Knowledge Management (2008)
7. Lavrenko, V., Croft, W.B.: Relevance based language models. In: SIGIR 2001: Proceedings of the 24th Annual International ACM SIGIR Conference on Research and Development in Information Retrieval (2001)
8. Yih, W.-T., Goodman, J., Carvalho, V.R.: Finding advertising keywords on web pages. In: WWW 2006: Proceedings of the 15th International Conference on World Wide Web (2006)
9. Atsuyoshi, M.L., Langheinrich, M., Abe, N., Kamba, T., Koseki, Y.: Unintrusive Customization Techniques for Web Advertising. Computer Networks (1999)
10. Jin, X., Li, Y., Mah, T., Tong, J.: Sensitive webpage classification for content advertising. In: ADKDD 2007: Proceedings of the 1st International Workshop on Data Mining and Audience Intelligence for Advertising (2007)
11. Spink, A., Building, R.I., Wolfram, D., Saracevic, T.: Searching the Web: the public and their queries. Journal of the American Society for Information Science and Technology (2001)
12. Xue, G.-R., Zeng, H.-J., Chen, Z., Yu, Y., Ma, W.-Y., Xi, W., Fan, W.: Optimizing web search using web click-through data. In: CIKM 2004: Proceedings of the Thirteenth ACM International Conference on Information and Knowledge Management (2004)
13. Bai, J., Song, D., Bruza, P., Nie, J.-Y., Cao, G.: Query expansion using term relationships in language models for information retrieval. In: CIKM 2005: Proceedings of the 14th ACM International Conference on Information and Knowledge Management (2005)

An Automatic Service Classification Approach

Haiyan Zhao and Qingkui Chen

School of Optical-Electrical and Computer Engineering, University of Shanghai for Science and Technology, Shanghai,200093, P.R. China
zhaohaiyan1992@sina.com.cn, chenqingk@sohu.com

Abstract. With the development of the service technology, more and more organizations are publishing their business function as service through Internet. Service classification is a key approach for service management. With the quick increase of service number, the cost of classifying these services through manual work is becoming more and more expensive. A service automatic classification approach based on WordNet by combining text mining, semantic technology and machine learning technology was given in the paper. The method only relies on text description of services so that it can classify different type services, such as WSDL Web Service, RESTfulWeb Service and traditional network based software component service. Though text mining and applying word sense disambiguation models, a service can be described as a sense vector with no ambiguous. Then a K-means algorithm is used to classify these services. Experimental evaluations show that our classification method has good precision and recall.

Keywords: Web Service, Text mining, WSD, Similarity Calculation.

1 Introduction

With the development of the service technology, more and more organizations are publishing their business function as service through Internet. Assigning a proper category to a service can be a tedious and error prone task due to the large number of services. Techniques for automatically classifying services will help to address this problem effortlessly. This paper describes an automatic classification approach based on text mining, machine learning techniques and semantic techniques.

During the past few years some efforts and research have been placed on assisting the developer to classify Web services. As a result, some semiautomatic and automatic methods have been proposed. These approaches are based on text mining and semantic annotations matching techniques. AWSC [1], METEROR-S [2] and Assam [3] are systems to classify web services based on text mining techniques. These systems generally consist of two parts, text mining module and web service classification module. Text mining module is responsible for extracting a feature vector from Web service descriptions. In this context, a feature vector is a term collection. Web service classification module is responsible for generating a feature vector for each classification

Y. Wang and T. Li (Eds.): Knowledge Engineering and Management, AISC 123, pp. 531–540.
springerlink.com

using machine learning techniques (such as Naive Bayes and Rocchio) and a training data-set. Then the similarity of each tested Web services and each classification is computed and the service is classified to the category with which it has the largest similarity. These methods have some limitations. The main limitation is that they need a group of classified Web services as the training data-set so that these methods are suitable for classifying newly added Web service after an initial classification framework has been set up. Secondly, most of these methods do not consider the semantic similarity and semantic conflict. For example, when a web service is descripted as *<advertising>* and another web service is descripted as *<campaign>*, these two services can probably be classified to different categories if without considering their semantic similarity.

An approach to add semantic annotations to Web services and classify these Web services by computing semantic similarity was proposed in [4]. The shortcoming of this approach is it relies on the domain ontology library. Although many domain experts are developing their domain ontology libraries, only ontology libraries of biology domain are widely accepted (such as TMO and Gene Ontology). Ontology libraries of other domains, such as finance, are far from complete. Therefore the classification approach based on semantic annotation and matching is difficult to implement.

In addition, current classifying methods mainly support web services classifying in terms of the WSDL documents. With the diversification of service types, more and more Web services adopt the RESTful style, such as Google AJAX Search API and Amazon Simple Storage Service (Amazon S3). Although currently the number of RESTful Web services is less than the number of WSDL Web services, this style of Web services has extensive application prospect. At the same time, traditional software components are also used in practical application development. But the research on how to classify RESTful web services and traditional software components is relatively less.

In order to overcome these limitations, a service automatic classification approach based on WordNet, which combines text mining, semantic technology and machine learning technology was given in the paper. The approach only relies on text description of services so that it can classify different type services, such as WSDL web service, RESTful web service and traditional software component service. This method does not need a training data set and is suitable for classifying large quantity services. This paper is organized as follows. Section 1 was an introduction. The framework of the system was given in Section 2. Section 3, 4, 5 and 6 discussed important steps of this approach respectively. Experiments and evaluations were given in Section 7. Section 8 concluded the whole paper.

2 The Framework for Service Automatic Classification

The framework of our system for service automatic classification consists of four parts, i.e., a text mining module, a word sense disambiguation module, an automatic web service classifier and a similarity calculation module.

The text mining module is used to extract interesting and trivial information and knowledge from description document of each service. Its outputs are a term vector.

Word semantic in the term vector is clarified using the word sense disambiguation module so that a matrix is obtained, where row represents each service and column represents a sense vector of the associated service. The automatic web service classifier based on K-Means algorithm is used to classify these services. The word sense disambiguation module and the automatic web service classifier both adopt a sense similarity algorithm based on WordNet to calculate the similarity of two senses.

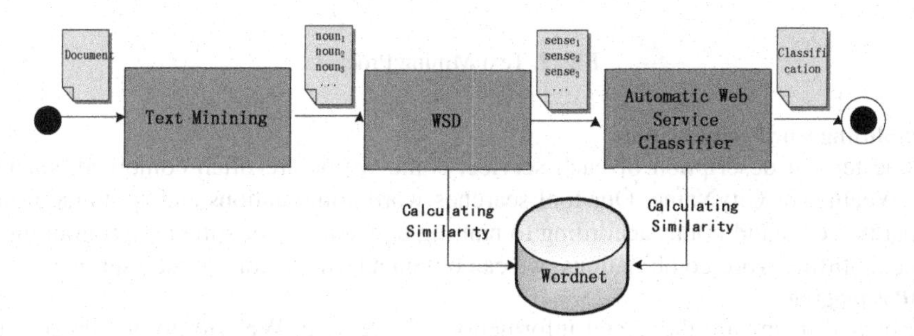

Fig. 1. System Framework for Service Automatic Classification

3 Text Mining

The text mining module is responsible for extracting interesting and trivial information from description document of each service. Its output is a term vector that is the input of the subsequent word sense disambiguation module. The quality of the output has direct impacts on the quality of service classification.

Current service classification methods extract trivial information from WSDL document of each service. WSDL is a well-structured and standard XML document format for describing Web services. Most methods rely on a parser based on WSDL4J, which can parses a WSDL document to extract out the service name, operations and arguments as the term vector of the service. This parser only deals with WSDL document and cannot process RESTful Web Service [5] and traditional software components. Our text mining module includes a more general parser which can extract trivial information from all kinds of description documents for each service by using text mining technology.

The traditional text mining approach often adopts TF/IDF algorithm. But in the description documents of a service the frequency of each term is very low. Therefore TF/IDF algorithm is no longer suitable. No matter what kind of services it is, it can be described as a tuple <*action, target, way*> formally, where an action is composed of verbs, target and way are composed of nouns. From this point of view, verbs and nouns are the trivial information of a service description. Our text mining module focuses on the verbs and nouns and ignores the else.

Fig.2 depicts the steps of text mining process and they are splitting word combinations, Pos tagging, reducing terms to their stems, removing noise and translating verbs to nouns.

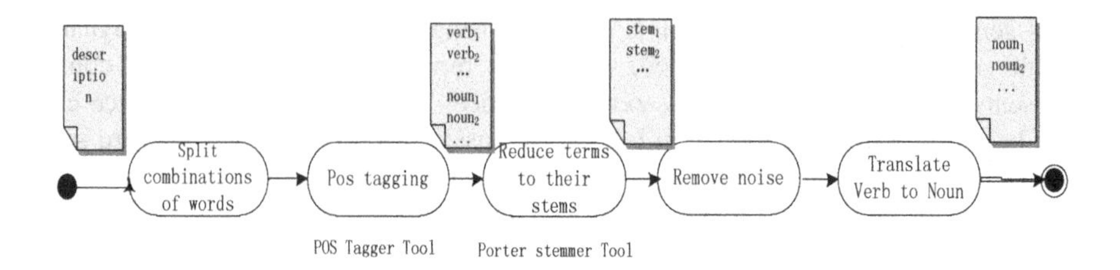

Fig. 2. Text Mining Process

(1)Splitting word combinations

In the textual description of each service, some words are often combined, such as GetWeather or CityName. Our tool searches word combinations and splits them into separate verbs and nouns according to naming conventions of software programming. After splitting word combinations, we can obtain a term vector for each service.

(2)Pos tagging

Verbs and nouns are the trivial information of a service. We will do pos tagging for each term of the term vector. We used Stanford POS Tagger [7] developed by the Stanford NLP (Natural Language Processing) Group to select verbs and nouns.

(3)Reducing terms to their stems

Considering the commoner morphological and inflectional ending, we also employed the Porter stemmer [6] to reduce words to their stems.

(4)Removing noise

Noise in service description will impact the quality of service classification. We removed noise from two aspects: one is common English stops words, the other is stops words which are related to services, such as request, response, soap, http, Result, platform, program, system, site, engine, database and so on.

(5)Translating verb to noun

Word sense disambiguation module and automatic service classifier adopt sense similarity algorithm based on WordNet to compute the similarity of two senses. These algorithms were given in [8][9][10]. The similarity of two senses will be zero when two senses are different parts of speech according to the algorithms. For example, although the first sense of *market* (as a verb) is similar to the second sense of *promotion*, the result of Lin's algorithm is zero because these two senses are different parts of speech. So we must translate all verbs to nouns such as *market* to *marketing* (as a noun). Then the similarity of *marketing* and *promotion* is 0.25 according to Lin's algorithm.

Table 1. Noun Vector for Service

Service Name	Noun Vector
Allegro	{auction marketplace}
AvantLink	{affiliate marketing}
123 Shop Pro	{shopping cart}
Allogarage	{marketplace car dealership review recommendation}
Cafe Press	{retail product}
...	...

We can obtain a noun vector for each service by text mining module. Table 1 depicts the noun vector for some services.

4 Word Sense Disambiguation

In WordNet, a word usually has multiple senses, for example, bus has four senses. When bus appears with station, its sense is automobile. But when bus appears with network, its sense is computer bus. Word sense disambiguation is used to find the correct meaning of a polysemous word in a given context.

As non-supervised WSD allows disambiguation of words without user interference, we use a variant of the SSI algorithm [11] to get the senses out of a set of words (Formula (1)). The algorithm disambiguates a word based on a previously disambiguated set of words and their related senses. For each sense of a word (s_j), a similarity with the senses from the context (sc_i) is calculated and the sense with the highest similarity is chosen. After that, the word and its chosen sense will be added to the context (I) and iteration will be done. This process continues until there are no ambiguous words left.

In Formula (1), senseSim (s_j, sc_i) is the similarity of two senses and it is shown in Formula (6).

$$selectedSense(word) = \arg \max_{s_j \in senses(word)} \sum_{sc_i \in I} senseSim(s_j, sc_i) \tag{1}$$

Table 1 will be changed into Table 2 after applying word sense disambiguation. For example, *auction#n#2*, *#n* means noun, *#2* means the 2th sense.

Table 2. Noun Sense Description of Service after Applying WSD

Service	Description	Sense Vector after WSD
Allegro	Online auction marketplace	{auction#n#2 marketplace#n#1}
AvantLink	Affiliate marketing network	{affiliate#n#2 marketing#n#2}
123 Shop Pro	Online shopping cart software	{shopping#n#1 cart#n#2}
Allogarage	marketplace car dealership reviews and recommendations	{marketplace#n#1car#n#1 dealership#n#1 review#n#1 recommendation#n#1}
Cafe Press	Customized retail product service	{retail#n#1 .product#n#1}
...	...	...

5 Automatic Service Classifier

After applying word sense disambiguation, we can obtain a sense vector for each service that is unambiguous. The next step is to classify by applying improved K-means clustering algorithm.

K-means clustering is a method of cluster analysis which aims to partition n observations into k clusters in which each observation belongs to the cluster with the nearest mean. This results into a partitioning of the data space into Voronoi cells.

In traditional K-means clustering algorithm, each observation is a d-dimensional real vector. But in our application, each observation is a d-dimensional sense vector and we need resolve two problems. One is how to calculate the new centroid of the observations in some cluster. The other is how to calculate the distance of the each observation to the cluster.

For the first problem, calculating the new centroid of the observations in some cluster can be treated as a problem of calculating the centroid of a group of sense vector. Since a cluster can be described by high frequency senses in this cluster, we retreat a group sense vector of all services belonging to some cluster as an input. Then TF-IDF algorithm is used to select the first m frequent senses $\{sense_1, sense_2, \ldots, sense_m\}$ as the centroid of this cluster. The Fig.3 depicts the detail.

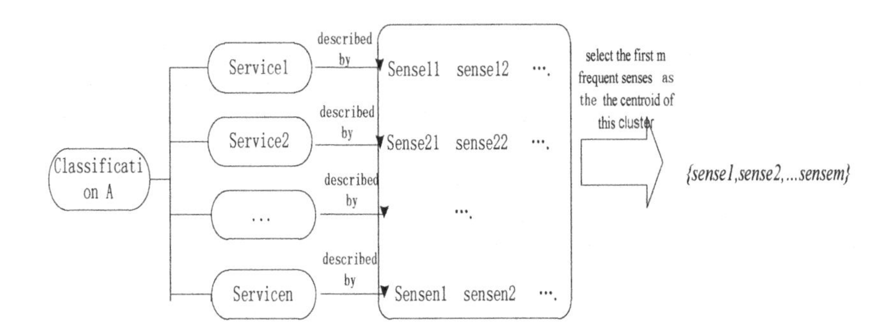

Fig. 3. Computing the Center of a Category

Formally, for each term t_i of a document d, $tf_{idfi}=tf_i*idf_i$, with:

$$tf_i = \frac{n_i}{\sum_{j=1}^{T_d} n_j} \tag{2}$$

Where the numerator (n_i) is the number of occurrences within d of the term being considered, and the denominator is the number of occurrences of all terms within d (T_d), and:

$$idf_i = \log \frac{|D|}{|\{d : t_i \in d\}|} \tag{3}$$

Where $|D|$ is the total number of documents in the corpus and $|\{d: t_i \in d\}|$ is the number of documents where the term ti appears.

The first problem also involves another problem of determining initial centroids. By experiments, we find the clustering quality is often poor if we randomly choose k observations as the initial centroids. For example, if we cluster all services in shopping catalogue and advertising catalogue in ProgrammableWeb site. If these services are classified into 2 sets and the initial two centroids which are randomly choose are both in shopping catalogues, the clustering quality is poor. To overcome this problem, we choose first k frequent sense as initial centroids by TF/IDF algorithm (k is the number of cluster). Although further experiment showed the clustering quality was improved, it is still not satisfied. The root cause is some senses often appear simultaneously. For above-mentioned example, shopping often appears with cart. In fact, these two senses stand for the same centroid. But by TF/IDF algorithm, shopping and cart should be treated as two initial centroids. In order to resolve this problem, the TF/IDF algorithm and association rules are combined in our approach. Firstly, TF/IDF algorithm is used to order each sense. Secondly association rules algorithm is applied to find senses sequence, which often appears simultaneously. Lastly, we choose first *k* frequent sense sequences as initial centroids

For the second problem, the distance of the each observation to the cluster can be calculated by computing the similarity of two sense vectors. Formula (4) shows how the similarity between sense vectors ssw and ss_u is computed. The average of similarity between any sense (s_u) of the sense vector (ss_u) and the sense vector (ss_w)is computed. The average of the similarity between any sense (s_w) of the sense vector (ss_w) and the sense vector (ss_u) is calculated. They are summed up to provide a symmetric match.

$$vectorVectorSim(ss_u, ss_w) = \sum_{s_u \in ss_u} \frac{senseVectorSim(s_u, ss_w)}{|ss_u| + |ss_w|} + \sum_{s_w \in ssw} \frac{senseVectorSim(s_w, ss_u)}{|ss_u| + |ss_w|} \quad (4)$$

In Formula (4), senseVectorSim(s_u, ss_w) or senseVectorSim (s_w, ss_u) means similarity of a sense and a sense vector. The similarity between a sense (s_a) and a sense vector (ss_b) is done by searching for the maximum similarity between the sense and each sense (s_b) of ss_b. Formula (5) shows how this is done.

$$senseVectorSim(s_a, ss_b) = \max_{s_b \in ss_b} senseSim(s_a, s_b) \quad (5)$$

In Formula (5), senseSim (s_a, s_b) is the similarity of two senses and its formula is shown in Formula (6).

6 Similarity Calculation

Many similarity measures have been proposed. Lexical comparison measure and semantic similarity measure based on WordNet are widely used. Semantic similarity refers to similarity between two concepts in taxonomy such as WordNet. Our word sense disambiguation module and automatic web service classifier both consider semantic similarity so that it is adopted to calculate the similarity of two concepts.

We adopted Lin's measure and used an open source java libraries:WordNet::Similarity to implement it.

Lin's measure is shown in Formula (6).

$$senseSim(C_1, C_2) = \frac{2\log^{-1} P(C_0)}{\log^{-1} P(C_1) + \log^{-1} P(C_2)} \quad (6)$$

C_1 and C_2 are two concepts of the taxonomy tree. C_0 is the most specific class that subsumes both C_1 and C_2. $P(C)$ is the probabilities that a randomly selected object belongs to C. In order to calculate $P(C)$, there must be a training corpus to get the probability of each concept. Here Semcor corpus is adopted which is developed by Princeton University.

In our experiment, shopping catalogue and travel catalogue in ProgrammableWeb site are used as test data sets. In these two catalogues, occurrence frequencies of shopping and travel are high. Let's consider the problem of determining the similarity between two concepts shopping and travel. Fig.4 depicts the fragment of the WordNet. The number attached to a node indicates its $IC(C)$ (here, $IC(C)=\log^{-1}P(C)$). From the Fig. 4 we can find that action is the concept that subsumes both shopping and travel. The similarity of shopping and travel is calculated as following:

$$senseSim(Shopping, Travel) = \frac{2 \times IC(Action)}{IC(Shopping) + IC(Travel)} = \frac{2 \times 3.7}{9.23 + 6.52} = 0.47$$

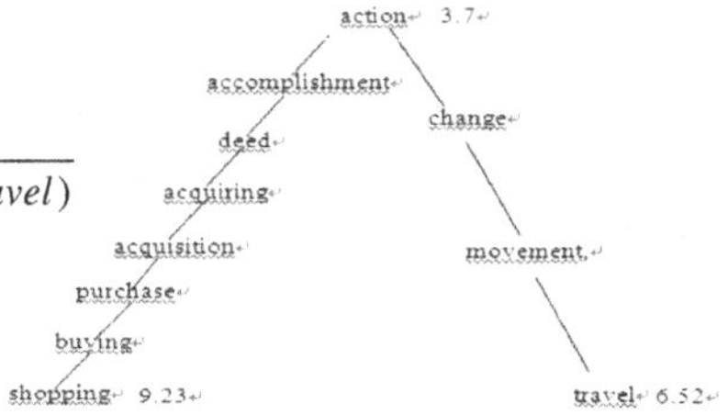

Fig. 4. A Fragment of WordNet

7 Experiment and Evaluation

We did the experiments on a test data set collected from Programmable Web Site. We extracted key information from service description documents and classified these services using our approach. The classification results were compared with the results from other approaches.

The precision and recall values are computed according to Formula (7), where n is the classification number of the testing data set, p_i is the precision of the *i-th* classification and r_i is the recall of the *i-th* classification.

$$p = \sum_{i=1}^{n} p_i \Big/ n \qquad r = \sum_{i=1}^{n} r_i \Big/ n \tag{7}$$

The *precision*(p_i) and *recall*(r_i) are calculated as follows: where m is the number of services which are classified to the i-th classification and n is the number of services which belong to the *i-th* classification in the test data set. If a service is classified correctly, then y_i is set 1, otherwise y_i is set 0.

$$p_i = \sum_{j=1}^{m} y_j \Big/ m \qquad r_i = \sum_{j=1}^{m} y_j \Big/ n \tag{8}$$

We did experiment six times and computed the precision and recall for each experiment. Fig.5 depicts the results.

From Fig.5, the average precision of our method is 74% and the average recall of our method is 71%. The result proves our method has satisfying quality of classification.

Comparing with Test 3, the precision and recall of Test 4 decrease greatly. In Test 4, we

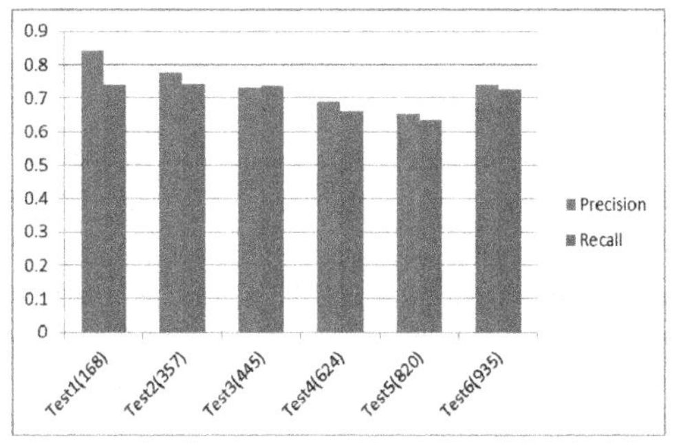

Fig. 5. The Experimental Result of Precision Rate and Recall Rate

added 179 APIs from shopping catalogue and advertising catalogue into the test data set. There are two reasons that lead to the quality decrease. Firstly, we find APIs from shopping catalogue and APIs from advertising catalogue are very similar. For example, their description documents both include *marketing* term and *advertising* term. Secondly, in Lin's method, the similarity is calculated in terms of the probability of each concept. In order to calculate the probability, we adopted Semcor corpus for training. Though Semcor corpus is the largest English corpus, in order to construct a semantic system with high accuracy, it is still necessary to expand this corpus. For example, in our experiment, some APIs are about retailer and should be classified to shopping catalogue. But by Lin's method, the similarity of (*retailer, shopping*) is 0.06934 and the similarity of (retailer, advertising) is 0.1192. Therefore these APIs are classified to advertising catalogue. In the subsequent research, we are going to expand Semcor corpus by adding the vocabularies coming from Programmable Web site.

Comparing with test 4, the precision and recall of test 5 decrease greatly. In test5, we added 196 APIs from travel catalogue and financial catalogue. The APIs from these two catalogues are very similar, which make the quality decrease.

Moreover, we found the imbalance of APIs number of different catalogue also had impact on classification accuracy. For example, in test 5, the APIs number of sports catalogue and weather catalogue is less than other catalogues'. Therefore in test 6, we replaced these APIs from sports catalogue and weather catalogue by the APIs from messaging catalogue and telephony catalogue whose numbers of APIs are proportional to other tested catalogues. Hence, the precision and recall of test 6 was improved greatly.

We compared our method with K-NN, Naive Bayes and Rocchiod method based on TF-IDF. The results are shown in Table 3. It shown that our method achieved better results than K-NN. But our method achieves worse results than Naive Bayes and Rocchiod method. The main reason was Naive Bayes and Rocchiod method based on TF-IDF were supervised machine learning algorithm, which need a training data set. Our method adopted K-means and it was a n unsupervised machine learning algorithm which don't need a training data set. We believe combing the supervised machine learning algorithm and unsupervised machine learning algorithm can achieve better result: firstly, using our method to classify the original set of services. Then using supervised machine learning algorithm to classify newly added services after having an initial classification framework.

Table 3. Comparison between Different Classifiers

Classifer	Average Precision
K-NN	39.59%
Naive Bayes	79.38%
Rocchiod method based on TF-IDF	85.08%
Our Approach	74%

8 Conclusions

A service automatic classification approach based on WordNet by combining text mining, semantic technology and machine learning technology was given in the paper. The method only relies on textual description of services so that it can classify different type services, such as WSDL web Service, RESTful web Service and traditional network based software component service. Through applying text mining and word sense disambiguation models, a service can be described as a sense vector with no ambiguous. Then a K-means algorithm is used to classify these services. Experimental evaluations show that our classification method has good precision and recall.

Our automatic classification apporach can be extended in several ways. For example, it could be extended with the ability of hierarchy clustering. Furthermore, distributed computing framework, such as Hadoop, can be applied to speed up the classification.

Acknowledgments. This work was supported by China NSF under Grant No. 60970012, 60873230 and 61073021, Shanghai Key Science and Technology Project in Information Technology Field (No.09511501000), Shanghai Key Science and Technology Project (No.09220502800) and Shanghai leading academic discipline project No.S30501.

References

1. Clifton, C., Leavens, G.T., Chambers, C., Millstein, T.: MultiJava: modular open classes and symmetric multiple dispatch for Java. ACM SIGPLAN Notices 35(10), 130–145 (2000)
2. Wegner, P., Zdonik, S.: Inheritance as an Incremental Modification Mechanism or What Like is and Isn't Like. In: Gjessing, S., Chepoi, V. (eds.) ECOOP 1988. LNCS, vol. 322, pp. 55–77. Springer, Heidelberg (1988)
3. Waxman, B.: Routing of multipoint connections. IEEE Journal on Selected Areas in Communications 6(9), 1617–1622 (1988)
4. Yonezawa, A.: ABCL: An Object-Oriented Concurrent System. MIT Press, Cambridge (1990)
5. Matsuoka, S., Yonezawa, A.: Analysis of inheritance anomaly in object-oriented concurrent programming languages. In: Agha, G., Wegner, P., Yonezawa, A. (eds.) Research Directions in Concurrent Object-Oriented Programming, pp. 107–150. MIT Press, Cambridge (1993)
6. Hemige, V.: Object-Oriented design of the groupware layer for the ecosystem information system. University of Montana (1995)
7. Rose, A., Perez, M., Clements, P.: Modechart toolset user's guide[R]. Technical Report, NML/MRL/5540-94-7427, University of Texas at Austin, Austin (1994)
8. Keene, S.: A Programmer's Guide to Object-Oriented Programming in Common LISP. Addison-Wesley Longman Publishing Co., Inc., Boston (1988)
9. Guo, L., Tang, Z.: Specification and verification of the triple-modular redundancy fault-tolerant system. Journal of Software 14(1), 28–35 (2003)
10. Schutze, H.: Dimensions of meaning. In: Whitelock, P. (ed.) Proc. of the Supercomputing, Los Alamitos, vol. 796, pp. 787–796 (1992), `ftp://parcftp.parc.xerox.com/pub/qca/papers/`
11. Sangers, J.: A Linguistic Approach for Semantic Web Service Discovery, Bachelor Thesis, Economics and Informatics.Erasmus University Rotterdam (July 2009)

Part VI

Data Simulation and Information Integration

A Method for Improving the Consistency of Fuzzy Complementary Judgement Matrices Based on Satisfaction Consistency and Compatibility

Yong Zeng and Zhoujing Wang

Department of Automation, Xiamen University, Xiamen, 361005, China
zy_lol@163.com, wangzj@xmu.edu.cn

Abstract. Focusing on the problem fuzzy analytic hierarchy process (FAHP), this paper investigates the inconsistency problems of preference information about alternatives expressed as a fuzzy complementary judgement matrix by a decision maker. Firstly, the average inconsistency index is selected to check whether the matrix meets the requirement of the consistency. Then, transform equation with a variable parameter is proposed after a fully consider of compatibility, and, a linear programming model is constructed under the constrain of both satisfaction consistency and compatibility to figure out the most suitable value of the variable parameter. Finally, the original matrix is transformed to an acceptable one, and the new method is illustrated with two numerical examples.

Keywords: Fuzzy complementary judgement matrices, satisfaction Consistency, compatibility.

1 Introduction

Since the fuzzy analytic hierarchy process (FAHP) was proposed, it has been attracting many scholars' attention; they also have made many great achievements. The FAHP has been widely used in many different areas, such as energy system analysis, city planning, economic management, scientific research and evaluation, social science, etc. However, the problems we are facing are always so complexity that the decision makers can not always give a perfect fuzzy judgment matrix. Therefore, how to improve the consistency of the judgment matrix has become a very important part of FAHP theory research and application.

At present, many scholars have been focusing on improving the consistency of fuzzy complementary judgment matrix. They also have made many achievements. Xu[4] offered two different consistency standards to inspect the consistency of the judgment matrix. Reference [5], [6], [7] proposed different ways to improve the satisfaction consistency based on the consistency standards proposed in Xu[4]. But all methods developed above are only keep the sequence information of the judgment matrix, and didn't take the acceptability into consideration. Reference [1], [8], [9],

Y. Wang and T. Li (Eds.): Knowledge Engineering and Management, AISC 123, pp. 543–552.
springerlink.com

[10], [11] presented some different consistency index, and brought in a predetermined deviation threshold value. If the consistency index is not bigger than the very deviation threshold value, we can say the judgment matrix has satisfactory consistency. Although these methods are all rational, there is not a standard way to get the deviation threshold value. Ma[3] derived two methods from graph theory to judge whether a fuzzy preference has weak transitivity or not, and a method was presented to repair its inconsistency. This method met the requirements of transitivity, but didn't take the compatibility into account.

In this paper, we chose the average inconsistency index to check whether the original fuzzy complementary judgment matrix meets the requirement of satisfaction consistency. Then, we propose a novel transform equation with a variable parameter for decision analysts to improve the consistency of original matrix. Following, with a fully consideration about the satisfaction consistency and the compatibility between the before and after repairing judgment matrix, a linear programming model is built and transformed to find out the most suitable value of variable parameter. Finally, we get the acceptable fuzzy complementary judgment matrix.

This paper is organized as follows. Section 2 gives some definitions to the problem discussed in this paper, such as additive consistency, satisfaction consistency and compatibility. In Section 3, a method is developed for repairing the inconsistency of a fuzzy preference relation, and the theoretical proofs are also presented. In Section 4, two numerical examples and the comparison are used to illustrate the usefulness of the proposed method. Our conclusions are presented in Section 5.

2 Preliminaries

Definition 1. Let $R=(r_{ij})_{n\times n}$ be a judgement matrix, if $1\ge r_{ij}\ge 0$, $r_{ij}+r_{ji}=1$, $r_{ii}=0.5$, $i,j=1,2,\dots,n$, then R is called a fuzzy complementary judgement matrix.

Definition 2. Let $R=(r_{ij})_{n\times n}$ be a fuzzy complementary judgement matrix, if $r_{ij}=r_{ik}+r_{kj}-0.5$ is fulfilled for all $i,j,k=1,2,\dots,n$, then R has the additive consistency.

Definition 3. Let $R=(r_{ij})_{n\times n}$ and $R'=(r'_{ij})_{n\times n}$ be two judgement matrices, $CI(R,R')=\frac{1}{n^2}\delta(R,R')$ is called the compatibility index between the R and R', where $\delta(R,R')=\|R-R'\|=\sum_{i=1}^{n}\sum_{j=1}^{n}|r_{ij}-r'_{ij}|$.

Supposing the R is the original judgement matrix, and the R' is the judgement matrix after the consistency improvement, then, the more the $CI(R,R')$ closes to 0, the better the compatibility between R and R' is ; On the contrary, the worse the compatibility is.

Definition 4. Let $R=(r_{ij})_{n\times n}$ be a fuzzy complementary judgement matrix, $\rho_R=\frac{1}{n(n-1)(n-2)}\sum_{i=1}^{n}\sum_{\substack{j=1\\ j\neq i}}^{n}\sum_{\substack{k=1\\ k\neq i,j}}^{n}|r_{ij}-(r_{ik}+r_{kj}-\frac{1}{2})|$ is used to describe the average inconsistency of R , if $\rho_R\leq\varepsilon$, where ε is a predetermined constant , then $R=(r_{ij})_{n\times n}$ has the satisfaction consistency.

Property 1. The necessary and sufficient condition for the $R=(r_{ij})_{n\times n}$ to be a fuzzy complementary judgement matrix has additive consistency is $\rho_R=0$.

LV[10] proposed a rational way to preset ε, it consider the average inconsistency index ρ_R should be a function of R 's order n, and the capability of matrix changes along with the n. After a large amount of simulations and compute, LV[10] figured out the value of ε.

Table 1. Value of ε for different orders.

n	3	4	5	6	7	8	9
ε	0.2	0.2	0.18	0.18	0.18	0.18	0.18

3 A Method for Improving Consistency of the Fuzzy Complementary Judgement Matrix

As the complexity of practical problems or the limitation of the decision makers' experience and knowledge, the original judgment matrix given by decision makers can not always meet the consistency requirements. There are two basic way to improve the consistency of fuzzy complementary judgment matrix: (1) decision makers(DMs) keep on revising their preference information on alternatives till an acceptable level of consistency is obtained; and, (2) decision analysts perform the medications if the DMs are unwilling or unable to refine their preference. This paper proposed a principle of modification based on the satisfaction consistency and compatibility.

For simplicity, let $R=(r_{ij})_{n\times n}$ be the original fuzzy complementary judgement matrix and $A=(a_{ij})_{n\times n}$ be the acceptable judgement matrix after the consistency improvement for the rest of paper. When R doesn't meet the requirement of the satisfaction consistency, i.e. $\rho_R>\varepsilon$. We can obtain $R'=(r'_{ij})_{n\times n}$ by the equation (1) :

$$r_{ij}^{'}=(1-t)r_{ij}+t\frac{\sum_{k=1,k\neq i,j}^{n}(r_{ik}+r_{kj}-0.5)}{n-2}\ ,\quad i,j\in\{1,...,n\}\ . \tag{1}$$

Where $t\in[0,1]$, if $\forall i,j\in\{1,...,n\}$, there is $r_{ij}^{'}\geq 0$, then we have $A=R^{'}$. If there exists one $r_{ij}^{'}<0,\ \ \forall i,j\in\{1,...,n\}$, then $R^{''}=(r_{ij}^{''})_{n\times n}$ can be obtained by equation (2) :

$$r_{ij}^{''}=\frac{r_{ij}^{'}+a}{1+2a},\quad \forall i,j\in\{1,...,n\}\ . \tag{2}$$

Where $a=\max\{|\,r_{ts}^{'}\,|\,\big|\,r_{ts}^{'}<0,t,s\in\{1,...,n\}\}$, and we have $A=R^{''}$.

Theorem 1. If $A=R^{''}$, then $R^{'}$ and $R^{''}$ contain the same amount of judgement information.

According to the equation (2), $R^{''}$ is obtained by an equal proportion zoom from $R^{'}$, so the transform doesn't change the amount of judgement information at all. Further more, no matter $A=R^{'}$ or $A=R^{''}$, we can always use the compatibility index $CI(R,R^{'})$ to measure the compatibility between R and A

Theorem 2. No matter $A=R^{'}$ or $A=R^{''}$, $A=(a_{ij})_{n\times n}$ is also a fuzzy complementary judgement matrix.

Proof. *Case* 1. If $A=R^{'}$, i.e. $\forall i,j\in\{1,...,n\}$, There is $r_{ij}^{'}\geq 0$. Obviously, there will be $r_{ij}^{'}+r_{ji}^{'}=1$, $0\leq r_{ij}^{'}\leq 1$, $i,j\in\{1,...,n\}$, so $A=(a_{ij})_{n\times n}$ is a fuzzy complementary judgement matrix.

Case 2. If $A=R^{''}$, there exists at least one $r_{ij}^{'}<0,\ \ \forall i,j\in\{1,...,n\}$, As

$$r_{ij}^{''}=\frac{r_{ij}^{'}+a}{1+2a},\quad \forall i,j\in\{1,...,n\}\ . \tag{3}$$

Hence

$$r_{ij}^{''}+r_{ij}^{''}=\frac{r_{ij}^{'}+a}{1+2a}+\frac{r_{ji}^{'}+a}{1+2a}=\frac{r_{ij}^{'}+r_{ji}^{'}+2a}{1+2a}\ . \tag{4}$$

We have known $r_{ij}^{'}+r_{ji}^{'}=1$ in *Case* 1. So, it's easy for us to get $r_{ij}^{''}+r_{ji}^{''}=1$ and $0\leq r_{ij}^{''}\leq 1$. Likely, $A=(a_{ij})_{n\times n}$ is still a fuzzy complementary judgement matrix.

Theorem 3. Let ρ_R and ρ_A are the average inconsistency index for R and A respectively, then we will have:

$$\rho_A = \begin{cases} \left|1-\dfrac{nt}{n-2}\right|\rho_R & A = R' \\ \dfrac{1}{1+2a}\bullet\left|1-\dfrac{nt}{n-2}\right|\rho_R & A = R^{*} \end{cases}.$$

Proof. *Case* 1. When $A = R'$, according to the *Definition* 4, we can get :

$$\rho_A = \rho_{R'} = \frac{1}{n(n-1)(n-2)}\sum_{i=1}^{n}\sum_{\substack{j=1 \\ j\neq i}}^{n}\sum_{\substack{k=1 \\ k\neq i,j}}^{n}\left| r'_{ij} - (r'_{ik} + r'_{kj} - \frac{1}{2})\right| . \tag{5}$$

For simplicity, we use r_{ij}^{k} to denote $r_{kj} - r_{ki} + 0.5$ for the rest of paper, then we can get :

$$\begin{aligned} &\left| r'_{ij} - (r'_{ik} + r'_{kj} - \frac{1}{2})\right| \\ &= \left|(1-t)[r_{ij} - (r_{ik} + r_{kj})] + \frac{t}{n-2}\left(\sum_{s=1,s\neq i,j}^{n} r_{ij}^{s} - \sum_{s=1,s\neq i,k}^{n} r_{ik}^{s} - \sum_{s=1,s\neq k,j}^{n} r_{kj}^{s}\right) + \frac{1}{2}\right| \\ &= \left|(1-t)\square[r_{ij} - (r_{ik} + r_{kj} - \frac{1}{2})] + \frac{t}{n-2}[\sum_{s=1}^{n}(r_{ij}^{s} - r_{ik}^{s} - r_{kj}^{s}) - 2(r_{ij} - r_{ik} - r_{kj})] + \right. \\ &= \left|(1 - t - \frac{2t}{n-2})\square[r_{ij} - (r_{ik} + r_{kj} - \frac{1}{2})] + \frac{t}{n-2}(1 - \frac{n}{2}) + \frac{t}{2}\right| \\ &= \left|1 - \frac{nt}{n-2}\right| \bullet \left| r_{ij} - (r_{ik} + r_{kj} - \frac{1}{2})\right| . \end{aligned} \tag{6}$$

So

$$\begin{aligned} \rho_A &= \left|1 - \frac{nt}{n-2}\right| \bullet \frac{1}{n(n-1)(n-2)}\sum_{i=1}^{n}\sum_{\substack{j=1 \\ j\neq i}}^{n}\sum_{\substack{k=1 \\ k\neq i,j}}^{n}\left| r_{ij} - (r_{ik} + r_{kj} - \frac{1}{2})\right| \\ &= \left|1 - \frac{nt}{n-2}\right|\rho_R . \end{aligned} \tag{7}$$

Case 2. When $A = R''$, we can get :

$$
\begin{aligned}
|r_{ij}'' - (r_{ik}'' + r_{kj}'' - \frac{1}{2})| &= |\frac{a + r_{ij}'}{1+2a} - (\frac{a + r_{ik}'}{1+2a} + \frac{a + r_{kj}'}{1+2a}) + \frac{1}{2}| \\
&= \frac{1}{1+2a} | r_{ij}' - (r_{ik}' + r_{kj}' - \frac{1}{2}) - \frac{1}{2(1+2a)} - \frac{a}{1+2a} + \frac{1}{2}| \\
&= \frac{1}{1+2a} | r_{ij}' - (r_{ik}' + r_{kj}' - \frac{1}{2})|.
\end{aligned}
\tag{8}
$$

Apparently, we can finally obtain $\rho_A = \frac{1}{1+2a} \bullet |1 - \frac{nt}{n-2}| \rho_R$.

Deduction 1. When $t = \frac{n-2}{n}$, $A = (a_{ij})_{n\times n}$ must be the fuzzy complementary judgement matrix has additive consistency.

Obviously, when $t = \frac{n-2}{n}$, ρ_A is equal to 0 for both Case 1 and Case 2 in Theorem 3. According to the Property 1, we can safely draw the conclusion that $A = (a_{ij})_{n\times n}$ is a fuzzy complementary judgement matrix has additive consistency.

Based on the Definition 3 and Theorem 1, the more the t close to 0, the smaller the $CI(R, R')$ is and the more original judgement information kept. Meanwhile, the more the t approaches $\frac{n-2}{n}$, the more the average inconsistency index ρ_A close to 0, and the better level of satisfaction consistency. Above all, a linear programming model (M-3.1) is built to minimize the compatibility index $CI(R, R')$ under the constraint of satisfaction consistency.

$$
\begin{aligned}
&\min \quad CI(R,R') = \frac{1}{n^2}\sum_{i=1}^{n}\sum_{j=1}^{n} | r_{ij} - (1-t)r_{ij} - t\frac{\sum_{k=1,k\neq i,j}^{n}(r_{ik} + r_{kj} - 0.5)}{n-2} | \\
&s.t. \quad \begin{cases} |1 - \frac{nt}{n-2}| \rho_R \le \varepsilon \\ 0 \le t \le 1 \end{cases}
\end{aligned}
\tag{M-3.1}
$$

Apparently

$$
| r_{ij} - (1-t)r_{ij} - t\frac{\sum_{k=1,k\neq i,j}^{n}(r_{ik} + r_{kj} - 0.5)}{n-2} | = t\, | r_{ij} - \frac{\sum_{k=1,k\neq i,j}^{n}(r_{ik} + r_{kj} - 0.5)}{n-2} |.
\tag{9}
$$

And

$$|1-\frac{nt}{n-2}|\rho_R \le \varepsilon \Leftrightarrow \frac{n-2}{n}(1-\frac{\varepsilon}{\rho_R}) \le t \le \frac{n-2}{n}(1+\frac{\varepsilon}{\rho_R}). \tag{10}$$

Hence, the model M-3.1 can be re-written as :

$$\min \ t$$
$$s.t. \ \begin{cases} \frac{n-2}{n}(1-\frac{\varepsilon}{\rho_R}) \le t \le \frac{n-2}{n}(1+\frac{\varepsilon}{\rho_R}) \\ 0 \le t \le 1 \end{cases} . \qquad \text{(M-3.2)}$$

By solving the model, we can finally get $t = \frac{n-2}{n}(1-\frac{\varepsilon}{\rho_R})$.

Algorithm:

Step1. Calculate the average inconsistency index ρ_R for $R=(r_{ij})_{n\times n}$, if $\rho_R \le \varepsilon$ go to step4 ; otherwise, go to step2.

Step2. Calculate t according $t=\frac{n-2}{n}(1-\frac{\varepsilon}{\rho_R})$, and construct $R^{'}=(r_{ij}^{'})_{n\times n}$ by equation (1), if every element in $R^{'}$ is not smaller than 0, then $A=R^{'}$, and go to step4 ; otherwise, go to step3.

Step3. Let $a=\max\{|r_{ts}^{'}| \, |r_{ts}^{'}<0, t,s\in\{1,...,n\}\}$, and construct $R^{''}=(r_{ij}^{''})_{n\times n}$ by equation (2), then $A=R^{''}$ and go to Step4.

Step4. Stop.

4 Numerical Examples and Methods Comparison

4.1 Numerical Examples

Example 1. A fuzzy complementary judgement matrix is given like following:

$$R=\begin{pmatrix} 0.5000 & 0.1000 & 0.6000 & 0.7000 \\ 0.9000 & 0.5000 & 0.8000 & 0.4000 \\ 0.4000 & 0.2000 & 0.5000 & 0.9000 \\ 0.3000 & 0.6000 & 0.1000 & 0.5000 \end{pmatrix}.$$

As $n=4$, So ε=0.2 , and the average inconsistency index ρ_R=0.5>0.2 , it obvious that R does not meet the requirement of satisfaction consistency. Then we can get $t=0.3$, and we construct $R^{'}=(r_{ij}^{'})_{n\times n}$ by equation (1) :

$$R' = \begin{pmatrix} 0.5000 & 0.2350 & 0.5250 & 0.6400 \\ 0.7650 & 0.5000 & 0.7100 & 0.6250 \\ 0.4750 & 0.2900 & 0.5000 & 0.7350 \\ 0.3600 & 0.3750 & 0.2650 & 0.5000 \end{pmatrix}.$$

Apparently, every elements in R' is bigger than 0. So, we can finally get $A = R'$, $\rho_A = 0.2$ and $CI(R, R') \approx 0.0938$.

Example 2. Another fuzzy complementary judgement matrix is given like following:

$$R = \begin{pmatrix} 0.5000 & 1.0000 & 1.0000 & 1.0000 \\ 0.0000 & 0.5000 & 0.6000 & 0.4000 \\ 0.0000 & 0.4000 & 0.5000 & 1.0000 \\ 0.0000 & 0.6000 & 0.0000 & 0.5000 \end{pmatrix}.$$

Likely, there is ε=0.2 , and the average inconsistency index ρ_R=0.35>0.2 . So, R doesn't meet the requirement of satisfaction consistency too. Then $t = 0.214$, and we can get $R' = (r'_{ij})_{n\times n}$ by equation (1):

$$R' = \begin{pmatrix} 0.5000 & 1.0000 & 0.9571 & 1.0429 \\ 0.0000 & 0.5000 & 0.5143 & 0.4857 \\ 0.0429 & 0.4857 & 0.5000 & 0.8714 \\ -0.0429 & 0.5143 & 0.1286 & 0.5000 \end{pmatrix}.$$

Though $\rho_{R'} = 0.2$, $R' = (r'_{ij})_{n\times n}$ is not a fuzzy complementary judgement matrix because of the $r'_{41} < 0$. We can work out $CI(R, R') \approx 0.0482$ and $a = 0.0429$, then, we can get $R'' = (r''_{ij})_{n\times n}$ by equation (2):

$$R'' = \begin{pmatrix} 0.5000 & 0.9605 & 0.9210 & 1.0000 \\ 0.0395 & 0.5000 & 0.5132 & 0.4868 \\ 0.0790 & 0.4868 & 0.5000 & 0.8421 \\ 0.0000 & 0.5132 & 0.1579 & 0.5000 \end{pmatrix}.$$

Finally, we have $A = R''$ and $\rho_A = \dfrac{1}{1+2a}\rho_{R'} = 0.1842$.

4.2 Methods Comparison

In order to have a relative comprehensive evaluation about the method proposed in this paper, we mainly compare it with the method proposed in Ma[3]. Table 2 will show us difference result by using different methods for both *Example* 1 and

Example 2. It includes the final acceptable judgement matrix A , the sort weight ω , the average inconsistency index ρ_A , the compatibility index $CI(R,R')$ and the total transformed times d .

Table 2. Comparison result

	Example 1				
	A	ω	ρ_A	$CI(R,R')$	d
Ma[3]	$A=\begin{pmatrix} 0.5000 & 0.1675 & 0.5625 & 0.6700 \\ 0.8325 & 0.5000 & 0.7550 & 0.5125 \\ 0.4375 & 0.2450 & 0.5000 & 0.8175 \\ 0.3300 & 0.4875 & 0.1825 & 0.5000 \end{pmatrix}$	$\begin{bmatrix} 0.2417 \\ 0.3000 \\ 0.2500 \\ 0.2081 \end{bmatrix}$	0.35	0.0469	3
This text	$A=\begin{pmatrix} 0.5000 & 0.2350 & 0.5250 & 0.6400 \\ 0.7650 & 0.5000 & 0.7100 & 0.6250 \\ 0.4750 & 0.2900 & 0.5000 & 0.7350 \\ 0.3600 & 0.3750 & 0.2650 & 0.5000 \end{pmatrix}$	$\begin{bmatrix} 0.2417 \\ 0.3000 \\ 0.2500 \\ 0.2081 \end{bmatrix}$	0.2	0.0938	1
	Example 2				
	A	ω	ρ_A	$CI(R,R')$	d
Ma[3]	$A=\begin{pmatrix} 0.5000 & 0.9500 & 0.9000 & 1.0000 \\ 0.0500 & 0.5000 & 0.4900 & 0.5100 \\ 0.1000 & 0.5100 & 0.5000 & 0.8000 \\ 0.0000 & 0.4900 & 0.2000 & 0.5000 \end{pmatrix}$	$\begin{pmatrix} 0.3625 \\ 0.2125 \\ 0.2425 \\ 0.1825 \end{pmatrix}$	0.1400	0.0713	6
This text	$A=\begin{pmatrix} 0.5000 & 0.9605 & 0.9210 & 1.0000 \\ 0.0395 & 0.5000 & 0.5132 & 0.4868 \\ 0.0790 & 0.4868 & 0.5000 & 0.8421 \\ 0.0000 & 0.5132 & 0.1579 & 0.5000 \end{pmatrix}$	$\begin{pmatrix} 0.3651 \\ 0.2116 \\ 0.2423 \\ 0.1809 \end{pmatrix}$	0.1842	0.0482	2

Based on the Table 2, we can safely draw the following conclusions:(1)Both method figure out the same sort order for Example 1 and Example 2, they are $x_2 \succ x_3 \succ x_1 \succ x_4$ and $x_1 \succ x_3 \succ x_2 \succ x_4$ respectively; (2)When comes to the transform times, the method proposed in this paper has absolutely advantage than it is in Ma[3], it needs only two times at most; (3)The method developed in this paper takes both satisfaction consistency and compatibility into account, while the method proposed in Ma[3] does not always meet the requirements of the consistency and compatibility

5 Concluding Remarks

In this paper, by a fully considering of the satisfaction consistency and the compatibility, a novel and simply method is developed to improve the consistency of the fuzzy complementary judgment matrix. Finally, by comparing with the method proposed in the Ma[3], on one hand, the method developed in this paper enhance the rationality of the consistency improving process; on the other hand, it also enhances the efficiency of the process. The new method enriches the fuzzy complementary judgment matrix consistency repair methods.

References

1. Wu, X.-H., Lv, Y.-J., Yang, F.: The verifier and adjustment of consistency for fuzzy complementary judgement matrix. Fuzzy Systems and Mathematics 24(2), 105–111 (2010)
2. Xiao, S.-H., Fan, Z.-P., Wang, M.-G.: Study on consistency of fuzzy judgement matrix. Journal of Systems Engineering 16(2), 142–145 (2001)
3. Ma, J., Fan, Z.-P., Jiang, Y.-P., Mao, J.-Y., Ma, L.: A method for repairing the inconsistency of fuzzy preference relations. Fuzzy Sets and Systems 157, 20–33 (2006)
4. Zeshui, X.: A improved method for constructing judgement matrix with fuzzy consistency. Communication on Applied Mathematics and Computation 12, 62–67 (1996)
5. Jiang, Y.-P., Fan, Z.-P.: A method for improving the consistency of fuzzy judgement matrix. Fuzzy Systems and Mathematics 16(2), 74–78 (2002)
6. Jiang, Y.-P., Fan, Z.-P.: A new method for regulating the consistency of fuzzy judgement matrix. Mathematics in Practice and Theory 33(12), 82–87 (2003)
7. Xu, Z.-S.: Research on compatibility and consistency of fuzzy complementary judgement matrices. Journal of PLA University of Science and Technology 3(2), 94–96 (2002)
8. Song, G.-X., Yang, D.-L.: Methods for identifying and improving the consistency of fuzzy judgement matrix. Systems Engineering 21(1), 110–116 (2003)
9. Lv, Y.-J., Xu, G.-L., Qin, J.-Y.: A method for adjusting the consistency of fuzzy complementary judgement matrices and its convergence. Fuzzy Systems and Mathematics 21(3), 86–92 (2007)
10. Lv, Y.-J., Guo, X.-R., Shi, W.-L.: A New Way for Improving Consistency of the Fuzzy Complementary Judgemetn Matrices. Operations Research and Management Science 16(2), 54–58 (2007)
11. Yang, J., Qiu, W.-H.: Research on Consistency Test and Modification Approach of fuzzy Judgement Matrix. Journal of Systems & Management 19(1), 14–18 (2010)

Study on EKF-Based Optimal Phase-Adjusted Pseudorange Algorithm for Tightly Coupled GPS/SINS Integrated System

Yingjian Ding and Jiancheng Fang

School of Instrumentation Science and Optoelctronics Engineering, Beihang University
Beijing 100191, China
dingyingjian@163.com

Abstract. An approach has been proposed to enhance the performance of tightly coupled GPS/SINS integrated system, with applying phase-adjusted pseudorange algorithm (PAPA), which is dedicated to decrease pseudorange noise with the aid of high precise carrier phase optimally, based on the extended Kalman filter (EKF). The theory of optimal PAPA is described first, followed by the construction of tightly coupled GPS/SINS integration system as well as the Kalman filter. Relevant experiments including both static and dynamic ones have been accomplished to verify the validity of the approach proposed. The experiment results demonstrated prominent promotion with the EKF-based PAPA, compared with conventional tightly coupled integrated algorithm.

Keywords: GPS/SINS integration, tightly coupled, phase-adjusted pseudorange, EKF.

1 Introduction

Global Positioning System (GPS)/Strap-down Inertial Navigation System (SINS) Integrated System has been applied to various domains that need precise navigation. This integration can be classified into three modes according to extent of coupling: loosely, tightly and ultra-tightly [1]. Compared with loosely coupled ones, the main advantage of tightly coupled GPS/INS integrated system is that ultimate navigation information when number of GPS satellites falls below four can still be computed [2].

Although both pseudorange and carrier phase measurements of GPS can be utilized to integrate with SINS, pseudorange is preferred to phase for it is free of cycle-slips and integer ambiguities that are inevitable in phase measurements. However, since phase measurements are much less noisy than pseudorange, several methods are proposed to improve the accuracy of pseudorange with the extremely precise carrier phase measurement. The classical Hatch smoothing algorithm [3] and phase-connected pseudorange algorithm [4] were introduced to do this work in the case of no cycle-slips, and hence they are often used with cycle-slips detection approaches. The optimal solution for this question was proposed by Teunissen in 1991, which is called phase-adjusted pseudorange algorithm (PAPA), where both the pseudorange and phase measurements are included in observation equations, and is proved statistically optimal

Y. Wang and T. Li (Eds.): Knowledge Engineering and Management, AISC 123, pp. 553–558.
springerlink.com

[4]. This paper is dedicated to apply optimal PAPA method to promote the performance of tightly coupled GPS/INS integration system.

In non-linear filtering methods, the Extended Kalman Filter (EKF) [5] and the Unscented Kalman Filter (UKF) have also been used in a wide range of integrated navigations. In practice, EKF is still used widely due to its good real-time performance ([6], [7]).

In this paper, an EKF combining with optimal phase-adjusted pseudorange algorithm is proposed, in which GPS provides both pseudorange and phase measurements, as observation vector, to integrate with SINS. Data analysis results show a considerable improvement.

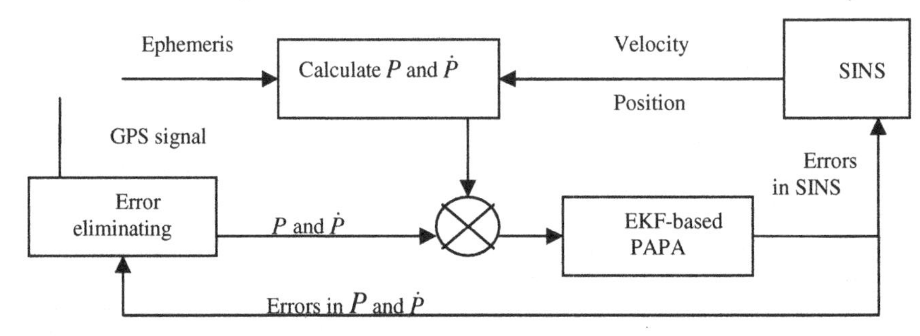

Fig. 1. Tightly coupled integration scheme

2 EKF-Based Phase-Adjusted Pseudorange Algorithm for Tightly Coupled GPS/SINS System

The Extended Kalman Filter (EKF) will be used to integrate GPS and SINS information, in which phase-adjust pseudorange algorithm (PAPA) serves as observation equations. The tightly coupled integration scheme is shown in figure 1.

2.1 The State Vector

Due to inertial navigation theory [7], the classical error model is used in this paper and the state vector contains navigation errors, inertial sensor errors, GPS errors and GPS carrier phase ambiguities.

$$X(t) = [\varphi_E, \varphi_N, \varphi_U, \delta V_E, \delta V_N, \delta V_U, \delta_L, \delta_\lambda, \delta_h, \varepsilon_{bx}, \varepsilon_{by}, \varepsilon_{bz}, \varepsilon_{rx}, \varepsilon_{ry}, \varepsilon_{rz}, \nabla_X, \nabla_Y, \nabla_Z, \delta t_U, \delta t_{ru}, \underline{N}]^T$$

Where $\varphi_E, \varphi_N, \varphi_U$ denotes the body angular error of pitch, roll and heading, $\delta V_E, \delta V_N, \delta V_U$ denotes the velocity error of eastern, northern and vertical direction in terrestrial frame, $\delta_L, \delta_\lambda, \delta_h$ is the error of latitude, longitude and altitude, $\varepsilon_{bx}, \varepsilon_{by}, \varepsilon_{bz}, \varepsilon_{rx}, \varepsilon_{ry}, \varepsilon_{rz}$ is the constant angle rate measurement error of gyroscope (gyro drift) and the 1 order gyro markov noise, $\nabla_X, \nabla_Y, \nabla_Z$ is the specific force measurement error of accelerometers (accelerometer bias), $\delta t_U, \delta t_{ru}$ are the error of pseudorange P

(receiver clock bias) and the error of pseudorange rate $\dot{P}$, and $\underline{N}$ is the carrier phase integer ambiguity vector .

2.2 The Measurement Equation

According to EKF-based PAPA method, the observation equation can be shown as:

$$Z_k = h(X_k) + V_k \tag{1}$$

in which:

$Z_k = [\delta P^1, \delta P^2 \cdots \delta P^m, \delta\phi^1, \delta\phi^2, \cdots, \delta\phi^m, \delta\dot{P}^1, \delta\dot{P}^2 \cdots, \delta\dot{P}^m]^T$, m is the total number of satellites in view, and $P^j, \phi^j, \dot{P}^j$ denotes the pseudorange ,the carrier phase measurement, and the pseudorange rate for the satellite j at the same epoch. And $\delta P^j, \delta\phi^j, \delta\dot{P}^j$ are the differences between the real value and the calculated value of the $P^j, \phi^j, \dot{P}^j$.

The non-linear relation between pseudorange and receiver's antenna position is as follow:

$$(\phi^j + N^j)\lambda_w = \sqrt{(x - x^j)^2 + (y - y^j)^2 + (z - z^j)^2} + \delta t_U + \varepsilon\phi \tag{2}$$

$$\dot{P}^j = e_1^j(\dot{x} - \dot{x}^j) + e_2^j(\dot{y} - \dot{y}^j) + e_3^j(\dot{z} - \dot{z}^j) + \delta t_{rU} + \varepsilon\dot{P} \tag{3}$$

$[e_1^j, e_2^j, e_3^j]$ is the direction cosine between satellite j and antenna in the WGS-84 frame.

The key point of EKF is the Jacobian matrix $\frac{\delta h(X_k)}{\delta X_k}$, and through simple but complicated calculation, the Jacobian matrix $\frac{\delta h(X_k)}{\delta X_k}$ is:

$$\begin{bmatrix} 0^{m\times 6} & H_{\rho 1}^{m\times 3} & 0^{m\times 9} & H_{\rho 2}^{m\times 2} & H_{N1}^{m\times m} \\ 0^{m\times 6} & H_{\phi}^{m\times 3} & 0^{m\times 9} & H_{\phi}^{m\times 2} & H_{N2}^{m\times m} \\ 0^{m\times 3} & H_{\dot{\rho} 1}^{m\times 3} & 0^{m\times 12} & H_{\dot{\rho} 2}^{m\times 2} & H_{N1}^{m\times m} \end{bmatrix}$$

Where,

$$H_{\rho 1}^{m\times 3} = \begin{bmatrix} a_{11} & a_{12} & a_{13} \\ & \vdots & \\ a_{m1} & a_{m2} & a_{m3} \end{bmatrix}, H_{\rho 2}^{m\times 2} = \begin{bmatrix} 1 & 0 \\ \vdots & \\ 1 & 0 \end{bmatrix}, \quad H_{N1}^{m\times m} = 0^{m\times m}$$

$$a_{i1} = (R_N + H)(-e_{i1}\sin L\cos\lambda - e_{i2}\sin L\sin\lambda) + [R_N(1-e)^2 + H]e_{i3}\cos L$$

$$a_{i2} = (R_N + H)(e_{i2}\cos L\cos\lambda - e_{i1}\cos L\sin\lambda)$$

$$a_{i3} = e_{i1}\cos L\cos\lambda + e_{i2}\cos L\sin\lambda + e_{i3}\sin L$$

$$H_{\phi}^{m\times 3} = \begin{bmatrix} b_{11} & b_{12} & b_{13} \\ & \vdots & \\ b_{m1} & b_{m2} & b_{m3} \end{bmatrix}, H_{N2}^{m\times m} = \begin{bmatrix} -1 & & \\ & \ddots & \\ & & -1 \end{bmatrix}, \quad H_{\phi}^{m\times 2} = \frac{1}{\lambda_w} H_{\rho 2}^{m\times 2}$$

Where,

$$H_{\dot{\rho}1}^{m\times 3} = \begin{bmatrix} c_{11} & c_{12} & c_{13} \\ & \vdots & \\ c_{m1} & c_{m2} & c_{m3} \end{bmatrix}, H_{\dot{\rho}2}^{m\times 2} = \begin{bmatrix} 0 & 1 \\ \vdots & \vdots \\ 0 & 1 \end{bmatrix}, b_{ij} = \frac{a_{ij}}{\lambda_w}, i = 1\ldots m, j = 1,2,3$$

$$c_{i1} = -e_{i1}\sin\lambda + e_{i2}\cos\lambda$$
$$c_{i2} = -e_{i1}\sin L\cos\lambda - e_{i2}\cos L\sin\lambda + e_{i3}\cos L$$
$$c_{i3} = e_{i1}\cos L\cos\lambda + e_{i2}\cos L\sin\lambda + e_{i3}\sin L$$

2.3 Kalman Filter Equation

The EKF algorithm is given as:

1) Time update

$$\Delta\hat{X}_{k+1,k} = \Phi\bullet\Delta\hat{X}_k \tag{4}$$

$$X_{k+1}^* = \Phi X_k^* \tag{5}$$

$$P_{k+1,k} = \Phi P_k \Phi^T + \Gamma Q_k \Gamma^T \tag{6}$$

in which the matrix Φ and Γ is given in section D, and $P_{k/k-1}$ is the *a priori* error covariance matrix.

2) Measurement update

$$K_{k+1} = P_{k+1,k}H_{k+1}^T(H_{k+1}P_{k+1}H_{k+1}^T + R_{k+1})^{-1} \tag{7}$$

$$\Delta\hat{X}_{k+1} = \Delta\hat{X}_{k+1,k} + K_{k+1,k}(Z_{k+1} - H_{k+1}\Delta\hat{X}_{k+1,k}) \tag{8}$$

$$P_{k+1} = (I - K_{k+1}H_{k+1})P_{k+1,k}(I - K_{k+1}H_{k+1})^T + K_{k+1}R_{k+1}K_{k+1}^T \tag{9}$$

$$\hat{X}_{k+1} = X_{k+1}^* + \Delta\hat{X}_{k+1} \tag{10}$$

Where K_k is the filter gain matrix at step *k*.

3 Experiment Verification and Result Analysis

To evaluate the performance of EKF-based PAPA for tightly coupled GPS/INS integrated system, both static and kinematic experiments have been conducted, in which real flight data is also post-processed using EKF-based PAPA.

A Static Experiment

Static test was performed in campus of Beihang University in Beijing, with IMU developed by Beihang University and GPS receiver of Novatel. The hardware configuration is specified in table 1.

Table 1. Hardware configuration of tightly coupled integrated system

	Specification
IMU	DTG-gyroscope: Random constant $0.1^{\circ}/h$, White noise $0.1^{\circ}/h$ Quartzose accelerometer: Constant Bias $100\mu g$, Random noise $50\mu g$,Frequency:100Hz
GPS	Novatel DL-V3 , Frequency:20Hz

The GPS/SINS integration results for static experiment are shown in figure 2. The static positioning standard deviance (STD) of latitude, longitude, and height for single pseudorange and PAPA are [0.49, 0.28, 0.86] (m) and [0.39, 0.22, 0.53] (m) in testing time of 350s respectively.

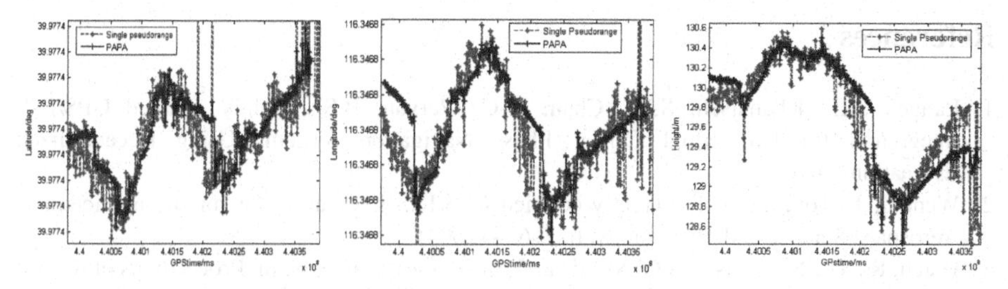

Fig. 2. Static experiments positioning results of GPS/SINS integration with PAPA compared with single pseudorange processing

B Kinematic Experiments

The kinematic experiments are performed on the 5th ring road of Beijing, also using the same system. The testing time is 1h, and the speed of the vehicle is limited in 50km/h.

In kinematic experiments, base and rover stations are both set to calculate the phase-differential GPS results which are highly precise and can be taken as reference. So the results described in positioning error are shown in figure 3. The RMS errors of latitude, longitude and height for single pseudorange integration and PAPA integration are [2.41, 4.32, 8.63] (m), and [2.21, 3.96, 7.69] (m). Meanwhile the STD of them are respectively [1.68, 1.36, 2.49] (m), and [0.82, 0.67, 1.02] (m).

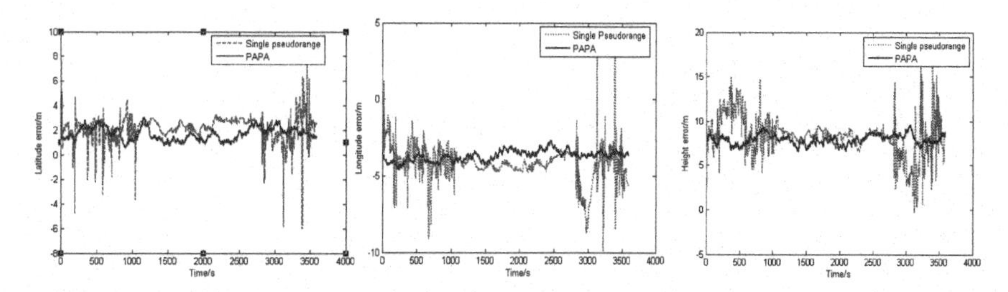

Fig. 3. Kinematic experiments results of GPS/SINS integration with PAPA compared with single pseudorange processing, in which the error calculating is taking the differential GPS (DPGS)/SINS integration results reference

4 Conclusion

The paper presents an approach to improve the performance of tightly coupled GPS/SINS integrated system. Through EKF-based PAPA, using high precise carrier phase to assist pseudorange in order to improve the positioning accuracy, the overall accuracy of tightly coupled integrated system has seen a prominent promotion. Static and dynamic experimental results demonstrate that the RMS which contains the constant bias does not reduce much, while STD which merely contains the random errors decreases evidently.

Acknowledgment. This work has been supported by National Basic Research Program of China (973) under the grant No.2009CB724002.

References

1. Langel, S.E., Khanafseh, S.M., Chan, F.-C., Pervan, B.S.: Tightly coupled GPS/INS Integration for Differential Carrier Phase Navigation Systems Using Decentralized Estimation (2010)
2. Wendal, J., Tommer, G.F.: Tightly coupled GPS/INS integration for missile applications. Aerospace Science and Technology, 627–634 (2004)
3. Hatch, R.: The Synergism of GPS code and carrier measurements. In: Proceedings of the 3rd International Geodetic Symposium on Satellite Doppler Positioning, Las Cruces - New Mexico, vol. 2, pp. 1213–1231 (1982)
4. Le, A.Q., Teunissen, P.J.G.: Recursive least-squares filtering of pseudorange measurements. In: European Navigation Conference, Manchester, U.K, May 7-10 (2006)
5. Brown, R.G., Hwang, P.Y.C.: Introduction to Random Signals and Applied Kalman Filtering. Wiley, New York (1997)
6. McGraw, G.A., Murphy, T., Brenner, M., Pullen, S., Van Dierendonck, A.J.: Development of the LAAS Accuracy Models. In: Proceedings of the Institute of Navigation's ION GPS 2000, Salt Lake City, UT, September 19-22 (2000)
7. Stakkeland, M., Prytz, G., Booij, W.E., Pedersen, S.T.: Characterization of accelerometers using nonlinear kalman filters and position feedback. IEEE Transactions on Instrumentation and Measurement 56(6), 2698–2704 (2007)

Study of Spatial Data Index Structure Based on Hybrid Tree

Yonghui Wang[1,2,3], Yunlong Zhu[1], and Huanliang Sun[3]

[1] Shenyang Inst . of Automation, Chinese Academy of Sciences, Shenyang 110016, China
[2] Graduate Sch. of Chinese Academy of Sciences, Beijing 100039, China
[3] School of Information and Control Engineering, Shenyang Jianzhu University, Shenyang 110168, China
yonghuiwang@188.com

Abstract. In order to improve the efficiency of spatial data access and retrieval performance, an index structure is designed, it solves the problem of low query efficiency of the single index structure when there are large amount of data. Through the establishment of correspondence between the logical records and physical records of the spatial data, the hybrid spatial data index structure is designed based on 2^K–tree and R-tree. The insertion, deletion and query algorithm are implemented based on the hybrid tree, and the accuracy and efficiency are verified. The experimental results show that the hybrid tree needs more storage space then R-tree, but with the data volume increasing the storage space needed declining relatively, and the hybrid tree is better than the R-tree in the retrieval efficiency, and with the data volume increasing the advantage is more obvious.

Keywords: Spatial indexing, Index structure, 2^K–tree, R-tree, hybrid tree.

1 Introduction

The spatial data index technology has been a key technology in spatial database systems, which directly influences the efficiency of accessing spatial data and the retrieval performance of the spatial index. A spatial data index describes spatial data stored in the media, establishing the logic records of spatial data and facilitating the correspondence among physical records, in order to improve the spatial efficiency of both data access and retrieval. The basic method of spatial data indexing is to divide the entire space into different search-areas, searching the spatial entities in these areas by a certain order. The usual spatial data indexing methods can be divided into two categories: one is a single index structure, such as the B tree , K-D tree , K-D-B trees, quad tree [1], R-tree[2-3] and its variant trees, the grid index, etc., in the case of large amounts of data, the retrieval efficiency of such indexes technology are relatively low; The other is to make use of hybrid index structure, such as the QR-tree [4], QR*-tree and PMR tree etc., which adopts the strategy of four equal divisions of spatial data or super nodes. Therefore, when data distribution is non-uniform, production of the spatial index quad-tree will form serious imbalances, in other words, the height of the tree will increase greatly, which will seriously affect the speed of the query. This paper aims to resolve this problem, integrating 2^K -tree and R-tree, in accordance with

Y. Wang and T. Li (Eds.): Knowledge Engineering and Management, AISC 123, pp. 559–565.
springerlink.com

the conditions of data distribution; meanwhile, the goal is to make the overlap among intermediate nodes is as small as possible, which establishes a kind of spatial data indexing structure based on a hybrid tree. In this structure, the segmentation of data space is based upon the conditions of the spatial data contained within; the height of the tree is log4n, which will serve to increase the speed of queries.

2 Quad-Tree and R*-Tree

2.1 Quad-Tree

A quad-tree [1] is a kind of important hierarchical data structure, mainly used to express spatial hierarchy in two-dimensional coordinates. In fact, a 1-dimensional binary tree extended in 2-dimensional space which essentially is a 2^K-tree (Unless otherwise indicated, all 2^K-tree are quad-trees) in which K is the dimension of spatial data. A quad tree with a root, in which each non-leaf node has four children (NE, NW, SW, SE), each node of quad tree corresponds to a rectangular area. Fig.1 shows an example of a quad-tree (a) and the corresponding spatial division (b).

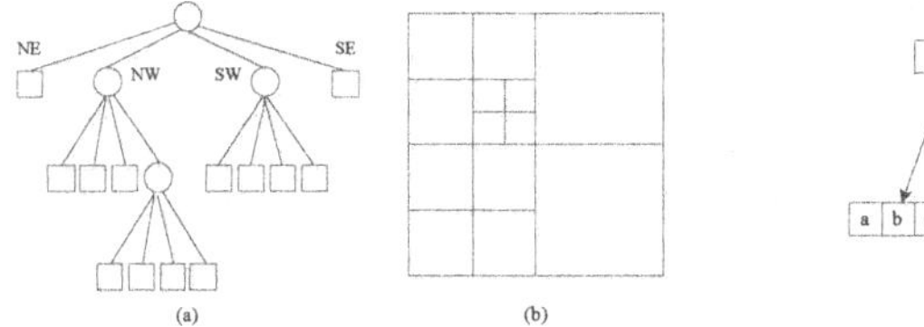

Fig. 1. Quad-tree (a) and its place division (b)

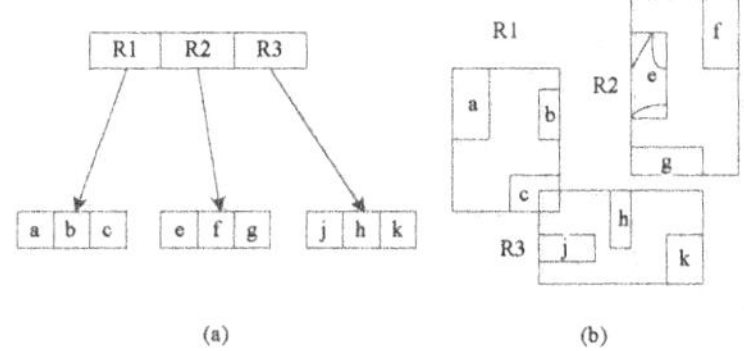

Fig. 2. R*- tree (a) and its Space division (b)

A quad-tree (2^K-tree) is a distinct data structure in which hierarchical data structure can be set up easily. Not only does it the ability to apply denotation to the spatial targets; but, because of its ability to aggregate spatial objects, it can improve the performance of spatial searches. The index structure of quad-trees (2^K-tree) have problems: The first occurs when there are large amounts of index data – if levels of the quad tree are too small it will lead to a decrease in the search properties. The second occurs if levels of the quad tree are so large that it will lead to the increase in duplicate storage, thereby increasing the consumption of space, which will affect the search performance.

2.2 R*-Tree

In 1990, Beckmann proposed the R*-tree [5-6] by optimizing the R-tree. R*- tree has the same data structure with the R-tree, it is a highly balanced tree, which is made up of the intermediate nodes and leaf nodes, the minimum bounding rectangle of the actual object is stored in the leaf nodes and the intermediate node is made up by gathering the index range of lower nodes (external rectangular). The number of child nodes hold by each node in R*- tree have the (M) under (m) limit, the (m) limit ensures the index to use the disk space effectively, the number of child nodes will be deleted if they are less than the (m) limit and the number of child nodes of nodes will be adjusted if they are greater than the (M) limit, if necessary, the original node will be divided into two new nodes. Fig.2 shows the R*-tree spatial division (a) and schematic (b).

When the amount of target data increase, the depth of the R*-tree and the overlap of spatial index will increase, and searching is always starting from the root, finding the number of branches visited, as a result with the number of nodes increase correspondingly and the finding performance will decline. And because of there is overlap among the regions, the spatial index may get the final result after searching several paths, which is inefficient in the case of large amount of data.

3 Spatial Data Based on Hybrid Tree Index Structure

Firstly, the quad dividing is carried out in the index space, according to the specific situation to determine the number of layers of quad division; then determining where the each objects included in index space is, constructing the R-tree corresponding to the index space for the objects included in index space [7-8]. A space object belongs to index space can fully contain the smallest space of index space. Therefore, the different ranges of space objects will be divided into different layers of the index space after a division, which is divided into different R-trees, these intermediate nodes within the R-tree overlap will be reduced as much as possible, which effectively reduces the multi-redundancy operation of R-tree structure to improve retrieval performance in the case of massive data[9-10].

3.1 Hybrid Tree Structure

A hybrid tree with depth of d from the macro view is a d-layer quad-tree, all the children space of the same floor where each of two disjoints and all of them together form the whole index space S0, the index space between father and son nodes is the relationship of inclusion, four children nodes corresponding to the index space is also each of two disjoints and they all together constitute index space of father nodes. Hybrid tree from the micro view totally contain the R-tree with the number ($n = \sum_{i=0}^{d-1} 4^i$), in which each node of quad-tree has an R-tree corresponded, the R-tree is compliant with the classic R-tree structure completely [2-3].

When constructing hybrid tree, which the R-tree should be inserted into by the space object, follow the rules: For the space target r, supposed to locate in all levels of sub-index space divided by "Quad-tree", the smallest index space completely surrounds MBRr is Si, so the r should be inserted into the R_i of R-tree corresponded to Si. Fig.3 shows an example of spatial distribution to illustrate hybrid tree structure of depth 2: the smallest index space surrounding r1, r5, r9 r10 is S1 (Note: Although the S0 can also completely surround r1, it is not the "minimum"), they are inserted into R-tree R1 corresponding to S1, in the same way, r3, r4, r8, corresponding to the index space are: S2, S4 and S3, respectively. Therefore they are inserted into the R2, R4 and R3, which can completely contain the minimum index space of r2, r7, r11, r6 is R0, as a result they correspond to the index space is R0, constructing a hybrid tree structure shown in Fig.4.

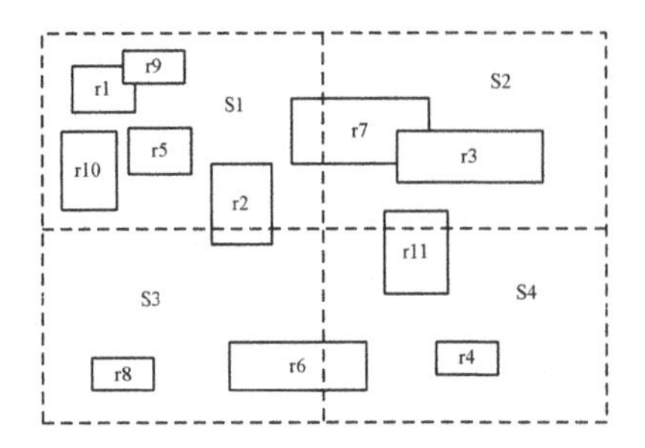

Fig. 3. The Space division of hybrid tree

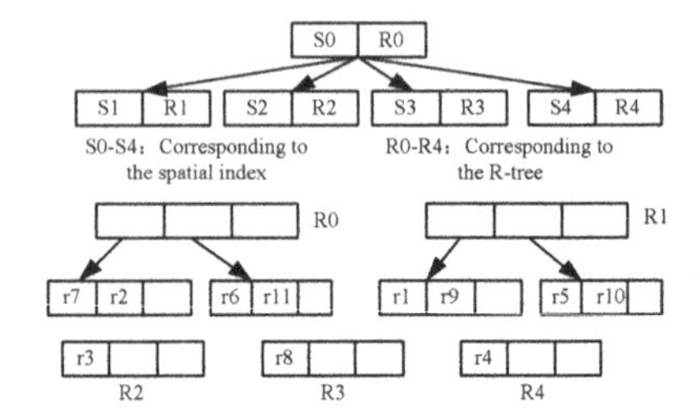

Fig. 4. The structure of hybrid tree

3.2 Hybrid Tree Algorithm Description

The insertion, deletion and search algorithms of Hybrid tree are to find the spatial goal at first, then call the corresponding R-tree algorithm operation. Supposed that the *Root* is the root node of hybrid tree, *MBRobj* is the minimum bound rectangle of the inserted object *obj*, *CS* as the given query space.

Algorithm 1: Hybrid Tree Insertion Algorithm
INSERT *(Root, MBRobj)*
Input: Root is the root of the hybrid tree, *MBRobj* is the minimum bound rectangle of the object to be inserted *obj*;
Output:
BEGIN
STEP 1: If the Root is leaf node, go to STEP 4;
STEP 2: Traverse the child nodes of Root, if the index range of all child nodes cannot surround *MBRobj* completely; Go to STEP 4;
STEP 3: A child node can fully contain *MBRobj* as the new *Root*, go to STEP 1;
STEP 4: Get the R-tree corresponding Root;
STEP 5: Call the R-tree insertion algorithm to the insert *MBRobj* into the hybrid tree.
END

Algorithm 2: Hybrid tree Deletion Algorithm
DELETE (*Root, Obj*)
BEGIN
STEP 1: If the Root is leaf node, go to STEP 4;
STEP 2: Traverse the child nodes of Root, if the index range of all child nodes cannot surround *MBRobj* completely; Switch to STEP 4;
STEP 3: A child node can fully contain *MBRobj* as the new Root, go to STEP 1;
STEP 4: Root access to the corresponding R-tree;
STEP 5: Call the R-tree deletion algorithm, the data will be removed from the hybrid data.
END

Algorithm 3: Hybrid Tree Search Algorithm
SEARCH (*Root, CS*)
BEGIN
STEP 1: If the Root index space corresponding to no overlap with the *CS*, go to STEP 4;
STEP 2: If the Root is leaf node, and receive the R-tree corresponding Root, call the search algorithm of R-tree to find it;
STEP 3: Traverse all the child nodes of Root, call SEARCH (*Root-> Children, CS*) followed by the recursive;
STEP 4: End query.
END

4 Experiment Analyses

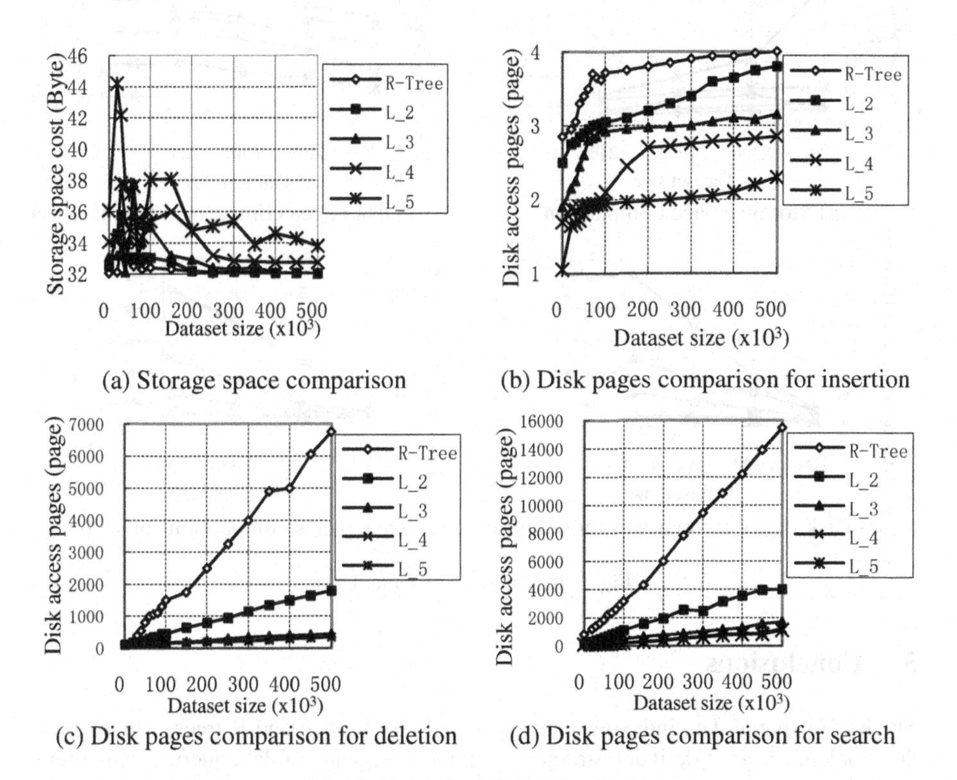

(a) Storage space comparison (b) Disk pages comparison for insertion

(c) Disk pages comparison for deletion (d) Disk pages comparison for search

Fig. 5. Result for 2D Random Data

In order to verify the correctness and efficiency of mixed trees, using Visual C + + 6.0 as a development tool to achieve the experiment process of R-tree and mixed tree, in the Windows Server 2003 Chinese operating system environment with a 2.4GHz Pentium Ⅳ CPU, 1GB RAM, using both random and real data to test space overhead and the inserting, deleting, and finding the visited disk number of pages of the mixed tree with different layers. The results are shown in Fig.5 & 6, which L_i ($2 \le i \le 5$) on behalf of hybrid tree with depth (layers) of *i*. The amount of data in the figure refers to the number of spatial objects. From the figure we can see the results: the consumed space of the hybrid tree is generally larger than that of the R-tree. The amount of data is smaller in the hybrid tree structure under the lower space utilization, which will result in a waste of storage space; With the increasing of the amount of hybrid data, the space utilization of the mixed tree structure increases gradually, in the case of a large amount of data, mixed tree has a little space utilization, and is equal to R-tree space utilization. Because of the efficiency in the search algorithm, the hybrid tree index structure is always better than the R-tree, and the larger the amount of data and the deeper the layers of hybrid tree, the more obvious the advantages become.

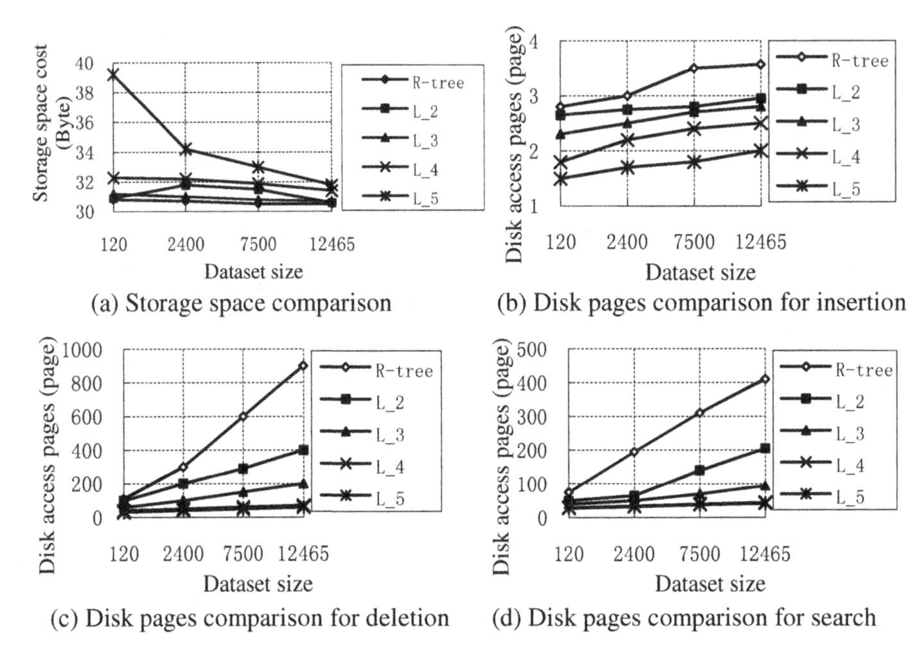

Fig. 6. Result for 2D Factual Data

5 Conclusions

The hybrid spatial data indexing structure based on 2K -tree and R-trees, which solves the problems of a single index structure with large amounts of data queried with a relatively low efficiency. Deletion and search algorithms have been designed based on the analysis of the key technologies of spatial database systems. The experiments show that the accuracy and efficiency of the structure and algorithm of a hybrid tree index structure can carry out a quick search of spatial data and achieve satisfactory results.

Acknowledgments. The work is supported by National Natural Science Foundation of China (61070024), The National Key Technology R&D Program (2008BAJ08B08-05) and Liaoning Provincial Natural Science Foundation of China under grant (20092057).

References

1. Samet, H.: The Quadtree and Related Hierarchical Data Structures. ACM Comp. Surveys, 47–57 (1984)
2. Guttman, A.: R-Tree: A Dynamic Index Structure for Spatial Searching. In: Proc ACM SIGMOD (June 1984)
3. Brakatsoulas, S., Pfoser, D., Theodoridis, Y.: Revisiting R-tree construction principles. In: Manolopoulos, Y., Návrat, P. (eds.) ADBIS 2002. LNCS, vol. 2435, p. 149. Springer, Heidelberg (2002)
4. Li, G., Li, L.: A Hybrid Structure of Spatial Index Based on Multi-Grid and QR-Tree. In: Proceedings of the Third International Symposium on Computer Science and Computational Technology, pp. 447–450 (August 2010)
5. Beckmann, N., Kriegel, H.P., Schnieider, R., et al.: The R*-tree: An Efficient and Robust Access Method for Points and Rectangles. In: Proc ACM SIGMOD, Atlantic City, USA, pp. 300–350 (1990)
6. Seeger, B.: A revised r*-tree in comparison with related index structures. In: Proceedings of the 35th SIGMOD International Conference on Management of Data, pp. 799–812. ACM (2009)
7. Gao, C., Jensen, C.S.: Efficient retrieval of the top-k most relevant spatial web objects. Proceedings of the VLDB Endowment 2(1), 337–348 (2009)
8. Luaces, M.R., Paramá, J.R., Pedreira, O., Seco, D.: An ontology-based index to retrieve documents with geographic information. In: Ludäscher, B., Mamoulis, N. (eds.) SSDBM 2008. LNCS, vol. 5069, pp. 384–400. Springer, Heidelberg (2008)
9. Luaces, M.R., Places, Á.S., Rodríguez, F.J., Seco, D.: Retrieving documents with geographic references using a spatial index structure based on ontologies. In: Song, I.-Y., Piattini, M., Chen, Y.-P.P., Hartmann, S., Grandi, F., Trujillo, J., Opdahl, A.L., Ferri, F., Grifoni, P., Caschera, M.C., Rolland, C., Woo, C., Salinesi, C., Zimányi, E., Claramunt, C., Frasincar, F., Houben, G.-J., Thiran, P. (eds.) ER Workshops 2008. LNCS, vol. 5232, pp. 395–404. Springer, Heidelberg (2008)
10. Shen, H.T., Zhou, X.: An adaptive and dynamic dimensionality reduction method for high-dimensional indexing. The International Journal on Very Large Data Bases 16(2), 219–234 (2007)

Research on an Improved GuTao Method for Building Domain Ontologies

Guoyan Xu and Tianshu Yu

College of Computer and Information, HoHai University, NanJing, China
gy_xu@126.com, yutianshu02@yahoo.com.cn

Abstract. Building domain ontologies is an important topic in semantic Web researches. The GuTao method that is a kind of the formal concept analysis is used in this paper for the problem. Because the GuTao method can only process single-value attributes, the interest rate is introduced as a basis of the conversion from multi-value attributes to single-value ones. An improved GuTao method is proposed to build domain ontologies. In this method domain ontologies is represented as derivative concept lattices in a formal background to make them complete and unique. Its building process is based on rigorous mathematical foundation.

Keywords: Domain Ontology, GuTao Method, Interest Rate, Concept Lattice.

1 Introduction

Ontology [1] is an explicit and formal description for a shared concept model, and a way to represent a concept model of information systems on semantic and knowledge levels. Ontologies have a good prospect in the semantic Web [2] research.

The formal concept analysis, as known as FCA, is an effective tool to support data analysis. In the FCA, its core data structure is concept lattices [3] that are formed by the concept hierarchy structure, and the concept is formed by the connotation and extension. The concept lattices are complete and unique with its building process based on rigorous mathematical theories. The formal concepts and ontologies have the same algebraic structure, so the domain ontologies can be represented as a form of derivative concept lattices in the formal background. So it is a feasible and efficient method to build domain ontologies with the FCA.

2 Building Domain Ontologies with the GuTao Method

The GuTao method [4] is proposed by GuTao based on the FCA for building domain ontologies. Compared with other ones, its self-developed plug-in FcaTab automatically gets formal backgrounds from domain concepts and relations, thus enabling semi-automatic domain ontology construction. It can eliminate the redundancy of concepts in the classification structures, and obtain required concepts. It also has

Y. Wang and T. Li (Eds.): Knowledge Engineering and Management, AISC 123, pp. 567–574.
springerlink.com

advantages in terms of the mechanisms of improvement and feedback loop of the formal background and concept lattice. Its building process flow is shown in Fig. 1.

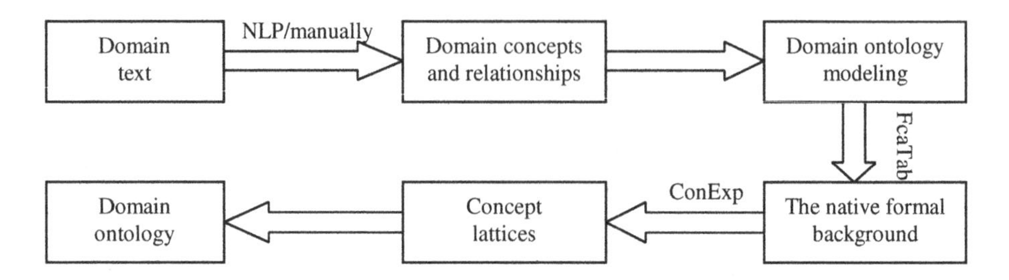

Fig. 1. Building Domain Ontologies Process Flow of the GuTao Method

Its building process are mainly 4 steps from Fig. 1: ①Get concepts, attributes and relationships between concepts from the domain text with the NLP methods or manually ; ②Build models with modeling tools. Use classes (domain concepts), slots (attributes of concepts), and facets (constraints on properties) to represent the domain ontologies; ③Get the formal background by the plug-in FcaTab and convert the formal background to the input format required by the ConExp; ④Build concept lattices by the ConExp.

3 An Improved GuTao Method

The GuTao method can build domain ontologies semi-automatically, generate the formal background and eliminate the redundancy between concepts automatically. But it cannot handle multi-value attributes. In this paper, the interest rate is introduced as the weight and as a basis of the conversion from multi-value attributes to single-value ones to get a composite interest rate formula and provide a method handling multi-value attributes.

3.1 Composite Interest Rate Formula

For the defect that the GuTao method cannot handle multi-value attributes, an interest rate is introduced in this paper as the weight and as a basis of the conversion from multi-value attributes to single-value ones, and based on a basic formula interest rate, a composite formula is derived to meet the needs of computing the weight.

(1) A basic interest rate formula

The basic formula of interest rate mainly includes the following three:

- $I(Ci)$: represents the interest rate on concept class Ci, as shown in Formula 1.

$$I(C_i) = \log \frac{time}{\log \log length} \cdot r(\mathrm{d}.\ C_i) \tag{1}$$

time represents the time of a user browsing a specific web page. *length* refers to the size of the page (bytes etc.). r(d, *Ci*) represents the matching degree between pages and concept classes, and it is a result of page characterization process.

- *Frequency*(*Ci*): reflects a user's interest on a concept, as shown in Formula 2.

$$Frequency(Ci)=Q(Ci)/Q \tag{2}$$

Frequency(*Ci*) reflects a user's interest rate on a concept Ci in the ontologies. *Q*(*Ci*) is the number of querying on the concept *Ci*, and *Q* represents the user's total number of querying.

- *Clarity*(*Ci*): reflects the accuracy of the user's interest, as shown in Formula 3.

$$Clarity(C_i) = (numAttr(C_i)+1).\frac{1}{numSubConcepts(C_i)+1} \tag{3}$$

numAttr(*Ci*) is the number of attributes of the concept *Ci*, and *numSubConcepts*(*Ci*) is the number of its sub-concepts.

(2) Composite Interest rate Formula

Formula 1 is hard to get and process in actual applications. So Formulas 2 and 3 are used in this paper with the idea of probability multiplication formula to propose a composite interest rate formula, as shown in Formula 4.

$$Irate(Ci) = Frequency(Ci) \times Clarity(Ci) \tag{4}$$

The querying frequency of the user is defined as the ratio between the number of URL querying of a certain web page and the total number of URL querying, as shown in Formula 5.

$$Frequency(C_i) = \sum_{i=1}^{n} Frequency(URL_i) = \sum_{i=1}^{n} \frac{Q(URL_i)}{Q(URL)} \tag{5}$$

3.2 An Improved GuTao Method

Based on the composite interest rate formula, an improved GuTao method is introduced in this paper for building domain ontologies. The method mainly has two following steps.

(1) Get Single-Value Formal Background

The one step classifies, extracts and gets object sets *URLS* and property sets *ATTRS* via user information. The object and attribute information are put into Formula 4 for the conversion, determining the Boolean relation *R*. It gets an initial formal background *Contology*, based on the object sets *O*, attribute sets *A* and the Boolean relation *R*. The algorithm in detail is shown in Fig. 2.

Algorithm 1 Get a single-value formal background

Input: user information collected by web spiders

Output: initial single-value formal background *Contology*=(*O*, *A*, *R*)

Main stpes

(1) for each $Ci \in CS(C)$ / /compute the clarification of concepts

(2) *Clarity*(*Ci*)=0

(3) *numAttr*(*Ci*)= *Count*(*A*) //number of *Ci* attributes

(4) *numSubConcepts*(*Ci*)=*Sum*(*SubConcepts*(*A*)) //number of *Ci* sub-conceptes

(5) *Clarity*(*Ci*)=(*numAttr*(*Ci*)+1) / (*numSubConcepts*(*Ci*)+1)

(6) *Frequency*(*Ci*)=0 //compute visiting frequency of concepts

(7) for each $URLi \in URLS$

(8) $$Frequency(C_i) = \sum_{i=1}^{n} Frequency(URL_i) = \sum_{i=1}^{n} \frac{Q(URL_i)}{Q(URL)}$$

(9) *Irate*(*Ci*) = *Clarity*(*Ci*)**Frequency*(*Ci*) //get the interest rate of single-value concepts

(10) *Sort*(*Irate*(*Ci*)) //order the results of interest rates

(11) If *Irate*(*Ci*) in *top*(5)

(12) Then $(O,A) \in R$ //get the Boolean relationship *R*

(13) *Contology*=(*O*,*A*,*R*) //get single-value formal background

Fig. 2. Algorithm of Getting Single-Value Formal Background

(2) Get Domain Ontology

The other step is used to get domain ontology. The algorithm in detail is shown in Fig. 3.

Algorithm 2 Get domain ontology

Input: initial single-value formal background *Contology*=(*O*, *A*, *R*)

Output: *Ontology*=(*CS*(*C*)), *Ships*)

Main steps:

(14) Build domain ontology with *URLS* objects and *ATTRS* as attributes. *URLS* is expressed as Classes and *ATTRS* as Slots in the ontology.

(15) Use the GuTao plug-in to improve and expand the formal background semi-automatically, and convert to the input format required by the concept lattice tool ConExp. Context, Object and Attribute in the FCA are corresponded to *Ontology*, Class and Slot respectively in the ontology

(16) Build concept lattice *Ontology*=(*CS*(*C*), *Ships*) with the ConExp, and store the information the domain ontology with the OWL.

Fig. 3. Algorithm of Getting Domain Ontology

3.3 Advantages of the Improved GuTao Method

The improved GuTao method have many advantages than the GuTao method. The differences between the GuTao method and the improved one in detail are shown in Table 1.

Table 1. Comparison between the GuTao method and the improved one

Item \ Name		GuTao Method	Improved GuTao Method
Automation		Semi	Semi
Formal Background	Object	Class	URL
	Attribute	Slot	Keyword
Handling Range		Single-value attributes only	Single- and multi-value attributes
Representation of ontology concept models		Protégé models	Concept lattices
Ontology formal degree		Higher than traditional methods	Higher than traditional methods
Ontology Visual degree		Based on concept lattices	Based on concept lattices

4 Build Domain Ontologies of Water Resources Based on the Improved GuTao Method

Based on the improved GuTao method to build a domain ontology of water resources, the improved algorithm is verified in terms of feasibility and practicability. Its steps are as the followings:

(1) Get initial information source of water resources

Initial water resources information is obtained from Baidu Baike when water resources are entered. Domain ontology of water resources is built based on initial water resources information .

(2) Get water resource keywords, attributes and relationships

Arrange water resource domain texts from Baidu Baike. Take a sentence as a unit and cut off the main domain keywords and phrases as the following:

{Fresh water, salt water; groundwater, surface water; solid water, liquid water, pneumatolytic water, natural water; river, lake, ocean; steam, glacier, swamp, accumulated snow; perched water, piestic water}.

Convert these domain keywords and their attributes to the "object-attribute" relationship, and the conversion is as the following:

{Lake---natural water; lake---fresh water; ocean---salt water, glacier--solid water; pneumatolytic water--steam; Submersible---groundwater}.

(3) Build domain ontologies with Protégé

Define two big classes: water resource and classification. There are 10 sub-classes in the water resource, upper thin glacier, glacier, piestic water, steam, swamp, river, ocean, lake, submersible and accumulated snow. And there are solid water, groundwater, surface water, pneumatolytic water, liquid water, fresh water, salt water and natural water in the classification.

(4) Generate initial formal background with the improved GuTao method

Build domain ontologies with improved GuTao method and generate the initial formal background as shown in Fig. 4. Each row means an object, and each column means an attribute. The X symbol means an object with the attribute.

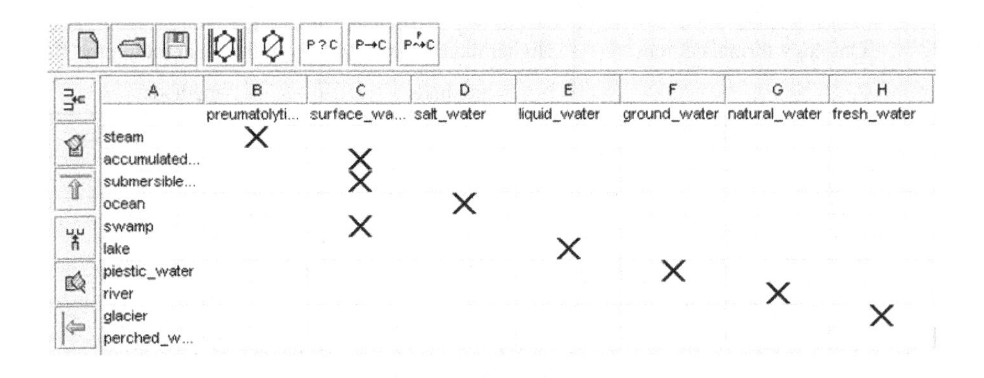

A	B	C	D	E	F	G	H
	preumatolyti...	surface_wa...	salt_water	liquid_water	ground_water	natural_water	fresh_water
steam	X						
accumulated...		X					
submersible...		X					
ocean			X				
swamp		X					
lake				X			
piestic_water					X		
river						X	
glacier							X
perched_w...							

Fig. 4. Generation of the initial formal background

(5) Improve the background with the common knowledge

After getting the initial formal background, expand and improve the background with the common knowledge. Accumulated snow, glacier and river also belong to fresh water, and perched water and submersible belong to surface water, and so forth. The improved formal background is as Fig. 5.

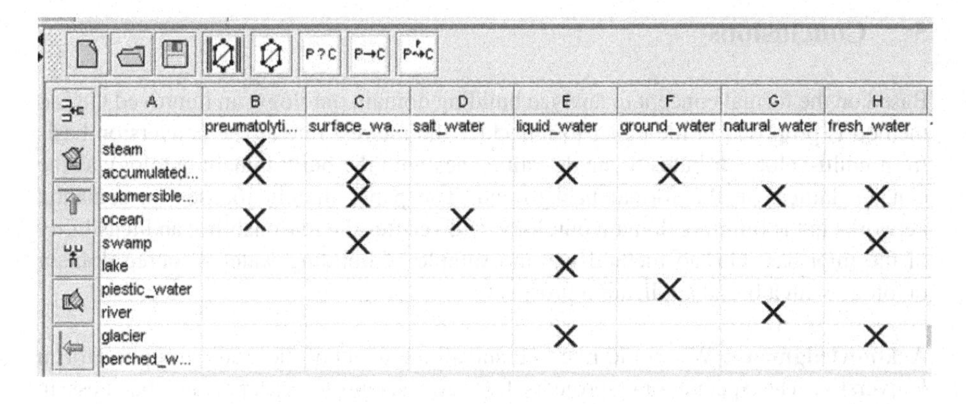

	A	B	C	D	E	F	G	H
		preumatolyti...	surface_wa...	salt_water	liquid_water	ground_water	natural_water	fresh_water
	steam	X						
	accumulated...	X	X		X	X		
	submersible...		X				X	X
	ocean	X		X				
	swamp		X					X
	lake				X			
	piestic_water					X		
	river						X	
	glacier				X			X
	perched_w...							

Fig. 5. Expanded Formal Background

(6) Get the concept lattices of the domain ontology

The concept lattices of the domain ontology are got based on Figure 5, as shown in Fig. 6. Each circle filled with colors means a formal concept. The small circles on the top mean the sets with all object attributes, and there is no such object in the real situation. The ones at the bottom mean the sets of all attributes, and obviously there is no such object. The sets formed by the 11 formal concepts in the figure are the required concept lattices.

- Concepts with all objects: ({river, lake, ocean, submersible, perched water, piestic water, steam, accumulated snow, glacier, swamp}{ }).
- Concept with all attributes: ({ }, {fresh water, groundwater, surface water, liquid water, solid water, pneumatolytic water, natural water}).
- Raw concept: ({glacier}, {groundwater}), ({submersible}, {groundwater}), ({sea water}, {salt water}), ({pneumatolytic water}, {steam}).
- Middle concepts: ({swamp}, {surface water}), ({piestic water}, {natural water}), ({glacier, submersible, perched water, accumulated snow}, {fresh water}), ({river, lake}, {fresh water}), ({river, lake, sea water}, {liquid water})

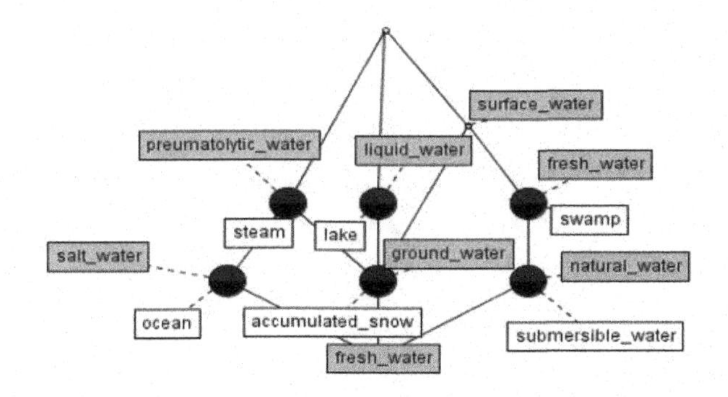

Fig. 6. Generated Concept Lattices

5 Conclusions

Based on the formal concept to analyze building domain ontology, an improved GuTao method is proposed in the paper to introduce the interest rate as the conversion basis from multi-value attributes to single-value ones with the built domain ontology as the concept lattices in the formal background. The paper mainly focuses on semantic networks in the building domain ontology, and verifies the reasonability and feasibility of the improved GuTao method via an example of building water resource domain ontology with a broad application prospect.

Acknowledgments. We would like to thank all members of the study group in HoHai University. The special fund projects for basic scientific research and business in central universities under the grant No. 2009B21214 supports our work.

References

1. Zhou, W., Liu, Z., Chen, H.: A Survey of the Research about Both FCA and Ontology. Computer Science 2, 8–12 (2006)
2. Huang, M.-L., Liu, Z.-T.: Research on Domain Ontology Building Methods Based on Formal Concept Analysis. Computer Science 1, 210–239 (2006)
3. Ganter, B., et al.: Formal concept analysis. Science press, Beijing (2007)
4. Gu, T.: Using formal concept analysis for ontology structuring and building. In: ICIS, Nanyang Technological University (2003)

Research and Application of Hybrid-Driven Data Warehouse Modeling Method

Junliang Xu[1], Boyi Xu[2], and Hongming Cai[1]

[1] School of Software, Shanghai Jiaotong University, Shanghai, 200240
rogerjimmy@sjtu.edu.cn, hmcai@sjtu.edu.cn
[2] Aetna College of Economics & Management, Shanghai Jiaotong University, Shanghai, 200052
byxu@sjtu.edu.cn

Abstract. In order to solve the problem of hardly keeping good performance on data presentation and usability of the data warehouse, a hybrid-driven data warehouse modeling method is proposed based on ontology. Firstly, a data-driven method is established using ontology to define the potential facts and dimensions. Furthermore, a demand-driven method is proposed based on the analysis of business process and specific business requirement to clear the fact and dimensions. Finally, with the analysis of medical domain and the specific business requirement, a data warehouse modeling about medical domain and data mining are used as a case study to verify the correctness and feasibility of the method. The result shows that the method can meet the medical domain requirements of data mining and provides a valuable reference to presenting data.

Keywords: data warehouse, ontology, data mining, multidimensional model modeling.

1 Introduction

In the early 1990s, the data warehouse is proposed as a solution to information management. It is the structured data environment of a decision support system and online analytical application data source. The main features of the data warehouse is subject oriented, integrated, stability and time variability. Currently mainstream data warehouse model are enterprise-class data warehouse proposed by Inmon model and dimensional model proposed by the Kimball [1] [2].

It is a very challenging problem to build a data warehouse. Data warehouse development has only the beginning, not the end. So it is a major difficulty to define the scope of data warehouse. Cleaning up dirty data affects the availability of the data warehouse. These two difficulties are the inevitable questions to build a data warehouse.

This paper presents a hybrid-driven data warehouse modeling method. We use ontology to describe the data source in order to achieve data cleaning. And we confirm the scope and boundaries of the data warehouse by specific business needs and business processes. This hybrid approach combines the data-driven method and demand-driven method to make the resulting data warehouse good features in practical application and describing the requirements.

Y. Wang and T. Li (Eds.): Knowledge Engineering and Management, AISC 123, pp. 575–583.
springerlink.com

The paper is organized as follows. Section 2 discusses the related work undertaking data warehouse designing. Section 3 outlets the framework for our modeling method. In Section 4 and 5 the detail of our method is discussed. A case study is presented in Section 6 to demonstrate the process of DW modeling using our method. Conclusions are drawn in Section 7.

2 Related Work

Current data Warehouse development methods can be divided into three different approaches such as: data-driven, goal-driven and user-driven [3].

(1) Supply-driven (Data-driven): This approach has been widely implemented by several researches [4] [5]. Data Warehouse design started with the analysis of the operational database. Bill Inmon, the founder of data warehousing argues that data warehouse environments are data driven. He states that requirements are the last thing to be considered in the decision support development life cycle. Romero and Abelló proposed a semi-automatable method aimed to find the business multidimensional concepts from domain ontology. [6]

(2) Demand-driven (User-driven): DW requirement for design are supplied directly from users. Next is the process of mapping the requirement with the available data. Golfarelli proposed a survey of the literature related to these design steps and points out pros and cons of the different techniques in order to help the reader to identify crucial choices and possible solutions more consciously.[7]

(3) Goal-driven: The corporate strategy and business objectives are considered as the company requirement for the Data Warehouse. Giorgini [8] adopt two type of modeling. Organizational modeling focus on stakeholder and decisional modeling relate to the company executive. Böhnlein and Ulbrich present an approach that is based on the SOM (Semantic Object Model) process modeling technique in order to derive the initial data warehouse structure. [9]

In summary, each of these three methods has distinct characteristics and advantages, in particular, supply-driven and demand-driven are the most commonly used and widely recognized. However, both methods have their deficiencies. Supply-driven relies too much on the database. Once the database is poorly designed, it will have a huge impact on the data warehouse. Demand-driven, in contrast, emphasizes too much on requirement and ignores the underlying data. It makes the resulting data warehouse be good description of business requirements, but out of touch with the underlying data in specific applications.

3 Framework for Hybrid-Driven Multidimensional DW Modeling

In this modeling framework, ontology is used to distinguish fact and dimension tables; business requirements are used to determine business process and granularity.

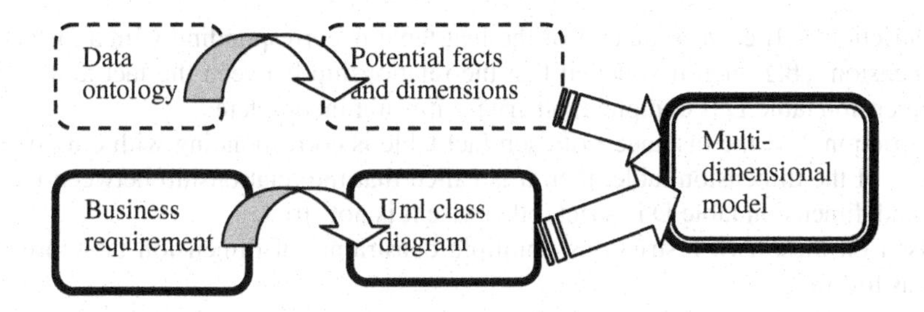

Fig. 1. The framework of hybrid-driven data warehouse modeling method

In Figure 1, the main framework of this method is divided into two parts. First the constraints of ontology concepts are analyzed to find the potential facts and dimensions in these ontology concepts; and then depending on the business needs, uml class diagram and business process diagrams are established.

4 Finding Potential Facts and Dimensions

According to the definition in The Data Warehouse Toolkit, the fact table is the main table to store measure of business; Dimension table reflects the different business analysis perspective. Combined with practical applications, the relationship between facts and dimensions can be described in Table 1.

Table 1. Multiple relationships of facts and dimensions

	The smallest multiple relationship		The largest multiple relationship	
	0	1	1	*
Fact table	Normal	Normal	Normal	Normal
Dimension table	Incomplete	Normal	Normal	Non-strict

First, the multiple constraints of the fact table should be analyzed (i.e. each instance of the dimension table corresponds to several instances of the fact table) to clear the multiple relationships between fact and dimension, so as to the fact and dimension could be identified by ontology constraints.

When the smallest multiple relationship of the fact is 0: This situation means the instances of dimension table are allowed to have no association with any instances of fact table; When the smallest multiple relationship of the fact is 1: This situation shows that each instance of dimension table must associate with at least one fact table instance; When the largest multiple relationship of fact table is 1: This situation means that each instance of dimension table has association with a maximum of one fact table instance; When the largest multiple relationship of the fact table is *: This situation shows that every instance of each dimension table can be associated with several instances of the fact table. All these relationships are common in practical applications.

Before the analysis of the multiple constraints of dimension table, two restrictions of the facts and dimensions are proposed to ensure the multi-dimensional model better practicality.

Restriction 1: If each instance f of the fact table is corresponding with a instance d of dimension table, then it is called that the relationship between the fact table F and the dimension table D is complete, otherwise it is not incomplete.

Restriction 2: If the instance f of each fact table is corresponding with only one instance d of the dimension table, then it is called that the relationship between fact table F and dimension table D is strict, otherwise it is not strict.

Based on these two restrictions, multiple constraints of dimension table are analyzed as follows:

(1) When minimum multiple relationship of the dimension table is 0: This situation means that an instance of fact table may have no association with any instance of dimension table, which is incomplete.

(2) When the minimum multiple relationship of the dimension table is 1: this situation shows that an instance of fact table has association with at least one instance of dimension table, it is common.

(3) When the largest multiple Relationship of the dimension table is 1: This situation means that an instance of fact table has association with at most one instance of the dimension table, which is common.

(4) When the largest multiple relationship of the dimension table is *: This situation shows that an instance of a fact table has association with several instances of the dimension table, which is not strict.

With the above analysis, we get the best relationship between the fact and the dimensions which should be 1-to-many relationship, shown in Figure 2.

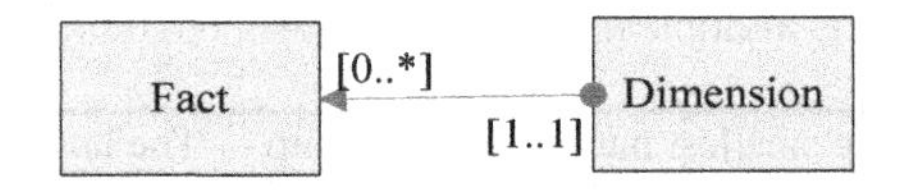

Fig. 2. The relationship between fact and dimension

Semantic formula is used to express this relationship:

$$\{\forall f, (f.P \in D) \cap (f.P \in D = 1) \rightarrow <F, D> \in R\} . \quad (1)$$

The formula shows that for a instance f of any concept F, if the value of a property P of f are all involved in the instances of concept D and each f has associate with one instance of D, then the relationship between F and D is considered the relationship between the fact and dimension, including the concept F is the fact and the concept D is the dimension.

Analyzed the above formula, the conditions of this formula have three parts:

(1) $\{\forall f , f.P \in D\}$ shows concept D is referenced by the property P of concept F. The property P can also be considered as a relationship between instance f and instance d. In this point, the expression can be transferred into

$$\{\forall d, <f, d> \in F.P \rightarrow d \in D\} . \quad (2)$$

(2) $\{\forall f , f.P \in D \geq 1\}$ shows one instance f of concept F has association with at least one instance of the concept D through the property P.

(3) $\{\forall f, f.P \in D \leq 1\}$ shows one instance f of concept F has association with at most one instance of the concept D through the property P.

Combined (2) and (3),

$$\{\forall f, f.P \in D = 1\}. \tag{3}$$

It shows the multiple relationships between dimension and fact is 1. On the other hand, the constraints of ontology concepts can also be described by semantic expression. Compared these two expressions, the following definition can be confirmed:

Definition 1: If a property R of the ontology concept F has constraints allValuesFrom(D) and maxCardinality(1), where D is another ontology concept, the relationship between D and F is considered as the relationship between dimension and fact.

5 Building Multi-dimensional Model by Business Requirement

In this part, the business concept is analyzed to clear business processes and granularity of the multi-dimensional model. The business concepts in this paper are mainly about the concepts in business process and the relationship between these concepts. The business concepts define the final facts and dimensions from the potential set which is confirmed in previous step.

First business requirements should be obtained. Here, a dual group called requirement set is defined to represent the business requirements:

Definition 2: R <C, P>, where C is the set of conditions, P is the prediction set that is the result set.

For example, requirement set expresses as < (jobs, working hours), income> to describe the relationship between work time and income of different jobs. Condition set is used to describe conditions of the business needs. Predict set is the data that business needs want to get. By using requirement set, every business needs can be uniformly described.

After accessing to specific business needs, a logical model can be created by uml. Here, how many use case are involved in the business requirements must be cleared first, then we can use the method combining uml class diagram and business process diagram:

Rule 1: When business needs involves only one uml use case, the logical model is determined only by using uml class diagram. For example when the business requirement 1 is <(position, gender, age), income>, this requirement only relates to the income of the staff use case, so the class diagram in Figure 3 can be described the needs.

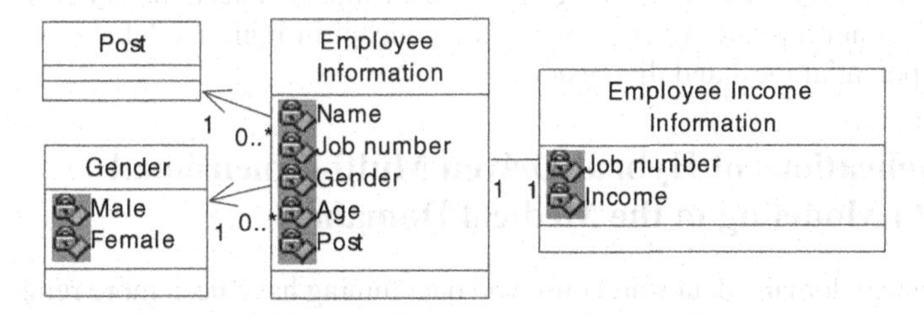

Fig. 3. Requirement 1 described by UML class diagram

Rule 2: When the business needs involves several uml use cases, the logical model is determined by combining class diagram and business process diagram. For example when the business requirement 2 is <(gender, age, type of disease), cost>, which involves the diagnosis use case and disease cost use case, so we should find the association and sequence between two use cases by using business process diagram. The business processes can be analyzed to get that diagnostic and cost information are different stages in a process, one diagnostic information is corresponding to one cost information. Figure 4 is a description of the needs by uml class diagram.

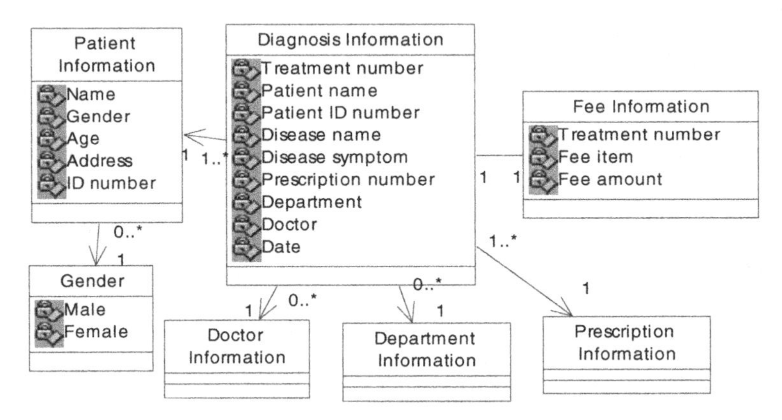

Fig. 4. Requirement 2 described by UML class diagram

Through the method above uml class diagrams can be get for different business needs, and mapping logical model by following rules.

Rule 3: The classes in class diagram are mapped to dimension tables or fact tables, where the fact table is in the many side in the relationship with other classes, the dimension table is in one side;

Rule 4: The class properties in class diagram are mapped to columns of dimension tables or fact tables;

Rule 5: The associations between classes in class diagram are mapped to relationships between the dimension tables and the fact tables.

A logic model cannot be used in practical applications, because the logic model is only based on business requirements and business processes, not the data source. So the logic model should be combined with the potential facts and dimensions obtained before to draw the final multi-dimensional data model. Facts and dimensions which are the most appropriate description for the concept in logic model should be chosen from the potential facts and dimensions.

6 Applications of Hybrid-Driven Multi-dimensional Data Modeling in the Medical Domain

In the medical domain, data warehouse and data mining have been more remarkable and have more applications. Medical field data has the following characteristics: large amount of data, complex amount of information and multiple correlations between data and data; heterogeneous medical business and uniform medical processes. These facts make building data warehouse difficult: large volumes of data, difficult to clean data; complex business concept and hard to integrate business into the data warehouse.

In order to effectively resolve these problems, hybrid-driven multi-dimensional modeling method is proposed to integrate and meet specific business needs. First the medical data ontology is gotten from the medical business database and evolved by medical standards. According to the medical ontology and definition 1, the ontology concepts are analyzed to getting the potential facts and dimensions, as shown in Table 2.

Table 2. Potential facts and dimensions

Potential Facts	Potential Dimensions
Patient Information	Age, gender, address, place of birth, occupation
Treatment process information	Patient information, fee information, surgery information, diagnosis information
Employee Information	Staff positions, staff titles, department
Diagnosis information	Disease information, patient information, doctor information
Fee information	Fee type, patient information, employee information

On this basis, business requirements are added to clear business particle size, facts and dimensions of ulti-dimensional model. For example the business requirement is the treatment to different patient's gender, age and departments. According to definition 2, the dual group should be: < (gender, age, departments), disease>.

After analysis of the dual group, uml class diagram can be gotten to be transformed into logical model according to mapping rules (rules 3, 4, 5). The concepts in logical model are from business processes and business requirements, but without instances. Then, the logical model is corresponded to the potential facts and dimensions in Table 2 to get the final multi-dimensional model shown in Figure 5.

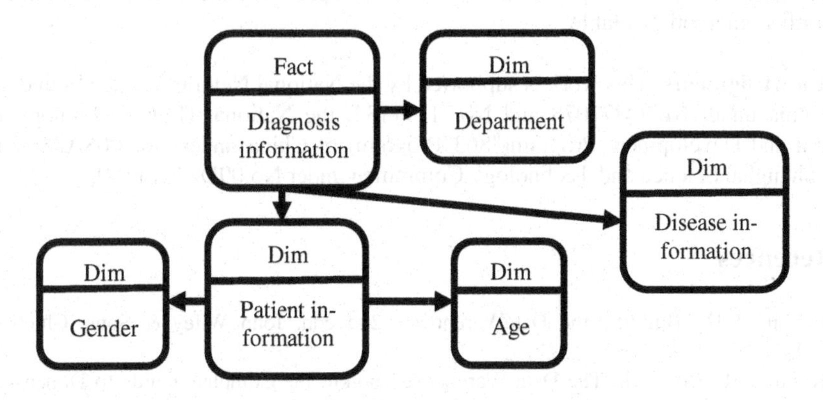

Fig. 5. Resulting multi-dimensional model

In Figure 5, multi-dimensional model is established for the specific needs <(gender, age, sections), the disease>, in which diagnostic information is the fact table, the others are the dimension tables. Then the multi-dimensional model should be mined to verify availability. Figure 6 is the ratio chart of treatment departments, age and sex obtained by data mining.

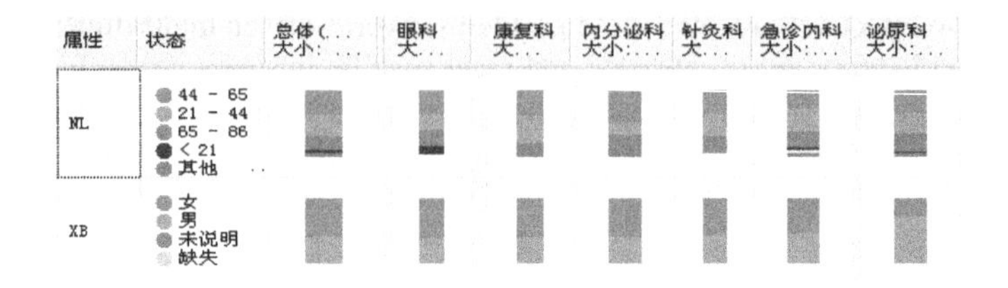

Fig. 6. Data mining - classification analysis

Figure 6 indicates the distribution of the number of patients on different departments with different age and gender. This figure can clearly reflect the relationship between age, gender, and departments.

7 Conclusion

This paper presents a hybrid-driven multi-dimensional data modeling method that combines features of data-driven and demand-driven method to achieve multi-dimensional model good data presenting and the usability of practical application. In allusion to Data-driven method which over relies on the underlying data, the method based on ontology makes data source more semantic more and more suitable for display. Faced with lack of usability for traditional demand-driven method, this paper describes the business requirements with the dual group, describes the requirements by uml class diagram, and finally associates these concepts with the underlying data. In the validation, analyzing the characteristics for the medical field, multi-dimensional model is completed according to our approach, and achieved data mining to confirm method available.

Acknowledgments. This work is supported by the National Natural Science Foundation of China under No.70871078 and No.71171132, the National High-Technology Research and Development Program("863"Program) of China under No.2008AA04Z126 and Shanghai Science and Technology Committee under No.09DZ1121500.

References

1. Inmon, W.H.: Building the DataWarehouse, 2nd edn. John Wiley & Sons, Chichester (1996)
2. Kimball, R., Ross, M.: The Data Warehouse Toolkit: the Complete Guide to Dimensional Modeling, 2nd edn. John Wiley & Sons, Chichester (2002)
3. Romero, O., Abelló, A.: A survey of multidimensional modeling methodologies. IJDWM 5(2), 1–23 (2009)

4. Moody, D., Kortink, M.: From enterprise models to dimensional models: A methodology for data warehouse and data mart design. In: Proc. 2nd DMDW, Stockholm, Sweden (2000)
5. Phipps, C., Davis, K.C.: Automating Data Warehouse Conceptual Schema Design and Evaluation. In: Proceedings of DMDW 2002: International Workshop on Design and Management of Data Warehouses (2002)
6. Romero, O., Abelló, A.: Automating multidimensional design from ontologies. In: Proceedings of the ACM Tenth International Workshop on Data Warehousing and OLAP, pp. 1–8 (2007)
7. Golfarelli, M.: From user requirements to conceptual design in data warehouse design – a survey[M]//L. In: Data Warehousing Design and Advanced Engineering Applications: Methods for Complex Construction, Bellatreche, pp. 1–16. IGI Global, Hershey (2010)
8. Giorgini, P., Rizzi, S., Garzetti, M.: GRAnD: A goal-oriented approach to requirement analysis in data warehouses. Decis. Support Syst. 45(1), 4–21 (2008)
9. Boehnlein, M., Ulbrich-vom Ende, A.: Business Process Oriented Development of Data Ware-house Structures[M]//R. In: Jung, R.W. (ed.) Data Warehousing 2000- Methoden, Anwendungen, Strategien, pp. 3–21. Physica Verlag, Heidelberg (2000)

Traditional Chinese Medicine Syndrome Knowledge Representation Model of Gastric Precancerous Lesion and Risk Assessment Based on Extenics

Weili Chen[1], Aizi Fang[2], and Wenhui Wang[3]

[1] Library of Guangdong University of Technology, Guangzhou, Guangdong, China
[2] Hospital of Guangdong University of Technology, Guangzhou, Guangdong, China
[3] Dongguan Hospital of Traditional Chinese Medincine, Dongguan, Guangdong

Abstract. To strengthen the research on Traditional Chinese Medicine theory and the knowledge representation of syndrome. Extension set is chosed to symbolize the syndromes of Gastric disease and Gastric Precancerous Lesion. After Extenics conjugate analysis of Model of Gastric Precancerous Lesion, overall evaluation elements is established and framework of risk assessment of gastric cancer in traditional Chinese medicine syndrome formed. Finally we establish Classical domain and section domain so as to calculate the exact risk for Gastric Precancerous Lesion changing into Gastric Cancer.

Keywords: extenics, extension set, TCM, syndrome, Gastric Precancerous Lesion.

1 Introduction

Gastric Precancerous Lesion has become the focus of the study, and early detection, prevention and treatment of precancerous diseases and precancerous lesions, become a effective way of lower incidence and mortality of gastric cancer. The application of Traditional Chinese Medicine(TCM) in the treatment of chronic atrophic gastritis has a distinct advantage. Gastric Precancerous lesions is belong to the category of "Stomachache", "fullness" in TCM, which root from a long-term improper diet, ill emotions, internal weary, inappropriate medication and chronic illness. TCM is one of the world's oldest medical systems that has made brilliant achievements and contributed to the health of the people, Traditional Chinese Medicine has been studied and applied in many countries all over the world. Researches on modernizing TCM become the significant event that attracts more and more attentions from the practitioners of TCM. First of all, automatic quantitative diagnosis of syndromes is one of the most important.

Syndrome is the key character and the essence of TCM which refers to the pathological and physiological characteristics in people who is under disease state, Syndrome Researches began in the late 1950s, and after 1980 further researches have carried on to reveal deeper meaning, stressing the objective index of the syndrome and the process micro-objective indicators, but it still cannot express the real meaning of syndromes. The common linear modeling methods to study syndrome are multiple

Y. Wang and T. Li (Eds.): Knowledge Engineering and Management, AISC 123, pp. 585–590.
springerlink.com

linear regression analysis, factor analysis, discriminant analysis and other multivariate statistical methods. Liufeng Bin [1] used Logistic regression analysis to model the diagnosis of stomach disease. However, due to the syndrome itself is a non-linear, multi-dimensional and multi-level system, intuitive, linear methods cannot meet the TCM research. Nonlinear modeling method can deepen the meaning of the syndrome, providing a new and more reliable modern tools, for example, Heqing Yong and Wang Jie[2] proposed a complex algorithm for standardization of TCM research. However, those methods cannot resolve the structure of the syndrome with deep TCM meaning, such as the role of balance between regulation and the constitution, the role of genetic factors TCM theory system stress the complex relationships that to some extent determines, what is we call systematic, overall and dynamic thinking.

2 Traditional Chinese Medicine Syndrome Knowledge Representation Model of Gastric Precancerous Lesion

Extension set introduce [things, features, value]as the basic element to describe things, as matter-element. It combines formal logic and dialectical logic advantages, with not only integrating with a formal language expression, but also considering the meaning of things and the extension of logic.

2.1 Extenics Model of Knowledge Representation

At very beginning of the year 1983, Prof. Cai Wen[3] published his creative paper "The Extension Set and non-compatible Problems" and later a book named "Analysis of Matter Element". Extension Set introduces a concept of matter elements. With an ordered three elements group $R = (N, c, v)$ to describe the basic unit of an object and it is named as a matter element. Here "N" stands for the objective, "c" the name of character of the object and "v" the quantity of "N" about "c". If an object has multiple characters, then a multi-dimensional (n-dimensions) matter element can be defined as:

$$R = \begin{bmatrix} N, & C_1, & V_1 \\ & C_2, & V_2 \\ & \cdots\cdots & \\ & C_n, & V_n \end{bmatrix}$$

2.2 Extenics Model of Gastric Precancerous Lesion Knowledge Representation

Changing patterns of Gastric Precancerous Lesion goes as follows: superficial gastritis —atrophic gastritis—intestinal metaplasia—dysplasia—gastric cancer, where exist a quantitative change from to qualitative change[4]. Extension study creates extenics degree of function and quality of the concept of domain section that a quantitative change to qualitative change. Traditional Chinese Medicine Syndrome information is closely related to disease pathogenesis , only after having comprehensive and systematic analysis can we express better the relationship between the characteristics of external syndrome and internal signs of pathogenesis. This paper using Extenics method to describe the characteristics of TCM Syndrome, establish a comprehensive representation model of TCM Syndrome, introduce the correlation analysis of the model, as well as prospect to realize intelligentize of quantify associated syndromes using extenics mathematical correlation function method.

We can use extenics to estable Gastric Disease for example:

$$R1 = \begin{bmatrix} \text{Gastric Disease Syndrome,} & \text{pathogeny} & V1n \\ & \text{Clinical symptom} & V2n \\ & \text{......} & \end{bmatrix}$$

$$R12 = \begin{bmatrix} \text{Gastric Disease Syndrome, Clinical symptom} & \text{stomach ache} \\ & \text{Gastric fullness} \\ & \text{Diet reduce} \\ & \text{Belch hiccups} \\ & \text{nausea and vomiting} \\ & \text{......} \end{bmatrix}$$

3 Extenics Conjugate Analysis of Model of Gastric Precancerous Lesion

The concept of the conjugated characters in Extenics Theory is used to analyze the character of content in four aspects. These four Characters are as follows: the explicit and the implicit; the positive and the negative; the actual and the virtual; the hard and the soft., by which we can describe the structure of Gastric Precancerous Lesion in full scale, as well as reveal deeply the nature of itself and its relationship with the outside world.

Syndrome reflect the disease state, pathology and illness and so on. In addition, the syndrome has physical, emotional and environmental characteristics because it also emphasized the physical, emotional status and geographical, climatic and environmental factors. We can use extenics conjugate analysis to describe the structure of Model of Gastric disease. Namely, the real part of N was reN, the imaginary part was imN, then $N = imN \otimes reN$, and $Rre = [\ reN, crei, vrei\ ]$ (i= 1,2,...n) ; $Rim = [\ imN, cimi, vimi\ ]$ (i= 1,2,...n). For example, vrei= (gastric cancer), vimi= (atrophic gastritis; intestinal metaplasia; dysplasia). Apparent part of N was apN , latent part was ltN, then $N = apN \otimes ltN$, and $Rap = [\ apN, capi, vapi\ ]$ (i= 1,2,...n) ; $Rlt = [\ ltN, clti, vlti\ ]$ (i= 1,2,...n), For example, vapi= (intestinal metaplasia; dysplasia; atrophy; stomach ache; Gastric fullness; Diet reduce; Belch hiccups; bitter taste of mouth and dry mouth; fatigue; nausea and vomiting; gastric upset; Chest tight; coated tongue; unsmooth pulse); vlti= (genetic family history; environmental pollution;industry pollution; food security). Positive part of N was psN,negative part was ngN, then $N = psN \otimes ngN$, and $Rps = [\ psN, cpsi, vpsi\ ]$ (i= 1,2,...n) ; $Rng = [\ ngN, cngi, vngi\ ]$ (i= 1,2,...n), For example, vngi= (ill habit ; night shift; stay up); vpsi= (emotional status; physical status).

Table 1. Factors of Risk Assessment of gastric cancer in Traditional Chinese Medicine Syndrome aspect

Target hiberarchy	Index	Elements	Contents	weight
pathology	I 1	intestinal metaplasia	The level of it	0.141
	I 2	dysplasia	The level of it	0.148
	I 3	atrophy	The level of it	0.106
Clinical symptom	I 4	stomach ache	The frequency and the degree of it	0.032
	I 5	Gastric fullness	The degree of it	0.009
	I 6	Diet reduce	The degree of it	0.006
	I 8	bitter taste and dry mouth	The degree of unsuitable feeling in mouth	0.003
	I 9	fatigue	The frequency and the degree of it	0.004
	I 10	Nausea and vomit	The frequency and the degree of it	0.009
	I 11	gastric upset	The frequency and the degree of it	0.007
	I 12	Chest tight	The frequency and the degree of it	0.008
	I 13	coated tongue	The frequency and the degree of it	0.052
	I 14	unsmooth pulse	The degree of it	0.030
environment	I 15	pollution	The degree of environmental pollution	0.048
	I 16	food security	financial capacity	0.021
habit	I 17	Ill habit	The habit of smoking	0.047
			The habit of drinking	
			The habit of having fried food	
			The habit of taking anodyne for long term	
	I 18	profession	The degree of work stress	0.031
			The frequency of night shift	
	I 19	Genetic history	The number of relative who had cancer	0.138
	I 20	physical status	Possible fatigue	0.017
			The degree of irregular bowel movement	
			The quality of sleeping	
	I 21	emotional status	The degree of depression	0.138

4 Risk Assessment System

4.1 The Establishment of Elements of Set

The risk of changing Gastric Precancerous Lesion to gastric cancer is determined by a variety of elements, Appropriate selection of which play a key role in the final judging result.We apply conjugate analysis above to find out 21 Syndrome item of gastric disease and define the these elements set is: I —{ I 1, I 2, I 3, I 4, ……, I 21}.

4.2 Evaluation Elements

After extenics conjugate analysis of Model of Gastric Precancerous Lesion, we dig out overall evaluation elements to establish risk assessment of cancer. These elements are as follows: (1) pathology: intestinal metaplasia; dysplasia; atrophy (2) Clinical symptom: stomach ache; gastric fullness; diet reduce; belch hiccups; bitter taste of mouth and dry mouth; fatigue; nausea and vomiting; gastric upset; chest tight; coated tongue; unsmooth pulse. (3) environment: pollution; food security. (4) habit: ill habit; profession. (5) Potential pathogeny: genetic history ; physical status; emotional status.

To determine the weight of these various elements, we selected 16 experts, who is either deputy chief Chinese physician or chief Chinese physician, and have been working in the area for more than 10 years, to provide a more comprehensive view and score for every elements, and then we used AHP measure to determine the weights. At last we formed the framework of risk assessment of gastric cancer in traditional Chinese medicine syndrome as follows (Table 1):

5 Classical Domain and Section Domain

According to Extension[5] theory, its classical domain is

$$\text{Roj} = (\text{Noj}, \text{C}, \text{Xoj}) = \begin{bmatrix} \text{Noj}, & \text{C1}, & \text{Xoj1} \\ & \text{C2}, & \text{Xoj2} \\ & \cdots\cdots & \\ & \text{Cn}, & \text{Xojn} \end{bmatrix} \quad (j=1,2,3,4)$$

The level of document resources allocation can be divided into 3 grades, namely A (High risk), B (Moderate risk), C(Low risk). then RAj= (NAj, C, XAj); RBj= (NBj, C, XBj); RCj= (NCj, C, XCj).We choose the 1-9 level score method, Classical domain is RAj= (NAj, C, <5, 9>); RBj= (NBj, C, <3, 7>) ; RCj= (NCj, C, <1, 4>).

6 Correlation Function Calculated

In extension theory, the formula $\rho(x, X_0) = \left| x - \frac{a+b}{2} \right| - \frac{1}{2}(b-a)$ is defined as the distance between the point x and the interval X_0. When $X_0 = (a,b)$,

$X=(c,d)$, $X_0 \subset X$ and there is not any common terminal points of intervals, the formula $K(x)=\frac{\rho(x,X_0)}{D(x,X_0,X)}$ is called also as an elementary incidence function of point x about the nested interval X_0 and X. When V_i is described by X_{o_i} and $X_i\left(X_{o_i} \in X_i\right)$, then $K\left(x_{ij}\right)=\frac{\rho\left(x_i,X_{o_{ji}}\right)}{\rho\left(x_i,X_{m_i}\right)-P\left(x_i,X_{o_{ji}}\right)}$.

With the equations above, we can calculate and find out which level the patient's risk belongs to of changing into gastric cancer.

7 Conclusions and Remarks

Extenics theory has been successfully applied in social and economical fields when there is not or is not possible to get an accurate and analytic model, people will be idealess. The writer believes that extenics theory and extension engineering methods are also very good choice of studying Traditional Chinese Medicine. The significance of this paper lies in the establishment of knowledge representation of Traditional Chinese Medicine syndrome, which is hard to be symbolized, and later provide the quantization criteria to Assess the Risk of changing into cancer.

References

1. Liu, F., Tao, H.: Logistic regression analysis in simulation experts dialectical diagnosis. Chinese Medicine Journal 2, 58–59 (2001)
2. Yong, H., Jie, W.: Research on TCM Syndromes and Diagnosis of Patients after Percutaneous Coronary Intervention based on Cluster Analysis. Traditional Chinese Medicine 10, 918–921 (2008)
3. C. Wen: Analysis of Matter-Element. Guangdong Higher Educational Publisher, Guangzhou (1987)
4. Liu, S.X., Tian, Y.: Medicine for reversal of precancerous lesions of gastric cancer. Jilin Traditional Chinese Medicine 10, 849–851 (2010)
5. Yang, C., Cai, W.: Extension engineering. Publisher of Science, Beijing (2007)

Theorem about Amplification of Relational Algebra by Recursive Objects

Veronika V. Sokolova and Oxana M. Zamyatina

Department of Control Systems Optimization, Institute of Cybernetics,
National Research Tomsk Polytechnic University, 634034, Tomsk, Russia
{veronica,zamyatina}@tpu.ru

Abstract. This paper discusses relational algebra extended by recursive relations (tables). The interpretation of the recursive table is proposed, the closure of the extended relational algebra is proved, and a new approach to modeling a physical database structure is offered that is suitable for representing complicated hierarchical data sets. It combines methods of set theory for the recursive relations within the framework of a single modeling paradigm, which allows users to define self-similar, partially self-similar, or hierarchical sets. The use of recursive relations in the definitions of self-similar objects yields representations that can be rendered at varying levels of detail or precision at run time.

Keywords: relational algebra, relational operations, recursive table, database, domain, tuple.

1 Introduction

Recently, relational approach has been widely used in modern database management systems such as Oracle, MS SQL Server, and etc. The relational approach to representation of the information model, developed by E. Codd in 1970, received huge popularity due to simplicity of the basic ideas and the strict formal theoretical base [1]. The relational model is described in concepts of classical algebra, mathematical logic, theory of formal systems, and graph theory. The given circumstance provides development of effective formal methods in database design. Since the time of appearance of Codd's first works on relational data model the range of tasks of database creation increased, which caused the development of new relational database modeling methods and improvements of classical relational data model [2].

Mathematically a relational database is a finite set of finite relations of any dimensions between predetermined sets of elementary data. In other words, a relational database (more precisely, any of its states) is a finite model in the sense of mathematical logic. It is possible to perform various algebraic operations with the model of relations. Thus, the relational database theory becomes the application domain of mathematical logic and modern algebra and is based on exact mathematical formalism. According to relational model, all data are considered as relations stored in flat tables in which each line has the same format. However users often have to deal with hierarchical data sets with recursion, such as a chemical (chemical elements), medical (model genome), economical, and etc.

Y. Wang and T. Li (Eds.): Knowledge Engineering and Management, AISC 123, pp. 591–596.
springerlink.com

In order to represent the hierarchical data structures there are various solutions: rejection of normalization rules [3], input additional attributes into relational table [4], or processing hierarchies with the help of special SQL operators [5]. Each of these solutions have restrictions. For example, to display a tree structure by using "adjacency list" it is necessary to know the number of nesting levels in the hierarchy. If you use the "subsets method", the data integrity is provided by the triggers that every time rewriting list of root and subordinate elements. The method "nested set" of Joe Selco [4] does not guarantee the integrity of the tree while editing the elements and requires processing multiple requests and time to refresh left and right values within a tree, at this time the advantage of "embedded sets" for fast tree generation is lost.

To guarantee the data integrity and nonredundancy of data the method for represent hierarchical structures with recursion in the form of nested relational tables is proposed.

2 Basic Definitions

Relation is a subset of the result of the Cartesian product operation under one or more domains:

$R(A_1, A_2, ..., A_n)$, where A_i – attributes of relation.

The relation has two important properties:

1) it does not contain coincide tuples,
2) the order of tuples is irrelevant.

We call the table a subset of the fixed domain D. The terms "relation" and "table" are synonymous, however, it makes sense to distinguish them, because not every relation can be represented by a two-dimensional table.

An informational model of a relational database it is a set of interconnected relations. Physical representation for relation (non-recursive) is a "flat" table, the header is defined by list of attributes, strings (tuples) – by the list of appropriate values, and attributes give names to the table columns.

In classical interpretation, the relational algebra represents a collection of operations upon relations (tables). In an article published in the Communication of the ACM magazine [6], Codd has determined the set of such operations and has proven that this set has the property of relational closure.

A set of operations of the relational algebra are divided into two classes: the set-theoretic and special relational operations. For the following, assume that T_1 and T_2 are relations. The union it is a binary operation that is written as $T_1 \cup T_2$ and is defined as the usual set union in set theory: $T_1 \cup T_2 = \{t: t \in T_1 \vee t \in T_2\}$. The result of the union is only defined when the two relations have the same headers. Since the result of any relational operation is some relation, it is possible to derive relational expressions in which instead of a relation-operand of some relational operation there is a nested relational expression.

Recursive object is partially defined by itself. Is permitted only explicit recursion in the sense of the following definition: a table $T(r)=r.(a_1{:}D_1, a_2{:}D_2, ..., a_n{:}D_n)$ is accessible if the syntactic equality $D_i \cong T$, defines the recursion is not running more than n-1 domains D. Note that the tuple can not refer to itself either directly or indirectly.

A recursive relation is also called a unary relation. It is a relation between instances of one entity type.

Recursive relation:

1. A recursive relation refers to a relation established within a table.
2. In other words, a table "refers to itself".
3. Recursive relations are usually established in situations where it is very useful to perform a self-join.

Self-referencing or recursive data may initially look like a hierarchy. The key difference is that all of the elements are stored in the same place. There are keys that relate one row in a table back to a different row in the same table.

Examples of recursive data:

a) Company organizational structure: president – director – employee; object type – employee.

b) Inventory: product – assembly – sub-assembly – component; object type – inventory item.

c) Project management: project – task – sub-task; object type – project entry.

In each of the above examples, the type of object (or node) type is the same at any level. For example, the table "Employee" includes the following columns: *id_employee*, *name*, and *id_manager* (Fig. 1 shows an example). The column *id_manager* keeps the *id_employee* (not the *name*) of the employee's manager. In order to prepare a report that lists employees and their managers by their names, a self-join is required. In this case, the column *id_manager* is typically made into a foreign key that refers to the primary key – field *id_employee*. Establishing this internal relation allows enforcing a referential integrity (e.g., you can not insert a manager *id_employee* into the column *id_manager* that does not exist in the employee column *id_employee*). This means that an employee can be managed by only one existing manager, and that each employee can manage more than one employee.

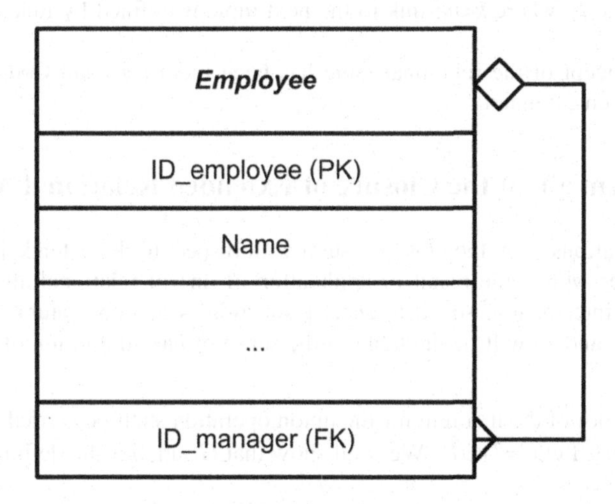

Fig. 1. Example of recursive data

Let us extend the relational algebra, having redefined the concept of table.

A *recursive table* it is a fixed (unlimited) set of tuples with unlimited length corresponding to the following conditions:

1. C=<a_1:D_1, a_2:D_2, ..., a_n:D_n> is a tuple-type where all attributes-names a_i – are mutually various; C being named a *main tuple* or a *basis*;
2. The table T, composed by a finite set of mutually various *main tuples-values*, is named the *main table*;
3. The *generalized table* of the *main table* T is the finite collection of recursive *tuples-values* C_r of the main table defined by the *main tuple-type* C of the table T in the following way:
   a) Tuples-values C_r of the generalized table are pair wise different.
   b) Each tuple C_{ri} is in essence a m_i-fold ($m \geq 1$) repeated set of attributes-values $<a_{11},a_{12},..,a_{1n},a_{21},a_{22},..,a_{2n},..,a_{m1},a_{m2},..,a_{mn}>$, such that $a_{ik} \in D_k$ ($1 \leq k \leq n$) of the main tuple C and for any $i \neq j$ any two subtuples $<a_{i1}, a_{i2},\ldots, a_{in}>$ and $<a_{j1}, a_{j2},\ldots, a_{jn}>$ do not coincide;
   c) In any tuple of the generalized table the reference attribute sometime receives the *null* value (an empty link).

Let the key set K and the index set I in the generalized table T be defined in a standard way, such that:

a) $D(K)=D(I)$, domain of keys match with the domain of indices;

b) $K \cap I = \varnothing$, key is not the index, thus avoiding an infinite recursion;

c) mapping is defined as $R{:}I \rightarrow K$, and, obviously, the fixed point of transitive closure R^* is the *null* value of the reference attribute.

Then the recursive natural join of the table T in relation to R is the operation, the result of which is such the generalized table T_R, that for each of its tuples-value C_{ri} consisting more than of two subtuples of the main table $< c_1, c_2,\ldots, c_m>$:

a) $I(c_m)$=*null*, a reference element of the last tuple must be empty;

b) $I(c_{k-1})=K(c_k)$, where $k<m$; link to the next tuple is defined by functional connectivity R.

So, the concept of the relational table has been redefined, and Codd's traditional algebra has been extended.

3 Theorem about the Closure of Extended Relational Algebra

Theorem: Relations form the closed system with respect to the extended relational algebra. In other words, the result of application of one of relational algebraic operations (union, intersection, set difference, projection, selection, natural join) to arbitrary tables T_1 and T_2 will be the table in the sense of base definition of the recursive table.

Proof: The proof of the theorem for the union operation shall be carried out. Consider tables T_1 and T_2. Let $G=T_1 \cup T_2$. We shall show that G satisfies the definition of the recursive table.

a) As T_1 and T_2 are tables in the sense of new definition of the recursive table, each of them consists of mutually various tuples C'_r $(1 \le r \le n)$ and C''_r $(1 \le r \le m)$ accordingly. Then according to the definition of the union operation table G will also consist of mutually various tuples C_r $(1 \le r \le k, k \le m+n$ – repeating tuples are eliminated).

b) For similar reasons in table G each tuple C_r will be an m-fold repeating set of attributes-values

$<a_{11},a_{12},\ldots,a_{1n},a_{21},a_{22},\ldots,a_{2n},\ldots,a_{m1},a_{m2},\ldots,a_{mk}>$, such that $a_{is} \in D_s$ $(1 \le s \le n)$ of the main tuple C and no two subtuples $<a_{i1}, a_{i2},\ldots, a_{ik}>$ and $<a_{j1}, a_{j2},\ldots, a_{jk}>$ do not coincide at $i \ne j$.

c) T_1 is a table in the sense of definition of the recursive table. Therefore, at least one of the attributes of the first subtuple has *null* value. T_2 is also a table in the sense of definition of the recursive table. Again it means that at least one of the attributes of its first subtuple has *null* value. Then table G will also contain at least one such subtuple (according to result of the union operation).

d) T_1 is a table in the sense of definition of the recursive table, therefore key set K' and index set I', such that $D(K')=D(I')$ and $K' \cup I' = \varnothing$ are preset in it. T_2 is also a table in the sense of definition of the recursive table, therefore key set K'' and index set I'', such that $D(K'')=D(I'')$ and $K'' \cup I'' = \varnothing$ are preset in it.

From the compatibility requirement of tables T_1 and T_2 on union and result of the union operation it follows that $D(K)=D(I)$ and $K \cup I = \varnothing$, where $K=K' \cup K''$, $I=I' \cup I''$.

e) T_1 is a table in the sense of definition of the recursive table, consequently, there exists mapping $R': I' \rightarrow K'$, the fixed point of such transitive closure R'^* being the *null* value. T_1 is a table in the sense of definition of the recursive table, therefore, there exists mapping $R'': I'' \rightarrow K''$, and the fixed point of such transitive closure R''^* is the *null* value.

In table G it is possible to determine mapping $R: I \rightarrow K$ which operates by the rule $R(i)=R'(i)$ at $i \in I'$ and $R(i)=R''(i)$ at $i \in I''$. Transitive closure R^* will have fixed points with *null* value.

f) For each tuple-value C_r of table G consisting of more than two tuples of the main table $<c_1, c_2,\ldots, c_m>$ the conditions $I(c_m)=null$ and $I(c_{k-1})=K(c_k)$, $k<m$ will be met.

Thus, convincing evidence is given that the result of the union operation is a table in the sense of definition of the recursive table. In the same manner, the result of other algebraic operations will also be a table in the sense of definition of the recursive table. The analysis of results of these operations with reference to the recursive tables shows that the constructed extension of Codd algebra is closed.

4 Conclusion

Thus, the relational algebra that uses sets of related tuples has been extended to include recursive relations. The technique includes formal set theoretic specifications of related sets of tuples that are related in any of the subrelations of a recursive relation. The approach covers both recursive one-to-many and recursive many-to-many relations. The method provides a set theoretic foundation for database management systems that allow using recursive relations.

The technique of displaying hierarchical data structures in the form of embedded relational tables is suggested. This idea of hierarchies improves the informativity of data models and provides opportunities for the development of formal methods of convolving and unfolding the hierarchies of any level of nesting.

It is proved, that result of any relational operation in itself is always another relation. The important consequence of relational algebra closure is the opportunity to make nested relational expressions. The process of transformation of each construction into attributes of relational tables described in the article guarantees that they will depend only on key attributes. Hence, each obtained recursive data model is guaranteed to have the fourth normal form. The developed technique can be used as theoretical base in construction of new database structures with use of modern design tools.

Acknowledgments. We would like to thank Prof. Dr. Vitaly B. Novoseltsev for his direction, assistance, and guidance. In particular, Prof. Novoseltsev's recommendations and suggestions have been invaluable for the project and for software development.

References

1. Codd, E.F.: Derivability, Redundancy, and Consistency of Relations Stored in Large Data Banks. IBM Research Report RJ599, vol. 4, 12–18 (1969)
2. Darwen, H., Date, C.J.: The Third Manifesto. ACM SIGMOD, 1. Record 24 (1995)
3. Chris, J.: Date: Introduction to Database Systems, 8th edn. Addison-Wesley (2003)
4. Celko, J.: Trees in SQL. Some Answers to Some Common Questions About SQL Trees and Hierarchies. Intelligent Enterprise Magazine (2010)
5. Kyte, T.: Expert Oracle Database Architecture: 9i and 10g Programming Techniques and Solutions. Apress (2005)
6. Codd, E.F.: A Relational Model of Data for Large Shared Data Banks. CACM 13(6), 30–42 (1970)

A Building Method of Virtual Knowledge Base Based on Ontology Mapping

Huayu Li[1] and Xiaoming Zhang[2]

[1] College of Computer & Communication Engineering, China University of Petroleum, Qingdao 266555, China
[2] School of Information Science & Engineering Hebei University of Science and Technology, Shijiazhuang 050018, China
lhyzj@upc.edu.cn, zxm1975@gmail.com

Abstract. For distributed relational data sources in oilfield domain, we propose a semantic integration method of building virtual knowledge base using ontology mapping technology. In this method, two kinds of mappings are created and two ontologies designed specially to record these mappings, by which semantic query can be rewritten into SQL statements to be executed on data sources, and results will be returned as knowledge instances. Moreover, a knowledge exchange center is provided to cache some results to improve query efficiency. Since query results directly come from data sources rather than by instances transformation, this query-driven integration system can provide a unified, virtual knowledge base which can be served to achieve decision support for oilfield production.

Keywords: semantic integration, virtual knowledge base, ontology mapping.

1 Introduction

At present, research of semantic integration mostly focuses on specific domains, and ontology mapping is used as main technology which can describe relationships between global and local concepts clearly, and they will be served as important reference for instances transformation or query rewriting. Instances transformation method includes two procedures: first, extraction rules are defined to map data source model into a local ontology and mappings are recorded; second, records of data sources are transformed into ontology instances to build Knowledge Base to support semantic query. This method can achieve complex query requirement and is applicable to those data sources with low update frequency, SOAM [1] and Stojanovic [2] belong to this kind of method. Another is query rewriting method and it can obtain results directly from data sources by rewriting algorithm, which can convert a semantic query into a group of SQL statements to be executed at data sources. This query-driven method is suitable for those data sources with such features as real-time update and moderate scale. For example, Dart Grid [3] and INDUS [4] all make use of this method.

For distributed relational data sources of Well-Engineering, we propose a semantic integration solution of building virtual KB by ontology mapping technology. This solution first establishes global and local ontologies to describe domain-concept models.

Y. Wang and T. Li (Eds.): Knowledge Engineering and Management, AISC 123, pp. 597–602.
springerlink.com

Second, a mapping ontology is created to record two kinds of mappings: one is between global and local ontologies; the other is between local ontology and data sources model. Finally, a semantic query rewriting process is executed to get results from data sources, which will be converted into ontology instances to populate KEC (Knowledge Exchange Center). KEC is used to cache part of instances to reduce consumption caused by data conversion and loading.

Reminder of this paper is organized as follows: Section 2 illustrates implementation framework; Section 3 describes building method of global and local ontologies; Section 4 describes realization mechanism and expression method of ontology mapping; Section 5 introduces execution process of semantic query and rewriting algorithm; Section 6 gives conclusions.

2 Implementation Framework

According to proposed solutions, we implement WeVKB (Well-Engineering Virtual Knowledge Base) system. As shown in Fig. 1, WeVKB involves three layers: User Layer, Virtual-KB Layer and Wrapper Layer.

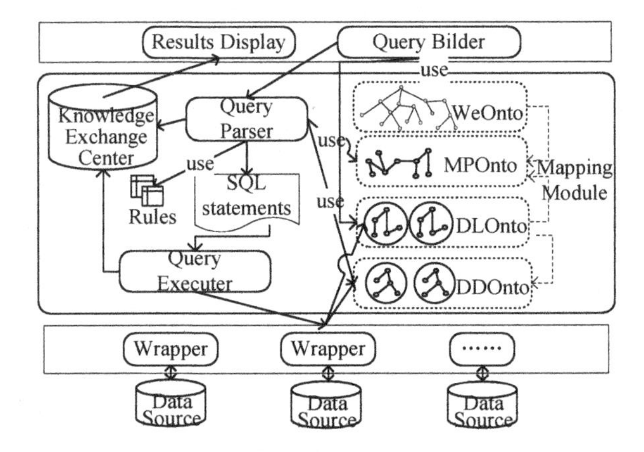

Fig. 1. Framework of WeVKB is consisted of three layers. By their respective functions, this system can provide an effective semantic integration platform.

(1) User Layer: it provides users query interfaces and contains two components: Query-Builder and Query-Display. The former takes charge of creating SPARQL query Q_S according to requirements set by users. The function of Query-Display is to reorganize the results sent by KEC and display them in certain format.

(2) Virtual-KB Layer: it is the core of WeVKB and completes two functions: receive Q_S from Query-Builder and rewrite it into a group of SQL statements to be executed by Wrapper-Layer; receive query results and returned them to KEC.

Four key components are included in Ontology-Mapping Module: WeOnto (Global Well-Engineering Ontology), DLOnto (Data Source Local Ontology), DDOnto (Data Source Description Ontology) and MPOnto (Mapping Ontology). WeOnto and DLOnto describe global and local domain concepts model, which are displayed by

Query-Builder as hierarchical structures for user to set query requirements. Two kinds of mappings need to be created for query rewriting: one is between WeOnto and DLOnto; the other is between DLOnto and data source model. MPOnto and DDOnto are defined to describe them respectively in standard format. KEC saves the latest query results. When a semantic query is presented, it is first delivered to KEC, if no results or insufficient results returned, the whole query process will be executed, and the returned results will firstly populate KEC before submitted to Query-Display.

(3) Wrapper Layer: this layer includes all distributed data sources and corresponding wrappers. Data sources involve relational database such as SQL Server, Oracle and Access. Wrappers are responsible for executing the dispatched SQL statements and submit the results to Query-Executer.

3 Ontology Construction

By referring to methods proposed in [5] and [6], we construct WeOnto under guidance of domain expert. According to application requirements, concepts range of WeOnto is determined in accordance with data model of Exploration-Production database which is widely used in most oilfields. At present, we define 52 classes and 246 properties relating to production, operation, measurement and cost. Part structure of WeOnto is shown as Fig.2.

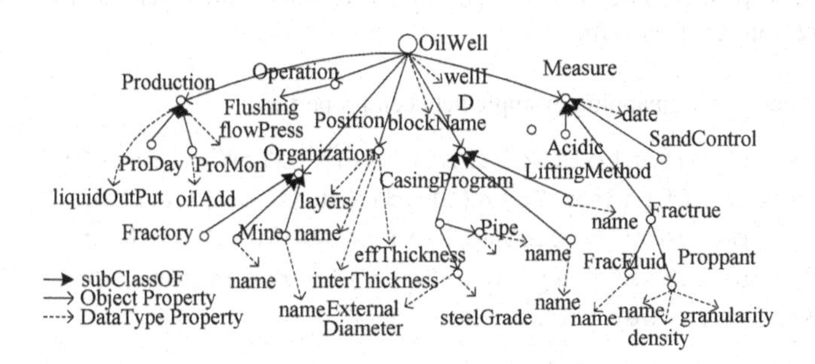

Fig. 2. *OilWell* is the root class of WeOnto. For each class, its subclass is denoted as a start node of a solid arrow line; its object property and data type property is denoted respective as an end node of a solid arrow line and a dotted arrow line.

With referential integrity constraints extracted from database schema, four procedures are defined to create DLOnto from data source by mapping table and attributes to ontology classes and properties: the first is to classify all user tables into three table Sets; the second is the definition of three extraction rules for each table Set; the third is to start a program to create initial local ontology; the fourth is the perfection process by ontology edit tool to construct the final local ontology.

4 Ontology Mapping

Two kinds of mappings exist in WeVKB. One is between DLOnto and data source model, named $M_{L\text{-}DS}$; the other is between WeOnto and DLOnto, named $M_{G\text{-}L}$.

We adopt a similar method with Relational.OWL [7] to express $M_{L\text{-}DS}$ by creating a data source discription ontology named DDOnto. Some classes and properties are defined in DDOnto to describe data source model and mappings of $M_{L\text{-}DS}$. Four objects of data source model including *database*, *table*, *column* and *primary key*, are expressed by *ddonto:Database*, *ddonto:Table*, *ddonto:Column and ddonto:PrimaryKey*; relationships among these classes are denoted by three object properties: *ddonto:hasTable, ddonto:hasColumn,* and *ddonto:hasPK*. Mappings of inter-classes and inter-properties are key information needed to be recorded in DDOnto, and they are described by two object properties: *ddonto:relatedClass* and *ddonto:relatedProperty*.

$M_{G\text{-}L}$ is expressed by MPOnto with which two kinds of mappings are expressed:

(1) Equivalence and sub-class mapping among classes: *EquivalentClass* and *subClassOf* are used as descriptors to denote this mapping.

(2) Equivalence and sub-property mapping among properties: two kinds of express forms need to be considered. For simple case, *subPropertyOf* and *equivalentProperty* can be directly used. While in complex case, a series of logical rules need to be used altogether to describe every semantic expression in RDF triples format. For instances, *oilIncrMon*, a property of $DLOnto_{DS1}$, can be defined equivalent to the property *oilAdd* in WeOnto by following rules:

Following rules gives a mapping example between properties

```
(?x rdf:type dlonto1:WellInfo)∧(?x dlonto1:meaEffect ?a)∧
(?a dlonto1:effIndex ?b)∧(?b dlonto1:oilIncrMon ?y)→
(?x rdf:type weonto:OilWell)∧(?x weonto:production ?c)∧
(?d rdfs:subClassOf weonto:?c)∧(?d rdf:type weon-
to:ProMon)∧(?d weonto:oilAdd ?y)
```

With above expressions method and *owl:imports* statement which is used to import WeOnto and DLOnto, MPOnto can distinctly describe mappings of $M_{G\text{-}L}$.

5 Implementation of Semantic Query

As a query-driven semantic integration system, WeVKB takes advantage of ontology mappings and query rewriting algorithm to achieve query process, and this process can be divided in four procedures:

(1) Q_S is first delivered to KEC to query cached instances. If returned results exists *null* values, proceeding to execute next procedure; otherwise, Query-Display presents query results to users in certain format.

(2) According to query scope of Q_S, Query-Parser reads mappings from MPOnto or DDOnto and executes query rewriting algorithm. Consequently, Q_S is rewritten into a group of SQL statements: Q_{DS1}, Q_{DS2}, … Q_{DSn}.

(3) These SQL statements are dispatched to corresponding wrappers to execute, and their returned results are denoted as R_{DS1}, R_{DS2}, … R_{Dn}.

(4) Query-Executer reorganizes all results and transforms them into ontology instances which will be submitted to KEC and Query-Display.

By constructing semantic query tree, we realize a rewriting algorithm and it involves five realization steps:

(1) Q_S is read into Query-Parser and analyzed to extract all classes, properties, operators and constants included in *Select*, *Where*, and *Filter* sub-statements. Meanwhile, relationships contained in every RDF triple of Q_S are also recorded.

(2) Basic information recorded in Step 1 are packaged into three Java objects named *NodeC*, *NodeP* and *NodeR*, with which a query tree named Q_S-Tree is build and its leaf nodes are *NodeP* or *NodeR*, branch nodes are *NodeC*.

(3) A traversal process is launched to access every node of Q_S-Tree, by mappings information, each node is mapped to one or several table object, column object or condition object named *NTab*, *NCol*, and *NCon*. According to their pre-mapped relationships in Q_S-Tree, these objects are then interconnected each other to create a SQL query graph called SQL-Graph.

(4) A search process is executed to find all sub-graphs of SQL-Graph which meets two conditions: each sub-graph is a tree including all *NCol* objects; each node in sub-graph has valid relationship conforming to the logical model of any data source. Returned sub-graphs are marked as SQL-Tree$_1$, SQL-Tree$_2$, …, SQL-Tree$_n$.

(5) For these SQL-Tree, a bottom-to-top traversal process is executed to generate a group of SQL statements named Q_{DS1}, Q_{DS2}, …, Q_{DSn}, each of which will be dispatched to its corresponding data source.

The following statement is an example of SPARQL query Q_S

```
Select ?x ?y Where { ?a rdf:type OilWell ?a weonto:
wellID ?x ?a weonto:production ?b ?b rdf:type Production
?b weonto:proMon ?y ?a weonto:blockName "Guan15-6" ?a
weonto:measure ?c ?c rdf:type Fracture ?c weonto:name
"yl" Filter (?y > 10)}
```

By this algorithm, Q_S-Tree, SQL-Graph and two SQL-Trees are build as Fig. 3. Two SQL statements are generated, and they are expressed as following statements.

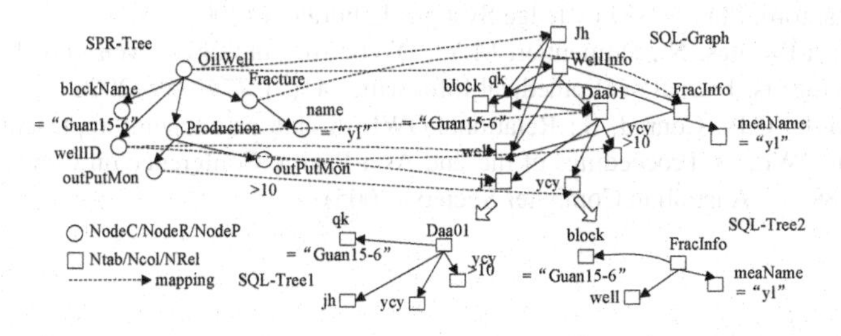

Fig. 3. The top part illustrates a process of step 3, which converts *SPR-Tree* into a *SQL-Graph*. Based on *SQL-Graph* and step 4, two *SQL-Trees* are generated and displayed at the under part.

The following two statements are the final SQL queries generated from Q_S

```
For DS1: Select fracInfo.well
         from fracInfo where fracInfo.block = "Guan15-6"
         and fracInfo.meaname = 'yl'
For DS2: Select daa01.jh, daa01.ycy from daa01
         where daa01.qk= "Guan15-6" and daa01.ycy>10.
```

6 Conclusions

In order to test proposed methods, we implement WeVKB system using Java-Struct development framework. By practical application, it can provide effective semantic integration services for oilfield production and decision makings. The future work of WeVKB is to realize automation discovery of mappings and increase efficiency of semantic query rewriting.

Acknowledgments. This work is supported by the "Fundamental Research Funds for the Central Universities": No. 27R1007007A.

References

1. Li, M., Du, X.-Y., Wang, S.: A semi-automatic ontology acquisition method for the semantic web. In: Fan, W., Wu, Z., Yang, J. (eds.) WAIM 2005. LNCS, vol. 3739, pp. 209–220. Springer, Heidelberg (2005)
2. Stojanovic, L., Stojanovic, N., Volz, R.: Migrating data-intensive web sites into the Semantic Web. In: Proceedings of the 2002 ACM Symposium on Applied Computing, pp. 1100–1107. ACM (2002)
3. Chen, H., Wang, Y., Wang, H.: Towards a Semantic Web of Relational Databases: a Practical Semantic Toolkit and an In-Use Case from Traditional Chinese Medicine. In: Proceedings of the 5th International Semantic Web Conference, pp. 750–763. Springer, Heidelberg (2006)
4. Caragea, D., Pathak, J., Bao, J.: Information Integration and Knowledge Acquisition from Semantically Heterogeneous Biological Data Sources. In: Proceedings of the Data Integration in Life Sciences, pp. 175–190. Springer, Heidelberg (2005)
5. Noy, N.F., Mcguinness, D.L.: Stanford Medical Informatics Technical Report SMI-2001-0880. Stanford: Stanford Knowledge Systems Laboratory (2001)
6. Villanueva-Rosales, N., Dumontier, M.Y.: OWL: An ontology-driven knowledge base for yeast biologists. Journal of Biomedical Informatics 41(5), 779–789 (2008)
7. De Laborda, C.P., Conrad, S.: Relational.OWL: a data and schema representation format based on OWL. In: Proceedings of the 2nd Asia-Pacific Conference on Conceptual Modeling, pp. 89–96. Australian Computer Society (2005)

Research on Computer Science Domain Ontology Construction and Information Retrieval

Dandan Li, Jianwei Du, and Shuzhen Yao

School of Computer, Beihang University, Beijing, 100191, China
lidandan04@163.com, djw9@163.com, szyao@buaa.edu.cn

Abstract. The paper combined seven-step method and the skeleton method of ontology constructions, and proposed computer domain ontology construction method. On the basis of computer domain ontology, the paper discussed semantic retrieving technology and finally, we design and implement a computer literature retrieval system prototype which has certain reasoning and semantic expansion capabilities.

Keywords: Domain ontology construction, Information retrieval, Semantic retrieval.

1 Introduction

Along with the rapid development of Internet, information on the web grow explosively, which lead to "information overload" and "information lost " getting more and more serious. Due to the present information retrieval has more information but less knowledge, experts and scholars had proposed semantic retrieval technology, also known as knowledge retrieval. Ontology can capture knowledge of specific domains by using object-oriented way, and express the relationships among concepts[1].By means of ontology, information retrieval is improved from a keyword-based retrieval to knowledge-based search that have the reasoning ability and help users to build semantic query, so as to improve service quality of information retrieval.

Taking computer science and technology domain for example, this paper discusses ontology construction and semantic retrieving technology. On this basis, we design and implement a computer literature retrieval system prototype which has reasoning and semantic expansion capabilities.

2 Computer Science Domain Ontology Construction

International research teams concluded a number of ontology construction methods, in which the more influential such as: skeleton method[2], enterprise modeling method[3], methontology method[4], seven-step method[5],etc.

Skeleton method provided guidelines for ontology construction. The goal of enterprise modeling method is to establish an integrated ontology for business and public enterprise modeling. Methontology method is to construct chemical ontology specially

Y. Wang and T. Li (Eds.): Knowledge Engineering and Management, AISC 123, pp. 603–608.
springerlink.com

and seven-step method has seven steps for constructing ontology. Although computer science knowledge and concepts are clear, the concepts are multiple and dispersed. Therefore, in the ontology construction, we need a guideline. The construction process of seven-step method is clear and universal. So we combined seven-step method and the skeleton method with the actual needs, then proposed computer domain ontology constructing process, showed in Fig. 1.

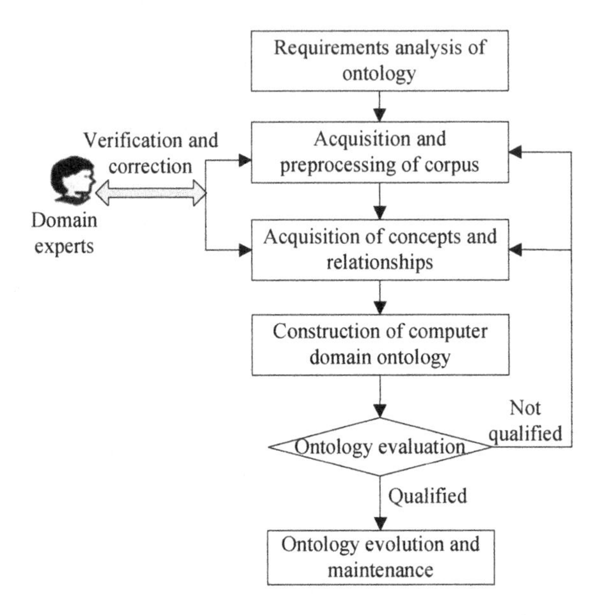

Fig. 1. Computer Science Domain Ontology Construction Process

1) Requirements analysis of ontology. The domain ontology describes concepts and relationships of the domain of computer science and technology.

2) Acquisition and preprocessing of corpus and books that can be referenced.

Currently, Department of Computer Information and Technology of Fudan University provided 20 domains of corpus. We selected and filtered out 100 representative documents from computer corpus as the original corpus. Then develop ICTCLAS [6] secondly to achieve word segmentation, removing stop words and other pretreatment.

Professional books and thesaurus of computer domain, such as" Encyclopedia of Computer Science and Technology", "Computer Science and Technology Chinese Thesaurus" which collected computer domain standard and authoritative core professional vocabulary, are the important source of computer domain.

3) Acquisition concepts and relationships of computer domain. Using feature words weight calculation algorithm TF-IDF calculates feature words' weights, extract the maximum weight of 10 feature words, constitute computer domain's candidate concepts set.Ultimately, 220 feature words were selected as core concepts, which formed a computer domain concepts hierarchical structure. Part of the structure showed in Fig. 2.

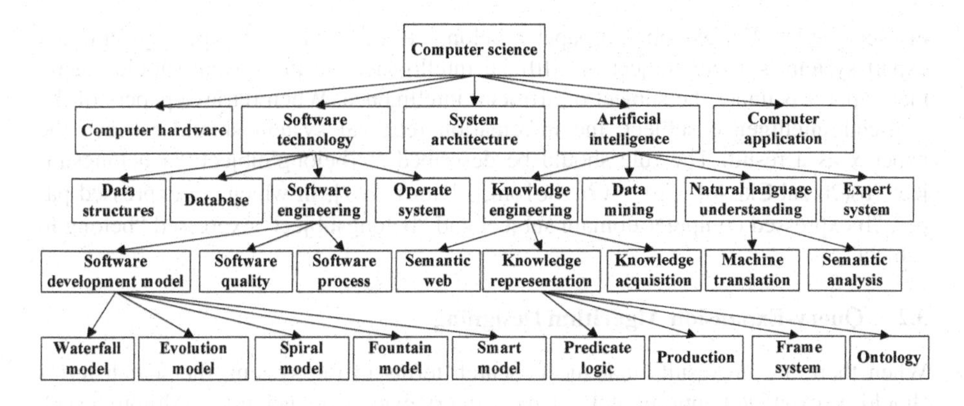

Fig. 2. Computer Science Domain Concepts Hierarchical Structure

4) Construction of computer domain ontology. We choose the ontology construction tool protégé3.4.4 to formalize the computer science knowledge.

5) Evaluation and evolution of ontology. Refer to criteria(clarity, coherence, extensibility, minimal encoding bias and minimal commitment) of ontology modeling to evaluate the computer domain ontology. Ontology evaluation and ontology construction is a cyclical, gradually improvement process.

3 Ontology-Based Information Retrieval

The recall and precision rate of current information retrieval system are not high. There are two main reasons. Firstly, the system does not have the reasoning capabilities and couldn't reason implicit information of user retrieval words. Secondly, systems based on keyword matching does not have the semantic analysis and automatic expansion capabilities. Therefore, we can improve the retrieval quality in two aspects:

1) Building domain reasoning rules to make the retrieval system reasonable.
2) Expanding search concepts by the relationships among concepts of ontology.

3.1 Establishing of Domain Reasoning Rules

Based on the concept level and attribute relationships of computer domain ontology, could reason the searching words. Usually consider object attributes such as parent class, subclass, symmetric, transitive, inverse, as well as the instance relationships between concepts and individuals. Through the establishment of certain rules, reason out of implicit knowledge data which have semantic relations. Rules described derivation relations among domain knowledge concepts. Domain reasoning rules formal format is:[rule-name:(X R_1 Y)(Y R_2 Z)→(X R_3 Z)], in which, rule-name is the name of the rule, R_1 is the concept of the relationship between X and Y, R_2 is the relationship between concept Y and Z, which can reason out the relationship between X and Z is R_3. Therefore, by establishing of domain rules could reason out potential semantic relations based on defined relations of concepts, so that achieve semantic understanding of user queries. For example, a paper x belongs to the subject of expert system, and expert system is a sub-subject of artificial intelligence, so the system should reason that paper x is under the subject of artificial intelligence. When retrieve papers of the artificial intelligence subject, the information retrieval system should retrieve the paper x as a result. The rule should be described as:[belongSubject:(?a belongsubject ?b)(?b subClassof ?c) → (?a belongsubject ?c)], in which, ?a expressed paper, ?b expressed computer domain subject and "belongsubject" expressed "belong to subject".

3.2 Query-Expansion Algorithm Designing

When there are no results that meets search terms in the system, the search terms should be expanded appropriately, namely query expansion technique. Although such queried results may not fully consistent with expected results, but can also be sufficiently relevant. The domain ontology hierarchy relationship can be used as a basis for the expansion. That commonly used to expand query relationships are: synonyms relation, hypernym(father concepts), hyponym(subclass concepts).

To ensure correlation between the retrieval results with query terms, the depth of query expansion should be limited. For example, based on hypernym and hyponym relationships, the expansion depth can be limited for 1.

In the Ontology-based literature retrieval system, query term is a collection of some key words. For the convenience of description, the following will just describe query expansion steps of one keyword K.

Step 1: Supposing the query term users entered is K, if K is empty, the query is end,or turn to Step 2;

Step 2: Search for the concept C could match query term K in computer domain ontology. If C does not exist, that expresses the concept haven't be added into the ontology, then the query term K will be directly used for building query and search literature indexing ontology. If C exists, turn to Step 3;

Step 3: Activate synonyms set of the concept C, so: Synonym={Syn_1,Syn_2,...,Syn_n};

Step 4: Activate the hypernym set of the concept set{C, Synonym},so: Sup={Sup_1,Sup_2,...,Sup_n}.Therefore, after synonyms words and hypernyms expansion, the query term set C_Set={C,Synonym,Sup},which is used as new query term to bulid queries to retrieve, if the retrieval result is null, turn Step5 or return search results;

Step 5: Activate the hyponym set Sub of concepts set{C, Synonym},then Sub={Sub_1,Sub_2,...,Sub_n},which is used as query term to build queries to retrieve, if the retrieval result is null, the retrieval is end; or return the search results.

4 Design and Implementation of Literature Retrieval System

The literature retrieval system designed in the paper is mainly used to provide efficient retrieval service quality for paper searchers. It should have query term expansion capabilities, literature retrieval results expansion capabilities and the fast response.

B/S three-tier system architecture is used for designing the system. And it is divided into: the presentation layer, business logic and data layer. There are two kinds of users: ontology maintenance personnel and literature retrieval personnel. Showed in Fig. 3.

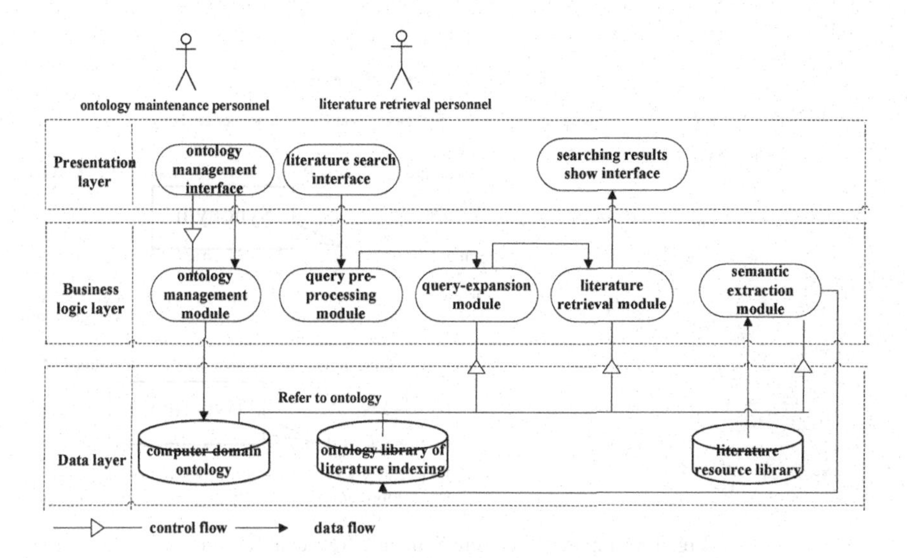

Fig. 3. Literature Retrieval System Overall Architecture

1) Presentation layer. It provides literature search interface and ontology management interface for retrieval personnel and ontology maintenance personnel respectively. Literature search interface is used to accept search request and display results.

2) Business logic layer. Ontology management module is responsible for evolution and maintenance of the ontology. Referring to ontology, semantic extraction module extract the title, keywords, authors, author affiliations, publication date and other metadata of a literature from literature library. Query pre-processing module accept input query terms and make some preliminary handling. Then the query terms will be transferred to the query-expansion module and they will be expanded referring to the synonyms, hypernym and hyponym relationships in computer domain ontology. Finally, according to the expanded query terms, the literature retrieval module build queries to search ontology indexing library, then return the search results.

3) Data layer. There are computer domain ontology library, ontology library of literature indexing and literature resource library. Ontology library of literature indexing is the index between computer domain knowledge and literature resource.

In the retrieval process, the proposed query-expansion algorithm is used to expand the search terms automatically then build new queries for searching literature indexing library. The retrieval system focuses on interaction with users and provides professional concepts such as synonyms, hypernyms and hyponyms in search results interface, hereby users can quickly find and be clear about retrieval needs. And the system also show the articles list of concepts set after expanded. Showed in Fig. 4.

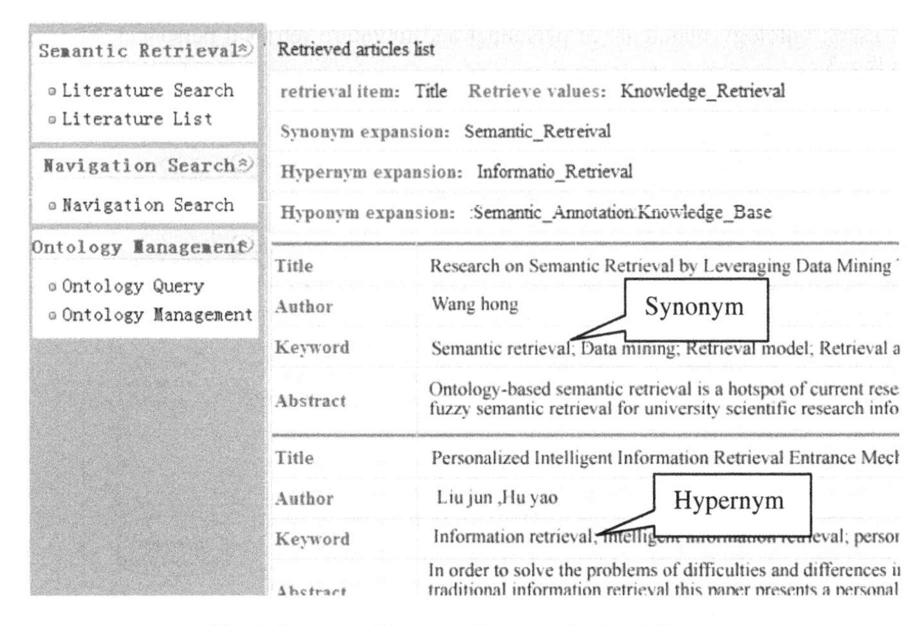

Fig. 4. Computer Literature Semantic Retrieval System

5 Conclusion

In the paper, we designed and implemented a computer literature retrieval system prototype based on domain ontology. After testing, the system have a certain reasoning and semantic expansion capabilities. Compared with the traditional keyword search, the system has better accuracy. In future studies, we will further achieve the intelligent semantic search for large-scale literature resources.

References

1. Studer, R., Benjamins, V.R., Fensel, D.: Knowledge Engineering, Principles and Methods. Data and Knowledgeing 25(122), 161–197 (1998)
2. Bruce, K.B., Cardelli, L., Pierce, B.C.: Comparing Object Encodings. In: Ito, T., Abadi, M. (eds.) TACS 1997. LNCS, vol. 1281, pp. 415–438. Springer, Heidelberg (1997)
3. Gruninger, M., Fox, M.S.: Methodology for the Design and Evaluation of Ontologies, Workshop on Basic Ontological Issues in Knowledge Sharing. In: IJCAI 1995, Montreal (1995)
4. Fernandez, M., Gomez-Perez, A., Juristo, N.: METHONTOLOGY: From Ontological Art Towards Ontological Engineering. In: AAAI 1997 Spring Symposium on Ontological Engineering, Stanford University, March 24-26 (1997)
5. Noy, N.F., McGuinness, D.L.: Ontology Development 101: A Guide to Creating Your First Ontology [EB/OL]
6. http://ictclas.org/

A New LED Screen Dynamic Display Technology

Wei Jin[1], Zhiquan Wu[1], Quan Liu[1], and Bocheng Sun[2]

[1] E Mei School, Southwest Jiaotong University Univeristy, Emei, 614202, China
[2] School of Information Science & Technology, Southwest Jiaotong University, Chengdu, 610031, China
{jinwei,wuzhiquan,liuquan,sunboch}@163.com

Abstract. In this paper, a new algorithm of dynamic display is introduced for design of the LED screen; it put forward drum type data organization way for any size display area LED screen. Illustrated the Corresponding relationship between LED display area X、Y coordinators and memory storage unit i、j byte address. Dynamic LED screen rolling display arithmetic and programming are completed. In order to verify the effectiveness of our new algorithm, a LED display system is designed and implemented with the single chip microcomputer VRS51L3074. The result shows that this algorithm improves the performance of dynamic display.

Keywords: LED screen, drum type data organization, dynamic display.

1 Introduction

For the LED panel is running for display data organization, the first problem to consideration is that all kinds of display effect are consider when the LED panel is displaying, for example, screen mobile and grayscale display. Any kind of screen movement can be divided into two basic mobile ways: moving horizontally and vertically mobile. Slash or curve move mobile can be regarded as a superposition of two basic mobile ways. N level grayscale display can be considered as the result of rapid alternate shows by pictures of n field different brightness. In order to achieve the fastest data output, best display effects, display data must serially arrange in memories by the order of unit board serial shift register group. In other words, arrangement of each line display data of the LED panel sequence is recorded in memory. If circumstance is possible, according to scan the scanning sequence, it will arrange every line of display data and then also arrange the data blocks.

In this paper, a new algorithm of dynamic display is introduced for design of the large LED screen. Paper put forward drum type data organization way on the LED screen, and illustrated the Corresponding relationship between display area X、Y coordinators and memory storage unit i、j byte address, then static and dynamic LED screen rolling display arithmetic and programming are completed. In order to verify the effectiveness of our new algorithm, a LED display system is designed and implemented with the single chip microcomputer VRS51L3074.

Y. Wang and T. Li (Eds.): Knowledge Engineering and Management, AISC 123, pp. 609–615.
springerlink.com

2 Data Organization of Static LED Panel

When the LED panel is analysis, the coordinate origin we use is the point upper left corner of the screen, Figure 1 shows the relationship of each unit board, screen coordinates of points on the X, Y, and the scan straight line. Any point only maps bytes of one in memory on LED screen when static displays.

For double color LED screen data display organization, red scan line can be corresponded to D0、D2、D4、D6, and green scan line correspond to D1、D3、D5、D7, that is to say , we use two bits in a byte to indicate one point in LED panel. The rest is the same to monochromatic LED display. When using an 8-bit microcontroller's 8 parallel port to drive two-color LED display screen having the same length. In vertical coordinate, two-color LED display has half points than monochromatic LED display. In order to show gray scale of eight levels, different levels of 8 display data are also in order. In routine 1, add a 8 level gray outer circle layer can realize M-8 grade grayscale display. But only adding auxiliary hardware and output rate must be greatly improved, single-game display data can realize grayscale display really.

Because of the 8-bit microcontroller, a parallel output has only 8, we can allow its each time-sharing homologous to the two or more scan lines. Each independent (SCK) shift signals control a scan line. According to share the display region of a shift signal, the organization for displaying data is blocked. The circuit diagram is in the Fig.1 (a) below.

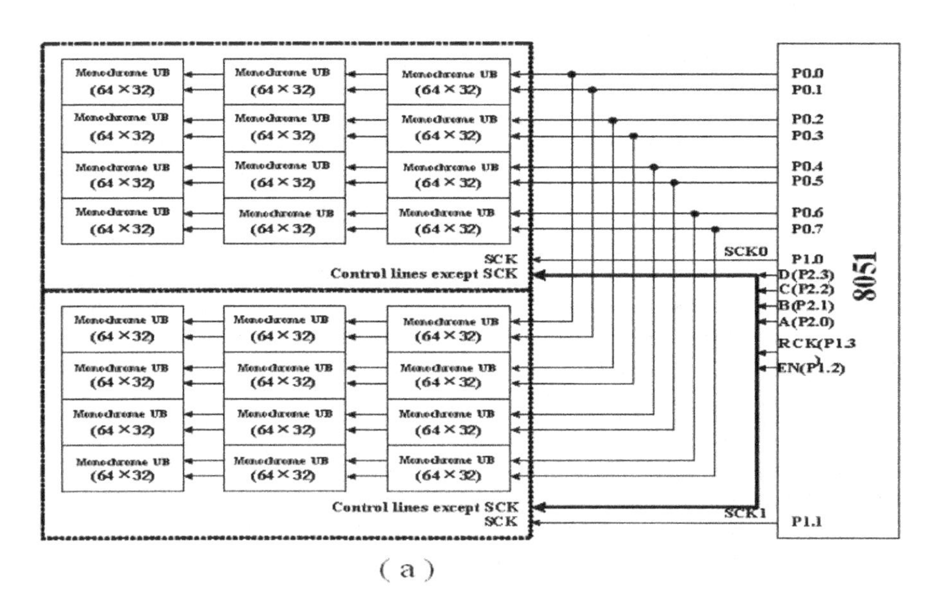

Fig. 1. The circuit diagram of increasing scan line

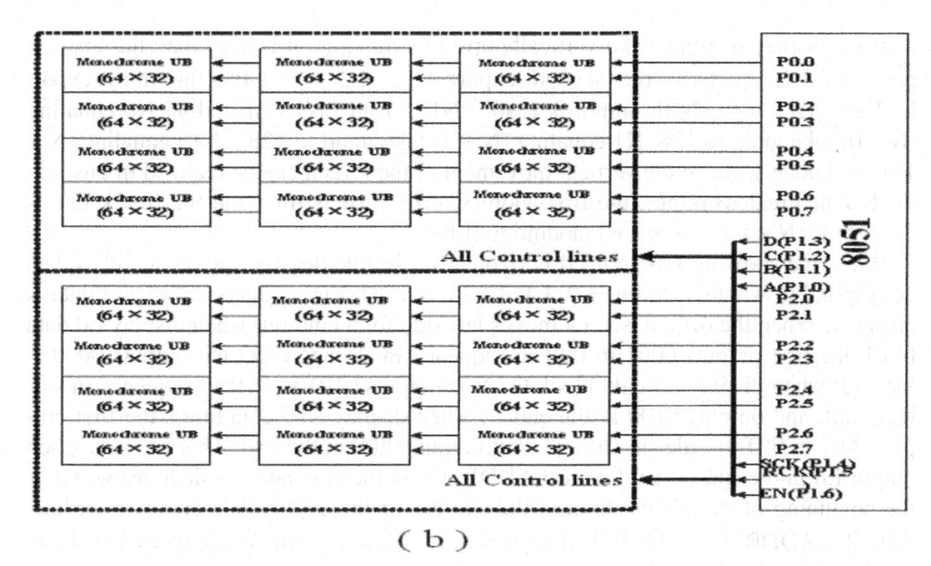

Fig. 1. *(continued)*

When the MCU is an 8-bit micro-controller, we also can use expanding hardware to realize, thus a parallel could send 16-bit or 32-bit data. For the system using 16 output and at the same time using 8-bit memory storage display data, allow low 8 bits display data stored in the even address storage unit , and high 8 bits display data stored in the odd address storage unit . So, the continuous arrange of display data can be sure. Circuit diagram shows in Fig.1 (b).

3 Static LED Screen Scrolling Display

Fig2 is other form of fig1, from fig1, fig2, D0 is only corresponding to Y = 0 to Y = 15. Figure 2 (b) shows three picture , the relative position of each scan line corresponds in three different cases: D0 to the 0 scan lines and LED screen Y = 0, D7 to the seventh scan lines and LED screen Y = 127 and D0 to the 0 scan lines and LED screen Y = 127.

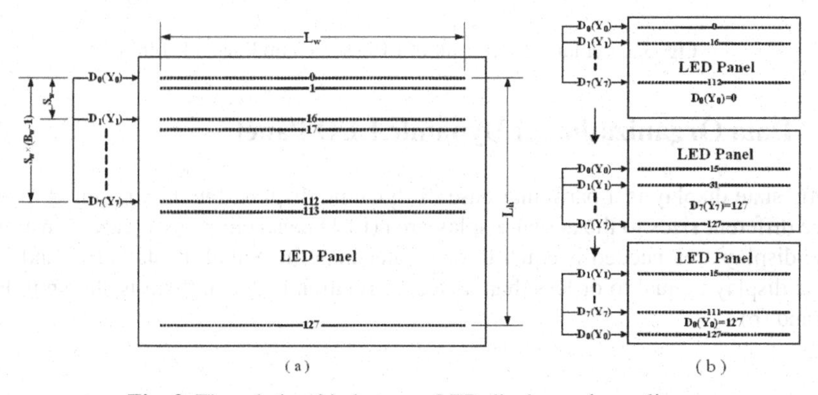

Fig. 2. The relationship between LED display and scan line

If the display screen need a vertically upwards moving , it is need that: the first export L0, L1, ..., L15; the second export L1 , L2 , …,L16; the third export L2,L3,…,L17; ...the N th output is LN0, LN1,... And LN15 (including N0 equaling (N + 0) take mold to 128, N1 equaling (N + 1) take mold to 128... N15 equaling (N + 15) take mode 128), so the vertical movement of the screen can be realized in this method. When real working , the first address of the first N time and the first k line is equal to K (N × Lw + K × Lw) modulo to (Lh).

When the showing screen need horizontal move, the display data area still is data correspond to which scan line L0, L1,... L15, but the origin address of export data is changed. When the display screen moves left side for a column: when display L0 data block the first outputs 0001 to 00ff in sequence in storage unit, and outputs 0000 at the end; when display L1 data block the first outputs 0101 to 01ff in sequence in storage unit, and outputs 0100 in the end…; when display L15 data block the first outputs 0f01 to 0fff in order in storage unit, outputs 0f00 in the end. That's to say, when output Ln in the line, if Ln line use ADDR_Ln as the first address data, using X0 as the beginning of the X-coordinates, the address of the export data can express like: ADDR = ADDR_Ln + （X0+i） [Lw] (i = 0, 1, ..., Lw-1). Among them [Lw] is meaning that (X0 + i) take modulo to Lw.

As above mentioned Vertical movement display can be regarded as connecting between the first line and last line of the screen; horizontal movement display can be regarded as connecting between the first row and last row of the screen. The relation corresponding with LED display is showed in Fig.3 (a), (b) below.

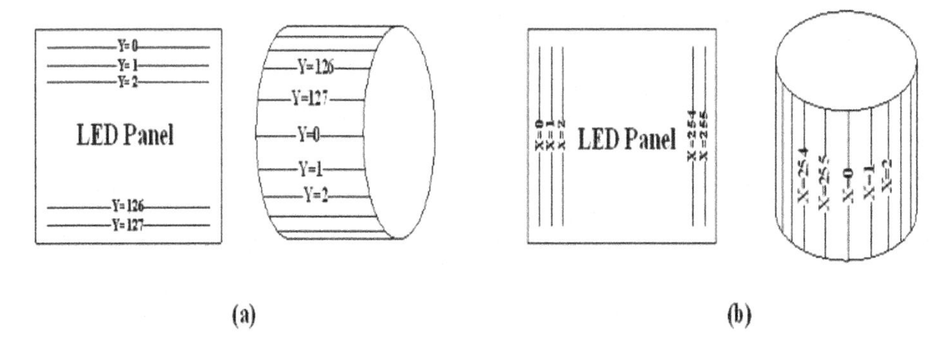

Fig. 3. Circular arrangement of LED screen up line and column

4 Data Organization of Dynamic LED Panel

Static state display is a particular case of dynamic display, the area required to show have different sizes to the actual display area (LED panel) in general case. Most of the case display area needed is equal to or greater than the actual display area, and when static display , equal to or less than area , the relation between them is shown in Fig.4 (a) and (b) .

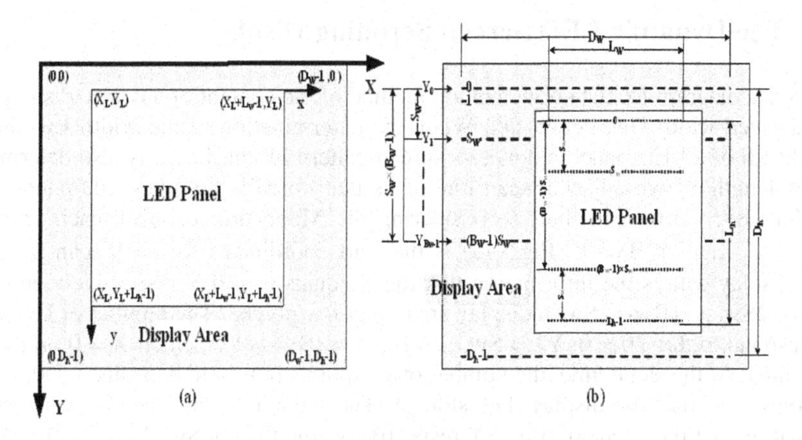

Fig. 4. Schematic of data organization methods

If need to show as in Fig4 display area, let XL, YL as the top-left corner of the LED panel origin coordinate, the width and the height are Lw, Lh, the contents of display area is shown by the LED panel, supposing the beginning address of display data block was RAM_BEGIN, we can organize data export by following steps:

1. Confirm XL, YL coordinates corresponding to origin memory unit address L0 of scan line Y0, then

$$L0=RAM\_BEGIN+(Dh+YL)\,[Dh]\times Dw+XL$$

Now Bw-bit in L0 memory unit each corresponding to (XL, YL), (XL, YL + 1 × Sw),... , (XL, YL + (Bw - 1) × Sw) ,there are totals Bw points on display area.

2. L0 corresponding to the Bw output lines: L0 corresponds Bw Y, needs to export XL, XL + 1..., X L + w - 1, total is Lw. On display area the X points can represented by below formula:

$$X=(XL+k)\,[Dw] \qquad (k=0,\ 1,\ \ldots,\ Lw-1.)$$

At the same time, consider the X cycle along horizontal directions, we can know that:

$$L0=RAM\_BEGIN+(Dh+YL)\,[Dh]\times Dw+(XL+k)\,[Dw]$$

3. outputting Sw×L0 lines complete export a complete display data: To complete a complete display, it must make the beginning points position at (XL, YL), (XL, YL + 1),... , (XL, YL + 2),... , (XL, YL + Sw - 1) which totals Sw points, in turn Sw times , and we will complete the display with the width and the height is each stand at Lw、Lh. and Lh equals Sw× Bw at his point. Due to origin point of the adjacent two lines display data differing Dw in storage, complete display data formula can be got as follows below :

$$i=RAM\_BEGIN+(Dh+YL+l)\,[Dh]\times DW+(XL+k)\,[Dw]$$

$l=0,\ 1,\ \ldots,\ Sw-1\ ;\ k=0,\ 1,\ \ldots,\ Lw-1$

Above formula shows the first address data in the 1st item display memory , the 2nd item shows starting address of data of two adjacent differs Dw, in the 3rd item , output the line display data in memory is continuous arranged.

5 The Dynamic LED Screen Scrolling Display

Fig.5 is a diagram for data organization method of the content of any size display area for a given width Dw, height Dh. When the determination of the width Lw and the height Lh of a LED panel is done, its spread pattern of unit board is also determined. If the length of two adjacent scan line of its unit board is Sw, the LED panel needs totaling Bw × Sw scan lines (considering the Monochrome only), each represent Y0、 Y1、 …、 Y Bw-1 . Fig.5 (a) is the start coordinates for each scan line. The first display data is the information that the X equals 0, and Y respectively equals to 0, Sw, 2Sw, ..., (Bw-1) Sw in display area, total Bw points . The number of Bw points correspond to the Y0 = 0, Y1 = Sw, ...,YBw-1 = (Bw-1) Sw which X = 0 on the display area. At the same time, the number of Bw points in turn to deposits in D0, D1,... , D Bw – 1 of the first display data storage; The second display data deposits the information that the X equals 1,and Y respectively equals to 0, Sw, 2Sw, ..., (Bw-1) Sw in display place, also total Bw points. According to above rules in turn it's over until X = Dw-1 total of Dw display data. Then letting Y0 = 1, Y1 = Sw + 1,... and YBW - 1 = (Bw - 1) Sw + 1 and X from 0 to Dw - 1 repeats the process.

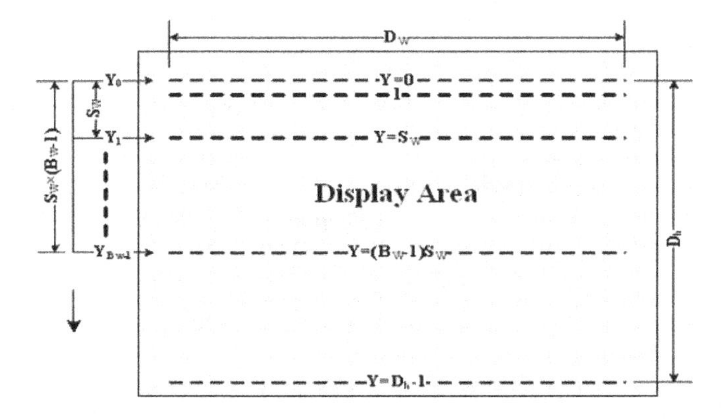

Fig. 5. Relation between scan and display area

In the Fig.6 the display area of a flat rectangular is bind up in a rotating drum along on the level, it can more graphically illustrate the data organization method. Y0, Y1, ..., YBw-1 Total Bw rigid link between scan lines and put them in the slide. When it starts, the Y0, Y1, ..., YBw-1 scan lines direct to the Y which is equal to 0, Sw, ..., (Bw-1) Sw line and raw position X = 0. The Bw scan lines move from left to right along the slide, that's to say , scanning from X = 0 to X = DW-1 can obtain the number of Dw display data and in turn to store in display memory. And the drum roll rotate by the direction of the arrow line. Let the Bw scan lines scanning from the X = 0 to X = DW-1 and from left to right along the slide repeat, and then get the DW display data number which were stored in the display memory after Previous lines' scanning data. Let the above process rotating drum rolls a circle repeat, totaling for Dh times, so Bw bytes totaling Dw × Dh Bw-bit units can be involved in the whole display data. This method of data organization and rotating drum moving along the slide has a vivid show: after Organization, display data is seriated in the X-axis direction; And series in Y directions, but in display memory Dw is a difference of positions, that's to say the distance of a line data.

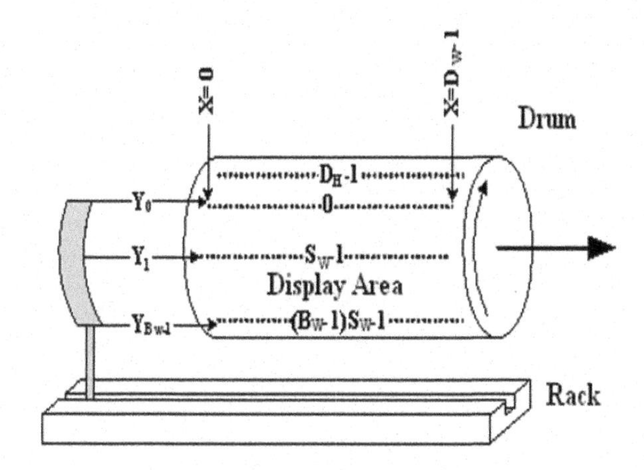

Fig. 6. Drum type data organization

6 Conclusion

Paper put forward drum type data organization way on the LED screen, and illustrated the Corresponding relationship between display area X、Y coordinators and memory storage unit i、j byte address, then static and dynamic LED screen rolling display arithmetic and programming are completed. In the follow study, relation between display result and size of data memory, the use of double RAM to reduce memory byte, are main problems we must solve.

References

1. Jin, W., Wu, Z.Q., Liu, Q.: LED Screen Developing Based On 51 Series Single Chip Microcomputer. M. Beihang University press (2009)
2. Jin, W., Wu, Z., Li, Q.: Data Organization Method Discussion Base on LED Screen of Muti-scanning-beam. Information Terminal & Display. J. 33(8), 36–39 (2009)
3. Svilainis, L.: LED Brightness Control Technology for Video Display Application. C. Displays 29(5), 506–511 (2008)
4. Liang, Y., Ma, X.: The design of LED dot-matrix graphic display system bu using the single chip computer. J. China Science and Technology Information, 99–100 (2009)

Redundant and Conflict Detection in Knowledge Management System

Kaixiang Wang and Yinglin Wang

Department of Computer Science and Engineering
Shanghai Jiao-Tong University
{seieewkx,ylwang}@sjtu.edu.cn

Abstract. In this paper, we shall first briefly introduce Knowledge Management System (KMS) and a background on the Recognizing Textual Entailment (RTE) that is the theoretical basis for our redundant and conflict detection (RCD). Based on the above concepts, a framework of redundant and conflict detection in KMS will be introduced, and some common information retrieval and RTE methods will be used in our framework. Finally, we implement our framework on an existing KMS and have an acceptable result.

Keywords: Redundant and Conflict Detection, Knowledge Management System, Recognizing Textual Entailment.

1 Introduction

In this section we shall briefly introduce KMS and the importance of redundant and conflict detection (RCD) in KMS, and then the RTE will be introduced in order to elaborate the common approach to RCD [2][6][7].

1.1 Knowledge Management System (KMS)

Knowledge Management (KM) refers to a multi-disciplined approach to achieve organizational objectives by making the best use of knowledge [9]. Building knowledge management systems to help employees of an enterprise to share their knowledge has become more and more important. However, the knowledge items provided by different employees may be redundant. Moreover, some conflicts may exist between the knowledge items. Redundant knowledge items in a knowledge base will increase the burden while knowledge items are retrieved, updated and maintained. Conflict knowledge items in KMS indicate that at least one knowledge item is not true, which is often fatal, as users of KMS will get wrong information. These conflicts may come from misunderstanding of knowledge or typing errors [8].

1.2 Recognizing Textual Entailment (RTE)

The concept "textual entailment" is used to indicate the state in which the semantics of a natural language written text can be inferred from the semantics of another one.

Y. Wang and T. Li (Eds.): Knowledge Engineering and Management, AISC 123, pp. 617–623.
springerlink.com

If the semantics of the one text (Hypothesis, H) can be inferred from the semantics of the other text (Hypothesis, T); then, it is said that textual entailment exists between both texts[5]. . Recognizing textual entailment (RTE) is a kind of difficult task. Here is an example from (RTE-4) challenge:

T: Barely six months after her marriage with the French President, supermodel Carla Bruni has admitted having problems with her "conservative" hubby Nicolas Sarkozy's "right- wing politics".

H: Carla Bruni is the French President.

From the above example, we can figure out that hypothesis is contradicted by text. In general, H and T can have the relations like "Entailment", "contradict" and "Unknown". There're some methods such as word overlap, semantic knowledge, probabilistic models or learning models are used for identifying the entailment relationship.

2 The Framework

In this chapter, we will first introduce the overall architecture of redundant and conflicts detection in KMS. It takes new knowledge items as input, and the output is the ranked existing knowledge items in the KMS in order to inform users the most likely redundant or conflict entry in the KMS, leave to the users to determine whether it is really the case.

The main goal of the system is to recommend to the user a ranked candidate list which contains the most likely redundant or conflict items in KMS, from the highest to the lowest degree of certainty. There are two points should be mainly concerned with. The first is the precision and recall; the system must give more accurate results than ordinary information search, and the accuracy here means not only the literally keyword match but also semantic similarity (which can be inferred based on some background knowledge. The second concern is the efficiency, RTE or more generally speaking the natural language processing (NLP) is a time-consuming and computing-intensive task. We must find a trade-off between accuracy and efficiency, in order to make the recommended results as accurate as possible within an acceptable time frame. Fig. 1 shows the overall architecture of our system that will be introduced in the following sections.

2.1 The Search Stage

When an employee in an enterprise input a new knowledge item, in the search stage of the system, we regard the knowledge item as a bag of words. We remove the stop words from the entry, and take the remaining words as keywords. We use Boolean search, and all the keywords are combined with the Boolean operator "OR". When the new knowledge item is very long, this Boolean OR search will return too many search results and it is inefficiency. If new knowledge item is longer than a given threshold, we just extract named entities in the new knowledge item as the keywords, and use top N search results (N is chosen based on the performance and other attributes).

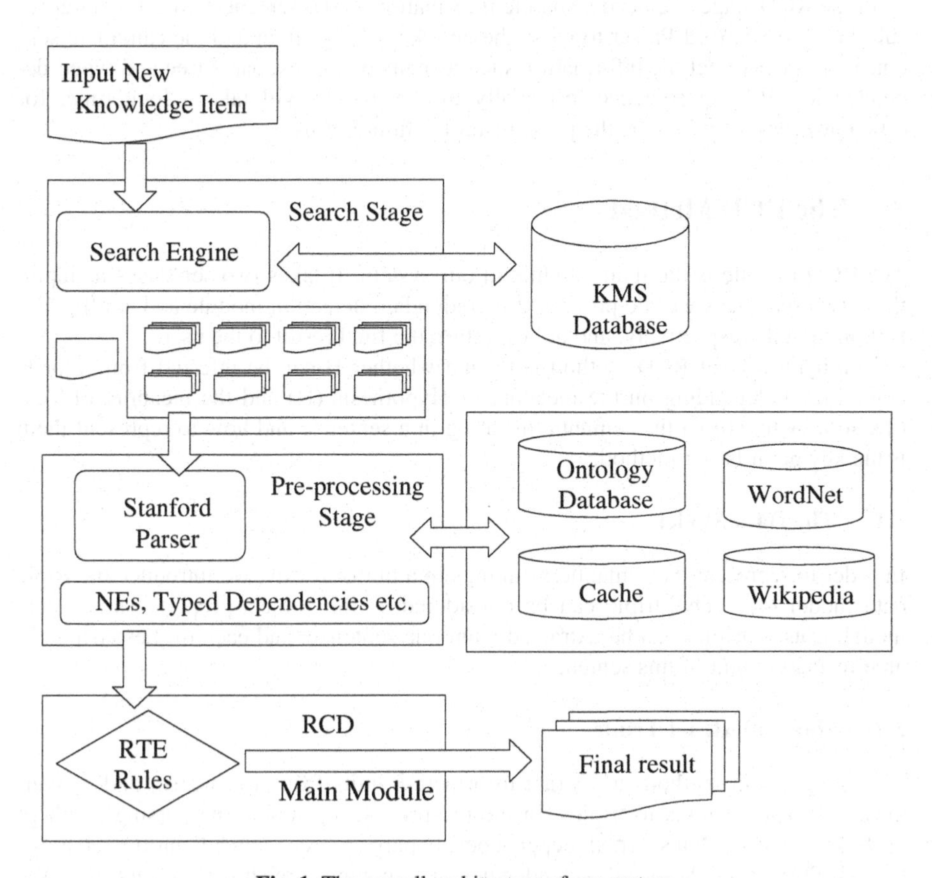

Fig. 1. The overall architecture of our system

2.2 The Pre-processing Stage

In order to improve the accuracy and performance of our system, before taking a new knowledge item and the searched candidates to the RCD stage, some additional steps are performed.

Because current RTE method can only handle the entailment between two sentences, one major task of the pre-processing stage is to select the most likely candidate sentences from the searched candidates as the input of the next stage (the RTE stage). We use a heuristic method for candidate sentence selection:

```
For each sentence (S) in the new knowledge item
    Find Named Entities (Nes) in sentence (S)
    For each sentence(S') in searched candidates
        IF S' contains at least one Ne in Nes
            Select S' as a candidate sentence
```

In the RCD stage, we need semantic information of the sentences to get a better result. We use Stanford-Parser to parse the new knowledge item and the candidate sentences in order to get the information such as named entities, parse tree and typed dependencies. Parsing sentence, especially the long one, will take a long time, for efficiency reasons we cache the parse result for future re-use.

3 The RCD Method

The RCD module is the main module in our system. It takes two sentences as input; these two sentences will be processed by redundant detection module and conflict detection module respectively, and at last, return the final result to the user.

The main task of RCD method is to find whether H can be inferred from T or H contradict T depending on the meaning of Hypothesis (H) and the meaning of Text (T), so how to extract the semantic meaning in a sentence and how to represent them is the key point in our method.

3.1 The Data Model

In order to represent the semantic meanings in a uniform way, we introduce the triple data model [4]. This triple can be considered as the "object-property-value" data model. Lots of triples can be extracted from one sentence, and each of these triples is an information unit of this sentence.

3.2 How to Extract Triple

We use different methods to extract information units from a sentence, and convert these information units to triples. In the pre-processing model, the parsing result of each sentence contains typed dependencies, parsing tree, named entities, Part of Speech(POS)tags. The typed dependencies of a sentence are shown in Table 1. We use these results to extract information units we need.

Table 1. The typed dependencies of a Chinese sentence "铝具有良好的防腐性能 (Lv Ju You Liang Hao De Fang Fu Xing Neng)"

Dependency Type	Governor	Dependent
Nsubj	具有(Ju You)	铝(Lv)
Rcmod	性能(Xing Neng)	良好(Liang Hao)
Cpm	良好(Liang Hao)	的(De)
Amod	性能(Xing Neng)	防腐(Fang Fu)
Dobj	具有(Ju You)	性能(Xing Neng)

It is extremely difficult to design a pervasive RCD system, but we can concentrate on a specific domain [3] to achieve satisfactory results. Take KMS for mechanical manufacture enterprises as an example, in the manufacture domain, numerical error tend to be a serious problem, hence, we can analyze the possible numerical descriptions in the domain, and find the methods for extracting numerical information units (triples) from a sentence to get a better performance on numerical error detection.

3.3 Redundant Detection

At this point, each of two sentences to be compared is represented as a collection of triples. To check whether a sentence H can be inferred from a sentence T, we calculate the percentage of triples contained in H that are also contained in T. We use the following formula to compute the redundant score:

$$Score_R(H,T) = \frac{\#\{triple \mid triple \in H \text{ and } triple \in T\}}{\#(triple \mid triple \in H)} \quad (1)$$

Here, we also use some extra alignment method to enhance the quality of redundant score. We use WordNet for the synonym match, and use the semantic similarity score, which computed from Wikipedia articles [1].

3.4 Conflict Detection

In this paper, for simplifying the problem, we only consider a simple situation of the conflicts: we assume that the two conflict sentences use similar words but one have negative description and the other one do not have negative description. The compute of conflict score is similar to the redundant score, because the two conflict sentences talk about the same thing but have opposite polarity. If two sentences have a relatively high $Score_R$, and have one and only one NEG triple in one sentence, we regard these two sentences conflict to each other. NEG triple is extracted from typed dependencies of parser result. We can also make different rules for different kinds of confliction. The common types of conflict are shown in table 2.

If two tuples have the same object and property, and have different numerical values, we can regard this as a confliction. A numeric conflict is shown in table 3. From the "Numeric" row, firstly, we parse these two sentences and get the typed dependencies. We extract tuple [展弦比, value range, 6 ~ 8] from the key dependency (the "range" dependency), and we can extract the same type of tuple [展弦比, value range, < 7] from the hypothesis。These two numeric ranges are not totally matched, so we regard this as a confliction and present these two conflict sentences to the user.

Table 2. The common types of conflict

Type	Text	Hypothesis
Negation	铝具有很好的防腐性。 Aluminum has good corrosion resistance.	铝的防腐性并不是很好。 Corrosion resistance of aluminum is not very good.
Antonym	纯铝的强度明显劣于铝合金。 Pure aluminum intensity was significantly inferior to the aluminum alloy.	纯铝在强度上优于铝合金。 Pure aluminum is superior in strength to the aluminum alloy.
Numeric	后掠翼的展弦比为6～8。 The gap/span ratio of Sweptback wing is 6~8.	机翼后掠翼的展弦比小于7即可。 The gap/span ratio of sweptback wing is less than 7.
Structure	固定翼飞机包括滑翔机、旋翼机。 Fixed-wing aircraft, including gliders, rotorcraft.	旋翼机不属于固定翼飞机范畴。 Rotorcraft category does not belong to fixed-wing aircraft,
Logic	当飞机遇到气流扰动时，机身会剧烈震荡。 When the plane encountered turbulence, the body will be turbulent.	飞机的震荡是因为气流的扰动。 The turbulence of aircraft is because of air turbulence disturbance.

Table 3. A numeric conflict example

Sample	后掠翼的展弦比为6～8。	机翼后掠翼的展弦比下于7即可。
Typed dependencies	assmod(展弦比-3, 后掠翼-1) assm(后掠翼-1, 的-2) **nsubj(为-4, 展弦比-3)** conj(～-6, 为-4) **attr(为-4, 6-5)** **range(～-6, 8-7)**	lobj(后-2, 机翼-1) loc(掠翼-3, 后-2) rcmod(展弦比-5, 掠翼-3) cpm(掠翼-3, 的-4) **nsubj(小于-6, 展弦比-5)** **range(小于-6, 7-7)** advmod(可-9, 即-8) conj(小于-6, 可-9)
Key typed dep.	nsubj(为-4, 展弦比-3) attr(为-4, 6-5) range(～-6, 8-7)	nsubj(小于-6, 展弦比-5) range(小于-6, 7-7)
Tuple	[展弦比, value range, 6~8]	[展弦比, value range, < 7]
Conclusion	Range Mismatch Conflict.	

4 Results and Conclusion

We implement the above framework in an ontology-based complex-product-design knowledge management system. All these design is under Chinese language. In this system, there is always tradeoff between result quality and system performance. For example, using synonyms from WordNet or Wiki will improve the accurate of final result, at other hand; it will lead to a lot of extra computational complexity. Quality corpus can greatly enhance the final results, especially named entity, synonyms and antonyms thesaurus. In some domain-specific area, domain ontology is a good enhancement for our system. As quantifying the final results need too many workload, we just give an intuitive result of our system. Overall, this system returns results within acceptable amount of time. And the top results can be a good representation of redundant and conflict entries in the KMS.

5 Discussion and Future Works

In the future, we will try to use concept hierarchy information of the domain ontology, for a better concept inference. For example, the derived class should contain the attributes in base Class. On the other hand, we will try to use distributed computing to reduce the overall computation time in the generation of the parse tree of each sentence in the knowledge base.

Acknowledgments. Thank the National High Technology Research and Development Program of China (Grant No. 2009AA04Z106) and the Natural Science Funds of China (Grant No. 60773088) for financially supporting this research.

References

1. Gabrilovich, E., Markovitch, S.: Computing Semantic Relatedness using Wikipedia-based Explicit Semantic Analysis. In: Proceedings of the 20th International Joint Conference on Artificial Intelligence, IJCAI 2007 (2007)
2. Li, F., et al.: THU QUANTA at TAC 2009 KBP and RTE Track. In: Proceedings of Text Analysis Conference (2009)
3. de Marneffe, M.-C., Rafferty, A.N., Manning, C.D.: Finding Contradictions in Text. In: Proceedings of ACL 2008: HLT, Columbus, Ohio, USA, June 2008, pp. 1039–1047 (2008)
4. Katz, B., Felshin, S., Yuret, D., Ibrahim, A., Lin, J., Marton, G., McFarland, A.J., Temelkuran, B.: Omnibase: Uniform Access to Heterogeneous Data for Question Answering. In: Andersson, B., Bergholtz, M., Johannesson, P. (eds.) NLDB 2002. LNCS, vol. 2553, pp. 230–234. Springer, Heidelberg (2002)
5. Main Task and Novelty Detection Subtask Task Guidelines. 6TH Textual Entailment Challenge @ TAC (2010)
6. Vanderwende, L., Menezes, A., Snow, R.: Microsoft Research at RTE-2: Syntactic Contributions in the Entailment Task: an implementation. Second Recognizing Textual Entailment Challenge (2006)
7. Cuza, A.I.: Textual Entailment (2009)
8. Knudsen, M.P.: Knowledge Redundancy: Suggesting and Testing an Empirical Construct for Knowledge Sharing. In: DRUID Summer Conference (June 2007)
9. http://www.unc.edu/~sunnyliu/inls258/Introduction_to_Knowledge_Management.html

Part VII
Formal Engineering & Reliability

Measuring Method of System Complexity

Guoning Zhang and Chunli Li

NanJing Army Command College, Nanjing, 210045, China
guoning_zh@yahoo.com.cn

Abstract. According to the idea of Kauffman's NK model, and considering the meaning of the measurement of complexity that a system possessing, a method of how to measure the complexity of a system was built. Here, a system was divided to three parts: itself, input and output. Each part had its elements, and among all the elements of the three parts there are relations, so the matrixes can be built based on the relations, therefore, with the idea of NK model, the complexity can be calculated through the matrixes.

Keywords: Method, Measurement, Complexity, NK model.

1 Introduction

At present, complexity of system is distinguish to structural complexity and evolutive complexity, and methods of measurement of the complexity consist of *Lyapunov* exponent method, associated dimensions method, *Lemple-Ziv* method, information entropy method and systematic entropy method and so on[1,2,3,4]. For two types of complexity of a system, the writers believe that they do not have essential differences at all, considering the relations among the system and between the system and its environments. As for these methods of measurement of complexity, the anterior three are used on the single data series generally, and the entropy methods, to the definition of it , the entropy means the out-of-order degree of a system, but for the complexity in the complexity science, its meaning do not equal the out-of-order degree of a complex system evidently, so we can say that to calculate the complexity of a system through the entropy, is irrelevant, although the method is useful under some conditions.

Stuart Kauffman, a researcher of the SFI, put forward a NK model to describe the fitness of evolving biology. Using its idea for reference, a new measuring method of system complexity was designed.

2 Briefs of the NK Model

In the NK model Kauffman uses Sewell Wright's idea of a fitness landscape in which agents seek to move from "fitness valleys" to higher "fitness peaks". At higher points in the landscape survival is more likely and the risk of extinction reduced.

The N parameter defines the number of characteristics each agent shall have, while K defines the degree of inter-dependence of each agent, and the effect is called epistasis. Each point in the landscape is identified by a kind of coordinate, and can be

Y. Wang and T. Li (Eds.): Knowledge Engineering and Management, AISC 123, pp. 627–630.
springerlink.com

viewed as a vertex in N -dimensional hypercube. The fitness f of a point d = d_1 , . . . , d_N is then:

$$f(d)=\frac{1}{N}\sum_{i=1}^{N}f_i(d_i,d_{i1},\ldots,d_{iK}) \quad (1)$$

Where the fitness contribution f_i of the gene at locus i depends on the allele (value of the gene) d_i and K other alleles d_{i1} , . . . , d_{iK} .

Kauffman's NK model was originally conceived as a means of exploring, through parameter variation, the correlation of fitness landscapes in biological evolution and speciation. In "The Origins of Order", Kauffman describes the NK Model in the biological context for which he designed it and not as the generic model he also claims it may be. However, in the years since, the NK model has become a common tool in the exploration of biological evolution, and more recently popular with business analysts applying it to anything from situated learning theory to manufacturing strategies [5].

3 Route of the Method

For a system, it can be divided into three parts: itself, input and output. The parameter N defines the number of elements the system have itself, while parameter S defines the outputs of the system, and parameter L defines the Inputs of the system. the elements, the outputs and the inputs of the system can be distinguished to several states, which can be denoted as $\theta_i \in \Theta$, $\theta_i^m \in \Theta^m$, $\theta_i^e \in \Theta^e$. According to the relations among the elements, between the elements and outputs, between the elements and inputs, build the relation matrixes K, M, and E, and calculate the complexity of a system, taking the numbers, states and relations into account.

4 Measurement Model of System Complexity

The measurement model consists of three sub-models.

4.1 Measurement Model of Complexity Derived from Inner Elements

The relation matrix of inner elements of a system as below:

Table 1. Matrix of relations among inner elements of a system

	$elment_1$	$elment_2$	$\cdots$	$elment_j$	$\cdots$	$elment_n$
$elment_1$	k_{11}	k_{12}	$\cdots$	k_{1j}	$\cdots$	k_{1n}
$elment_2$	k_{21}	k_{22}	$\cdots$	k_{2j}	$\cdots$	k_{2n}
$\vdots$	$\vdots$	$\vdots$	$\vdots$	$\vdots$	$\vdots$	$\vdots$
$elment_i$	k_{i1}	k_{i2}	$\cdots$	k_{ij}	$\cdots$	k_{in}
$\vdots$	$\vdots$	$\vdots$	$\vdots$	$\vdots$	$\vdots$	$\vdots$
$elment_n$	k_{n1}	k_{n2}	$\cdots$	k_{nj}	$\cdots$	k_{nn}

The symbol k_{ij} denote whether the element i has influence to element j, if j carry out its function, if it does, k_{ij}=1, or k_{ij}=0. So the complexity derived from inner elements system C^k can be calculated as formula 2:

$$C^k = \frac{1}{N}\sum_{i=1}^{N}\left[\prod_{j=1}^{n}\theta_j^{k_{ij}} \cdot \left(\sum_{j=1}^{n} k_{ij}\right)\right] \cdot \quad (2)$$

4.2 Measurement Model of Complexity Derived from Relations between System and Outputs

An output of a system is corresponded to some elements of the system but not all of the elements generally, so according the action relations, a matrix can be built as table 2 shows:

Table 2. Matrix of elements between elements of a system and outputs

	$output_1$	$output_2$	$\cdots$	$output_j$	$\cdots$	$output_s$
$element_1$	m_{11}	m_{12}	$\cdots$	m_{1j}	$\cdots$	m_{1s}
$element_2$	m_{21}	m_{22}	$\cdots$	k_{2j}	$\cdots$	k_{2s}
$\vdots$	$\vdots$	$\vdots$	$\vdots$	$\vdots$	$\vdots$	$\vdots$
$element_i$	m_{i1}	m_{i2}	$\cdots$	m_{ij}	$\cdots$	m_{is}
$\vdots$	$\vdots$	$\vdots$	$\vdots$	$\vdots$	$\vdots$	$\vdots$
$element_n$	m_{n1}	m_{n2}	$\cdots$	m_{nj}	$\cdots$	m_{ns}

The symbol m_{ij} in the matrix denote whether the element i has influence to output j, if j carry out its function, if it does, m_{ij}=1, or m_{ij}=0. So the complexity derived from relations between system and outputs C^m can be calculated as formula 3:

$$C^m = \frac{1}{S}\sum_{j=1}^{S}\left[\prod_{i=1}^{N}\theta_i^{m_{ij}} \cdot \left(\sum_{i=1}^{N} m_{ij}\right)\right] \cdot \quad (3)$$

4.3 Measurement Model of Complexity Derived from Relations between System and Inputs

If a system wants to keep its functions and to remain dynamic stability, it must have many kinds of continuous supplies; usually different element of the system needs different supply combinations, so according the supplying relations, a matrix can be built as table 3 shows:

Table 3. Matrix of elements between elements of a system and inputs

	$element_1$	$element_2$	$\cdots$	$element_j$	$\cdots$	$element_n$
$input_1$	e_{11}	e_{12}	$\cdots$	e_{1j}	$\cdots$	e_{1n}
$input_2$	e_{21}	e_{22}	$\cdots$	e_{2j}	$\cdots$	e_{2n}
$\vdots$	$\vdots$	$\vdots$	$\vdots$	$\vdots$	$\vdots$	$\vdots$
$input_i$	e_{i1}	e_{i2}	$\cdots$	e_{ij}	$\cdots$	e_{in}
$\vdots$	$\vdots$	$\vdots$	$\vdots$	$\vdots$	$\vdots$	$\vdots$
$input_l$	e_{l1}	e_{l2}	$\cdots$	e_{lj}	$\cdots$	e_{ln}

The symbol e_{ij} in the matrix denote whether the element i has influence to output j, if j carry out its function, if it does, e_{ij}=1, or e_{ij}=0. So the complexity derived from relations between system and outputs C^e can be calculated as formula 4:

$$C^e = \frac{1}{L}\sum_{i=1}^{L}\left[\prod_{j=1}^{n}\theta_j^{e_{ij}}\cdot\left(\sum_{j=1}^{n}e_{ij}\right)\right]. \tag{4}$$

4.4 Combination Measurement Model of System Complexity

The total complexity of system C can be attained from $C^k \cdot C^m \cdot C^e$, i.e.:

$$C = \frac{1}{NSL}\cdot\sum_{i=1}^{N}\left[\prod_{j=1}^{N}\theta_j^{k_{ij}}\cdot\left(\sum_{j=1}^{N}k_{ij}\right)\right]\cdot\sum_{j=1}^{S}\left[\prod_{i=1}^{N}\theta_i^{m_{ij}}\cdot\left(\sum_{i=1}^{N}m_{ij}\right)\right]\cdot\sum_{i=1}^{L}\left[\prod_{j=1}^{N}\theta_j^{e_{ij}}\cdot\left(\sum_{j=1}^{N}e_{ij}\right)\right]. \tag{5}$$

5 Conclusions

The measurement model here is an instantaneous description of a system; the dynamic description can be attained from a series of states that sampled during the evolutive process of the system. So the method here is not only a way to calculate a static system, but also a heuristic model to explore more felicitous description of the complexity of a complex system.

References

1. Xu, G.: System Science, pp. 415–438. Shanghai Technology Education Press, Shanghai (2000)
2. Lin, Y.-B., Du, G., Lu, J.: Structure Complexity Measurement and Complex Structure Organization. Journal of Beijing Institute of Technology (Social Sciences Edition) 11(3), 20–23 (2009)
3. Jiao, F., Wang, L., Hou, J.: Dependency Matrix Based Metric Model for Complexity of Component Based Software. Computer Applications and Software 26(5), 55–56 (2009)
4. Deshmukh, A.V., Talavage, J.J., Barash, M.M.: Complexity in Manufacturing Systems, Part 1: Analysis of Static Complexity. IIE Transactions 30(10), 645–655 (1998)
5. Kauffman, S.: The Origins of Order: Self-Organization and Selection in Evolution, pp. 183–219. Oxford University Press, London (1993)

Research on Channel Estimation of Uplink for MC-CDMA Based on OFDM

Qiuxia Zhang, Yuanjie Deng, and Chao Song

HuangHe Science and Technology College, morden education technical center
qiuxiang6@sina.com.cn

Abstract. On the uplink MC-CDMA channel estimation of multi-carrier, the thesis proposes a new channel estimation scheme with the aided of the estimated channel parameters. Compared to conventional uplink MC-CDMA channel estimation, the proposed scheme can estimate channel paratemeters accurately, including estimation path numbers and time delays. The proposed scheme successfully resolves the robust and reduce the computational complexity when the conventional scheme determines the important taps by threshold. The simulation result proved the proposed scheme.

Keywords: measurement, MC-CDMA, OFDM, computational complexity.

1 Introduction

MC-CDMA technology is combined with the OFDM technology of multipath fading, and combined with the characteristics of the high frequency spectrum characteristics of CDMA efficiency, therefore it is considered to be the key technology about following evolution of 3G technology [1]. MC-CDMA system in order to achieve coherent detection receiver, channel estimation is a key part of the problem.

For MC-CDMA system uplink channel estimation [2], Sanguinetti, who made use of to estimate the equilibrium uplink channel, proposed two schemes, one operating in the frequency domain, deal with the adjacent carrier as a separate unknown parameters; the similarity between reference [3] and reference [4] is also by setting important path to select the threshold number, such estimates are still rough, not exact. Reference [5] provides scheme about an accurate estimation of OFDM systems in effective channel parameters, which can be used to overcome not robustness of the above-described threshold-based scheme. However, the program need to use the eigenvalue decomposition and the estimation of Signal parameters using Rotational Invariance Technique (ESPRIT) to estimate the channel parameters, so the program has been proved to be high complexity, and low signal-to-noise ratio (SNR)under poor performance.

Based on the above discussion, this paper proposes a scheme improved for uplink MC-CDMA system channel estimation. The program, through the use of information theory criteria --- Hannan-Quinn (HQ) criterion can accurately estimate the channel parameters, channel parameters including the number of channel path and path delay. There are two main contribution of this paper: first, successfully overcome the threshold constraints of the proposed scheme in reference[4], and the performance has

Y. Wang and T. Li (Eds.): Knowledge Engineering and Management, AISC 123, pp. 631–636.
springerlink.com

been proved to be robust; second, the scheme greatly reduced the complexity, and got much better performance than reference [5] under the low signal-to-noise ratio.

2 3G Technology Standard

Communication as a means of information transmission and exchange, for the construction of the information society has made a huge and positive contribution. Mobile communications began from the 1980s, just 20 years, has experienced the first-generation analog technology and the second-generation narrowband digital technology, the current mobile communications technology has begun to enter the third stage - the third generation mobile communication technology (3G).

This standard adopts the evolution strategy of GSM (2G)-GPRS-EDGE-WCDMA (3G), Although the above three evolutionary lines varies, but they have a common, the physical layer the two kinds of effective transmission technology are orthogonal frequency division multiplexing (OFDM) technology and multiple input multiple output (MIMO) technology. These two kinds of technology can greatly improve the spectrum utilization of next generation wireless communication system, and supporting high data transfer rates, therefore it has been established as the core technology of the physical layer of next-generation wireless communications.

3 Based on the Parametric Estimation of Uplink Channel Estimation Strategy of MC-CDMA

MC-CDMA uplink channel estimation algorithm can be divided into two steps, as shown in Fig.1. The first step, use the HQ number of criteria to estimate the channel path and path delay. The second step, using the channel number and time delay got on the step 1 are used to estimate the channel coefficients more accurately.

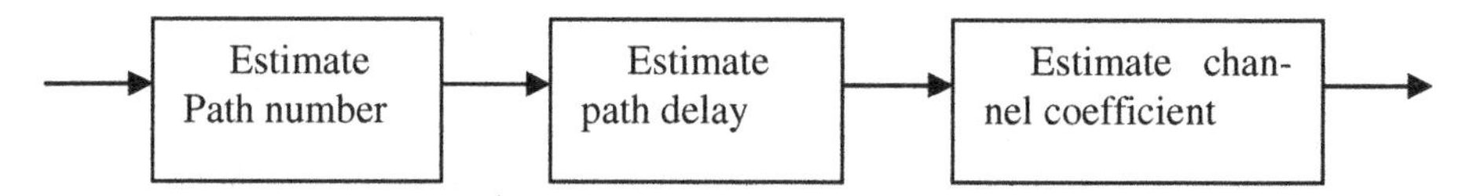

Fig. 1. The basic strategy process proposed

3.1 Estimated the Number of Channel Path and Path Delay

According to the pilot strategy, written time domain channel $\hat{h}_i(\tilde{n},m)$ to vector form

$$\hat{h}_i(\tilde{n}) = \begin{bmatrix} \hat{h}_i(\tilde{n},0) \\ \hat{h}_i(\tilde{n},1) \\ \vdots \\ \hat{h}_i(\tilde{n},G-1) \end{bmatrix} \tag{1}$$

The value of the average frequency of N_p pilot channel symbols

$$R_i(m)=\frac{1}{N_p}\sum_{n\in p_1}\left\|(\hat{h}_i,m)\right\|^2 \tag{2}$$

Given a power vector $R_i=[R_i(0),R_i(1),\cdots,R_i(G-1)]^T$, then order the vector by descending, get a new vector

$$\eta_i=[\eta_i(0),\eta_i(1),\cdots\eta_i(G-1)]^T \tag{3}$$

Combined with (3) and HQ criteria, get

$$HQ_i(l)=N_p((G-l)\ln(\frac{\sum_{m=l}^{G}\eta_i(m)}{G-l})-\sum_{m=l}^{G}\ln\eta_i(m))$$
$$+\frac{1}{2}l(2G-l)\ln\ln N_p \tag{4}$$

Then, by searching for feasible values l, the number of paths can be identified as

$$\hat{L}_i=\arg\min_{l\in\{0,1,\cdots,G-1\}}HQ_i(l) \tag{5}$$

Using the number of paths to determine the path delay, as following:

Step 1: choose the index corresponding to the largest of $\hat{L}_i$ values in the power vector R_i, and order the $\hat{L}_i$ indexes by ascending.

Step 2: the index value after ordering are obtained corresponding to the path delay $\hat{\xi}_i(l)(l=0,1,\cdots,\hat{L}_i-1)$

In summary, the general steps of using HQ criteria to estimate the number of channel path and the path delay.

Step 1: According to (2) calculate the value of the average power $R_i(m)$ of N_p pilot channel symbols.

Step 2: get a new vector η_i based on the power vector

$R_i=[R_i(0),R_i(1),\cdots,R_i(G-1)]^T$, get the new vector η_i by ordering the power vector R_i by descending.

Step 3: according to the (4) calculate decision vector $HQ_i(l)$.

Step 4: according to the (5) to judge the decision vector, and then select the most appropriate number $\hat{L}_i$ of paths.

Step 5: according to the path $\hat{L}_i$ has been estimated, choose the index corresponding to the largest of $\hat{L}_i$ values in the power vector R_i

Step 6: order the index values got from step 5 by ascending, the new index values after ordering are the path delay $\hat{\xi}_i(l)(l=0,1,\cdots,\hat{L}_i-1)$ required to estimate.

3.2 Estimated Channel Impulse Effect

Has been based on the number of paths and path delay, the time-domain channel $\hat{h}_i(\tilde{n},m)$ at the pilot position update to

$$\tilde{h}_i(\tilde{n})=[\tilde{h}_i(\tilde{n},0),\tilde{h}_i(\tilde{n},1),\cdots\tilde{h}_i(\tilde{n},\hat{L}_i-1)]^T \tag{6}$$

The next major work is to estimate all the channel coefficients

$$h_i(n)=[h_i(n,0),h_i(n,1),\cdots h_i(n,\hat{L}_i-1)]^T \quad (n=0,1,\cdots N-1)$$

According to the correlation of l th channel time-domain each element of the whole channel coefficient can be obtained as

$$\hat{h}_i(n,l)=\Psi_i^{-1}\tilde{h}_i(l) \tag{7}$$

$$\Psi_i=E\{\tilde{h}_i(n,l)\tilde{h}_i^H(l)\}(E\{\tilde{h}_i(l)\tilde{h}_i^H(l)\}) \tag{8}$$

In which, $h_i(l)=[\tilde{h}_i(\tilde{n}_0,l),\tilde{h}_i(\tilde{n}_1,l),\cdots,\tilde{h}_i(\tilde{n}_{Np-1},l)]^T$

The first type of Bessel function which instead of time domain, we used the robust time-domain correlation function under the ideal bandwidth-limited power spectrum, its expression as that

$$\gamma_l=E\{h(n_1,l)h^*(n_2,l)\}=\frac{\sin(w_d(n_2-n_1))}{w_d(n_2-n_1)} \tag{9}$$

In which, $w_d=2\pi T_s(K+G)f_d$, f_d is the Doppler frequency.

Finally, the frequency domain channel get (10),

$$\hat{H}_i(n)=\tilde{w}_i\hat{h}_i(n) \tag{10}$$

In which, $\hat{H}_i(n)=[\hat{H}_i(n,0),\hat{H}_i(n,1),\cdots,\hat{H}_i(n,K-1)]^T$,

$\hat{h}_i(n)=[\hat{h}_i(n,0),\hat{h}_i(n,1),\cdots,\hat{h}_i(n,K-1)]^T$,

$\tilde{w}_i$ is the $\hat{\xi}_i(0),\hat{\xi}_i(1),\cdots,\hat{\xi}_i(\hat{L}_i-1)$ column of the matrix $\sqrt{K}w$

In summary, The proposed algorithm is summarized into the following steps:

Step 1: According to (2) calculate the value of the average power $R_i(m)$ of N_p pilot channel symbols.

Step 2: Get sorted vector η_i.

Step 3: using HQ Criteria to determine the number of paths

Step 4: find the largest $\hat{L}_i$ of index from the vector R_i, and the corresponding delay path $\hat{\xi}_i(l)$ can be got.

Step 5: estimate the channel coefficient from (2)-(6).

4 The Simulation Results

We assume a MC-CDMA system which the user number $I = 16$, using QPSK modulation, the length of spreading code $D = 16$, the bandwidth of signal is 5MHz, sampling period is $0.2\mu s$, the total number of the carrier is 512, Cyclic prefix is128, and the most user number of each group is $I_g = k / G = 4, U = I / I_g$. In order to facilitate the simulation, we set $S_p = U = 4$, Assume the frame length $N = N_d S_d + N_p S_p = 160$, in which N_p=20, N_d=20, S_d=4,Doppler frequency $f_d = 277.78Hz$, each user through the different channels.

The simulation 1: HQ criterion to MDL criteria

In order to prove the HQ criterion performance, we compared the error probability of correct detection between HQ criteria and MDL criteria which the path number is estimated. Figure2-2. It can be seen from the figure both can correctly detect the number of paths in the case of high SNR. In the low SNR, the number of low-power path will be covered up by the noise, so it is difficult to detect. However, the HQ criteria mentioned is better than the MDL criterion in the case of low SNR.

The simulation 2: comparison the algorithm proposed with the algorithm [5] about the complexity

Consider [5] scheme is for the single-user, so the complexity of the proposed scheme also takes the complexity of one user for example. And only consider the numbers of Multiplication and division, such as the number of other sort are ignored. Moreover, considering the complexity of the estimated channel impulse response in the second part is the same, we only need to Calculate the complexity of estimating number of channel path and path delay. Indicate a complex multiplication and division requires 6 floating operations, a complex addition and subtraction take two floating operations Therefore, proposed algorithm is

$$87G^3+6(N_p+1)G^2+6(N_p+8)G+42 \quad (11)$$

The times of general floating operations of reference [13] is

$$36G^3+84L^3+6(N_p+1)G^2+18(G-1)L^2+6(7+N_p)G+6L+42 \quad (12)$$

As figure 2 shows, obviously, the total number of floating operations of the algorithm proposed is less than reference [5].

5 Conclusion

This paper proposes an improved uplink MC-CDMA channel estimation scheme, The scheme by estimating the channel parameters. Channel estimation by the aid of accurate channel parameters can achieve more robust performance than the traditional channel estimation DFT-based. Channel parameters include the number of channel path and multipath delay. We use the information theory HQ criterion to accurately estimate the channel parameters.

References

1. Li, Y., Cimini Jr., L.J., Sollenberger, N.R.: Robust ehannel estimation for OFDM systems with rapid dispersive fading channels. IEEE Trans. Commun. 46, 902–914 (1998)
2. Sanguinetti, L., Morelli, M.: Channel acquisition and tracking for MC-CDMA uplink transmissions. IEEE Trans. on Vehicular Tech. 49(4m), 1207–1215 (2000)
3. Zhang, H., Li, Y., Yi, Y.-W.: Practical considerations on channel estimation for uplink MC-CDMA systems. IEEE Trans. on Wireless Commun. 7(11) (November 2008)
4. Zheng, K., Zeng, G., Wang, W.: A novel uPlink channel estimation in OFDM-CDMA systems. IEEE Trans. on consummer Electronics 50, 125–129 (2004)
5. Yang, B., Letaief, K.B., Clieng, R.S., Cao, Z.: Channel estimation for OFDM transmission in multiPathfading channels based on Parametric channel modeling. IEEE Trans. Commun. 49, 467–478 (2001)

Properties of Linguistic Truth-Valued Intuitionistic Fuzzy Lattice

Fengmei Zhang[1], Yunguo Hong[2], and Li Zou[1]

[1] School of Computer and Information Technology, Liaoning Normal University, Dalian 116029, P.R. China
zoulicn@163.com

[2] Dalian Vocational and Technical College, Dalian 116035, China

Abstract. A kind of linguistic truth-valued intuitionistic fuzzy lattice algebra based on the point view of intuitionistic fuzzy set and linguistic truth-valued lattice implication algebra is discussed. As a fundament of linguistic truth-valued intuitionistic fuzzy, some algebra properties are obtained. The result shows that linguistic truth-valued intuitionistic fuzzy lattice algebra is a residual lattice, but it is not MTL-algebra and R_0-algebra.

Keywords: Lattice implication algebra,Linguistic truth-valued intuitionistic fuzzy lattice, residual lattice, Intuitionistic fuzzy operation.

1 Introduction

Formal fuzzy logics are generalizations of classical logic that allow us to reason gradually. Formally, the semantics of formulas in these logics can be assigned not only 0 and 1 as truth values, but also elements of $[0,1]$, or, more generally, of a bounded lattice L. In L, the partial ordering of L can serve to compare the truth values of formulas which can be true to some extent. In order to research the many-valued logical system whose propositional value is given in a lattice, Xu and Qin proposed the notion of lattice implication algebras and investigated many useful properties [9,11]. Since then this logical algebra has been extensively investigated by several researchers, and some interesting results have been obtained.

Information processing corresponding linguistic values is translated to their semantics, and fuzzy sets theory becomes main tool for CWW. Due to some drawbacks in linguistic approaches based on fuzzy sets, there exist many alternative methods for processing linguistic information, *e.g.*, Huynh proposed a new model for parametric representation of linguistic truth-values [5,6,7]. Turksen studied the formalization and inference of descriptive words, substantive words and declarative sentence based on type-2 fuzzy sets [8]. Ho discussed the ordering structure of linguistic hedges, and proposed hedge algebra to deal with CWW [3,4]. Xu, et al. proposed linguistic truth-valued lattice implication algebra to deal with linguistic truth inference [10,12]. Zou [13,14,15,16] proposed a framework of linguistic truth-valued propositional logic and developed the reasoning method of six-element linguistic truth-valued logic system.

Intuitionistic fuzzy sets (A-IFSs) introduced by Atanassov is a powerful tool to deal with uncertainty. Formally, an A-IFS is $A = \{(x, \mu_A(x), \nu_A(x))|x \in X\}$, where $\mu_A(x)$

Y. Wang and T. Li (Eds.): Knowledge Engineering and Management, AISC 123, pp. 637–641.
springerlink.com

and $\nu_A(x)$ are the degree of membership and degree of non-membership of the element $x \in X$ to the set $A \subseteq X$ such that for every $x \in X$, $0 \leq \mu_A(x) + \nu_A(x) \leq 1$, the value $\pi_A(x) = 1 - \mu_A(x) - \nu_A(x) (\forall x \in X)$ is called the degree of indeterminacy of x to A. A-IFSs concentrate on expressing advantages and disadvantages, pros and cons. A-IFSs has been widely researched and applied [1,2,9]

Inspired by linguistic truth-valued lattice implication algebra and A-IFSs, we discuss the properties of linguistic truth-valued intuitionisitic fuzzy Lattice. We analyze the relationship between linguistic truth-valued intuitionisitic fuzzy Lattice and other logical algebras. The organization of this paper is as following: In Section 2, we introduce the basic concept of linguistic truth-valued intuitionistic fuzzy lattice. Then linguistic truth-valued intuitionistic fuzzy lattice algebra is compared with residual lattice, MTL-algebra and R_0-algebra in Section 3. We conclude in Section 4.

2 Linguistic Truth-Valued Intuitionisitic Fuzzy Lattice

Denote all the linguistic truth-valued intuitionisitic fuzzy pairs based on $2n - element$ linguistic truth-valued lattice implication algebra as:

$$LI_{2n} = \{((h_i,t),(h_j,f)) | (h_i,t),(h_j,f) \in \mathscr{L}_{V(n\times 2)}, i \leq j\}. \tag{1}$$

For any $((h_i,t),(h_j,f)),((h_k,t),(h_l,f)) \in LI_{2n}$, define the operation "$\cup$","$\cap$" and "$\neg$" as follows:

$$((h_i,t),(h_j,f)) \cup ((h_k,t),(h_l,f)) = ((h_i,t) \vee (h_k,t),(h_j,f) \wedge (h_l,f)), \tag{2}$$
$$((h_i,t),(h_j,f)) \cap ((h_k,t),(h_l,f)) = ((h_i,t) \wedge (h_k,t),(h_j,f) \vee (h_l,f)). \tag{3}$$

Where "$\vee$" and "$\wedge$" are operations of $\mathscr{L}_{V(n\times 2)}$.

Based on $2n$ linguistic truth-valued lattice implication algebra $\mathscr{L}_{V(n\times 2)}$, we can construct linguistic truth-valued intuitionistic fuzzy lattice. Formally, denote

$$\mathscr{L}\mathscr{I}_{2n} = (LI_{2n}, \cup, \cap)$$

as a linguistic truth-valued intuitionistic fuzzy lattice where $((h_n,t),(h_n,f))$ and $((h_1,t),(h_1,f))$ are the greatest element and the least element of $\mathscr{L}\mathscr{I}_{2n}$, respectively.

Definition 1. In the linguistic truth-valued intuitionistic fuzzy lattice $\mathscr{L}\mathscr{I}_{2n} = (LI_{2n}, \cup, \cap)$ (Figure.1), for any $((h_i,t),(h_j,f)),((h_k,t),(h_l,f)) \in LI_{2n}$, $((h_i,t),(h_j,f)) \leq ((h_k,t),(h_l,f))$ if and only if $i \leq k$ and $j \leq l$, also

$$((h_i,t),(h_j,f)) \cup ((h_k,t),(h_l,f)) = ((h_{max(i,k)},t),(h_{max(j,l)},f)), \tag{4}$$
$$((h_i,t),(h_j,f)) \cap ((h_k,t),(h_l,f)) = ((h_{min(i,k)},t),(h_{min(j,l)},f)). \tag{5}$$

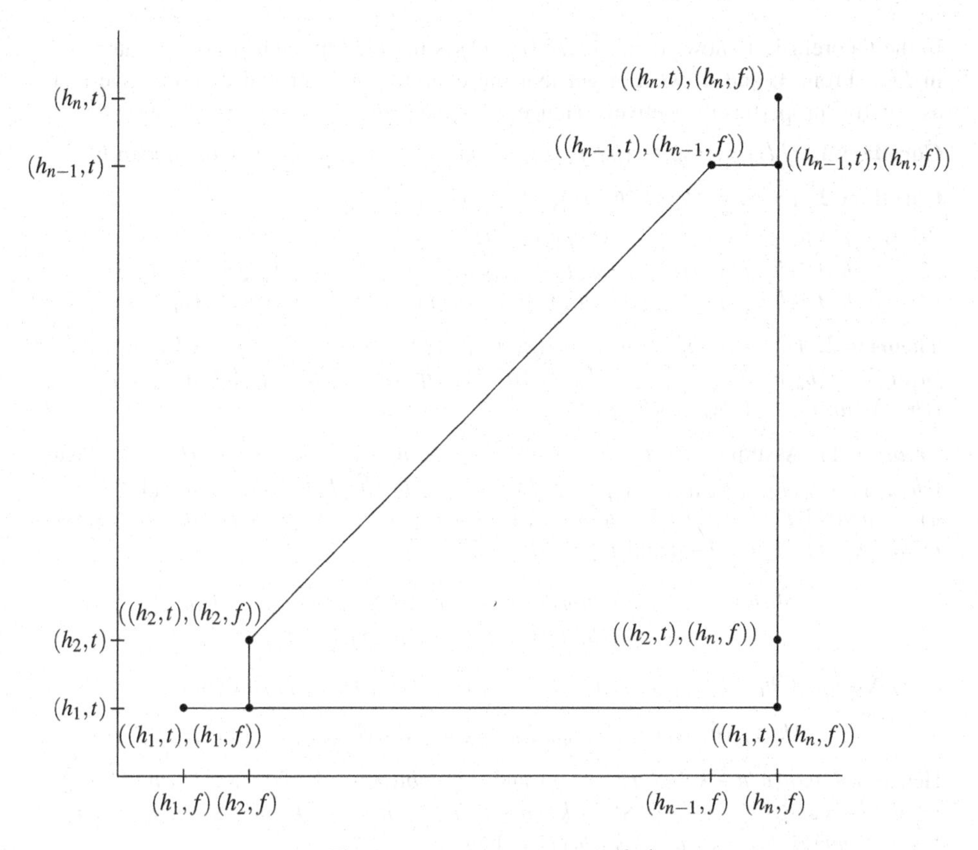

Fig. 1. Structure Diagrams of $\mathscr{L}\mathscr{I}_{2n}$

3 The Relationship between $\mathscr{L}\mathscr{I}_{2n}$ and Other Logic Algebra

we compare linguistic truth-valued intuitionistic fuzzy lattice algebra $\mathscr{L}\mathscr{I}_{2n} = (LI_{2n}, \cup, \cap, \rightarrow, ((h_n,t),(h_n,f)), ((h_1,t),(h_1,f)))$ with residual lattice, MTL-algebra, BL-algebra, MV-algebra, lattice implication algebra and R_0-algebra. For any $((h_i,t),(h_j,f)), ((h_k,t),(h_l,f)) \in LI_{2n}$, define the opeator "*" as follow:

$$((h_i,t),(h_j,f)) * ((h_k,t),(h_l,f)) = ((h_{max(1,i+k-n)},t),(h_{max(1,i+l-n,j+k-n)},f)). \quad (6)$$

Theorem 1. *For any* $((h_i,t),(h_j,f)), ((h_k,t),(h_l,f)), ((h_m,t),(h_s,f)) \in LI_{2n}$,

1. $((h_i,t),(h_j,f)) * ((h_n,t),(h_n,f)) = ((h_i,t),(h_j,f))$;
2. $((h_i,t),(h_j,f)) * ((h_1,t),(h_1,f)) = ((h_1,t),(h_1,f))$;
3. $((h_1,t),(h_n,f)) * ((h_1,t),(h_n,f)) = ((h_1,t),(h_1,f))$;
4. $((h_i,t),(h_j,f)) * ((h_k,t),(h_l,f)) = ((h_k,t),(h_l,f)) * ((h_i,t),(h_j,f))$;
5. $(((h_i,t),(h_j,f)) * ((h_k,t),(h_l,f))) * ((h_m,t),(h_s,f)) = ((h_i,t),(h_j,f)) * (((h_k,t),(h_l,f)) * ((h_m,t),(h_s,f)))$.

In the theorem 1, 1.shows that $((h_n,t),(h_n,f))$ is the identity element of operator "$*$" in LI_{2n}. From 3. and 4. we can get that the operator "$*$" satisfied commutative and associative properties respectively. Hence, we have some corollaries as follows:

Corollary 1. *$(LI_{2n},*,((h_n,t),(h_n,f)))$ is a commutative semigroup with an identity.*

Corollary 2. *For any $((h_i,t),(h_j,f)),((h_k,t),(h_l,f)) \in LI_{2n}$,*

1. $((h_i,t),(h_j,f)) * (\neg((h_i,t),(h_j,f))) = ((h_1,t),(h_1,f))$;
2. $\neg(((h_i,t),(h_j,f)) * ((h_k,t),(h_l,f))) = ((h_i,t),(h_j,f)) \rightarrow (\neg((h_k,t),(h_l,f)))$;
3. $\neg(((h_i,t),(h_j,f)) \rightarrow ((h_k,t),(h_l,f))) = ((h_i,t),(h_j,f)) * (\neg((h_k,t),(h_l,f)))$.

Theorem 2. *For any $((h_i,t),(h_j,f)),((h_k,t),(h_l,f)),((h_m,t),(h_s,f)) \in LI_{2n}$, $((h_i,t),(h_j,f)) * ((h_k,t),(h_l,f)) \leq ((h_m,t),(h_s,f))$ if and only if $((h_i,t),(h_j,f)) \leq ((h_k,t),(h_l,f)) \rightarrow ((h_m,t),(h_s,f))$.*

Proof. 1) Assume $((h_i,t),(h_j,f)) * ((h_k,t),(h_l,f)) \leq ((h_m,t),(h_s,f))$, then $((h_{max(1,i+k-n)},t),(h_{max(1,i+l-n,j+k-n)},f)) \leq ((h_m,t),(h_s,f))$. Hence, $max(1,i+k-n) \leq mmax(1,i+l-n,j+k-n) \leq s,$, $i+k-n \leq m$, $i+l-n \leq sj+k-n \leq s$. So $i \leq n-k+m$, $i \leq n-l+s$ and $j \leq n-k+s$, i.e.,

$$\begin{aligned}((h_i,t),(h_j,f)) &\leq ((h_{min(n,n-k+m,n-l+s)},t),(h_{min(n,n-k+s)},f)) \\ &= ((h_k,t),(h_l,f)) \rightarrow ((h_m,t),(h_s,f)).\end{aligned}$$

2) Assume $((h_i,t),(h_j,f)) \leq ((h_k,t),(h_l,f)) \rightarrow ((h_m,t),(h_s,f))$, then

$$((h_i,t),(h_j,f)) \leq ((h_{min(n,n-k+m,n-l+s)},t),(h_{min(n,n-k+s)},f)).$$

Hence, $i \leq min(n,n-k+m,n-l+s)$ and $j \leq min(n,n-k+s)$, i.e., $i \leq n-k+m$, $i \leq n-l+s$ and $j \leq n-k+s$. So $i+k-n \leq m$, $i+l-n \leq sj+k-n \leq s$, i.e., $max(1,i+k-n) \leq mmax(1,i+l-n,j+k-n) \leq s$, then,

$$\begin{aligned}((h_m,t),(h_s,f)) &\geq ((h_{max(1,i+k-n)},t),(h_{max(1,i+l-n,j+k-n)},f)) \\ &= ((h_i,t),(h_j,f)) * ((h_k,t),(h_l,f)).\end{aligned}$$

□

From the theorem 2, $(*,\rightarrow)$ is company pair on LI_{2n}. According to the corollary 1 and the theorem 2, we have some corollary as follows:

Corollary 3. *$\mathscr{L}\mathscr{I}_{2n} = (LI_{2n},\cup,\cap,*,\rightarrow,((h_1,t),(h_1,f)),((h_n,t),(h_n,f)))$ is a residual lattice.*

4 Conclusions

The linguistic truth-valued intuitionistic fuzzy lattice is the algebra fundament for the intuitionistic fuzzy logic. We got some algebra properties, especially, the relationship between residual lattice, MTL-algebra and R_0-algebra. It can better express and handle both comparable and incomparable information during the information processing. The proposed method can deal with both positive evidence and negative evidence together. The properties of the implication operators are good for the reasoning. The further work is to apply the theory into the linguistic truth-valued intuitionistic fuzzy logic and to find the method for automatic reasoning.

Acknowledgements. This work is partly supported by national nature science foundation of China (Grant No.61105059,61175055), the research fund of Sichuan key laboratory of intelligent network information processing (SGXZD1002-10) and the key laboratory of the radio signals intelligent processing (Xihua university) (XZD0818-09).

References

1. Atanassov, K.T., Answer to Dubois, D., Gottwald, S., Hajek, P., Kacprzyk, J., Prade's, H. paper : Terminological difficulties in fuzzy set theory the case of 'Intuitionistic Fuzzy Sets'. Fuzzy Sets and Systems 156, 496–499 (2005)
2. Dubois, D., et al.: Terminological difficulties in fuzzy set theory the case of intuitionistic fuzzy sets. Fuzzy Sets and Systems 156, 485–491 (2005)
3. Ho, N.C.: A topological completion of refined hedge algebras and a model of fuzziness of linguistic terms and hedges. Fuzzy Sets and Systems 158, 436–451 (2007)
4. Ho, N.C., Long, N.V.: Fuzziness measure on complete hedge algebras and quantifying semantics of terms in linear hedge algebras. Fuzzy Sets and Systems 158, 452–471 (2007)
5. Huynh, V.N., Ho, T.B., Nakamori, Y.: A parametric representation of linguistic hedges in Zadeh's fuzzy logic. International Journal of Approximate Reasoning 30, 203–223 (2002)
6. Martinez, L., Ruan, D., Herrera, F.: Computing with Words in Decision support Systems: An overview on Models and Applications. International Journal of Computational Intelligence Systems 3(4), 382–395 (2010)
7. Nguyen, C.H., Huynh, V.N.: An algebraic approach to linguistic hedges in Zadeh's fuzzy logic. Fuzzy Set and Systems 129, 229–254 (2002)
8. Turksen, I.B.: Meta-linguistic axioms as a foundation for computing with words. Information Sciences 177, 332–359 (2007)
9. Pei, Z., Liu, X., Zou, L.: Extracting association rules based on intuitionistic fuzzy sets. International Journal of Innovative Computing, Information and Control 6(6), 2567–2580 (2010)
10. Pei, Z., Xu, Y., Ruan, D., Qin, K.: Extracting complex linguistic data summaries from personnel database via simple linguistic aggregations. Information Sciences 179, 2325–2332 (2009)
11. Xu, Y., Ruan, D., Qin, K.Y., Liu, J.: Lattice-valued Logic. Springer (2004)
12. Xu, Y., Ruan, D., Kerre, E.E., Liu, J.: α-resolution principle based on lattice-valued propositional logic LP(X). Information Sciences 130, 195–223 (2000)
13. Zou, L., Ruan, D., Pei, Z., Xu, Y.: A linguistic truth-valued reasoning approach in decision making with incomparable information. Journal of Intelligent and Fuzzy Systems 19(4-5), 335–343 (2008)
14. Zou, L., Pei, Z., Liu, X., Xu, Y.: Semantics of linguistic truth-valued intuitionistic fuzzy proposition calculus, International Journal of Innovative Computing. Information and Control 5(12), 4745–4752 (2009)
15. Zou, L., Liu, X., Wu, Z., Xu, Y.: A uniform approach of linguistic truth values in sensor evaluation. International Journal of Fuzzy Optimization and Decision Making 7(4), 387–397 (2008)
16. Zou, L., Ruan, D., Pei, Z., Xu, Y.: A Linguistic-Valued Lattice Implication Algebra Approach for Risk Analysis. Journal of Multi-Valued Logic and Soft Computing 17, 293–303 (2011)

Formal Engineering with Fuzzy Logic

Victoria Lopez

Dep. Computer Architecture and Automatics, Complutense University, Madrid, Spain
vlopez@fdi.ucm.es

Abstract. Formal engineering constitutes a very important issue in software engineering projects in real life. The developing of software does not always reach the desired level of reliability and performance even the life cycle of the project used to be controlled by methodologies and specific tools as Formal Languages and Formal Methods. Despite the efforts the question is that even in the best cases, the final product has a lot of errors and in some cases these errors produce catastrophic disasters (ARIANE 5, for example). This paper shows a new proposal to formalize the life cycle according to the worker at each stage and the importance of using not only the classical logics (propositional and first order), but also fuzzy logic in order to formalize the certain and uncertain information involved in natural language.

Keywords: Formal Engineering, Fuzzy Logic, Software Engineering, Formal Specification, Industrial Software, Critical Systems.

1 Introduction

In any real engineering project, it would be unbelievable to make a large system without a rigorous documentation of what is going to be built. However, in computer science developments, this is not usual. Each day, software developers and companies create and send to the markets a lot of software tools full of errors. On one hand, there are not too many complaints about small software errors and users do not put too much attention into the real cause of these problems. The truth is in a few seconds the problem is solved and the system is running again. Each day we need to restart our computer, systems delay and operative systems fall down without an evident reason. The fact is all of us lose time in regarding this kind of software errors but we do not a real complaint about them. On the other hand, there is a list of software engineering disasters and failures, such as the Therac-25 and the Ariane5 [1], which understanding is now incorporated into engineering teaching and learning.

Software engineering faces the challenge of creating very large systems in which descriptions and correctness arguments are required. Software engineering is a process that is vulnerable to cost overruns, poor communication and increasing complexity. Despite the efforts of industry to mitigate these risks, engineers have not able to solve it completely.

Y. Wang and T. Li (Eds.): Knowledge Engineering and Management, AISC 123, pp. 643–652.
springerlink.com

2 Software Engineering

Software is regarded not only as a set of programs but as a combination of documentation and programs. Documentation is part of software project that may contain the user's requirements, the design of the program, the goal to be achieved by the programs as well as the manual for using the resulting software tool. A good structure and writing of the document is important for ensuring the quality and for avoiding misunderstanding and the risk of producing a loss of money and time.

A typical software life cycle [2] fits with the well known waterfall model. Although the real picture of the software life cycle may be much more complicated, waterfall model depicts the main stages and almost every other model is based on this idea. Life cycle models comprise the following five stages: 1. Analysis of Requirement and Specification; 2. Design; 3. Implementation and Code Development; 4. Testing; 5. Deliver and Maintenance. Figure 1 shows the life cycle stages. It is shown as a cycle to remember a real software project must be dynamic, adaptive and open to the business plan model.

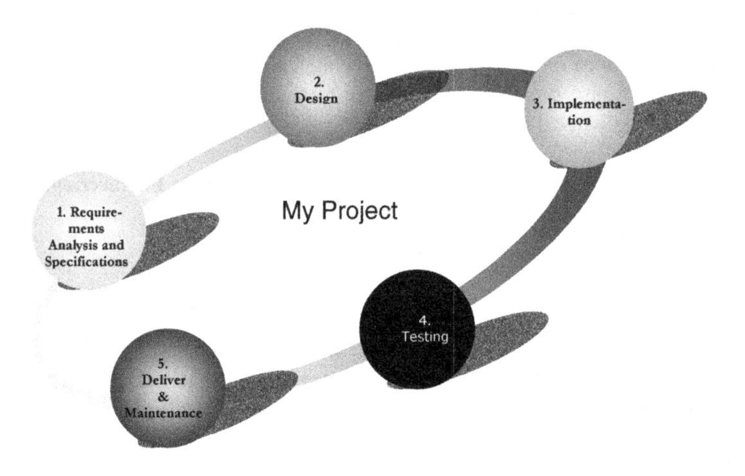

Fig. 1. Basic life cycle stages

Stage 1 is a study aiming to discover and document the exact requirements for the software system to be constructed. To this aim, the system in the real world may need modeling so that all necessary requirements can be explored. Requirements tell what is to be done by the system in a document which is usually called a requirements specification document.

Stage 2 refers to the design of the project, i.e. the definition and shape of all its functionalities and properties. Design is an activity to construct a system, to meet the system requirements and it is concerned with how to provide a solution for the problem reflected in the requirements. It can be done in two phases: *abstract design*, that builds the architecture of the entire system defining the relation between software modules and components, and *detailed design*, that focuses on the definition of data structures and the construction of algorithms. In this second phase the designer choose the imperative or functional paradigm for developing the code as well as details about complexity of the algorithms.

Stage 3 is about implementation, where the design specification is transformed into a program written in a specific programming language (Java, C++, ...). The primary concerns in implementation are the functional correctness of the program against its design and requirements specification by means of formal verification techniques. It is important to note here that a good design could prevent errors in the developing of codes and decrease the responsibility of programmers.

Testing is in State 4. This is a way to detect potential faults in the program by running the program with test cases. There are many ways to introduce faults during the software development process therefore detecting and removing faults are necessary.

Finally, Stage 5 is deliver and maintenance. Maintenance of a system requires a thought understanding by maintenance engineers, and that is because well documented requirements specification and design specification are needed to enhance the reliability and efficiency of this stage.

The main conclusion of this review of software engineering phases is the importance of using mathematical modeling of natural language. Software engineering is a set of human resources, reasoning and knowledge. The use of first order logic to model reasoning and knowledge is well known and accepted by researchers. Uncertainty is also a component of the knowledge at any software project. That is why we propose the use of fuzzy logic in order to model reasoning and knowledge over all in early stages of the life cycle: Specification and design. This proposal also includes a new organization of the stages as Figure 2 shows. In this figure, stage 1 refers to the first phase of any project in which the human participation is mandatory and natural language with uncertainty and fuzzy information are the base of the information and data collected. Stage 2 corresponds with formal specification and requirements already done by using not only first order logics but also fuzzy logic. At this stage the human participation is also mandatory. At stage 3, formal design and complexity considerations are done. Human participation is not mandatory here but high recommended in order to optimize the complexity of the algorithms. Stage 4 is devoted to detection of critical algorithms. This issue is regarded at this point due to many algorithms became critical at working time as a part of the project, and not as isolated algorithms. In this case also the human participation is highly recommended. Automatic development and verification or formal methods implement are in stage 5. Stage 6 refers to evaluation of the software, despite of after all these stages; reliable software must be reached. Testing techniques are also due; actually, some testing techniques could be implemented at stages 2-6.

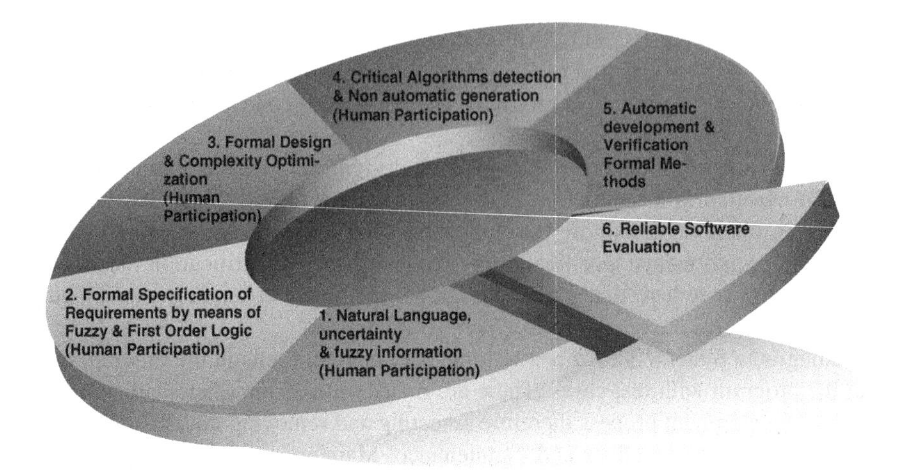

Fig. 2. New life cycle stages

3 Formal Engineering

3.1 Natural Language

By natural language we mean a language that is used for everyday communication by humans. In contrast to artificial languages such as programming languages and mathematical notations, natural languages have evolved in ambiguity and uncertainty. Natural Language Processing (NLP) [3] covers any kind of computer manipulation of the natural language. Within industry, it is important to have a working knowledge of NLP. This includes people in human-computer interaction, business information analysis and web software development. One of the most commonly researched tasks in NLP is the identification of the intended semantic form the multiple possible semantics which can be derived from a natural language expression. Within the software engineering process, we aim to convert chunks of text into formal representations such as first-order-logic structures or formulae that are easier for computer programs to manipulate.

Due to the way of human thinking, some statements from natural language cannot be formalized by only firs-order-logic structures. Fuzzy Logic, Computer with Words and Computational Theory of Perception [4] have been conceived as a merger of natural language processing and fuzzy logic for solving how to compute and reason with perceptual information.

3.2 Development of Rigorous Software

Formal Engineering deals with the formalization of all stages at life cycle in software engineering projects when developing complex software systems. Formal Methods [5] have made significant contributions to the establishment of theoretical foundations and rigorous approach for software development; however, engineers used to feel frustrated by using them in practice. The fact is they need some experience not only in programming but also in discrete mathematics, algorithmic and formal logics. Moreover, Formal Methods in industry as SOLF [6], VDM, Z and B are developed for ensuring verification of the code in Object Oriented Languages, and sometimes they focus on the pre/post conditions of data structures, testing the good behavior when working with list or queues, for example. By applying Formal Methods, the work becomes so heavy that engineers may forget the main issue: a good development (writing) of the critical algorithms guarantees the reliability of the entire system. That means a development of correct algorithms, and therefore their formal specification, formal design and formal verification [7].

Design by Contract (DbC) [8] is another approach to design reliable software. It prescribes that software designers should define formal, precise and verifiable specification for software systems as an extension of abstract data types with preconditions, postconditions and invariants. This formalization is done by means of assertions which test the good behavior of the codes and some rules in inheritance hierarchy (subclasses are allowed to weaken preconditions but not strengthen them, and are allowed to strengthen postconditions and invariants but not weaken them) so that behavior of subtypes is well established.

On the other hand, Design for Reliability (DFR) [9] is a process that describes the entire set of tools that support the effort to increase a product's reliability (software, hardware or both). The success of a DFR process is directly related to the selection of the appropriate reliability tools for each phase of the product development life cycle and the correct implementation of those tools. Mean Time Between Failures (MTBF) is an example of a reliability requirement that is a standard in many industries. The use of an MTBF as a sole reliability requirement implies a constant failure rate, which is not the case for most systems or components. Even if the assumption of a constant failure rate is not made, the MTBF does not tell the whole story about reliability. Research in this area is a challenge for software engineers.

In the last years, some researchers as Jim Woodcock (University of York), Tony Hoare (Microsoft Research Ltd.), Peter O'Hearn (University of London) and Cliff Jones (University of Newcastle) have been involved in the 'International Grand Challenge Project' to construct a program verifier that would use logical proof to give an automatic check of the correctness of programs. This project has a long-term vision, around 20-50 years from 2005, as it is pointed in [10,11].

4 Functional Modeling under Fuzziness

4.1 Antecedents

Several works are already done in modeling specification under fuzziness. Starting with [7] in which is developed an approach to formal specification and verification by means of fuzzy logic, and continuing with some works on formal evaluation as [12, 13]. The FFOLE software tool [14], which stands for Fuzzy First Order Logic Evaluation, was developed in 2006. The main goal of this tool is the evaluation on fuzzy logic assertions. Evaluation of logic assertions is very well known and used in Formal Engineering. The new challenge is to cover the uncertainty of the natural languages by using fuzzy logic and its formalization and evaluation. The FFOLE tool [14] is useful to assist an expert to decide about the convenience of different logics. It was developed in Haskell which allows considering functions as parameters, producing a general and efficient tool. Nevertheless it is still needed that the user writes down the whole expression to be evaluated.

Algorithm design and software specification are also very important issues in decision making problems. [13, 15, 16] are examples in which application of fuzzy specification of algorithms is shown. In all of them the system is based in using fuzzy logic to formalize natural language and information. The result is a set of first order logic formulae in which elements of fuzzy logic are part of the system.

Uncertain risk analysis and uncertain reliability analysis in the sense of modeling human uncertainty (as a branch of axiomatic mathematics) is recently studied by B. Liu [17]. In his works, Liu has applied uncertainty theory and uncertain programming to industrial engineering successfully.

4.2 Formal Specification under Fuzziness

In this subsection we show by means of an example, how fuzzy logic can be useful to write specifications as a previous step in the process of analysis and design of new software. From algorithms theory point of view, a fuzzy specification of pre and postconditions is introduced. Thanks to fuzzy inference rules techniques, these kind of specifications allow the correctness of algorithms within a fuzzy framework by means of predicate evaluations. In the same way used by formal methods in software engineering and more specifically, specifications focus on the consistency between the process and the related invariant. The invariant must not violated before and after the execution process according to its formal specification. All these constraints are formalized as predicate expressions or assertions by means of well designed software tools. The examples provided will show how fuzzy logic can improve the meaning of the predicates as fuzzy assertions even about testing which is usually performed dynamically, with the need to evaluate specifications.

We should realize that quite often some parts of the specifications are expressed in natural language just because the variety and non-technical background of people involved in the project. We speak then about informal specifications. Hence, any computation will require a previous formalization. Natural language uses a lot of fuzzy relations and perceptions that we can not avoid, either in the input data and their restrictions (precondition) nor the output data (postcondition).

The following definition generalizes the concept of formal specification given by Tony Hoare in 1969 of an algorithm in a fuzzy framework. All details about fuzzy logic elements and fuzzy formulae can be consulted in [7].

Definition 1. [12] Let A be an algorithm. The 'fuzzy formal specification' of A is given by a triple *(ED, fPre, fPost)* where

- *ED* is an explanatory database in the sense of [4] that is, a collection of relations (fuzzy and crisp) in terms of which the meaning of *fPre* and *fPost* are defined. It contains also the universe of discourse (as programming types) and whatever function that the goals need to be explained;

- *fPre* (fuzzy precondition) is the canonical form in the sense of [4] of propositions and fuzzy first order formulas that expresses the initial constraints and knowledge about input data;
- *fPost* (fuzzy postcondition) is the canonical form after execution of the algorithm (fuzzy constraint propagation and inference is due, and the postcondition is the formalized goal of the algorithm or the process).

The following examples shows the difference between evaluations in fuzzy and crisp framework, see [12].

Example 1. Specification and code with crisp relations and first order logic. The next code and specification corresponds to an algorithm (procedure Majority1) in which the aim is to check if the majority of the members of the array has the property P. Property P is informally defined as 'P(x) if and only if x has the property P' in order to generalized the problem to other environments as well as the type (Type1) of the members into the array. In this case, the specification is also gotten by crisp relations and first order logic. Majority is understood as half plus one (Ndiv2+1).

Code of a crisp definition of majority.

```
procedure Majority1 (Input v:array[1..N] of Type1;
Output c: natural; b: boolean);
{Pre: P(x)=defP}

ASSERT(Pre)

i=1;

c=0;

{INV: c=#j∈{1..i-1}·P(v[j])}

ASSERT(INV)

while (i≤N),

          if P(v[i])then c++;

          i++;

          ASSERT(INV)

end;

{Post: c=#j∈{1..N}·P(v[j])∧ b=(c≥Ndiv2+1)}

ASSERT(Post)
```

Example 2. Specification and code with fuzzy relations, fuzzy logic and first order logic. The procedure Majority2 shows the fuzzy generalization of the procedure Majority1. At the same time, and for improving the understanding of the example, Majority2 is based on the following items:

- The property P becomes here the fuzzy μ_{sweet} (which set the grade of sweetness of an element in the array), to illustrate the sense of the fuzziness although any other property could be used instead. An example of this function can be found in [12], and there, is fixed level.
- An aggregator function (Aggr) to define majority is used as well. Any aggregator can be used instead the average, mean and median are suitable as well as some OWA operators already used in [12].

Code of fuzzy relations by means of first order logic and fuzzy logic.

```
function Majority2 (Input v: array[1..N] of Type1;
Aggr:[0,1]^N →[0,1]; μ_sweet:Type1 →[0,1]; Output b: [0,1])
{Pre: μ_sweet=defμ_sweet =(2) ∧ Aggr=(3)}
ASSERT(Pre)
i=1;
b=0;
{INV: b=Aggr j∈{1..i-1}·μ_sweet (v[j])}
ASSERT(INV)
while (i≤N),
          b= Aggr(b, μ_sweet(v[i]);
          i++;
          ASSERT(INV)
end;
{Post: b= Aggr i∈{1..N}·μ_sweet (v[j])}
ASSERT(Post)
```

Using the input v = [175, 176, 180, 160, 160, 160, 174, 174, 174, 174, 173] and P(x)=(x>175), and after some considerations in the Majority code, Majority1 returns 0 (***false***), which means the majority of the array has not the property P) while Majority2 returns 0.5027 using the average as Aggr (which could be translated in a fuzzy framework as majority aims in a grade of around 50%) and 0.63 using the median (which means the majority is reached in a grade of 63%). At a glance, a simple view of the array data and the sweetness fuzzy function output ([0.70, 0.77, 1.0, 0, 0, 0, 0.63, 0.63, 0.63, 0.63, 0.54]) drives us towards the Majority2, closer to the natural intuition and human perception. Once here must take into account that the crisp version say that a member is sweet if its level is greater than 175. Despite of many members are very close to this level, the number of then below 175 is greater than above.

A well use of fuzzy logic items (relations, sets and properties) is necessary in order to aim good formalizations. From a computational point of view, behavior of computing states in a fuzzy framework is an issue to be studied carefully and where human participation is nowadays a necessity in formal engineering.

A suitable using of aggregators in formal engineering can be found in [16]. In this paper a good classification of aggregators are done from a computational point of view. This classification permits not only to choose a suitable aggregator but also to get good designs for developing correct codes.

5 Conclusions and Future Work

At this point, the question is whether we can create software 100% reliable. The optimistic answer is yes, we can, but only for programming in the small. However, real life and the projects we do in it are not so easy. Multiple factors prevent the outcome of a reliable project, even featuring techniques. Software development projects are faced with a big problem: the insufficient level of training of programmers. As shown in the examples in the previous section, specifications and codes appear together. However, the idea is to develop an automated program code from specifications (Task 1) or a tester statements that determines the veracity of the assertions described in terms of computational logic, above all, the invariant (Task 2). Unfortunately, looking at the codes and specifications of simple examples there is perceived serious difficulties to perform both Task 1 and Task 2. The most optimistic engineers believe that software development tool will automatically perform both tasks without error better than any non-specialist programmer (keep in mind that developers specializing in logical and algorithmic derivations are a very low percentage). So far, this idea is just a wish of many [10, 11] and for the moment, we must accept that software is not only unreliable but it fails constantly.

Users are used to restart the computer, close windows, and even send details of errors to the company that caused the error. Therefore a good political awareness must be developed if we aim the Grand Challenge. Millions of working-hours are lost every day to restore systems. Then, what to do to improve the reliability of software? We propose three ideas:

- Using techniques of error detection and calculation of reliability in the sense of Model Checking, Testing and especially the reliability theory [9]. Among the works that are already developing the CDRISK team of Complutense University has developed some work [15].
- Improving the preparation of systems engineers and professionals involved in the early stages of project life cycle (with respect to the proposal shown in Figure 2) and the using of fuzzy logic to formalize natural languages together with the first order logic.
- Raise awareness among governments, institutions and users in general not to accept buggy software. Thus a positive response is expected by the software development companies and a positive attitude to the effort required creating higher quality products in smaller amounts. Too bad the software creation is a fact that is already affecting investors [18], making it a good time to start with political awareness. A marketing plan could be helpful.

Acknowledgments. We appreciate the support and funding of CAM research group and project GR58/08 TIN2009-07 901 of Spain.

References

1. Lions, J.L., et al.: ARIANE 5. Flight 501 Failure. Report by the Inquiry Board (1996), `http://www.ima.umn.edu/~arnold/disasters/ariane5rep.html`
2. Sommerville, I.: Software Engineering. Addison-Wesley (2004)
3. Jurafsky, D., Martin, J.H.: Speech and Language Processing: An Introduction to Natural Language Processing, Speech Recognition, and Computational Linguistics, 2nd edn. Prentice-Hall (2008)
4. Zadeh, L.A.: From Computing with Numbers to Computing with Words: From manipulation of measurements to manipulation of perceptions. In: Ito, T., Abadi, M. (eds.) TACS 1997. LNCS, vol. 1281, pp. 35–68. Springer, Heidelberg (1997)
5. Hall, A.: Realizing the Benefits of Formal Methods. Journal of Universal Computer Science 13, 669–675 (2007)
6. Liu, S.: Formal Engineering for Industrial Software Development, 3rd edn. Springer, Heidelberg (2004)
7. Lopez, V., et al.: Specification and Computing States in Fuzzy Algorithms. International Journal of Uncertainty, Fuzziness and Knowledge-Based System 16, 301–336 (2008)
8. Mitchell, R., McKim, J.: Design by Contract: by example. Addison-Wesley, Reading (2002)
9. ReliaSoft Team.: Avoiding Common Mistakes and Misapplications in Design for Reliability. Reliability Edge: The Newsletter for Reliability Professionals, vol. 11 (2011)
10. Jones, C., et al.: Verified Software: A Grand Challenge. Computer, pp. 93–95. IEEE Computer Society (2006)
11. Hoare, T., Misra, J.: Vision of a Grand Challenge Project. In: Proceedings of Verified Software: Theories, Tools, Experiments Conference, pp. 1–11 (2005)
12. Lopez, V., Montero, J.: Formal Specification under Fuzziness. Journal of Mult-Valued Logic & Soft Computing 15, 209–228 (2009)
13. Lopez, V., et al.: Fuzzy Specification in Real Estate Market Decision Making. Int. Journal of Computational Intelligence Systems 3, 8–20 (2010)
14. Lopez, V., et al.: A Functional Tool for Fuzzy First Order Logic Evaluation. In: Applied Intelligence: Proceedings of FLINS Conference, pp. 19–26. Word Scientific (2006)
15. Lopez, V., et al.: Improving reliability and performance in computer systems by means of fuzzy specifications. In: Proceedings ISKE Conference, Series on Computer Engineering and Information Science, vol. 2, pp. 351–356. World Scientific (2009)
16. Lopez, V., et al.: Formal Specification and Implementation of Computational Aggregation Functions. In: Applied Intelligence: Proceedings of FLINS Conference, vol. 4, pp. 523–528 (2010)
17. Liu, B.: Uncertain Risk Analysis and Uncertain Reliability Analysis. Journal of Uncertain Systems 4, 163–170 (2010)
18. Hoffman, D.: The Software Bubble is Going to End Badly for Investors. Wall St. Cheat Sheet, April, 12 (2010)

An Approach to the Analysis of Performance, Reliability and Risk in Computer Systems

Victoria López, Matilde Santos, and Toribio Rodríguez

Facultad de Informática, Universidad Complutense de Madrid,
28040-Madrid, Spain
vlopez@fdi.ucm.es, msantos@dacya.ucm.es, trodriguez@gmail.com

Abstract. A computer system is a network of devices working together. Devices are in constant change during the life cycle of the system. Replacing devices is one of the main aims of systems performance optimization. Assuming a system contains several uncertain variables and taking into account some system evaluation indexes such as loss function, risk, uncertainty distribution, etc., this article deals with the problem of measuring performance, risk and reliability of computer systems simultaneously. A tool for the evaluation of these three factors has been developed, named EMSI (Evaluation and Modelling of Systems). It is based on an uncertainty multicriteria decision-making algorithm. Several implemented functionalities allow making comparisons between different devices, evaluating the decision of including or not a new unit into the system, measuring the reliability/risk and performance of an isolated device and of the whole system. This tool is already working at Complutense University of Madrid.

Keywords: Uncertainty, Performance, Reliability, Risk, Computer system, Fuzziness.

1 Introduction

In a computer system any device acts like any other component in any machine (car, robot, etc.), regarding performance and reliability. These features of each unit should be evaluated both individually and globally, i.e., within the system in which it interacts. The aim of increasing the performance of a system or device requires an analysis of the functionality of every part. It is important to calculate the relative and absolute speedup as well as to analyze the capacity of each of the devices. A good reference on computing performance can be found at [3].

The fundamental problem that arises when dealing with maintenance is to determine which devices should be replaced to optimize the overall system performance. When the devices are evaluated in isolation, solutions can be obtained by modeling a multicriteria decision problem. However, it is quite exceptional to find isolated devices in computer systems.

Therefore, as it is said in [3], a Computer System is a set of interrelated parts, hardware, software and human resources, with the objective of making a sequence of data operations. The system presents a lot of dependences between components:

Y. Wang and T. Li (Eds.): Knowledge Engineering and Management, AISC 123, pp. 653–658.
springerlink.com

usually, the component performance is directly related with other performances in the system and other devices.

Performance analysis is used in computer science and engineering fields. It could be carried out by a combination of measurements of a computer system's characteristics. The analysis will depend on the specific situation and the features, interests, and facilities involved in the system.

This article deals with the problem of measuring performance, risk and reliability of computer systems simultaneously. A software tool that includes several functionalities for making comparisons between different devices, between the decision of including or not a new unit into the system, measuring the reliability/risk of an isolated device, assessing the reliability/risk of the whole system itself, and evaluating performance of each component and of the whole system, is presented.

2 Uncertainty and Reliability

Uncertainty theory is a branch of axiomatic mathematics for modeling human uncertainty. The uncertainty can be considered in Computer Systems as unpredictable variations in the performance of current tasks.

An uncertain process is essentially a sequence of uncertain variables indexed by time or space. The uncertainly theory have three foundations stones. The first one is uncertain measure that is used to measure the truth degree of an uncertain event. The second one is uncertainty variable that is used to represent imprecise quantities. The third one is uncertainty distribution that is used to describe uncertain variables in incomplete but easy to use way.

There are others fundamentals concepts in uncertainty: independence, operational law, expected value, variance, moments, entropy, distance, convergence, and conditional uncertainty. There are also many methods of working with uncertainty [4]. Some of them are shown within the following subsection following.

3 EMSI Software Tool

The queuing network models provide a rough idea of the correct operation between devices. There are several studies that develop successfully this idea. Sharpe [12], Symbolic Hierarchical Automated Reliability and Performance Evaluator, is a tool for specifying and analyzing performance and reliability in Computer Systems. It is a toolkit that provides a specification language and solution methods for most of the commonly used model types for performance, reliability and modeling. It facilitates the hierarchical combination of different model types.

EMSI [7, 9] has been also developed as a toolkit but it evaluates the model (Computer System) by means of the aggregation of data and then by applying decision making techniques under uncertainty. In addition, in this work we propose a new way of evaluation by considering that the optimization of the computer system performance involves not only the performance analysis but also the relative and global reliability, the security of devices and the stability. This new feature is developed through two functionalities: Fuzzy MCDM and Warranty Analysis that are explained in detail in next sections.

The previous and new features of the tool are summarized below. They are implemented in modules o new features can be added.

- *Computer System Configuration.* This module facilitates the creation of a new system and it can be also used to update the current system as well as to recovery previously stored data. In this module user can select the new devices that will become part of the new system and configure their own characteristics.
- *Performance Evaluation:* This module allows to measure the performance of the system as a whole and for each component.
- *System's Reliability.* This module calculates the reliability of each component and the whole system from the manufacturer's data, using reliability functions associated with the variable time between failures for each component
- *System Activity Reporter:* This module is used for monitoring and analyzing results associated with operating systems, specifically, with well known Unix SAR.
- *Comparative Analysis:* It makes a comparative analysis of the devices as part of the network of queues of the global system.
- *Fuzzy MCDM.* Here the problem of multiple criteria decision is approached by fuzzy logic [8, 11]. Some devices are upgraded in order to optimize computer network devices performance and reliability. The technique of optimal points [13] is applied.
- *Network.* This module allows the entry of transaction data on the network tasks and analyzes their behavior, both open and close net cases.
- *Warranty Analysis.* This module analyzes the security of the devices on a given system as a problem with uncertainty, using the technique of Hurwitz [1]. It also analyzes the security of the whole system.

3.1 Uncertainty and Guarantees Analysis

In order to study the guarantees of a certain component as part of the computer system, uncertainty and multicriteria decision making techniques are applied. This analysis is important in prevention of future failure of any of devices of a computer.

Security analysis is discussed extensively in the literature. There are commercial applications for companies that offer guarantees on an analysis of components such as Weibull++ [14] or [10]. These applications are distributed by quite expensive licenses. For this reason the development of a separate module to calculate this values, that is integrated into the EMSI tool, is a significant added value especially for teaching, where free software and open source are the best options.

The functionality presented in this article is based on the analysis of a dimension [2]. To determine the guarantees some comparisons are made based on two dependent variables: the number of units launched and the number of members returned, over a given period of time. As it is known, because of the probability of occurrence of different events that can affect the problem, it was decided to solve it with a model of decision problem under uncertainty. When purchasing a new component to be integrated within a computer system, besides taking into account issues such as cost or efficacy, we should not forget the reliability of the component. This reliability depends on several factors such as the date of manufacture, time of use, or reliability of plant distribution. A variable of particular interest in this study is the age of the component. *A priori* it seems logical that a hard disk (for example) is worse than another if it is older. The justification may be due to advances in technology, improved materials, etc. However, experience dictates that the reliability of components depends on factors that are not always easy to control.

Decision problems under uncertainty involve situations in which the probabilities associated with potential outcomes are not known or can not be estimated by classical methods. For these situations there are several criteria that can be analyzed:

- *Maximax:* this criterion chooses the alternative that provides a maximum possible profit. Used by optimistic, it can lead to achieve the maximum benefit or the maximum loss.
- *Maximin:* choose the alternative that provides a possible minimum benefit, which is characteristic of pessimists who seek the minimum risk in terms of losses.
- *Minimax:* indicates that we choose the alternative in which the maximum possible loss that may occur is as small as possible, trying to minimize the maximum potential loss.
- *Hurwitz:* it is the criterion used in this tool, and it is between *maximax* and *maximin*. It is also known as the criterion of the realism because it leaves the decision maker the option of setting a coefficient of optimism. It is based on that, when making decisions, people often look at the benefits or consequences of extreme and do not take into account that the results found between the two extremes, of course, will be the most realistic since it is what normally happens in reality.

The Hurwitz criterion weights the outliers so that the weights assigned reflect the importance that the decision maker gives to them (a level of pessimism or optimism). By the coefficient of optimism (chosen by the user), the weighted average profit of each alternative is calculated. The best alternative will be the one A* that maximizes the value obtained by the following function for each alternative A_i and a coefficient of optimism α, where x_{ij} is the value of alternative A_i regarding criteria C_j.

$$\begin{gathered} K(A_i,\alpha) = \alpha \cdot \min\{x_{ij}\} + (1-\alpha)\max\{x_{ij}\} \\ K(A^*,\alpha) = \max k(A_i,\alpha) \end{gathered} \tag{1}$$

3.2 Example

We study the case of a computer system that has two components: a processor and a hard disk. By means of windows forms we can see the number of returns that have been occurred because they are defective in a period of 3 months. In these windows, you may enter the data and the system generates a window with the corresponding records for each of the devices of the system under study. Then the application proceeds to perform the security analysis of the components regarding the coefficient of optimism that user had chosen. This ratio is also introduced as an input because it depends on the user.

For example, with an optimistic coefficient of 0.8 (the closer to 1 the more optimistic it is), analysis yields to the results of Figure 1 (left). However, if the user wants to analyze it from a pessimistic point of view, for example, with a coefficient $\alpha = 0.2$, we obtain the results shown in Figure 1 (right).

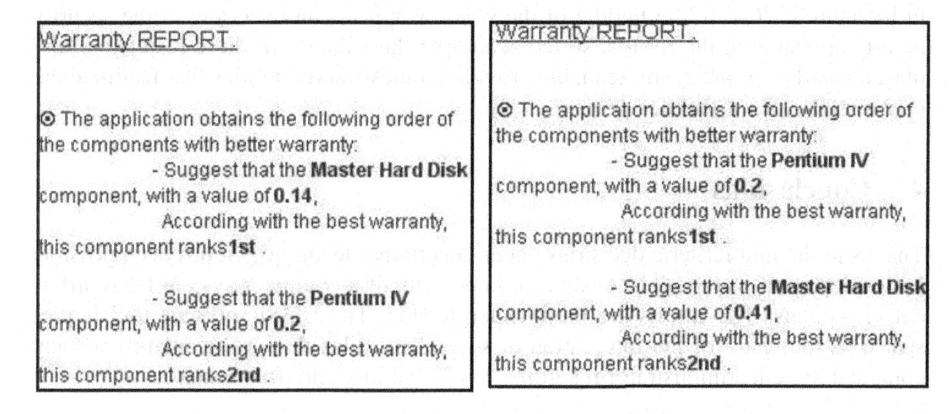

Fig. 1. Optimistic and pessimistic warranty report

The selection of a larger optimistic or pessimistic coefficient in the range of [0, 1] allows some flexibility in the process and it incorporates the user perceptions, which is essential when the tool is comparing a large number of devices.

Finally we discuss the results obtained in the respective optimistic and pessimistic reports. On the one hand, we had the same number of processors than hard drives, and therefore the total number of components returned is the same in both cases (60) during the 3 months. The optimistic analysis keeps the hard disk that offers better security component, considering bad luck the month with a high percentage of defectives. However, a pessimistic analysis returned the processor component, as it tends to avoid a bad result.

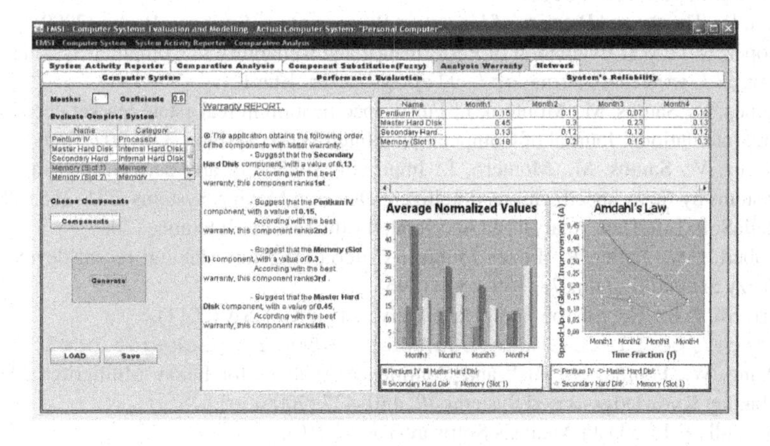

Fig. 2. Analysis Warranty Report

EMSI tool does not only return those reports but, as shown in Figure 2, it also reproduces with bar charts and graphs standard statistical measures the average of the acceleration of the devices under study, calculated by the Amdah formula widely used in literature [5, 6]]. In the middle of the window appears an overview of the security report summarizing the results, sorted according the values. All these results are displayed clearly and are easily readable providing quick interpretations that facilitate the user to make an optimal decision.

4 Conclusions

Thanks to the multicriteria decisions under uncertainty techniques it has been possible to develop a software tool to evaluate components of a computer system from different viewpoints. The results are easily interpretable. The EMSI software tool is presented as the result of the integration of several modules that are of interest for any computer system administrator regarding risk, reliability and performance.

Acknowledgments. We appreciate the support and funding of CAM research group GR58/08 and project TIN2009-07901 from the Government of Spain.

References

1. Arnold, B.F., Grössl, I., Stahlecker, P.: The minimax, minimin and the Hurwicz adjustment principle. Theory and Decision 52(3) (2002)
2. Blischke, W.: Warranty Cost Analysis. CRC Press (1993)
3. Lilja, D.J.: Measuring computer performance. Cambridge University Press (2000)
4. Liu, B.: Uncertainty Theory: A Branch of Mathematics for Modeling Human Uncertainty. Springer, Berlin (2010a)
5. Liu, B.: Uncertain Risk Analysis and Uncertain Reliability Analysis. J. Uncertain Systems 4(3), 163–170 (2010b)
6. Liu, B.: Theory and Practice of Uncertain Programming. Springer, Berlin (2009)
7. Lopez, V.: EMSI Evaluation of Performance and Reliability in Computer Systems (2011), http://www.mat.ucm.es/~vlopez/mcdm.html
8. López, V., Santos, M., Montero, J.: Fuzzy specification in real estate market decision making. International Journal of Computational Intelligence Systems 3, 8–20 (2010)
9. Lopez, V., Santos, M., Montero, J.: Improving reliability and performance in computer systems by fuzzy specifications. Intelligent Decision Making Systems, 351–356 (2009)
10. ReliaSoft. Life Data Analysis & Accelerated Life Testing Reference (2007)
11. Robert, C.C., Fuller, R.: Fuzzy Multiple Criteria Decision Making: recent developments. Fuzzy Set and Systems 78, 139–153 (1996)
12. Trivedi, K.S.: The Sharpe Tool, Duke University, NC, USA (2011), http://people.ee.duke.edu/~kst/software_packages.html
13. Wang, W., Fenton, N.: Risk and Confidence Analysis for Fuzzy Multicriteria Decision Making. Knowledge-Based Systems 19, 430–437 (2006)
14. Weibull++. Life Data Analysis Software Tool (2010), http://www.reliasoft.com/Weibull/

Empirical Bayes Estimation in Exponential Model Based on Record Values under Asymmetric Loss

Haiping Ren

School of Software, Jiangxi University of Science and Technology,
330013 Nanchang, China
chinarhp@163.com

Abstract. In this paper, the empirical estimates is derived for the parameter of the exponential model based on record values by taking quasi-prior and inverse prior Gamma distributions using the LINEX loss and entropy loss functions. These estimates are compared with the corresponding maximum likelihood (ML) and Bayes estimates, and also compared with the corresponding Bayes estimates under squared error loss function. A Monte Carlo simulation is used to investigate the accuracy of estimation.

Keywords: Bayes estimate, Empirical Bayes estimate, LINEX loss, entropy loss, record value.

1 Introduction

The exponential distribution is one of the most important distributions in life-testing and reliability studies. Inference procedures for the exponential model have been discussed by many authors.See Cohen and Helm(1973), Sinha and Gutman(1976), Basubramanian and Balakrishnan(1992), Balakrishnan et al(2005),and reference therein.

Record values is first proposed Chandler(1952), and it has drawn great attention by many researchers due to the fact that the extensive application in clued industrial stress testing, meteorological analysis, sporting and athletic events, oil and mining surveys using record values. Let $X_1, X_2, \ldots$ be a sequence of independent and identically distributed random variables with cumulative distribution function (cdf) F(x) and probability density function (pdf) f(x).Set

$$Y_n = \max\{X_1, X_2, \ldots, X_n\}, n \geq 1.$$

We say X_j is an upper record value of this sequence if $Y_j > Y_{j-1}, j > 1$.For more details on record values,see Arnold et al(1998) ,Raqab (2002), Jaheen(2004), and Soliman et al(2010).

We say X be a random variable from exponential, if it has a pdf

$$f(x;\theta) = \frac{1}{\theta} e^{-\frac{1}{\theta}x}, \quad x > 0, \theta > 0 \tag{1}$$

Y. Wang and T. Li (Eds.): Knowledge Engineering and Management, AISC 123, pp. 659–667.
springerlink.com

And cdf

$$F(x;\theta)=1-e^{-\frac{1}{\theta}x}, \quad x>0,\theta>0 \tag{2}$$

In this paper, the empirical Bayes estimate is derived for the parametert of the exponential model(1) based on record values. The rest of this paper is organized as follows. In Section2, we will give some preliminaries about the prior knowledge and the asymmetric loss functions. The MLE and Bayes estimates will deriven in Section3. In Section4, we will propose a empirical Bayes method to obtained the empirical Bayes estimates of the parameter. Finally, In Section5, we will give a Monte Carlo numerical example to illustrate the result.

2 Preliminaries

Suppose we observe n upper record values $X_{U(1)}, X_{U(2)}, \ldots, X_{U(n)}$ from the exponential distribution(1), and $x=(x_1, x_2, \ldots, x_n)$ is the observation of $X=(X_{U(1)}, X_{U(2)}, \ldots, X_{U(n)})$. The likelihood function is given by (see Arnold et al (1998))

$$l(\theta;x)=\prod_{i=1}^{n-1} H(x_i) f(x_n)=\theta^{-n}e^{-\frac{x_n}{\theta}} \tag{3}$$

Where $H(\cdot)$ is the harzard function corresponding to the pdf $f(\cdot)$.

Then the MLE of θ can be shown to be of the form

$$\hat{\theta}_{ML}=\frac{x_n}{n} \tag{4}$$

2.1 Prior Distribution

In the Bayes approach, we further assume that some prior knowledge about the parameter θ is available to the investigate from past experience with the exponential model. The prior knowledge can often be summarized in terms of the so-called prior densities on parameter space of θ. In the following discussion, we assume the following priors:

(i) The quasi-prior: For the situation where the experimenter has no prior information about the parameter θ, one may use the quasi density as given by

$$\pi_1(\theta;d)\propto\frac{1}{\theta^d}, \quad \theta>0, d>0 \tag{5}$$

Hence, $d=0$ leads to a diffuse prior and $d=1$ to a non-informative prior.

(ii) The Inverse Gamma prior: The most widely used prior distribution of θ is the Inverse Gamma prior distribution with parameters α and β (>0) , denoted by $I\Gamma(\alpha,\beta)$,with pdf given by

$$\pi_2(\theta;\alpha,\beta)=\frac{\beta^{\alpha}}{\Gamma(\alpha)}\theta^{-(\alpha+1)}e^{-\beta/\theta},\quad \theta>0,\alpha,\beta>0 \tag{6}$$

2.2 Loss Function

In Bayesian estimation, the loss function plays an important role, and the squared error loss as the most common symmetric loss function is widely used due to its great analysis properties. And the squared error loss function (SELF) $L(\hat{\theta},\theta)=(\hat{\theta}-\theta)^2$, which is a symmetrical loss function that assigns equal losses to overestimation and underestimation. However, in many practical problems, overestimation and underestimation will make different consequents.

(a) The LINEX Loss Function
Thus using of the symmetric loss function may be inappropriate, Varian(1975) and Zellner (1986) proposed an asymmetric loss function known as the LINEX function, which draws great attention by many researchers,see more details about LINEX in Al-Aboud(2009),also . Basu and Ebrahimi(1991) proposed the LINEX loss is expressed as:

$$L(\Delta)=b[e^{a\Delta}-a\Delta-1],\quad a\neq 0,b>0 \tag{7}$$

where $\Delta=\frac{\hat{\theta}}{\theta}-1$, and $\hat{\theta}$ is an estimator of θ, and studied Bayes estimation under this asymmetric loss function for an exponential lifetime distribution. This loss function is suitable for situations where overestimation is more costly than underestimation.

The Bayes estimator of θ, denoted by $\hat{\theta}_{BL}$ under the LINEX loss function is the solution of the following equation

$$E_{\pi}[\frac{1}{\theta}\exp(\frac{\hat{\theta}_{BL}}{\theta})]=e^{a}E_{\pi}[\frac{1}{\theta}] \tag{8}$$

Provide that the two posterior expectations in equation (8) $E_{\pi}(\cdot)$ exist and are finite.

(b)The Entropy Loss Function
In many practical situations, it appears to be more realisticto express the loss in terms of the ratio $\hat{\theta}/\theta$. In this case, Calabra and Pulcini(1994) point out that a useful asymmetric loss function is entropy loss function:

$$L(\hat{\theta},\theta)\propto(\frac{\hat{\theta}}{\theta})^{q}-q\ln\frac{\hat{\theta}}{\theta}-1 \tag{9}$$

Whose minimum occurs at $\hat{\theta}=\theta$. Also, this loss function (9) has been used in Dey et al(1987) , Dey and Lin(1992), in the original form having $q=1$ .Thus $L(\hat{\theta},\theta)=b(\frac{\hat{\theta}}{\theta}-\ln\frac{\hat{\theta}}{\theta}-1), b>0$. The Bayes estimator under the entropy loss function is denoted by $\hat{\theta}_{BE}$, given by

$$\hat{\theta}_{BE}=[E_{\pi}(\theta^{-1})] \tag{10}$$

3 Bayes Estimation

3.1 Bayes Estimation under Quasi-prior

Combining (3) with quasi-prior(5) using Bayes theorem, the posterior pdf of θ is given by

$$h_1(\theta \mid x)=\frac{x_n^{n+d-1}}{\Gamma(n+d-1)}\theta^{-(n+d)}e^{-\frac{x_n}{\theta}} \tag{11}$$

Then (i)the Bayes estimate under the squared error loss function is given by

$$\hat{\theta}_{S1}=\frac{x_n}{n+d-2}, \quad n+d>2 \tag{12}$$

(ii) Using (8) and (10), the Bayes estimate under the LINEX loss function is come out to be

$$\hat{\theta}_{L1}=\frac{1-\exp(-\frac{a}{n+d})}{a}x_n \tag{13}$$

(iii) Using (10) and (11), the Bayes estimate under the entropy loss function is obtained as

$$\hat{\theta}_{E1}=\frac{x_n}{n+d-1} \tag{14}$$

3.2 Bayes Estimation under Inverse Gamma Prior

Combining (3) with Inverse Gamma prior (6) and using Bayes theorem, the posterior pdf of θ is given by

$$h_2(\theta \mid x)=\frac{(\beta+x_n)^{n+\alpha}}{\Gamma(n+\alpha)}\theta^{-(n+\alpha+1)}e^{-\frac{\beta+x_n}{\theta}}, \quad \theta>0 \tag{15}$$

which is also an inverse Gamma family with parameter $(\beta+x_n, n+\alpha)$.

Thus , (i) the Bayes estimate under the squared error loss function is given by

$$\hat{\theta}_{S2} = \frac{\beta + x_n}{n + \alpha - 1} \tag{16}$$

(ii) the Bayes estimate under the LINEX loss function is come out to be

$$\hat{\theta}_{L2} = \frac{1 - \exp(-\frac{a}{n+\alpha+1})}{a}(\beta + x_n) \tag{17}$$

(iii) the Bayes estimate under the entropy loss function is obtained as

$$\hat{\theta}_{E2} = \frac{\beta + x_n}{n + \alpha} \tag{18}$$

4 Empirical Bayes Estimation

When the prior parameters α and β are unknown, we may use the empirical Bayesian approach to get their estimates. Since the prior density (3) belongs to a parametric family with unknown parameters, such parameters are to be estimated using past samples. Applying these estimates in (11) and (12), we obtain the empirical Bayes estimates of the parameter θ based on squared error loss and LINEX loss functions, respectively. For more details on the empirical Bayes approach, see Maritz and Lwin (1989).

For a random variable X, suppose that $X_{N+1,U(1)}, X_{N+1,U(2)}, \ldots, X_{N+1,U(n)}$ is the current sample when the random variable θ has the value θ_{N+1}. When the current sample is observed, there are available past samples $X_{l,U(1)}, X_{l,U(2)}, \ldots, X_{l,U(n)}$, $l = 1,2,\ldots,N$ .Each sample is supposed to follow the exponential distribution (1). The LF of the jth sample is given by (3) with x_n being replaced by $x_{n:j}$. For a random sample $l, l = 1,2,\ldots,N$, the past MLE of θ_l is given by(4) and can be rewritten as

$$\hat{\theta}_l \equiv Y_l = \frac{x_{n:j}}{n} \tag{19}$$

For a given θ_l, the conditional pdf of $x_{n:j}$ is gamma and then Y_l has the following inverted gamma pdf:

$$h(y_l \mid \theta_l) = \frac{1}{\Gamma(n)(\theta_l / n)^n} y_l^{n-1} e^{-\frac{y_l}{\theta_j / n}} \tag{20}$$

Then the marginal pdf of $Y_l, l = 1,2,\ldots,N$, can be shown to be

$$h(y_l) = \int_0^\infty f(y_l \mid \theta_l) g(\theta_l; \alpha, \beta) d\theta_l \tag{21}$$

where $g(\theta_l; \alpha, \beta)$ is the gamma prior density of θ_l with density(4).

Then, from(14) and (15), the random variable $Y_l, l = 1,2,\ldots,N$ has the inverted beta pdf in the form

$$h(y_l) = \frac{n^n \beta^\alpha}{B(\alpha,n)} \frac{y_l^{\,n-1}}{(\beta + n y_l)^{n+\alpha}}, \quad y_l > 0 \tag{22}$$

Therefore, the moments estimates of the parameters α and β may be obtained by using (16), to be of the forms

$$\hat{\alpha} = \frac{(n+1)S_1^2}{nS_2 - (n+1)S_1^2} + 2 \qquad \text{and} \qquad \hat{\beta} = S_1(\hat{\alpha} - 1) \tag{23}$$

where

$$S_1 = \frac{1}{N}\sum_{l=1}^{N} \hat{\theta}_l \qquad \text{and} \qquad s_2 = \frac{1}{N}\sum_{l=1}^{N} \hat{\theta}_l^2 \tag{24}$$

Therefore, the empirical Bayes estimates of the parameter θ under the squared error and LINEX loss and entropy loss functions are given, respectively,

$$\hat{\theta}_{ES} = \frac{\hat{\beta} + x_n}{n + \hat{\alpha} - 1} \tag{25}$$

$$\hat{\theta}_{EL} = \frac{1 - \exp(-\frac{a}{n + \hat{\alpha} + 1})}{a} (\hat{\beta} + x_n) \tag{26}$$

$$\hat{\theta}_{EE} = \frac{\hat{\beta} + x_n}{n + \hat{\alpha}} \tag{27}$$

5 Simulation Study and Comparisons

5.1 Numerical Example

In this section, the MLE, Bayes, and empirical Bayes estimates of the parameter θ are computed according to the following steps.

(1) For given values of $\alpha = 3.5$ and $\beta = 1.1$, we generate $\theta = 0.8803$ from the inverse gamma prior density (6).

(2) Based on the generated value θ, an upper record sample of size $n = 10$ is then generated from the density of the exponential distribution(1), which is considered to be the informative sample. This sample is

0.1325, 0.3430, 0.9750, 1.7531, 2.5116, 3.6330, 5.2406, 6.0390, 8.5970, 10.3065

(3) The ML estimate of θ is given from (9),by $\hat{\theta}_{ML} = 1.0307$

(4) For given values of c, the Bayes estimates of θ are computed from (11) and (12), based on squared error and LINEX loss(a=1) and entropy loss functions, respectively.

$$\hat{\theta}_{S2} = 0.9125,\ \hat{\theta}_{EL} = 0.7601,\ \hat{\theta}_{EE} = 0.9048$$

(5) For the given values of the prior parameters $\alpha = 3.5$ and $\beta = 1.0$, the past estimates $\hat{\theta}_l, l = 1,2,\ldots,N$ are generated from the inverted beta pdf, given by (16), using suitable transformations on the beta random variables, and the sample is given by

0.3785, 0.2630, 1.5464, 0.2699, 0.3837, 0.0863, 0.6605, 0.7990, 0.3691, 0.3174, 0.3383, 0.4750, 0.2071, 0.2748, 0.2751, 0.2515, 0.3648, 1.4693, 0.2314, 0.1614

The estimates of α and β are respectively given by $\hat{\alpha} = 3.8058$, $\hat{\beta} = 1.2798$.

Then the empirical Bayes estimates are respectively given by

$$\theta_{ES} = 0.9048,\ \hat{\theta}_{EL} = 0.7567,\ \hat{\theta}_{EE} = 0.8392.$$

5.2 Monte Carlo Simulation

In this section, we present a Monte Carlo simulation study to illustrate the presented next.

The MLE, Bayes, and empirical Bayes estimates of the parameter θ are compared based on Monte Carlo simulation as follows. Assume d=1,a=1.

(1) For given values of α and β, we generate $\theta = 0.8803$ from the inverse gamma prior density (6).

(2) Based on the generated value $\theta = 0.8803$, a random sample of size n is then generated from the density of the exponential distribution(1), which is considered to be the informative sample.

(3) The ML estimate of θ is computed from (4).

(4) For given values of c, the Bayes estimates of θ are computed, under squared error and LINEX loss functions, respectively.

(5) For the given values of the prior parameters α and β, the past estimates $\theta_l, l = 1,2,\ldots,N$ are generated from the inverted beta pdf, given by (16), using suitable transformations on the beta random variables.

(6) The estimates of the prior parameters α and β are evaluated by using the past estimates obtained in step (5), substituting these estimates in (25),(26) and (27) yields the empirical Bayes estimate of θ.

(7) Steps (1) to (6) are repeated N=1000 times and the risks under squared-error loss of the estimates are computed by using

$$ER(\hat{\theta}) = \frac{1}{N}\sum_{i=1}^{N}(\hat{\theta}_i - \theta)^2$$

where $\hat{\theta}_i$ is the estimate at the ith run.

Table 1. Estimated Risk(ER) of various estimates for different sample sizes (n, N)

(n,N)	$ER(\hat{\theta}_{ML})$	$ER(\hat{\theta}_{S1})$	$ER(\hat{\theta}_{L1})$	$ER(\hat{\theta}_{E1})$	$ER(\hat{\theta}_{S2})$
(10,10)	0.6348	0.6252	0.6546	0.6384	0.5282
(20,10)	0.6990	0.6952	0.7040	0.6990	0.6273
(20,15)	0.6999	0.6962	0.7049	0.6999	0.6281
(30,15)	0.7241	0.7224	0.7265	0.7241	0.6713
(30,20)	0.72569	0.7253	0.7291	0.7261	0.6738
(40,25)	0.7373	0.7363	0.7386	0.7373	0.6956
(50,30)	0.7454	0.7448	0.7463	0.7454	0.7110

Table 2. Estimated Risk(ER) of various estimates for different sample sizes (n, N)

(n,N)	$ER(\hat{\theta}_{L2})$	$ER(\hat{\theta}_{E2})$	$ER(\hat{\theta}_{ES})$	$ER(\hat{\theta}_{EL})$	$ER(\hat{\theta}_{EE})$
(10,10)	0.5653	0.5445	0.4529	0.4879	0.4680
(20,10)	0.6412	0.6323	0.5705	0.5865	0.5772
(20,15)	0.6419	0.6351	0.5860	0.6018	0.5927
(30,15)	0.6784	0.6758	0.6388	0.6474	0.6424
(30,20)	0.6808	0.6757	0.6463	0.6546	0.6497
(40,25)	0.6999	0.6988	0.6769	0.6819	0.6789
(50,30)	0.7139	0.7110	0.6974	0.7008	0.6988

Acknowledgments. This work is partially supported by Natural Science Foundation of Jiangxi Province of China,NO. 2010GZS0190, and Foundation of Jiang'xi Educational Committee, NO. GJJ116 00, GJJ11479.

References

1. Cohen, A.C., Helm, F.R.: Estimators in the exponential distribution. Technometrics 15, 415–418 (1973)
2. Sinha, S.K., Kim, J.S.: Ranking and subset selection procedures for exponential populations with type I and type II censored data. In: Frontiers of Modern Statistical Inference Procedures, pp. 425–448. American Science Press, New York (1985)
3. Basubramanian, K., Balakrishnan, N.: Estimation for one and two parameter exponential distributions under multiple type II censoring. Statistische Hefte 33, 203–216 (1992)
4. Balakrishnan, N., Lin, C.T., Chan, P.S.: A comparison of two simple prediction intervals for exponential distribution. IEEE Transactions on Reliability 54(1), 27–33 (2005)
5. Chandler, K.N.: The distribution and frequency of record values. J. Royal Statistics. Society B14, 220–228 (1952)
6. Arnold, B.C., Balakrishnan, N., Nagaraja, H.N.: Records. John Wiley and Sons, New York (1998)
7. Gulati, S., Padgett, W.J.: Parametric and Nonparametric Inference from Record-Breaking Data. Springer, New York (2003)
8. Raqab, M.Z.: Inferences for generalized exponential distribution based on recors statistics. J. Statist. Plan. Inference, 339–350 (2002)
9. Jaheen, Z.F.: Empirical Bayes analysis of record statistics based on LINEX and quadratic loss functions. Computers and Mathematics with Applications 47, 947–954 (2004)

10. Soliman, A., Amin, E.A., Abd-ElAziz, A.A.: Estimation and prediction from inverse Rayleigh distribution based on lower record values. Applied Mathematical Sciences 4(62), 3057–3066 (2010)
11. Varian, H.R.: A Bayesian approach to real estate assessment. In: Fienberg, S.E., Zellner, A. (eds.) Studies in Bayesian Econometrics and Statistics in Honor of Leonard J. Savage, pp. 195–208. North Holland, Amsterdam (1975)
12. Zellner, A.: Bayesian estimation and prediction using asymmetric loss functions. J. Amer. Statist. Assoc. 81, 446–451 (1986)
13. Al-Aboud, F.M.: Bayesian estimations for the extreme value distribution using progressive censored data and asymmetric loss. International Mathematical Forum 4(33), 1603–1622 (2009)
14. Pandey, H., Rao, A.K.: Bayesian estimation of the shape parameter of a generalized pareto distribution under asymmetric loss functions. Hacettepe Journal of Mathematics and Statistics 38(1), 69–83 (2009)
15. Basu, A.P., Ebrahimi, N.: Bayesian approach to life testing and reliability estimation using asymmetric loss function. Journal of Statistical Planning and Inferences 29, 21–31 (1991)
16. Calabria, R., Pulcini, G.: An engineering approach to Bayes estimation for the Weibull distribution. Micro-Electron. Reliab. 34(5), 789–802 (1994)
17. Dey, D.K., Ghosh, M., Srinivasan, C.: Simultanous estimation of parameters under entropy loss. J. Statist. Plan. and Infer., 347–363 (1987)
18. Dey, D.K., Liu, P.-S.L.: On comparison of estimators in a generalized life model. Micro-Electron. Reliab. 32(1), 207–221 (1992)
19. Martz, J.L., Lwin, T.: Empirical Bayes Methods, 2nd edn. Chapman & Hall, London (1989)

Image Registration Algorithm Based on Regional Histogram Divergence and Wavelet Decomposition

Cheng Xin[1] and Bin Niu[2]

[1] Department of Electrical and Information Engineering, Liaoning Univeristy, Shenyang 110036, China
[2] College of Information Science & Technology, Liaoning University, Shenyang, 110036, China
chengxin_1985@126.com.cn, niub@lnu.edu.cn

Abstract. This paper presents new methods that have been developed for registration of gray scale image, these methods are based on improve the registration precision and the ability of anti-noise, the image registration process was based on analyzed the situation of the point set of histogram, and defined the formula of the histogram divergence. To speed up searching the registration parameters, all were done in the wavelet field and a hybrid algorithm based on genetic algorithm and Powell's method was used to optimize this parameters. Experimental results proved the algorithm can apply wider optimization methods and have better anti-noise robustness performance.

Keywords: image registration, histogram divergence, wavelet decomposition, Powell optimization algorithm, weighted summation.

1 Introduction

Image registration is the process of overlaying two or more images of the same scene taken at different times, from different viewpoints, and/or by different sensors. It is widely used in automatic mapping aerial images, three-dimensional image reconstruction, computer vision, fusion, pattern recognition and many other fields [1].

At present, mutual information is commonly used in image registration method, which is the similarity measure based on gray scale statistics[3], so the inconsistency distribution of image gray may affect registration results of mutual information. This paper analyzes the situation of the point set of the joint histogram, and proposed a new similarity measure - histogram divergence (HD). And on this basis, proposed an automatic image registration method based on the wavelet domain. Because the low-frequency image contains the main contours of the original image information, we used this as a histogram of gray-scale divergence, and weighted slice to the low-frequency image, using the method of a weighted sum of local histogram divergence as the similarity measure function of image registration. The result shows that the registration algorithm is more accurate and stronger anti-noise.

Y. Wang and T. Li (Eds.): Knowledge Engineering and Management, AISC 123, pp. 669–675.
springerlink.com

2 HD-Based Similarity Measure

2.1 Histogram Divergence Measure

Joint histogram is a new concept which is proposed by Greg Pass and Ramin Zabih in 1999[2], which uses color information and the pixel location of different images. It actually counts the frequency of different gray-scale combination of corresponding pixels in two images.

Supposed that A and B represent a pair of images to register, their probability distributions are respectively denoted by M and N , Where M and N are two gray-scale images A and B series. The defined is as follows:

1) Define a matrix which size is M×N named HIST[M,N] ,M and N are the gray-scale of image A and B.

2) For each pixel $i \in A \cap B$, command HIST[A(i),B(i)]+1, In which A(i) and B(i) are the gray of A and B, respectively, for the image pixel i in the gray.

In this way, the final HIST [M, N] matrix that statistics out of is the joint histogram of image A and image B. Figure 1 is the joint histogram that two images translate different sizes in the x-direction.

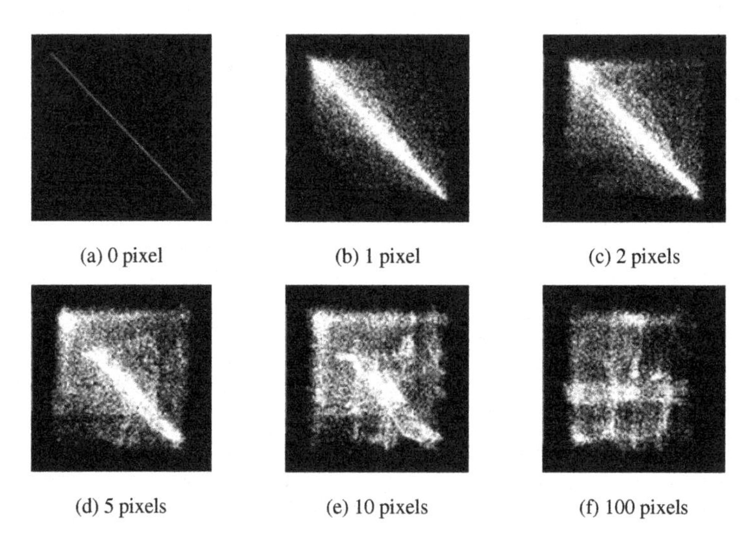

Fig. 1. (a) ~ (f) are the joint histograms which generated after two 200 × 200 size images translate 0,1,2,5,10,100 pixels. The vertical and horizontal coordinates of images are the grayscales of two registration images, and each coordinate represents the frequency of pixel pairs of gray combination.

From these several figures, it is clear that as the amount of shift increases, the non-zero in the histogram get into bulk distribution as the center of 45°line. This phenomenon can be explained by the definition of the joint histogram: When the translation difference is 0, that is, complete registration of two images, for the two images of the same imaging mode, the corresponding position of the pixel gray is the same or similar .

Therefore, we can use this divergence as a similarity measure of two images. The little different of the image, the little of the value of the measure. When the two images are complete registration, the measure reaches the minimum value of the function. The measure can be defined as follows:

$$HD = \frac{\sum_{(x,y)} |y - x|^2 HIST[x, y]}{\sum_{(x,y)} HIST[x, y]} \tag{1}$$

The geometry of molecules in formula 1 is the sum of each point on the 45° line to the weighted square of the distance. And the denominator is the pixel number of overlap part of two images. Its function is to remove the similarity measure and the two images overlap correlation. The value of formula 1 is increasing the size of the relationship with the translation and rotation, and it will achieve the minimum value when the two images are complete registration.

2.2 Local Histogram Divergence

Because the pixels of intersection part between two registration images are in the center of the images, this paper choice the weighted method which let the center area of the image as the most interesting region. The measure can be defined as follows:

- Defined fragment of the parity

The weight value of center point is 4, and the weight value of points at up, down, left and right direction are 2, and the weight value of remaining four points are 1.

- Calculate the similarity measure HD_L value of each slice and put it into the formula 1.

Therefore, the locally weighted histogram divergence is defined as:

$$HD_L = \sum_i HD_{L_i} = \sum_i \frac{\sum_{(x,y)} |y - x|^2 HIST[x, y, i]}{\sum_{(x,y)} HIST[x, y, i]} \times P \tag{2}$$

Where x, y are the gray of two images A and B in the corresponding position, and i is the fragment sequence number. Therefore, the min sum value of local histogram divergence is the best registration results. It can improve the registration accuracy of interesting region and it does not affect the final value of the histogram divergence.

3 Algorithm Implementation

First, wavelet decomposes the reference image and the registration image. In the most advanced decomposition use formula 2 as the standard of measure registration. And then parametric search algorithms use genetic algorithms combined with the Powell method to optimize.

3.1 Image of Wavelet Decomposition

The choice of wavelet is decided by the function of decomposed and decomposition requires. Taking into account the Daubechies wavelet functions have good frequency domain localization nature and choice flexible etc. For choose the frequency channel of the wavelet image, due to the diagonal direction of the wavelet image contains more noise, it is not treated as objects generally. Jacqueline LeMoigne et al select the horizontal and vertical information relevant measure at the literature [4], but this paper is choose in the low-frequency wavelet image processing.

3.2 Parameters Optimization Algorithm

Two image registration parameters of the solution process, is actually finding the greatest similarity to measure the value of registration parameters that optimize the process parameters. In this paper, we choose the search strategies to optimize the solution process which combines genetic algorithm (genetic algorithm, GA) [5] and Powell (Powell) algorithm [6]. GA is a general optimization algorithm and the coding techniques and genetic manipulation is relatively simple to optimize. It is not be bounded by the restrictive conditions, and the most notable feature is the implicit parallelism and global solution space search. Powell method does not require compute derivatives, it use iterative search using the Brent algorithm in each dimension, so the search speed is fast, the local optimization ability is strong, and its accuracy is higher than other algorithm in local search optimization. But the Powell method has a defect that it depends on the initial point, so it is very easy to fall into local optimum in the registration process, the final optimization results depend on the initial point. We can get a solution where is near global optimum using genetic algorithm. Powell method has strong local optimization ability, so it can be GA's global search ability and Powell method of local optimization ability to combine and taking advantage of the advantages of two algorithms.

First, we use wavelet decomposition to the floating image and reference registration image to get the smaller sub-image. In the optimization process, the population size is 50, the image encoding parameter is the rigid transformation parameters (t_x, t_y, θ), the algorithm uses one-point crossover, crossover probability is 0.80, and mutation probability is 0.05. In the optimization process, the objective function of the algorithm is local histogram divergence between the two images.

Second, the starting point is the optimize parameters which are get by using GA. Then register in the high-resolution images and get registration optimal solution.

Because the GA registration results provides a very effective initial point for the Powell, the optimization time is much shorter than using default initial point or random initial point to search. GA optimizes in low resolution images, so the optimization process can be completed within a relatively short period of time, the total running time of the algorithm is only slightly longer than the ordinary method of Powell.

3.3 Transformation Model and the Interpolation Model

The purpose of image interpolation is to estimate the floating point pixel map values Non-grid point. In this paper three interpolation methods interpolate will be used.

Four-neighborhood of current pixels, the kernel function of image convolution operation as follows:

$$Cubic(X) = \begin{cases} X^3 - 2X^2 + 1 & , \ X \le 1 \\ -X^3 + 5X^2 - 8X + 4, & 1 < X \le 2 \end{cases} \quad (3)$$

Where, X is the distance from the pixel in the neighborhood to the center pixel.

4 Experimental Results and Discussion

4.1 Registration Results of This Algorithm

We uses two group images to verify. One group is image registration using the original image , which generates by floating image rotating clockwise 15°, and the other group is intercepted original [0,0,185,186] rectangular section and [15,15,200,201] rectangular section shifted from the x-direction to registration.

(a) Original image

(b) Rotate the image 15 degrees

(c) Registration result

Fig. 2. Registration result of Image Rotation

(a) [0,0,185,186] matrix part of original image

(b) [15,15,200,201] matrix part of original image

(c) Registration result

Fig. 3. Registration result of image translation

Figure 2 , 3 shows that there is a very precisely fusion between the two images and the specific experimental data is shown in Table 1.We do wavelet-4 transform in the image to generate one sub-image, and the image size is determined by (200 200) pixels into a (25 25) pixels. And calculate the similarity (HD_L) parameter values. It can be seen from Table I that the search accuracy in the sub-pixel level.

Table 1. Registration experimental data

Test Image	Actual parameter $(t_x/t_y/\theta)$	Measurement parameter $(t_x/t_y/\theta)$	Measurement error $(\Delta t_x/\Delta t_y/\Delta\theta)$
Rotation Image	(0.000/0.000/0.2616)	(0.0479/0.0756/ 0.2645)	(0.0479/0.0756/ 0.0029)
Translation Image	(15.000/16.000/0.00 0)	(14.9956/16.1616/ 0.0034)	(0.0044/0.1616/ 0.0034)

4.2 Comparison between Histogram Divergence and Mutual Information

(1) The comparison between HD_L and MI in Search trajectory sharp

Theory can be proved that the more intense the search trajectory curve, the more the search of the parameters process is not easy to fall into local extreme value. In the experiment, select two same pictures, and then let one of them turn a translation along the t_x direction 0-50 pixels and calculate HD_L and MI. In order to observe the two curves visually, now they are normalized to [0,1] range. Shown in Figure 4 that dotted line is MI curve, and solid line is HD_L curve. From Figure 4 it is clear that HD_L curve is sharper than MI curve, but the convergence are different. So comparing with MI, HD_L can use more search parameters optimization strategy.

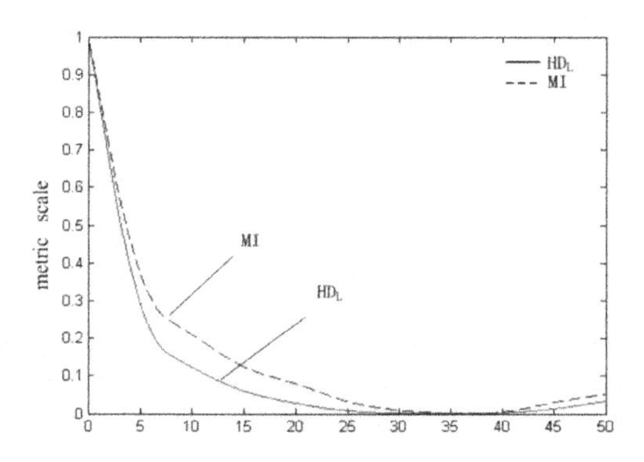

Fig. 4. Search trajectory of HD_L and MI

(2) Noise robustness comparison

In order to test the anti-noise performance of HD_L and MI, adding different image noise to the previous experiment images. The noise include Gaussian noise, salt and pepper noise, multiplicative noise. After adding four kinds noise to the original image, the registration error results of X-direction a pixel image shown in Table 3.

Table 3. HD_L and MI error comparison in different noise levels

Noise level	Gaussian(0.02)	Gaussian(0.2)	salt & pepper(0.02)	Speckle(0.02)
HD_L $(\Delta t_x / \Delta t_y / \Delta\theta)$	(0.3298/0.2218/0.0015)	(-0.4009/1.5891/-0.0007)	(0.0000/0.3433/0.0059)	(0.7519/0.1281/0.0012)
MI $(\Delta t_x / \Delta t_y / \Delta\theta)$	(-0.6177/1.5321/0.0006)	(0.6531/1.7109/0.0005)	(-0.4404/0.0933/0.0016)	(0.1731/-1.9275/0.0015)

(3) Run time comparison

Compared with MI, although the computation amounts about joint histogram are same, but the computation amount of HD_L is much less than MI in the next. For two Gray-level images of 256, HD_L only need 2×256×256+1+9 multiplications and 256×256 addition operations, but MI need 2×256×256+2×256+1 multiplications and 3×256×256 +2 addition operations. So in the time consumption, HD_L has more advantages.

5 Conclusion

Use Matlab7.8.0 AMD 4200+2.20GHz, 2.00GB Win7 computer .This paper bases on analyzing joint histogram of two images, which can achieve sub-pixel accuracy that proved by the image registration experiments of pan and rotate images. Comparing with the mutual information, it can be concluded that the algorithm has certain advantages in application and noise immunity.

References

1. Zitova, B., Flusser, J.: Image registration methods:a survey. Image and Vision Computing 21, 977–1000 (2003)
2. Pass, G., Zabih, R.: Comparing images using joint histograms (3) (1999)
3. West, J., Fitzpatrick, J.M., Wang, M.Y., et al.: Comparison and evaluation of retrospective inter modality brain image registration techniques. Journal of Computer Assisted Tomography 21(4), 554–566 (1997)
4. LeMoigne, J., Campbell, W.J., Cromp, R.F.: An automated parallel image registration technique based on the correlation of wavelet freatures. IEEE Transactions on 40(8), 1849–1864 (2002)
5. Chen, G., Wang, X., Zhuang, Z., et al.: Genetic Algorithms and Applications. People Post Press, Beijing (1996)
6. Tang, H., Qin, X.: Practical Methods of Optimization. Dalian University of Technology Press, Dalian (2001)

Some Preservation Results of Location Independent Riskier Order

Bo-de Li and Jie Zhi

School of Information Engineering Lanzhou University of Finance and Economics
Lanzhou 730020 People's Republic of China

Abstract. The preservations of location independent riskier order under certain monotone transformations are investigated. It is also showed that location independent riskier order can preserved formation of order statistics.

Keywords: Location independent riskier order, Monotone transformation, Order statistics, Preservation, Random minima, Series systems Preservation.

1 Introduction

It is of interest in comparing variability of two random variables by means of stochastic orders. For this purpose, lots of stochastic orders have been introduced in literature. For example, disperse order, increasing convex order, increasing concave order, right spread order and total time on test transform order, etc. Readers can refer to Barlow and Proschan(1981), Shaked and Shanthikumar(1994), Shaked and Shanthikumar(1998) and Kochar, Li and Shaked(2002) for more details on these stochastic orders.

Besides all above stochastic order, basing upon the Rothschild and Stiglitz's comparative definition of risk (see Scarsini, 1994),Jewitt(1989) defined another criterion to compare different random assets in risk analysis. With respect to the standard way to compare risky prospects, the criterion defined in Jewitt(1989) dispenses with the constraint that the variables under comparison must have the same expectation. In particular, this stochastic order is location-free, and for this reason it is called the Jewitt's location independent riskier order.

Formally, given two random variables X and Y, we say that X is less location independent riskier than Y (denoted by $X \leq_{lir} Y$) if for all positive number c, $E[v(Y)] = E[v(X-c)]$) implies $E[u(Y)] \leq E[u(X-c)]$) whenever u and v are increasing concave functions with u being more risk averse than v,that is $u = h(v)$ with h increasing and concave.

Jewitt(1989) proved that if X and Y have distribution functions F and G respectively, then $X \leq_{lir} Y$ if and only if,

$$\int_{-\infty}^{F^{-1}(p)} F(x)dx \leq \int_{-\infty}^{G^{-1}(p)} G(x)dx \quad for\ all\ p \in (0,1) \tag{1}$$

Y. Wang and T. Li (Eds.): Knowledge Engineering and Management, AISC 123, pp. 677–684.
springerlink.com

Another necessary and sufficient condition for $X \leq_{lir} Y$ is, for all $0 < p_1 \leq p_2 \leq 1$,

$$\frac{1}{p_1}\int_0^{p_1}[F^{-1}(u)-G^{-1}(u)]du \geq \frac{1}{p_2}\int_0^{p_2}[F^{-1}(u)-G^{-1}(u)]du \tag{2}$$

The location independent riskier order has good interpretation in auction theory. Let v be the utility function of a risk averse agent (that is, v is an increasing concave function), and let X and Y be two random asserts, c be the premium that the agent is willing to pay in order to replace the assert Y by the assert X . If anybody who is more risk averse than the agent is ready to pay more than c in order to replace Y by X , then $X \leq_{lir} Y$. So, the location independent riskier order has potential valuable applications in this area. Landsberger and Meilijson (1994) investigated some of the properties of the location independent riskier order. The application of location independent riskier order can be seen in Scarsini(1994). Afterwards, Fagiouli, Pellerey and Shaked(1999) built some relationships between location independent riskier order and other stochastic orders. The location independent risk order has been used to compare different random assets in risk analysis without the requirement of equal means(2006). Location-independent riskier order and its dual version, excess wealth order, compare random variables in terms of dispersion,derived the relationship of both orders to the usual stochastic order.and some new properties of these orders are obtained(2009). A class of location-independent variability orders, with applications has been studied(2010). Some new properties of this stochastic order were showed there as well.

In Section 2 of this paper, it is showed that location independent riskier order can be preserved under formation of certain monotone transformations. The preservations and reversed preservations of location independent riskier order under order statistics are proved in section 3.

2 Preservations under Monotone Transformations

In this section, we first exhibit a sufficient condition for location independent riskier order.

Theorem 1. If X and Y are non-negative absolutely continuous random variables and $G^{-1}F(x)-x$ is increasing, then $X \leq_{lir} Y$.

Proof. Denote $f(x)=[F^{-1}(x)-G^{-1}(x)]x-\int_0^x[F^{-1}(u)-G^{-1}(u)]du$, then we have

$$\begin{aligned} f'(x) &= [F^{-1}(x)-G^{-1}(x)]'x \\ &= \left[\frac{1}{f[F^{-1}(x)}-\frac{1}{g[G^{-1}(x)]}\right]x. \end{aligned}$$

Where f and g are respective density function of X and Y .

Note that $G^{-1}F(x)-x$ is increasing is equivalent to $f[F^{-1}(x)] \geq g[G^{-1}(x)]$ for all $0<x<1$, so $f'(x) \leq 0$, then $f(x)$ is decreasing, and hence $f(x) \leq f(0)=0$ for all $0<x<1$. Now for all $0<x<1$,

$$\left[\frac{\int_0^x [F^{-1}(u)-G^{-1}(u)]du}{x}\right]'$$

$$=\frac{[F^{-1}(x)-G^{-1}(x)]x-\int_0^x [F^{-1}(u)-G^{-1}(u)]du}{x^2}=\frac{f(x)}{x^2} \leq 0.$$

Thus, $\int_0^x [F^{-1}(u)-G^{-1}(u)]du/x$ is decreasing in $x \in (0,1)$, (2) holds for all $0<p_1 \leq p_2 \leq 1$. That is, $X \leq_{lir} Y$.

Remark. *In consideration that* X *is less than* Y *in dispersive order (denoted by* $X \leq_{disp} Y$ *) if* $G^{-1}F(x)-x$ *is increasing in* $x \geq 0$, *Theorem 1 shows that if* X *and* Y *are non-negative absolutely continuous random variables, then* $X \leq_{disp} Y$ *is the sufficient condition for* $X \leq_{lir} Y$.

Recall that X is less than Y in convex order (denoted by $X \leq_c Y$) if $G^{-1}F(x)$ is convex for all $x>0$; X is less than Y in star order (denoted by $X \leq_* Y$) if $G^{-1}F(x)/x$ is increasing for all $x>0$ and $X \leq_c Y$ implies $X \leq_* Y$. For more details on convex order and star order, readers can refer to Barlow and Proschan(1981) and Shaked and Shanthikumar(1994). Using Theorem 1, we obtain the following corollary directly.

Corollary. If X and Y are non-negative absolutely continuous random variables and $X \leq_c Y$ ($X \leq_* Y$) and $\lim_{x\to 0} G^{-1}F(x)/x \geq 1$, then $X \leq_{lir} Y$.

Proof. It suffices to prove that Corollary holds for the case of $X \leq_* Y$.

Note that $X \leq_* Y$ is equivalent to say $[G^{-1}F(x)/x]' \geq 0$, that is, for all $x>0$,

$$[G^{-1}F(x)]' \geq \frac{G^{-1}F(x)}{x},$$

Additionally, by assumption, we have

$$\frac{G^{-1}F(x)}{x} \geq \lim_{x\to 0}\frac{G^{-1}F(x)}{x} \geq 1.$$

Thus, $[G^{-1}F(x)]' \geq 1, G^{-1}F(x)-x$ is increasing in $x>0$, and hence $X \leq_{lir} Y$ by Theorem 1.

We now consider the stochastic comparison between baseline distribution and transformed distribution.

Let $\phi(x)$ be an increasing function, the reverse of ϕ is φ, and then the distribution of $\phi(X)$ is

$$H(x) = P(\phi(X) \leq x) = P(X \leq \varphi(x)) = F(\varphi(x))$$

and

$$H^{-1}F(x) = \phi(x).$$

So, Theorem 1 shows that the increasing property of $\phi(x)-x$ is the sufficient condition for $X \leq_{lir} \phi(X)$ if both X and $\phi(X)$ are non-negative absolutely continuous random variables. This theorem is useful in judging whether baseline distribution is less than transformed distribution in location independent riskier order. However, the non-negative absolutely continuous constraint in Theorem 1 is somewhat strong. One may wonder whether this constraint can be omitted. The following Theorem 2 gives an answer for this question.

Theorem 2. If $\phi(x)-x$ is an increasing concave function such that $\phi(0)=0$, then $X \leq_{lir} \phi(X)$.

Proof. Note that for all $x \geq 0$,

$$\int_{-\infty}^{x} F(u)d[H^{-1}F(u)-u] = \int_{-\infty}^{x} F(u)d[\phi(u)-u]$$

$$= \int_{-\infty}^{x} F(u)[\phi(u)-u]'du$$

By assumption, $[\phi(u)-u]'$ is non-negative decreasing. Thus, according to Lemma 7.1(b)(Barlow and Proschan1981, p.120), for all $x \geq 0$, we have

$$\int_{-\infty}^{x} F(u)d[H^{-1}F(u)-u] = \int_{-\infty}^{x} F(u)[\phi(u)-u]'du \geq 0$$

or equivalently, for all $p = F(x) \in (0,1)$,

$$\int_{-\infty}^{F^{-1}(p)} F(u)du \leq \int_{-\infty}^{H^{-1}(p)} H(u)du.$$

By (1) $X \leq_{lir} \phi(X)$.

3 Preservations under Order Statistics

Suppose that $X_1, X_2, \cdots, X_n$ is a sample with size n from parent distribution X, $X_{1:n} \leq X_{2:n} \leq \cdots \leq X_{n:n}$ is the order statistics according to this sample. Similarly, $Y_1, Y_2, \cdots, Y_n$ is another sample with size n from parent distribution Y,

$Y_{1:n} \leq Y_{2:n} \leq \cdots \leq Y_{n:n}$ is the order statistics according to $Y_1, Y_2, \cdots, Y_n$. If the distribution functions of $X_{i:n}$ and $Y_{i:n}$ are denoted by $F_{i:n}(x)$ and $G_{i:n}(x)$ respectively, then

$$F_{i:n}(x) = \frac{n!}{(i-1)!(n-i)!} \int_0^{F(x)} u^{i-1}(1-u)^{n-i} du = B_{i,n-i+1}(F(x)),$$

$$G_{i:n}(x) = \frac{n!}{(i-1)!(n-i)!} \int_0^{G(x)} u^{i-1}(1-u)^{n-i} du = B_{i,n-i+1}(G(x)),$$

where $B_{\alpha,\beta}(x)$ is the distribution function of a beta distribution with parameter $\alpha > 0$ and $\beta > 0$ with the density function given by

$$b_{\alpha,\beta}(x) = \frac{\Gamma(\alpha+\beta)}{\Gamma(\alpha)\Gamma(\beta)} u^{\alpha-1}(1-u)^{\beta-1}, \quad 0 < u < 1.$$

So it is obvious that, for all $1 \leq i \leq n$,

$$G_{i:n}^{-1}(x) F_{i:n}(x) = G^{-1} B_{i,n-i+1}^{-1} B_{i,n-i+1} F(x) = G^{-1} F(x) \tag{3}$$

Order statistics has good interpretation in reliability theory. For example, $i-th$ order statistics represents the life of a $n-i+1-out-of-n$ systems. In particular, $n-th$ order statistics and 1-th order statistics represent the life time of a series systems and a parallel systems respectively. Readers can refer to Balakrishnan and Rao (1989a,b), Kim and David (1990) for more details on order statistics.

As the first theorem of this section, the following Theorem 2.1 reveals that location independent riskier order can be preserved under formation of series systems.

Theorem 3. If $X \leq_{lir} Y$, then $X_{1:n} \leq_{lir} Y_{1:n}$.

Proof. By (1), for all $x \geq 0$, we have

$$\int_{-\infty}^{x} F(u) d[G^{-1}F(u) - u] \geq 0.$$

Since $\bar{F}^{n-1}(u) + \bar{F}^{n-2}(u) + \cdots + 1$ is decreasing in u, it follows from (3) and Lemma 7.1(b) in Barlow and Proschan (1981, p.120) that, for all $x \geq 0$,

$$\begin{aligned}
&\int_{-\infty}^{x} F_{1;n}(u) d[G_{1:n}^{-1} F_{1:n}(u) - u] \\
&= \int_{-\infty}^{x} F_{1;n}(u) d[G^{-1}F(u) - u] \\
&= \int_{-\infty}^{x} [1 - \bar{F}^{n}(u)] d[G^{-1}F(u) - u] \\
&= \int_{-\infty}^{x} F(x)[\bar{F}^{n-1}(u) + \bar{F}^{n-2}(u) + \cdots + 1] d[G^{-1}F(u) - u] \geq 0.
\end{aligned}$$

That is, $X_{1:n} \leq_{lir} Y_{1:n}$.

In order to show the reversed preservations of location independent riskier order, we need the following lemma.

Lemma. For all $1 \leq i < j \leq n$, $\dfrac{F_{j:n}(x)}{F_{i:n}(x)}$ is increasing in $x \geq 0$.

Proof. Note that $\dfrac{F_{j:n}(x)}{F_{i:n}(x)} = \dfrac{n-i}{i} h(F(x))$,

where $h(p) = \int_0^p u^i (1-u)^{n-i-1} du \Big/ \int_0^p u^{i-1}(1-u)^{n-i} du,\ p \in (0,1)$.

The numerator of $h'(p)$ is

$$p^{i-1}(1-p)^{n-i-1}[p\int_0^p u^{i-1}(1-u)^{n-i} du - (1-p)\int_0^p u^i(1-u)^{n-i-1} du$$
$$= p^{i-1}(1-p)^{n-i-1}k(p),$$

where $k(p) = p\int_0^p u^{i-1}(1-u)^{n-i} du - (1-p)\int_0^p u^i (1-u)^{n-i-1} du$.

$$k'(p) = \int_0^p u^{i-1}(1-u)^{n-i} du + \int_0^p u^i (1-u)^{n-i-1} du \geq 0.$$

Then, for all $p \in (0,1), k(p) \geq k(0) = 0$,and hence $h'(p) \geq 0, h(p)$ is increasing in p.

Now, $\dfrac{F_{(i+1):n}(x)}{F_{i:n}(x)}$ is increasing in x for all $i \geq 1$,and desired result follows from the fact

$$\frac{F_{j:n}(x)}{F_{i:n}(x)} = \frac{F_{j:n}(x)}{F_{(j-1):n}(x)} \cdot \frac{F_{(j-1):n}(x)}{F_{(j-2):n}(x)} \cdot \frac{F_{(j-2):n}(x)}{F_{(j-3):n}(x)} \cdots\cdots \frac{F_{(i+1):n}(x)}{F_{i:n}(x)}.$$

Theorem 4. If $X_{n:n} \leq_{lir} Y_{n:n}$, then $X_{j:n} \leq_{lir} Y_{j:n}$ for all $1 \leq j < n$.

Proof. By (1), for all $x \geq 0$, we have

$$\int_{-\infty}^x F_{n:n}(u) d[G_{n:n}^{-1} F_{n:n}(u) - u] \geq 0.$$

According to (3), Lemma 3.2 and Lemma 7.1 (b) in Barlow and Proschan (1981, p.120),for all $x \geq 0$, we further have

$$\int_{-\infty}^x F_{j:n}(u) d[G_{j:n}^{-1} F_{j:n}(u) - u] = \int_{-\infty}^x F_{j:n}(u) d[G_{j:n}^{-1} F_{j:n}(u) - u]$$
$$= \int_{-\infty}^x \frac{F_{j:n}(u)}{F_{n:n}(u)} F_{n:n}(u) d[G_{j:n}^{-1} F_{j:n}(u) - u] \geq 0 .$$

That is, $X_{j:n} \leq_{lir} Y_{j:n}$ for all $1 \leq j < n$.

Theorem 5. If $X_{n:n} \leq_{lir} Y_{n:n}$, then $X \leq_{lir} Y$.

Proof. By (1), for all $x \geq 0$, we have

$$\int_{-\infty}^{x} F_{n:n}(u)d[G_{n:n}^{-1}F_{n:n}(u)-u] \geq 0.$$

Since that $1/[F(x)]^{n-1}$ is decreasing in x, it follows from Lemma 7.1 (b) in Barlow and Proschan (1981, p.120) that, for all $x \geq 0$,

$$\int_{-\infty}^{x} F(x)d[G^{-1}F(x)-x] = \int_{-\infty}^{x} [F(x)]^{n} \frac{1}{[F(x)]^{n-1}} d[G^{-1}F(x)-x]$$

$$= \int_{-\infty}^{x} F_{n:n}(x) \frac{1}{[F(x)]^{n-1}} d[G_{n:n}^{-1}F_{n:n}(x)-x] \geq 0.$$

That is, $X \leq_{lir} Y$.

The following theorem generalize the Theorem 2.1 in the sense that fixed sample can be replaced by random sample.

Theorem 6. If $X \leq_{lir} Y$, N is a non-negative integer-valued random variable which is independent of X and Y, then $X_{1:N} \leq_{lir} Y_{1:N}$.

Proof. Denote the probability mass of N by $P_N(n) = P(N=n)$, then the distribution function of $X_{1:N}$ is $F_{1:N}(x) = 1 - \sum_{n=1}^{\infty} \overline{F}^{n}(x) P_N(n)$,

and by (8) in Bartoszewicz (2001), for all $1 \leq i \leq N$,

$$G_{i:N}^{-1}F_{i:N}(x) = G^{-1}F(x), \tag{4}$$

where $G_{i:N}(x)$ is the distribution function of $Y_{i:N}$.

Now, by (1), for all $x \geq 0$, we have

$$\int_{-\infty}^{x} F(u)d[G^{-1}F(u)-u] \geq 0.$$

Since $\sum_{n=1}^{\infty} [\overline{F}^{n-1}(u) + \overline{F}^{n-2}(u) + \cdots + 1] P_N(n)$ is decreasing in u, it follows from (4) and Lemma 7.1 (b) in Barlow and Proschan (1981, p.120) that, for all $x \geq 0$,

$$\int_{-\infty}^{x} F_{1:N}(u)d[G_{1:N}^{-1}F_{1:N}(u)-u]$$

$$= \int_{-\infty}^{x} F_{1:N}(u)d[G^{-1}F(u)-u]$$

$$= \int_{-\infty}^{x} [1 - \sum_{n=1}^{\infty} \bar{F}^{n}(u) P_N(n)] d[G^{-1}F(u) - u]$$

$$= \int_{-\infty}^{x} \sum_{n=1}^{\infty} [1 - \bar{F}^{n}(u)] P_N(n) d[G^{-1}F(u) - u]$$

$$= \int_{-\infty}^{x} \sum_{n=1}^{\infty} F(u)[\bar{F}^{n-1}(u) + \bar{F}^{n-2}(u) + \cdots + 1] P_N(n) d[G^{-1}F(u) - u]$$

$$= \int_{-\infty}^{x} F(u) \sum_{n=1}^{\infty} [\bar{F}^{n-1}(u) + \bar{F}^{n-2}(u) + \cdots + 1] P_N(n) d[G^{-1}F(u) - u] \geq 0.$$

That is, $X_{1:N} \leq_{lir} Y_{1:N}$.

References

1. Barlow, R., Proschan, F.: Statistical Theory of Reliability and Life Testing, Probability Model. Holt, Rinehart and Winnston, New York (1981)
2. Fagiouli, E., Pellerey, F., Shaked, M.: A characterization of dilation order and their applications. Statistical Papers 40, 393–406 (1999)
3. Jewitt, I.: Choosing between risky prospects: The characterization of comparative statics results, and location independent risk. Management Science 35, 60–70 (1989)
4. Li, K.S., Shaked, M.: The total time on test transform and the excess wealth stochastic orders of distribution. Advance in Probability 34, 826–845 (2002)
5. Landsberger, M., Meilijson, I.: The generating process and an extension of Jewitt's location independent risk concept. Management Science 40, 662–669 (1994)
6. Shaked, M., Shanthikumar, J.: Stochastic Orders and their Applications. Academic Press, Boston (1994)
7. Shaked, M., Shanthikumar, J.: Two variable orders. Probability in Engineering and Informational Science 12, 1–23 (1998)
8. Scarini, M.: Comparing risk and risk aversion. In: Shaked, M., Shanthikumar, J. (eds.) Stochastic Orders and their Applications, pp. 351–378. Academic Press, New York (1994)
9. Hu, T., Chen, J., Yao, J.: Preservation of the location independent risk order under convolution. Insurance: Mathematics and Economics 38, 406–412 (2006)
10. Sordo, M.A., Ramos, H.M.: Characterizations of stochastic orders by L-functionals. Statist. Papers 48, 249–263 (2007)
11. Sordo, M.A., Ramos, H.M., Ramos, C.D.: Poverty measures and poverty orderings. SORT 31, 169–180 (2007)
12. Sordo, M.A.: On the relationship of location-independent riskier order to the usual stochastic order. Statistics & Probability Letters 79, 155–157 (2009)
13. Sordo, M.A., Alfonso, S.-L.: A class of location-independent variability orders, with applications. Journal of Applied Probability 47, 407–425 (2010)
14. Shaked, M., Shanthikumar, J.G.: Stochastic Orders. Mathematical Reviews (MathSciNet): MR2265633. Springer, New York (2007)
15. Sordo, M.A., Ramos, H.M., Ramos, C.D.: Poverty measures and poverty orderings. SORT 31, 169–180 (2007)
16. Wang, F.: Some Results on the Generalized TTT Transform Order for Order Statistics. Journal of Applied Probability and Statistics 25(1), 27–37 (2009)

A Method Based on AHP to Define the Quality Model of QuEF

M. Espinilla[1], F.J. Domínguez-Mayo[2], M.J. Escalona[2], M. Mejías[2], M. Ross[3,4], and G. Staples[4]

[1] University of Jaén,
Campus Las Lagunillas s/n 23071, Jaén, Spain
mestevez@ujaen.es
[2] University of Seville,
Avda. Reina Mercedes s/n. 41012 Seville
fjdominguez@us.es, mjescalona@us.es, risoto@us.es
[3] Southampton Solent University,
East Park Terrace, Southampton, Hampshire, UK
margaret.ross@solent.ac.uk
[4] British Computer Society (BCS)

Abstract. QuEF is a framework to analyze and evaluate the quality of approaches based on Model-Driven Web Engineering (MDWE). In this framework, the evaluation of an approach is calculated in terms of a set of information needs and a set of quality characteristics. The information needs are requirements demanded by users of approaches. On the other hand, the quality characteristics are specific aspects that the approaches provide to their users. In these lines, there is a gap in the importance of each quality characteristic in the QuEF and the degree of coverage of each information need regarding the quality characteristics. In this contribution, we propose a method to define the Quality Model within QuEF. This method is based on the Analytic Hierarchy Process in order to establish the importance of the quality characteristics and the degree of coverage of each requirement of the information needs regarding the set of quality characteristics. Furthermore, a software application that develops the proposed method is presented.

Keywords: Model-Driven Web Engineering, Quality Assurance, Analytic Hierarchy Process, Information Needs, Quality Characteristics.

1 Introduction

A good strategy to manage quality is essential to obtain good results on the improvement of the quality. In these lines, it is important to define what are the goals and the set of steps to achieve the goals. The web development is currently being an important task to take into account in the sense that more and more web applications are developed every day. So, it is important to define the right steps to manage the development of this kind of products. In this context, Model-Driven Engineering (MDE) [1] paradigm plays a key role because it pertains to software development; MDE refers to

Y. Wang and T. Li (Eds.): Knowledge Engineering and Management, AISC 123, pp. 685–694.
springerlink.com

a range of development approaches that are based on the use of software modeling as a primary form of expression. Model-Driven Web Engineering (MDWE) is a specific domain of the Model-Driven Engineering (MDE) paradigm [1] which focuses on Web environment.

The growing interest on the Internet has led to generate several MDWE approaches, which offer a frame of reference for the Web environment. Therefore, it is necessary to analyze and evaluate the quality of different approaches in order to choose the most appropriate taking into account the user requirements, i.e., users' information need. However, the process of quality evaluation of an approach is a complex task because there are lot of MDWE approaches without standard consensus [2][3][4] and an important gaps between the users' information needs and quality characteristics [5][6].

Recently, in our previous research, we have defined QuEF (Quality Evaluation Framework) [7][8][9], an approach to establish an environment in order to analyze and evaluate MDWE (Model-Driven Web Engineering) approaches under quality characteristics. The framework has been structured and organized with four different components: *Thesaurus & Glossary component*, *Quality Model component*, *Approach Features Template component* and *Quality Evaluation Process component.*

The most important component is the *Quality Model,* which defines and describes the set of quality criteria, its weight and its relations with the users' information needs. Each approach will be assessed in this evaluation framework. Therefore, a process of decision among users of approaches, decision-makers, on the definition of this evaluation framework is crucial in the success of QuEF.

A common approach in decision making is the *Analytic Hierarchy Process* (AHP) introduced by Saaty [10][11], which is a widely accepted as a multi-criteria decision-making methodology. In this process, the criteria are structured in several levels and then, different decision alternatives are evaluated and prioritized, taking into account the preferences of a set of decision makers. In this contribution, the essence of the AHP is used to propose a method to define the *Quality Model* in QuEF in order to establish the importance of the set of quality characteristics and the degree of coverage of each information need regarding the set of quality characteristics. Furthermore, the proposed method is developed in a software application in order to support the process of definition of the Quality Model. In this way, this definition process can lay the bases of a framework for quality assessment of approaches on Model-Driven Web Engineering (MDWE) approaches.

The outline of this contribution is as follows. Section 2 describes Quality Evaluation Framework. Section 3 reviews the Analytic Hierarchy Process. Then, Section 4 proposes an AHP-based method to build a quality model. Section 5 presents a software application that develops the proposed method. Finally, the conclusion and future works are described in the Section 6.

2 QuEF and the Quality Management

QuEF is a framework to analyze and evaluate the quality of approaches based on Model-Driven Web Engineering (MDWE), although it could be extended to other areas or domains. The main objective of QuEF is to provide a set of guidelines, techniques, processes and tools for the structuring of specifications, which are expressed as models.

The framework is a basic conceptual structure composed of a set of four elements used to evaluate, in this case, MDWE approaches (see figure 1). These elements are based on existing literature for the evaluation of the quality of MDWE methodologies and they are described as follow:

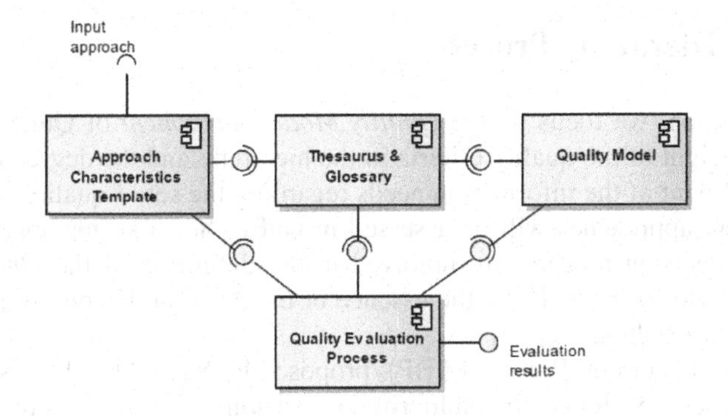

Fig. 1. Basic Conceptual Structure of QuEF

1. *Quality Model Component.* This component includes a set of information needs and a hierarchy of quality criteria composed by quality subcharacteristics and characteristics which provide the basis for specifying quality requirements and evaluating quality in a specific domain (in this case, MDWE). Furthermore, the model contains association links between the information needs and the quality subcharacteristics. These links represented the degree of coverage of each requirement of the information needs regarding the set of quality subcharacteristics. In this contribution, we focus in this component in order to establish the weight of the quality criteria in the hierarchy and the degree of coverage of each requirement of the information needs regarding the set of quality subcharacteristics.
2. *Thesaurus & Glossary component.* An important element for QuEF is the thesaurus component. A thesaurus is a list containing the "terms" used to represent concepts, themes or contents of documents in order to standardize the terminology which improves the access channel and communication between users of different MDWE methodologies. This component is necessary to carry out a standardization of terminology to improve the access channel for communication on MDWE. A set of concepts for MDWE methodologies is currently being described and related.
3. *The Approach Characteristic Template Component.* Templates with users' information needs based on the Quality Model are used to describe an input methodology. These templates are used as input to QuEF. They are analyzed in the evaluation process and compared with the model quality of the Quality Model component. Templates for MDE, Web Modelling, Tool Support and Maturity have already been developed.

4. *The Quality Evaluation Process Component.* The Quality Evaluation Process component contrasts the information from each input approach template with information from the Quality Model. The main purpose of this evaluation is to identify tradeoffs and sensitivity points of the methodology under study. The idea is to determine which aspect needs to be improved on MDWE methodology.

3 Analytic Hierarchy Process

In this contribution, we focus on the *Quality Model component* of QuEF in order to establish the weight of the quality criteria in the hierarchy and the degree of coverage of each requirement of the information needs regarding the set of quality subcharacteristics. Different approaches will be assessed in QuEF. So, it seems appropriate that several users, decision makers, are involved in the definition of the *Quality Model component.* To do so, we will use the essence of the Analytic Hierarchy Process that is reviewed in this section.

The Analytic Hierarchy Process (AHP), proposed by Saaty [10][11], is systematic analysis technique developed for multicriteria decision by means of creating a ratio scale corresponding to the main alternatives. The output of AHP is a ranking indicating the overall preference for alternative.

This process is based on three axioms: (1) breaking down the problem; (2) pairwise comparison of the various alternatives and (3) synthesis of the preferences. Conventional AHP includes four steps: modelling, valuation, priorization and synthesis, which are detailed below.

1. *Modelling*: The first step builds a hierarchy where the goal is at the top, criteria and sub-criteria are respectively placed at levels and sub-levels, and decision alternatives appear at the bottom of the hierarchy.
2. *Valuation*: The second step analyzes the elements of the problem by means of reciprocal pairwise comparison matrices, which are provided for each decision maker. This step involves two states:
   a) *Weight of the criteria*: group pairwise comparisons are performed to determine the relative scores for the weights of the criteria and sub-criteria in the hierarchy.
   b) *Judgments.* The assessment is conducted by means of reciprocal pairwise comparison matrices against a third element. In this way, the process obtains the preferences of the individuals regarding the different components of the model (criteria, sub-criteria, alternatives).

   Each individual provides a preference in terms of importance, preference or probability, assigning a numerical value, which measures the intensity of their preference. So, Saaty suggested a nominal scale with 9 points, the so-called "Saaty's Fundamental Scale", shown in Table 1, in order to provide judgments of each individual.

Table 1. The rate of importance of criterion Y over X

Numerical rating	Linguistic judgments
1	X is equally preferred to Y
2	X is equally to moderately preferred over Y
3	X is moderately preferred over Y
4	X is moderately to strongly preferred over Y
5	X is strongly preferred over Y
6	X is strongly to very strongly preferred over Y
7	X is very strongly preferred over Y
8	X is very strongly to extremely preferred over Y
9	X is extremely preferred over Y

3. *Prioritation:* In this step, the local and global priorities are obtained by using, respectively, any of the existing prioritation procedures. The eigenvector method and the row geometric mean method are the most widely used.
4. *Synthesis.* In this step, the total priorities are derived by applying any aggregation procedure. It can be additive or multiplicative.

One of the main characteristics of AHP is the ability to assess the degree of inconsistency present in the judgments expressed by the decision makers in the reciprocal pairwise comparison. The consistency ratio is obtained by comparing the consistency index with the random index [9][10] which is an average random consistency index derived from a sample of randomly generated reciprocal matrices using the scale in Table 1. Saaty defined the consistency ratio (CR) as:

Consistency ratio= Consistence index / Random index (1)

Where the consistency index (CI) is in the form:

$$CI= (\lambda_{max} - n) / (n-1) \quad (2)$$

λ_{max} is a principal eigenvalue of a pairwise matrix such that $\lambda_{max} \geq n$. This method is good for measuring consistency by using the eigen system method. A value of the consistency ratio $CR \leq 0.1$ is considered acceptable. Larger values of CR require that the decision-maker revise his judgments.

Table 2. Random consistency index (RI)

N	1	2	3	4	5	6	7	8	9	10
RI	0	0	.52	.89	1.11	1.25	1.35	1.40	1.45	1.49

4 AHP-Based Method to Build a Quality Model

QuEF represents a valid framework for the quality management of MDWE approaches. In this framework is necessary to fix in the Quality Model the importance of the criteria in the hierarchy and the degree of coverage of each requirement of the information need regarding the set of quality criteria. Furthermore, in QuEF is important that the definition of the Quality Model is performed by a group of users, decision makers, in order to fix a consensus quality model.

In this section, we present a method to define the Quality Model of QuEF based on the essence of the AHP, which takes into account the opinions of different decision makers. Our method contains the following steps that are described in detail in the following sections.

4.1 Hierarchy Design

In this step, all the elements in the Quality Model are identified: 1) the objetive, 2) criteria and sub-criteria and 3) alternatives.

In our case, the aim is the establishment of the degree of coverage of each information need in a quality framework.

The criteria and sub-criteria composed the quality hierarchy; they are the factors that affect to the objective. In the Quality Model, they are the set of quality characteristics and quality sub-characteristics that are shown in Table 3.

Table 3. Hierarchy of quality criteria

Quality Characteristics	Quality Sub-Characteristics
Q_1=Usability	
	q_{11} = Learnability
	q_{12} = Understandability
	q_{13} = Simplicity
	q_{14} = Interpretability
	q_{15} = Operability
	q_{16} = Attractiveness
Q_2=Functionality	
	q_{21} = Suitability
	q_{22} = Accuracy
	q_{23} = Interoperability
	q_{24} = Compliance
	q_{25} = Interactivity
	q_{26} = Applicability
	q_{27} = Accessibility
	q_{28} = Flexibility
Q_3=Maintainability	
	q_{31} = Stability
	q_{32} = Analyzability
	q_{33} = Changeability
Q_4= Reliability	
	q_{41} = Maturity
	q_{42} = Recoverability case of failure
	q_{43} = Fault Tolerance
	q_{44} = Availability
	q_{45} = Currently
	q_{46} = Compactness
	q_{47} = Relevancy

Finally, the alternatives are proposes that can achieve the goal. In our case, they are the set of information needs that are represented as features or requirements. These information needs can group as shown in Table 4.

Table 4. Alternatives represent as information needs

	Information needs
F_1 = MDE	
	f_{11} = Standard Definition
	f_{12} = Model–Based Testing
	f_{13} = Traces
	f_{14} = Level of Abstraction
	f_{15} = Transformations
F_2=Web Modelling	f_{21} = Web Conceptual Levels
	f_{22} = Interfaces
	f_{23} = Content Modelling
	f_{24} = Presentation Modelling
	f_{25} = Navigation Modelling
	f_{26} = Business Modelling
	f_{27} = Development Process
F_3=Tool Support	f_{31} = Analysis Tool Support
	f_{32} = Code Generation and Specific Tool Support
	f_{33} = Team Work Tool Support
	f_{34} = Creation, Edition and Composition Tool Support
	f_{35} = Transformation Tool Support
	f_{36} = Trace Tool Support
F4= Maturity	
	f_{41} = Modelling Examples
	f_{42} = Publications
	f_{43} = Topicality
	f_{44} = Application in Real-World Projects
	f_{45} = External Web References
	f_{11} = Standard Definition

4.2 Development of Judgment Matrices

In this step, the information is obtained from decision makers in order to establish the Quality Model.

At the beginning, each individual provides his preferences about the weight of the set of criteria in the quality hierarchy, i.e., the importance that each quality characteristics and quality sub-characteristics has in the Quality Model. These weights are provided by means of reciprocal pairwise comparison matrices, using the scale shown in Table 1.

When the priority of the quality hierarchy has been provided, the evaluation of information need takes place. Each decision maker provides his preferences about the information needs, i.e, the degree of coverage of each information need regarding the quality sub-characteristics. This involves yet another set of pairwise comparisons, this time between each alternative against each quality sub-characteristics.

Finally, the consistency ratio for each judgment matrix is checked. If CR ≥ 0.1, the pairwise matrix is not consistent, then the comparisons should be revised by the decision maker.

4.3 Compute the Weights of the Quality Criteria

Once judgment matrices have been obtained by decision-makers, the proposed method computes a collective matrix for the weights of the set of quality characteristic and quality sub-characteristic which summarizes preferences of the group decision. This aggregation can be carried out applying some types of OWA operator [13]. To use an OWA operator is recommended because it reorders arguments according to the magnitude of their respective values. Furthermore, they satisfy some interesting properties such as compensativeness, idempotency, symmetry and monotonicity.

4.4 Compute the Degree of Coverage of Each Information Need

In this step, the proposed method computes the collective preferences to obtain a final vector of degree of coverage for each information need. This is done by following a path from the top of the hierarchy down to each alternative at the lowest. The outcome of this aggregation is a normalized eigenvector of the overall weights of the options [10][11] level, and multiplying the weights along each segment of the path.

5 A Software Application of the AHP-Based Method

In this section, we present an application that develops the proposed method. The objectives of the application are two. The first objective is to obtain the preferences of decision makers in a quick and simple way. The second objective is to automate the computations to obtain the importance of the criteria in the quality hierarchy and the degree of coverage of each information need.

This application generates a Web application with the set of surveys that consider all the elements in the Quality Model in order to carry out the proposed method. The Quality Model has been defined using the Enterprise Architect tool support as is shown in Fig. 2. This tool support can generate an XML file of the model defined in the tool.

On the other hand a Windows form application has been implemented using the Visual Studio .NET environment to generate automatically all the code of a Web Application that include all the Surveys for the elements which have already defined in the Quality Model. If the Quality Model or the set of questions to carry out the AHP method is changed the Web application can be generated automatically. This program can be used to carry out the AHP method for any other domain.

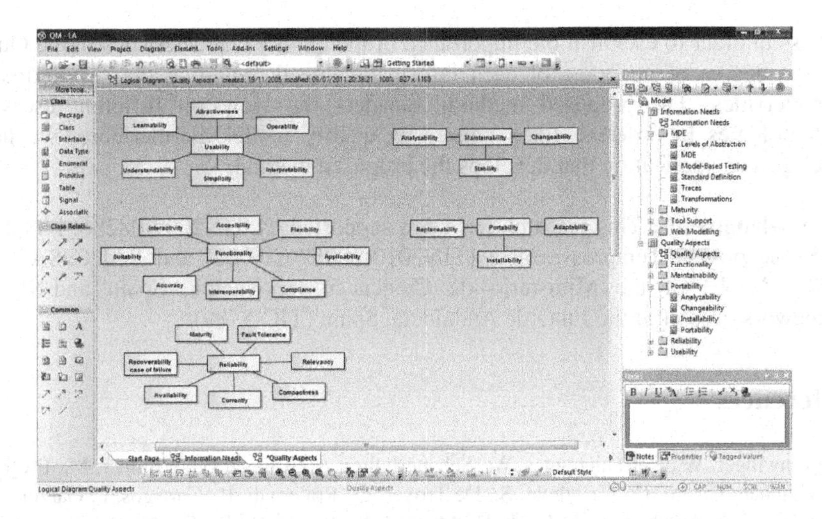

Fig. 2. Application of the method

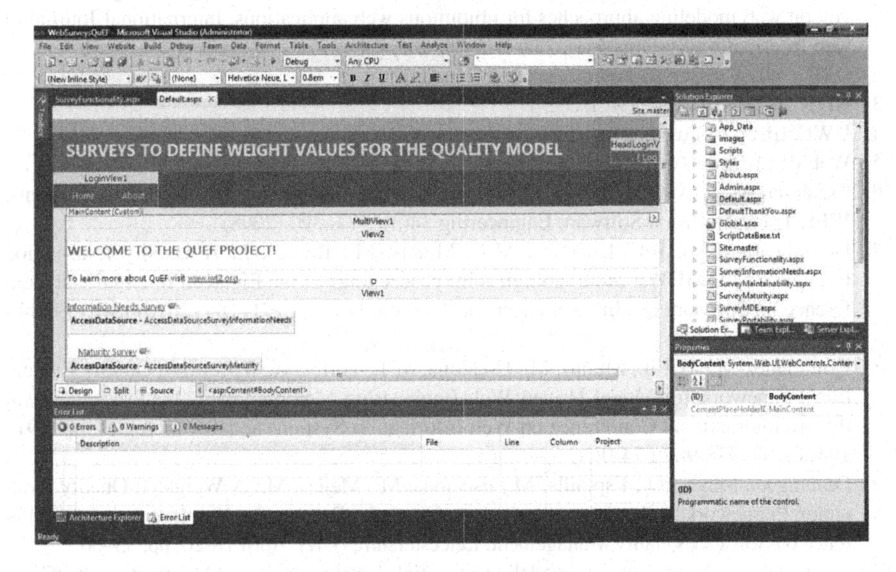

Fig. 3. Web application

Finally, the Web Application is generated. The Visual Studio .NET has been used to implement this environment as is shown in Fig. 3.

6 Conclusions

The *Quality Model* of QuEF defines and describes an hierarchy of quality aspects and the relationships between them and the information needs. In this contribution, we have presented a method to define the Quality Model based on the Analytic Hierarchy

Process in order to establish the importance of the quality characteristics in the QuEF and the degree of coverage of each information need regarding the set of quality characteristics. The proposed method considers the views of different users of methodologies in order to fix a consensus quality model. Furthermore, we have developed an application that develops the proposed method.

Acknowledgments. This research has been supported by TIN2009-08286, P08-TIC-3548, the project Tempros project (TIN2010-20057-C03-02) and Red CaSA (TIN 2010-12312-E) of the Ministerio de Ciencia e Innovación, Spain and NDTQ-Framework project of the Junta de Andalucia, Spain (TIC-5789).

References

1. Schwinger, W., Retschitzegger, W., Schauerhuber, A., Kappel, G., Wimmer, M., Pröll, B., Ca-chero Castro, C., Casteleyn, S., De Troyer, O., Fraternali, P., Garrigos, I., Garzotto, F., Gi-nige, A., Houben, G.-J., Koch, N., Moreno, N., Pastor, O., Paolini, P., Pelechano Ferra-gud, V., Rossi, G., Schwabe, D., Tisi, M., Vallecillo, A., van der Sluijs, Zhang, G.: A survey on web modeling approaches for ubiquitous web applications. International Journal of web Information Systems 4(3), 234–305 (2008)
2. OMG: MDA, `http://www.omg.org/mda/`
3. NDT, `http://www.iwt2.org`
4. UWE, `http://uwe.pst.ifi.lmu.de`
5. WebML, `http://www.webml.org`
6. Escalona, M.J., Aragón, G.: NDT. A Model-Driven Approach for Web Requirements. IEEE Transactions on Software Engineering 34(3), 377–390 (2008)
7. Domínguez-Mayo, F.J., Escalona, M.J., Mejías, M., Ramos, I.: A Quality Eval-uation Framework for MDWE Methodologies. In: Proceedings of the Eighteen International Conference on Software Quality Management, London, UK, pp. 171–184 (2010) ISBN: 978-0-9557300
8. Domínguez-Mayo, F.J., Mejías, M., Escalona, M.J., Torres, A.: Towards a Quality Evaluation Framework for Model-Driven Web Engineering Methodologies. In: Proceedings of the 6th International Conference on Web Information Systems and Technologies, pp. 191–194, ISBN: 978-989-674-0
9. Domínguez-Mayo, F.J., Espinilla, M., Escalona, M., Mejías, M.: A Weighted Quality Evaluation Framework by Applying the Analytic Hierarchy Process. In: 19th International Conference on Software Quality Management, Leicestershire (UK), April 18-20, pp. 55–66
10. Saaty, T.L.: Introduction to a modeling of social decision process. Mathematics and Computers in Simulation 25, 105–107 (1983)
11. Saaty, T.L.: The analytic hierarchy process. MacGraw-Hill, New York (1980)
12. Yaguer, R.R.: On Ordered Weighted Averaging Operators in Multicriteria Decision Makin. IEEE Transactions on Systems, Man, and Cybernetics 18, 183–190 (1988)

Patterns for Activities on Formalization Based Requirements Reuse

Zheying Zhang[1], Jyrki Nummenmaa[1], Jianmei Guo[2], Jing Ma[1], and Yinglin Wang[2]

[1] School of Information Sciences, University of Tampere, Tampere, Finland
[2] Department of Computer Science & Engineering, Shanghai Jiao Tong Univeristy, Shanghai 200240, China
{zheying.zhang,jyrki.nummenmaa,jing.ma}@uta.fi,
{guojianmei,ylwang}@sjtu.edu.cn

Abstract. This paper aims at specifying procedures and patterns for developing high-quality reusable requirements and for engineering new requirements documents with reuse. Formalization of the requirements is a promising approach to validating requirements in the reuse process. It improves the consistency, correctness, and completeness of reusable requirements. This paper reviews the current research on reuse and requirements formalization. Based on a strategy that introduces formalization into systematic requirements reuse, we present patterns for activities on formalization-based requirements reuse.

Keywords: requirements reuse, formalization, process, patterns.

1 Introduction

When variants of software applications are developed, the variants share a part of their functionalities, and, therefore, a part of the requirements. Different variants' requirements documents hence should have many similarities. Consequently, requirements can be reused when developing a new variant of an application. Studies have shown that similar applications can reuse up to 80% of the requirements [1]. Instead of redefining the same information over and over again among a variant of similar applications, it is important to alleviate redundant requirements documentation, and find ways of reusing existing requirements.

Requirements reuse reduces the development cost by specifying new applications from predefined reusable requirements rather than from scratch. Since the reusable requirements have been validated and verified in the previous software applications and are less prone to error, they have a better chance of being understood by all stakeholders, and can further lead to reuse in upcoming engineering activities [1, 2]. The reuse process, however, neither occurs by accident, nor comes as a goal. It is a consequence of good design. Because requirements documents facilitate communication among stakeholders, they are commonly documented in natural language. The individual statements, however, often inherit ambiguity from natural language (NL). When reusing them, the vagueness may introduce potential risks of misunderstanding, incompleteness, inconsistency, conflicts, etc. Studies have addressed that ad-hoc recycling between requirements documents of product variants is a major source of

Y. Wang and T. Li (Eds.): Knowledge Engineering and Management, AISC 123, pp. 695–707.
springerlink.com

requirements defects [3]. In order to benefit from the reuse activities, we need a set of high quality reusable requirements, and a well-defined process to be integrated into the requirements engineering process.

Formalization of the requirements means representing requirements in notations with a set of rules which define their syntax and semantics. Because of a higher degree of precision in the formulation of statements, precise rules for their interpretation and more sophisticated forms of validation and verification that can be automated by software tool [4, 5, 6, 7], consistency, correctness, completeness, etc. can be analyzed and inspected in a formal specification. Formalization is a practical strategy to ensure requirements quality. As formal or semi-formal specifications often refer to state variables designating objects, the temporal sequences of object's states, the transitions, etc. [7], functional requirements are often addressed in requirements formalism. Therefore, our discussion in this paper mainly refers to functional requirements. We study how formalism can be integrated in developing requirements for and with reuse.

2 Requirements Reuse

Adopting a well-defined repeatable reuse process is an effective way for organizations to obtain benefits from reuse. The adaptation process, however, is not a trivial task. A lot of research has been endeavored on the framework of systematic reuse, such as Draco [10], RSEB [8], FORM [11], and frameworks for product line engineering [12, 13]. These frameworks show that a typical systematic reuse process is divided into two essential phases, i.e. the development for reuse and the development with reuse. During these two phases, the following activities are commonly performed [9, 14]:

- Abstraction is the process of specifying reusable software components. They are classified and stored in the library.
- Selection is the process of retrieving reusable components and choosing appropriate ones for reuse.
- Adaptation is the process of changing the selected component for use in a new application.
- Integration is the process of constructing applications by putting together the reusable component and adding new application specific artifacts.
- Maintenance is the process comprising a set of sub-activities for managing reusable component storage, retrieval, version control, and change control when new components are added into the library and old ones.

Despite the reuse processes, most requirements reuse practice occurs in an ad hoc way by engineers who can develop new requirements documents quicker because they have developed similar products before[15]. However, ad hoc reuse of requirements between application variants form a major source of requirements defects [3]. Besides, systematic requirements reuse bas been studied from different perspectives. Maiden and Sutcliffe [16] applied techniques based on artificial intelligence to support the structural and semantic matching when retrieving requirements, and presented a paradigm for reuse of analogous specifications during requirements analysis. This paradigm focuses on requirements at a high abstraction level, which represent common knowledge among different domains. Similar research has been pursued by

Massonet and Lamsweerde [7] on reuse of requirements framework. Meanwhile, case-based reasoning [17], metamodels [18] have been applied as higher abstraction structures for requirements reuse. A variety of approaches [11, 13, 19, 20, 21] have been explored to identify, analyze and model variable requirements on different abstraction levels.

Although different paradigms and approaches have been explored for requirements reuse, researchers argue that the diversity of representations and the existence of different levels of requirements description make reuse difficult [4, 22]. Hence it is important to ensure requirements reusability before storing them for a systematic way of reuse.

3 Reusable Requirements

Requirements are often reused in groups rather than as a monolithic document or an individual statement. A reusable component comprises a set of software requirements which are at the same abstraction level, and together representing a specific functionality [15]. Requirements reusability is an attribute that refers to the expected reuse potential of a requirements component. Requirements are commonly documented in NL. An ideal requirements document shall be correct, comprehensible, complete, consistent, precise, modifiable, verifiable, and traceable [23]. In practice, however, it is hard to create and maintain a complete and consistent requirements document throughout the project in an ever-changing environment. Existing requirements form a key enabler of the reuse process. Poor quality requirements not only decrease their reusability, but also bring potential risk in software development. Taking into account the difficulties in developing a perfect requirements document, we shall develop a meaningful way of supporting existing requirements in a refineable and reusable form. Taking into account the reuse process, the following characteristics are particularly expected in the context of requirements reuse.

Unambiguity and Comprehensibility. A reusable requirements component shall be specified in a way that is understood by all stakeholders with only one possible interpretation [23]. Requirements documentation allows different kinds of representations, being more or less formal [26]. Textual requirements have the most expressive power and form the most common way of requirements documentation. It, however, often inherits ambiguity from NL [25, 26]. The vagueness may lead to misunderstanding and affect the follow-up design and implementation [24], as well as later requirements reuse. On the contrary, formal representation provides precise syntax and rich semantics and thus, enables reasoning and validation. Even though the formal representations improve the clarity of requirements documentation, they are developer-oriented techniques. Without additional information, it is hard to understand and reuse them.

Meanwhile, high-level requirements, e.g. user requirements, hardly bring straightforward vision about the practical operation of the application. Often a lack of well-documented exemplar application in a domain can inhibit analysis at a sufficient abstraction level. Low-level requirements, e.g. software requirements, are often tied in with design, more detailed in nature, and difficult to reuse [24]. Requirements are reusable when they specify what an application must do without placing any constraints on how that behavior is achieved. A requirement that enforces a technique puts unnecessary limitations, and is hard to understand.

Completeness. A set of reusable requirements shall be self-contained with clearly defined boundaries. Requirements are related to each other. Without enough knowledge about requirements and their interdependency relationships, it is difficult to ensure that all requirements related to the reused ones are included [3, 27]. At the same time, it is a problem that too many requirements are included in the new requirements document [25]. Furthermore, it is also hard to analyze the interdependencies arising between reused requirements and the new ones [27].

Modifiability and validity. 'As-is' reuse is extremely rare. In general, reusable requirements need to be changed in some way to match the new product. Thus a reusable requirement shall be modifiable, i.e. the structure and the statement allow the adaptation of changes in an easy, consistent and complete way.

Taking into account formalized representation, precision, conciseness, consistency and completeness of a requirement can be improved [6, 7]. On the downside, formalized specifications have limited expressiveness. They are not easily accessible to practitioners, and thus less widely applicable [8]. In order to benefit from the formalism in a requirements reuse process, problems of integrating formalization into requirements reuse shall be addressed. This short paper reviews the set of problems abstractions on formalization based requirements reuse. It ends with the process model that exploits the problem abstractions in different requirements reuse activities.

4 Issues in Formalization of the Requirements

Formal specifications have a mathematical basis and employ a formal notation to model requirements [5, 7]. Unlike statements in NL, formal specifications are less inclined to some of the requirements defects discussed in the prior section. Using the formal notation, precision and conciseness of specifications can be achieved [5]. Meanwhile, the mathematical proof procedures can be used to validate the internal consistency and syntactic correctness of the requirements [28]. Furthermore, the completeness of the specification can be checked in the sense that all enumerated options and elements have been specified [29, 30]. Formal specifications are often grounded on some logic and refer to state variables designating objects involved in an application, the temporal sequences of states of objects, the event-driven transitions between objects at different states, etc. Therefore, functional requirements are often addressed in formal representation.

On the downside, formal specifications have limited expressiveness. They are not easily accessible to practitioners, and thus less widely applicable [7]. There have been studies on formal requirements representations and requirements formalization [7]. In order to benefit from the formalism in a requirements reuse process, approaches to bridging the gap between formal and informal requirements representations form a key driver.

4.1 Formalizing Requirements

In order to derive formalized requirements from the informal statement, analyzing NL statements to identify basic constructs and their semantic relationships forms an indispensable step in requirements formalization.

Natural language processing. A common way to deal with textual requirements specifications using restricted or unrestricted NL is linguistic analysis. Some researchers use linguistic analysis tools to automatically analyze NL requirements documents. The tools incorporate individual or combined natural language processing (NLP) techniques, originated in artificial intelligence [31]. As discussed in [33], there are different types of NLP techniques, including lexical analysis, syntactic analysis, semantic analysis, categorization, and pragmatic analysis.

In general, textual requirements have to be pre-processed using NLP techniques before being used as the input to generate formalized requirements specification. Mu et al. [34] applied Stanford Parser [35] to syntactically analyze textual functional requirements and then transformed them to a formal representation based on Fillmore's case grammar [36]. A NL parser is a program that works out the grammatical structure of sentences, for instance, which groups of words go together (as "phrases") and which words are the subject or object of a verb. Probabilistic parsers use knowledge of language gained from manually parsed sentences to try to produce the most likely analysis of new sentences. However, Mu's experiments [34] showed that pure syntactical analysis based on Stanford Parser did not fully extract various semantic cases defined in the case grammar because of the native gap between syntax and semantics.

Semantic Role Labeling (SRL) [37] could provide a promising solution for the above issue. Semantic parsing of sentences is believed to be an important task on the road to NL understanding, and has immediate applications in tasks such as information extraction and question answering. SRL is a shallow semantic parsing task, in which for each predicate in a sentence, the goal is to identify all constituents that fill in a semantic role, and to determine their roles (e.g. Agent, Patient, Instrument, etc.) and their adjuncts (e.g Locative, Temporal, Manner etc.). However, the impact of SRL to functional requirements extraction and analysis is still an open issue.

In addition, grammatical framework (GF) [38] provides another approach to creating multilingual machine translation applications. It is a computation linguistic method that analyzes NL statements. This approach provides a syntactic analysis of texts. Its output forms the ground to generate grammatical structure with respect to a given formal grammar.

Using pre-defined requirements templates. Pre-processing the NL requirements to generate formalized specification is a challenge task. Besides, defining requirements models or templates to collect needed requirements constructs and to derive formalized specification form an alternative to NLP. Yue et al. [33] examined current research on transformation of textual requirements into analysis models using formal or semi-formal representations. They suggested that a better requirements source for deriving a formal representation is use case models with or without restricted NL and/or glossaries instead of pure textural requirements specifications. They also recommended the formal representation transformed from textual requirements should adopt the model techniques that are most often used in practice.

Many existing studies adopted use cases as input to generate various formal or semi-formal representations. Diaz et al. [39] presented an approach to transforming use case models into interaction models using role-driven patterns. They used roles to capture the linguistic abstractions on the use case text as well as to describe the corresponding interaction. Smialek et al. [40] proposed a metamodel and concrete syntax for three complementary representations of use case scenarios. Their approach essentially accomplished the transformation from ordinary use cases into extended sequence and extended activity diagrams, which facilitates better comprehension by different participants in a software project.

4.2 Visualizing Formalized Specification

Stakeholders find it very difficult to visualize how a written requirement statement will be translated into an executable application. When the statement is formalized, it is even hard to understand and imagine the behavior of an application fulfilling the requirements.

The primary purpose of formalism is to facilitate automated validation and verification rather than to communicate requirements among stakeholders. The higher degree of precision in the formulation of statements and precise rules enables the automated process, and can further improve the information provided to stakeholders for feedback through prototypes, simulation, animation, etc.

In most of the cases prototyping [25, 41] means building software to mimic the behavior and functionality of the application to be built. It provide stakeholders the most straightforward and actual context of the application to be built and demonstrates the realization of the requirements specification at the implementation level. Besides prototyping, simulation and animation are promising techniques to visualizing formalized specification and to checking whether the specifications meet stakeholder's real intents and expectations [4, 42]. Much work has been devoted in this area since it was originally proposed [43]. For example, Magee et al. [44] describe how behavior models specified by Labeled Transition Systems (LTS) can drive graphical animations. The semantic model of the approach is based on timed automata. Van et al. [42] presents a tool aiming at animating goal-oriented requirements models. The tool reduces the gap between user requirements and operational models by automatically generating parallel state machines from goal operationalizations. Furthermore, the work by Nummenmaa et al. [30] utilizes directly an action system [45] which simulates the dynamic behavior of an expected application. The action-based specification is derived from pre-processed functional requirements.

Research has been devoted to explicit event-based descriptions and state-machine models [42]. It is easy to see that behavior forms the primary focus of visualization of requirements specification. This provides a rich support of validating formalized functional requirements for reuse.

4.3 Managing Requirements Interdependencies

Requirements are interdependent, and affect each other in complex ways. Interdependency between requirements affects various decisions and activities in an application development process. Besides the reuse process [3, 27], it also supports the decision making in activities such as release planning, requirements management, software design, etc. In particular, requirements interdependencies can express the temporal transition relationships among functional requirements, which provide essential information in formalizing NL requirements to a formal representation. Despite this, little research has addressed the nature of requirements interdependency, and studies on requirements interdependency remain on the analysis from the theoretical point of view.

5 Integrating Requirements Formalization into Reuse

Considerable amount of research has been elaborated on formalization of the requirements in different perspectives. It forms the ground for the possibility to enhance validation and verification of reusable requirements. Integrating requirements formalization in the reuse process seems a promising strategy in requirements engineering. Consequently, requirements engineering, systematic reuse, and requirements formalization form our major concerns of improving systematic reuse at the requirements stage. As a reuse process occurs at the requirements stage, it shall be inextricably intertwined with the requirements engineering activities. Meanwhile, formalization of the requirements adds a complementary dimension to improve the validity and quality of requirements to be reused.

In an attempt to clarify our reuse strategy, Fig. 1 illustrates the process of integrating requirements formalization activities into requirements reuse. The basic idea is that reusable requirements are indispensable enabler of a systematic reuse process. They should be of high quality, which can be ensured through formalization of the requirements. Circles represent processes, and round-edged rectangles represent files or databases which provide or store data in the reuse process. As our purpose is to illustrate the strategy of formalizing requirements for reuse, we emphasize the stores or externals which are closely related to the process and ignore the others. For example, requirements templates or models are also an input in the validation process, as changes or modification in requirements shall follow the pre-defined template. However, we did not present the flow between requirements templates and the adaptation process, as they are not directly related to requirements formalization.

Processes in this diagram are presented in two rows. The lower part represents the basic reuse activities in a systematic reuse process, i.e. “abstract”, “select”, “adapt”, “integrate”, and “maintain”. The upper row represents the formalization and validation of reusable and reused requirements, i.e. “unify requirement statements”, “formalize requirements”, and “validate requirements”. They interact with requirements reuse at three stages in the process.

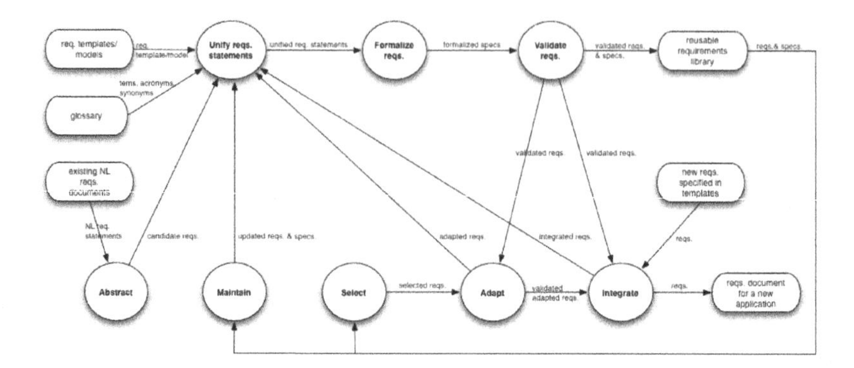

Fig. 1. A data flow diagram of formalization-based requirements reuse

- When a set of textual requirements are selected to be further specified as a reusable component at the abstraction stage, the characteristics of reusable requirements shall be taken into account. That is to say, requirements' completeness, unambiguity, and consistency shall be analyzed through a set of activities, including unification of requirements representation, formalization of the requirements, and validation of the specifications.
- When any changes have been done in reusable requirements or its specification at the stages of adaptation or maintenance, they shall be unified, formalized, and validated. The interdependencies and impacts of changes inside the component shall be inspected and analyzed.
- When reusable components are integrated with other requirements at the integration stage, the dependency relationship between reusable components and other requirements, the completeness, and the consistency shall be further analyzed.

Formalizing requirements is not the only solution to requirements validation. Requirements review meeting [25] is a common way of validating requirements. The approach, however, is participants oriented, and hardly provides an objective evaluation of requirements. Therefore, formalization forms a systematic way of validating the completeness, correctness and consistency of requirements. With the help of automated validation and verification tool to simulate the formalized specification, the comprehensiveness of requirements is greatly improved.

6 Patterns Supporting Systematic Reuse through Requirements Formalization

Based on the above discussion, we have identified some key activities in formalization-based requirements reuse. We describe these activities as patterns. This, in particular, means that we have identified the need for these activities, and the tension that poses the difficulty for these activities, where the tension, in turn, is a result of conflicting forces. The first pattern addresses the initial stage of documenting the requirements, where the tension is caused by the need to use NL, however some level of unification of expressions is needed for formalization.

Pattern 1: Use templates to document the requirements

Task: Unifying requirements statements
Quality goal: Unified systematic format of requirements
Forces, F→: Requirements are documented using NL, which is hard to formalize in the general form.
Forces, F←: Requirements need to be represented in a unified way for successful formalization.
Action: Create templates for systematic requirements documentation.

Templates help out in presenting the functional requirements using a unified structure. They can be used as forms to be filled in to describe the requirements. Thus templates can have separate fields for different parts of functional requirements definition, e.g. the condition to enable the functional requirements execution may have a field of its own, and the formation of the condition may be restricted using a grammar for well-formed conditions. However, templates mainly guide structure of the functional requirements. In terms of content, they do not help equally. A major problem comes with terminology and variance in expressions. Our second pattern tackles this problem.

Pattern 2: Unify the expression and terminology of requirement statements

Task: Unifying requirements statements
Quality goal: Linguistic unification of requirements
Forces, F→: Requirements are documented using NL, which has multitude of homonyms and synonyms and a variance of expressions, which may be used.
Forces, F←: Automated formalization of requirements calls for linguistic compatibility and consistency.
Action: Define homonyms, synonyms and alternative expressions for the domain, choose the priority expressions, and unify the expressions.

A glossary describing and classifying all the domain-specific terms used in the requirements is a necessary input for unification of requirements. It documents the key concepts and the vocabulary of an application domain. Besides, selecting the vocabulary from synonyms forms an important part in unification of the expression and terminology. For this, we need to construct for our domain a synonym table to indicate, which words are synonyms. For each synonym class, we need to pick up the recommended form. Using the table, synonyms are replaced by their recommended form.

Once the linguistic expressions are unified by format and content, it is possible to generate a formal specification from them, based on the conditions and actions represented in the functional requirements.

Pattern 3: Generate (semi-)formal specifications from requirements specified in template

Task: Formalizing requirements
Quality goal: To obtain a formal representation of the requirements, that can be used to analyze the behavior of the required system.
Forces, F→: Requirements are presented in a uniform way, but they are still a set of separate statements of what the system should do.

Forces, F←: The analysis of the required system calls for a unified set of requirements that forms a set that describes the collective behavior of the required system.
Action: Combine the information from the requirements to find out a collective description of all the data included in the required system and all the actions needed to represent the functionalities of the system.

One part of formalization of the requirements may be the identification of interdependencies or other logic relationships between functional requirements. Conditions and triggers of a functional requirement can be defined as a part of a requirement. Besides, structural interdependencies such as "explains" or "influences" shall be clarified to represent the operation of a set of requirement. Meanwhile, domain models describing the various concepts and their relationships are usually represented as a class diagram to provide domain information for the formalization process.

Once the requirements have been formalized, the formalization needs to be utilized to analyze the collective behavior of a system fulfilling the set of requirements.

Pattern 4: Analyze the behavior of the application fulfilling the requirements

Task: Validating requirements
Quality goal: To study the collective behavior of the required system.
Forces, F→: It is easy to understand the behavior of each single functional requirement.
Forces, F←: The complete set of functional requirements typically represents a system, which has a behavior too complicated to be completely understood
Action: Use different methods, such as simulations, animations, and mathematical analysis methods, to obtain information of the behavior of the system.

Analysis of the behavior can be done in various different forms. These include random and probabilistic simulations and their animation and analysis using e.g. statistical and data mining methods. Using stepwise user-controlled execution is a further possibility. In principle, the state models generated from the execution system can be used to analyze the mathematical behavior of the specified system, and to check for e.g. deadlocks and liveness properties.

Notably, as the patterns of this section are on activity level, they are on a fairly high abstraction level and their application require careful consideration of different approaches, methods and tools.

7 Conclusions

Promoting reuse at the requirements stage still remains a major challenge in software development. The aim in this work has been to collect problem abstractions as the starting point for outline our research on utilizing formalized specification in requirements engineering. In this paper, we have discussed the needs and possibilities for formalization of the requirements in a reuse process, and present procedures that guide the process. The procedure differs from other requirements reuse processes in that it integrates formalization of the textual requirement for validation into the process in order to define high-quality requirements for reuse and with reuse. Developing a set of 'best' practices in requirements reuse involves application and validation across a variety of domains and organizations. From this perspective, the issues we have proposed in this paper are limited, and we would seek to develop our ideas in different contexts on an on-going basis.

References

1. McClure, C.: Software Reuse: A Standards-based Guide. IEEE Computer Society, Los Alamitos (2001)
2. Gibbs, W.W.: Software Chronic Crisis. Scientific American 271(3), 86–95 (1994)
3. Knethen, V.A., Peach, B., Kiedaisch, F., Houdek, F.: Systematic requirements recycling through abstraction and traceability. In: Proceedings of IEEE Joint International Conference on Requirements Engineering, Essen, Germany, pp. 273–281 (2002)
4. Cheng, B.H.C., Atlee, J.M.: Research directions in requirements engineering. In: Proceedings of Future of Software Engineering, FOSE 2007 (2007)
5. Fraser, D.M., Kumar, K., Vaishnavi, V.K.: Informal and Formal Requirements Specification Languages: Bridging the Gap. IEEE Transactions On Software Engineering 17(5) (1991)
6. Dulac, N., Viguier, T., Leveson, N., Storey, M.: On the Use of Visualization in Formal Requirements Specification. In: Proceedings of the 10th Anniversary. IEEE Joint International Conference on Requirements Engineering (2002)
7. van Lamsweerde, A.: Requirements Engineering – From system goals to UML models to software specifications. Wiley (2009)
8. Jacobson, I., Griss, M.L., Jonsson, P.: Reuse-driven Software Engineering Business (RSEB). Addison-Wesley (1997)
9. Zhang, Z.: Model Component Reuse: Conceptual Foundations and Application in the Metamodeling Based Systems Analysis and Design Environment, Doctoral thesis, Department of Computer Science and Information Systems, Univ. of Jyväskylä, Jyväskylä Studies in Computing (39) (2004)
10. Neighbors, J.M.: Draco: A method for engineering reusable software systems. In: Software Reusability. Concepts and Models. ACM Frontier Series, vol. I, pp. 295–219. Addison-Wesley, NY (1989)
11. Kang, K.C., et al.: FORM: A feature-oriented reuse method with domain-specific reference architectures. Annals of Software Engineering 5(1), 143–168 (1998)
12. Clements, P., Northrop, L.: Software Product Lines. Addison-Wesley, Reading (2002)
13. Pohl, K., Bockle, G., van der Linden, F.: Software Product Line Engineering: Foundations, Principles, and Techniques. Springer, Heidelberg (2005)
14. Karlsson, E.: Software Reuse: A Holistic Approach. Wiley Series in Software Based Systems. John Wiley and Sons Ltd (1995)
15. Shehata, M., Eberlein, A., Hoover, J.: Requirements Reuse and Feature Interaction Management. In: Proceedings of the 15th International Conference on Software and Systems Engineering and their Applications (ICSSEA 2002), Paris, France (2002)
16. Maiden, N.A.M., Sutcliffe, A.G.: Reuse of Analogous Specificaitons during Requirements Analysis. In: Proceedings of the Sixth International Workshop on Software Specification and Design, pp. 220–223 (1991)
17. Lam, W.: Reasoning about requirements from past cases, PhD thesis. Kings College, University of London (1994)
18. Lopez, O., Laguna, M.A., Garcia, F.J.: Metamodeling for Requirements Reuse. In: Anais do WER 2002 - Workshop em Engenharia de Requisitos, Valencia, Spain (2002)

19. Moon, M., Chae, H.S., Yeom, K.: An approach to developing domain requirements as a core asset based on commonality and variability analysis in a product line. IEEE Trans. Softw. 31(7), 551–569 (2005)
20. Liaskos, S., Lapouchnian, A., Yu, Y.E., Mylopoulos, J.: On goal-based variability acquisition and analysis. In: Proc. RE 2006, Minneapolis, USA (2006)
21. Niu, N., Easterbrook, S.: Extracting and modeling product line functional requirements. In: Proc. RE 2008, Barcelona, Spain, September 8-12, pp. 155–164 (2008)
22. Nuseibeh, B., Easterbrook, S.: Requirements Engineering: A Roadmap. In: Proceedings of 22nd International Conference on Software Engineering, Future of Software Engineering Track, Limerick, Ireland, pp. 35–46 (2000)
23. IEEE Recommended practice for software requirements specification. IEEE Standard 8301998 (1998)
24. Lam, W., Whittle, B.R.: A Taxonomy of Domain-Specific Reuse Problems and their Resolutions - Version 1.0. Software Engineering Notes 21(5) (1996)
25. Kotonya, G., Somerville, I.: Requirements Engineering. John Wiley & Sons (1998)
26. Pohl, K.: The three dimensions of requirements engineering: a framework and its applications. Information Systems 19(3) (1994)
27. Dahlstedt, Å., Persson, A.: Requirements Interdependencies: State of the Art and Future Challenges. In: Engineering and Managing Software Requirements, pp. 95–116 (2005)
28. Méry, D., Singh, N.K.: Trustable Formal Specification for Software Certification. In: Proceedings of ISoLA 2010, Heraklion, Crete, Greece, pp. 312–326 (2010)
29. Lee, S.W., Rine, D.C.: Missing Requirements and Relationship Discovery through Proxy Viewpoints Model. Studia Informatica Universalis 3(3), 315–342 (2004)
30. Nummenmaa, J., Zhang, Z., Nummenmaa, T., Berki, E., Guo, J., Wang, Y.: Generating Action-Based Executable Specifications from Functional Requirements. University report, University of Tampere (2010)
31. Luisa, M., Mariangela, F., Pierluigi, I.: Market research for requirements analysis using linguistic tools. Requirements Engineering 9(1), 40–56 (2004)
32. Mich, L., Garigliano, R.: Ambiguity measures in requirements engineering. In: Proceedings of ICS 2000 16th IFIP WCC, Beijing, China, pp. 39–48 (2000)
33. Yue, T., Briand, L., Labiche, Y.: A Systematic Review of Transformation Approaches between User Requirements and Analysis Models. In: Requirements Engineering. Springer (2010)
34. Mu, Y., Wang, Y., Guo, J.: Extracting Software Functional Requirements from Free Text Documents. In: Proceedings of ICIMT 2009, Jeju Island, South Korea (2009)
35. The Stanford Parser, `http://nlp.stanford.edu/index.shtml` (access date: May, 2011)
36. Fillmore, C.: The case for case. In: Bach, E., Harms, R. (eds.) Universals in Linguistic Theory, Holt, Rinehart and Winston, pp. 1–88. Holt, Rinehart and Winston (1968)
37. Semantic Role Labeling, `http://www.lsi.upc.edu/~srlconll/home.html` (access date: March 2011)
38. Ranta, A.: Grammatical Framework: Programming with Multilingual Grammars. CSLI Publications, Stanford (2011)
39. Díaz, I., Pastor, O., Matteo, A.: Modeling Interactions using Role-Driven Patterns. In: Proceedings of RE 2005, Paris, France, pp. 209–220 (2005)
40. Śmiałek, M., Bojarski, J., Nowakowski, W., Ambroziewicz, A., Straszak, T.: Complementary Use Case Scenario Representations Based on Domain Vocabularies. In: Engels, G., Opdyke, B., Schmidt, D.C., Weil, F. (eds.) MODELS 2007. LNCS, vol. 4735, pp. 544–558. Springer, Heidelberg (2007)

41. Zhang, Z., Arvela, M., Berki, E., Muhonen, M., Nummenmaa, J., Poranen, T.: Towards Lightweight Requirements Documentation. Journal of Software Engineering and Applications 3(9), 882–889 (2010)
42. Van, H.T., van Lamsweerde, A., Massonet, P., Ponsard, C.: Gpaö-oriented requirements animation. In: Proc. Of the IEEE Int. Reg. Eng. Conf., pp. 218–228 (2004)
43. Balzer, R.M., Goldman, N.M., Wile, D.S.: Operational Specification as the Basis for Rapid Prototyping. ACM SIGSOFT Softw. Eng. Notes 7(5), 3–16 (1982)
44. Magee, J., Pryce, N., Giannakopoulou, D., Kramer, J.: Graphical animation of behavior models. In: Proc. Of the IEEE Int. Conf. on Soft. Eng (ICSE), pp. 499–508 (2000)
45. Kurki-Suonio, R.: A Practical Theory of Reactive Systems: Incremental Modeling of Dynamic Behaviors. Springer (2005)

Author Index